Foundation Engineering in Difficult Ground

Foundation Engineering in Difficult Ground

Edited by

F. G. BELL

With 17 specialist contributors

NEWNES-BUTTERWORTHS
LONDON - BOSTON
Sydney - Wellington - Durban - Toronto

The Butterworth Group

United Kingdom **Butterworth & Co (Publishers) Ltd.**
London: 88 Kingsway, WC2B 6AB

Australia **Butterworths Pty Ltd**
Sydney: 586 Pacific Highway, Chatswood, NSW 2067
Also at Melbourne, Brisbane, Adelaide and Perth

Canada **Butterworth & Co. (Canada) Ltd**
Toronto: 2265 Midland Avenue, Scarborough,
Ontario M1P 4S1

New Zealand **Butterworths of New Zealand Ltd**
Wellington: T & W Young Building, 77–85 Customhouse Quay,
1, CPO 472

South Africa **Butterworth & Co (South Africa) (Pty) Ltd**
Durban: 152–154 Gale Street

USA **Butterworth (Publishers) Inc**
Boston: 19 Cummings Park, Woburn, Mass. 01801

First published 1978

British Library Cataloguing in Publication Data

Foundation engineering in difficult ground.
1. Foundations
1. Bell, Frederic Gladstone
624′.15 TA775 77–30301

ISBN 0-408-00311-1

Typeset and produced by **Scribe Design** · Medway · Kent

Printed in England by Billing & Sons Ltd.,
Guildford, London and Worcester

Preface

The basis of this book was a conference on Foundation Engineering which was held in Sheffield in September 1976. However, rather than publish the papers in the form of proceedings it was decided to produce a text, which would be of value to the practicing foundation engineer on the one hand and the teacher and student on the other. This involved expanding the number of topics considered in order to provide a more balanced treatment of the subject. The topics were suggested to the individual authors, many of whom have international reputations, who, of course, developed their own contributions in their own way. Their style and philosophy has been retained as much as possible, within the concept of the text.

Although it is assumed that the reader has a basic knowledge of soil mechanics, the major emphasis of the book is placed on ground conditions, especially difficult ground conditions. In this way, it differs from most other works on foundation engineering. Poor ground conditions arise from lack of strength, excess ground water or appreciable ground movement, generally in some combination rather than in isolation. It is poor ground conditions which present the engineer with problems. In addition it must be borne in mind that the heavy demand placed on land previously considered unsuitable for development means that sites are now used for building purposes. In order to overcome these problems at the design stage the engineer must first investigate the subsurface conditions in order that some attempt at understanding their likely behaviour may be made. With such information the engineer can now choose either to enhance the performance of the ground or to adopt special foundation structures, or to combine the two. This method of attack should allow the structure, which is to be erected, to perform satisfactorily.

The book deals with foundation engineering in the context outlined above. It starts with an examination of ground conditions, focusing attention on geology, soils, rocks and ground water, and their influence on foundations; foundations of all types, from those of small dimensions such as necessary for houses, to those required for concrete dams. Consideration is then given to how ground conditions are investigated. Then follow chapters on ground movements; settlement, subsidence and, worse still, earthquakes, all provide appreciable problems for the foundation engineer. If ground conditions prove difficult then they can be subjected to some form of treatment to improve their engineering performance. Ground treatment is therefore dealt with in chapter 14. The problem of slope stability, whether in soils or rocks, is present during excavation. As a consequence slopes in soils and rocks are considered before open excavation.

The latter part of the book deals with foundation structures, particularly piles, and failures.

Time involved in production and the size of the work place restrictions on the number of topics which can be covered. No treatment can ever be exhaustive and the topics are accordingly selected for their intrinsic importance. One is conscious that certain topics may have been included by others, for instance, retaining walls springs to mind. But this is not a textbook of design methods, which a treatment of retaining walls mainly would have involved. For the same reason, no detailed attempt is made to cover the various aspects of design of foundations structures. These are included in most texts of foundation engineering, indeed, one of the authors, M.J. Tomlinson, is responsible for a first class textbook on design and construction of foundations.

Finally, the editor would like to express his warm appreciation to the authors for helping to make this text a reality, and in particular acknowledges the kind help given by Dr. Ian Farmer who commented upon chapter 9.

F.G. Bell
Blyth, Notts

The design on the jacket is a diagram of the Shell building, London showing raft on piles. The piles extend beneath the London Underground Bakerloo line. (Reproduced by courtesy of the Institution of Civil Engineers)

Contents

Chapter 1

Introduction

SOIL CLASSIFICATION

Soil is an unconsolidated assemblage of solid particles which may or may not contain organic matter, the voids between the particles being occupied by air and/or water. It derives mainly from rock material broken down by physical or chemical weathering* and by water, wind and ice which are responsible for the varying amounts of transport which particles may undergo prior to deposition. Gravity likewise has a hand in particle transport. The type of breakdown and amount of transport have a significant influence on the character of a deposit (Table 1.1), as has the parental material. In addition changes occur in a deposit after it has been laid down. Accordingly different types of soil evolve, with different grain size distributions, with differing degrees of sorting and packing, and with differently shaped particles.

Classification of soils is made on a basis of certain fundamental properties and provides an ordered framework for their systematic description. Such a classification ideally should indicate the engineering performance of the soil type and should provide a quick means of identification.

The fundamental property upon which most engineering classifications of soils are based is particle size distribution, since it is readily measurable and has an important influence on soil behaviour. Boulders, cobbles, gravels, sands, silts and clays are distinguished as individual groups, each group being given the following symbol and size range:

1. Boulders (B), over 200 mm
2. Cobbles (Cb), 60 to 200 mm
3. Gravel (G), 2 to 60 mm (20 to 60 mm, coarse; 6 to 20 mm, medium; 2 to 6 mm, fine).
4. Sand (S), 0.06 to 2 mm (0.6 to 2 mm, coarse; 0.2 to 0.6 mm, medium; 0.06 to 0.2 mm, fine)
5. Silt (M), 0.002 to 0.06 mm (0.02 to 0.06 mm, coarse; 0.006 to 0.02 m, medium; 0.002 to 0.006 mm, fine)
6. Clay (C), less than 0.002 mm.

The gravel, sand and silt size ranges are further divided up into coarse, medium and fine categories. Sands and gravels are granular materials ideally possessing .10

*Peat, being an organic soil, does not fall within this context

Table 1.1 EFFECTS OF TRANSPORTATION ON SEDIMENTS

	Gravity	*Ice*	*Water*	*Air*
Size	Various	Varies from clay to boulders	Various sizes from boulder gravel to muds	Sand size and less less
Sorting	Unsorted	Generally unsorted	Sorting takes place both laterally and vertically. Marine deposits often uniformly sorted. River deposits may be well sorted	Uniformly sorted
Shape	Angular	Angular	From angular to well rounded polished surface	Well rounded
Surface texture	Striated surfaces	Striated surfaces	Gravel: rugose surfaces. Sand: smooth, polished surfaces. Silt: little effect	Impact produces frosted surfaces

cohesion whereas silts and clays are cohesive materials. Mixed soil types can be indicated as follows:

1. sandy GRAVEL gravelly SAND — With sand sized or gravel sized material as an important second constituent of the coarse fraction
2. clean GRAVEL/SAND — With under 5% fines (G; S)
3. GRAVEL/SAND with some fines (F) — With 5 to 20% fines (can be distinguished as silt or clay: G-F, G-M, G-C; S-F, S-M, S-C)
4. GRAVEL/SAND with much fines — With 20 to 50% fines (can be distinguished as silt or clay: GF, GM, GC; SF, SM, SC)
5. SILT/CLAY with gravel or with sand — With 30–50% gravel or sand (MG, MS; CG, CS)

These major groups are divided into subgroups on a basis of grading and the plasticity of the fine material (Table 1.2). In the proposed British Soil Classification[1], which is based upon the Casagrande classification[2], granular soils are described as well graded (W) or poorly graded (P). Two further types of poorly graded granular soils are recognised, namely, uniformly graded (Pu) and gap-graded (Pg). Silts and clays are generally subdivided according to their liquid limits (LL) into low (LL = less than 30. L), medium (LL = 35 to 50. M) and high (LL = over 50. H) subgroups. Very high (LL = 70 to 90. V) and extremely high (LL = over 90. E) categories have also been recognised. Each subgroup is given a combined symbol in which the letter describing the predominant size fraction is written first (e.g. GW = well graded gravels; CH = clay with high liquid limit). Silty soil has a restricted plastic range in relation to its liquid limit whilst clay is fully plastic. The plasticity chart for use in soil classification (Table 1.2c) was introduced by Casagrande[2]. Silts tend to plot below, and clays above the A-line. Organic soils also plot below the A-line.

Any group may be referred to as organic if it contains a significant proportion of organic matter, in which case the letter O is suffixed to the group symbol (e.g. CVSO = organic clay of very high liquid limit with sand). The symbol Pt is given to peat.

In many soil classifications boulders and cobbles are removed before an attempt is made at classification, for example, their proportions are recorded separately in the proposed British Soil Classification. Their presence should be recorded in the soil description, a plus sign being used in symbols for soil mixtures, for example, G + Cb for gravel with cobbles. The British Soil Classification has proposed that very coarse deposits should be classified as follows:

1. BOULDERS	Over half of the very coarse material is of boulder size (over 200 mm). May be described as cobbly BOULDERS if cobbles are an important second constituent in the very coarse fraction.
2. COBBLES	Over half of the very coarse material is of cobble size (200–60 mm). May be described as bouldery COBBLES if boulders are an important second constituent in the very coarse fraction.

Mixtures of very coarse material and soil can be described by combining the terms for the very coarse constituent and the soil constituent as follows:

1. clean BOULDERS/COBBLES ,	up to 5% soil
2. BOULDERS/COBBLES with some SOIL *	5–20% soil
3. BOULDERS/COBBLES with much SOIL *	20–50% soil
4. SOIL* with many BOULDERS/COBBLES	50–20% boulders
5. SOIL* with some BOULDERS/COBBLES	20–5% boulders
6. SOIL* with occasional BOULDERS/COBBLES	up to 5% boulders

*Give soil name in brackets, e.g. Cobbly BOULDERS with some SOIL (sand with some fines)

The British Soil Classification also has proposed that classification can be made either by rapid assessment or by full laboratory procedure (Tables 1.2a and 1.2b). It was recommended that if classification was by means of rapid assessment, brackets should enclose the group symbol indicating a lower degree of accuracy, whereas if full laboratory procedure was used for classification the group symbol should not be bracketed. For a comparison with the Unified Soil Classification which is used in the United States see Table 1.2c. A note on the engineering uses of these soils is provided in Table 1.2d.

COARSE GRAINED SOILS

The composition of a gravel deposit reflects not only the source rocks of the area from which it was derived but is also influenced by the agents responsible for its formation and the climatic règime in which it was or is being deposited. The latter two factors have a varying tendency to reduce the proportion of unstable material. Relief also influences the nature of a gravel deposit, for example, gravel production under low relief is small and the pebbles tend to be chemically inert residues such as vein quartz, quartzite, chert and flint. By contrast high relief and rapid erosion yield coarse, immature gravels.

Sands consist of a loose mixture of mineral grains and rock fragments. Generally they tend to be dominated by a few minerals, the chief of which is quartz. There is a presumed dearth of material in those grades transitional to gravel on the one hand and silt on the other (see Glossop and Skempton[3]). Sands vary appreciably in their textural maturity.

The engineering behaviour of a soil is a function of its structure or fabric, which in turn is a result of the geological conditions governing deposition and the subsequent stress history. The macro-structure of a soil includes its bedding, laminations, fissures, joints and tension cracks, all of which can exert a dominant influence on the shear strength and drainage characteristics of a soil mass.

Table 1.2a THE BRITISH SOIL CLASSIFICATION SYSTEM FOR ENGINEERING PURPOSES (BSCS). FIXED OR RAPID IDENTIFICATION
[First remove material coarser than 60 mm and record as *COBBLES* (60–200 mm) or *BOULDERS* (over 200 mm)]

	SOIL GROUPS					*Remoulded at suitable moisture content*	
	GRAVEL and SAND may be qualified sandy GRAVEL and gravelly SAND where appropriate		*SYMBOL*		*TEXTURE*	*COHESION* Ability to stick together	*PLASTICITY* Ability to deform without rupture
COARSE SOILS More than half coarser than 60 μ (i.e. visible to eye or gritty to feel)	*GRAVELS* More than half of coarse material is of gravel size (coarser than 2 mm	Clean *GRAVEL*	G	GW GP	Little or no fines. GW has wide range of grain sizes, well distributed. GP has one size predominating, GPu, or missing, GPg	None	None
		GRAVEL with some silt *GRAVEL* with some clay	G-F	G-M G-C	Gravel with some silt Gravel with some clay	None to low Low to medium	None None
		GRAVEL with much silt *GRAVEL* with much clay	GF	GM GC	Gravel with much silt Gravel with much clay	None to low Medium to high	None Low to medium
	SANDS More than half of coarse material is of sand size (finer than 2 mm)	Clean *SAND*	S	SW SP	Little or no fines. SW has wide range of grain sizes, well distributed. SP has one size predominating, SPu, or missing SPg	Low	None
		SAND with some silt *SAND* with some clay	S-F	S-M S-C	Sand with some silt Sand with some clay	Low Low to medium	None None to low
		SAND with much silt *SAND* with much clay	SF	SM SC	Sand with much silt Sand with much clay	Low to medium Medium to very high	Low to medium Medium to very high

FINE SOILS less than half coarser than 60 μ (i.e. visible to eye or gritty to feel	*SILTS* and *CLAYS* with gravel or sand 50–70% fines	*SILT* with gravel *CLAY* with gravel	FC	MG CG	Coarse material mainly over 2 mm	Similar to *SILT* and *CLAY* groups (see below), but with properties modified by a considerable proportion of coarse material		
		SILT with sand *CLAY* with sand	FS	MS CS	Coarse material mainly under 2 mm			
						First remove coarser particles by hand		
						DRY STRENGTH Dried from plastic limit	*TOUGHNESS* Consistency at plastic limit	*DILATANCY*
	SILTS and *CLAYS* 70–100% fine (finer than 60 μ)	*SILT*	M	ML	Dries moderately quickly and can be brushed off the fingers. Inorganic	None to low	None to low	High to moderate
		CLAY of low liquid limit	CI	CL		Medium	Medium	Low
		of intermediate liquid limit		CI		High	Medium	None
		of high liquid limit		CH	Sticks to fingers and dries slowly. Shrinks appreciably on drying, usually showing cracks	High	High	None
		of very high liquid limit		CV		Very high	High	None
ORGANIC SOILS	'Organic' prefixed to any group name and letter O, suffixed to symbol, e.g. CO				Organic matter suspected to be a significant constituent. Dark colour, distinctive odour, moisture content may be very high			
PEAT	Pt	Peat soil consists predominantly of plant resins, which may be fibrous or amorphous. Dark colour, distinctive odour low bulk density, moisture content may be very high						

Table 1.2b BRITISH SOIL CLASSIFICATION SYSTEM FOR ENGINEERING PURPOSES (BSCS)
[First remove material coarser than 60 mm and record as *COBBLES* (60 mm–200 mm) or *BOULDERS* (over 200 mm)]

Soil Groups			*Sub-Groups and laboratory identification*				
GRAVEL and SAND may be qualified Sandy GRAVEL and Gravelly SAND where appropriate			*Group Symbol*		*Sub-Group Symbol*	*Fines % less than 60 μ*	*Sub-Group name*
COARSE SOILS More than 50% coarser than 60 μ	*GRAVELS* More than 50% of coarse material is of gravel size (coarser than 2 mm)	Clean *GRAVEL*	G	GW GP	GW GP GPu GPg	0–5	Well-graded Poorly/uniformly/gap graded — clean *GRAVEL*
		GRAVEL with some silt *GRAVEL* with some clay	G-F	G-M G-C	GWM GPM GWC GPC	5–20	Well/poorly graded *GRAVEL* with some silt / clay
		GRAVEL with much silt *GRAVEL* with much clay	GF	GM GC	GM GCL GCI GCH GCV	20–50	*GRAVEL* with much silt: sub-divide like GC *GRAVEL* with much clay of low/intermediate/high/very high plasticity
	SANDS More than 50% of coarse material is of sand size (finer than 2 mm)	Clean *SAND*	S	SW SP	SW SP SPu SPg	0–5	Well graded Poorly/uniformly/gap graded — clean *SAND*
		SAND with some silt *SAND* with some clay	S-F	S-M S-C	SWM SPM SWC SPC	5–20	Well/poorly graded *SAND* with some silt / clay
		SAND with much silt *SAND* with much clay	SF	SM SC	SM SCL SCI SCH SCV	20–50	*SAND* with much silt: sub-divide like SC *SAND* with much clay of low/intermediate/high/very high plasticity

						Liquid Limit %	
FINE SOILS more than 50% finer than 60 μ	*SILTS* and *CLAYS* with gravel or sand 50–70% fines	*SILT* with gravel	FG	MG	MG		*SILT* (divide like CG) with gravel
		CLAY with gravel		CG	CLG	< 35	*CLAY* of low liquid limit with gravel
					CIG	35–50	of intermediate liquid limit with gravel
					CHG	50–70	of high liquid limit with gravel
					CVG	70–90	of very high liquid limit with gravel
		SILT with sand	FS	MS	MS		*SILT* with sand
		CLAY with sand		CS	CLS etc		*CLAY* with sand sub-divide like GC
	SILTS and *CLAYS* 70–100% fines	*SILT*	F	M	M		*SILT*. sub-divide like C
		CLAY		C	CL	< 35	*CLAY* of low liquid limit
					CI	35–50	of intermediate liquid limit
					CH	50–70	of high liquid limit
					CV	70–90	of very high liquid limit
ORGANIC SOILS		Descriptive letter O suffixed to any group or sub-group symbol					Organic matter suspected to be a significant constituent. Example: *MHO, ORGANIC SILT of HIGH LIQUID LIMIT*
PEAT		Pt					Peat soils consist predominantly of plant remains, which may be fibrous or amorphous.

Table 1.2c UNIFIED SOIL CLASSIFICATION (After Wagner, 1957)*

Field identification procedures (excluding particles larger than 3 in, and basing fractions on estimated weights)				*Group Symbols*	*Typical names*	*Information required for describing soils*	*Laboratory classification criteria*		
Coarse-grained soils More than half of material is *larger* than No. 200 sieve size[b]	Gravels More than half of coarse fraction is larger than No. 7 sieve size*	Clean gravels (little or no fines)	Wide range in grain size and substantial amounts of all intermediate particle sizes	*GW*	Well graded gravels, gravel-sand mixtures, little or no fine	Give typical name; indicate approximate percentages of sand and gravel; maximum size; angularity, surface condition, and hardness of the coarse grains; local or geologic name and other pertinent descriptive information: and symbols in parentheses	Use grain size curve in identifying the fractions as given under field identification	Determine percentages of gravel and sand from grain size curve. Depending on percentage of fines (fraction smaller than No. 200 sieve size) coarse grained soils are classified as follows: Less than 5%: *GW, GP, SW, SP*. More than 12%: *GM, GC, SM, SC*. 5% to 12%: *Borderline* cases requiring use of dual symbols	$C_u = \frac{D_{60}}{D_{10}}$ Greater than 4; $C_c = \frac{(D_{30})^2}{D_{10} \times D_{60}}$ Between 1 & 3
			Predominantly one size or a range of sizes with some intermediate sizes missing	*GP*	Poorly graded gravels, gravel-sand mixtures, little or no fine				Not meeting all gradation requirements for *GW*
		Gravels with fines (appreciable amount of fines)	Nonplastic fines (for identification procedures see *ML* below)	*GM*	Silty gravels, poorly graded gravel-sand-silt mixtures	For undisturbed soils add information on strataification, degree of compactness, cementation, moisture conditions and drainage characteristics			Atterberg limits below "A" line, or *PI* less than 4. Above "A" line with *PI* between 4 and 7 are *borderline* cases requiring use of dual symbols
			Plastic fines (for identification procedures, see *CL* below)	*GC*	Clayey gravels, poorly graded gravel-sand-clay mixtures				Atterberg limits above "A" line, with *PI* greater than 7
	Sands More than half of coarse fraction is smaller than No. 7 sieve size*	Clean sands (little or no fines)	Wide range in grain sizes and substantial amounts of all intermediate particle sizes	*SW*	Well graded sands, gravelly sands, little or no fines	Example: *Silty sand*, gravelly; about 20% hard, angular gravel particles ½-in maximum size; rounded and subangular sand grains coarse to fine, about 15% nonplastic fines with low dry strength; well compacted and moist in place; alluvial sand; (*SM*)			$C_u = \frac{D_{60}}{D_{10}}$ Greater than 6; $C_c = \frac{(D_{30})^2}{D_{10} \times D_{60}}$ Between 1 & 3
			Predominantly one size or a range of sizes with some intermediate sizes missing	*SP*	Poorly graded sands, gravelly sands, little or no fines				Not meeting all gradation requirement for *SW*
		Sands with fines (appreciable amount of fines)	Nonplastic fines (for identification procedures, see *ML* below)	*SM*	Silty sands, poorly graded sand-silt mixtures				Atterberg limits below 'A' line or *PI* less than 5. Above "A" line with *PI* between 4 and 7 are *borderline* cases requiring use of dual symbols
			Plastic fines (for identification procedures, see *CL* below	*SC*	Clayey sands, poorly graded sand-clay mixtures				Atterberg limits below 'A' line with *PI* greater than 7

Fine-grained soils More than half of material is *smaller* than No. 200 sieve size[b]	Identification Procedures on Fraction smaller than No. 40 Sieve Size								
	Silts and clays liquid limit less than 50	*DRY STRENGTH (crushing character-istics)*	*DILA-TANCY (reaction to shaking)*	*TOUGH-NESS (consis-tency near plas-tic limit)*					
		None to slight	Quick to slow	None	*ML*	Inorganic silts and very fine sands, rock flour, silty or clayey fine sands with slight plasticity	Give typical name; indicate degree and character of plasticity, amount and maximum size of coarse grains; colour in wet condition, odour if any, local or geologic name, and other pertinent descriptive information, and symbol in parentheses	Use grain size curve in identifying the functions as given under field identification	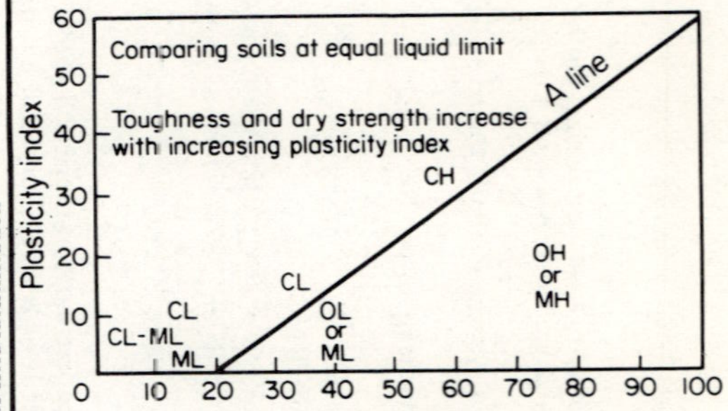 *Plasticity chart for laboratory classification of fine grained soils*
		Medium to high	None to very slow	Medium	*CL*	Inorganic clays of low to medium plasticity, gravelly clays, sandy clays, silty clays, lean clays	For undisturbed soils add information on structure, stratification, consistency in undisturbed and remoulded states, moisture and drainage conditions		
		Slight to medium	Slow	Slight	*OL*	Organic silts & organic silt-clays of low plasticity			
	Silts and clays liquid limit greater than 50	Slight to medium	Slow to none	Slight to medium	*MH*	Inorganic silts, micaceous or diatomaceous fine sandy or silty soils, elastic silts	Example: *Clayey silt*, brown: slightly plastic; small percentage of fine sand; numerous vertical root holes; firm and dry in place; loess; (*ML*)		
		High to very high	None	High	*CH*	Inorganic clays of high plasticity, fat clays			
		Medium to high	None to very slow	Slight to medium	*OH*	Organic clays of medium to high plasticity			
Highly organic soils		Readily identified by colour, odour, spongy feel and frequently by fibrous texture			*Pt*	Peat and other highly organic soils			

* Wagner, A.A. (1957), 'The use of the Unified Soil Classification System for the Bureau of Reclamation', *Proc.* 4th Int. Conf. Soil. Mech. & Found. Engng. London, 1, pp. 125–134

Footnotes to Table 1.2c

[a]*Boundary classifications.* Soils possessing characteristics of two groups are designated by combinations of group symbols. For example *GW–GC*, well graded gravel-sand mixture with clay binder.
[b]All sieve sizes on this chart are US standard. The No 200 sieve size is about the smallest particle visible to the naked eye.
*For visual classification, the ¼ in size may be used as equivalent to the No 7 sieve size.

Field Identification Procedure for Fine Grained Soils or Fractions

These procedures are to be performed on the minus No. 40 sieve size particles, approximately 1/64 in. For field classification purposes, screening is not intended, simply remove by hand the coarse particles that interfere with the tests.

Dilatancy (reaction to shaking):

After removing particles larger than No. 40 sieve size, prepare a pat of moist soil with a volume of about one-half cubic inch. Add enough water if necessary to make the soil soft but not sticky.
Place the pat in the open palm of one hand and shake horizontally, striking vigorously against the other hand several times. A positive reaction consists of the appearance of water on the surface of the pat which changes to a livery consistency and becomes glossy. When the sample is squeezed between the fingers, the water and gloss disappear from the surface, the pat stiffens and finally it cracks and crumbles. The rapidity of appearance of water during shaking and of its disappearance during squeezing assist in identifying the character of the fines in a soil.
Very fine clean sands give the quickest and most distinct reaction whereas a plastic clay has no reaction. Inorganic silts, such as a typical rock flour, show a moderately quick reaction.

Dry Strength (Crushing characteristics):

After removing particles larger than No. 40 sieve size, mould a part of soil to the consistency of putty, adding water if necessary. Allow the pat to dry completely by oven, sun or air drying, and then test its strength by breaking and crumbling between the fingers. This strength is a measure of the character and quantity of the colloidal fraction contained in the soil. The dry strength increases with increasing plasticity.
High dry strength is characteristic for clays of the CH group. A typical inorganic silt possesses only very slight dry strength. Silty fine sands and silts have about the same slight dry strength, but can be distinguished by the feel when powdering the dried specimen. Fine sand feels gritty whereas a typical silt has the smooth feel of flour.

Toughness (Consistency near plastic limit):

After removing particles larger than the No. 40 sieve size, a specimen of soil about one-half inch cube in size, is moulded to the consistency of putty. If too dry, water must be added and if sticky, the specimen should be spread out in a thin layer and allowed to lose some moisture by evaporation. Then the specimen is rolled out by hand on a smooth surface or between the palms into a thread about one-eighth inch in diameter. The thread is then folded and re-rolled repeatedly. During this manipulation the moisture content is gradually reduced and the specimen stiffens, finally loses its plasticity, and crumbles when the plastic limit is reached.
After the thread crumbles, the pieces should be lumped together and a slight kneading action continued until the lump crumbles.
The tougher the thread near the plastic limit and the stiffer the lump when it finally crumbles, the more potent is the colloidal clay fraction in the soil. Weakness of the thread at the plastic limit and quick loss of coherence of the lump below the plastic limit indicate either inorganic clay of low plasticity, or materials such as kaolin-type clays and organic clays which occur below the A-line.
Highly organic clays have a very weak and spongy feel at the plastic limit.

The micro-structure of a sand or gravel refers to its particle arrangement which in turn involves the concept of packing – in other words the spacial density of particles in the aggregate (see Kahn[4]).

The conceptual treatment of packing begins with a consideration of the arrangement of spherical particles of equal size. These can be packed either in a disorderly or systematic fashion. The closest type of systematic packing is rhombohedral packing whereas the most open type is cubic packing, the porosities approximating to 26% and 48% respectively. Put another way, the void ratio of a well sorted and perfectly cohesionless aggregate of equidimensional grains can range between extreme values of about 0.35 and 1.00. If the void ratio is more than unity the micro-structure will be collapsable or metastable. If a large number of spheres of equal size is arranged in any systematic packing pattern then there is a certain diameter ratio for smaller spheres which can just pass through the throats between the larger spheres into the interstices, for example, in rhombohedral packing this critical diameter is 0.154 D (D being the diameter of the larger spheres). However, a considerable amount of disorder occurs in most coarse grained deposits and, according to Graton and Frazer[5], there are colonies of tighter and looser packing within any deposit.

In a single grain structure individual particles are bulky and pore passages have average diameters of the same order of magnitude as smaller particle diameters. There is virtually no effective combination of particles to form aggregates. Each particle functions individually in the soil framework, and particles are in contact with one another, so that the movement of any individual grain is influenced by the position of adjacent grains. For most equilibrium conditions in coarse grained soil the soil framework serves exclusively as the stressed member.

Size and sorting have a significant influence on the engineering behaviour of granular soils. Generally speaking the larger the particles, the higher the strength. Deposits consisting of a mixture of different sized particles are usually stronger than those which are uniformly graded. However, the mechanical properties of such sediments depends mainly on their relative density which in turn depends on packing. For instance, densely packed sands are almost incompressible whereas loosely packed deposits, located above the water table, are relatively compressible but otherwise stable. If the relative density of a sand varies erratically this can give rise to differential settlement. Generally settlement is relatively rapid. However, when the stresses are large enough to produce appreciable grain fracturing, there is a significant time lag.

Greater settlement is likely to be experienced in granular soils where foundation level is below the water than above. Additional settlement may occur if the water table fluctuates or the ground is subject to vibrations. Although relative density may decrease in a general manner with decreasing grain size there is ample evidence to show, for example, that water deposited sands with similar grain size can vary between wide limits. Hence factors other than grain size, such as rate of deposition and particle shape, influence relative density.

DEFORMATION OF GRANULAR SOIL

Two basic mechanisms contribute towards the deformation of granular soil, namely, distortion of the particles, and the relative motion between them.

Table 1.2d ENGINEERING USE CHART (after Wagner, 1957)

		Important properties				*Relative desirability for various uses (Graded from 1 (highest) to 14 (lowest)*									
						Rolled earth dams			*Canal sections*		*Foundations*		*Roadways*		
													Fills		
Typical names of soil groups	*Group symbols*	*Permeability when compacted*	*Shearing strength when compacted and saturated*	*Compressibility when compacted and saturated*	*Workability as a construction material*	*Homogeneous embankment*	*Core*	*Shell*	*Erosion resistance*	*Compacted earth lining*	*Seepage important*	*Seepage not important*	*Frost heave not possible*	*Frost heave possible*	*Surfacing*
Well-graded gravels, gravel-sand mixtures, little or no fines	*GW*	pervious	excellent	negligible	excellent	–	–	1	1	–	–	1	1	1	1
Poorly graded gravels, gravel-sand mixtures, little or no fines	*GP*	very pervious	good	negligible	good	–	–	2	2	–	–	3	3	3	–
Silty gravels, poorly graded gravel-sand-silt mixtures	*GM*	semi-pervious to impervious	good	negligible	good	2	4	–	4	4	1	4	4	9	5
Clayey gravels, poorly graded gravel-sand-clay mixtures	*GC*	impervious	good to fair	very low	good	1	1	–	3	1	2	6	5	5	1
Well-graded sands, gravelly sands, little or no fines	*SW*	pervious	excellent	negligible	excellent	–	–	3 if gravelly	6	–	–	2	2	2	4
Poorly graded sands, gravelly sands, little or no fines	*SP*	pervious	good	very low	fair	–	–	4 if gravelly	7 if gravelly	–	–	5	6	4	–

Silty sands, poorly graded sand-silt mixtures	*SM*	semi-pervious to impervious	good	low	fair	4	5	–	8 if gravelly	5 erosion critical	3	7	8	10	6
Clayey sands, poorly graded sand-clay mixtures	*SC*	impervious	good to fair	low	good	3	2	–	5	2	4	8	7	6	2
Inorganic silts and very fine sands, rock flour, silty or clayey fine sands with slight plasticity	*ML*	semi-pervious to impervious	fair	medium	fair	6	6	–	–	6 erosion critical	6	9	10	11	–
Inorganic clays of low to medium plasticity, gravelly clays, sandy clays, silty clays, lean clays	*CL*	impervious	fair	medium	good to fair	5	3	–	9	3	5	10	9	7	7
Organic silts and organic silt-clays of low plasticity	*OL*	semi-pervious to impervious	poor	medium	fair	8	8	–	–	7 erosion critical	7	11	11	12	–
Inorganic silts, micaceous or diatomaceous fine sandy or silty soils, elastic silts	*MH*	semi-pervious to impervious	fair to poor	high	poor	9	9	–	–	–	8	12	12	13	–
Inorganic clays of high plasticity, fat clays	*CH*	impervious	poor	high	poor	7	7	–	10	8 volume change critical	9	13	13	8	–
Organic clays of medium to high plasticity	*OH*	impervious	poor	high	poor	10	10	–	–	–	10	14	14	14	
Peat and other highly organic soils	*Pt*	–	–	–	–	–		–	–	–	–	–		–	–

These mechanisms are usually interdependent. At any instant during the deformation process different mechanisms may be acting in different parts of the soil and these may change as deformation continues. Interparticle sliding can occur at all stress levels, the stress required for its initiation increasing with initial stress and decreasing void ratio. Crushing and fracturing of particles begins in a minor way at small stresses, becoming increasingly important when some critical stress is reached. This critical stress is smallest when the soil is loosely packed and uniformly graded, and consists of large, angular particles with a low strength. Usually fracturing only becomes important when the stress level exceeds 3.5 MN/m^2.

The internal shearing resistance of a granular soil is generated by friction *generated* when grains in the zone of shearing are caused to slide, roll and rotate against each other. At the commencement of shearing in a sand some grains are moved into new positions with little difficulty. The normal stress acting in the direction of movement is small but eventually these grains occupy positions in which further sliding is more difficult. By contrast other grains are so arranged in relation to the grains around them that sliding is difficult. They are moved without sliding by the movements of other grains. The frictional resistance of the former is developed as the grains become impeded whereas in the latter case it is developed immediately. The resistance to rolling represents the sum of the behaviour of all the particles, and the resistance to sliding is essentially attributable to friction which, in turn, is proportional to the confining stress.

Frictional resistance is built up gradually and consists of establishing normal stresses in the intergranular structure as the grains push or slide along (see Cornforth[6]). At the same time sliding allows the structure to loosen in dilatant soils which reduces normal stress. The maximum shearing resistance is a function of these two factors. The packing and external stress conditions govern the amount of sliding by individual grains in mobilizing shearing resistance. According to Cornforth the latter factor is the more important and in fact is really a strain condition. He therefore concluded that the strain condition during shear is a major factor contributing to the strength of sand.

STRENGTH AND DISTORTION

Interlocking grains contribute a large proportion of the strength in densely packed granular soils and shear failure occurs by overcoming the frictional resistance at the grain contacts. Conversely, interlocking has little or no effect on the strength of very loosely packed coarse grained soils in which the mobility of the grains is greater (see Borowicka[7]). Figure 1.1 shows that dense sand has a high peak strength and that, when it is subjected to shear stress, it expands up to the point of failure, after which a slight decrease in volume may occur. Loose sand, on the other hand, compacts under shearing stress and its residual strength may be similar to that of dense sand, and tends to remain so. Hence a constant void ratio is obtained, that is, the critical volume condition which has a critical angle of friction and a critical void ratio. These are independent of initial density being a function of the normal effective stress at which shearing occurs. For a discussion of the shear strength of sands see Rowe[8].

Both curves in Figure 1.1 indicate strains which are approximately proportional to stress at low stress levels, suggesting a large component of elastic distortion. If the stress is reduced the unloading stress-strain curve indicates that not all the strain is recovered on unloading. The hysteresis loss represents the energy lost in crushing and repositioning of grains. At higher shear stresses the strains are proportionally greater indicating greater crushing and reorientation. Indeed Arnold and Mitchell[9] showed that as a sample of sand is subjected to cyclic loading, the unloading response involves an increasing degree of hysteresis; in other words they found that on unloading recoverable deformation in

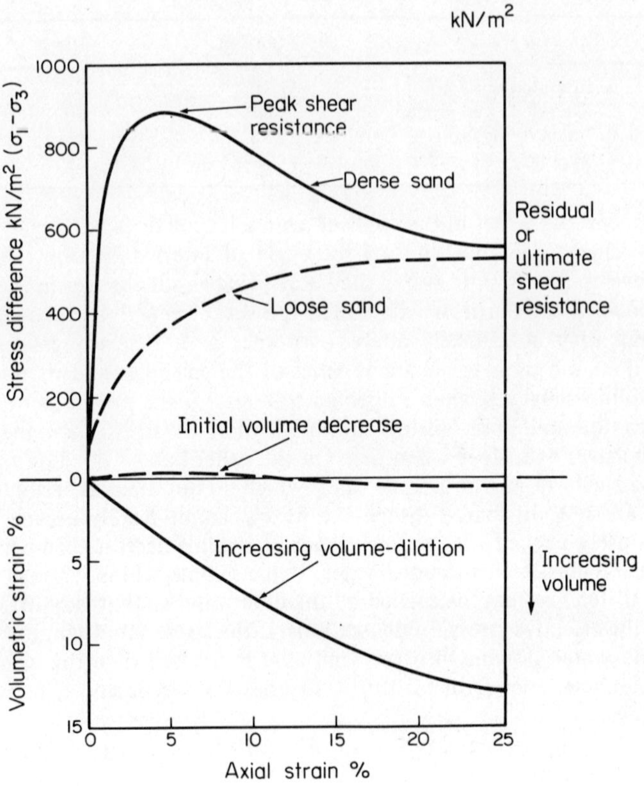

Figure 1.1 Stress-strain curves for dense and loose sand

sand under triaxial conditions was small. Because irrecoverable strains were larger than the elastic strains this led them to suggest that total strains could, in fact, be regarded as irrecoverable. As would be expected loose sand with larger voids and fewer points of contact exhibits greater strains and less recovery when unloaded than dense sand (see Chapter 3).

The angle of shearing resistance is also influenced by the grain size distribution and grain shape (see Holtz and Gibbs[10]); the larger the grains the wider the zone affected, the more angular the grains the greater the frictional resistance to their relative movement since they interlock more thoroughly than do rounded ones and they therefore produce a larger angle of shearing resistance (Table 1.3). A well graded granular soil experiences less breakdown on loading than a uniformly sorted soil of the same mean particle size since in the former type there are more interparticle contacts and hence the load per contact is less than in the latter. In gravels the effect of angularity is less because of particle crushing.

Kirkpatrick[12]; Sutherland and Mesdary[13] and Arnold and Mitchell[9] all studied the failure state of sand in a three-dimensional stress system. They found that generally the Mohr-Coulomb law for sand based on triaxial tests under-predicts the failure strength of the material in other stress conditions.

Table 1.3 EFFECT OF GRAIN SHAPE AND GRADING ON THE PEAK FRICTION ANGLE OF COHESIONLESS SOIL (after Terzaghi[11])

Shape and grading	*Loose*	*Dense*
1. Rounded, uniform	30°	37°
2. Rounded, well graded	34°	40°
3. Angular, uniform	35°	43°
4. Angular, well graded	39°	45°

The presence of water in the voids of a granular soil does not usually produce significant changes in the value of the angle of internal friction. However, if stresses develop in the pore water they may bring about changes in the effective stresses between the particles whereupon the shear strength and the stress-strain relationships may be radically altered. Whether or not pore pressures develop depends upon the drainage characteristics of the soil mass and its tendency to undergo volume changes when subjected to stress. If the pore water can readily drain from the soil mass during the application of stress then the granular material behaves as it does when dry. On the other hand if loading takes place rapidly, particularly in fine grained sands which do not drain as easily, then pore pressures are not dissipated. Since the water cannot readily escape from the voids of loosely packed, fine grained sands no volume decrease can occur and so the pressure increases in the pore water. If the sample is loose enough nearly all the stress difference may be carried by the pore water so that very little increase occurs in the effective stress. In dense sands, if the stress and drainage conditions prevent the water flowing into the sand as it is stressed then the usual volume increase characteristic of dense dry sand does not occur and a negative pore pressure develops.

The relationship between unit load on a potential surface of sliding (p) and the shearing resistance per unit area (s) can be approximately expressed by the equation:

$$s = (p - u) \tan \phi$$

where u is the hydrostatic pressure of the pore liquid prior to the application of the shearing force.

In the light of what has been said above concerning changing pore pressures this expression can be modified as follows:

$$s = (p - u - \Delta u) \tan \phi$$

where Δu represents the change in pore pressure.

SILTS AND LOESS

Silts are Elastic sediments, i.e. they are derived from pre-existing rock types chiefly by mechanical breakdown processes. They are mainly composed of fine quartz material. Silts may occur in residual soil horizons but in such instances they are usually not important. However, silts are often found in alluvial, lacustrine and marine deposits. As far as alluvial sediments are concerned silts are typically present in flood plain deposits, they may also occur on terraces which border such plains. These silts tend to interdigitate with deposits of sand and clay. Silts are also present with sands and clays in estuarine and deltaic sediments.

Lacustrine silts are often banded and may be associated with varved clays, which themselves contain a significant proportion of particles of silt size. Marine silts are also frequently banded and have high moisture contents. Wind blown silts are usually uniformly sorted. Grains of silt are often rounded with smooth outlines which influence their degree of packing. The latter, however, is more dependent on the grain size distribution within a silt deposit, uniformly sorted deposits not being able to achieve such close packing as those in which there is a range of grain size. This, in turn, influences the porosity and void ratio values as well as the bulk and dry densities (Table 1.4).

Table 1.4 SOME VALUES OF THE COMMON PROPERTIES OF SOILS

A. COHESIONLESS SOILS		
	Gravels	*Sands*
Specific gravity	2.5–2.8	2.6–2.7
Bulk density	1450–2300 kg/m³	1400–2150 kg/m³
Dry density	1400–2100 kg/m³	1350–1900 kg/m³
Porosity	20–50%	25–35%
Shear strength	200–600 kN/m²	100–400 kN/m²
Angle of friction	35–45%	32–42°
B. COHESIVE SOILS		
	Silts	*Clays*
Specific gravity	2.64–2.66	2.55–2.75
Bulk density	1820–2150 kg/m³	1500–2150 kg/m³
Dry density	1450–1950 kg/m³	1200–1750 kg/m³
Void ratio	0.35–0.85	0.42–0.96
Liquid limit	24–35%	Over 25
Plastic limit	14–25%	Over 20
Coefficient of consolidation	12.2 m³/yr	5–20 m²/yr
Effective cohesion	75 kN/m²	20–200 kN/m²
Effective angle of friction	32–36°	
C. ORGANIC SOILS AND FILL		
	Peat	*Coarse discard*
Moisture content	650–1100%	6–14%
Specific gravity	1.3–1.7	1.8–2.7
Bulk density	910–1050 kg/m³	1200–2400 kg/m³
Dry density	70–110 kg/m³	1050–2000 kg/m³
Void ratio	12.7–14.9	0.35–Over 1
Liquid limit		23–45%
Plastic limit		Non-plastic–35%
Effective angle of friction	5°	28°–40°
Effective cohesion	20 kN/m²	20–50 kN/m²

Schultze and Kotzias[14] showed that consolidation of silt was influenced by grain size, particularly the size of the clay fraction, porosity and natural moisture content. Primary consolidation accounted for 76% of the total consolidation exhibited by the Rhine silts tested by Schultze and Kotzias, secondary consolidation contributing the remainder. It was noted that unlike many American silts, which are often unstable when saturated, and undergo significant settlements when loaded, the Rhine silts in such a condition were usually stable. The difference no doubt lies in the respective soil structures. Most American silts are, in fact, loess soils which have a more open structure than the reworked river silts of the Rhine with a void ratio of less than 0.85. Nonetheless, in many silts, settlement continues to take place after construction has been completed and may exceed 100 mm. Settlement may continue for several months after completion because the rate at which water can drain from the voids under the influence of applied stress is slow.

Schultze and Horn[15] found that the direct shear test proved unsuitable for the determination of the shear strength of silt, this had to be obtained by triaxial testing. They demonstrated that the true cohesion of silt was a logarithmic function of the water content and that the latter and the effective normal stress determined the shear strength. The angle of friction was dependent upon the plasticity index.

In a series of triaxial tests carried out on silt Penman[16] showed that in drained tests with increasing strain, the volume of the sample first decreased, then increased at a uniform rate and ultimately reached a stage where there was no further change. The magnitude of the dilatancy which occurred when the silt was sheared, and was responsible for these volume changes, increased with increasing density as it does in sands. The expansion was caused by the grains riding over each other during shearing. The strength of the silt was attributed mainly to the friction between the grains and the force required to cause dilatancy against the applied pressures.

These drained tests indicated that the angle of shearing resistance (ϕ) decreased with increasing void ratio and with increasing lateral pressure. Grain interlocking was responsible for the principal increase in the angle of shearing resistance and increased with increasing density. A fall in pore water pressure occurred in the undrained tests during shearing, and there was an approximately linear relation between the maximum fall in pore pressure and the void ratio. Provided the applied pressures were sufficiently high the drop in pore pressure governed the ultimate strength. The fall in the pore pressure was dependent on the density of the silt, the greater the density, the greater the fall in pore pressure. At a given density the ultimate strength was independent of applied pressure (above a critical pressure) and so the silt behaved as a cohesive material ($\phi = 0$). Below this critical pressure silt behaved as a cohesive and frictional material. An exceptional condition occurred when a highly dilatable sample was placed under low cell pressure. When the pore pressure fell below atmospheric pressure gas was liberated by the pore water and the sample expanded.

Frost heave (see below) is commonly associated with silty soils and loosely packed silts can exhibit quick conditions.

LOESS

Loess is a wind-blown deposit which is mainly of silt size and consists mostly of quartz particles with lesser amounts of feldspar and clay minerals. It is characterised by a lack of stratification and uniform sorting and occurs as blanket deposits in central and western Europe, USA, Russia and China. Loess deposits are of Pleistocene age and because they show a close resemblance to fine grained glacial debris their origin has customarily been assigned a glacial association. In other words winds blowing from the arid interiors of the northern continents during glacial times picked up fine glacial outwash material and carried it many hundreds of kilometres before deposition took place. Deposition is presumed to have occurred over steppe lands, the grasses having left behind fossil root-holes which typify loess. The lengthy transport accounts for the uniform sorting of loess.

As can be inferred from the previous paragraph, loess owes its engineering characteristics largely to the way in which it was deposited since this gave it a metastable structure, in that initially the particles were loosely packed. The porosity of the structure is enhanced by the presence of fossil root holes, but the latter have been subsequently lined with carbonate cement, which also helps bind the grains together. This has meant that the initial metastable structure has been preserved and the carbonate cement provides the bonding strength of loess. It must be pointed out, however, that the chief binder is usually the clay matrix. In a detailed examination of the micro-structure of loess soils Larionov[17] found that the coarser grains were never in contact with each other, being carried in a fine granular dispersed mass. Hence the strength of the soil is largely determined by the character of this fine mass. The ratio of coarser grains to fine dispersed fraction varies not only quantitatively but morphologically. Consequently three micro-structures can be recognised, namely, granular, where a filmy distribution of the fine dispersed fraction predominates; aggregate, consisting mainly of aggregates; and granular-aggregate, having an intermediate character. Larionov suggested that generally loess soils with granular micro-structure have less water resistance than aggregate types, they also have lower cohesion and higher permeability. They are therefore more likely to collapse on wetting than the aggregate types.

Loess deposits generally possess a uniform texture consisting of 50–90% particles of silt size. Their liquid limit averages about 30. Liquid limits as high as 45 have been recorded, and the plasticity index ranges from about 4 to 9, but averages 6. As far as their angle of shearing resistance is concerned this usually varies from 30–34°. Loess deposits are better drained (their permeability ranges from 10^{-5} to 10^{-7} m/sec) than true silts because of their pattern of vertical root-holes. As would be expected the coefficient of permeability is appreciably higher in the vertical than horizontal direction.

In the unweathered state above the water table the unconfined compressive strength of loess may amount to several hundred kilonewtons a square metre. On the other hand, if loess is permanently submerged the metastable structure breaks down so that loess then becomes a slurry. The presence of vertical root-holes explains why vertical slopes are characteristic of loess landscapes. These may remain stable for long periods and when failure occurs it generally does so in the form of a vertical slice. By contrast an inclined slope is subject to rapid erosion.

Unlike silt, loess does not appear to be frost susceptible, this being due to its more permeable character. However, like silt, it can exhibit quick conditions and it is difficult, if not impossible, to compact. Because of its porous structure a 'shrinkage' factor must be taken into account when estimating earthwork. A full account of loess as a foundation material is provided by Clevenger[18].

CLAY DEPOSITS

Clay deposits are principally composed of fine quartz and clay minerals. The latter represent the commonest breakdown products of most of the chief rock forming silicate minerals. Clay minerals are phyllosilicates and their atomic structure, which significantly affects their engineering behaviour, can be regarded as consisting of two fundamental units. One of these units is composed of two sheets of closely packed oxygens or hydroxyls in which atoms of aluminium, magnesium or iron are arranged in octohedral coordination (Figure 1.2a). The other unit is formed of linked SiO_4 tetrahedrons which are arranged in layers (Figure 1.2d). In the common clay minerals these fundamental units are arranged

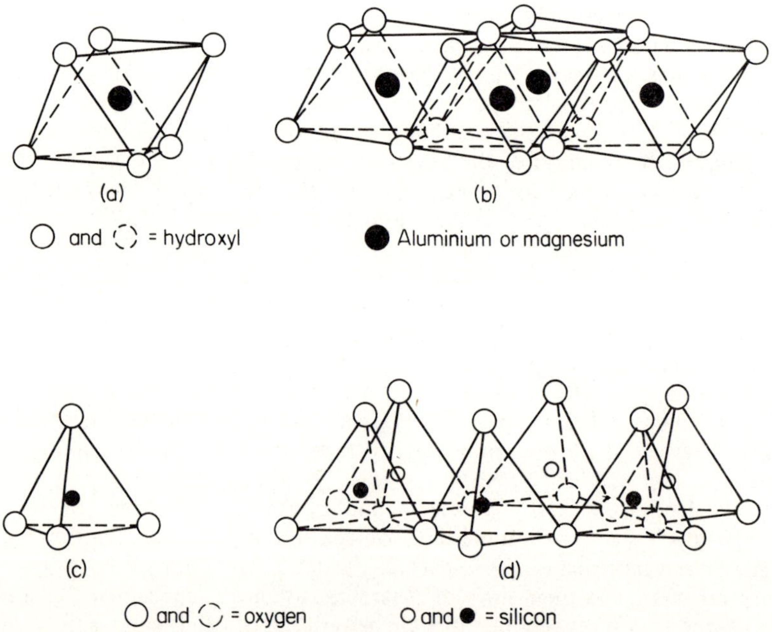

Figure 1.2 Fundamental units comprising the structure of clay minerals. (a) A single octahedral unit, in gibbite A1 is surrounded by 6 oxygens whereas in brucite they surround Mg in six-fold coordination; (b) the sheet structure of the octahedral units; (c) the silica tetrahedron; (d) the sheet structure of silica tetrahedrons arranged in a hexagonal network (After Grim[26])

in the respective atomic lattices shown in Figures 1.3a, b and c. The chemical composition of the clay minerals varies according to the amount of aluminium which is substituted for silicon in the atomic structure and also with the replacement of magnesium by other ions. Nevertheless they are basically hydrated aluminium silicates.

The three major clay minerals are kaolinite, illite and montmorillonite. Kaolinite is principally formed as an alteration product of feldspars, feldspathoids and muscovite as a result of weathering under acidic conditions. It is the most important clay mineral in china clays, ball clays and fireclays as well as most residual and transported clay deposits. Illite is a common mineral in most clays and shales, and is present in various amounts in tills and loess, but less common in soils. It develops due to the weathering of feldspars, micas and ferromagnesium silicates, or may form from other clay minerals upon diagenesis. Irrespective of the process responsible for its formation, this appears to be favoured by an alkaline environment.

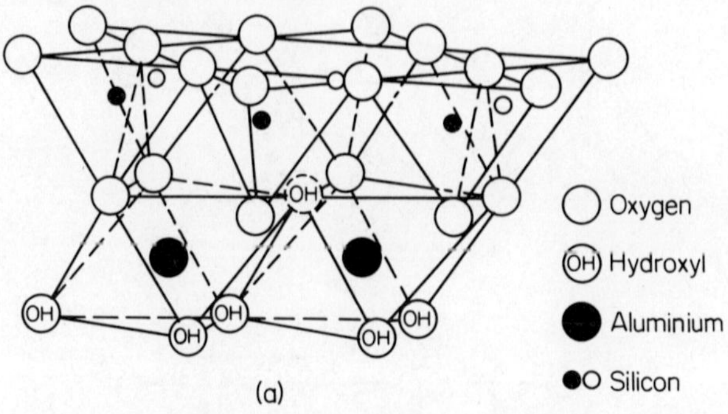

Figure 1.3(a) Diagrammatic sketch of the kaolinite structure

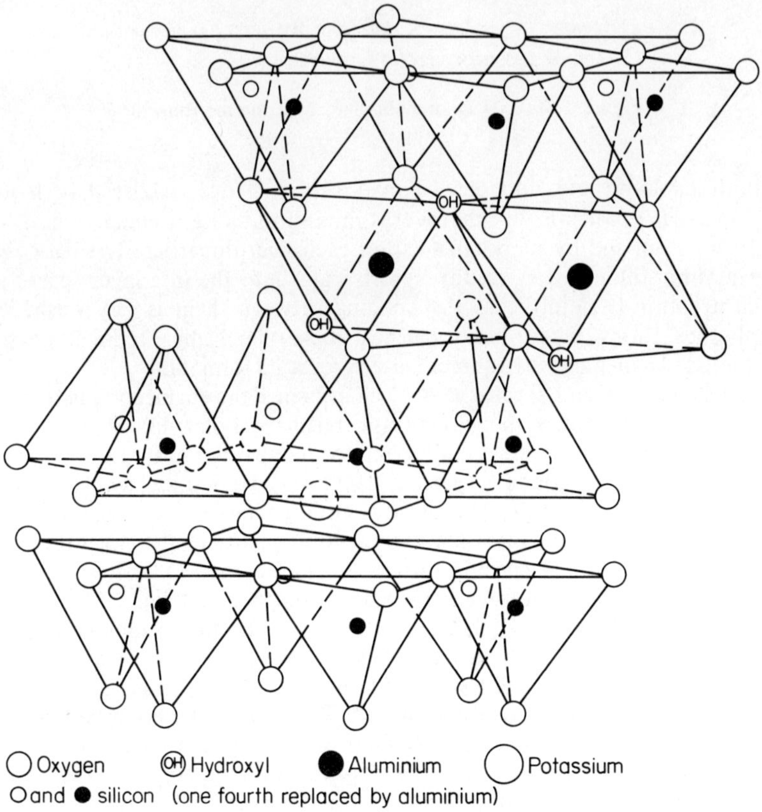

Figure 1.3(b) The above structure is that of muscovite which is regarded as essentially the same as that of illite

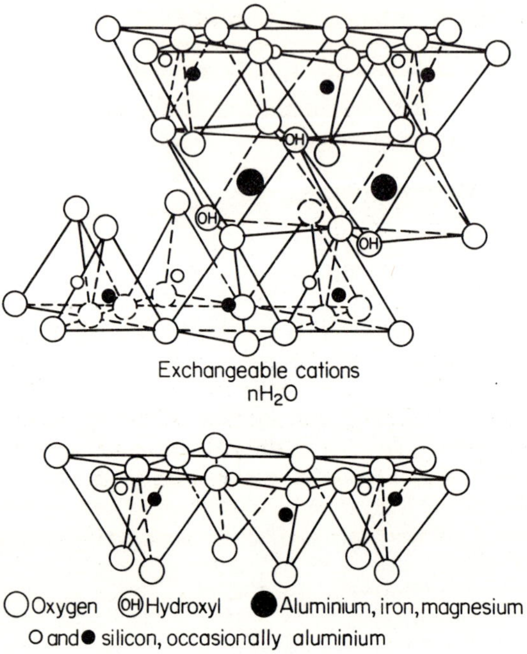

Figure 1.3(c) Diagrammatic sketch of the montmorillonite structure

Both kaolinite and illite have non-expansive lattices whilst that of montmorillonite is expansive. In other words montmorillonite is characterised by its ability to swell and by its notable cation exchange properties. The basic reason why montmorillonite can readily absorb water into the interlayer spaces in its sheet structure is simply that the bonding between them is very weak. Montmorillonite forms when basic igneous rocks, in badly drained areas, are weathered. An alkaline environment also favours its formation.

The shape, size and specific surface all influence the engineering behaviour of clay minerals. As noted above, clay minerals have a plate-like shape. They are very small in size, being measured in angstrom units. For example, an individual particle of montmorillonite is typically 1000 Å by 10 Å thick, whilst kaolinite is 10000 Å by 1000 Å thick. The specific surface refers to the magnitude of the surface in relation to the mass and the smaller the particle, the larger the specific surface (Table 1.5). The specific surface provides a good indication of the relative influence of electrical forces on the behaviour of a particle.

The surface of a clay particle has a nett charge. This means that a clay particle is surrounded by a strongly attracted layer of water, but as the diapolar water molecules do not satisfy the electrostatic balance at the surface of the clay particle, some metal cations are also adsorbed. The ions are usually weakly held and therefore can be readily replaced by others. Consequently they are referred to as exchangeable ions. The ion exchange capacity of soils normally ranges up to 40 milli-equivalents per 100 g. However, for some clay soils it may be greater as can be inferred from the ion exchange capacity of kaolinite, illite and montmorillonite (Table 1.5).

Table 1.5 SIZE AND SPECIFIC SURFACE OF SOIL PARTICLES

Soil particle	*Size* (mm)	*Specific surface* (m^2/g)	*Ion exchange/capacity* (Milli-equivalents/100 g)
Sand grain	1	0.002	
Kaolinite	d = 0.3 to 3^{-3} thickness = 0.3 to 0.1d	10–20	3–15
Illite	d = 0.1 to 2^{-3} thickness = 0.1d	80–100	20–40
Montmorillo-nite	d = 0.1 to 1^{-3} thickness = 0.01d	800	60–100

The type of adsorbed cations influences the behaviour of the soil in that the greater their valency, the better the mechanical properties. For instance, clay soils containing montmorillonite with sodium cations are characterised by high water absorption and considerable swelling. If these are replaced by calcium, a cation with a higher valency, both these properties are appreciably reduced. The thickness of the adsorbed layer influences the soil permeability, that is, the thicker the layer, the lower the permeability since a greater proportion of the pore space is occupied by strongly held adsorbed water. As the ion exchange capacity of a cohesive soil increases so does its plasticity index, the relationship between the two being almost linear.

MICRO-STRUCTURES

The micro-structure of cohesive soils is largely governed by the clay minerals present and the forces acting between them. Because of the complex electrochemistry of clay minerals the spatial arrangement of newly sedimented particles is very much influenced by the composition of the water in which deposition takes place. Single clay mineral platelets may associate in an edge-to-edge (EE), edge-to-face (EF) face-to-face (FF) or random type of arrangement depending on the interparticle balance between the forces of attraction and repulsion, and the amount or absence of turbulence in the water in which deposition occurs. Since sea water represents electrolyte rich conditions it causes clay particles to flocculate (see Lambe[19]). In other words the particles are attracted to one another in a loose, haphazard arrangement and considerable free water is trapped in the large voids (Figure 1.4a). Lambe[20] suggested that the clay particles were typically arranged in edge-to-edge or edge-to-face associations.

Flocculent soils are light in weight and very compressible but are relatively strong and insensitive to vibration because the particles are lightly bound by their edge-to-face attraction. They are sensitive to remoulding which destroys the bond between the particles so that the free water is released to add to the adsorbed layers at the former points of contact. By contrast, flocculation does not occur amongst clay particles deposited in fresh water. In this case they assume a more or less parallel, close-packed type of orientation. This has been referred to as a dispersed micro-structure (Figure 1.4b). The bulky grains are distributed throughout the mass and cause localised departures from the pattern. Soils having a dispersed structure are usually dense and watertight. Typical void ratios are often as low as 0.5. Estuarine clays, because they have been deposited in marine through brackish to freshwater conditions can contain a mixture of flocculated and dispersed micro-structures. Van Olphen[21] showed that the micro-structure of clay deposits was somewhat more complex than that proposed by Lambe[19]. The aggregate micro-structure (Figure 1.4c) was also proposed by Van Olphen[21].

The original micro-structure of a clay deposit is subsequently modified by overburden pressures due to burial, which bring about consolidation. Consolidation tends to produce a preferred orientation with the degree of reorientation

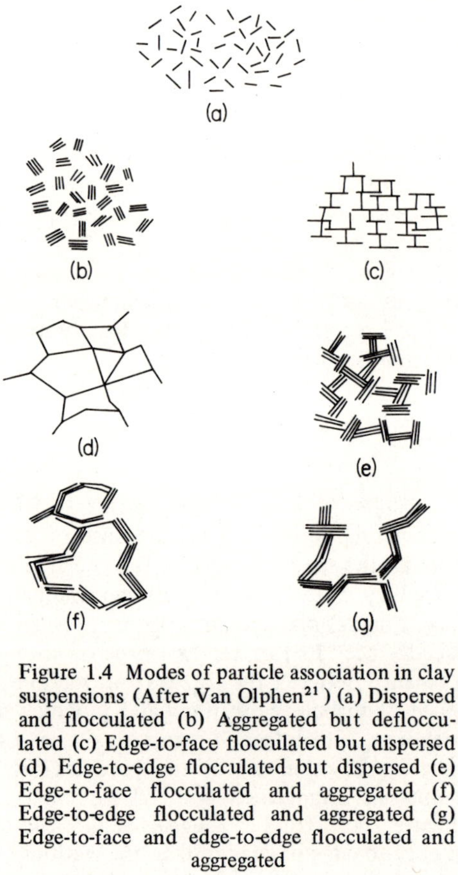

Figure 1.4 Modes of particle association in clay suspensions (After Van Olphen[21]) (a) Dispersed and flocculated (b) Aggregated but deflocculated (c) Edge-to-face flocculated but dispersed (d) Edge-to-edge flocculated but dispersed (e) Edge-to-face flocculated and aggregated (f) Edge-to-edge flocculated and aggregated (g) Edge-to-face and edge-to-edge flocculated and aggregated

of the clay particles being related to both the intensity of stress and the electrochemical environment, dispersion encouraging and flocculation discouraging clay particle parallelism. For instance, Barden[22] maintained that lightly consolidated marine clay (e.g. Romerike Clay) retains a random open structure, that medium consolidated brackish water clay (e.g. Boston Blue Clay) has a very high degree of orientation, and that extremely heavily consolidated marine clay (e.g. London Clay) develops a fair degree of orientation.

It is generally accepted that turbostratic groups or domains occur in consolidated clays although it has been suggested that ill-defined domains are present in unconsolidated clays (see Aylmore and Quirk[23]; Burnham[24]). These domains consist of aggregations of clay particles which have a preferred orientation but between the aggregates the orientation is random. With increasing overburden pressure it appears that the number of clay particles in each domain increases and that there is an increase in domain orientation. For example, it has been shown that most of the Oxford Clay is made up of small domains, up to 50^{-3} mm across, these showing some alignment with the bedding.

PERFORMANCE OF CLAY DEPOSITS

The engineering performance of clay deposits is very much affected by the total water content and by the energy with which this moisture is held. For example, the moisture content influences their density, consistency and strength, and the energy with which moisture is held influences their volume change characteristics since swelling, shrinkage and consolidation are affected by permeability and moisture migration. Furthermore moisture migration may give rise to differential movement in clay soils. The gradients which generate moisture migration in clays may arise from variations in temperature, extent of saturation, and chemical composition or concentration of pore solutions. In order to minimise the deleterious effects of moisture movements in cohesive soils footings can be placed at depths which are unaffected by seasonal fluctuations of moisture content.

The capillary potential or soil water potential is the force required to pull a unit mass of water away from a unit mass of soil, its magnitude indicating the force with which moisture is held. Moisture moves from wet to dry clay and will move upwards under the influence of capillary action against the force of gravity until equilibrium is established. At equilibrium the pore water pressures decrease linearly with height above the water table.

Fully saturated clay soils often behave as incompressible materials when subjected to rapid loading. The amount of elasticity increases continuously as the water content is decreased. Elastic recovery of the original size or shape may be immediate or may take place slowly. The linear relationship between stress and strain only applies to clays at low stresses.

PLASTICITY

There is no particular value of plastic limit that is characteristic of an individual clay mineral, indeed the range of values for montmorillonite is large. This is due to the inherent variations of structure and composition within the crystal lattice and the variations in exchangeable-cation composition. Generally the plastic limits for the three clay minerals noted decrease in the order montmorillonite, illite and kaolinite. As far as montmorillonite is concerned if the exchangeable ions are Na and Li then these give rise to high plastic limits. In the case of the other two clay minerals the exchangeable cations produce relatively insignificant variation in the plastic limit. On the other hand poorly crystalline kaolinite of small particle size has a substantially higher plasticity than that of relatively coarse, well-organised particles.

Similarly there is no single liquid limit which is characteristic of a particular clay mineral; indeed the range of limits is much greater than that of the plastic limits. Again the highest liquid limits are obtained with Li and Na montmorillonite; then follow, in decreasing order, Ca, Mg, K, Al montmorillonite; illite; poorly crystalline kaolinite; well crystallised kaolinite. Indeed the liquid limit of Li and Na montmorillonite cannot be determined accurately because of their high degree of thixotropy. The character of the cation is not the sole factor influencing the liquid limit, the structure and composition of the silicate lattice are also important. The liquid limits for illites fall in the range 60 to 90 whilst those for kaolinites vary from about 30 to 75. Again the crystallinity of the lattice and particle size are the controlling factors, for instance, poorly crystallised, fine grained samples may be over 100. The presence of 10% montmorillonite in an illitic or kaolinitic clay can cause a substantial increase in their liquid limits.

The plasticity indices of Na and Li montmorillonite clays have exceedingly high values, ranging between 300 and 600. Montmorillonites with other cations have values varying from about 50 to 300 with most of them in the range of 75 to 125. As far as the latter are concerned there is no systematic variation with cation composition. In the case of illitic clays the plasticity indices range from 25 to 50. The values for well crystallised illite are extremely low, indeed they are almost non-plastic. The presence of montmorillonite in these clays substantially increases the index. The range of plasticity indices for kaolinitic clays varies from about 1 to 40, generally being around 25. As can be inferred from above the limit values increase with a decrease in particle size and the liquid limit tends to increase somewhat more than the plastic limit.

SWELLING AND SHRINKAGE

One of the most notable characteristics of clays from the engineering point of view is their susceptibility to slow volume changes which can occur independently of loading due to swelling or shrinkage. The ability of a clay to imbibe water leads to it swelling, and when it dries out it shrinks. Such movements in heavy clays in south-east England have been responsible for appreciable damage to buildings (see Chapter 18).

Differences in the periods and magnitude of precipitation and evaporation are the major factors influencing the swell-shrink response of an active clay beneath a structure. Poor surface drainage and leakage from underground pipes can produce concentrations of moisture in clay. Trees with high water demand and uninsulated hot-process foundations may dry out a clay causing shrinkage. The density of a clay soil also influences the amount of swelling it is likely to undergo. Expansive clay minerals absorb moisture into their lattice structure (see below), tending to expand into adjacent zones of looser soil before volume increase occurs. In a densely packed soil having small void space, the soil mass has to swell to accommodate the volume change of the expansive clay particles.

Grim[25] distinguished two modes of swelling in clay soils, namely, intercrystalline and intracrystalline swelling. Interparticle swelling takes place in any type of clay deposit irrespective of its mineralogical composition; the process is reversible. In relatively dry clays, the particles are held together by relict water under tension from capillary forces. On wetting, the capillary force is relaxed and the clay expands. In other words intercrystalline swelling takes place when the uptake of moisture is restricted to the external crystal surfaces and the void spaces between the crystals.

Intracrystalline swelling, on the other hand, is characteristic of the smeitite family of clay minerals, and of montmorillonite in particular. The individual molecular layers which make up a crystal of montmorillonite are weakly bonded so that on wetting, water enters not only between the crystals but also between these unit layers which comprise the crystals. Swelling in Na montmorillonite is the most notable and can amount up to 1000% of the original volume, the clay then having formed a gel. Hydration volume changes are frequently assessed in terms of the free-swell capacity. Mielenz and King[26] showed that generally kaolinite has the smallest swelling capacity of the clay minerals and that nearly all of its swelling is of the interparticle type. Illite may swell by up to 15% but intermixed illite and montmorillonite may swell some 60–100%. Swelling in Ca montmorillonite is very much less than in the Na variety, it ranges from about 50–100%. The large swelling capacity of montmorillonite means that they give the most trouble in foundation work.

Cycles of wetting and drying are responsible for slaking in argillaceous sediments which can bring about an increase in their plasticity index and augment their ability to swell. The air pressure in the pore spaces helps the development of the swell potential under cyclic wetting and drying conditions. On wetting, the pore air pressure in a dry clay increases and it can become large enough to cause breakdown, which at times can be virtually explosive. The rate of wetting is important, slow wetting allowing the air to diffuse through the soil water so that the pressure does not become large enough to disrupt the soil. In weakly bonded clay soils, cyclic wetting and drying brings about a change in the swell potential as a result of the breakdown of the bonds between clay minerals and the alteration of the soil structure.

Freeze-thaw action and osmotic swelling also affect the swell potential. Freezing can give rise to large internal pressures at the freezing front in fine grained soils and if this front advances slowly enough the soil immediately beneath it can become quite desiccated. On melting the desiccation zone is saturated. Hence the swell potential is increased by the freeze pressures. Bjerrum[27] suggested that clay soils with high salt contents and a network of cracks undergo an increase in swell potential due to slaking consequent upon osmotic pressures being developed as rain water infiltrates into the cracks.

According to Schmertmann[28] some clays increase their swell behaviour when they undergo repeated large shear strains due to mechanical remoulding. He introduced the term swell sensitivity for the ratio of the remoulded swelling index to the undisturbed swelling index and suggested that such a phenomenon may occur in unweathered highly overconsolidated clay when the bonds, of various origins, which hold clay particles in bent positions have not been broken. When these bonds are broken by remoulding, the clays exhibit significant swell sensitivity.

An internal swelling pressure will reduce the effective stress in a clay soil and therefore will reduce its shearing strength. As a consequence Hardy[29] suggested that the Coulomb equation for swelling soils should take the swelling pressure as well as pore pressure into account, it becoming

$$\tau = c + (\sigma - u - p_s) \tan \phi'$$

where

τ is the shearing strength;

c is the cohesion;

σ is the total stress;
u is the pore water pressure;
p_s is the swelling pressure and
ϕ' is the angle of effective internal friction.

Holtz and Gibbs[30] showed that expansive clays can be recognised from their plasticity characteristics and Holtz[31] suggested a relationship between colloid content, plasticity index and shrinkage limit to indicate the degree of expansion of certain clay soils. However, the most widely used soil property to predict swell potential is the activity of a clay (see Van der Merwe[32]).

The plasticity of a cohesive soil is influenced by the amount of its clay fraction since clay minerals greatly influence the amount of attracted water held

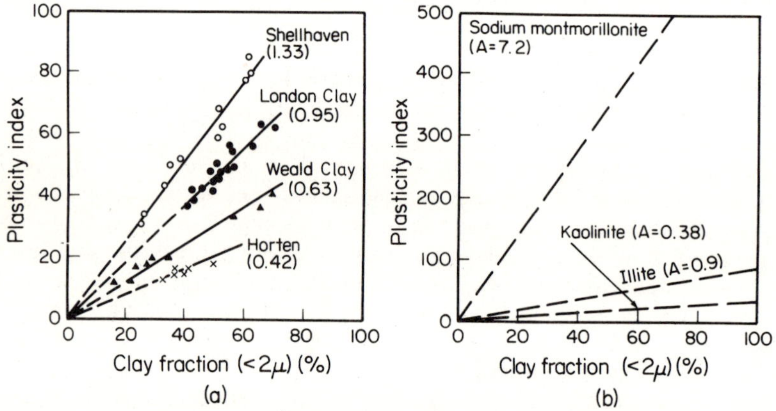

Figure 1.5 Relation between plasticity index and clay fraction. Figures in brackets represent the activities of the clays. (After Skempton[33])

in a soil. With this in mind Skempton[33] defined the activity of a clay deposit as the ratio of the plasticity index to the percentage of the clay fraction:

$$\text{Activity} = \frac{\text{Plasticity Index}}{\text{\% by weight finer than 0.002 mm}}$$

Skempton suggested three classes of activity, namely, active, normal and inactive which he further subdivided into five groups as follows:

1. Inactive with activity less than 0.5
2. Inactive with activity range 0.5 to 0.75
3. Normal with activity range 0.75 to 1.25
4. Active with activity range 1.25 to 2
5. Active with activity greater than 2.

The activity of many British soils varies between 0.75 and 1.25, the exceptions being lacustrine and estuarine clays which tend to be lower and higher respectively. It would appear that there is only a general correlation between the clay mineral composition of a deposit and its activity, that is, kaolinitic and illitic clays are usually inactive whilst montmorillonitic clays range from inactive to active (Figure 1.5). Usually active clays have a relatively high water holding capacity and a high cation exchange capacity. They are also highly thixotropic, have a low permeability and a low resistance to shear.

HEAVE POTENTIAL

Volume changes in clays also occur as a result of loading and unloading which bring about consolidation and heave respectively. When clay is first deposited in water, its water content is very high and it may have void ratios exceeding 2. As sedimentation continues overburden pressure increases and according to Hedberg[34] the particles are rearranged to produce a new packing mode of greater stability as most of the water is expelled from the deposit. He suggested that the expulsion of free and adsorbed water took place until a porosity of about 30% was reached. Any further reduction in pore volume he attributed to mechanical deformation of particles or diagenetic processes.

As material is removed from a deposit by erosion the effective overburden stress is reduced and elastic rebound begins. Part of the rebound or heave results from an increase in the water content of the clay. Cyclic deposition and erosion has resulted in multiple loading and unloading of many clay deposits.

The heave potential arising from stress release depends upon the nature of the diagenetic bonds within the soil, that is, the post-depositional changes such as precipitation of cement and recrystallization which have occurred (see Bjerrum[27]). It would appear that significant time dependent vertical swelling may arise, at least in part, from either of two fundamentally different sources, namely, localised shear stress failures or localised tensile stress failures. Localised shear stress failures are associated with long term deformations of soils having weak diagenetic bonds. Localised tensile stress failures are probably associated with long term deformations of soils having well developed diagenetic bonds such as clay shales.

When an excavation is made in a clay with weak diagenetic bonds, elastic rebound will cause immediate dissipation of some stored strain energy in the soil. However, part of the strain energy will be retained due to the restriction on lateral straining in the plane parallel to the ground surface. The lateral effective stresses will either remain constant or decrease as a result of plastic deformation of the clay as time passes (see Bjerrum[35]). These plastic deformations can result in significant time dependent vertical heaving. However, creep of weakly bonded soils is not a common cause of heaving in excavations.

The relationships between the stresses, failure mechanisms and time dependent heaving are complex in clay soils with well developed diagenetic bonds. According to Obermeier[36] heaving is in part related to crack development, cracking giving rise to an increase in volume, the rate of crack growth being of particular significance. The initial rate of heaving is probably controlled by cracking due to tensile failures and to plastic deformations arising from shear failures. Furthermore Obermeier maintained that because of the breakdown of diagenetic bonds there is an increase in the lateral stresses parallel to the ground surface.

When a load is applied to a clay soil its volume is reduced, this being due principally to a reduction in the void ratio. If such a soil is saturated then the load is initially carried by the pore water which causes a pressure, termed the hydrostatic excess pressure, to develop. The excess pressure of the pore water is dissipated at a rate which depends upon the permeability of the soil mass and the load is eventually transferred to the soil structure. The change in volume during consolidation is equal to the volume of the pore water expelled and corresponds to the change in void ratio of the soil. In clay soils, because of their low permeability, the rate of consolidation is slow. Consolidation brought about by a reduction in the void ratio is termed primary consolidation. Further consolidation may occur due to a rearrangement of the soil particles. This is much less significant and is referred to as secondary consolidation. The various factors which influence the compressibility of a clay soil have been reviewed by Wahls[37].

SETTLEMENT AND BEARING CAPACITY OF CLAY SOILS

For all types of foundation structures on clays the factors of safety must be adequate against bearing capacity failures. Experience has indicated that it is desirable to use a factor of safety of 3, yet although this will eliminate complete failure, settlement may still be excessive. It is, therefore, necessary to give consideration to the settlement problem if bearing capacity is to be viewed correctly (see chapter 11). More particularly it is important to make a reliable estimate of the amount of differential settlement that may be experienced by the structure. If estimated differential settlement is excessive it may be necessary to change the layout or type of foundation structure.

During the construction period the net settlement is comprised of immediate settlement due to deformation of the clay without a change in water content and consolidation settlement brought about by pore water being squeezed from the clay (see Chang, Broms and Peck[38]). As mentioned above the rate of consolidation is generally very slow because of the low permeability of clays so that the former type of settlement will have been the greater of the two by the end of the construction period. In the course of time consolidation becomes important, giving rise to long continued settlement, although at a decreasing rate for years or decades after the completion of construction. Accordingly the principal objects of a settlement analysis are, firstly, to obtain a reasonable estimate of the net final settlement corresponding to a time when consolidation is virtually complete and, secondly, to estimate the progress of settlement with time. It should be borne in mind that settlement depends primarily on the compressibility of the clay which is, in turn, intimately related to its geological history, that is, to whether it is normally consolidated or overconsolidated (see below). Clays which have undergone volume increase due to swelling or heave are liable to suffer significantly increased gross settlement when they are subsequently built upon.

A normally consolidated clay is that which, at no time in its geological history, has been subject to pressures greater than its existing overburden pressure, whereas an overconsolidated clay has been so subjected. The major factor in overconsolidation is removal of material that once existed above a clay deposit by erosion. Berre and Bjerrum[39] carried out a series of triaxial and shear tests on normally consolidated clay, the confining conditions simulating the overburden pressures in the field. They demonstrated that the clay could sustain a shear stress in addition to the *in situ* value, undergoing relatively small deformation as long as the shear did not exceed a given critical value. This critical shear value represents the maximum shear stress which can be mobilised under undrained conditions, and governs the bearing pressure such a clay can carry with limited amount of settlement.

Generally this critical shear value varies with plasticity and the rate at which load is applied. An overconsolidated clay is considerably stronger at a given pressure than a normally consolidated one, and it tends to dilate during shear whereas a normally consolidated clay consolidates. In both normally consolidated and overconsolidated clays the shear strength reaches a peak value and then, as displacements increase, decreases to the residual strength. The development of residual strength is therefore a continuous process.

The ultimate bearing capacity of foundations on clay soil depends on the shear strength of the soil and the shape and depth at which the foundation structure is placed (see Skempton[40]). Although there is a small decrease in the moisture content of a clay beneath a foundation structure, which gives rise to a small increase in soil strength, this is of no importance as far as estimation of the factor of safety against shear is concerned. In relation to applied stress saturated clays behave as purely cohesive materials provided that no change of moisture content occurs. Thus when a load is applied to saturated clay it produces excess pore pressures which are not quickly dissipated. In other words the angle of shearing resistance (ϕ) is equal to zero. The assumption that $\phi = 0$ forms the basis of all normal calculations of ultimate bearing capacity in clays (see Skempton[40]). The strength may then be taken as the undrained shear strength or one half the unconfined compressive strength. To the extent that consolidation does occur, the results of analyses based on the premise that $\phi = 0$ are on the safe side. Only in special cases, with prolonged loading periods or with very silty clays, is the assumption sufficiently far from the truth to justify a more elaborate analysis.

Lambe's[41] concept of the shear strength in clay postulated the existence of forces of attraction and repulsion between the particles with a net repulsive force in accordance with physico-chemical principles. Hence the equilibrium of internal stresses in a clay soil can be expressed as:

$$\sigma' = \sigma - u = R - A$$

where
- σ' is the effective stress;
- σ is the total stress;
- u is the pore water pressure;
- R is the total force of repulsion and
- A is the total force of attraction per unit area between the particles.

Unfortunately the R and A forces cannot be measured. The relationship is demonstrated by the fact that clay behaviour differs when immersed in solutions with different electrolyte concentrations.

SENSITIVITY OF CLAYS

The shear strength of an undisturbed clay is generally found to be greater than that obtained when it is remoulded and tested under the same conditions and at the same water content. The ratio of the undisturbed to the remoulded strength at the same moisture content was defined by Terzaghi[42] as the sensitivity of a clay. Subsequently Skempton and Northey[43] proposed the following grades of sensitivity:

1. Insensitive clays, under 1
2. Low sensitive clays, 1 to 2
3. Medium sensitive clays, 2 to 4
4. Sensitive clays, 4 to 8
5. Extra-sensitive clays, 8 to 16
6. Quick clay, over 16

Clays with high sensitivity values have little or no strength after being disturbed. Indeed if they suffer slight disturbance this may cause an initially fairly strong material to behave as a viscous fluid. High sensitivity seems to result from the metastable arrangement of equidimensional particles. The strength of the undisturbed clay is chiefly due to the strength of the framework developed by

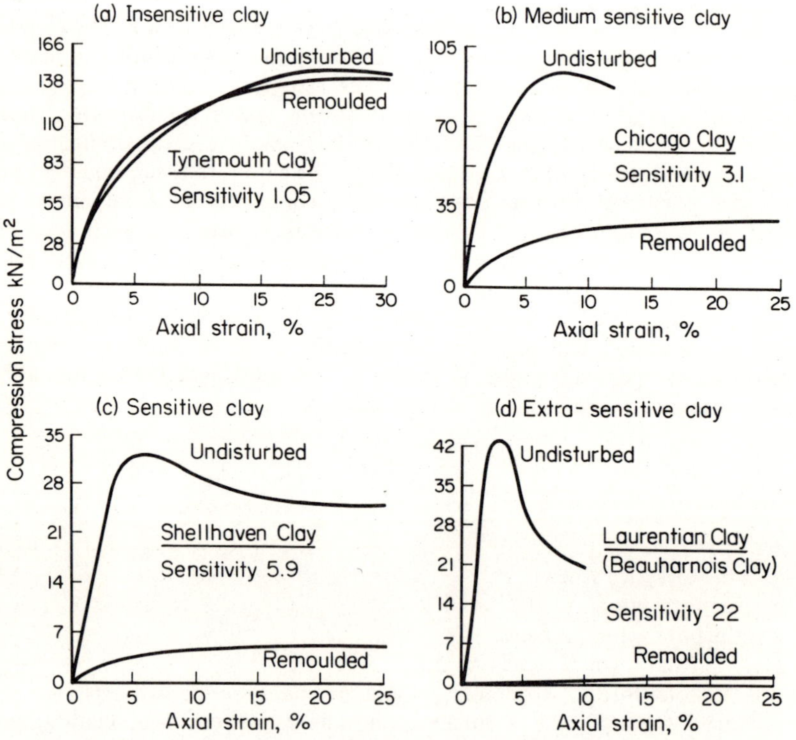

Figure 1.6 Stress-strain curves of different sensitivities (After Skempton, A.W., Soil Mechanics in Relation to Geology, *Proc. Yorks. Geol. Soc.* 29 (1953))

these particles and the bonds between their points of contact. If the framework is destroyed by remoulding, the clay loses most of its strength and any subsequent gain in strength due to thixotropic hardening does not exceed a small fraction of its original value. Sensitive clays generally possess high moisture contents, frequently with liquidity indices well in excess of unity. A sharp increase in moisture content may cause a great increase in sensitivity, sometimes with disastrous results. Heavily overconsolidated clays are insensitive. The effect of remoulding on clays of various sensitivities is illustrated in Figure 1.6.

FISSURES IN CLAY

Many clays are weakened by a network of fissures. Terzaghi[44] provided the first quantitative data relating to the influence of fissures and joints on the strength of clays, pointing out that such features are characteristic of overconsolidated clays. He[45] maintained that fissures in normally consolidated clays have no significant practical consequences and that they appeared to be due to a process comparable with syneresis. On the other hand they can have a decisive influence on the engineering performance of an overconsolidated clay, in that the overall strength of such fissured clay can be as low as one tenth that of the intact clay.

Skempton[46] attributed the reduction in strength in the London Clay exposed in cuttings to softening along fissures which opened as a result of small movements consequent upon the removal of lateral support on excavation. In addition to allowing clay to soften, fissures and joints allow concentrations of shear stress, which locally exceed the peak strength of the clay, thereby giving rise to progressive failure. Under stress, the fissures in clay seem to propagate and coalesce in a complex manner. In most instances slope failures in cuttings in stiff fissured clay take place several years or decades after excavation but where fissures are closely spaced failure can occur during excavation. In fact it is not unusual for slopes in fissured clay to fail even though they were designed on strengths lower than the undrained strength parameters for intact clay. It appears that the average shearing resistance of stiff fissured clay at the moment of sliding usually ranges between 15 and 30 kN/m^2 whereas the initial shearing resistance of such clays ranges between 100 and 300 kN/m^2. Slides in such clay may occur with little or no warning. Perhaps the best way to take care of this is to apply a higher factor of safety to undrained stability analyses. With emphasis on the probable area of fissures along a potential failure plane it should be possible to establish the total resistance to sliding offered by a fissured material by distinguishing those fissures likely to be open and offering no resistance from those closed and offering residual strength.

Skempton, Schuster and Petley[47] noted that the joints in London Clay were planar and that their orientation was predominantly normal to the horizontal bedding. They range up to 2.6 m in height and up to 6.0 m long. On the other hand the fissures may be curved or planar, or rarely more than 150 mm in size and exhibit no preferred orientation although they tend to be inclined at low angles to bedding planes. The mean size of the fissures decreases and the number per unit volume correspondingly increases as the upper surface of the clay is approached, suggesting that stress release and weathering play an important role in fissure formation. Similarly Ward, Marsland and Samuels[48] found that at Ashford the more weathered brown London Clay was more fissured than the blue clay below.

However, this concept has been disputed by some workers who have assumed that fissures are original features which are opened up on exposure. Fookes and Denness[49] suggested that the intensity of fissuring was influenced by near surface desiccation cracks and that the bedding planes appeared to be the major factor governing the fissure patterns which developed. They claimed that vertical stress release seemed to have little influence except when almost parallel to the bedding direction. But when this does happen they supposed that dilation occurs within the upper metre or so of the clay mass due to elastic expansion of the clay material and separation of individual lithological units. Fookes and Denness maintained that fissuring sympathetic to the bedding planes might have developed as a result of differential settlement between neighbouring particles shortly after deposition, leading to their separation with the formation of discontinuities approximately normal to the bedding. Minor changes in lithology do not appear to influence the attitude of fissuring but affect the size of fissures and the intensity of fissuring. Generally with increasing time after exposure, the proportion of non-planar to planar fissures increases since the former can develop from the latter by extension from their extremities.

Skempton, Schuster and Petley[47] summarised the shear strength parameters of the London Clay in terms of effective stress as follows:

Peak strength of intact clay

$$\sigma' = 31\ \text{kN/m}^2, \phi' = 20°$$

'Peak' strength on fissure and joint surfaces

$$\sigma' = 6.9\ \text{kN/m}^2, \phi' = 18.5°$$

Residual strength

$$\sigma' = 1.4\ \text{kN/m}^2, \phi'_r = 16°$$

Thus the strength along joints or fissures in clay is only slightly higher than its residual strength.

Similar results previously had been obtained by Marsland and Butler[50] from the stiff fissured Barton Clay. Hence the upper limit of the strength of fissured clay is represented by its intact strength whilst the lower limit is the strength along the fissures. The operational strength which is somewhere between the two is, however, often significantly higher than the fissure strength (see Lo[51]).

TILLS AND OTHER GLACIAL DEPOSITS

Till is regarded as being synonymous with boulder clay. It is deposited directly by ice whilst stratified drift or tillite is deposited in melt waters associated with glaciers. An extensive review of the various types of glacial deposits and their engineering properties has been provided by Fookes *et al.*[52]

The character of a till deposit depends on the lithology of the material from which it was derived, on the position in which it was transported in the glacier, and on the mode of deposition (see Boulton[53]; and Boulton and Paul[54]). The underlying bedrock usually constitutes up to about 80% of basal tills, depending on its resistance to abrasion and plucking. Argillaceous rocks such as shales, and mudstones are more easily abraded and produce fine grained tills which are presumably richer in clay minerals and therefore more plastic than other tills. Mineral composition also influences the natural moisture content which is slightly higher in tills containing appreciable quantities of clay minerals or mica. Upper tills have a high proportion of far-travelled material and may not contain any of the local bedrock.

Deposits of till consist of a variable assortment of rock debris ranging from fine rock flour to boulders. On the one hand they may consist essentially of sand and gravel with very little binder, alternatively they may have an excess of clay. Lenses and pockets of sand, gravel and highly plastic slickensided clay are frequently encountered in some tills. Most tills contain a significant amount of quartz in their silt-clay fractions. Frequently the larger, elongated fragments in till possess a general orientation in the path of ice movement.

The shape of the rock fragments found in till varies but is largely conditioned by the initial shape of the fragment at the moment of incorporation into the ice. Angular boulders are common, their irregular sharp edges resulting from crushing. Crushing or grinding of a rock fragment occurs when it comes in contact with another fragment or the rock floor.

Distinction has been made between tills derived from rock debris which was carried along at the base of a glacier and those deposits which were transported within and on the ice. The former is referred to as lodgement till whereas the latter is known as ablation till. Lodgement till is thought to be plastered onto the ground beneath the moving glacier in small increments as the basal ice melts. Because of the overlying weight of ice such deposits are overconsolidated. Ablation till accumulates on the surface of the ice when englacial debris melts out, and as the glacier decays the ablation till is slowly lowered to the ground. It is therefore normally consolidated. Lodgement till contains fewer, smaller stones (they generally possess a preferred orientation) than ablation till and they are rounded and striated. Due to abrasion and grinding, the proportion of silt and clay size material is relatively high in lodgement till (e.g. the clay fraction varies from 15 to 40%). Lodgement till is commonly very compact and fissile, and is practically impermeable. It oxidises very slowly so that it is usually grey. Because it has not been subjected to much abrasion, ablation till is characterised by abundant large stones that are angular and not striated, the proportion of sand and gravel is high and clay is present only in small amounts (usually less than 10%). Because the texture is loose, ablation till oxidises rapidly and commonly is brown or yellowish brown. Since ablation till consists of the load carried at the time of ablation it usually forms a thinner deposit than lodgement till. Accordingly till sheets may comprise one or more layers of different materials, not all of which are likely to be found at any one locality. Shrinking and reconstitution of an ice sheet can complicate the sequence. In fact Elson[55] went further and divided tills into four categories, namely, superglacial ablation till, subglacial ablation till, comminution till and deformation till. Only superglacial ablation till corresponds with the definition of ablation till given above. It generally consists of thin lenses of sand and gravel of irregular distribution.

Englacial debris occurs mainly in the lower 30 to 60 m of a glacier where rock detritus may comprise as much as 10 or 20% of its volume. Consequently an appreciable thickness of englacial drift can be melted out from the base of a glacier although this suffers a reduction in volume of anything up to 90%. This subglacial ablation till may be precompressed by the overlying ice and may be sliced by thrust planes, dipping upstream near the glacier margin. Elongate stones may possess preferred orientations. The average grain size is much smaller than that of superglacial till. At the base of subglacial till deposits striated boulder pavements or thin irregular lenses of sand and pebbles may occur as relicts of the last erosive movement of the ice.

Comminution till occurs beneath subglacial ablation till and is formed by the shearing action of the ice at the base of the glacier. Elson suggested that this action generates enough heat to produce a sufficient quantity of melt water to bring about compaction to maximum density. Particles are oriented in the position of least resistance by the shearing action of moving ice and pebbles show a preferred orientation. Silt-sized tabular particles lie approximately horizontal or parallel to surfaces of larger particles, giving rise to microfoliation. Stones are surrounded by a compact matrix containing a high concentration of fines formed by abrasion of their surfaces.

Deformation or soft till may form from sandstone, siltstone, soft shale or weathered mantle. These materials cannot form dense comminution tills because the surfaces of the particles are not created by the glacier. These materials are porous and some initially contain more water than required for compaction to maximum density. Hence they yield soft tills which tend to be deformed rather than crushed.

STRUCTURE OF TILL

Till may possess several structures that affect its strength and permeability. These include microfoliation and preferred orientation of the long axes of the larger rock fragments. The random occurrence in a till of boulders has been referred to as *raisin cake* structure, however, more frequently fragments are crudely aligned in rows rather than in a haphazard manner. Large irregular masses of sand and gravel occur in unpredictable profusion, some of these are interconnected, as a result of meltwater action, but many are isolated. Small distorted pockets of sand and silt have been termed *flame* structures.

Tills are often fissile, tending to split into irregular lenticular flakes, ranging from less than a millimetre thick in clay tills up to several tens of millimetres in sandy tills. Some tills are separated into beds by compact layers of sand a metre or two thick and silt several millimetres thick. Joints commonly extend obliquely through massive, clay tills; some may be caused by shearing and others by desiccation after deposition. Some joints contain thin layers of compact sand.

GRAIN SIZE DISTRIBUTION

After an investigation of the lithology of tills Krumbein[56] wrote that they possess irregular grain size distributions, being polymodal in aspect. He also concluded that tills were perhaps not so heterogeneous as had previously been supposed and that an individual ice sheet tended to produce till with a fairly well defined frequency distribution. This view was subsequently confirmed by Chryssafopoulos[57].

The grain size distribution has been used to classify tills (Figures 1.7a and b). According to McGown[58] tills are frequently gap graded, the gap generally occurring in the sand fraction. He also noted that large, often very local, variations can occur in the gradings of till which reflect local variations in the formation processes, particularly the comminution processes. The clast size consists principally of rock fragments and composite grains, and presumably was formed by frost action and crushing by ice. Single grains predominate in the matrix. The range in the proportions of coarse and fine fractions in tills dictates the degree to which the properties of the fine fraction influence the properties of the composite soil.

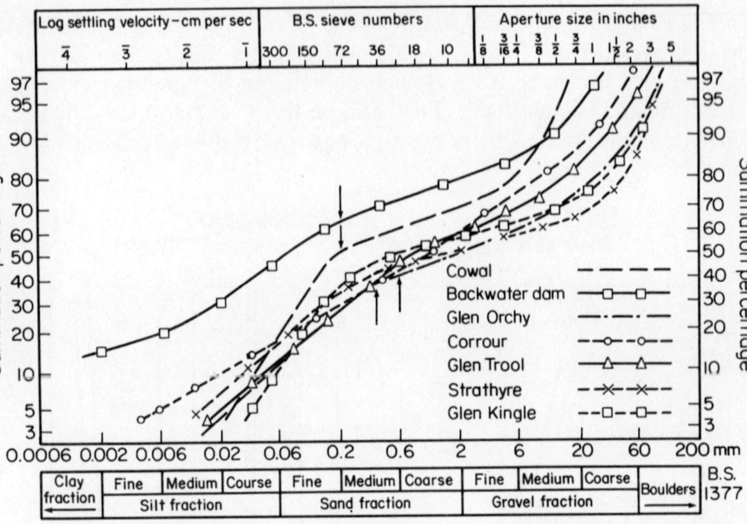

Figure 1.7(a) Typical gradings of some Scottish morainic soils (After McGown[58])

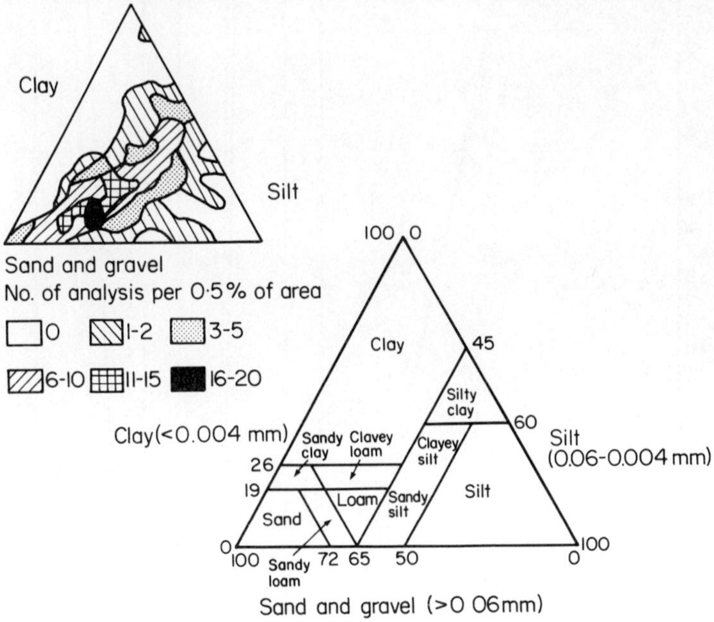

Figure 1.7(b) Grain size distribution for about 500 tills (After Elson[55])

The variation in the engineering properties of the fine soil fraction is greater than that of the coarse fraction, and this often tends to dominate the engineering behaviour of the till.

One of the earliest engineering studies of glacial material was made by Legget[59]. The site, near Fergus, Ontario, consisted of fluvio-glacial sands and gravels, and till deposits. Legget found that the specific gravity of these latter deposits was remarkably uniform, varying from 2.77 to 2.78. These values suggest the presence of fresh minerals in the fine fraction, that is, rock flour rather than clay minerals. Rock flour behaves more like granular material than cohesive and has a low plasticity. Furthermore the investigation revealed that the angle of internal friction of these deposits was always above 30° with a low value of cohesion.

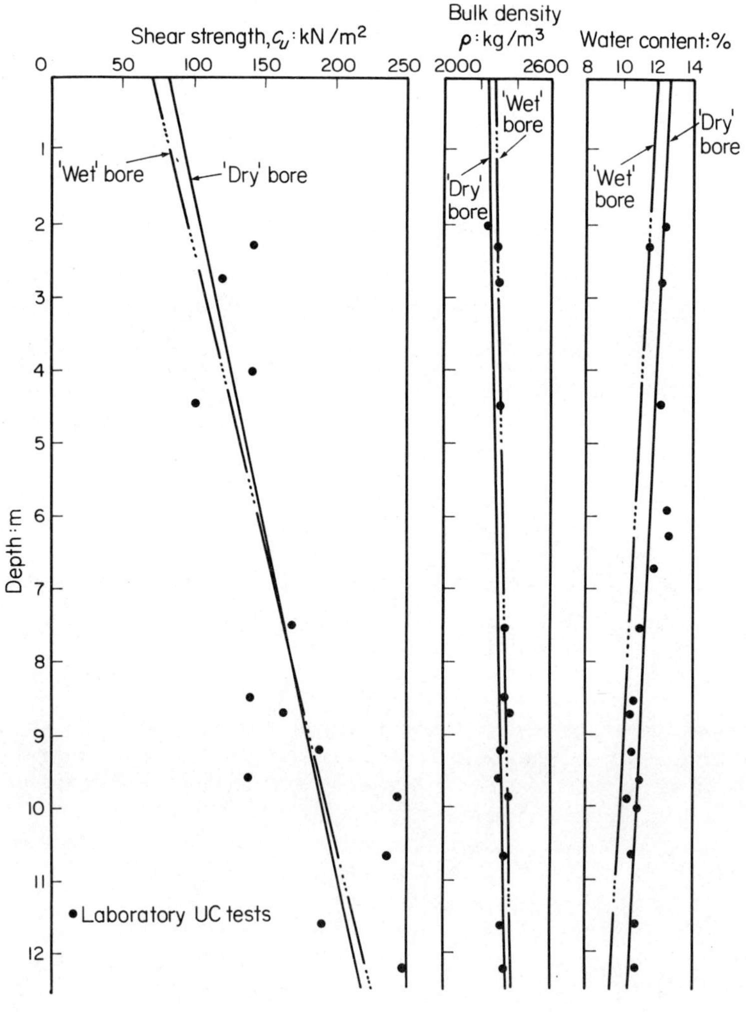

Figure 1.8 'Same day' tests on unconfined compression specimens (samples from 'dry' boreholes) (After McKinley, Tomlinson and Anderson[61])

The consistency limits of tills are dependent upon water content, grain size distribution and the properties of the fine grained fraction. Generally, however, their plasticity index is small and Bernell[60] has shown that the liquid limit of tills decreases with increasing grain size. He added that because of the difficulty of obtaining the plastic limits of Swedish tills by standard methods it had been found more useful to use the liquid and shrinkage limits in classification. The range of water content between these two limits was referred to as the LS difference and Bernell noted that the liquid limit increases linearly with the LS difference. However, McGown[58] dismissed the idea of using the LS difference for classification of tills. McKinley, Tomlinson and Anderson[61] in an investigation of lodgement till showed that plasticity tests on the matrix material placed this till in the CL group, just above the A line on the plasticity chart (e.g. plastic limit = 16%, liquid limit = 28%); the till appeared to be fully saturated. The

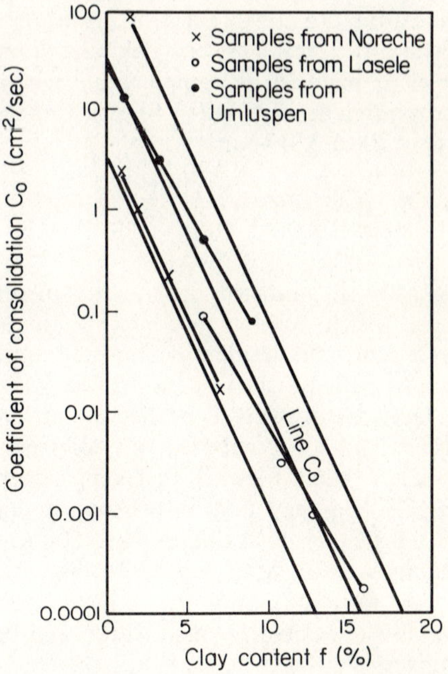

Figure 1.9 Relationship between clay content of till and the coefficient of consolidation (After Bernell[60])

values of moisture content, bulk density and shear strength derived from unconfined compression tests for this till are given in Figure 1.8. This shows that the values of moisture content and bulk density are relatively constant and that the shear strength generally increases with depth.

COMPRESSABILITY AND ELASTICITY

The compressibility and consolidation of tills are principally determined by the clay content. For example, the value of compressibility index tends to increase linearly with increasing clay content whilst for moraines of very low clay content, less than 2%, this index remains about constant (C_c = 0.01). The relationship between the coefficient of consolidation and clay content is given in Figure 1.9.

Klohn[62] noted that dense, heavily overconsolidated till is relatively incompressible and that when loaded undergoes very little settlement, most of which is elastic. For the average structure such elastic compressions are too small to consider and can therefore be ignored. However, for certain structures they are critical and their magnitude must be estimated prior to construction. Klohn carried out a series of plate loading tests which indicated that the modulus of elasticity of the till deposit concerned was very high, being of the order 1500 MN/m^2. Observations subsequently, taken on all major structures, indicated that settlement occurred almost instantaneously on application of load.

In another survey of dense till, Radhakrishna and Klym[63] found the undrained shear strength, as obtained by pressuremeter and plate loading tests, to average around 1.6 MN/m^2, while the values from triaxial tests ranged between 0.75 and 1.3 MN/m^2. The average values of the initial modulus of deformation were around 215 MN/m^2 which was approximately twice the laboratory value. These differences between field and laboratory results were attributed to stress relief of material on sampling and sampling disturbance. Much lower values of shear strength were found for the Cromer Till by Kazi and Knill[64]; these ranged from 170 to 220 kN/m^2.

FLUVIO-GLACIAL DEPOSITS

Deposits of stratified drift are often subdivided into two categories, namely, those which develop in contact with the ice – the ice contact deposits; and those which accumulate beyond the limits of the ice, forming in streams, lakes or seas – the proglacial deposits (see Flint[65]).

Outwash fans are deposited by streams which emerge from the snout of a glacier and are composed of sediments ranging in size from coarse sands to boulders. When they are first deposited their porosity may be anything from 25 to 50% and they tend to be very pervious. The finer silt-clay fraction is transported further downstream. Several outwash fans may initially extend from the terminus of a glacier but often they gradually merge to form one deposit. On retreat they may bury the terminal moraine. Kames, kame terraces and eskers are also deposited by melt waters and usually consist of sands and gravels (see Fookes *et al.*[52]).

The most familiar proglacial deposits are varved clays. These deposits accumulated in proglacial lakes and are generally characterised by alternating laminae of finer and coarser grain size, each conplet being termed a varve. The thickness of the individual varve is frequently less than 2 mm although much thicker varves have been noted in a few deposits. Generally the coarser layer is of silt size and the finer of clay size. Usually very finely comminuted quartz, feldspar and mica form the major part of varved clays rather than clay minerals. For example, the clay mineral content may be as low as 10%, although instances where it has been as high as 70% have been recorded. Montmorillonite clay has also been found in varved clays. Varved clays tend to be normally consolidated or lightly overconsolidated, although it is usually difficult to make the distinction. In many cases the precompression may have been due to ice loading.

The two normally discrete layers formed during the deposition of the varve present an unusual problem in that it may invalidate the normal soil mechanics analyses based on homogeneous soils from being used. As far as the Atteberg limits are concerned, assessment of the liquid and plastic limits of a bulk sample may not yield a representative result. However, Metcalf and Townsend[66] suggested that the maximum possible liquid limit obtained for any particular varved deposit must be that of the clayey portion, whereas the minimum value must be that of the silty portion. Hence they assumed that the maximum and minimum values recorded for any one deposit approximate to the properties of the individual layers. The range of liquid limits for varved clays tends to vary between 30 and 80 whilst that of plastic limit often varies between 15 and 30. These limits, obtained from varved clays in Ontario, allow the material to be classified as inorganic silty clay of medium to high plasticity or compressibility.

In some varved clays in Ontario the natural moisture content would appear to be near the liquid limit. They are consequently soft and have sensitivities generally of the order of 4. Since triaxial and unconfined compression tests tend to give very low strains at failure, around 3%, Metcalf and Townsend presumed that this indicates a structural effect in the varved clays. The average strength reported was about 5.9 MN/m^2, with a range of 3.4 to 7 MN/m^2. The effective stress parameters of apparent cohesion and angle of shearing resistance range from 0.7 to 2.8 MN/m^2, and 22° to 25° respectively.

QUICK CLAYS

Quick clays are composed of material which is predominantly smaller than 0.002 mm but many deposits seem very poor in actual clay minerals, containing a high proportion of ground down, fine quartz. For example, it has been shown that quick clay from St. Jean Vienney consists of very fine quartz and plagioclase. Indeed examination of quick clays with the scanning electron microscope has revealed that they do not possess clay based structures, although such work has not lent unequivocal support to the view that non-clay particles govern the physical properties.

Quick clays generally exhibit little plasticity, their plasticity index generally varying between 8 and 12. The most extraordinary property possessed by quick clays is their very high sensitivity. In other words a large proportion of their undisturbed strength is permanently lost following shear. The small fraction of the original strength gained after remoulding may be attributable to the development of some differnt form of interparticle bonding. The reason why only a small fraction of the original strength can ever be recovered is because the rate at which it develops is so slow. As an example the Leda Clay is characterised by exceptionally high sensitivity, commonly between 20 and 50, and a high natural moisture content and void ratio, the latter is commonly about 2. It has a low permeability in the range 10^{-10} m/sec. The plastic limit is around 25%, with a liquid limit about 60%, and undrained shear strength of 700 kN/m^2. When subjected to sustained load, an undrained triaxial specimen of Leda Clay exhibits a steady time dependent increase in both pore pressure and axial strain. Continuing undrained creep may often result in a collapse of the sample after long periods of time have elapsed (see Walker[67]).

Quick clays can liquefy on sudden shock. This has been explained by the fact that if quartz particles are small enough, having a very low settling velocity, and if the soil has a high water content, then the solid-liquid transition can be achieved.

The peculiar behaviour and properties of Norwegian quick clays were first explained by Rosenqvist[68,69] who suggested that they were developed by leaching of clay minerals by fresh water after initial deposition in a marine environment. He proposed that the decrease in salt content of the pore water due to this leaching and diffusion caused the original flocculated structure to become metastable since the change in electrolyte concentration would give rise to particle repulsion rather than attraction. Originally the strength of the soil fabric was dependent on a contribution from the ions in the pore water but when they are leached out by fresh water the strength of the soil depends mainly on the integrity of the fabric. However, Pusch and Arnold[70] disputed the Rosenqvist theory, they having attempted to produce quick behaviour in a specially prepared soil composed largely of illite, following the stages outlined by Rosenqvist, and failed.

The fact that it appears that clay minerals are not quantitatively important in quick clays has led to the development of other theories to explain their peculiar properties. For example, Cabrera and Smalley[71] suggested that these deposits owe their distinctive properties to the predominance of short range interparticle bonding forces which they maintained were characteristic of deposits in which there was an abundance of glacially produced, fine non-clay minerals. In other words they contended that the ice sheets supplied abundant ground quartz in the form of rock flour for the formation of quick clays. Certainly quick clays have a restricted geographical distribution, occurring in certain parts of the northern hemisphere which were subjected to glaciation during Pleistocene times.

ORGANIC SOILS: PEAT

Peat is an accumulation of partially decomposed and disintegrated plant remains which have been fossilized under conditions of incomplete aeration and high water content. Physico-chemical and biochemical processes cause this organic material to remain in a state of preservation over a long period of time.

All present day surface deposits of peat have accumulated since the last ice age and therefore have formed during the last 20 000 years. On the other hand, some buried peats may have been developed during inter-glacial periods. Peats have also accumulated in post-glacial lakes and marshes where they are interbedded with silts and muds. Similarly they may be associated with salt marshes. Fen deposits are thought to have developed in relation to the eustatic changes in sea level which occurred after the retreat of the last ice sheets from Britain. The most notable fen deposits in the UK are found south of the Wash. Similar deposits are also found in Suffolk and Somerset. These are areas where expanses of peat interdigitate with wedges of estuarine silt and clay. However, the most familiar type of peat deposit in the UK is probably the blanket bog. These deposits are found on the cool, wet uplands.

Macroscopically peaty material can be divided into three basic groups, namely, amorphous granular, coarse fibrous and fine fibrous peat (see Radforth[72]). The amorphous granular peats have a high colloidal fraction, holding most of their water in an adsorbed rather than a free state, the adsorption occurring around the grain structure. In the other two types the peat is composed of fibres, these usually being woody. In the coarse variety a mesh of second order size exists within the interstices of the first order network whilst in fine fibrous peat the interstices are very small and contain colloidal matter. Most of the water, which these two types of peat contain, is free water. Generally peat deposits are acidic in character, the pH values often varying between 5.5 and 6.5, although some fen peats are neutral or even alkaline.

The ash percentage of peat consists of the mineral residue remaining after its ignition, which is expressed as a fraction of the total dry weight. Ash contents may be as low as 2% in some highly organic peats, or may be as high as 50% as in some peats found on the Yorkshire moors (see Bell[73]). The mineral material is usually quartz sand and silt. In many peat deposits the mineral content increases with depth.

The void ratio of peat ranges between 9, for dense amorphous granular peat, up to 25, for fibrous types with a high content of sphagnum. It usually tends to decrease with depth within a peat deposit. Such high void ratios give rise to phenomenally high water contents. The latter is the most distinctive characteristic of peat. Indeed most of the differences in the physical characteristics of peat are attributable to the amount of moisture present. This varies according to the type of peat, it may be as low as 500% in some amorphous granular varieties whilst, by contrast, percentages exceeding 3000 have been recorded from coarse fibrous varieties. The volumetric shrinkage of peat increases up to a maximum and then remains constant, the volume being reduced almost to the point of complete dehydration. The amount of shrinkage which can occur ranges between 10 and 75% of the original volume of the peat and it can involve reductions in void ratio from over 12 down to about 2.

As would be expected, amorphous granular peat has a higher bulk density than the fibrous types. For instance, in the former it can range up to 1.2 Mg/m^3, whilst in woody fibrous peats it may be half this figure. However, the dry density is a more important engineering property of peat, influencing its behaviour under load. Hanrahan[74] recorded dry densities of drained peat within the range 65 to 120 kg/m^3. The dry density is influenced by the mineral content and higher values than those quoted can be obtained when peats possess high mineral residues. The specific gravity of peat has been found to range from as low as 1.1 up to about 1.8, again being influenced by the content of mineral matter.

Because of the variability of peat in the field the value of its permeability as tested in the laboratory can be misleading. Nevertheless Hanrahan[75] showed that the permeability of peat, as determined during consolidation testing, varied according to the loading and length of time involved as follows:

1. Before test: void ratio = 12; permeability = 4×10^{-6} m/sec.
2. After seven months loading at 55 kN/m^2: void ratio = 4.5; permeability = 8×10^{-11} m/sec.

Thus after seven months of loading, the permeability of the peat was 50 000 times less than it was originally. Miyahawa[76] and Adams[77] have also shown that there is a marked change in the permeability of peat as its volume is reduced under compression. The magnitude of construction pore water pressure is particularly significant in determining the stability of peat. Adams showed that the development of pore pressures in peat beneath embankments was appreciable, in one instance it approached the vertical unit weight of the embankment.

When loaded, peat deposits undergo high deformations but their modulus of deformation tends to increase with increasing load. If peat is very fibrous it appears to suffer indefinite deformation without planes of failure developing. On the other hand failure planes nearly always form in dense amorphous granular peats. Hanrahan and Walsh[78] found that the strain characteristics of peat were independent of the rate of strain and that flow deformation, in their tests, was negligible. Strain often takes place in an erratic fashion in a fibrous peat. This may be due to the different fibres reaching their ultimate strengths at different strain values, the more brittle, woody fibres failing at low strain whilst the non-woody types maintain the overall cohesion of the mass up to much higher strains. The viscous behaviour of peat is generally recognised as being non-Newtonian and the relationship between stress and strain is a function of the void ratio. As the void ratio decreases so the effective viscosity increases and hence a certain value of stress produces a correspondingly smaller value of strain rate.

Apart from its moisture content and dry density the shear strength of a peat deposit appears to be influenced, firstly, by its degree of humification and, secondly, by its mineral content. As both these factors increase so does the shear strength. Conversely the higher the moisture content of peat the lower its shear strength. The dry density is influenced by the effective load to which a deposit of peat has been subjected. As the effective weight of 1 m^3 of drained peat is approximately 45 times that of 1 m^3 of undrained peat the reason for the negligible strength of the latter becomes apparent. Due to its extremely low submerged density, which may be between 15 and 35 kg/m^3, peat is especially prone to rotational failure or failure by spreading, particularly under the action of horizontal seepage forces.

In an undrained bog the unconfined compressive strength is negligible, the peat possessing a consistency approximating to that of a liquid. The strength is increased by drainage to values between 20 and 30 kN/m^2 and the modulus of elasticity to between 100 and 140 kN/m^2. According to Hanrahan[79] unconfined compressive strengths of up to 70 kN/m^2 are not uncommon in peats consolidated under pavements, a typical modulus of elasticity being 700 kN/m^2.

SETTLEMENT AND CONSOLIDATION

Differential and excessive settlement is the principal problem confronting the engineer working on a peaty soil. When a load is applied to peat, settlement occurs because of the low lateral resistance offered by the adjacent unloaded peat. Serious shearing stresses are induced even by moderate loads. Worse still, should the loads exceed a given minimum, then settlement may be accompanied by creep, lateral spread, or in extreme cases by rotational slip and upheaval of adjacent ground. At any given time the total settlement in peat due to loading involves settlement with and without volume change. Settlement without volume change is the more serious for it can give rise to the types of failure mentioned. What is more it does not enhance the strength of peat.

Creep does not take place in peat at a constant rate. This is probably due to the increase in density consequent upon consolidation. A good example of the long term behaviour of peat was given by Buisman[80] who cited examples of embankments on peat in the Netherlands in which continuous settlement, linear with the logarithm of time, was recorded for more than 80 years.

When peat is compressed the free pore water is expelled under excess hydrostatic pressure. Since the peat is initially quite pervious and the percentage of pore water is high, the magnitude of settlement is large and this period of initial settelment is short (a matter of days in the field). Adams[77] showed that the magnitude of initial settlement was directly related to peat thickness and applied load. The original void ratio of a peat soil also influences the rate of initial settlement. Excess pore pressure is almost entirely dissipated during this period. Settlement subsequently continues at a much slower rate which is approximately linear with the logarithm of time. This is because the permeability of the peat is significantly reduced due to the large decrease in volume. During this

Figure 1.10 Settlement on peat at Benwick, Hunts. The entrance to the house has settled by about 1 m below pavement level

period the effective consolidating pressure is transferred from the pore water to the solid peat fabric. The latter is compressible and will only sustain a certain proportion of the total effective stress, depending on the thickness of the peat mass.

Adams[81] maintained that the macro- and micro-structure of fibrous peat influences its consolidation. He considered that primary consolidation of such peats took place due to a drainage of water from the macro-structure whilst secondary consolidation was due to the extremely slow drainage of water from the micro-pores into the macro-structure. Because of its higher permeability the rate of primary consolidation of a fine fibrous peat is higher than that of an amorphous granular peat.

Due to the highly viscous water adsorbed around soil particles, amorphous granular peat exhibits a plastic structural resistance to compression and hence has a similar rheological behaviour to that of clay. In this case secondary consolidation is believed to occur as a result of the gradual readjustment of the soil structure to a more stable configuration following the breakdown which occurs during the primary phase due to dissipation of pore pressure. The rate at which this process takes place is controlled by the highly viscous adsorbed water surrounding each soil particle; the colloidal material which the former contains tending to plug the interstices and thereby reduce permeability.

Wilson *et al*[82] suggested that amorphous granular peats exhibit considerable secondary consolidation and therefore settlement. Because of the highly complex structure of such peat they also suggested that it may also exhibit phases of tertiary and quaternary consolidation.

One-dimensional consolidation theories have been developed by Berry and Poskitt[83] for both amorphous granular and fibrous peat. These consider finite strain, decreasing permeability, compressibility and the influence of secondary compression with time. The different mechanisms involved in secondary compression of these two types of peat were found to give similar non-linear rheological models but their effective creep equations were fundamentally different. That for amorphous granular peat predicts an exponential increase in strain with incremental loading whilst that for fibrous peat predicts a linear increase.

To summarise, the factors which account for the complex manner in which peat consolidates and which therefore mitigate against a precise settlement analysis based on the Terzaghi theory of consolidation, include the abnormally large decrease in permeability which accompanies loading, the decreasing coefficient of compressibility and thixotropy, as well as the surface activity of organic material. What is more it must be remembered that primary and secondary consolidation are empirical divisions of a continuous compression process, both of which occur simultaneously during part of that process.

The quantity of water removed from peat in the later stages of consolidation results in an increase in strength considerably greater than that following the removal of the same quantity during the early stages. What happens to a peat is therefore very largely a function of the structure of the material since this affects the retention and expulsion of water and affords it its strength.

With few exceptions improved drainage has no beneficial effect on the rate of consolidation. This is because efficient drainage only accelerates the completion of primary consolidation which is anyhow completed rapidly.

There is an extremely small increase in the void ratio which follows the reduction of the load on a peat deposit, in other words the voids are not restored to their original value, and the compressibility of preconsolidated peat is greatly reduced. This can be illustrated from the following figures:

Coefficient of compressibility of peat for a range of loading from 13.4 to 26.8 kN/m^2.

1. Normally loaded, $m_v = 12.214\ m^2/MN$
2. Preconsolidated, $m_v = 0.599\ m^2/MN$

FILLS AND COARSE COLLIERY DISCARDS

Because suitable building sites are becoming scarce in urban areas the construction of buildings on fill or made-up ground recently has assumed a greater importance. A wide variety of materials is used for fills including domestic refuse, ashes, slag, clinker, building waste, chemical waste, quarry waste and all types of soils. The extent to which an existing fill will be suitable as a foundation depends largely on its composition and uniformity. In the past the control exercised in placing fill has frequently been insufficient to ensure an adequate and uniform support for structures immediately after placement. Consequently a time interval had to be allowed prior to building so that the material could consolidate under its own weight. Although this may be suitable for small, lightly loaded buildings it is unsatisfactory for more heavily loaded structures which can give rise to substantial settlement.

The stability and potential settlement of foundation structures on fill are largely governed by its density. Therefore random end tipping in thick layers, giving low densities, produces an unsatisfactory condition since excessive and non-uniform settlements are likely to occur under the load of the structure erected. The thickness of fill and site conditions also affect the amount and rate of settlement, the greater the thickness the larger the likely deformation and the increased length of time over which it is likely to take place.

The time taken for a fill to reach a sufficient degree of natural consolidation so that it becomes suitable for a foundation depends on the nature and thickness of the fill, the method of placing and the nature of the underlying ground, especially the ground water conditions. The best materials in this respect are obviously well graded, hard and granular. By contrast, fills containing a large

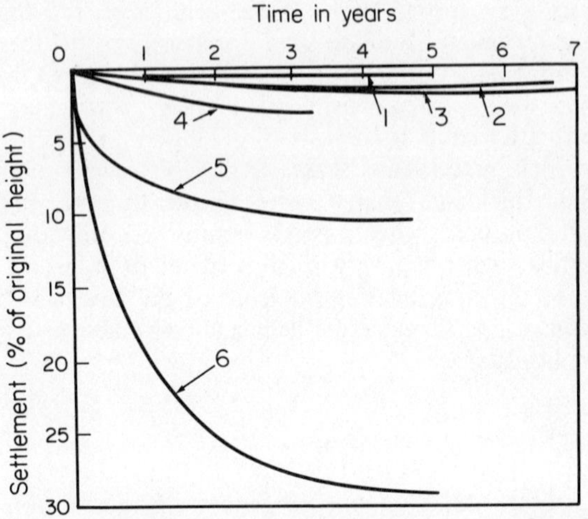

Figure 1.11 Observations of the settlement of various types of fill due to consolidation under its own weight (After Meyerhof[84]). Description of curves: 1. Well-graded sand, well compacted. 2. Rockfill, medium state of compaction. 3. Clay and chalk, lightly compacted. 4. Sand, uncompacted. 5. Clay, uncompacted. 6. Mixed refuse, well compacted

proportion of fine material may take a long while to settle. Generally rock fills will settle 2.5% of their thickness, sandy fills about 5% and cohesive material around 10%. The rate of settlement decreases with time but in some cases it may take 10 to 20 years before movements are reduced within tolerable limits for building foundations. In coarse grained soils the larger part of movement generally occurs within the first two years after the construction of the fill and after five years settlements are usually very small. Indeed Meyerhof[84] showed that most settlement in a fill occurs within the first year of placement and after two years is relatively small (Figure 1.11). Therefore two years generally may be regarded as the lower limit before buildings are constructed on fills consolidating under their own weight.

Fills placed over low lying areas of compressible or weak strata should be considered unsuitable unless tests demonstrate otherwise or the structure can be designed for low bearing capacity and irregular settlement. For example, clay soils beneath fills may undergo a prolonged period of consolidation. Frequently poorly compacted old fills continue to settle for years due to secondary consolidation. Mixed fills which contain materials liable to decay, which may leave voids or involve a risk of spontaneous combustion, afford very variable support and such sites should again, in general, be avoided.

Consequently the material of new fills should be spread uniformly in thin layers and compacted at the optimum moisture content required to produce the maximum density. Properly compacted fills on sound ground can be as good as, of better than, virgin soil (see BRS Digest No. 9[85]). According to Meyerhof[84] the support afforded foundations is generally inadequate if the density of a cohesionless fill is less than 90% of the maximum value given by the standard compaction test. For cohesive material a somewhat higher degree of compaction may be necessary (say around 95%). It has been suggested that if rockfill is compacted to greater than 85% of the solid dry density of the rock then building can commence immediately. For instance, Kilkenny[86] quoted the Expo '67 site in Montreal where a compacted density of 2.3 Mg/m^3 was achieved for the shale fill, as compared with a solid dry density of 2.7 Mg/m^3, and proved satisfactory. Well compacted rock, gravel, sand, shale and clay fills have shown settlements of only 0.5% of their thickness. Quarry wastes frequently have proved satisfactory fills for foundation purposes when properly compacted. Meyerhof indicated that hydraulic sand fills underwent very small settlements if placed above ground water level due to the consolidating influence of the downward perculation of the water. By contrast settlements are large and continue over a long time when clay fill is hydraulically placed.

BUILDING ON FILL

Settlement distribution over the loaded area of a fill may be very irregular, even under uniform loading conditions. It is necessary therefore to design a structure so that the total, as well as the differential, movements are restricted or can be withstood without damage. Hence, the structure should either be sufficiently rigid to redistribute the loads and thereby reduce relative settlement, or should be relatively flexible to accomodate them without cracks appearing in the structure. Long continuous structures should be avoided; it is best to divide them into sections. Ordinary pad and strip footings are rarely adequate. Wide reinforced strip footings may be adequate where good bearing can be obtained, otherwise a reinforced raft is necessary. If a fill is relatively thin and overlies a firm stratum then a beam and pier or beam and pile foundation may prove an alternative. Settlement observations on structures have revealed that a bearing pressure of 55 kN/m^2 is very conservative except on poorly compacted fine grained soils or industrial and domestic wastes consolidating under their own weight.

Waste disposal or sanitary land fills are usually very mixed in composition (Table 1.6) and suffer from continuing organic decomposition and physicochemical breakdown. Methane and hydrogen sulphide are often produced in the process and accumulations of these gases in pockets in fills have led to explosions. The production of leachate is another problem. Some material such as ashes and industrial wastes may contain sulphates and other products which are

Table 1.6 MUNICIPAL WASTE MATERIALS INCORPORATED IN FILLS (AFTER SOWERS, 1973[87])

Material	*Characteristics as fill*
1. Garbage: Food, waste	Wet. Ferments and decays readily. Compressible, weak
2. Paper, cloth	Dry to damp. Decays and burns. Compressible
3. Graden refuse	Damp. Ferments, decays, burns. Compressible
4. Plastic	Dry. Decay resistant, may burn. Compressible
5. Hollow metal, e.g. drums	Dry. Corrodable and crushable
6. Massive metal	Dry. Slightly corrodable. Rigid
7. Rubber, e.g. tyres	Dry. Resilient, burns, decay resistant. Compressible
8. Glass	Dry. Decay resistant. Crushable and compressible
9. Demolition timber	Dry. Decays and burns. Crushable
10. Building rubble	Damp. Decay resistant. Crushable and erodable
11. Ashes, clinker and chemical wastes	Damp. Compressible, active chemically and partially soluble

potentially injurious as far as concrete is concerned. The density of waste disposal fills varies from about 120 to 300 kg/m^3 when tipped. After compaction the density may exceed 600 kg/m^3. Moisture contents range from 10 to 50% and the average specific gravity of the solids from 1.7 to 2.5. Settlements are likely to be large and irregular.

According to Sowers[87] the mechanisms responsible for the settlement which occurs in waste disposal fills include: mechanical distortion by the bending, crushing and reorientation of materials which cause a reduction in the void ratio; ravelling, that is, the transfer of fines into the voids; physico-chemical and biochemical changes such as corrosion, combustion and fermentation; and the interactions of these various mechanisms. The initial mechanical settlement of waste disposal fills is rapid and is due to a reduction in the initial void ratio. It takes place with no build up of pore water pressure. Settlement continues due to a combination of secondary compression (material disturbance) and physico-chemical and biochemical action, and Sowers has shown that the settlement-log time relationship is more or less linear. However, the rate of settlement produced by ravelling and combustion is erratic.

Meyerhof[84] referred to a site where industrial and domestic waste had been dumped for some thirty years. When the site was developed only very small movements were recorded where the fill had been in place for many years but on newly filled areas settlements up to 38 mm per month were noted during the early stages of construction. Most of the buildings were of light steel frame construction and sufficiently flexible to accommodate differential movements. Provision was made to jack columns back to their original position if they were subjected to excessive differential settlement. Where large settlements were anticipated pile foundations were used.

Where urban renewal schemes are undertaken it may be necessary to construct buildings on areas covered by rubble fill. In most cases such fills have not been compacted to any appreciable extent and where the rubble has collapsed into old cellars large voids may be present. However, demolition rubble fill is usually comparatively shallow and the most economical method of constructing foundations is either to cut a trench through the fill and backfill it with lean concrete or to clear all the fill beneath the structure and replace it with compacted layers. Deep vibration techniques may prove economical in areas where old cellars make it difficult to operate backacter excavators.

Penman and Godwin[88] noted that maximum rates of settlement occurred immediately after the construction of houses on an old open cast site at Corby which had been backfilled. These settlements decreased to small rates after about four years. The authors suggested that two of the causes of settlement in this fill were creep, which is proportional to log time, and partial inundation. Similar conclusions were reached by Sowers *et al*[89] for a similar situation. The houses at Corby were constructed twelve years after the fill was placed. The amount of damage which they have suffered is relatively small and is attributable to differential settlement. It is not related to the type of foundation structure used.

Where opencast fills exceed 30 m in depth, because greater settlements may occur, Kilkenny[86] recommended that the minimum time which should elapse before development takes place should be twelve years after restoration is complete. He noted that settlement of opencast backfill appeared to be complete within five to ten years after the operation. For example, comprehensive observations of the opencast restored area at Chibburn, Northumberland, 23 to 38 m in depth, revealed that the ultimate settlement amounted to approximately 1.2% of the fill thickness and that some 50% of the settlement was complete after two years and 75% within five years. In shallow opencast fills, up to 20 m deep, settlements of up to 75 mm have been observed.

COLLIERY SPOIL

There are two types of colliery discard, namely, coarse and fine. Coarse discard consists of run-of-mine material and reflects the various rock types which are extracted during mining operations. It contains varying amounts of coal which have not been separated by the preparation process. Fine discard consists of either slurry or tailings from the washery, which is pumped into lagoons. Some tips, particularly those with relatively high coal contents, may be partly burnt or burning and this affects their composition and therefore their engineering behaviour.

The moisture content of coarse discard would appear to increase with increasing content of fines, and generally falls within the range 5 to 15%. The range of specific gravity depends on the relative proportions of coal, shale, mudstone and sandstone (Table 1.4). The proportion of coal is of particular importance; the higher the content, the lower the specific gravity. Tip material also shows a wide variation in bulk density and may, in fact, vary within a tip. Low densities are mainly a function of low specific gravity.

The majority of tip material is essentially granular. Often most of it falls within the sand range, but significant proportions of gravel and cobble range may also be present. Due to breakdown, older and surface materials tend to contain a higher proportion of fines than that which occurs within a tip. In coarse discard the liquid and plastic limits are only representative of that fraction passing the 425 μm BS sieve, which frequently is less than 40% of the sample concerned. Nevertheless, the results of these consistency tests suggest a low to medium plasticity whilst in certain instances spoil has proved virtually non-plastic.

As far as effective shear strength of coarse discard is concerned, ϕ' usually varies from 25° to 45°. The angle of shearing resistance, and therefore the strength, increases in spoil which has been burnt. With increasing content of fine coal, on the other hand, the angle of shearing resistance is reduced. The shear strength of colliery spoil, and therefore its stability, is dependent upon the pore pressures developed within it. These are likely to be developed where there is a high proportion of fine material which reduces the permeability below 5×10^{-7} m/sec.

Oxidation of pyrite within tip waste is governed by access of air. However, the highly acidic oxidation products which result may be neutralised by alkaline materials in the waste; when this does not happen these chemical changes may give rise to pollution of drainage water. The sulphate content of weathered, unburnt colliery waste is usually high enough to warrant special precautions in the design of concrete structures which may be in contact with the discard or water issuing from it.

Spontaneous combustion of carbonaceous material, frequently aggravated by the oxidation of pyrite, is the most common cause of burning spoil. The problem of combustion has sometimes to be faced when reclaiming old tips (see Bell[90]). The NCB[91] recommend digging out, trenching, blanketing, injection with non-combustible material and water, and water spraying as methods by which spontaneous combustion in spoil can be controlled. Spontaneous combustion may give rise to subsurface cavities in spoil heaps and burnt ashes may also cover zones which are red hot to appreciable depths. When steam comes in contact with red hot carbonaceous material watergas is formed and when the latter is mixed with air it becomes potentially explosive. Explosions may occur when burning spoil heaps are being reworked and a cloud of coal dust is formed near the heat surface. If the mixture of coal dust and air is ignited it may explode violently.

Noxious gases are emitted from burning spoil. These include carbon monoxide, carbon dioxide, sulphur dioxide and less frequently hydrogen sulphide. Carbon monoxide is the most dangerous since it cannot be detected by taste, smell or irritation and may be present in potentially lethal concentrations. By contrast, sulphur gases are readily detectable in the ways mentioned above and are usually not present in high concentrations.

THE MECHANICAL BEHAVIOUR OF ROCK

The factors which influence the deformation characteristics and failure of rock masses can be divided into internal and external categories. The internal factors include the inherent properties of the rock itself whilst the external factors are those of its environment at a particular period in time. Regarding internal factors the mineralogical composition and texture are obviously important. However, even more important are the planes of weakness within a rock mass, i.e. the discontinuities and the degree of mineral alteration.

The composition and texture of a rock are governed by its origin. For instance, the olivines, pyroxenes, amphiboles, micas, feldspars and silica minerals are the principal components in igneous rocks. These rocks have solidified from a magma. Solidification involves a varying degree of crystallisation; the greater the length of time involved the greater the development of crystallisation. Hence glassy, fine, medium and coarse grained types of igneous rocks can be distinguished. In metamorphic rocks either partial or complete recrystallisation has been brought about by changing temperature-pressure conditions. Not only are new minerals formed in the solid state but the rocks may develop certain lineation structures, notably cleavage and schistosity. A varying amount of crystallisation is found within the sedimentary rocks. This ranges from almost complete crystallisation as in the case of certain chemical precipitates, to slight such as diagenetic crystallisation in the pores of, for example, certain sandstones.

Few rocks are composed of only one mineral species and even if this happens, the properties of that species vary slightly from mineral to mineral. This in turn is reflected in their physical behaviour. As a consequence few rocks can be regarded as homogeneous, isotropic substances. The size and shape relationships of the component minerals also influence physical behaviour, generally the smaller the grain size the stronger the rock.

One of the most important features of texture as far as physical behaviour, particularly strength, is concerned, is the degree of interlocking of the component grains. Fracture is more likely to take place along grain boundaries than through grains and therefore irregular boundaries make fracture more difficult. The bond between grains in sedimentary rocks is usually provided by the cement and/or matrix, rather than by grains interlocking. The amount and, to a lesser extent, the type of cement/matrix is important, not only influencing strength and elasticity but density, porosity and permeability.

Grain orientation in a particular direction facilitates breakage along that direction. This applies to all fissile rocks whether they are cleaved, schistose, foliated, laminated or thinly bedded. For example, Donath[92] demonstrated that cores cut in Martinsburg Slate at 90° to the cleavage possessed the highest breaking strength whilst those cores cut at 30° exhibited the lowest. Similar experiments were carried out on slate by Hoek[93] who orientated the cleavage planes so as to minimise and maximise their influence. He found that the uniaxial compressive strength in these two directions varied by a factor of approximately four.

Brown *et al*[94] showed that the compressive strength of the Delabole Slate is highly directional, indeed it varies continuously with the angle made by the cleavage planes with the direction of loading (Figure 1.12). They found that even where the cleavage makes high or low angles with the major principal stress direction the mode of failure is mainly influenced by the cleavage. The water content and surface roughness are the principal factors governing shear strength along the cleavage planes. For example, the average friction angle of smooth wet surfaces was determined as 20.5° which was 9° less than that obtained for the same surfaces when dry. What is more it was found that surface roughness could add up to 40° to the basic friction angles. The degree of surface roughness was shown to vary appreciably according to direction along and character of the cleavage plane concerned, which was reflected in the range of shear strength.

Turning to external factors Griggs[95] noted the changes in rock behaviour with increasing pressure-temperature conditions, simulating increasing depth. He showed that the ultimate strength of the Solenhofen Limestone was increased by 360% under 10 000 atm. High temperatures tend to aid plastic deformation

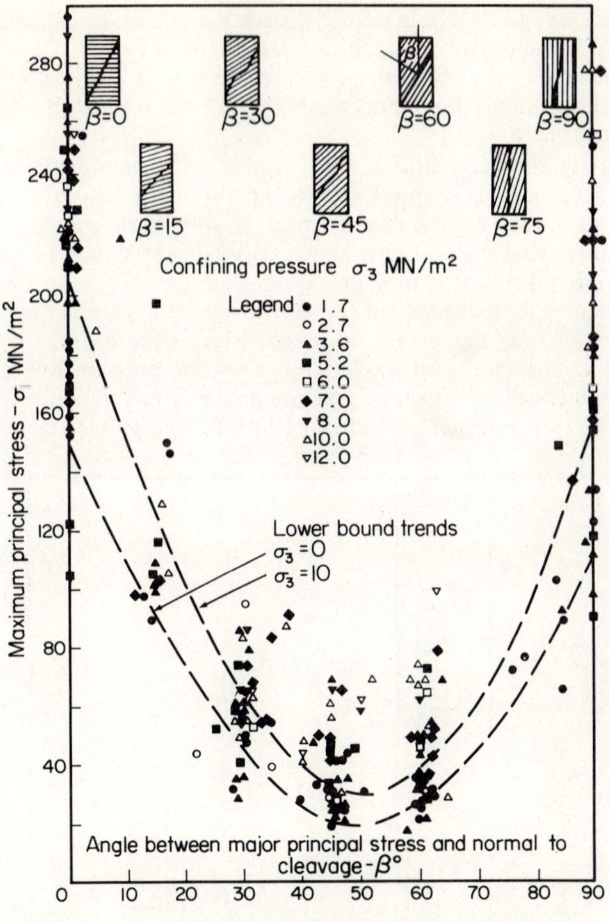

Figure 1.12 Compressive strength amisotrophy in Delabole slate (After Brown, Richards and Barr[94])

so that it becomes increasingly important with depth. Surface rocks are usually brittle and only some, such as salt, possess plasticity.

Colback and Wiid[96] carried out a number of uniaxial and triaxial compression tests at 8 different moisture contents on quartzitic shale and quartzitic sandstone with porosities of 0.28 and 15% respectively. The tests indicated that the compressive strengths of both rocks under saturated conditions were approximately half what they were under dry conditions (Figure 1.13a). From Figure 1.13b it will be noted that the slopes of the Mohr envelopes are not sensibly different indicating that the coefficient of internal friction is not significantly affected by changes in moisture content. Colback and Wiid therefore tentatively concluded that this reduction in strength was primarily due to a lowering of the tensile strength which is a function of the molecular cohesive strength of the material.

Tests on specimens of quartzitic sandstone showed that their uniaxial compressive strength was inversely proportional to the surface tension of the different liquids into which they were placed. As the surface free energy of a solid submerged in a liquid is a function of the surface tension of the liquid and since the uniaxial compressive strength is directly related to the uniaxial tensile strength and this to the molecular cohesive strength, it was postulated that the influence of the immersion liquid was to reduce the surface free energy of the rock and hence its strength.

Most strong rocks, such as granite, exhibit little time dependent strain or creep although creep in evaporitic rocks, notably, salt, may greatly exceed the instantaneous elastic deformations. Creep in soft chalk should also be given consideration in civil engineering practice. Although creep deformation usually is limited at low pressures, it may greatly exceed normal plastic flow when the pressures approach the limit of rupture. High temperatures also favour an increase in the rate and extent of creep. From experiments carried out on Solenhofen Limestone under moderate hydrostatic pressure Robertson[97] concluded that increased hydrostatic pressure caused a very significant decrease in the primary creep rate per unit stress; and that megascopic fracturing, chiefly

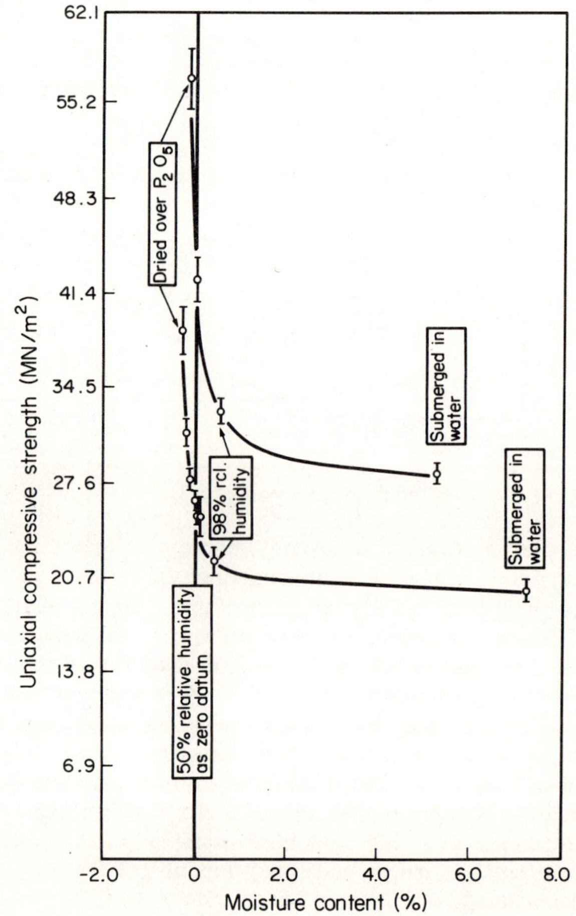

Figure 1.13(a) Relationship between uniaxial compressive strength and moisture content for quartzitic sandstone

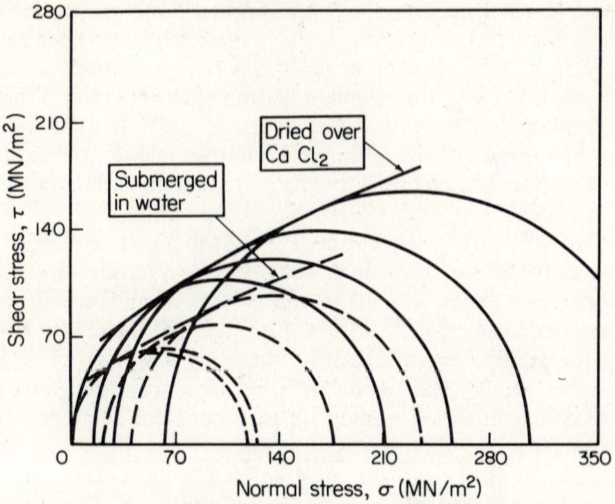

Figure 1.13(b) Mohr envelopes for quartzitic shale at two moisture contents (After Colback and Wild[96])

intragranular, decreased rapidly as the confining pressure increased. This conclusion was based upon bulk density measurements and suggested that in limestone under a constant load, fracturing is an important process of primary creep. It also seems likely that confining pressure might decrease the size, number and propagation of fractures during creep and that it may facilitate their healing when complete unloading has taken place.

STRENGTH OF ROCK

One of the most popular theories which was proposed to explain shear fractures was advanced by Coulomb[98]. The Coulomb criterion of brittle failure is based upon the idea that shear failure occurs along a surface if the shear stress acting in that plane is high enough to overcome the cohesive strength of the material and the resistance to movement. The latter is equal to the stress normal to the shear surface multiplied by the coefficient of internal friction of the material, whilst the cohesive strength is its inherent shear strength when the stress normal to the shear surface is zero. The Coulomb criterion has been shown to agree with experimental data for rocks in which the relationship between the principal stresses at rupture is, to all intents, linear. If the relationship, however, is non-linear then it is assumed that either the angle of internal shearing resistance is dependent upon pressure or, which is more probable, that the area of the grains in frictional contact increases as the normal pressure increases.

Coulomb's concept was subsequently modified by Mohr[99]. Mohr's hypothesis states that when a rock is subjected to compressive stress shear fracturing occurs parallel to those two equivalent planes for which shearing stress is as large as possible whilst the normal pressure is as small as possible. This assumes that a triaxial state of external stress is applied to a substance and that the maximum external stress is resolved into shear and normal components for any inlined potential shear planes existing in the stressed material. It has been suggested that shear fractures usually enclose an angle of less than 90° about the axis of maximum compression because the normal stresses which act across a shear plane are also involved in shear fracturing. The shearing angle is a constant which is governed by the ratio of the ultimate compressional to the ultimate tensile strength. It is therefore characteristic of a particular material at a given temperature and pressure.

In 1920 Griffith[100] claimed that because of the presence of minute cracks or flaws, particularly in surface layers, the measured tensile strengths of most brittle materials are much less than those which would be inferred from the values of the molecular cohesive forces. Although the mean stress throughout a body may be relatively low, local stresses developed in the vicinity of the flaws were assumed to attain values equal to the theoretical strength. Under tensile stress the stress magnification around a flaw is concentrated where the radius of curvature is smallest, in other words at its ends. Hence the tensile stresses which develop around the flaw have the most influence when the tensile stress zone coincides with the zone of minimum radius of curvature. The concentration of stress at the ends of flaws causes them to be enlarged and presumably with time they develop into fractures. Brace[101] showed that the fracture in hard rock was usually initiated in grain boundaries which could be regarded as the inherent flaws required by the Griffith theory. He wrote that as stress was increased prior to fracture grain boundaries at numerous sites in a rock became loosened and that at the instant before fracture it was filled with loosened sections along grain boundaries which had various lengths and orientations. Cracks grew in such sections and ultimately gave rise to fracture.

Griffith's original theory was concerned with brittle fracture produced by applied tensile stress and he based his calculations upon the assumption that the inherent flaw, from which fracture is initiated, could be regarded as an elliptical opening. This simple assumption, however, is not valid in conditions of compressive stress. Accordingly McClintock and Walsh[102] modified the Griffith theory to include the closing of flaws and the development of frictional forces across their surfaces as presumably occurs in compression. Brace[101] noted that in compression tests grain boundaries, although loosened, were in frictional contact at the moment of fracture, and thereby lent support to McClintock and Walsh's modification of the Griffith theory.

Although there is an encouraging agreement between experimental and theoretical results the Griffith theory does not provide a complete description of the mechanism of rock failure for it is only strictly correct when applied to fracture initiation under conditions of static stress. Indeed Hoek[103] contended that it was largely fortuitous that it could be so successfully applied to the prediction of fracture in rock since, once fracture is initiated, its propagation and ultimate failure is a relatively complex process. He further stated that the Griffith theory, when expressed in terms of the stresses at fracture, contributed little to the understanding of rock fracture under dynamic stress conditions, the energy changes associated with fracture or the deformation process of rock.

When the envelope fitted to a set of Mohr circles obtained from low pressure triaxial compression tests on brittle rocks is observed, it is usually represented by a straight line as suggested by the Coulomb equation. As most engineering undertakings involve low confining pressures it is sufficient to assume that the coefficient of friction, which is derived from the slope of the envelope, is a constant. At high confining pressures, however, this assumption is usually erroneous. More importantly it is misleading to assume that the coefficient of friction of 'soft rocks', like shales and mudstones, even at low confining pressures, is a constant. In fact their envelopes are curved. Accordingly it has been concluded that the coefficient of internal friction varies with normal compressive stress. The reason for this has been suggested by Hoek[104] who supposed that it was connected with the interlocking nature of the asperities along the shear plane. He contended that the interlocking depended upon the intimacy of their contact which, in turn, depended upon the amount of normal stress.

One of the most recent empirical laws advanced to describe the shear strength of intact rock has been given by Barton[105] and is as follows:

$$\tau/\sigma_n = \tan\left[50\log_{10}\left(\frac{\sigma_1-\sigma_3}{\sigma_n}\right)+\phi_c\right]$$

where

σ_n is the normal stress;
σ_1 is the axial stress at failure;
σ_3 is the effective confining pressure, and
ϕ_c is equal to 26.6°.

Barton argued that the the effective normal stress mobilised on conjugate shear surfaces was equal to the differential stress ($\sigma_1 - \sigma_3$) at a critical state. This happens to be the limiting value of the dimensionless ratio $(\sigma_1-\sigma_3)/\sigma_n$ used in formulating the empirical laws of friction and fracture strength. The critical state for intact rock was defined as the stress condition under which the Mohr envelope of peak shear strength reaches a point of zero gradient. This represents the maximum possible shear strength of the rock. There is a critical effective confining pressure for each rock above which strength cannot increase. The one-dimensional dilation associated with shearing is completely suppressed if the applied stress reaches the level of critical effective confining pressure.

THE INFLUENCE OF DISCONTINUITIES UPON THE ENGINEERING BEHAVIOUR OF ROCK MASSES

A discontinuity is simply a fracture within a rock mass, the most common discontinuities being joints and bedding planes. Other important discontinuities are cleavage planes, planes of schistosity and fissure zones. In this section the major consideration is given to joints, other discontinuities are dealt with in subsequent sections. Joints are fractures along which little or no displacement has occurred and are present within all rock types. If joints are planar and parallel they may be described as systematic, conversely, when they are irregular they are referred to as non-systematic.

A group of joints which run parallel to each other is termed a joint set whilst two or more joint sets which intersect more or less at a constant angle are referred to as a joint system. When one set of joints is dominant they are called primary joints, the other set being termed secondary. Joints may be classified, on a basis of their size, into master joints, which penetrate several rock horizons and persist for hundreds of metres; smaller joints, which are still well defined structures, are termed major joints; whilst minor joints are related to local structures and do not transcend bedding planes. Lastly, minute fractures occasionally occur in finely bedded sediments and such micro-joints are usually only a few millimetres in size.

Joints may be formed in a number of ways. For example, joints develop within igneous rocks when they cool down, and in wet sediments when they dry out. The most familiar of these are the columnar joints in lava flows, sills and some dykes. Cross joints, longitudinal joints, diagonal joints and flat-lying joints are associated with large granitic intrusions. Sheet or mural joints have a similar orientation to flat-lying joints. When they are closely spaced and well developed they impart a pseudo-stratification to the host rock. It has been noticed that the frequency of sheet jointing is related to depth of overburden, in other words the thinner the rock cover the more pronounced the sheeting. This suggests a connection between removal of overburden by denudation and the development of sheeting. Chapman[106] maintained that other sets of joints developed after the formation of sheet joints and that they were contained between the sheeting surfaces. He showed that the orientation and degree of development of these joints was related to topography.

Price[107], however, postulated that most joints formed after major episodes of folding, that is, they were post-compressional structures. Some spatially restricted small joints associated with folds, such as radial tension joints, are however, probably initiated during folding. He suggested that, after folding, rocks retained residual strain energy and that the associated residual stresses were modified during the succeeding stage of orogenic uplift in such a way as to give rise to tension and shear joints. Indeed joints are frequently associated with folds and faults, but they do not appear to form parallel to other planes of shear failure such as normal and thrust faults. The orientation of joint sets in relation to folds depends upon their size, the type and size of the fold and the thickness and competence of the rocks involved.

The shear strength of a rock mass and its deformability are very much influenced by the discontinuity pattern, its geometry and how well it is developed. Observation of discontinuity spacing, whether in a core or field exposure, aids the appraisal of rock mass structure. As joints represent surfaces of weakness, the larger and more closely spaced they are, the more influential they become in reducing the effective strength of the rock mass. The nature of the opposing discontinuity surfaces also influences rock mass behaviour as the smoother they are, the more easily can movement take place along them. However, joint surfaces are usually rough and may be slickensided.

Furthermore joints may be open or closed. The degree of opening is of importance in relation to the overall strength and permeability of the rock mass and this often depends largely on the amount of weathering which the rocks have suffered. Some joints may be partially or completely filled. The type and amount of the filling not only influences the effectiveness with which the opposing joint surfaces are bound together, thereby affecting the strength of the rock mass, but it also influences permeability. Mineralisation is frequently associated with joints. This may effectively cement a joint, however, in other cases the mineralising agent may have altered and weakened the rocks along the joint conduit.

Discontinuities in a rock mass reduce its effective shear strength at least in a direction parallel with them. Hence the strength of bedded and jointed rocks is highly anisotropic. Obviously where a load is applied in a direction parallel or subparallel to the discontinuity direction the shear strength depends upon the shearing resistance along that surface. At low normal pressures shearing stresses along a surface with relatively smooth asperities produces a tendency for one block to ride up onto the asperities of the other whereas at high normal pressures shearing takes place through the asperities. When a jointed rock mass undergoes shearing this may be accompanied by dilation especially at low pressures and small shear displacements probably occur as shear stress builds up. Joints offer no resistance to tension whereas they offer high resistance to compression. Nevertheless they may deform under compression if there are crushable asperities, compressible filling material or apertures along the joint surfaces or if the rock wall is altered.

Goodman, Taylor and Brekke[108] considered the application of the finite element method to the problem of jointed rocks. They maintained that for a realistic analysis the jointed rock mass should be treated as an aggregate of massive rock blocks separated by joints rather than as a continuum. In their analysis they distinguished three distinct joint parameters, namely, the unit stiffness across the joint (k_n), the unit stiffness along the joint (k_s) and the shear strength along the joint (S). The value of k_n depends upon the contact area ratio between the joint surfaces, the perpendicular aperture distribution and amplitude, and the relevant properties of the filling materials when present. The controlling factors of the k_s value include the roughness of the joint surfaces which is determined by the distribution, size and inclination of the asperities; and the tangential aperture distribution and amplitude. Lastly the factors which govern the value of S are the friction along the joint, the cohesion due to interlocking, and the strength of the filling material when present.

The moisture content influences all three parameters, however, and the authors regarded the joint water pressures as a separate variable whose effect was analogous to that of pore pressures in soil. Joints with a high stiffness have negligible joint displacements compared with the elastic displacements of the rock blocks whilst those which have a low stiffness value undergo displacements of very much greater magnitude than the elastic displacements of the blocks. A moderate joint stiffness corresponds to displacements of joints to the same extent as the elastic displacements in the blocks. The joint strength was considered as high, moderate or low in accordance with whether the joints played a negligible, participating or dominant role in the strength of the rock mass.

Barton[105] proposed the following expression for deriving the shear strength (τ) along joint surfaces:

$$\tau = \sigma_n \tan \left(JRC \log_{10} \left(\frac{JCS}{\sigma_n} \right) + \phi_b \right)$$

where

σ_n is the effective normal stress;
JRC is the joint roughness coefficient;
JCS is the joint wall compressive strength, and
ϕ_b is the basic friction angle.

According to Barton the values of the joint roughness coefficient range from 0 to 20, from the smoothest to the roughest surface. The joint wall compressive strength is equal to the unconfined compressive strength of the rock if the joint is unweathered whereas this may be reduced by up to 75% when the walls of the joints are weathered. Both these factors are related since smooth walled joints are less affected by the value of *JCS* since failure of asperities plays a less important role. The smoother the walls of the joints the more significant is the part played by its mineralogy (ϕ_b). The experience gained from rock mechanics indicates that under low effective normal stress levels, such as occur in engineering, the shear strength of joints can vary within relatively wide limits (Figure 1.14) and the maximum effective normal stress acting across joints considered

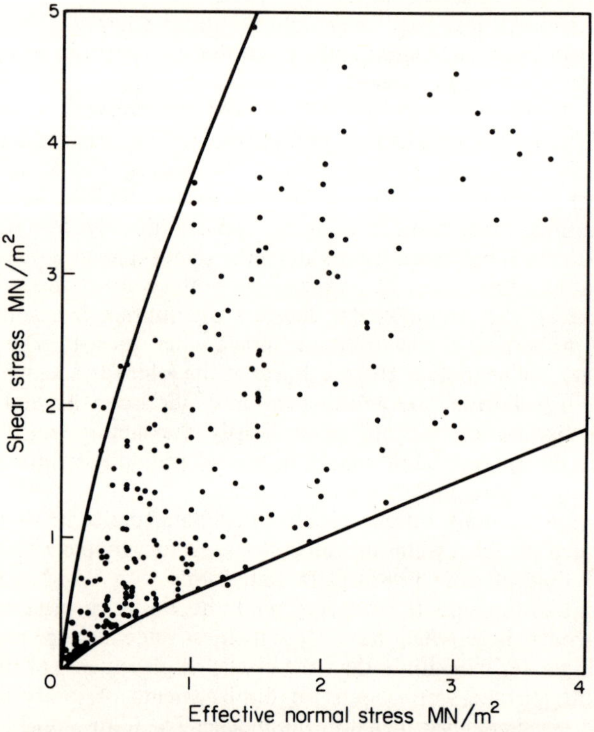

Figure 1.14 The peak shear strength of unfilled joints in rock from the published results of direct shear tests performed both in the laboratory and *in situ* (After Barton[105])

critical for stability lies, according to Barton, in the range 0.1 to 2.0 MN/m^2. A further account of shear strength along joints is given in chapters 5 and 16.

ROCK QUALITY INDICES

Several attempts have been made to relate the numerical intensity of fractures to the quality of unweathered rock masses and to quantify their effect on deformability. For example, the effect of discontinuities in a rock mass can be estimated by comparing the *in situ* compressional wave velocity with the laboratory sonic velocity of an intact core obtained from the same rock mass. The difference in these two velocities is caused by the structural discontinuities which exist in the field. The velocity ratio, V_f/V_l, where V_f and V_l are the compressional wave velocities of the rock mass *in situ* and of the intact specimen respectively, was first proposed as a quality index by Onodera[109]. For a high quality massive rock with only a few tight joints, the velocity ratio approaches unity. As the degree of jointing and fracturing becomes more severe, the velocity ratio is reduced. The sonic velocity is determined for the core in the laboratory under an axial stress equal to the computed overburden stress at the depth from which the sample was taken, and at a moisture content equivalent to that assumed for the *in situ* rock. The field seismic velocity is preferably determined by uphole or crosshole seismic measurements in drillholes or test adits, since by these measurements it is possible to explore individual homogeneous zones more precisely than by surface refraction surveys.

ROCK QUALITY DESIGNATION (RQD)

The concept of rock quality designation (*RQD*) was introduced by Deere *et al*[110]. However, it does not take account of the joint opening and condition, a further drawback being that with fracture spacings greater than 100 mm the quality is excellent irrespective of the actual spacing. This particular difficulty can be overcome by using the fracture spacing index as suggested by Franklin, Broch and Walton[111]. This simply refers to the frequency with which fractures occur within a rock unit.

The concept of the rock mass factor (j) has been recently introduced by Hobbs[112]. He defined the rock mass factor as the ratio of the deformability of a rock mass within any readily identifiable lithological and structural component to that of the deformability of the intact rock comprising the component. Consequently it reflects the effect of discontinuities on the expected performance of intact rock. The value of j depends upon the method of assessing the deformability of the rock mass, and the value beneath an actual foundation will not necessarily be the same as that determined from even a large scale field test, particularly on the more massive rocks. Generally the plate loading test, suitably scaled to the discontinuity spacing, is the best method of determining the field modulus of deformation. An assessment of rock quality and deformability can be made by using water injection tests between packers in drillholes. High hydraulic conductivity in relation to the fracture frequency means correspondingly low j-values and vice versa. In simple joint systems, it is theoretically possible to relate joint opening to hydraulic conductivity. According to Hobbs

Table 1.7 CLASSIFICATION OF ROCK QUALITY IN RELATION TO THE INCIDENCE OF DISCONTINUITIES

Quality classification	*RQD* %	*Fracture frequency* per metre	*Velocity ratio* V_f/V_l	*Mass factor* (j)
Very poor	0–25	Over 15	0–0.2	
Poor	25–50	15–8	0.2–0.4	Less than 0.2
Fair	50–75	8–5	0.4–0.6	0.2–0.5
Good	75–90	5–1	0.6–0.8	0.5–0.8
Excellent	90–100	Less than 1	0.8–1.0	0.8–1.0

the greatest difficulties which occur in jointed rock masses in relation to foundation design are experienced in those where the fracture spacing falls within a range of about 100 to 500 mm, in as much as small variations in fracture spacing and condition result in exceptionally large changes in the j-value. Some of the classifications of rock quality in relation to discontinuities are given in Table 1.7.

FAULTS AND FOLDS

The two most important features which are produced when strata are deformed are faults and folds. Such deformation principally takes place due to movements along shearing planes. When these are few and large they cause faulting, whilst if they are small and numerous flexuring and folding results.

Faults are fractures along which adjacent strata have been displaced. The amount of displacement may vary from only a few millimetres to several hundred kilometres, for example, the lateral displacement along the San Andreas fault in California is said to total over 650 km. In some faults the fracture is a clean break but in others the displacement is not restricted to a simple fracture but is developed throughout a fault zone, differential movement having occurred along innumerable closely spaced fractures. If a fault surface is irregular it is usually grooved in the direction of slip. At the time of their initiation faults are shear fractures unless they follow pre-existing planes of weakness. A recent review of the mechanics of fault formation has been given by Murrell[113].

If the movement along a fault has been severe then the rocks involved may have been crushed, sheared or pulverised (see Sibson[114]). Where shales or clays have been faulted the fault zone may be occupied by clay gouge (Figure 1.15).

Figure 1.15 Clay gouge occupying the fault zone of the Howick fault in the Limestone Group, Lower Carboniferous, near Howick, Northumberland

Fault breccias, which consist of a jumbled mass of angular fragments containing a high proportion of voids, come into being when more competent rocks are faulted. They are formed by the continued opening of tension gashes where the normal stresses across the fault are low. Accordingly fault breccias are confined to the uppermost region of the crust and are associated with normal faults.

Crush breccias develop when rocks are sheared by a regular pattern of fractures, the individual fragments being bounded by intersecting shear surfaces. If the movements are prolonged then the corners of the fragments may be rounded with the result that crush conglomerates are formed. The material between the lenses forms a fine grained matrix. Crush conglomerates and breccias indicate that strong normal stresses were operational across the fault. These rocks are squeezed as well as stretched and so they are elongated along a plane of flattening which bisects intersecting shear surfaces. Although crush conglomerates may occur in association with any type of fault they most commonly accompany reverse faults.

Movements of even greater intensity are responsible for the occurrence of mylonite along a fault zone, this may be regarded as a micro-breccia and shearing planes may be observed within it. If, because of the severity of the crushing, no individual fragments can be distinguished within the material then it is described as an ultramylonite. Mylonites are principally found in association with low angle thrusts and are thought to be developed at depths where the confining pressures are sufficient to maintain rock coherence.

The ultimate stage in the intensity of movement is reached with the formation of pseudotachylite. This looks like glass and many authorities have suggested that the movements responsible for its formation must have been so severe as to melt the rock material involved and that pseudotachylite was formed upon cooling.

Intraformational shears, that is, zones of shearing parallel to bedding are associated with faulting. Salehy *et al*[115] noted their presence in clays, mudstones and shales of Carboniferous age and that they generally occurred at the contact of such beds with overlying sandstones. They pointed out that these shear zones tended to die out when traced away from the faults concerned and suggested that such zones are probably formed as a result of flexuring of strata adjacent to faults. A shear zone may consist of a single polished or slickensided shear plane, a more complex shear zone may be up to 300 mm in thickness (see Skempton[116]). Intraformational shear zones are not restricted to argillaceous rocks, for instance, they occur in the Chalk. Their presence means that the strength of the rock along the shear zone has been reduced to its residual value.

Shear and tension joints and cleavage are frequently associated with major faults (see Louderback[117]). Shear and tension joints formed along a fault are frequently referred to as feather joints because of their barb-like appearance. Where pinnate shear planes are closely spaced and involve some displacement then fracture cleavage is developed. Slickensides are polished striated surfaces which occur on a fault plane and are caused by the frictional effects generated by its movement.

As a fault is approached the strata involved frequently exhibit flexures which suggest that the beds have been dragged into the fault plane by the frictional resistance generated along it, indeed along some large dip-slip faults the beds may be vertically inclined. A related effect is seen in faulted gneisses and schists where a pre-existing foliation is strongly turned into the fault zone and a secondary foliation results.

In almost all known instances of historic fault breaks the fracturing has occurred along a pre-existing fault. Whilst it seems probable that a given fault would break again at the same location as the last break this cannot be concluded with certainty. However, the likelihood of a new fault interfering with an engineering structure is so remote that it can be reasonably neglected except in unusual situations such as near the tip of the wedge of an active thrust fault.

There is a rough relationship between the length of a fault break and the amount of displacement involved, and both are related to the magnitude of the resultant earthquake. Displacements range from a few millimetres up to 11 m, which was the vertical displacement on the Chedrang fault during the Assam earthquake (1897; M = 8.7). The largest horizontal displacement was on the Bogdo fault after the Mongolian earthquake (1957; M = 8.3). As expected the longer fault breaks have the greater displacements and generate the larger earthquakes. On the other hand it has been shown by Ambraseys[118] and Bonilla[119] that the smaller the fault displacement the greater the number of observed fault breaks. They also found that for the great majority of fault breaks the maximum displacement was less than 6 m and that the average displacement along the length of the fault was less than 50% of the maximum.

The length of the fault break during a particular earthquake is generally only a fraction of the true length of the fault. Individual fault breaks during simple earthquakes have ranged in length from less than a kilometre to several hundred kilometres. What is more fault breaks do not only occur in association with large and infrequent earthquakes but also occur in association with small shocks and continuous slippage known as fault creep (maximum slip rate on the San Andreas fault is 20 mm per annum).

There is little information on the frequency of breaking along active faults, all that can be said is that some master faults have suffered repeated movements, in some cases recurring in less than 100 years. By contrast much longer intervals totalling many thousands of years have occurred between successive breaks. Therefore because movement has not been recorded in association with a particular fault in an active area it cannot be concluded that the fault is inactive. Earthquakes resulting from displacement and energy release on one fault can

Figure 1.16 Fracture cleavage developed in highly folded Horton Flags, Silurian, near Stainforth, Yorkshire. The inclination of the fracture cleavage is indicated by the near vertical hammer. The other hammer represents the dip of the bedding

sometimes trigger small displacements on other unrelated faults many kilometres distant. Breaks on subsidiary faults have occurred at distances as great as 25 km from the main fault, but with increasing distances from the main fault the amount of displacement decreases. For example, displacements on branch and subsidiary faults located more than 3 km from the main fault break are generally less than 20% of the main fault displacement.

Compressive forces acting in opposed directions cause strata to fold by minute adjustments taking place within and between constituent grains so giving rise to a permanent change in form. The positions of the folds occur where irregularities cause the resolutes of stress to act normal to them. The shape of a fold is influenced by the relative strength of opposing compressive forces and the competency or incompetency of the strata involved. Competent beds transmit stresses whereas incompetent beds behave passively. In fact it is the beds which do not yield to mass flowage which largely determine the shape of a fold. In folding a certain amount of bedding slip takes place. Where competent and incompetent strata are interbedded the movements of the former relative to each other subject the incompetent strata to shearing stress, which may cause small folds, termed drag folds, to develop. Fracture cleavage may be associated with folded competent strata (Figure 1.16).

Shear folding results when minute differential displacements occur along closely spaced shear planes. In shear folding the width of the individual beds along the direction of shear is constant although the beds themselves have been thinned. The apices of the fold are thicker than the flanks if measured normal to the bedding planes. Shear folding may affect folds which have been formed previously.

ENGINEERING ASPECTS OF ROCK WEATHERING

Weathering of rocks is brought about by physical disintegration, chemical decomposition and biological activity. The type of weathering which predominates in a region is largely dependent upon climate, which also affects the rate at which weathering proceeds. The latter is also influenced by the stability of the rock mass concerned, which in turn depends upon its mineral composition, texture and porosity, and the incidence of discontinuities within it.

Many rocks were originally formed at high temperatures and pressures and a large part of the weathering process consists of an attempt to reach a new stability under atmospheric conditions. High temperature minerals occur in the ultrabasic and basic igneous rocks. Hence such rocks tend to offer less resistance to weathering than the acid igneous rocks which are largely composed of soda and potash feldspar, quartz and, to a lesser extent, mica. The latter two minerals are particularly stable. Generally coarse grained rocks weather more rapidly than do fine grained types of similar mineral composition.

The degree of interlocking between the constituent minerals is a particularly important textural factor. The closeness of the grains governs the porosity of a rock which in turn determines the amount of water it can hold. Not only are the more porous rocks more susceptible to chemical attack but they are also more prone to frost action. Weathering action is concentrated along discontinuities. For example, the karstic features which occur in thick sequences of limestone are due to solutioning and enlargement of joints and bedding planes; spheroidal weathering, commonly seen in basalts and dolerites, is again caused by weathering along joints (Figures 1.17a and b). The bedding plane spacing in sedimentary rocks affects the rate at which they weather. Intense weathering may be associated with fault zones.

Physical weathering is particularly effective in climatic regions which experience significant changes of diurnal temperature. This does not necessarily imply a wide range of temperature as frost and thaw action can proceed where the range is limited. Alternate freeze-thaw action causes discontinuities to be enlarged and eventually leads to rock shattering. In hot deserts where the daily range of temperature is large the outer layers of rock masses undergo alternate expansion and contraction which eventually leads to them being exfoliated.

Figure 1.17(a) Gaping Ghyll pothole which descends about 150 m in the Carboniferous Limestone, near Clapham, Yorkshire

Figure 1.17(b) Spheroidal weathering of basalt lava flows, Tideswell, Derbyshire. In places these lavas are completely decomposed

Chemical weathering leads to mineral alteration and the solution of rocks. Alteration is principally effected by oxidation, hydration, hydrolysis and carbonation whilst solutioning is brought about by acidic or alkaline waters. In dry air, rock decay takes place slowly. However, the presence of moisture hastens the rate tremendously, firstly, because water is itself an effective agent of weathering and, secondly, because it holds in solution substances which react with rock forming minerals. The most important of these substances are free oxygen, carbon dioxide, organic acids and nitrogenous acids. Chemical weathering also aids rock disintegration by weakening the rock fabric and emphasizing any structural weaknesses. When decomposition occurs within a rock the altered material generally occupies a greater volume than the original parental rock material and in the process internal stresses are generated. If swelling occurs in the outer layers of a rock then it causes them to peel off.

Plants and animals play an important role in the breakdown and decay of rocks. Indeed their part in soil formation is of major significance.

The agents of weathering, unlike those of erosion, do not provide transport for the debris from the rock surface. Unless this rock waste is otherwise removed it will eventually act as a protective blanket and so prevent further weathering taking place.

Certain rocks are, of course, particularly susceptible to breakdown; shales and mudstones fall into this category. The disintegration of shales was studied by Badger, Cummings and Whitmore[120] who concluded that their breakdown was due to the dispersion of colloid material, which appeared to be a general cause of disintegration, and, to a lesser extent, due to air breakage, which only occurs in mechanically weak shales. They found that the variation in disintegration of different shales in water was not usually connected with their total amount of clay colloid or the variation in the types of clay mineral present. It was rather controlled by the type of exchangeable cations associated with clay particles, and by the accessibility of the latter to attack by water which in turn, depended upon the porosity of the shale. Air breakage could assist this process by presenting new surfaces of shale to water. The fracture pattern within mudstones and the lamination of shale aids their disintegration.

Nakano[121] attempted to assess the weatherability of some Japanese mudstones. On being immersed in water these mudstones swelled and underwent a reduction in strength but they did not disintegrate. However, when dried and then wetted again they disintegrated rapidly into small pieces.

Fireclays may disintegrate very rapidly. For instance, Taylor and Spears[122] showed that the Brooch and Park seatearths, of the Sheffield area, after desiccation, broke down in water in less than 30 min, in fact the former was literally explosive. Although the expandable clay content of these two rocks is high and leads to intra-particle swelling, these authors maintained that this alone was not responsible for rapid disintegration. They demonstrated that breakdown could be arrested by removal of air from samples. Thus they concluded that air breakage was a fundamental disintegration mechanism in such rocks. It was suggested that during dry periods evaporation from their surface gives rise to high suction pressures which result in increased shearing resistance. With extreme desiccation most of the voids in a sample of seatearth are filled with air, which, on immersion in water, becomes pressurised by capillary pressures developed in the outer pore spaces. The mineral fabric may then fail along its weakest plane exposing a new surface to the same process.

Swelling of rocks is associated with weathering, clays, shales, mudstones and marls being especially prone. The clay mineral content plays a significant role in these rocks, for example, kaolinite is non-expansive whilst montmorillonite is. Small amounts of swelling have been recorded in some sandstones. The swelling is due principally to the ingress of water therefore the rock must be porous or fractured. Rocks with intact unconfined compressive strengths in excess of 40 MN/m^2 are not subject to swelling. When rocks swell they may exert considerable pressures against confining structures, drainage is therefore important. However, it would be better to prevent the ingress of water in the first place. On some sites in swelling rocks, because of the drastic changes which may take place in their character upon wetting, resort to covering the offending strata may be necessary.

The effects of weathering on strengths and deformation moduli are dramatic. For example, a small increase in moisture content or porosity due to weathering can cause large reductions in these parameters, however, the modulus ratio may be only marginally affected. An important consequence of the reduction in modulus is that the joints and fractures have a smaller and smaller influence on the mass factor as the modules of the intact rock declines, that is, the j-value increases with the increase in the degree of weathering.

Table 1.8 ENGINEERING GRADE CLASSIFICATION OF WEATHERED ROCK (After Dearman, Fookes and Franklin[128])

Grade	Degree of decomposition	Field recognition (after Fookes and Horswill[129]) — Soils (i.e. soft rocks)	Field recognition — Rocks (i.e. hard rocks)	Engineering properties — After Little[130]	Engineering properties — After Hobbs[131]
VI	Soil	The original soil is completely changed to one of new structure and composition in harmony with existing ground surface conditions	The rock is discoloured and is completely changed to a soil in which the original fabric of the rock is completely destroyed. There is a large volume change	Unsuitable for important foundations. Unstable on slopes when vegetation cover is destroyed and may erode easily unless a hard cap is present. Requires selection before use as fill	In completely weathered rock and residual soil it may be possible to obtain fair quality samples depending upon the parent rock type and the consistency of the product Generally the samples will tend to be less disturbed than when taken in the same rock in the highly weathered state. The bearing capacity and settlement characteristics of rock in these extreme states can be assessed using the usual method for testing soils
V	Completely weathered	The soil is discoloured and altered with no trace of original structures	The rock is discoloured and is wholly decomposed and friable, but the original fabric is mainly preserved. The properties of the rock mass depend in part on the nature of of the parent rock	Can be excavated by hand or ripping without use of explosives. Unsuitable for foundations of concrete dams or large structures. May be suitable for foundations of earth dams and for fill. Unstable in high cuttings at steep angles. New joint patterns may have formed. Requires erosion protection	
IV	Highly weathered*	The soil is mainly altered with occasional small lithorelicts of original soil. Little or no trace of original structures	The rock is discoloured; discontinuities may be open and have discoloured surfaces and the original fabric of the rock near the discontinuities is altered; alteration penetrates	Similar to Grade V. Unlikely to be suitable for foundations of concrete dams. Erratic presence of boulders makes it an unreliable foundation for large structures	In highly weathered rock difficulties will generally be encountered in obtaining undisturbed samples for testing. If samples are obtained the strength and modulus will generally be underestimated, frequently by

			deeply inwards, but corestones are still present. The rock mass is partially friable		large margins, even with apparently undisturbed samples. In such rocks *in-situ* tests with either the Menard pressuremeter or the plate should be carried out to determine the bearing capacity and settlement characteristics. The greatest difficulties in assessing bearing capacity and settlement are likely to be encountered in highly weathered rocks, in which the rock fabric becomes increasingly disintegrated or increasingly more plastic
III	Moderately weathered*	The soil is composed of large discoloured lithorelicts of original soil separated by altered material. Alteration penetrates inwards from the surfaces of discontinuities	The rock is discoloured; discontinuities may be open and surfaces will have greater discolouration with the alteration penetrating inwards; the intact rock is noticeably weaker, as determined in the field, than the fresh rock. The rock mass is not friable	Excavated with difficulty without the use of explosives. Mostly crushes under bulldozer tracks. Suitable for foundations of small concrete structures and rock fill dams. May be suitable for semipervious fill. Stability in cuttings depends on structural features, especially joint attitudes.	In moderately weathered rock the intact modulus, and strength, can be very much lower than in the fresh rock and thus the *j*-value will be higher than in the fresh state, unless the joints and fractures have been opened by erosion or softened by the accumulation of weathering products. The intact modulus and strength can be measured in the laboratory, and the bearing capacity assessed, in the same way as for fresh rock. Triaxial tests may be more appropriate than uniaxial tests, and it would be advisable to adopt conservative values for the factor of safety

*The ratio of the original soil or rock to altered material should be estimated where possible.

Table 1.8 (*continued*)

		Field recognition (after Fookes and Horswill)		*Engineering properties*	
Grade	*Degree of decomposition*	*Soils (i.e. soft rocks)*	*Rocks (i.e. hard rocks)*	*After Little*[130]	*After Hobbs*[131]
II	Slightly weathered	The material is composed of angular blocks of fresh soil, which may or may not be discoloured. Some altered material starting to penetrate inwards from discontinuities separating blocks	The rock may be slightly discoloured; particularly adjacent to discontinuities which may be open and have slightly discoloured surfaces; the intact rock is not noticeably weaker than the fresh rock	Requires explosives for excavation. Suitable for concrete-dam foundations. Highly permeable through open joints. Often more permeable than the zones above or below. Questionable as concrete aggregate	In faintly and slightly weathered rock it is possible that the *j*-value, owing to the reduction in stiffness of the joints as a result of penetrative weathering alone, will show a fairly sharp decrease compared with that of the same rock in the fresh state. The intact modulus, by definition, is unaffected by penetrative weathering. The safe bearing capacity is not therefore affected by faint weathering, and may be only slightly affected by slight weathering
I	Fresh rock	The parent soil shows no discolouration, loss of strength or other effects due to weathering	The parent rock shows no discolouration, loss of strength or any other effects due to weathering	Staining indicates water percolation along joints; individual pieces may be loosened by blasting or stress relief and support may be required in tunnels and shafts	

A number of tests have been designed to assess certain aspects of weatherability, perhaps one of the most familar being the freeze-thaw test. Freeze-thaw action can quickly break down rocks such as shales and soft chalk. More recently the slake-durability test has been introduced. This test estimates the resistance to wetting and drying of mudstones and shales in particular. Franklin and Chandra[123] found a general qualitative correlation between the slake-durability index, the rate of weathering and the stable slope angle.

Several attempts have been made to devise an engineering classification of weathered rocks. Those methods which rely on a simple index test to provide a quantitative answer such as that devised by Hamrol[124], or indeed the slake-durability test, have not proved particularly useful. The most successful attempts have been based on a simple description of the geological character of the rock as seen in the field, the description embodying different grades of weathering related to engineering performance (see Moye[125]; Kiersch and Treasher[126]; Knill and Jones[127]). The most useful of these classifications is given in Table 1.8.

ENGINEERING CLASSIFICATION OF ROCK MASSES

An engineering classification attempts to assess the suitability of a rock mass for a given project and consequently the selection of the parameters for such a classification is of special importance. One of the first attempts at classifying rock masses for engineering purposes was made by Terzaghi[132]. However, this classification has been criticised on the grounds that it was too general to permit an objective evaluation of rock quality and that it provided no quantitative information on the properties of rock masses. Subsequently Deere *et al*[110] introduced the concept of rock quality designation (RQD) as a means of classifying rock masses. The RQD is a more general measure than fracture frequency and is based indirectly on both the degree of fracturing and the amount of weathering in the rock mass.

One of the most recent classifications has been advanced by Bieniawski[133]. This initially incorporated the RQD; the unconfined compressive strength; the degree of weathering; the spacing, orientation, separation and continuity of the discontinuities; as well as the ground water flow. The grades of rock quality related to RQD are given in Table 1.7. Unfortunately, however, the RQD takes no account of the spacing, orientation, tightness, roughness of the surface or continuity of discontinuities, or the presence and character of infilling material. Furthermore in certain circumstances, for instance, when discontinuities are widely spaced or the rock mass is weak, the intact strength of the rock mass has an important bearing on its engineering performance.

Because the unconfined compressive strength and the degree of weathering are two interdependent factors Bieniawski[133] subsequently revised his views and suggested that both factors should be regarded as one parameter, namely, the strength of rock material. He chose a somewhat modified version of Deere's[135] classification of intact strength for his classification, which is given at the top of page 72.

Special caution should be exercised in using this classification in the case of shales and other swelling materials. For instance, Bieniawski suggested that their behaviour under conditions of alternate wetting and drying should be assessed by a slake-durability test (see Franklin and Chandra[123]).

Very low strength	Under 25 MN/m^2
Low strength	25 to 50 MN/m^2
Medium strength	50 to 100 MN/m^2
High strength	100 to 200 MN/m^2
Very high strength	Over 200 MN/m^2

(cf. Strength classification proposed by Geological Society[134] which provides further subdivision of those rocks with strengths less than 25 MN/m^2 and tends to be used in Britain. Bieniawski argued that when assessing importance rating (see below) the strength of such rocks does not contribute to the overall mobility of the rock mass.)

The presence of discontinuities reduces the overall strength of a rock mass and their spacing and orientation govern the degree of such reduction. Hence the spacing and orientation of the discontinuities are of paramount importance as far as the stability of structures in jointed rock masses is concerned. Bieniawski accepted the classification of discontinuity spacing proposed by Deere[137], i.e.

Description	*Spacing of discontinuities*	*Rock mass grading*
Very wide	Over 3 m	Solid
Wide	1 to 3 m	Massive
Moderately close	0.3 to 1 m	Blocky/seamy
Close	50 to 300 mm	Fractured
Very close	Under 50 mm	Crushed

Another revision of Bieniawski's ideas included the continuity and separation of discontinuities which he later[133] grouped together under the heading, condition of discontinuities. This parameter also took account of their surface roughness and the quality of the wall rock. The condition of discontinuities is as important as the discontinuity spacing, for example, tight discontinuities with rough surfaces and no infill have a high strength, whilst, by contrast, open continuous discontinuities facilitate unrestricted flow of ground water. Obviously the condition of discontinuities influences the extent to which the rock material affects the behaviour of a rock mass.

Ground water has an important effect on the behaviour of a jointed rock mass. However, as pore water pressures are of greater significance in foundations than ground water inflow Pells[138] suggested that the pore pressure ratio (r_u), where r_u is defined as the ratio of the pore pressure to the major principal stress, should be included within the classification. This was duly done in the 1974 classification. Bieniawski grouped each of the chosen rock mass parameters into five classes (Table 1.9a).

Because these parameters vary in their relative importance from rock mass to rock mass and can contribute individually or collectively to its engineering performance Bienawski used a rating system. In other words a weighted numerical value was given to each parameter, the total rock mass rating being the sum of the weighted values of the individual parameters, the higher the total rating, the better the rock mass conditions (Table 1.9b). Bieniawski used a simplified version of the rating system advanced by Wickham *et al*[139]. The ratings given for discontinuity spacings apply to rock masses having three sets of discontinuities. Thus when only one or two sets are present a conservative assessment is attained. In describing a rock mass the class rating should be quoted with the class number, for example, Class 3, rating 68.

Name of Project:

Site of survey:

Conducted by:

Date:

Structural region	Rock type and origin
	....

Drill core quality R.Q.D.			*Weathering*
Very good quality	90–100%	....	Unweathered ....
Good quality	75–90%	....	Slightly weathered ....
Fair quality	50–75%	....	Moderately weathered ....
Poor quality	25–50%	....	Highly weathered ....
Very poor quality	< 25%	....	Completely weathered ....

NOTE: RQD Rock Quality Designation in accordance with the method of

Ground Water

Inflow per 10m length litres/min.................
or
Water Pressure kPa ....
or
General conditions (completely dry, moist only, water under pressure, severe problems)

....

Strength of intact rock material

Designation	Uniaxial compressive strength	Point-load strength index
Very high:	Over 200MN/m^2	>8MN/m^2
High:	100–200MN/m^2	4–8MN/M^2
Medium:	50–100MN/m^2	2–4MN/m^2
Low:	25–50MN/m^2	1–2MN/m^2
Very low:	10–25MN/m^2	<1MN/m^2
	3–10MN/m^2	
	1–3MN/m^2	

Figure 1.18
Input Data Form: Engineering classification of jointed rock masses (after Bieniawski[133])

(cont)

Spacing of joints

		Set 1	Set 2	Set 3	Set 4
Very wide	Over 3m	....	....	....	....
Wide	1–3m	....	....	....	....
Moderately close	0.3–1m	....	....	....	....
Very close	50mm	....	....	....	....

NOTE: These values are obtained from a joint survey and not from borehole logs. Provide data for each joint set.

Strike and dip orientations

Set 1 Strike:............................ (average) (from .... to ....) Dip:............................ (Angle) (Direction)

Set 2 Strike:............................ (from .... to ....) Dip:............................

Set 3 Strike:............................ (from .... to ....) Dip:............................

Set 4 Strike:............................ (from .... to ....) Dip:............................

NOTE: Provide data for each joint set. Refer all directions to magnetic north

Condition of joints

Continuity		Set 1	Set 2	Set 3	Set 4
Not continuous,	no gouge	....	....	....	....
	with gouge	....	....	....	....
Continuous,	no gouge	....	....	....	....
	with gouge	....	....	....	....

Separation				
Very tight joints: Less than 0.1mm	....	....	....	....
Tight joints 0–1.1mm	....	....	....	....
Moderately open joints 1–5mm	....	....	....	....
Open joints more than 5 mm	....	....	....	....
Roughness				
Very rough surfaces	....	....	....	....
Rough surfaces	....	....	....	....
Slightly rough surfaces	....	....	....	....
Smooth surfaces	....	....	....	....
Slickensided surfaces	....	....	....	....
Joint wall rock				
Hardrock	....	....	....	....
Medium hard rock	....	....	....	....
Soft rock	....	....	....	....

NOTE. Provide data for each joint set

Figure 1.18 (continued) *(cont)*

Figure 1.18 (continued)

Major faults or folds
Describe major faults and folds specifying their locality, nature and orientations
General remarks and additional data
If gouge is present specify its type, thickness, continuity and consistency. Describe waviness of joints. Assess regional stresses. NOTE: the data on this form constitute the minimum required for engineering design. The geologist, should, however, supply any further information which he considers relevant.

Bienawski provided a summary of the steps which were to be taken when using his classification, they are:

1. The rock mass is divided into structural regions.
2. The input data form (Figure 1.18) is completed for each region and an assessment is made of the input data.
3. Importance ratings are allocated to each parameter as per Table 1.9a.
4. The ratings are summed to establish rock mass rating and its class (Table 1.9b).
5. The meaning of each rock mass is considered in terms of Table 1.9c, as applicable to the given projects.

ENGINEERING PERFORMANCE OF ROCKS

IGNEOUS AND METAMORPHIC ROCKS

The plutonic igneous rocks are characterised by granular texture, massive structure and relatively homogeneous composition. In their unaltered state they are essentially sound and durable with adequate strength for any engineering

Table 1.9 ENGINEERING CLASSIFICATION OF JOINTED ROCK MASSES (After Bieniawski[133])
(a) CLASSIFICATION PARAMETERS AND THEIR RATINGS

1	Strength of intact rock material	Point load strength index (MN/m^2)	> 8	4–8	2–4	1–2	Use of uniaxial compressive test preferred
		Uniaxial compressive strength (MN/m^2)	> 200	100–200	50–100	25–50	< 25
		Rating	10	5	2	1	0
2	Drill core quality RQD		90%–100%	75%–90%	50%–75%	25%–50%	< 25%
		Rating	20	17	14	8	3
3	Spacing of discontinuities		> 3 m	1–3 m	0.3–1 m	50–300 mm	< 50 mm
		Rating	30	25	20	10	5
4	Orientations of discontinuities		Very favourable	Favourable	Fair	Unfavourable	Very unfavourable
		Rating	15	13	10	6	3
5	Condition of discontinuities		Extremely tight. Very rough surfaces Not continuous No separation Hard joint wall rock	Very tight Slightly rough surfaces Separation < 0.1 mm Hard joint wall rock Not continuous	Tight Slightly rough surface Separation < 1 mm No gauge Soft joint wall rock	Open slickensided surfaces OR Gouge < 5 mm OR Joints open 1–5 mm Continuous joints	Very open Soft gouge > 5 mm thick OR Joints open > 5 mm Continuous joints
		Rating	20	15	10	5	0

6	Ground Water	Inflow per 10 m tunnel length	None	None	< 25 litres/min	25–125 litres/min	> 125 litres/min
					Slight	Moderate	Heavy
		OR Ratio $\frac{\text{joint water pressure}}{\text{major principal stress}}$		0	0.0–0.2	0.2–0.5	> 0.5
		OR General conditions		Completely dry	Moist only (interstitial water)	Water under moderate pressure	Severe water problems
		Rating	10	10	8	5	2

(b) ROCK MASS CLASSES AND THEIR RATINGS

Class No.	I	II	III	IV	V
Description	Very good rock	Good rock	Fair rock	Poor rock	Very poor rock
Rating	100 ← 90	90 ← 70	70 ← 50	50 ← 25	< 25

(c) MEANING OF ROCK MASS CLASSES

Class No.	I	II	III	IV	V
Average stand up time	10 years for 5 m span	6 months for 4 m span	1 week for 3 m span	5 hours for 1.5 m span	10 minutes for 0.5 m span
Cohesion of the rock mass	> 300 kPa	200–300 kPa	150–200 kPa	100–150 kPa	< 100 kPa
Friction angle of the rock mass	< 45°	40°–45°	35°–40°	30°–35°	< 30°
Caveability of ore	Very poor	Will not cave readily. Large fragments	Fair	Will cave readily Good fragmentation	Very good

requirement (Table 1.10). In some instances, however, intrusives may be highly altered, by weathering or hydrothermal attack. Furthermore fissure zones are by no means uncommon in granites. The rock mass may be very much fragmented along such zones, indeed it may be reduced to sand size material (see Terzaghi[132]) and it may have undergone varying degrees of kaolinisation.

Table 1.10 SOME PHYSICAL PROPERTIES OF IGNEOUS AND METAMORPHIC ROCKS

	Specific gravity	*Unconfined compressive strength* (MN/m²)	*Point load strength* (MN/m²)	*Shore scleroscope hardness*	*Schmidt hammer hardness*	*Youngs modulus* (× 10³ MN/m²)
Mount Sorrel Granite	2.68	176.4	11.3	77	54	60.6
Eskdale Granite	2.65	198.3	12.0	80	50	56.6
Dalbeattie Granite	2.67	147.8	10.3	74	69	41.1
Markfieldite	2.68	185.2	11.3	78	66	56.2
Granophyre (Cumbria)	2.65	204.7	14.0	85	52	84.3
Andesite (Somerset)	2.79	204.3	14.8	82	67	77.0
Basalt (Derbyshire)	2.91	321.0	16.9	86	61	93.6
Slate* (North Wales)	2.67	96.4	7.9	41	42	31.2
Slate† (North Wales)		72.3	4.2			
Schist* (Aberdeenshire)	2.66	82.7	7.2	47	31	35.5
Schist†		71.9	5.7			
Gneiss	2.66	162.0	12.7	68	49	46.0
Hornfels (Cumbria)	2.68	303.1	20.8	79	61	109.3

*Tested normal to cleavage or schistocity
†Tested parallel to cleavage or schistocity.

In humid regions valleys carved in granite may be covered with residual soils which extend to depths often in excess of 30 m. Fresh rock may only be exposed in valley bottoms which have actively degrading streams. At such sites it is necessary to determine the extent of weathering and the engineering properties of the weathered products. Generally the weathered product of plutonic rocks has a large clay content although that of granitic rocks is sometimes porous with a permeability comparable to that of medium grained sand.

Joints in plutonic rocks are often quite regular, steeply dipping structures in two or more intersecting sets. Sheet joints tend to be approximately parallel to the topographic surface. The sheet joints introduce a dangerous element of weakness into valley slopes. For example, in a consideration of Mammoth Pool Dam foundations on sheeted granite Terzaghi[140] observed that the most objectionable feature was the sheet joints orientated parallel to the rock surface. In the case of dam foundations such joints, if they remain untreated, may allow the escape of large quantities of water from the reservoir, and this may lead to the development of hydrostatic pressures in the rock downstream which are high enough to dislodge sheets of granite.

Generally speaking the older volcanic deposits do not prove a problem in foundation engineering, ancient lavas having strengths frequently in excess of 200 MN/m^2. But volcanic deposits of geologically recent age at times prove treacherous, particularly if they have to carry heavy loads such as concrete dams. This is because they often represent markedly anisotropic sequences in which lavas, pyroclasts and mud flows are interbedded. Hence foundation problems in volcanic sequences arise because weak beds of ash, tuff and mudstone occur within lava piles which give rise to problems of differential settlement and sliding. In addition weathering during periods of volcanic inactivity may have produced fossil soils, these being of much lower strength. The individual lava flows may be thin and transected by a polygonal pattern of cooling joints. They also may be vesicular or contain pipes, cavities or even tunnels.

Pyroclastics usually give rise to extremely variable foundation conditions due to wide variations in strength, durability and permeability. Their behaviour very much depends upon their degree of induration, for example, many agglomerates have a high enough strength to support heavy loads such as concrete dams and also have a low permeability. By contrast ashes are invariably weak and often highly permeable. One particular hazard concerns ashes, not previously wetted, which are metastable and exhibit a significant decrease in their void ratio on saturation. Tuffs and ashes are frequently prone to sliding. Montmorillonite is not an uncommon constituent in the weathered products of basic ashes.

Slates, phyllites and schists are characterised by textures which have a marked preferred orientation. Platey minerals such as chlorite and mica tend to segregate into almost parallel or subparallel bands alternating with granular minerals such as quartz and feldspar. This preferred alignment of platey minerals accounts for the cleavage and schistocity which typify these metamorphic rocks and means that slate, in particular, is notably fissile. Obviously such rocks are appreciably stronger across, than along the lineation (Table 1.10). The orientation of the planes of cleavage or schistocity in relation to the foundation structure can be significant. Not only does cleavage and schistocity adversely affect the strength of metamorphic rocks, it also makes them more susceptible to decay.

Generally speaking, however, slates, phyllites and schists weather slowly but the areas of regional metamorphism in which they occur have suffered extensive folding so that in places rocks may be fractured and deformed. Some schists, slates and phyllites are variable in quality, some being excellent foundations for heavy structures; others, regardless of the degree of their deformation or weathering, are so poor as to be wholly undesirable. For instance, talc, chlorite and sericite schists are weak rocks containing planes of schistocity only a millimetre or so apart. Some schists become slippery upon weathering and therefore fail under a moderately light load.

The engineering performance of gneiss is usually similar to that of granite. However, some gneisses are strongly foliated which means that they possess a texture with a preferred orientation. Generally this will not significantly affect their engineering behaviour. They may, however, be fissured in places and this can mean trouble. For example, it would appear that fissures opened in the gneiss under the heel of the Malpasset Dam, which eventually led to its failure (see Jaeger[141, 142]).

Fresh, thermally metamorphosed rocks such as quartzite and hornfels are very strong and afford excellent foundations. Marble has the same advantages and disadvantages as other carbonate rocks.

SANDSTONE

Simple defined a sandstone is an indurated sand, the latter being classified in terms of particle size. In other words a sandstone is a Elastic sediment in which mineral grains or rock fragments are bound together with cement and/or matrix. However, there are several fundamental types of sandstone depending on their composition, more particularly the proportions of feldspar, quartz and detrital matrix they contain. For example, quartz arenites contain over 95% quartz whereas greywackes contain 15 to 25% detrital matrix with little or no cement (see Pettijohn, Potter and Siever[143]).

Sandstones may vary from thinly laminated micaceous types to very thickly bedded varieties. Moreover they may be cross-bedded and are invariably jointed. With the exception of shaley sandstone, sandstone is not subject to rapid surface deterioration on exposure.

The dry density and especially the porosity of a sandstone are influenced by the amount of cement and/or matrix material occupying the pores. Usually the density of a sandstone tends to increase with increasing depth below the surface (see Bell[144]).

The compressive strength of a sandstone is influenced by its porosity, amount and type of cement and/or matrix material as well as the composition of the individual grains. Price[145, 146] showed that the strength of sandstones with a low porosity (less than 3.5%) was controlled by their quartz content and degree of compaction. In those sandstones with a porosity in excess of 6% he found that there was a reasonably linear relationship between dry compressive strength and porosity, for every 1% increase in porosity the strength decreased by approximately 4%. If cement binds the grains together then a stronger rock is produced than one in which a similar amount of detrital matrix performs the same function. However, the amount of cementing material is more important than the type of cement although if two sandstones are equally well cemented, one having a siliceous, the other a calcareous cement, then the former is the stronger. For example, ancient quartz arenites in which the voids are almost completely occupied with siliceous material are extremely strong with crushing strengths exceeding 240 MN/m^2. By contrast poorly cemented sandstones may possess crushing strengths less than 3.5 MN/m^2.

The pore water plays a very significant role as far as the compressive strength and deformation characteristics of a sandstone are concerned. For example, the Fell Sandstone and Bunter Sandstone may suffer a reduction of dry compressive strength on saturation of nearly 30 and 60% respectively (Table 1.11).

Moore derived a value of Young's modulus of 1100 MN/m^2 for the Bunter Sandstone from long term (up to 18 months) plate load testing. The modulus was found to increase with depth. At the highest loading, 5.6 MN/m^2, settlement did not exceed 4 mm. Creep accounted for 20 to 30% of the total settlement at loads varying between 0.3 and 1.5 MN/m^2, but at 3.0 and 5.6 MN/m^2 it was lower. Moore and Jones[148] concluded that at fairly low stresses the Bunter Sandstone, even though weathered near the surface provided a sound foundation. Moreover the rapid reduction in settlement with depth presumably means

Table 1.11 SOME PHYSICAL PROPERTIES OF ARENACEOUS SEDIMENTARY ROCKS

	Fell Sandstone (Rothbury)	*Chatsworth Grit (Stanton in the Peak)*	*Bunter Sandstone (Edwinstowe)*	*Keuper Waterstones (Edwinstowe)*	*Horton Flags (Helwith Bridge)*	*Bronllwyn Grit (Llanberis)*
Specific gravity	2.69	2.69	2.68	2.73	2.70	2.71
Dry density (Mg/m^3)	2.25	2.11	1.87	2.26	2.62	2.63
Porosity	9.8	14.6	25.7	10.1	2.9	1.8
Dry unconfined compressive strength (MN/m^2)	74.1	39.2	11.6	42.0	194.8	197.5
Saturated unconfined compressive strength (MN/m^2)	52.8	24.3	4.8	28.6	179.6	190.7
Point load strength (MN/m^2)	4.4	2.2	0.7	2.3	10.1	7.4
Scleroscope hardness	42	34	18	28	67	88
Schmidt hardness	37	28	10	21	62	54
Young modulus ($\times 10^3$ MN/m^2)	32.7	25.8	6.4	21.3	67.4	51.1
Permeability ($\times 10^{-6}$ mm/s)	1740	1960	3500	22.4	–	–

that simple spread foundation structures may be suitable even for sensitive buildings.

Many sandstones in the valleys excavated in the Millstone Grit series (Namurian) have been fractured by valley bulging or cambering. For example, spectacular valley bulges were recorded in the foundations of the Howden, Derwent and Ladybower Dams. In the latter the folding was present to a depth of almost 60 m (see Hill[149]). A further consequence of valley bulges is the opening up of tension fissures in sandstones forming the valley sides. These fissures run parallel to the valley and may be up to 250 mm wide close to the valley side, but they become progressively narrower and finally disappear when followed into the hillside. Valley bottom and valley side disturbance appear to be the result of a number of factors including stress relief, moisture up-take by underlying shales, artesian water pressures, valley notch concentration of stresses or frozen ground conditions.

Frequently thin beds of sandstone and shale are interbedded. Foundations on such sequences may give rise to problems of shear, settlement, and rebound, the magnitude of these factors depending upon the character of the shales. In some cases this even accentuates the undesirable properties of the shale by permitting access of water to the shale-sandstone contacts. Contact seepage may weaken shale surfaces and cause slides in dipping formations.

CARBONATE ROCKS

Carbonate rocks contain more than 50% of carbonate minerals, amongst which calcite and/or dolomite predominate. Normally the term limestone is used for those rocks in which the carbonate fraction is composed principally of calcite and the term dolostone is reserved for those rocks in which dolomite accounts for more than half the carbonate fraction. Chalk is a rather peculiar type of soft, remarkably pure limestone which is characteristically developed in the Upper Cretaceous.

Representative values of some physical properties of carbonate rocks are listed in Table 1.12. It can be seen that, generally, the density of these rocks

Table 1.12 SOME PHYSICAL PROPERTIES OF CARBONATE ROCKS

	Carboniferous limestone (Buxton)	*Magnesium limestone (Anston)*	*Ancaster freestone (Ancaster)*	*Bath stone (Corsham)*	*Middle chalk (Hillington)*	*Upper chalk (Northfleet)*
Specific gravity	2.71	2.83	3.70	2.71	2.70	2.69
Dry density (Mg/m^3)	2.58	2.51	2.27	2.30	2.16	1.49
Porosity (%)	2.9	10.4	14.1	15.6	19.8	41.7
Dry unconfined compressive strength (MN/m^2)	106.2	54.6	28.4	15.6	27.2	5.5
Saturated unconfined compressive strength (MN/m^2)	83.9	36.6	16.8	9.3	12.3	1.7
Point load strength	3.5	2.7	1.9	0.9	0.4	–
Scleroscope hardness	53	43	38	23	17	6
Schmidt hardness	51	35	30	15	20	9
Youngs modulus ($\times 10^3$ MN/m^2)	66.9	41.3	19.5	16.1	30.0	4.4
Permeability ($\times 10^{-9}$ m/s)	0.3	40.9	125.4	160.5	1.4	13.9

increases with age, whilst the porosity is reduced. Diagenetic processes mainly account for the lower porosities of the Carboniferous and Magnesion Limestones in particular. On the other hand the high porosity values of the Upper Chalk may be due to the presence of hollow tests and the complex shapes of the constituent particles. What is more the Upper Chalk is poorly cemented and has not suffered the same degree of pre-consolidation loading as the Middle and Lower Chalk.

Age often has an influence on the strength and deformation characteristics of carbonate and other sedimentary rocks. From Table 1.12 it can be seen that Carboniferous Limestone is generally very strong, conversely the Bath Stone (Great Oolite, Jurassic) is only just moderately strong. Similarly the oldest limestones tend to have the highest values of Young's modulus.

Thick-bedded, horizontally lying limestones relatively free from solution cavities afford excellent foundations. For example, intact samples of Carboniferous Limestone may have unconfined compressive strengths greater than 100 MN/m^2. On the other hand thin bedded, highly folded or cavernous limestones are likely to present serious foundation problems. A possibility of sliding may exist in thinly bedded, folded sequences. Similarly if beds are separated by layers of clay or shale these, especially when inclined, may serve as sliding planes and result in failure.

Limestones are commonly transected by joints. These have generally been subjected to varying degrees of solutioning so that some may gape. Sink holes may develop where joints intersect and these may lead to subterranean galleries and caverns. The latter are characteristic of thick, massive limestones. Sometimes solutioning produces a highly irregular, pinnacled surface on limestone pavements. The size, form, abundance, and downward extent of the aforementioned features depends upon the geological structure and the presence of interbedded impervious layers. Individual cavities may be open; they may be partially or completely filled with clay, silt, sand or gravel mixtures, or they may be water-filled conduits. Solution cavities present numerous problems in the construction of large foundations such as for dams, among which bearing strength and water-tightness are paramount.

Few sites are so bad that it is impossible to construct safe and successful structures upon them, but the cost of the necessary remedial treatment may be prohibitive. Dam sites should be abandoned where the cavities are large and numerous and extend to considerable depths.

As already pointed out, the removal of limestone in solution by ground water leads to the progressive opening of discontinuities and consequently to an increase in mass permeability. Loss of support to overlying rock or superficial material may on rare occasions take place and this can lead to collapse, resulting in subsidence. Subsidences in Liassic Limestone in Bridgend have been described by North[148]. Significant solution can occur where an adequate flow of unsaturated ground water takes place through a limestone mass but this occurs over an extensive time period which only exceptionally would be less than the life expectancy of the engineering works concerned. However, it could be important where extensive solutioning has been proceeding. For instance, the rate of solutioning at Hales Bar Dam, Tennessee, was very much higher than the normally accepted rates, 0.254 to 0.016 mm per year as compared with 0.041 to 0.099 mm per year (in water at 25°C the solubility of $CaCO_3$ ranges from 0.01 to 0.05 g/litre).

With the exception of certain horizons in the Lower Chalk of SE England which possesses an appreciable mud content, the Chalk is a remarkably pure, soft limestone usually containing over 95% calcium carbonate. Generally chalk can be divided into coarse and fine fractions. The coarse fraction, which may constitute 20 to 30%, falls within the 10 to 100 micron range. This contains material derived from the mechanical breakdown of large shelled organisms and, to a lesser extent, from foraminifera. The fine fraction, which takes the form of calcite particles about one micron or less in size, is almost entirely composed of coccoliths and may form up to, and sometimes over, 80% of certain horizons.

Bell[151] found a notable range in the dry density of chalk, which substantiated the work of Higginbottom[152]. For example, low values have been recorded from the Upper Chalk of Kent (1.35 to 1.64 Mg/m^3) whilst those from the Middle Chalk of Norfolk and the Lower Chalk of Yorkshire frequently exceed 2.0 Mg/m^3. In other words the density of the Chalk tends to increase with depth, presumably due to increasing overburden pressures, the void ratio decreasing. The porosity of chalk tends to range between 30 and 50%.

Carter and Mallard[153] found that chalk compressed elastically up to a critical pressure, the apparent preconsolidation pressure. Marked breakdown and substantial consolidation occurs at higher pressures. The apparent preconsolidation pressure is influenced by consolidation, cementation and possibly creep. They obtained coefficients of consolidation (c_V) and compressibility (m_V) similar to those found by Meigh and Early[154] and Wakeling[155]; c_V = 1135 m^2/yr, m_V = 0.019 m^2/MN. The unconfined strength of the Chalk ranges from moderately weak (much of the Upper Chalk) to moderately strong (much of the Lower Chalk of Yorkshire and the Middle Chalk of Norfolk) according to the strength classification recommended by the Geological Society[136]. However, the unconfined compressive strength of chalk undergoes a marked reduction when it is saturated. For example, according to Bell, some samples of Upper Chalk from Kent suffer a dramatic loss on saturation amounting to approximately 70%.

Samples from the Lower and Middle Chalk may show a reduction in strength averaging some 50%. The pore water also has a critical influence on the triaxial strengths of chalk and obviously its modulus of deformation (see Meigh and Early[154]).

Bell[151] noted that the mode of failure of the Upper Chalk when tested in triaxial conditions was influenced by the confining pressure. Diagonal shear failure occurred at the lower confining pressures but at 4.9 MN/m^2 confining pressure and above plastic deformation took place, giving rise to barrel shaped failures in which numerous small, inclined shear planes were developed. Meigh and Early[154] suggested that at high cell pressures the sample underwent disaggregation, which they demonstrated by wetting and drying failed samples, this causing a collapse of their structure.

Chalk is a non-elastic rock (E_i = less than 5×10^4 MN/m^2), and the Upper Chalk from Kent is particularly deformable, a typical value of Young's modulus being 5×10^3 MN/m^2. In fact the Upper Chalk exhibits elastic-plastic deformation, with perhaps incipient creep, prior to failure. The deformation properties of chalk in the field depend upon its hardness, and the spacing, tightness and orientation of its discontinuities. These values are also influenced by the amount of weathering it has undergone. Using these factors, Ward *et al*[156] classified the Middle Chalk at Mundford, Norfolk, into five grades, and showed that the value of Young's modulus varies with grade (Table 1.13). They pointed out that grades IV and V were largely the result of weathering and were therefore independent of lithology whereas grades I and II were completely unweathered so that the difference between them was governed by their lithological character. Accordingly grades V, IV and III occur in succession from the surface down whilst grade I may overlie grade II or vice versa.

Burland and Lord[159] observed that both the full-scale tank test and plate load

Table 1.13 CORRELATION BETWEEN GRADES AND THE MECHANICAL PROPERTIES OF MIDDLE CHALK AT MUNDFORD (after Ward *et al.**)

Grade	*Description*	*Approximate range of E (MN/m^2)*	*Bearing pressure causing yield (kN/m^2)*	*Creep properties*	*SPT N Value*† *(after Wakeling[155])*	*Rock mass factor (after Burland and Lord[154])*
V	Structureless melange. Unweathered and partly weathered angular chalk blocks and fragments set in a matrix if deeply weathered remoulded chalk. Bedding and jointing are absent	Below 500	Below 200	Exhibits significant creep	Below 15	0.1
IV	Friable to rubbly chalk. Unweathered or partially weathered chalk with bedding and jointing present. Joints and small fractures closely spaced, ranging from 10–60 mm apart	500–1000	200–400	Exhibits significant creep	15–20	0.1 to 0.2
III	Rubbly to blocky chalk. Unweathered medium to hard chalk with joints 60–200 mm apart. Joints open up to 8 mm, sometimes with secondary staining and fragmentary infillings	1000–2000	400–600	For pressures not exceeding 400 kN/m^2 creep is small and terminates in a few months	20–25	0.2 to 0.4
II	Medium hard chalk with widely spaced, closed joints. Joints more than 200 mm apart. Fractures irregularly when excavated, does not break along joints. Unweathered	2000–5000	Over 100	Negligible creep for pressure of at least 400 kN/m^2	25–35	0.6 to 0.8
I	Hard, brittle chalk with widely spaced, closed joints. Unweathered	Over 5000	Over 1000	Negligible creep for pressure of at least 400 kN/m^2	Over 35	Over 0.8

*Ward *et al.*[156] emphasized that their *N* classification was specifically developed for the site at Mundford and hence its application elsewhere should be made with caution

†The correlation between SPT *N* value and grade may be different in the Upper Chalk (see Dennehy[158])

tests at Mundford indicated that at low applied pressures even grade IV chalk behaves elastically. At higher pressures chalk exhibits yielding behaviour. They pointed out that for stresses up to 1.0 MN/m^2 the plate load tests showed that grades IV and V exhibit significant creep, and in the long term creep deflections may be considerably larger than immediate deflections, creep in grade III is smaller and terminates more rapidly whilst grades II and I undergo negligible creep.

Burland *et al*[160] found that settlements of a five-storey building founded in soft low grade chalk at Reading were very small. Their findings agreed favourably with those previously obtained at Mundford, as did the results of an investigation carried out at Basingstoke by Kee *et al*[161].

As in limestone, the discontinuities are the fundamental factors governing the mass permeability of chalk (see Ineson[162]). Chalk is also subject to solutioning along discontinuities. However, subterranean solution features tend not to develop in chalk since it is usually softer than limestone and so collapses as solution occurs. Nevertheless solution pipes and swallow holes are present in the Chalk, being commonly found near the contact of the Chalk with the overlying Tertiary and Drift deposits. West and Dumbleton[163] suggested that high concentrations of water, for example, run-off from roads, can lead to the re-activation of swallow holes and the formation of small pipes within a few years. They also recorded that new swallow holes often appear at the surface without warning after a period of heavy rain or following the passage of plant across a site. Moreover they found that voids can gradually migrate upwards through chalk due to material collapsing. Lowering of the chalk surface beneath overlying deposits due to solution can occur, disturbing the latter deposits and lowering their degree of packing. Hence the chalk surface may be extremely irregular in places.

Chalk during cold weather may suffer frost heave (see Lewis and Croney[164]), ice lenses up to 25 mm in thickness being developed along bedding planes. Higginbottom[152] suggested that a probable volume increase of some 20 to 30% of the original thickness of the ground may ultimately result.

Solifluction deposits known as head or coombe deposits, and which were formed under periglacial conditions during Pleistocene times, are commonly found along valley bottoms carved in the Chalk of southern England. Head is a poorly stratified, poorly sorted deposit of angular chalk fragments, frequently set in a matrix of remoulded, pasty, fine chalk detritus. Coombe deposits are to a varying degree cemented with secondary carbonate. Frost shattering also took place during Pleistocene times and its effects in the Chalk of southern England may extend to depths of several metres.

EVAPORITIC ROCKS

Representative specific gravities and dry densities of gypsum, anhydrite, rock salt and potash are given in Table 1.14, as are the porosity values. Anhydrite according to the classification of unconfined compressive strength (Geological Society[136]) is a strong rock, gypsum and potash are moderately strong, whilst rock salt is moderately weak (Table 1.14). Values of Young's modulus are also given in Table 1.14, from which it can be ascertained that gypsum and anhydrite have high values of modulus ratio whilst potash and rock salt have medium values. Evaporitic rocks exhibit varying degrees of plastic deformation prior to failure, for example, in rock salt the yield strength may be as little as one tenth the ultimate compressive strength, whereas anhydrite undergoes comparatively

Table 1.14 SOME PHYSICAL PROPERTIES OF EVAPORITIC ROCKS

	Gypsum (Sherburn in Elmet)	*Anhydrite (Sandwith)*	*Rock salt (Winsford)*	*Potash (Loftus)*
Specific gravity	2.36	2.93	2.2	2.05
Dry density (Mg/m^3)	2.19	2.82	2.09	1.98
Porosity (%)	4.6	2.9	4.8	5.1
Unconfined compressive strength (MN/m^2)	27.5	97.5	11.7	25.8
Point load strength (MN/m^2)	2.1	3.7	0.3	0.6
Schleroscope hardness	27	38	12	9
Schmidt hardness	25	40	8	11
Young's modulus ($\times 10^3$ MN/m^2)	24.8	63.9	3.8	7.9
Permeability ($\times 10^{-10}$ m/s)	6.2	0.3	–	–

little plastic deformation. Creep may account for anything between 20 and 60% of the strain at failure when these evaporitic rocks are subjected to incremental creep tests. Rock salt is most prone to creep. However, Justo and Zapico[165] recorded that the amount of settlement which occurred when gypsum was subjected to plate load testing, the maximum load being 1.2 MN/m^2, was negligible.

Gypsum is more readily soluble than limestone, for example, 2100 ppm can be dissolved in non-saline waters as compared with 400 ppm. Sink holes and caverns can therefore develop in thick beds of gypsum (see Eck and Redfield[166]) more rapidly than they can in limestone. Indeed in the United States they have been known to form within a few years where beds of gypsum are located beneath dams. Extensive surface cracking and subsidence has occurred in parts of Oklahoma and New Mexico due to the collapse of cavernous gypsum (see Brune[167]). The problem is accentuated by the fact that gypsum is weaker than limestone and therefore collapses more readily.

Kendal and Wroot[168] quoted vivid accounts of subsidences which occurred in the Ripon area in the eighteenth and nineteenth centuries due to the solution of gypsum. They wrote that wherever beds of gypsum approach the surface craters have been formed by the collapse of overlying rocks into areas from which gypsum has been removed by solution. However, where gypsum is effectively sealed from the ingress of water by overlying impermeable strata such as marl, solutioning does not occur (see Redfield[169]). The solution of gypsum gives rise to sulphate bearing ground waters which mean that normal concrete may suffer accordingly.

Uplift is a problem associated with anhydrite. This takes place when anhydrite is hydrated to form gypsum; in so doing there is a volume increase of between 30 and 58% which exerts pressures that have been variously estimated between 2 and 69 MN/m^2. It is thought that no great length of time is required to bring about such hydration. When it occurs at shallow depths it causes expansion but the process is gradual and is usually accompanied by the removal of gypsum in solution. At greater depths anhydrite is effectively confined during the process. This results in a gradual build-up of pressure and the stress is finally liberated as an explosive force. According to Brune[167] such uplifts in the United States have taken place beneath reservoirs, these bodies of water providing a constant supply for the hydration process, perculation taking place via cracks and fissures. Examples are known of the ground surface being elevated by about 6 m. The rapid, explosive movement causes strata to fold, buckle and shear which further facilitates access of water into the ground.

SILTSTONES

Siltstones may be massive or laminated, the individual laminae being picked out by mica and/or carbonaceous material. Micro-cross bedding is frequently developed and in some siltstones the laminations may be convoluted. Siltstones have a high quartz content with a predominantly siliceous cement. They therefore tend to be hard, tough rocks (Table 1.15). Frequently siltstones are interbedded with shales or fine grained sandstones, the siltstones occurring as thin

Table 1.15 ENGINEERING PROPERTIES OF SOME COAL MEASURES ROCKS

	Mudstone	*Siltstone*	*Shale*	*Barnsley Hards Coal*	*Deep Duffryn Coal*
Specific gravity	2.69	2.67	2.71	1.5	1.2
Dry density (Mg/m^3)	2.32	2.43	2.35	–	–
Dry unconfined compressive strength (MN/m^2)	45.5	83.1	20.2	54.0	18.1
Saturated unconfined compressive strength (MN/m^2)	21.3	64.8	–	–	–
Point load strength (MN/m^2)	3.8	6.2	–	4.1	0.9
Scleroscope hardness	32	49	–	–	–
Schmidt hardness	27	39	–	–	–
Youngs modulus ($\times 10^3$ MN/m^2)	25	45	5.2	26.5	–

ribs. Like sandstones, their disintegration is governed by their fracture pattern. After several months of weathering debris in excess of cobble size may be produced. Subsequent degradation down to component grain size takes place at a very slow rate.

SHALES

Shales are the most abundant sediments accounting for approximately half the stratigraphical column (see Kuenen[170]). There is no sharp distinction between shales and mudstones, one grading into the other. Shales, however, are characterised by their lamination. Shale differs from residual clay in grain size distribution, the average shale consisting of about two thirds silty material and one third clay fraction. There are, of course, exceptions. Non-silty siliceous shales are usually diatom rich or derived from volcanic ash. Bauxitic shales, black pyritous shales and sideritic shales are also usually fine grained. Shales may also contain appreciable quantities of lime or gypsum, indeed calcareous shales frequently grade into shaley limestone. Carbonaceous shales are those which accumulated slowly under anaerobic conditions and are rich in sulphur compounds.

Quartz usually accounts for approximately one third of a normal shale, clay minerals, including micas and chlorite, for another third and other minerals such as feldspar, calcite, dolomite, pyrite, hematite and limonite, together with some carbonaceous matter, make up the remainder. The mineral content of shales influences their geotechnical properties, the most important factor in this respect being the quartz-clay minerals ratio. For example, the liquid limit of clay shales increases with increasing clay mineral content, the amount of montmorillonite, if present, being especially important. Mineralogy also affects the activity of an argillaceous material, again this increases with clay mineral content, particularly with increasing content of montmorillonite. Activity influences the slaking characteristics of a shale.

Compaction with concomitant recrystallisation on the one hand and the parallel orientation of platey minerals, notably micas, on the other give rise to the fissility of shales. An increasing content of siliceous or calcareous material gives a less fissile shale whilst carbonaceous shales are exceptionally fissile. Lamination effects an important control on the breakdown of shales (see Taylor and Spears[122]). Other controls on the breakdown of shaley materials include air breakage and dispersal of colloidal material (see Badger *et al*[120]). A feature of the breakdown of shales and mudstones is their disintegration to produce silty clays (see Grice[171]).

Shales vary widely in their engineering behaviour, which to a large extent depends upon their degree of compaction and cementation, indeed Mead[172] divided shales into compaction and cementation types in his classification (Tables 1.16a and 1.16b). The cemented shales are invariably stronger and more durable. Marine shales may be impregnated with carbonate cement which makes them appreciably stronger and indeed they may grade into impure limestone. Carbonaceous shales contain a significant proportion of organic matter and are therefore softer.

Low grade compaction shales undergo complete disintegration after several cycles of drying and wetting, whilst well cemented shales are resistant. Indeed De Graft-Johnson *et al*[174] found that the compacted variety of Accra Shale could be distinguished from the cemented variety by wetting and drying tests. The compacted shales generally crumbled to fine material after 2 or 3 cycles whilst the cemented samples withstood 6 cycles, none of the samples exceeding a loss of 8%. The primary problem attributable to slaking of shale during construction is that upon exposure it becomes coated with mud when wetted, which prevents the development of bond between concrete and rock. This can be prevented by coating the surface with a protective material, or by pouring a protective concrete cover immediately after exposure. Slaking of shales after construction causes ravelling and spalling of cut slopes and is sometimes the cause of the undermining and collapse of more competent beds, but this is rarely a serious problem. Mudstones tend to break down along irregular fracture patterns, which when well developed, can mean that these rocks disintegrate within one or two cycles of wetting and drying.

The degree of packing, and hence the porosity, void ratio and density of a shale, depend on its mineral composition and grain size distribution, its mode of sedimentation, its subsequent depth of burial and tectonic history, and the effects of diagenesis. The porosity of shale may range from slightly under 5 to just over 50% with natural moisture contents of 3 to 35%. Argillaceous materials are capable of undergoing appreciable suction before pore water is removed, drainage commencing when the necessary air-entry suction is achieved (about pF 2). Under increasing suction pressure the incoming air drives out water from a shale and some shrinkage takes place in the fabric before air can offer support. Generally as the natural moisture content and liquid limit increase so the effectiveness of soil suction declines.

It would appear that the strength of compacted shales decreases exponentially with increasing void ratio and moisture content. However, in cemented shales the amount and strength of the cementing material are the important factors influencing intact strength. Unconfined compressive strength tests on Accra Shales carried out by De Graft-Johnson *et al*[174] indicated that the samples usually failed at strains between 1.5 and 3.5%. The compressive strengths varied from 200 kN/m^2 to 20 MN/m^2, with the values of the modulus of elasticity ranging from 6.5 to 1400 MN/m^2. Those samples which exhibited the high strengths were generally cemented types. In fact values of Young's modulus as high as 15000 MN/m^2 have been recorded from well cemented shales. The higher the degree of fissility possessed by a shale the greater the anisotropy with regard to strength, deformation and permeability. For instance, the influence of fissibility on Young's modulus can be illustrated by two values quoted by

Table 1.16a CLASSIFICATION OF SHALE (After Mead[173])

SHALE

COMPACTION SHALE	
CLAY SHALE	50% or more clay particles which may or may not be true minerals
SILTY SHALE	25% to 45% silt sized particles. Silt may be in thin layers between clayey shale bands
SANDY SHALE	25% to 45% sand sized particles. Sand may be in thin layers between clayey shale bands
BLACK SHALE	Organic rich, splits into thin semi-flexible sheets

CEMENTED SHALE	
CALCEROUS SHALE	20% to 35% $CaCo_3$ (Marls and shaly chalk 35% to 65% $CaCo_3$)
SILICEOUS SHALE	70%–85% amorphous silica often highly siliceous volcanic ash (quartzose shale – detrital quartz)
FERRUGINOUS SHALE	(25%–35% Fe_2O_3) (Potassic shale – 5%–10% potash)
CARBONACEOUS SHALE (Oil shale, Bone coal)	Carbonaceous matter (3% to 15%) tends to bond constituents together and imparts a certain degree of toughness
CLAY BONDED SHALE	Welded by recrystallization of clay minerals, or by other diagenetic bonds

Table 1.16b ENGINEERING EVALUATION OF SHALES (After Underwood[173])

Physical properties			*Probable* in situ *behaviour*						
Laboratory tests and in situ observations	*Average range of values* *Unfavourable*	*Favourable*	*High pore pressure*	*Low bearing capacity*	*Tendency to rebound*	*Slope stability problems*	*Rapid sinking*	*Rapid erosion*	*Tunnel support problems*
Compressive strength (kN/m²)	350–2070	2070–3500	√	√					
Modulus of elasticity (MN/m²)	140–1400	1400–14000		√					√
Cohesive strength (kN/m²)	35–700	700–>10500			√	√			√
Angle of internal friction, degrees	10–20	20–65			√	√			√
Dry density Mg/m³	1.12–1.78	1.78–2.56	√					√(?)	
Potential swell (%)	3–15	1–3			√	√		√	√
Natural moisture content (%)	20–35	5–15	√			√			
Coefficient of permeability m/s	10^{-7}–10^{-12}	$>10^{-7}$	√			√	√		
Predominant clay minerals	Montmorillonite or illite	Kaolinite and chlorite	√			√			
Activity ratio	0.75–>20	0.35–0.75				√			
Wetting and drying cycles	Reduces to grain sizes	Reduces to flakes					√	√	
Spacing of rock defects	Closely spaced	Widely spaced		√		√		√(?)	√
Orientation of rock defects	Adversely oriented	Favourable oriented		√		√			√
State of stress	> Existing overburden load	≅ Overburden load			√	√			√

Note: According to S. Irmay (Israel Journal of Technology, Vol 6, No. 4, pp 165–172, 1968), the maximum possible $\phi = 47.4°$

Chappell[175], 6000 and 7250 MN/m^2, for cemented shale tested parallel and normal to the lamination respectively.

According to Burwell[176] well-cemented shales, under structurally sound conditions, present few problems for large structures such as dams, though their strength limitations and elastic properties may be factors of importance in the design of concrete dams of appreciable height. They, however, have lower moduli of elasticity and generally lower shear values than concrete and therefore in general are unsatisfactory foundation materials for arch dams.

The problem of settlement in shales generally resolves itself into one of reducing the unit bearing load by widening the base of structures or using spread footings. In some cases appreciable differential settlements are provided for by designing articulated structures capable of taking differential movements of individual sections without damage to the structure. Severe settlements may take place in low-grade compaction shales. Uplift frequently occurs in excavations in shales and is attributable to swelling and heave (see below). Differential rebound movements in the foundations require special design provisions.

The stability of slopes in excavations can be a major problem in shale both during and after construction. This problem becomes particularly acute in dipping formations and in formations containing expansive clay minerals.

Sulphur compounds are frequently present in shales, clays, mudstones and marls. An expansion in volume large enough to cause structural damage can occur when sulphide minerals such as pyrite and marcasite suffer oxidation to give anhydrous and hydrous sulphates.

According to Fasiska, Wagenblast and Dougherty[177] the pyrite structure may be regarded as a stacking of almost close packed hexagonal sheets of sulphide ions with iron ions in the interstices between the sulphide layers. The packing density is related to the radius of the sulphide ion which is 1.85 Å (volume = 26.1 Å). In the sulphate structure each atom of sulphur is surrounded by four atoms of oxygen in tetrahedral coordination. The packing density is related to the radius of the sulphate ion which is 2.8 Å, giving a volume of 92.4 Å. This represents an increase in volume per packing unit of approximately 35%. Hydration involves a further increase in volume. In fact such a reaction is electrolytic, that is, water is required and the sulphate ion exists in solution. Any cation in the system may cause the precipitation of sulphate crystals. If calcium carbonate is present, gypsum may be formed, which may give rise to an eightfold increase in volume over the original sulphide, exerting pressures of up to about 0.5 MN/m^2. This leads to further disruption and weakening of the rocks involved.

Penner *et al*[178] quoted a case of heave in a black shale of Ordovician age in Ottawa which caused displacement of the basement floor of a three storey building. The maximum movement totalled some 107 mm, the heave rate being almost 2 mm per month. When examined the shale in the heaved zone was found to have been altered to a depth of between 0.7 and 1 m. Beneath, the unaltered shale contained numerous veins of pyrite, indeed the sulphur content was as high as 1.6%. The heave was attributable to the alteration of pyrite to gypsum and jarosite, these forming in the fissures and between the laminae of the shales in the altered zone. Measurements of the pH gave values ranging from 2.8 to 4.4, and it was therefore concluded that the alteration of the shales was the result of biochemical weathering brought about by autotrophic bacteria. The heaving was therefore arrested by creating conditions unfavourable for bacterial growth. This was done by neutralising the altered zone by introducing a potassium hydroxide solution into the examination pits. The water table in the altered zone was also kept artificially high so that the acids would be diffused and washed away, and to reduce air entry.

Similarly sulphuric acid and sulphate are produced when gypsum is subjected to weathering. Aqueous solutions of sulphate and sulphuric acid react with tricalcium aluminate in Portland cement to form calcium sulphoaluminate or ettringite. This reaction is accompanied by expansion. The rate of attack is very much influenced by the permeability of the concrete or mortar and the position of the water table. For example, sulphates can only continue to reach cement by movement of their solutions in water. Thus if a structure is permanently above the water table it is unlikely to be attacked.

By contrast below the water table movement of water may replenish the sulphates removed by reaction with cement, thereby continuing the reaction. Concrete with a low permeability is essential to resist sulphate attack, hence it should be fully compacted. Sulphate resistant cements, that is, those in which the tricalcium aluminate is low can also be used for this purpose (Table 1.17). Foundations used to be made larger than necessary to counteract sulphate attack but it is now more economical to protect them by impermeable membranes or bituminous coatings.

MARL

Marl is a term which has been assigned various meanings, although it has recently been defined by Pettijohn[180] as a rock with 35 to 65% carbonate and a complementary content of clay. However, this definition cannot be applied to many of the rocks in Britain which are referred to as marls, for example, most of the marls of the Keuper series contain less than 20% carbonate material. Such rocks according to the classification of clay-lime carbonate mixtures, after Barth, Correns and Eskola[181], are marly clays or mudstones. In the account which follows the term marl is used in the latter context.

Marls are particularly well developed in the Old Red Sandstone, Permian and Trias systems. It is assumed that many of these marls were formed by the deposition of vast quantities of fine sediment in shallow inland lakes or seas in regions of high evaporation, others may represent fine grained aeolian deposits. These rocks are extensively distributed in England and Wales, for example, the Red Marls of the Lower Old Red Sandstone in South Wales outcrop from Pembrokeshire to Brecon Beacons, the Middle and Upper Marls of the Permian occur, with a north-south outcrop, in Yorkshire. But the Keuper Marl (Trias) is the most extensively distributed occurring in Yorkshire, Nottinghamshire, Lancashire, Cheshire, the Midlands, Gloucestershire and south west England. Because the marls of the Keuper series are extensively distributed and because they have been more thoroughly investigated as far as their geotechnical properties are concerned, attention will be focused on them alone.

The marls of the Keuper series consist of between 50 and 90% clay minerals. The marl may contain thin veins or beds of gypsum, in such cases the ground waters contain sulphates. According to Dumbleton[182] illite accounts for 28 to 56% of the clay minerals and chlorite may total some 39%. Usually more than half the chlorite is of the swelling type. Quartz tends to vary between 5 and 35%. The other minerals include calcite and dolomite (which usually comprise

Table 1.17 REQUIREMENTS FOR CONCRETE EXPOSED TO SULPHATE ATTACK

Class	*Concentration of sulphates expressed as SO_3: In soil, total SO_3*	*In ground water*	*Type of cement*	*Minimum cement content, Nominal maximum size of aggregate: 40 mm*	*20 mm*	*10 mm*	*Maximum free water/ cement ratio*
	%	*parts per 100 000*		*kg/m³*	*kg/m³*	*kg/m³*	
1	less than 0.2	less than 30	Ordinary Portland or Portland-blastfurnace	240	280	330	0.55
2	0.2–0.5	30–120	Ordinary Portland or Portland-blastfurnace	290	330	280	0.50
			Sulphate-resisting Portland	240	280	330	0.55
			Supersulphated	270	310	360	0.50
3	0.5–1.0	120–250	Sulphate-resisting Portland or supersulphated	290	330	380	0.50
4	1.0–2.0	250–500	Sulphate-resisting Portland or supersulphated	330	370	420	0.45
5	over 2	over 500	As for Class 4, but with the addition of adequate protective coatings of inert material such as asphalt or bituminous emulsions reinforced with fibreglass membranes				

Notes. This table applies only to concrete made with aggregates complying with the requirements of BS 882 or BS 1047 placed in near-neutral groundwaters of pH 6 to pH 9, containing naturally occurring sulphates but not contaminants such as ammonium salts. Concrete prepared from ordinary Portland cement would not be recommended in acidic conditions (pH 6 or less); sulphate-resisting Portland cement is slightly more acid-resistant but no experience of large-scale use in these conditions is currently available. Supersulphated cement has given an acceptable cement, provided that the concrete is dense and prepared with a free water/cement ratio of 0.40 or less in mineral acids down to pH 3.5.

The cement contents given in Class 2 are the minima recommended by the manufacturers. For SO_3 contents near the upper limit of Class 2, cement contents above these minima are advised.

For severe conditions, e.g. thin sections, sections under hydrostatic pressure on one side only and sections partly immersed, consideration should be given to a further reduction of water/cement ratio and, if necessary, an increase in cement content to ensure the degree of workability needed for full compaction and thus minimum permeability.

(*After BRE Digest 174. Crown Copyright.* Reproduced by permission of the Controller of HMSO and by courtesy of the Director of the Building Research Establishment, UK)

less than 20%) and hematite -- one or two per cent. Occasionally other clay minerals such as sepiolite and palygorskite have been found.

The clay particles tend to be aggregated mainly into silt-size units, the aggregated structure being extremely variable (see Davis[186]). Particles composing the aggregate are held together by cement. Sherwood[184] suggested that silica might be the cementing agent whilst Lees[185] assumed that the clay particles were bound together by physical forces. The former explanation seems the more likely. Aggregation leads to the lack of correlation between consistency limits and shear strength on the one hand and clay content on the other. Because engineering behaviour is controlled by the aggregates, rather than the individual clay minerals, the plasticity, according to Davis[186], is lower than would be expected. He consequently proposed the aggregation ratio as a means of assessing the degree of aggregation. The aggregation ratio was defined as the percentage weight of clay as determined by mineralogical analysis, expressed as a ratio of the percentage weight of clay particles determined by sedimentation techniques. On the other hand Sherwood[184] maintained that aggregation did not give anomalous plasticity values for Keuper marls and that they can generally be classified as materials of low to medium plasticity (Table 1.18). The activity of the Keuper marls increases as the degree of aggregation increases.

Table 1.18 SOIL CLASSIFICATION TESTS ON KEUPER MARL (From Sherwood[184])

	Clay content by sedimentation (%)	*Clay content by X-ray analysis* (%)	*Liquid limit* (%)	*Plastic limit* (%)	*Plasticity index* (%)
1	26	94	71	40	31
2	36	58	33	19	14
3	30	87	46	28	25
4	12	77	48	29	19

These marls are very often fissured, weathering and water penetrating the fissures and thereby further reducing the strength of the material. Certain marls exhibit rapid softening when exposed to wet conditions. The fissures close with increasing depth. Skempton and Davis (see Chandler[187]) proposed the following classification of weathered Keuper marl:

Zone	*Description*
V Fully weathered	Matrix only
IV Highly weathered	Matrix with occasional pellets < 3 mm diameter
III Moderately weathered	Matrix with frequent lithorelicts up to 25 mm
II Slightly weathered	Angular blocks of unweathered marl with virtually no matrix
I Unweathered	Marl (often fissured)

Weathering first develops along fissures in zone II material, the weathered product consisting of a thin veneer of silt. A significant proportion of the marl is weathered in zone III, the unweathered material occurring as angular fragments set in a weathered matrix which is predominantly silty. The water content of the matrix exceeds that of the lithorelicts. In zone IV the lithorelicts are mainly of coarse sand size, and the marl has lost much of its silty texture, indeed up to 50% may be composed of clay-sized particles. This indicates that particle aggregation is broken down upon weathering. Little or no trace of the original structure now remains and the material has a lower permeability than has zone III, 5×10^{-9} to 5×10^{-10} m/sec as compared with 1×10^{-8} to 1×10^{-9} m/sec respectively. Finally the marl is completely weathered, becoming a plastic, slightly silty, clay. Chandler[187] showed that highly and fully weathered marl could be distinguished from material from zones I, II and III by their particle size distribution and plasticity index (Table 1.19).

Table 1.19 SOME INDEX PROPERTIES OF KEUPER MARL (AFTER CHANDLER[187])

Index property	*Weathering zone*		
	I and II	*III*	*IV*
Bulk density, Mg/m^3	2.5–2.3	2.3–2.1	2.2–1.8
Dry density, Mg/m^3	2.4–1.9	2.1–1.8	1.8–1.4
Natural moisture content (%)	5–15	12–20	18–35
Liquid limit (%)	25–35	25–40	35–60
Plastic limit (%)	17–25	17–27	17–33
Plasticity index	10–15	10–18	17–35
% clay size (BS 1377)	10–35	10–35	30–50
Aggregation ratio (A_r)	10–2.5	10–2.5	2.5
c' (kN/m^2)	$\geqslant$ 27.6	$\leqslant$ 17.2	$\leqslant$ 17.2
ϕ'	40°	42°–32°	32°–25°
ϕ'_r	32°–23°	29°–22°	24°–18°
$\dfrac{\tau_{max} - \tau_{res}^*}{\tau_{max}}$ (%)	55	55–30	35–20

*Percentage reduction from peak to residual strength

An extensive review of the Keuper Marl as a foundation material has been provided by Meigh[188].

SEATEARTHS AND COAL SEAMS

Seatearths are almost invariably found beneath coal seams. Indeed they represent fossil soils and as such are characterised by the presence of fossil rootlets. These tend to destroy the lamination and bedding. The character of a seatearth depends on the type of deposit which was laid down immediately before the establishment of plant growth. If this was mud then a fireclay underlies the coal, whereas if silts and sands were deposited then a gannister was subsequently formed. Many gannisters are pure siltstones and because they are usually well cemented they are hard and strong. Fireclays with a low quartz content are typically highly slickensided and break easily along randomly orientated listric surfaces. The presence of listric surfaces may mean that a fireclay will disintegrate within a few wetting and drying cycles.

Most coal seams are composite in character. At the base the coal is often softer and is sometimes simply referred to as 'bottom coal'. Bright coal is often of most importance in the centre of a seam whilst dull coal may predominate in the upper part of a seam. Coal generally breaks into blocks which have three faces approximately perpendicular to each other. These surfaces are referred to as cleat. The cleat direction is usually fairly constant and is best developed in bright coal. Cleat perhaps may be coated with films of mineral matter, commonly calcite, ankerite and pyrite. Coal seams may split or be replaced, totally or partially, by washouts. Coal is more suspect to mechanical than chemical weathering. Its crushing strength varies but generally it is less than 20 MN/m^2 (Table 1.15).

SUBSURFACE WATER AND FOUNDATIONS

The principal source of ground water is precipitation or meteoric water. The amount of water which perculates into the ground depends upon how that precipitation is dispersed, that is, on what proportions are assigned to immediate run-off and to evapotranspiration, the remainder constituting the proportion allotted to perculation. Infiltration is the movement of surface water into the ground, perculation being its subsequent movement to the zone of saturation, but in reality one cannot be separated from the other.

The retention of water in a soil depends upon the capillary force and the molecular attraction of the particles. When the pores in a soil become thoroughly wetted the capillary force declines so that gravity becomes more effective. In this way downward perculation can continue after infiltration has ceased, but as the soil dries capillarity increases in importance. No further perculation occurs after the capillarity and gravity forces are balanced. Thus water perculates into the zone of saturation when the retention capacity is satisfied. The pores within the zone of saturation are filled with water, frequently referred to as phreatic water. The upper surface of this zone is therefore known as the phreatic surface but is more commonly termed the water table. Above the zone of saturation is the zone of aeration in which both air and water occupy the pores.

Capillary movement in a soil refers to the movement of moisture through the minute pores between the soil particles which act as capillaries. It takes place as a consequence of surface tension, therefore moisture can rise from the water table. This movement, however, can occur in any direction, not just vertically. It occurs whenever evaporation takes place from the surface of the soil, thus exerting a 'surface tension pull' on the moisture, the forces of surface tension increasing as evaporation proceeds. Accordingly capillary moisture is in hydraulic continuity with the water table and is raised against the force of gravity, the degree of saturation decreasing from the water table upwards. Equilibrium is attained when the forces of gravity and surface tension are balanced.

The boundary separating capillary moisture from the gravitational water in the zone of saturation is, as would be expected, ill defined and cannot be determined accurately. That zone immediately above the water table which is saturated with capillary moisture is referred to as the closed capillary fringe, whilst above this air and capillary moisture exist together in the pores of the open capillary fringe. The depth of the capillary fringe is largely dependent upon the

Table 1.20 CAPILLARY RISES AND PRESSURES IN SOILS (After Jumikis[189])

Soil	*Capillary rise* (mm)	*Capillary pressure* (kN/m^2)
Fine gravel	Up to 100	Up to 1.0
Coarse sand	100–150	1.0–1.5
Medium sand	150–300	1.5–3.0
Fine sand	300–1000	3.0–10.0
Silt	1000–10000	10.0–100.0
Clay	Over 10000	Over 100.0

particle size distribution and density of the soil mass, which in turn influence pore size. In other words the smaller the pore size, the greater is the depth. For example, capillary moisture can rise to great heights in clay soils (Table 1.20) but the movement is very slow. In soils which are poorly graded the height of the capillary fringe generally varies, whereas in uniformly textured soils it attains roughly the same height.

Where the water table is at shallow depth and the maximum capillary rise is large, moisture is continually attracted from the water table due to evaporation from the surface so that the uppermost soil is near saturation. For instance, under normal conditions peat deposits may be assumed to be within the zone of capillary saturation. In other words, the height to which water can rise in peat by capillary action is greater than the depth below ground to which the water table is reduced by drainage. The coarse fibrous type of peat, containing appreciable sphagnum, may be an exception.

Drainage of capillary moisture cannot be effected by the installation of a drainage system within the capillary fringe as only that moisture in excess of that retained by surface tension can be removed, but it can be lowered by lowering the water table. The capillary ascent, however, can be interrupted by the installation of impermeable membranes or layers of coarse aggregate. These two methods can be used in the construction of embankments, or more simply the height of the fill can be raised.

At each point where moisture minisci are in contact with soil particles the forces of surface tension are responsible for the development of capillary or suction pressure (Table 1.20). This tends to force the soil particles together and these compressive stresses contribute towards the strength and stability of the soil. There is a particular suction pressure for a particular moisture content in a given soil, the magnitude of which is governed by whether it is becoming wetter or drier. In fact as a clay soil dries out the soil suction may increase to the order of several thousands of kilonewtons per square metre. However, the strength of a soil attributable to soil suction is only temporary and may be destroyed upon saturation.

Below the water table the water contained in the pores is under normal hydrostatic load, the pressure increasing with depth. Because these pressures exceed atmospheric pressure they are designated positive pressures. On the other hand the pressures existing in the capillary zone are less than atmospheric and so are termed negative pressures. Thus the water table is usually regarded as a datum of zero pressure between the positive pore pressure below and the negative above.

POROSITY AND PERMEABILITY

Porosity and permeability are the two most important factors governing the accumulation, migration and distribution of water in soils and rocks. However, porosity and permeability are not necessarily as closely related as would be expected, for instance, very fine textured sandstones frequently have a higher porosity than coarser ones though the latter are more permeable. In other words the size of the pores is all important as far as the permeability of a formation is concerned. It is not uncommon to find variations in both porosity and permeability throughout a formation. The factors affecting the porosity of a rock include particle size distribution, grain shape, fabric, degree of compaction and cementation, solution effects, and lastly mineralogical composition, particularly the presence of clay particles. The highest porosity is commonly obtained when the grains are all of the same size.

In ordinary hydraulic usage a substance is termed permeable when it permits the passage of a measurable quantity of fluid in a finite period of time and impermeable when the rate at which it transmits that fluid is slow enough to be

Table 1.21 VALUES OF SOIL PERMEABILITIES

Degree of permeability	*Range of Coefficient of permeability (k)* m/sec	*Soil type*
High	10^{-3}	Medium and coarse gravel
Medium	10^{-3} to 10^{-5}	Fine gravel; coarse, medium and fine sand; dune sand; clean sand-gravel mixtures
Low	10^{-5} to 10^{-7}	Very fine sand, silty sand, loose silt, loess, well fissured clays
Very low	10^{-7} to 10^{-9}	Dense silt, dense loess clayey silt, poorly fissured clays
Impermeable	10^{-9}	Unfissured clays

negligible under existing temperature-pressure conditions (Table 1.21). The flow through a unit cross section of material is modified by temperature, hydraulic gradient and the coefficient of permeability. The latter is affected by the range of grain size and shape, stratification, consolidation and cementation of the material. Temperature changes affect the flow rate of a fluid by changing its viscosity. The rate of flow is commonly assumed to be directly proportional to the hydraulic gradient but this is not always so in practice.

Determination of the permeability of a sample in the laboratory does not necessarily bear any relation to the permeability of the formation concerned in the field. This is because most soil masses are anisotropic in character and in rock masses the discontinuities are the most important conduits for water movements (see Ineson[162]; Bell[190]). Discontinuities allow water to perculate through rocks with extremely low values of porosity. Indeed the frictional resistance to flow through discontinuities is frequently much lower than that offered by a porous medium, hence appreciable quantities of water may be transmitted. However, discontinuities tend to close with depth. Similarly the permeability of soil, even if fairly uniform in character, tends to decrease with increasing depth because the increasing weight of overburden leads to its densification. Field tests therefore provide much more reliable results (see chapter 9). In a sequence of layered strata the permeability of the individual beds will vary, usually the average permeability in the vertical direction will be less than that in the horizontal.

Subsurface water is normally under pressure which increases with increasing depth below the water table to very high values. Such water pressures have a significant influence on the engineering behaviour of most rock and soil masses and their variations are responsible for changes in the stresses in these masses which affect their deformation characteristics and failure. The influence of interstitial water on rock has been briefly mentioned above, a fuller account is given by Serafim[191].

PORE PRESSURES

Piezometers are installed in the ground in order to monitor and obtain accurate measurements of pore water pressures. Observations should be made regularly so that changes due to external factors such as excessive precipitation, tides, the seasons, etc are noted, it being most important to record the maximum pressures whicn have occurred. Standpipe piezometers allow the determination of the position of the water table and the permeability (see Sherrell[192]). Hydraulic piezometers connected to a manometer board record the changes in pore water pressure. Usually simpler types of piezometer are used in the more permeable soils. The response to piezometers in rock masses can be very much influenced by the incidence and geometry of the discontinuities so that the values of water pressure obtained may be misleading if due regard is not given to these structures. The efficiency of a soil in supporting a structure is influenced by effective or intergranular pressure, that is, the pressure between the particles of the soil which develops resistance to applied load. Because the moisture in the pores offers no resistance to shear it is ineffective or neutral and therefore pore water pressure has also been referred to as neutral pressure. Since the pore or neutral pressure plus the effective pressure equals the total pressure, reduction in pore pressure increases the effective pressure. Reduction of the pore pressure by drainage consequently affords better conditions for carrying a proposed structure. The effective pressure at a particular depth is simply obtained by multiplying the unit weight of the soil by the depth in question and subtracting the pore pressure for that depth.

In a layered sequence the individual layers may have different unit weights. The unit weight of each should then be multiplied by its respective thickness and the pore pressure involved subtracted. The effective pressure for the total thickness concerned is then obtained by summing the effective pressures of the individual layers. Water held in the capillary fringe by soil suction does not affect the values of pore pressure below the water table. However, the weight of water held in the capillary fringe does increase the weight of overburden and so the effective pressure.

Volume changes brought about by loading compressive soils depend upon the level of effective stress and are not affected by the area of contact. The latter may also be neglected in saturated or near saturated soils.

There is some evidence which suggests that the law of effective stresses as used in soil mechanics, in which the pore pressure is subtracted from all direct stress components, holds true for some rocks, those with low porosity may at times prove the exception. However, Serafim[191] suggested that it appeared that pore pressures have no influence in brittle rocks. This is probably because the strength of such rocks is mainly attributable to the strength of the bonds between the component crystals or grains.

COEFFICIENT OF PERMEABILITY

If at two different points within a continuous mass of water there are different amounts of energy, then flow takes place towards the point of lesser energy and the difference in head is expended in maintaining that flow. Other things being equal the velocity of flow between two points is directly proportional to the difference in head between them. The hydraulic gradient refers to the loss of head or energy of water flowing through the ground. This loss of energy by the water is due to the friction resistance of the ground material, and this is greater in fine than coarse grained soils. Thus, in a given engineering project, there is no guarantee that the rate of flow will be uniform and indeed this is the exception. However, if it is assumed that the resistance to flow is constant, then for a given difference in head the flow velocity is directly proportional to the flow path.

The permeability of a particular material is defined by its coefficient of permeability (k). The basic law concerned with flow is that enunciated by Darcy[193] which states that the rate of flow (v) per unit area is proportional to the gradient of the potential head (i) measured in the direction of flow:

$$v = ki$$

and for a particular rock or soil or part of it, of area (A):

$$Q = vA = Aki$$

where Q is the quantity in a given time. The ratio of the cross sectional area of the pore spaces in a soil to that of the whole soil is given by $e/(1 + e)$. Hence a truer velocity of flow, that is, the seepage velocity (v_s) is:

$$v_s = \left(\frac{1 + e}{e}\right) ki$$

Darcy's law is valid as long as a laminar flow exists; departures from this law therefore occur when the flow is turbulent. They also occur when the velocity of flow is high. Such conditions exist in very permeable media, normally when the Reynolds number can attain values above four. In other words, it is usually accepted that this law can be applied to those soils which have finer textures than gravels. Furthermore Darcy's law probably does not accurately represent the flow of water through a porous medium of extremely low permeability, because of the influence of surface and ionic phenomena and the presence of gases.

Darcy omitted to recognise that permeability also depends upon the density (ρ) and viscosity of the fluid(μ) and the average pore size of the porous medium (d) as well as the pore shape. In fact permeability is directly proportional to the unit weight of the fluid concerned and is inversely proportional to its viscosity. The latter is very much influenced by temperature. The hydraulic conductivity (k_c) is an attempt to take these factors into account and is expressed as follows:

$$k_c = Cd^2 \frac{\rho}{\mu}$$

where C is a dimensionless constant or shape factor which takes note of the effects of stratification, packing, particle size distribution and porosity.

It is assumed in the above expression that both the porous medium and the water are mechanically and physically stable, but this may never be true. For example, ion exchange on clay and colloid surfaces may bring about changes in mineral volume which in turn affect the shape and size of the pores. Moderate to high ground water velocities tend to move colloids and clay particles. Solution and deposition may occur from the pore fluids. Small changes in temperature and/or pressure may cause gas to come out of solution which may block pore spaces.

Generally it is the interconnected systems of discontinuities which determine the permeability of a particular rock mass. Indeed the permeability of a jointed rock mass is usually several orders higher than that of intact rock. According to Serafim[189] the following expression can be used to derive the filtration through a rock mass intersected by a system of parallel sided joints with a given opening (e) separated by a given distance (d):

$$k = \frac{e^3 \gamma_W}{12 d \mu}$$

where γ_W is the unit weight of water. The velocity of flow through a single joint of constant gape is expressed by:

$$v = \left(\frac{e^2 \gamma_W}{12 \mu}\right) i$$

where i is the hydraulic gradient.

Subsequently Wittke[194] suggested that where the spacing between discontinuities is small in comparison with the dimensions of the rock mass it is often admissible to replace the fissured rock, with regard to its permeability, by a continuous anisotropic medium. The permeability of this can be described by means of Darcy's law. Wittke also provided a resume of procedures by which three dimensional problems of flow through rocks under complex boundary conditions could be solved.

FLOW NETS

Flow nets provide a graphical representation of the flow of water through the ground and indicate the loss of head involved. They also provide data relating to the changes in head velocity and effective pressure which occur in a foundation subjected to flowing ground water conditions. For instance, when the flow lines of a flow net move closer together this indicates that the flow increases, although their principal function is to indicate the direction of flow. The equipotential lines indicate equal losses in head or energy as the water flows through the ground, so that the closer they are the more rapid is the loss in head. Hence a flow net can provide quantitative data related to the flow problem in question, for example, seepage pressures can be determined at individual points within the net (Figure 1.19).

CRITICAL GRADIENT AND QUICK CONDITIONS

As water flows through the soil and loses head, its energy is transferred to the particles past which it is moving, which in turn creates a drag effect on the particles. If the drag effect is in the same direction as is the force of gravity, then the effective pressure is increased and the soil is stable. Indeed the soil tends to become more dense. Conversely if water flows towards the surface then the drag effect is counter to gravity thereby reducing the effective pressure between particles. If the velocity of upward flow is sufficient it can buoy up the particles so that the effective pressure is reduced to zero. This represents a critical condition where the weight of the submerged soil is balanced by the upward acting seepage force. The critical hydraulic gradient (i_c) involved can be calculated from the following expression:

$$i_c = \frac{G_s - 1}{1 + e}$$

where G_s is the specific gravity of the particles and e is the void ratio.

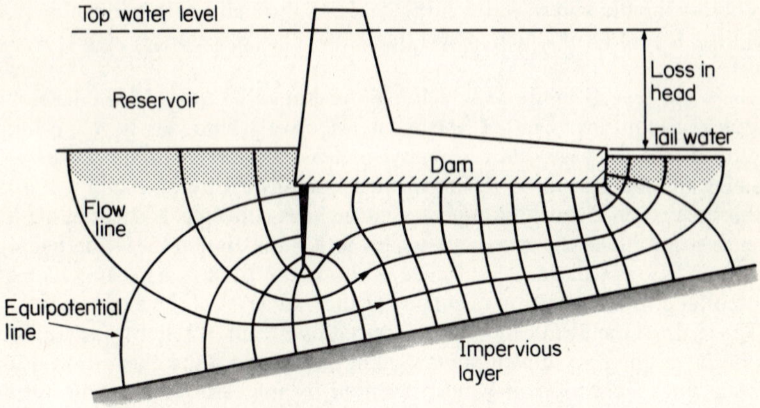

Figure 1.19 Flow net beneath concrete gravity dam, with cut-off at the heel, showing seventeen equipotential drops and four flow channels

A critical condition sometimes occurs in loosely packed, saturated silts and sands, and if the upward velocity of flow increases beyond the critical hydraulic gradient a quick condition develops.

Quicksands, if subjected to deformation or disturbance, can suffer a spontaneous loss of shear strength. This loss of strength causes them to flow like viscous liquids. Terzaghi[195] explained the quicksand phenomenon in the following terms. Firstly, the sand must be saturated and loosely packed. Secondly, on disturbance the constituent grains become more closely packed, but this leads to an increase in pore water pressure which reduces the forces acting between the grains. This brings about the reduction in strength. If the pore water can escape very rapidly the loss in strength is momentary. Hence the third condition requires that pore water cannot readily escape. This is fulfilled if the sand has a low permeability and/or the seepage path is long.

Casagrande[196] demonstrated that a critical porosity existed above which a quick condition could be developed. He maintained that many coarse grained sands, even when loosely packed, have porosities approximately equal to the critical condition whilst medium and fine grained sands, particularly if uniformly graded, exist well above the critical porosity when loosely packed. Accordingly fine sands tend to be more unstable than coarse grained varieties. It must also be remembered that their permeability is lower.

Quick conditions brought about by seepage forces are frequently encountered in excavations made in fine sands which are below the water table as, for example, in cofferdam work. As the velocity of the upward seepage force increases further from the critical gradient the soil begins to boil more and more violently. At such a point structures fail by sinking into the quicksand. Liquefaction of potential quicksands may be caused by sudden shocks such as the action of heavy machinery (notably pile driving), blasting and earthquakes. Such shocks increase the stress carried by the water, the neutral stress, and give rise to a decrease in the effective stress and shear strength of the soil. There is also a possibility of a quick condition developing in a layered soil sequence where the individual beds have different permeabilities. Hydraulic conditions are particularly unfavourable where water initially flows through a very permeable horizon with little loss of head which means that flow takes place under a great hydraulic gradient.

There are several methods which may be employed to avoid the development of quick conditions. One of the most effective techniques is to prolong the length of the seepage path thereby increasing the frictional losses and so reducing the seepage force. This can be accomplished by placing a clay blanket at the base of an excavation where seepage lines converge. If sheet piling is used in excavation then the depth to which it is sunk determines whether or not quick conditions will develop. Hence it should be sunk to a depth which avoids a potential critical condition occurring at the base level of the excavation.

The hydrostatic head can also be reduced by means of relief wells and seepage can be intercepted by a well point system placed about the excavation. Furthermore a quick condition may be prevented by increasing the downward acting force. This may be brought about by placing a load on the surface of the soil where seepage is discharging. Gravel filter beds may be used for this purpose. Suspect soils can also be densified, treated with stabilizing grouts or frozen.

When water perculates through heterogeneous soil masses it moves preferentially through the most permeable zones and it issues from the ground as springs. Piping refers to the erosive action of some such springs, where sediments are removed by seepage forces so forming subsurface cavities and tunnels. In order that erosion tunnels may form the soil must have some cohesion, the greater the cohesion the wider the tunnel. In fact fine sands and silts are most susceptible to piping failures. Obviously the danger of piping occurs when the hydraulic gradient is high, that is, when there is a rapid loss of head over a short distance. This may be indicated on a flow net by a close network of squares where the flow is upward. As the pipe develops by backward erosion it nears the source of water supply so that eventually the water breaks into and rushes through the pipe. Ultimately the hole, so produced, collapses from lack of support. Piping has been most frequently noted downstream of dams, the reservoir providing the water source, but leaking drains can also give rise to piping.

Subsurface structures should be designed to be stable with regard to the highest ground water level that is likely to occur. Structures below ground water level are acted upon by uplift pressures. If the structure is weak this pressure can break it and, for example, cause a blow-out of a basement floor or collapse of a basement wall. If the structure is strong but light it may be lifted. Uplift can be taken care of by adequate drainage or by resisting the upward seepage force. Continuous drainage blankets are effective but should be designed with filters to function without clogging. The entire weight of the structure can resist uplift if a raft foundation is used. Anchors grouted into bedrock can also provide resistance to uplift.

If water flowing under pressure through the ground is confined between two impermeable horizons then it is termed artesian water. Artesian conditions are commonly developed in synclinal structures where an aquifer is sandwiched between two impermeable layers but outcrops at a higher elevation than the position at which the pressure is measured. Artesian conditions can cause serious trouble in excavations and both the position of the water table and the piezometric pressures should be determined before work commences. Otherwise excavations which extend close to strata under artesian pressure may be severaly damaged due to blow-outs taking place in their floors. Slopes may also fail. Indeed such sites may have to be abandoned.

FROST ACTION IN SOIL

Obviously frost action in a soil is influenced by the initial soil temperature, as well as the air temperature, the intensity and duration of the freeze period, the depth of frost penetration, the depth of the water table, and types of ground and exposure cover. If frost penetrates down to the capillary fringe in fine grained soils, especially silts, then, under certain conditions, lenses of ice may be developed. The formation of such ice lenses may, in turn, cause frost heave and frost boil which may lead to the break up of roads, the lifting of structures or the failure of slopes.

Frost action not only causes local heaving but is followed by a loss in the bearing capacity of a soil when thawing takes place. When the thaw occurs the water liberated greatly exceeds that originally present in the soil. Moreover a soil usually thaws out from the top downwards, however, it is not unknown for some thawing to take place from the bottom. As the soil thaws downwards the upper layers become saturated, and since water cannot drain through the frozen soil beneath, they may suffer a complete loss of strength.

Jumikis[200] listed the following factors as necessary for the occurrence of frost heave, namely, capillary saturation at the beginning of and during the freezing of the soil, a plentiful supply of subsoil water and a soil possessing fairly high capillarity together with moderate permeability. However, the essential factor governing frost heave appears to be grain size since this influences moisture movement. For example, well sorted soils in which less than 30% of the particles are of silt size are non-frost heaving (see Glossop and Skempton[3]). Moreover frost heave does not occur in clays because of their low permeability. Indeed Taber[197] gave an upper size limit of 0.007 mm, above which, he maintained, layers of ice do not develop, although Casagrande[198] suggested that the particle size critical to heave formation is 0.02 mm. If the quantity of such particles in a soil is less than 1%, no heave is expected, but considerable heaving may take place if this amount is over 3% in non-uniform soils and over 10% in very uniform soils. In silts the moisture of the upward capillary rise and/or film flow, if frost penetrates downward into the capillary fringe, forms ice lenses under prolonged and severe freezing conditions. This is because silt particles are small enough to provide a comparatively high capillary rise.

Croney and Jacobs[199] suggested that under the climatic conditions experienced in Britain well-drained cohesive soils with a plasticity index exceeding 15%

could be looked upon as non-frost susceptible. Where the drainage is poor and the water table is within 0.6 m of formation level they suggested that the limiting value of plasticity index should be increased to 20%. In experiments with sand they noted that as the amount of silt added was increased up to 55% or the clay fraction up to 30%, increase in permeability in the freezing front was the overriding factor and heave tended to increase. Beyond these values the decreasing permeability below the freezing zone became dominant and progressively reduced the heave. In other words the permeability below the frozen zone was principally responsible for controlling heave. These two authors also suggested that the permeability of soft chalk is sufficiently high to permit very serious frost heave but in the harder varieties the lower permeabilities minimize or prevent heaving.

Maximum heaving, according to Jumikis[200], does not necessarily occur at the time of maximum depth of the 0°C line, there being a lag between the minimum air temperature prevailing and the maximum penetration of the freeze front. In fact soil freezes at temperatures slightly lower than 0°C. As heaves amounting to 30% of the thickness of the frozen layer have frequently been recorded, moisture, other than that initially present in the frozen layer, must be drawn from below, since water increases in volume by only 9% when frozen.

In fact when a soil freezes there is an upward transfer of heat from the ground water towards the area in which freezing is occurring. The thermal energy, in its turn, initiates an upward migration of moisture within the soil. The moisture in the soil can be translocated upwards either in the vapour or liquid phase or by a combination of both. Moisture diffusion by the vapour phase occurs more readily in soils with larger void spaces than in fine grained soils and if a soil is saturated migration in the vapour phase cannot take place.

In a very dense, closely packed soil where the moisture forms uninterrupted films throughout the soil mass, down to the water table, then, depending upon the texture of the soil, the film transport mechanism becomes more effective than the vapour mechanism. The upward movement of moisture due to the film mechanism, in a freezing soil mass, is slow. Nonetheless a considerable amount of moisture can move upwards as a result of this mechanism during the winter. What is more in the film transport mechanism the water table is linked by the films of moisture to the ice lenses.

Before freezing, soil particles develop films of moisture about them due to capillary action. This moisture is drawn from the water table. As the ice lens grows, the suction pressure it develops exceeds that of the capillary attraction of moisture by the soil particles. Hence moisture moves from the soil to the ice lens. But the capillary force continues to draw moisture from the water table and so the process continues.

The amount of segregated ice in a frozen mass of soil depends largely upon the intensity and rate of freezing. When freezing takes place quickly no layers of ice are visible whereas slow freezing produces visible layers of ice of various thicknesses. Ice segregation in soil also takes place under cyclic freezing and thawing conditions.

Where there is a likelihood of frost heave occurring it is necessary to estimate the depth of frost penetration (see Jumikis[200]). Once this has been done provision can be made for the installation of adequate insulation or drainage within the soil. Alternatively the amount by which the water table needs to be lowered so that it is not affected by frost penetration can be determined. The base of footings should be placed below the estimated depth of frost penetration, which in the UK is generally at a depth of 0.5 m below ground level.

Frost susceptible soils may be replaced by free draining gravels. The addition of certain chemicals to soil can reduce its capacity for water absorption and so can influence frost susceptibility. For example, Croney and Jacobs[199] noted that the addition of calcium lignosulphate and sodium tripolyphosphate to silty soils were both effective in reducing frost heave. The freezing point of the soil may be lowered by mixing in solutions of calcium chloride or sodium chloride, in concentrations of 0.5 to 3.0% by weight of the soil mixture. The heave of non-cohesive soils containing appreciable quantities of fines can be reduced or prevented by the addition of cement or bituminous binders. Addition of cement both reduces the permeability of a soil mass and gives it sufficient tensile strength to prevent the formation of small ice lenses as the freezing isotherm passes through.

PERMAFROST

Perennially frozen ground or permafrost is characteristic of tundra environments; in fact permafrost covers 20% of the Earth's land surface. During Pleistocene times it was developed over an even larger area, evidence for its former presence being found in the form of fossil ground-ice wedges, solifluction deposits and stone polygons. The temperature of perennially frozen ground below the depth of seasonal change ranges from slightly less than 0°C to –12°C. Generally the depth of thaw is less the higher the latitude. It is at a minimum in peat or highly organic sediments and increases in clay, silt and sand to a maximum in gravel where it may extend to 2 m depth.

Because permafrost represents an impervious horizon it means that water is prevented from entering the ground so that soils are commonly supersaturated. Hence solifluction deposits and mud flows are typical, indeed soil material on all slopes greater than 1° to 3° is on the move in summer. Permafrost, by maintaining saturated or supersaturated conditions in surface soils, aids frost stirring, frost splitting and mass wasting processes. Frost action and gravity movements result in certain characteristic features, for example, involutions, frost boils, hummocks, antiplanation terraces, terracettes and stone stripes (see Black[201]). Annual freezing in permafrost areas brings about changes in surface and ground water movements which may result in the development of frost blisters, ice mounds, icings or even pingos.

According to Muller[202] there are two methods of construction in permafrost, namely, the passive and the active methods. In the former the frozen ground is not disturbed or is provided with additional insulation so that heat from the structure does not bring about thawing in the ground below, thereby reducing its stability. By contrast the ground is thawed prior to construction in the active method and it is either kept thawed or removed and replaced by materials not affected by frost action.

Permafrost is an important characteristic, although it is not essential to the definition of periglacial conditions, the latter referring to conditions under which frost action is the predominant weathering agent. A broad survey of periglacial features in Britain and their engineering significance has been provided by Higginbottom and Fookes[203]. Superficial structural disturbances have several causes. The results of frost shattering are particularly noteworthy in the Chalk (see Ward *et al*[156]). Ice wedges may originate as thermal contraction cracks which are enlarged by the growth of ice and subsequently infilled, and in plan may form part of a polygonal network (see Shotton[204]). Surface cracking may be associated with the formation of stone polygons. Involutions are pockets or tongues of highly disturbed material, generally possessing inferior geotechnical properties, which have been intruded from below into overlying layers due to hydrostatic uplift of water trapped beneath a refreezing surface layer (see West[205]). Frost mounds and pingos are subject to solifluction during their life-time consequently a depression develops when the ice lens melts. This then may be filled with deposits of saturated, normally consolidated fine sediments, which alternate with layers of peat. Mass movements in former periglacial areas in Britain included cambering and valley bulging, landsliding and mudflows (see chapter 13).

References

1. Code Drafting Committee, 'Proposed British soil classification system for use in the revised Code of Practice on site investigations, *Inst. C. E.* (1976).
2. Casagrande, A. 'Classification and identification of soils, *Trans. Am. Soc. C. E.*, **113**, 901–992 (1948).
3. Glossop, R. and Skempton, A.W., 'Particle size in silts and sands, *Jour. Inst. C. E.*, **25**, Paper No. 5492, 81–105 (1945).
4. Kahn, J.S., 'The analysis and distribution of the properties of packing in sand size particles', *Jour. Geol.*, **64**, 385–395, 578–606 (1956).
5. Graton, L.C. and Frazer, H.J., 'Systematic packing of spheres with particular relation to porosity and permeability', *Jour. Geol.*, **43**, 785–909 (1935).
6. Cornforth, D.H., 'Some experiments on the influence of strain conditions on the strength of sand,' *Geotechnique*, **14**, 143–167 (1964).
7. Borowicka, H., 'Rearrangement of grains by shear tests with sand,' *Proc. 8th Int. Conf. Soil Mech. Found. Eng., Moscow*, **1**, 71–77 (1973).
8. Rowe, P.W., 'The relation between the shear strength of sands in triaxial compression, plane strain and direct shear', *Geotechnique*, **19**, 75–86 (1969).
9. Arnold, M. and Mitchell, P.W., 'Sand deformation in three-dimensional stress state', *Proc. 8th Int. Conf. Soil Mech. Found. Eng., Moscow*, **1**, 11–18 (1973).
10. Holtz, W.G. and Gibbs, H.J., 'Shear strength of pervious gravelly soils', *Proc. Am. Soc. Civ. Eng.*, **82**, Paper No. 867, (1956).
11. Terzaghi, K., 'The influence of geological factors in the engineering properties of sediments,' *Econ. Geol.*, 50th Ann. Vol., 557–618 (1955).
12. Kirkpatrick, W.M., 'The condition of failure of sands,' *Proc. 4th Int. Conf. Soil Mech. Found. Eng., London*, **1**, 172–185 (1957).
13. Sutherland, H. and Mesdary, M., 'The influence of the intermediate principal stress on the strength of sand,' *Proc. 7th Int. Conf. Soil Mech. Found. Eng.*, **1**, 391–399 (1969).
14. Schultze, E. and Kotzias, A.B., 'Geotechnical properties of Lower Rhine Silts,' *Proc. 5th Int. Conf. Soil Mech. Found. Eng.*, **1**, 329–333 (1961).
15. Schultze, E. and Horn, A., 'The shear strength of silt,' *Proc. 5th Int. Conf. Soil Mech. Found. Eng.*, **1**, 350–358 (1961).
16. Penman, A.D.M., 'Shear characteristics of a saturated silt in triaxial compression,' *Geotechnique*, **3**, 312–315 (1953).
17. Larionov, A.K. 'Structural characteristics of loess soils for evaluating their constructional properties,' *Proc. 6th Int. Conf. Soil Mech. Found. Eng., Montreal*, 64–68 (1965).
18. Clevenger, M.A., 'Experiences with loess as a foundation material,' *Trans. Am. Soc. Civ. Eng.*, **123**, 151–180 (Paper No. 2961), (1958).
19. Lambe, T.W., 'The structure of inorganic soil,' *Proc. Am. Soc. Civ. Eng.*, **79**, Separate No. 315 (1953).
20. Lambe, T.W., 'Structure of compacted clays,' *A.S.C.E. Soil Mech. Found. Eng. Div.*, **SM4**, 85–106 (1958).

21. Van Olphen, H., *An introduction to clay colloid chemistry*, Wiley, New York, (1963).
22. Barden, L., 'The relation of soil structure to the engineering geology of clay soil,' *Quar. Jour. Eng. Geol.*, **5**, 85–102 (1972).
23. Aylmore, L.A.G. and Quirk, J.P., 'Domain of turbostratic structures of clays,' *Nature*, **187**, 1046–1048 (1960).
24. Burnham, C.P., 'Micromorphology of argillaceous sediments: particularly calcareous clays and siltstones,' article in *Micromorphological Techniques and Applications*, Ed. by Osmond, D.A. and Bullock, P. *Soil Surv. Tech. Mono.*, *2*, 83–96 (1970).
25. Grim, R.E., *Applied clay mineralogy*, McGraw-Hill, New York (1962).
26. Mielenz, R.C. and King, M.E., 'Physical-chemical properties and engineering performance of clays (article in *Clays and clay technology*, Ed. by Pask, J.A. and Turner, M.D.) California Division of Mines, Bulletin 169, 196–254 (1955).
27. Bjerrum, L., 'Progressive failure in slopes of overconsolidated plastic clay and clay shales,' *A.S.C.E. Soil Mech. Found. Eng. Div.*, **SM5**, 93, 2–49 (1967).
28. Schmertmann, J.H., 'Swell sensitivity,' *Geotechnique*, **19**, 530–533 (1969).
29. Hardy, R.M., 'Identification and performance of swelling soil types,' *Canadian Geot. Jour.*, **2**, 141–153 (1966).
30. Holtz, W.G. and Gibbs, H.J., 'Engineering properties of expansive clays,' *Trans. Am. Soc. Civ. Eng.*, **121**, 641–663 (1956).
31. Holtz, W.G., 'Expansive clays – properties and problems (article in *Theoretical and practical treatment of expansive soils*) *Colorado School of Mines Quarterly*, **54**, 89–117 (1959).
32. Van der Merwe, D.H., 'The prediction of heave from the plasticity index and the percentage clay fraction of soils. The Civil Engineer,' *S. Af. Inst. Civ. Eng.*, **6**, 103 *et seq* (1964).
33. Skempton, A.W., 'The colloidal activity of clays,' *Proc. 3rd Int. Conf. Soil Mech. Found. Enf., Zurich*, **1**, 57–61 (1953).
34. Hedburg, H.D., 'Gravitational compaction of clays and shales,' *Am. Jour. Sci. (5th series)*, **31**, 241–287 (1936).
35. Bjerrum, L., 'Embankments on soft ground (article in *Performance of earth and earth supported structures, Vol. II*) *A.S.C.E. Proc. Specialty Conf.*, Purdue Univ., Lafayette, Indiana, 32–33 (1972).
36. Obermeier, S.F., 'Evaluation of laboratory techniques for measurement of swell potential of clays.' *Bull. Ass. Eng. Geol.*, **11**, 293–314 (1974).
37. Wahls, H.E., 'Analysis of primary and secondary consolidation,' *ASCE Jour. Soil Mech. Found. Eng. Div.*, **88**, SM6, 207–231 (1962).
38. Chang, Y.C.E., Broms, B. and Peck, R.B., 'Relationship between the settlement of soft clays and excess pore pressure due to improved loads,' *Proc. 8th Int. Conf. Soil Mech. Found. Eng., Moscow*, **1**, 93–96 (1973).
39. Berre, T. and Bjerrum, L., 'The shear strength of normally consolidated clays,' *Proc. 8th Int. Conf. Soil Mech. Found. Eng., Moscow*, **1**, 39–49 (1973).
40. Skempton, A.W., 'The bearing capacity of clays,' *Building Research Congress, Div. 1*, 180–189 (1951).
41. Lambe, T.W., 'A mechanistic picture of shear strength in clay,' *ASCE Proc. Res. Conf. Shear Strength Cohesive Soils*, Boulder, Colorado, 555–580 (1960).
42. Terzaghi, K., 'Ends and means in soil mechanics,' *Engineering Jour., (Canada)*, **27**, 608–615 (1944).
43. Skempton, A.W. and Northey, R.D., 'The sensitivity of clays,' *Geotechnique*, **2**, 30–53 (1952).
44. Terzaghi, K., 'Stability of slopes of natural clay,' *Proc. 1st Int. Conf. Soil Mech. Found. Eng.*, Cambridge, Mass., **1**, 161–165 (1936).
45. Terzaghi, K., 'The influence of modern soil studies on the design and construction of foundations,' *Building Research Congress, Div. 1*, 139–145 (1951).
46. Skempton, A.W., 'The rate of softening in stiff fissured clay with special reference to the London Clay,' *Proc. 2nd Ind. Conf. Soil Mech. Found. Eng.* (Rotterdam), **2**, 50–53 (1948).
47. Skempton, A.W., Schuster, R.L. and Petley, D.J., 'Joints and fractures in the London Clay at Wraysbury and Edgware,' *Geotechnique*, **19**, 205–217 (1969).
48. Ward, W.H., Marsland, A. and Samuels, S.G., 'Properties of the London Clay at the Ashford Common Shaft: *in situ* and undrained strength tests,' *Geotechnique*, **15**, 321–344 (1965).

49. Fookes, P.G. and Denness, B. 'Observational studies on fissure patterns in Cretaceous sediments of South East England,' *Geotechnique*, **19**, 453–477 (1959).
50. Marsland, A. and Butler, M.E., 'Strength measurements on stiff fissured Barton Clay from Fawley, Hampshire,' *Proc. Geot. Conf. Oslo*, **1**, 139–146 (1967).
51. Lo, K.Y., 'The operational strength of fissured clay,' *Geotechnique*, **20**, 57–74 (1970).
52. Fookes, P.G., Gordon, D.L. and Higginbottom, I.E., 'Glacial landforms, their deposits and engineering characteristics,' (article in *The engineering behaviour of glacial materials*) *Proc. Symp. Midland Soil Mech. Found. Eng. Soc.*, Birmingham Univ., 18–51 (1975).
53. Boulton, G.S., 'The genesis of glacial tills, a framework for geotechnical interpretation,' (article in *The engineering behaviour of glacial materials*) *Proc. Symp. Midland Soil Mech. Found. Eng Soc.*, Birmingham Univ., 52–59 (1975).
54. Boulton, G.S. and Paul, M.A., 'The influence of genetic processes on some geotechnical properties of glacial tills,' *Quar. Jour. Eng. Geol.*, **9**, 159–194 (1976).
55. Elson, J.A., 'Geology of tills,' *Proc. 14th Canadian Conf. Soil Mech.*, Section 3, 5–17 (1961).
56. Krumbein, W.C., 'Textural and lithologic variations in glacial till,' *Jour. Geol.*, **41**, 382–408 (1933).
57. Chryssafopoulos, H.W.S., 'An example of the use of engineering properties for differentiation of young glacial till sheets,' *Proc. 2nd Pan. Conf. Soil Mech. Found. Eng.*, **2**, 35–43 (1963).
58. McGown, A., 'The classification for engineering purposes of tills from moraines and associated landforms,' *Quar. Jour. Eng. Geol.*, **4**, 115–130 (1971).
59. Leggett, R., 'An engineering study of glacial drift for an earth dam near Fergus, Ontario,' *Econ. Geol.*, **37**, 531–556 (1942).
60. Bernell, L., 'Properties of moraines,' *Proc. 4th Int. Conf. Soil Mech. Found. Eng.*, **2**, 286–290 (1957).
61. McKinley, D.G. Tomlinson, M.J. and Anderson, W.F., 'Observations on the undrained strength of glacial till,' *Geotechnique*, **24**, 503–516 (1974).
62. Klohn, E.J., 'The elastic properties of a dense glacial till deposit', *Canadian Geot. Jour.*, **2**, 116–128 (1965).
63. Radhakrishna, H.S. and Klym, T.W., 'Geotechnical properties of a very dense glacial till,' *Canadian Geot. Jour.*, **11**, 396–408 (1974).
64. Kazi, A. and Knill, J.L., 'The sedimentation and geotechnical properties of the Cromer till between Happisburgh and Cromer, Norfolk,' *Quart. Jour. Eng. Geol.*, **2**, 63–86 (1969).
65. Flint, R.F., *Glacial and Pleistocene geology*, Wiley, New York (1967).
66. Metcalf, J.B. and Townsend, D.L., 'A preliminary study of the geotechnical properties of varved clays as reported in Canadian engineering core records', *Proc. 14th Canadian Conf. Soil Mech.*, Section 13, 203–225 (1961).
67. Walker, L.K., 'Undrained creep in a sensitive clay,' *Geotechnique*, **19**, 515–529 (1969).
68. Rosenqvist, I.Th., *'Om lieres Kvikkaktighet (on the sensitivity of clays), Meddelelser fra Vegdirektoren*, No. 3, 24–36 (1946).
69. Rosenqvist, I. Th., 'Considerations on the sensitivity of Norwegian quick clays,' *Geotechnique*, **3**, 195–200 (1953).
70. Pusch, R. and Arnold, M., 'The sensitivity of artificially sedimented organic free illitic clay,' *Eng. Geol.*, **3**, 135–148 (1969).
71. Cabrera, J.G. and Smalley, I.J., 'Quick clays as products of glacial action: a new approach to their nature, geology, distribution and geotechnical properties,' *Eng. Geol.*, 7, 115–133 (1973).
72. Radforth, N.W., 'Suggested classifications of muskeg for the engineer,' *Engineering Jour. (Canada)*, **35**, 1194–1210 (1952).
73. Bell, F.G., 'Peat: a note and its geotechnical properties,' *Civil Engineering* (1977).
74. Hanrahan, E.T., 'The mechanical properties of peat with special reference to road construction,' *Trans. Inst. Civ. Eng.*, Ireland, **78**, 179–215 (1952).
75. Hanrahan, E.T., 'An investigation of some physical properties of peat,' *Geotechnique*, **4**, 108–123 (1954).
76. Miyahawa, I., 'Some aspects of road construction over peaty or marshy areas in Hokkaido, with particular reference to filling methods,' *Civ. Eng. Res. Inst.*, Sapporo, Japan (1960).

77. Adams, J.I., 'The engineering behaviour of Canadian muskeg,' *Proc. 6th Int. Conf. Soil Mech. Found. Enf., Montreal,* **1**, 3–7 (1965).
78. Hanrahan, E.T. and Walsh, J.A., 'Investigations of the behaviour of peat under varying conditions of stress and strain,' *Proc. 6th Int. Conf. Soil Mech. Found. Eng., Montreal,* **1**, 226–230 (1965).
79. Hanrahan, E.T., 'A road failure on peat,' *Geotechnique,* **14**, 185–203 (1964).
80. Buisman, A.S.K., 'Results of long duration settlement tests,' *Proc. 1st Int. Conf. Soil Mech. Found. Eng.*, Cambridge, Mass., **1**, 103–105 (1936).
81. Adams, J.I., 'A comparison of field and laboratory measurement of peat,' *Proc. 9th Muskeg Res. Conf.*, NRC-ACSSM Tech. Memo. 81, 117–135 (1963).
82. Wilson, N.E., Radforth, N.W., Macfarlane, I.C. and Lo, M.B., 'The rates of consolidation for peat,' *Proc. 6th Int. Conf. Soil Mech. Found. Eng.*, Montreal, **1**, 407–412 (1965).
83. Berry, P.L. and Poskitt, T.J., 'The consolidation of peat,' *Geotechnique,* **22**, 27–52 (1972).
84. Meyerhof, G.G., 'Building on fill with special reference to the settlement of a large factory,' *Struct. Eng.*, **29**, No. 11, 297–305 (1951).
85. BRS Digest No. 9, *Building on made-up Ground*, Building Research Station, HMSO, Garston, Watford (1967).
86. Kilkenny, W.M., 'A study for the settlement of restored opencast coal sites and their suitability for building development,' *Dept. Civ. Eng., Newcastle Univ., Bull. No. 38* (1968).
87. Sowers, G.E., 'Settlement of waste disposal fills,' *Proc. 8th Int. Conf. Soil Mech. Found. Eng., Moscow,* **2**, 207–212 (1973).
88. Penman, A.D.M. and Godwin, E.W., 'Settlement of experimental houses on land left by opencast mining at Corby (article in *Settlement of Structures*) British Geotechnical Society, Pentech Press, London, 53–61 (1975).
89. Sowers, G.F., Williams, R.C. and Wallace, T.S., 'Compressibility of broken rock and settlement of rockfills,' *Proc. 6th Int. Conf. Soil Mech. Found. Eng., Montreal,* **2**, 561–565 (1965).
90. Bell, F.G., 'Coarse discard from mines', *Civil Engineering,* 37–39 (March 1977).
91. National Coal Board, *Spoil heaps and lagoons*, Technical Handbook, National Coal Board, London (1973).
92. Donath, E.A., 'Experimental study of shear failure in anisotropic rocks,' *Bull. Geol. Soc. Am.,* **72**, 985–991 (1961).
93. Hoek, E., 'Fracture of anisotropic rock,' *Jour. S. Af. Inst. Min. Met.,* **64**, 510–518 (1964).
94. Brown, E.T., Richards, L.R. and Barr, M.V., 'Shear strength characteristics of the Delabole States,' *Proc. Conf. Rock Eng.*, Newcastle Univ., **1**, 33–51 (1977).
95. Griggs, D.T., 'Deformation of rocks under high confining pressures,' *Jour. Geol.,* 44 (1936).
96. Colback, P.S.B., and Wiid, B.L., 'Influence of moisture content on the compressive strength of rock,' *Symp. Canadian Dept. Min. Tech. Survey.*, Ottawa, 65–83 (1965).
97. Robertson, E.C., 'Creep of Solenhofen Limestone under moderate hydrostatic pressure, in *Rock Deformation,' Geol. Soc. Am.*, Memoir 79, 227–244 (1960).
98. Coulomb, G.A., 'Sur une application des regles de maximus et minimus a quelques problemes de statique relatifs a l'architecture,' *Acad. Roy. des Sci.,* Mem. de Math. et de Phys. par divers Savans, 7, 343–382 (1773).
99. Mohr, O., *Abhandlungen aus dem gebiete der technische mechanik*, Ernst und Sohn, Berlin (1882).
100. Griffith, A.A., 'The theory of rupture,' *Proc. 1st Cong. Appl. Mech., Delft,* 55–70 (1920).
101. Brace, W.F., 'Brittle fracture of rocks (article in *Symp. State of Stress in the Earth's Crust*, Ed. by Judd, W.R.) Elsevier, Santa Monica, 111–180 (1964).
102. McClintock, F.A. and Walsh, J.B., 'Friction on Griffith cracks in rocks under pressure,' *Proc. 4th Cong. Appl. Mech.*, 1015–1021 (1962).
103. Hoek, E., 'Rock mechanics – an introduction for the practical engineer', *Min. Mag.*, **114**, 236–255 (1966).
104. Hoek, E., 'Brittle fracture of rock' (article in *Rock mechanics in engineering practice*, Ed. By Stagg, K.G. and Zienkiewicz, O.C.) Wiley, London, 99–124 (1968).

105. Barton, N., 'The shear strength of rock and rock joints,' *Int. Jour. Rock Mech. Min. Sci.*, **13**, 255–279 (1976).
106. Chapman, C.A., 'Control of jointing by topography,' *Jour. Geol.*, **66**, 522–533 (1958).
107. Price, N.J., *Fault and joint development in brittle and semi-brittle rock*, Pergamon, London (1966).
108. Goodman, R.E., Taylor, R.L. and Brekke, T.L., 'A model for the mechanics of jointed rocks,' *ASCE Soil Mech. Found. Eng. Div.*, **94**, 637–659 (1968).
109. Onodera, T.F., 'Dynamic investigation of foundation rocks,' *Proc. 5th Symp. Rock Mech. Minnesota*, Pergamon, New York, 517–533 (1963).
110. Deere, D.U., Hendron, A.J., Patton, F.D. and Cording, E.J., 'Design of surface and near-surface construction in rock'. *Proc. 8th Symp. Rock Mech., Minnesota*, A.I.M.E., 237–302 (1967).
111. Franklin, J.A., Broch, E. and Walton, G., 'Logging the mechanical character of rock,' *Trans. Inst. Min. Met.*, **81**, Mining Section, A1–9 (1971).
112. Hobbs, N.B., 'Factors affecting the prediction of settlement of structures on rocks with particular reference to the Chalk and Trias' (article in *Settlement of Structures*) British Geotechnical Society, Pentech Press, London, 579–610 (1975).
113. Murrell, S.A.F., 'Natural faulting and the mechanics of brittle shear failure,' *Jl. Geol. Soc.*, **133**, 175–190 (1977).
114. Sibson, R.H., 'Fault rocks and fault mechanisms', *Jour. Geol. Soc.*, **133**, 191–214 (1977).
115. Salehy, M.R., Money, M.S. and Dearman, W.R., 'The occurrence and engineering properties of intraformational shears in Carboniferous rocks,' *Proc. Conf. Rock. Eng.*, Newcastle Univ., **1**, 311–328 (1977).
116. Skempton, A.W., 'Some observations on tectonic shear zones,' *Proc. 1st Int. Cong. Rock. Mech.*, Lisbon, **1**, 329–355 (1966).
117. Louderback, G.D., 'Faults and engineering geology' (article in *Application of geology to engineering practice*, Berkey Volume) Am. Geol. Soc., 125–150 (1950).
118. Ambraseys, N.N., 'Maximum intensity of ground movements caused by faulting,' *Proc. 4th World Conf. Earthquake Eng.*, Chile, **1**, 154–162 (1969).
119. Bonilla, M.G., 'Surface faulting and related effects' (article in *Earthquake Engineering*, Ed. by Weigel, R.L.) Prentice-Hall, Englewood Cliffs, New Jersey, 47–74 (1970).
120. Badger, C.W., Cummings, A.D. and Whitmore, R.L., 'The disintegration of shale,' *Jour. Inst. Fuel*, **29**, 417–423 (1956).
121. Nakano, R., 'On weathering and change of properties of Tertiary mudstone related to landslide,' *Soil and Found*, 7, 1–14 (1967).
122. Taylor, R.K. and Spears, D.A., 'The breakdown of British Coal Measures rocks,' *Int. Jour. Rock Mech. Min. Sci.*, 7, 481–501 (1970).
123. Franklin, J.A. and Chandra, R., 'The slake durability test,' *Int. Jour. Rock Mech. Min. Sci.*, **9**, 325–341 (1972).
124. Hamrol, A., 'A quantitative classification of weathering and weatherability of rocks,' *Proc. 5th Int. Conf. Soil Mech. Found. Eng.*, **2**, 771–773 (1961).
125. Moye, D.G., 'Engineering geology for the Snowy Mountain scheme,' *Jour. Inst. Engrs. Aust.*, **27**, 287–298 (1955).
126. Kiersch, G.A. and Treasher, R.C., 'Investigations, areal and engineering geology – Folsam Dam Project, Central California,' *Econ. Geol.*, **50**, 271–310 (1955).
127. Knill, J.L. and Jones, K.S., 'The recording and interpretation of geological conditions in the foundations of the Rosieres, Kariba and Latiyan Dams, *Geotechnique*, **15**, 94–124 (1965).
128. Dearman, W.R., Fookes, P.G. and Franklin, J.A., 'Some engineering aspects of weathering with field examples from Dartmoor and elsewhere,' *Quar. Jour. Eng. Geol.*, **3**, 1–24 (1972).
129. Fookes, P.G. and Horswill, P., Discussion on 'The load deformation behaviour of the Middle Chalk at Mundford, Norfolk' (article in *In situ investigations in soils and rocks*) British Geotechnical Society, London, 53–57 (1970).
130. Little, A.L., 'The engineering classification of residual tropical soils,' *Proc. 7th Int. Conf. Soil Mech. Found. Eng., Mexico*, **1**, 1–10 (1969).
131. Hobbs, N.B., *Foundations on rock*, Soil Mechanics, Bracknell (1975).
132. Terzaghi, K., 'Introduction to tunnel geology' (article in *Rock tunnelling with steel supports* by Procter, R. and White, T.) Commercial Shearing and Stamping Co., Youngstown, Ohio, 17–99 (1946).

133. Bieniawski, Z.T., 'Engineering classification of jointed rock masses,' *Trans. S. Af. Inst. Civ. Engrs.*, **15**, 335–343 (1973).
134. Bieniawski, Z.T., 'Geomechanics classification of rock masses and its application in tunnelling,' *Proc. 3rd Int. Cong. Rock Mech., Denver*, **2**, 27–32 (1974).
135. Deere, D.U., 'Technical description of cores for engineering purposes,' *Rock. Mech. Eng. Geol.*, **1**, 17–22 (1964).
136. Geological Society Engineering Group, 'Working party report on the logging of cores for wngineering purposes,' *Quar. Jour. Geol. Soc.*, **3**, 1–24 (1970).
137. Deere, D.U., 'Geological considerations' (article in *Rock mechanics in engineering practice*, Ed. by Stagg, M.G. and Zienkiewicz, O.C.) Wiley, London, 1–19 (1968).
138. Pells, P.J.H., 'Discussion: engineering classification of rock jointed masses,' *Trans. S. Af. Inst. Civ. Engrs.*, **16**, 242 (1974).
139. Wickham, G.E., Tiedemann, H.R. and Skinner, E.H., 'Support determinations based on geological predictions,' *Proc. 1st N. Am. Tunnelling Conf.*, AIME, New York, 43–64 (1972).
140. Terzaghi, K., 'Dam foundations on sheeted rock,' *Geotechnique*, **12**, 199–208 (1962).
141. Jaeger, C., 'The Malpasset report,' *Water Power*, **15**, 55–61 (1963).
142. Jaeger, C., 'The stability of partly immersed fissured rock masses and the Vajont Slide,' *Civ. Eng. Pub. Works Rev.*, **64**, 1204–1207 (1969).
143. Pettijohn, P.J., Potter, P.E. and Siever, R., *Sands and sandstones*, Springer-Verlag, Berlin (1975).
144. Bell, F.G., 'The physical and mechanical properties of the Fell Sandstone,' *Eng. Geol.*, (1977).
145. Price, N.J., 'The compressive strength of Coal Measures rocks,' *Colliery Guardian*, 283–292 (1960).
146. Price, N.J., 'The influence of geological factors on the strength of Coal Measures rocks,' *Geol. Mag.*, **100**, 428–443 (1963).
147. Moore, J.F.A., 'A long term plate test on Bunter Sandstone,' *Proc. 3rd Int. Cong. Rock. Mech., Denver*, **2**, 724–732 (1974).
148. Moore, J.F.A. and Jones, C.W., '*In situ* deformation of Bunter Sandstone,' (article in *Settlement of Structures*) British Geotechnical Society, Pentech Press, London, 311–319 (1975).
149. Hill, H.P., 'The Ladybower Reservoir,' *Jour. Inst. Water Engrs.*, **3**, 414–433 (1949).
150. North, F.J., 'Some geological aspects of subsidence not due to mining,' *Proc. S. Wales Inst. Engrs.*, 127–158 (1951).
151. Bell, F.G., 'A note on the geotechnical properties of chalk,' *Eng. Geol.*, **11**, (1977).
152. Higgi.ıbottom, I.E., 'The engineering geology of the Chalk,' *Proc. Symp. on Chalk in Earthworks, I.C.E.*, London, 1–14 (1965).
153. Carter, P.G. and Mallard, D.J., 'A study of the strength, compressibility and density trends within the Chalk of South East England,' *Quart. Jour. Eng. Geol.*, **7**, 43–56 (1974).
154. Meigh, A.C., and Early, K.R., 'Some physical and engineering properties of chalk,' *Proc. 4th Int. Conf. Soil Mech. Found. Eng.*, **1**, 68–73 (1957).
155. Wakeling, T.R.M., 'Foundations on chalk,' *Proc. Symp. on Chalk in Earthworks, I.C.E.*, London, 15–23 (1965).
156. Ward, W.H., Burland, J.B. and Gallois, R.W., 'Geotechnical assessment of a site at Mundsford, Norfolk for a large Proton accelerator,' *Geotechnique*, **18**, 399–431 (1968).
157. Wakeling, T.R.M., 'A comparison of the results of standard site investigation methods against the results of a detailed geotechnical investigation in Middle Chalk at Mundford, Norfolk (article in *In situ investigations in soils and rocks*) British Geotechnical Society, London, 17–22 (1970).
158. Dennehy, J.P., 'Correlating the SPT N value with chalk grade from some zones of the Upper Chalk,' *Geotechnique*, **26**, 610–614 (1976).
159. Burland, J.B. and Lord, J.A., 'The load deformation behaviour of Middle Chalk at Mundford, Norfolk: a comparison between full-scale performance and *in situ* and laboratory measurements' (article in *In situ investigations in soils and rocks*) British Geotechnical Society, London, 3–16 (1969).
160. Burland, J.B., Kee, R. and Burford, D., 'Short term settlement of a five-storey building on soft chalk (article in *Settlement of structures*) British Geotechnical Society, Pentech Press, London, 259–265 (1974).

161. Kee, R., Parker, A.S. and Wehale, J.E.C., 'Settlement of a twelve-storey building on piled foundations in chalk at Basingstoke' (article in *Settlement of structures*) British Geotechnical Society, Pentech Press, London, 275–282 (1974).
162. Ineson, J., 'A hydrogeological study of the permeability of chalk,' *Jour. Inst. Water Eng.*, **16**, 255–286 (1962).
163. West, G. and Dumbleton, M.J., 'Some observations on swallow holes and mines in the Chalk,' *Quar. Jour. Eng. Geol.*, **5**, 171–178 (1972).
164. Lewis, W.A. and Croney, D., 'The properties of chalk in relation to road foundations and pavements,' *Proc. Symp. on Chalk in Earthworks, I.C.E.*, London, 27–42 (1965).
165. Justo, J.L. and Zapico, L., 'Compression between measured and estimated settlements at two Spanish aqueducts on gypsum rock (article in *Settlement of structures*) British Geotechnical Society, Pentech Press, London, 266–274 (1975).
166. Eck, W. and Redfield, R.C., 'Engineering geology problems at Sanford Dam, Texas,' *Bull. Ass. Eng. Geol.*, **3**, 15–25 (1965).
167. Brune, G., 'Anhydrite and gypsum problems in engineering geology,' *Bull. Ass. Eng. Geol.*, **3**, 26–38 (1965).
168. Kendal, P.F. and Wroot, H.E., 'The geology of Yorkshire,' *Printed privately* (1924).
169. Redfield, R.C., 'Brantley Reservoir site – an investigation of evaporiate and carbonate facies,' *Bull. Am. Eng. Geol.*, **6**, 14–30 (1968).
170. Kuenen, Ph.H., 'Geotechnical calculations concerning the total mass of sediments of the Earth,' *Am. Jour. Sci.*, **239**, 161–190 (1941).
171. Grice, R.H., 'Test procedures for the susceptibility of shale to weathering,' *Proc. 7th Int. Conf. Soil Mech. Found. Eng., Mexico City*, **3**, 884–889 (1969).
172. Mead, W.J., 'Engineering geology of dam sites,' *Trans. 2nd Int. Congr. Large Dams, Washington, D.C.*, **4**, 183–198 (1936).
173. Underwood, L.B., 'Classification and identification of shales,' *ASCE Soil Mech. Found. Eng. Div.*, **93**, No. SM6, 97–116 (1967).
174. De Graft-Johnson, J.W.S., Bhatia, H.S. and Yeboa, S.L., 'Geotechnical properties of Accra Shales,' *Proc. 8th Int. Conf. Soil Mech. Found. Eng., Moscow*, **2**, 97–104 (1973).
175. Chappell, B.A., 'Deformational response of differently-shaped and sized test pieces of shale rock,' *Int. Jour. Rock Mech. Min. Sci.*, **11**, 21–28 (1974).
176. Burwell, E.B., 'Geology in dam construction, Part I' (article in *Applications of geology to engineering practice, Berkey Volume*) Geol. Soc. Am., 11–31 (1950).
177. Fasiska, E., Wagenblast, N. and Dougherty, M.T., 'The oxidation mechanism of sulphide minerals,' *Bull. Ass. Eng. Geol.*, **11**, No. 1, 75–82 (1974).
178. Penner, E., Eden, W.J. and Gillott, J.E., 'Floor heave due to biochemical weathering of shale,' *Proc. 8th Int. Conf. Soil Mech. Found. Eng., Moscow*, **2**, 151–158 (1973).
179. BRS Digest 90, *Concrete in sulphate bearing soils and ground waters*, Building Research Station Establishment, Garston, Watford, HMSO (1975).
180. Pettijohn, F.J., *Sedimentary rocks*, Harper & Row, New York (1975).
181. Barth, T.F.W., Correns, C.W. and Eskola, P., *Die enstehung der gestiene*, Springer, Berlin (1939).
182. Dumbleton, M.J., 'Origin and mineralogy of African Red Clays and Keuper Marl,' *Quart. Jour. Eng. Geol.*, **1**, 39–46 (1967).
183. Davis, A.G., 'The structure of Keuper Marl.,' *Quar. Jour. Eng. Geol.*, **1**, 145–153 (1968).
184. Sherwood, P.T., 'Classification tests on African Red Clays and Keuper Marl,' *Quar. Jour. Eng. Geol.*, **1**, 47–56 (1967).
185. Lees, G., 'Geology of the Keuper Marl,' *Proc. Geol. Soc. Lond.*, No. 1621, 46 (1965).
186. Davis, A.G., 'On the mineralogy and phase equilibrium of Keuper Marl,' *Quar. Jour. Eng. Geol.*, **1**, 25–46 (1967).
187. Chandler, R.J., 'The effects of weathering on the shear strength properties of Keuper Marl,' *Geotechnique*, **19**, 321–334.
188. Meigh, A.C., 'The Triassic Rocks, with particular reference to predicted and observed performance of some major foundations,' *Geotechnique*, **26**, 391–452 (1976).
189. Jumikis, A.R., *Soil mechanics*, Van Nostrand, Princeton, N.J. (1968).
190. Bell, F.G., 'Some petrographic factors relating to porosity and permeability in the Fell Sandstones of Northumberland,' *Quar. Jour. Eng. Geol.*, **10** (1977).
191. Serafim, J.L., 'Influence of interstitial water on rock masses' (article in *Rock mechanics in engineering practice*, Ed. by Stagg, K.G. and Zienkiewicz, O.C.) Wiley, London, 55–97 (1968).

192. Sherrell, F.W., 'Engineering geology and ground water,' *Ground Engineering*, 9, No. 4, 21–27 (1976).
193. Darcy, H., *Les Fontaines Publiques de la Ville de Dijon*, Dalmont, Paris (1856).
194. Wittke, W., 'Perculation through fissured rock,' *Bull. Int. Ass. Eng. Geol.*, No. 7, 3–28 (1973).
195. Terzaghi, K., *Erdbaumechanik auf bodenphysikalischer grundlage*, Deuticke, Vienna (1925).
196. Casagrande, A., 'Characteristics of cohesionless soils affecting the stability of slopes and earth fills,' *Jour. Boston Soc. Civ. Engrs.*, **23**, 3–32 (1936).
197. Taber, S., 'Mechanics of frost heaving,' *Jour. Geol.*, **38**, 303–317 (1930).
198. Casagrande, A., 'Discussion on frost heaving,' *Proc. Highways Res. Board*, **12**, 169, Washington D.C., (1932).
199. Croney, D. and Jacobs, J.C., 'The frost susceptibility of soils and road materials,' *Trans. Road Res. Lab., Rept. LR90*, Crowthorne (1967).
200. Jumikis, A.R., 'The soil freezing experiment, *Highway Res. Board Bull. No. 135*, Factors influencing ground freezing,' *Nat. Acad. Sci.*, Nat. Res. Council Pub. 425, Washington D.C. (1956).
201. Black, R.F., Permafrost, *in Applied Sedimentation* (Ed. by Trask, P.D.) Wiley, New York, 247–275 (1950).
202. Muller, S.W., 'Permafrost or permanently frozen ground and related engineering problems,' *U.S. Geol. Surv., Spec. Rept., Strategic Engineering Study 62*, Military Intelligence Division Office, Chief of Engineers, U.S. Army (1945). Also published by Edwards Bros., Ann Arbor, Mich. (1947).
203. Higginbottom, I.E. and Fookes, P.G., 'Engineering aspects of periglacial features in Britain,' *Quar. Jour. Eng. Geol.*, 3, 85–118 (1970).
204. Shotton, F.W., 'Large scale patterned ground in the valley of the Worcestershire Avon,' *Geol. Mag.*, 97, 407–408 (1960).
205. West, R.G., *Pleistocene geology and biology*, Longmans, London (1968).

Chapter 2

Geology and Foundations

GENERAL

All engineering structures, however simple or sophisticated they may be are founded on geological materials. An appreciation of geology is therefore fundamental to the understanding of foundation-structure interactions. In many instances the foundation materials approach the engineering ideal of predictable uniformity, thereby making design and construction procedures more straightforward. The degree of difficulty of a foundation can, therefore, be a direct consequence of the degree of deviation of the geological realities from this engineering ideal but also may be the result of other factors – the scale, cost or sensitivity of the structure, public awareness or failure during investigation to anticipate fully the ground conditions.

In this chapter it is the geological factors which can give rise to particular difficulties in foundation design which will be highlighted and in summary these include the presence of the following foundation conditions:

Compressible, weak or variable materials.
Deep overburden.
Irregular rockhead.
High permeability associated with high water table.
Ground movements { subsidence, seismicity, mass movement }
Deleterious physical and chemical changes.

Returning to the simple engineering ideals for the present it is clear that soil and rock can be simulated by essentially uniform materials. Taking clay as illustrative of the ideal engineering soil, then a mass of isotropic clay (Figure 2.1, A), being composed of finegrained particles is as close to homogeneity as can be achieved. However, in practical terms clays can never achieve such a degree of uniformity for with burial and consequential dissipation of gravitationally-induced pore pressures the clay will increase in strength and decrease in deformability progressively with depth (Figure 2.1, C). During this same process fissuring will develop within the clay (Figure 2.1, C) thereby creating discontinuities within the clay and increasing the deviation from uniformity. Sedimentary variations will also give rise to lateral and vertical changes in grain size and mineralogy thus contributing to internal variations in material properties (Figure 2.1, D).

Similarly, the ideal rock mass can be simulated as a brick wall-like structure with individual joint-defined blocks of uniform materials stacked together in a regular pattern as illustrated in B, Figure 2.1. Three particular factors can contribute to deviations from this ideal – the occurrence of sedimentary layering and associated variations in material properties (Figure 2.1, E), less regular, consistent patterns of discontinuities (Figure 2.1, E) and the consequential effects of weathering and stress release (Figure 2.1, F). Combined

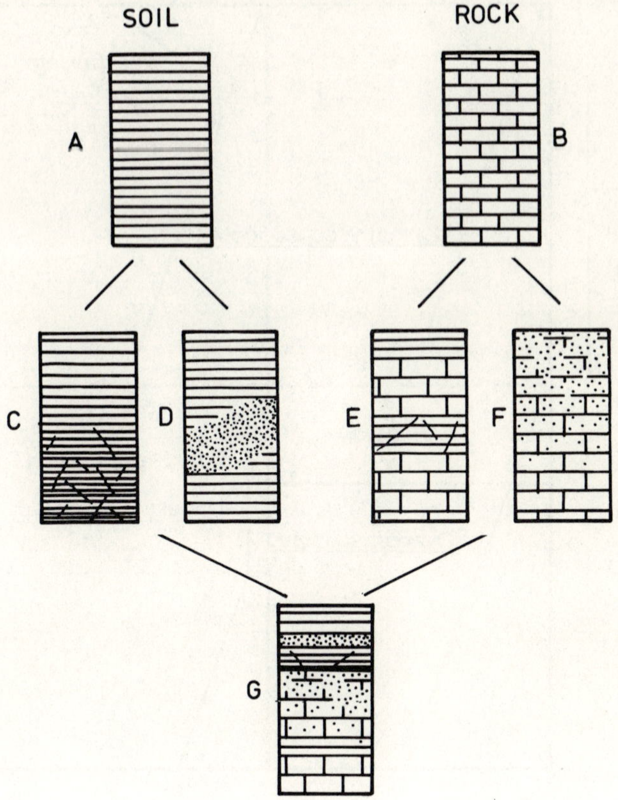

Figure 2.1 Diagrammatic representation of the geological factors in soils and rocks which influence their foundation behaviour. A and B are idealized representations; C, D, E and F introduce the geological realities of variability; G combines soil as overburden in contact at rockhead with underlying bedrock

together (Figure 2.1, G) these various factors introduce a potentially complex pattern of material and mass properties combined with the additional component of rockhead forming the interface between overburden and bedrock.

ENGINEERING SOILS

Clays are possibly the most quoted example of potentially uniform foundation materials even though the geological history of clays as sediments can contribute to significant variations in their material properties. Clay of uniform classification properties deposited in water, and progressively buried, will develop a linear relationship between undrained shear strength and depth such materials being described as normally consolidated sediments (Figure 2.2). If the rate of sedimentation is rapid, for example in excess of a few cm/yr in the case of marine

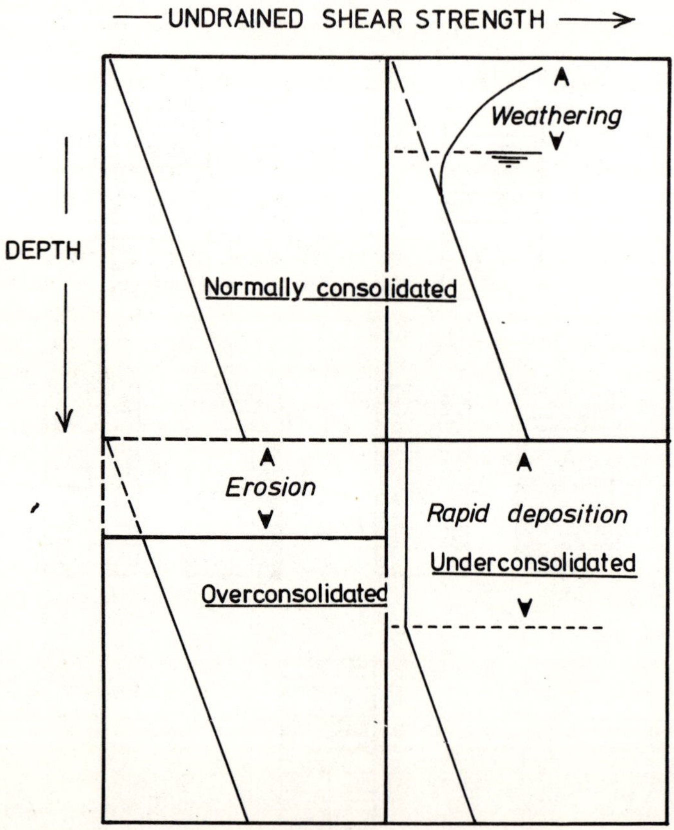

Figure 2.2 Distribution of undrained shear strength with depth for clay sediments

clays, there will be inadequate dissipation of pore water pressures and the clay will not increase in strength with depth. Such underconsolidated materials occur in areas of rapid deposition such as deltas, or lagoons used for the disposal of materials such as fly ash or mine tailings.

If the normally consolidated sediment is exposed to subaerial weathering, possibly as a consequence of a fall in sea level, there will almost certainly be a drop in the water table, partial desiccation of the sediment and a consequential increase in strength thereby forming a crust resting on top of weaker materials (Figure 2.2). If the overlying crust together with the underlying clay is removed by erosion then the clay is described as being in an overconsolidated state. Fresh, overconsolidated clay is likely to become softened as a consequence of weathering. Complex situations can develop where there are fluctuations in sea, or lake level giving rise to rhythmic sequences of both normally and overconsolidated clays interleaved with stiff fossil soils.

One of the best documented examples of marine, normally consolidated clays are in the late-glacial and post-glacial deposits of southern Scandinavia. These sediments were deposited in the sea when the local Earth's surface had been depressed by the weight of glacial ice. With subsequent rebound of the Earth's surface, the sea retreated to the present position and around Oslofjord, for example, the vertical rise has been in excess of 220 m. As a consequence normally consolidated sediments occur both below the sea as well as covering parts of the surrounding land areas.

These clays demonstrate a progressive increase in strength with depth, variations in the strength gradient being a consequence of grain size variations which can also be correlated with past climatic changes. There is a surface crust of 2 to 5 m thickness and this has been used as a structural foundation, thereby enabling loads to be spread over the underlying softer and more deformable clays. A factor contributing to soil instability has been the creation of 'quick' clays resulting from the leaching of the salt content from the pore water originating from the sea water in which the clay was originally deposited. A clay laid down in sea water develops a flocculated structure as a consequence of the electrolyte contained in the pore water. Removal of the electrolyte creates a metastable situation so that the clay particles can only achieve stability by collapse of the flocculated structure to a more compact, dispersed form together with the release of excess pore water. Piles driven into such clays thereby rework the clay causing a catastrophic reduction in shear strength. As a result end-bearing, rather than friction, piles founded on bedrock have been commonly used for deep foundations in Oslo. Even though the Oslo clays are relatively young and of low strength, consistent fissure patterns are present both within the soft clays and the overlying crust.

The influence of pore water chemistry on clay particle orientation, as reflected by the deposition of clays in fresh or marine waters, continues to be reflected in the engineering behaviour of clays, and shales, at even relatively advanced states of lithification. Thus, in the case of the London Clay, which has been buried to depths of the order of 600–800 m, the stress-strain curves of samples orientated horizontally and vertically (as related to the flat-lying bedding) are markedly different. The vertically-orientated specimens are, in relative terms, more deformable and weaker, and this implies that a horizontal parallelism of clay particles has been impressed on the primary random flocculated structure. Original differences in the salinity of pore water can also contribute to differences in the primary clay fabric. In the case of the Leda Clay (a post-glacial marine clay) and the Seven Sisters Clay (a glacial lake clay) from Canada the variations in fabric can be demonstrated by changes in response during shrinkage and subsequent swelling[1]. For example, the Seven Sisters Clay shrinks significantly more at right angles to the bedding (Figure 2.3) than parallel to the bedding; this is a consequence of the preferred clay particle orientation parallel to the bedding formed during deposition in fresh water. In contrast (Figure 2.3) the Leda Clay demonstrates the same difference, but far less markedly, thereby reflecting only a slight parallelism imposed on the original flocculated clay structure formed in sea water. Apart from such alterations in fabric resulting from gravitational compaction, the development of fissures can have important implications on the subsequent engineering behaviour of clay. Fissures can be

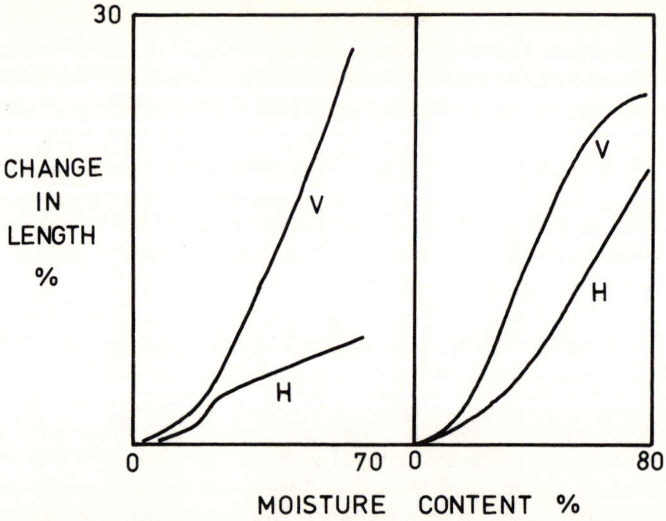

Figure 2.3 Comparative shrinkage curves for Seven Sisters (left) and Leda Clays (right) for vertically and horizontally orientated samples

formed by alternative processes and these include imposed gravitational or tectonic stresses, release of load associated with erosion, and as pore water drainage channels generated during consolidation.

ROCKS

In contrast to clay, sandstone can be adopted as a rock material which in the simplest terms can be composed of quartz grains bonded together. It is, however, apparent that even the simplest even-grained sandstone will be a multi-component system being composed of quartz grains, matrix (which may not be quartz) and voids.

More complex sandstones may involve an assemblage of different mineral grains (e.g. quartz, felspars, micas, lithic fragments) of varying grain size and with a range of alternative cements. This multi-component micro-structure sandstones can be reflected in their response to uniaxial loading. For example, Bunter Sandstone when loaded to failure has a sigmoidal stress-strain curve (Figure 2.4) with greater deformation at low stress levels thereby reflecting the closure of voids and the response of more deformable cementing materials. Repeated loading (Figure 2.4) results in the progressive closure of voids, the development of a permanent strain and a consequential decrease in deformability. Thus, even small scale rock specimens reflect, by their behaviour, the same responses to loading as do rock masses.

In adopting an example of a sandstone foundation, the Hartlepool Power Station, in north-east England, has been selected as representing a site with essentially uniform bedrock conditions. The site is underlain by 25 m or more of recent and glacial sediments resting on top of flat-lying Bunter Sandstone bedrock. The upper part of the overburden is composed of soft post-glacial clays

(Figure 2.5) resting on top of a complex of glacial outwash sandy gravels. As a consequence, there were alternative choices in foundation selection with the glacial sequence and the bedrock forming logical possibilities. In the event[2], the lighter parts of the structure including the Turbine House were founded on driven cast *in situ* piles, whereas the Reactor Halls were founded on 2.2 m diameter concrete piles formed in shafts excavated from the ground surface into bedrock (Figure 2.6).

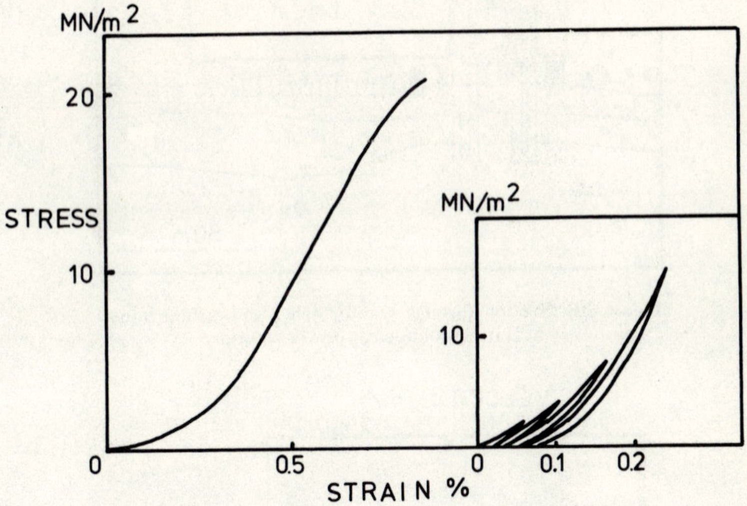

Figure 2.4 Larger graph is full uniaxial stress-strain curve to failure for Bunter Sandstone core; smaller graph is cyclic loading test for similar core

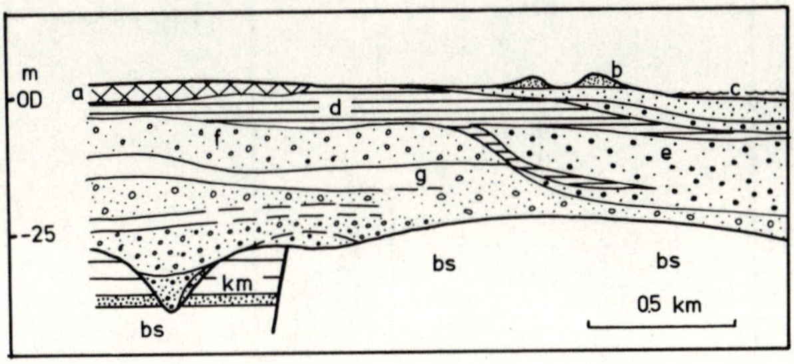

Figure 2.5 Longitudinal sketch cross-section through Hartlepool Nuclear Power Station site. (a) fill, (b) sand dunes, (c) marine sands and gravels, (d) soft marine silty clays, (e) sands and gravels, (f) stony clay (till), (g) laminated clays, (bs) Bunter Sandstone, (km) Keuper Marl

The investigations for the foundations in rock were carried out by drilling supplemented by cross-hole and down-hole seismic velocity measurements. In this situation the major problem which arises relates to the extent to which the precise settlement characteristics of the rock mass essential to the design of the structures can be anticipated and, particularly, the influence of weathering effects immediately below rockhead identified. Drilling yielded adequate

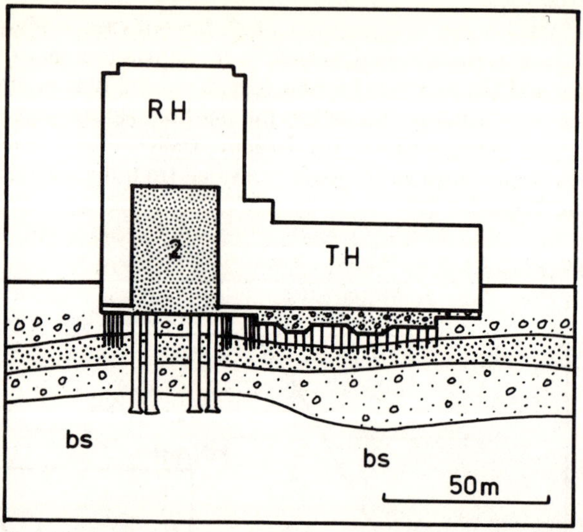

Figure 2.6 Section through reactor hall 2 and turbine house at Hartlepool nuclear power station

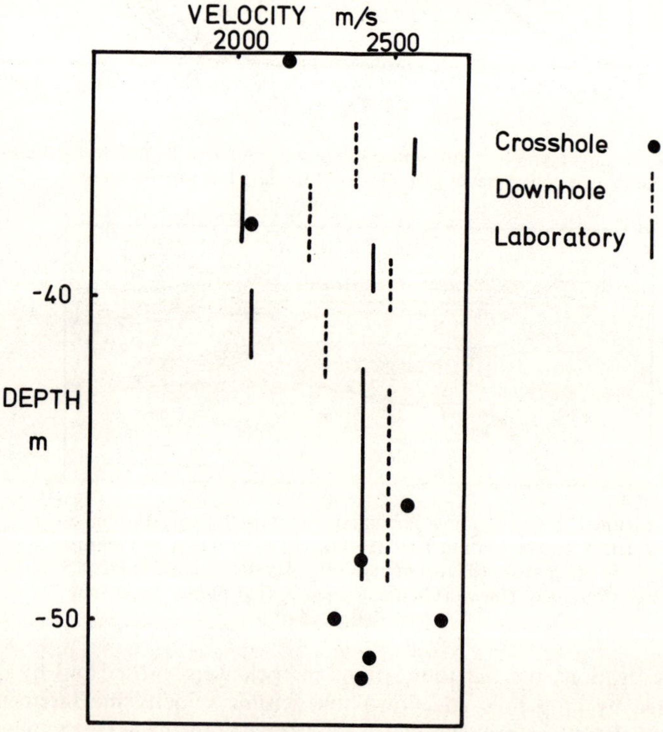

Figure 2.7 *In situ* and laboratory longitudinal wave velocity measurements in Bunter Sandstone plotted with depth for Hartlepool nuclear power station site

samples for conventional laboratory testing and this was supplemented by cyclical loading tests (Figure 2.4) intended to establish the extent to which non-recoverable strains generated by such means could be equated to strains resulting from long-term creep. The *in situ* velocity measurements provided an indication of the extent to which the rock mass was fractured.

It was of particular interest (Figure 2.7) to recognise that, in the main, the *in situ* velocity was greater than that measured on saturated specimens in the laboratory. In these circumstances it was apparent that the confined rock mass at depth was not significantly fractured. Settlements recorded[2] for the Reactor Halls were less than 1 cm during construction, well within that anticipated from the laboratory and related tests. It is apparent, therefore, that for an essentially homogenous, confined rock mass in which the effect of fracturing is not significant programmes of laboratory-scale testing can lead to practical conclusions.

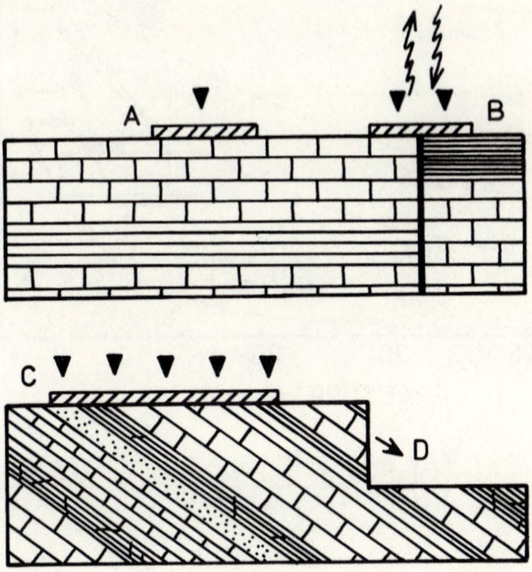

Figure 2.8 Diagrammatic representation of influence of rock structure on alternative foundation situations

However, in the more general case discontinuities within the rock mass and variations in rock material properties can influence foundation behaviour. Four examples reflecting the most common situations which arise are set out in Figure 2.8:

A. Essentially uniform foundations.

B. Foundation based on materials with distinct properties giving rise to risks of differential settlement; possible seismic displacements on fault.

C. Risk of differential settlement arising from dipping strata with significantly different mass/material properties.

D. Instability arising from undercutting of dipping strata.

Although these four situations identify those which arise most commonly in practice, other factors, such as the influence of scale, time or water, can be of considerable importance.

Load transmitted through a rock mass is primarily carried in two ways, by shear transference across joints or by the spanning effect of joint-defined blocks. In consequence, the discontinuity pattern, and thus the slope of joint blocks, has a critical bearing on the manner in which a rock mass will respond as a foundation. Within recent years attempts have been made to relate joint separation, as a readily measurable geological parameter, to quantitative figures for the deformability of rock mass. For example, Hobbs[3] has established relationships between the rock mass factor (ratio of the deformability of the rock mass *in situ* to the deformability of intact rock) and fracture separation, lower rock mass factors being associated with more closely spaced fracturing.

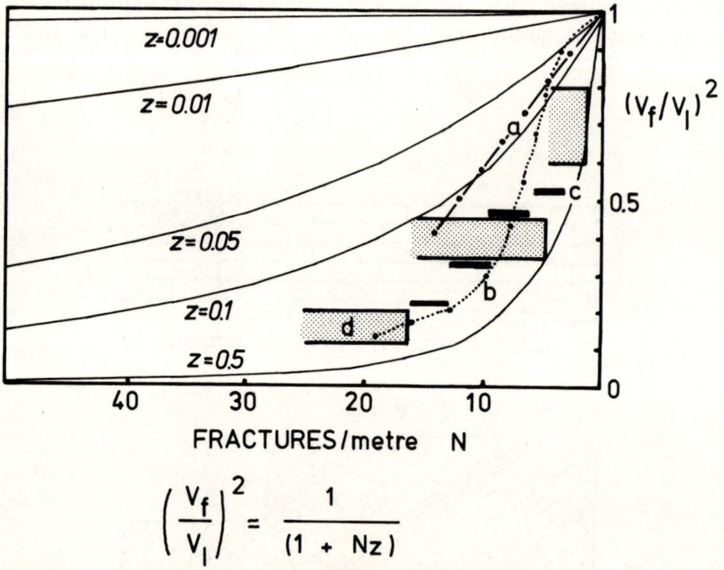

Figure 2.9 Relationship between velocity ratio, $(v_f/v_1)^2$ and fracture separation. (a and b) results of model experiments with different fracture spacings, (c) quartzitic sandstones at Farahnaz Pahlavi Dam, Iran, (d) chalk in south England

A similar approach can be adopted using the velocity ratio (ratio of *in situ* to laboratory longitudinal wave velocity squared) and this is illustrated by Figure 2.9; Malone[4] established a series of theoretical curves from the formula:

$$\left(\frac{V_f}{V_l}\right)^2 = \frac{1}{(1 + N_z)}$$

where
N = number of fractures per metre;
$z = V_l t'$ where t' is the time loss for each fracture.

Where velocity measurements are used to assess foundation deformability it is important to bear in mind the effect of confining conditions. This is clearly illustrated by Figure 2.10 which is based on laboratory tests on intact rock and a simulated fractured rock with $N = 15$. It will be noted that the measured velocity is sensitive to the applied stress which results in a closure of discontinuities and thus a higher measured velocity. It is certain that this situation has contributed to the observations made at Hartlepool Power Station (Figure 2.7).

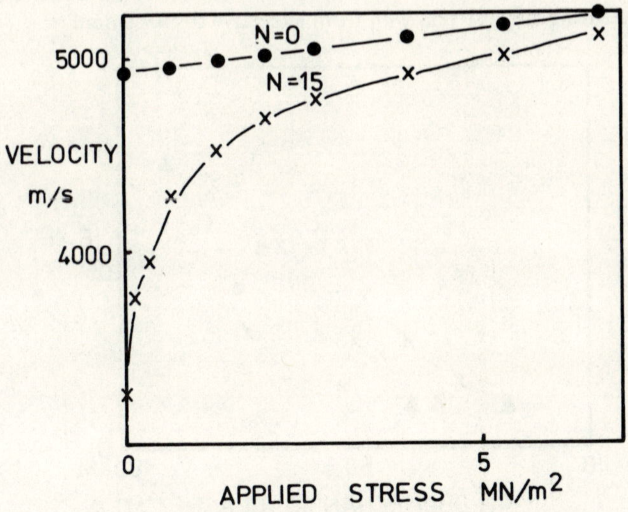

Figure 2.10 Influence of applied normal stress on longitudinal wave velocity for unfractured rock (spots) and rock with 15 fractures/m (crosses)

An alternative approach is that illustrated by the application of classifications which provide a numerical rating to different geological characteristics which can influence the engineering behaviour of rock masses. Of these, the Geomechanics Classification[5] (see Chapter 1) allocates points according to the following scales:

	Points available for allocation
Strength of intact rock	0–15
RQD	3–20
Joint spacing	5–30
Condition of joints	0–25
Groundwater	0–10
	100 maximum

The ratings range from Very Good rock (> 90) to Very Poor rock (< 25). It is of interest to note that there is a broad relationship between the rock mass factor and the points awarded under the Geomechanics Classification (Figure 2.11). This provides a method for estimating the relative deformability of rock masses from essentially geological information.

One of the major questions which arises in foundation evaluation occurs when the foundation spans rock materials of radically different properties, thereby inducing the possibilities of differential settlement. In the case of the

Farahnaz Pahlavi Dam, a 112 m buttress dam in northern Iran, the right bank buttresses founded on a steep hillside span a steep zone of shales sandwiched between quartzitic sandstones. In these circumstances there is a risk that shear stresses transmitted along the foot of the buttress webs will not be carried by the poorer quality rock but will be shed downstream, thereby risking the overstressing of the dam. Passive thrust blocks were constructed downstream of the dam thus securing the buttress web from excessive displacement.

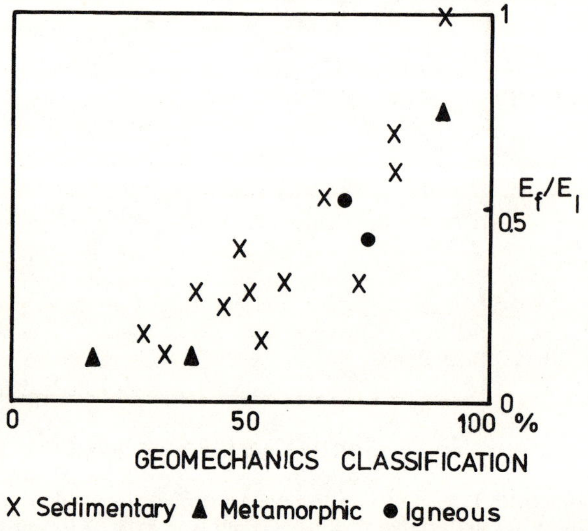

Figure 2.11 Relationship between ratings based on Geomechanics Classification for various sites and rock mass factor, (E_f/E_1)

Undercutting of dipping strata can result in potential instability and may require special support measures particularly in those cases where the critical area will form a part of the foundations. During the construction of the Hallamshire Hospital in Sheffield during the mid 1960s a slope failure occurred during excavation of Coal Measures shales dipping at 16–18° in a downhill direction. Subsequent shear box tests on bedding planes yielded a peak angle of shearing resistance of about 17°. Although in this example the slip interfered only slightly with construction, as allowance had been made for anchored supports to excavated faces, the event highlighted somewhat greater risks in the Bristol Royal Infirmary development. In the latter case[6], an alternating sequence of quartzitic sandstones, siltstones and shales (including clay shales) dipped in a downhill direction at 27°.

The excavation plans allowed for undercutting of the dip, the undercut benches being designed to carry foundation loads. It was apparent that limited failure, at the minimum, was almost inevitable. For these reasons, the foundations of the Outpatients Building were excavated in trench and formed as a series of concrete load-bearing walls and retaining walls. Bulk excavation was then carried out, the rock strata being fully supported by the pre-existing structural walls. Small-scale slips occurred on bedding planes within the trench excavations confirming that the measures adopted had been fully realistic of the ground conditions. Similar problems occur in those situations where inclined loads from arch or other kinds of dams are transmitted onto unfavourably orientated discontinuities. This condition was a primary factor in the failure of the Malpasset Dam, in southern France, where the left bank abutted against a potentially unstable structural wedge, one component of which outcropped on the valley side downstream of the dam.

SUBSIDENCE

Major foundation problems do arise in those cases where there is a likelihood of subsidence and such circumstances arise, in particular, where there are natural or artificial cavities below the ground surface. Removal of solid materials associated with mining will create voids and these may be associated with displacement of the overlying rocks. Imposition of load onto areas which have been undermined at shallow depth can give rise to large total and differential settlements. Upward propagation of mining-induced voids through rocks can be induced by ground disturbance, loading or modification to the groundwater flow pattern. The effects of such mining activity and the consequences of foundation design are now generally recognised[7] but there is less wide recognition of the effect of undermining on the subsiding rock.

It has been argued[8] that subsidence creates permanent void-openings which give rise to an enhanced secondary permeability. However, investigations[9] in association with the design of the Liège Metro in Belgium have demonstrated that subsidence can give rise to a deterioration in the physical properties of both rock materials and rock mass. Samples of rock were collected from boreholes drilled in the Coal Measures of the Liège area, these samples being tested for unconfined strength under uniform conditions, and also for quartz content. By separating out material from sites which have, and have not, been undermined it is apparent (Figure 2.12) that the rock material collected from undermined areas is significantly weaker than where there has been no mining. Possibly somewhat less surprising is the fact the pressuremeter tests carried out in the boreholes revealed that the mean deformability is consistently greater in those rocks which are in undermined areas as set out in Table 2.1.

Table 2.1

	Modulus of deformation MN/m²		*Modulus of elasticity MN/m²*	
	Undermined	*Not undermined*	*Undermined*	*Not undermined*
Coal	15	330	720	900
Silty Mudstone	700	5200	2150	8300
Sandy Mudstone	980	5800	4700	7000
Sandstone	1050	5000	4700	10000

The moduli quoted are mean values for all results obtained, the modulus of deformation being based on a single cycle test whereas the modulus of elasticity is derived after the non-elastic strain has been eliminated by cyclic loading. Although it has been recognised that subsidence fracturing can occur and may, in certain circumstances, be of massive proportions the only numerical evaluation of the consequences of undermining on foundations are those which can be derived from the Liège Metro studies.

The solution of carbonates and evaporites will result in the generation of voids which can cause, or contribute to, subsidence. In the case of limestone cavities modification to the groundwater flow pattern by groundwater extraction could result in internal erosion of overlying materials into the cavities, thereby resulting in upward, and often catastrophic, transference of voids to the ground surface. If there is a depression in the rockhead associated with a thickening of the overburden differential settlement may occur as a result of a reduction in groundwater pressures and consequential increased consolidation of the thicker cover.

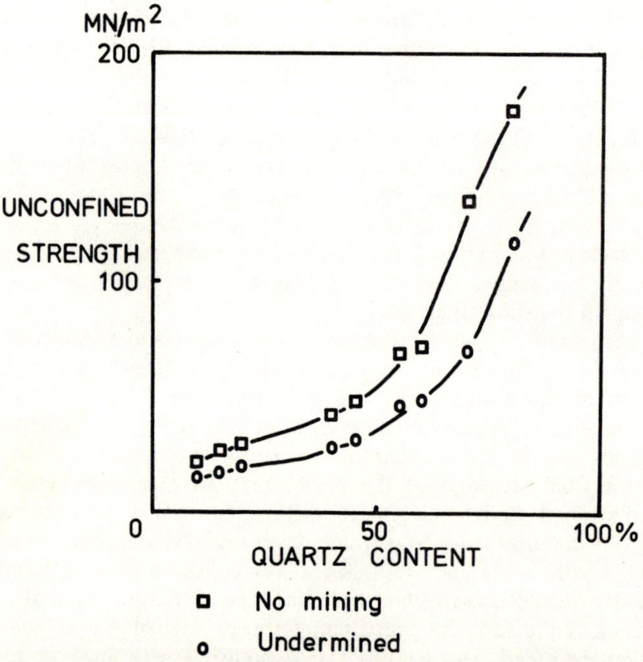

Figure 2.12 Relationship between uniaxial strength and quartz content for Coal Measure rocks from Liège Metro investigations, a distinction being drawn between those from sites which have, and have not, been undermined by coal extraction

Diversity in subsidence processes can be illustrated by the collapses occurring on the northern flanks of the Western Rand[10] where the underlying dolomitic rocks are dewatered from below by the underlying gold mining operations. Both catastrophic subsidences, associated with no surface evidence of underlying instability, and areas of settlement, possibly related to deeper areas of overburden have been recognised. Mineral extraction of fluids (e.g. water, brine, oil) and gas from the ground will, similarly, give rise to potentially uncontrolled settlement[11]. Possibly the most classic situation which has occurred in Britain has been the wild brine pumping in Cheshire[12] and Staffordshire resulting in extensive total and differential settlements.

In the case of the Baldwin Hills reservoir in California, the failure of the asphaltic lining to the reservoir appears to have arisen as a consequence of ground displacements around the Inglewood oilfield associated with extraction of hydrocarbons[13]. As subsidence took place the ground above the oilfield subsided with the ground surface being put into compression. Surrounding this zone the ground surface was in tension and this was reflected by ground ruptures on

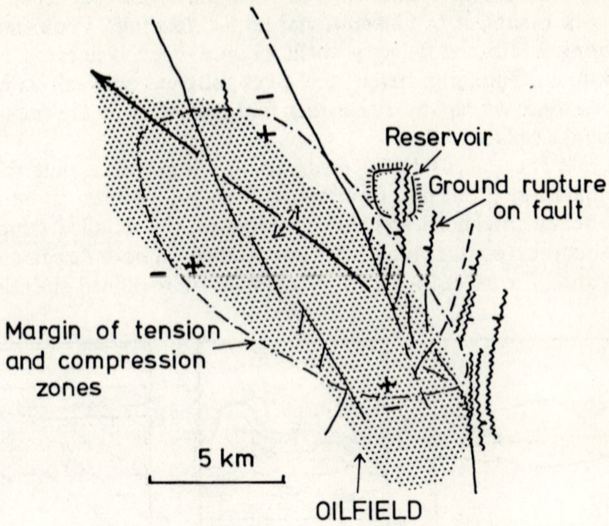

Figure 2.13 Sketch map of Baldwin Hills reservoir

faults (Figure 2.13). Two of these faults passed below the reservoir and eventually differential movements cracked the lining causing loss of water and partial washing out of the surrounding embankment.

OTHER FACTORS

The influence of groundwater on foundation conditions can be significant. The control of groundwater flows and pressures may be an important factor requiring careful evaluation and monitoring during design and construction. However, groundwater can also influence, to a major degree, the engineering properties of certain soils, and some rocks, particularly in the cases where metastable materials or soils subject to internal erosion are present. Granular soils laid down under partially saturated or dry conditions are liable to be deposited in a porous state and this can give rise to uncontrolled collapse when saturated or subject to vibration.

The removal of fines, by groundwater flow, from coarser grained soils can result in increases in permeability and, in the case of gravels, the development of an openwork texture. This situation has been identified in High Wycombe, in southern England, where permeability increases in alluvial gravels appear to have been induced by heavy groundwater extraction during trench excavation in the last century. Fines were apparently washed out of the gravels at this stage, subsequently giving rise to high groundwater velocities and consequential erosion damage to cast *in situ* piles installed several decades afterwards.

A variety of physical and chemical changes can contribute to deleterious effects on foundation materials. Unstable or reactive minerals such as sulphate or pyrites, or the presence of swelling clay minerals, can result in a deterioration in foundation properties and, in appropriate circumstances, aggressive responses in concretes. Mechanical breakdown of rock materials can occur particularly where the rock is subject to conventional impact loadings. For example chalk is readily re-worked into a soft putty chalk as pore water is released during mechanical breakdown. Similarly driven steel piles founded on weak slates have been found to lose their set rapidly suggesting that softening of the fragmented rock material has taken place.

Finally, reference needs to be made to the importance that rockhead can have on foundation problems. Commonly foundations have to be constructed on, or just below, rockhead and, in consequence, the detailed structure of the overburden-bedrock contact may have importance on both design and construction procedures. Figure 2.14 identifies a range of rockhead situations each of

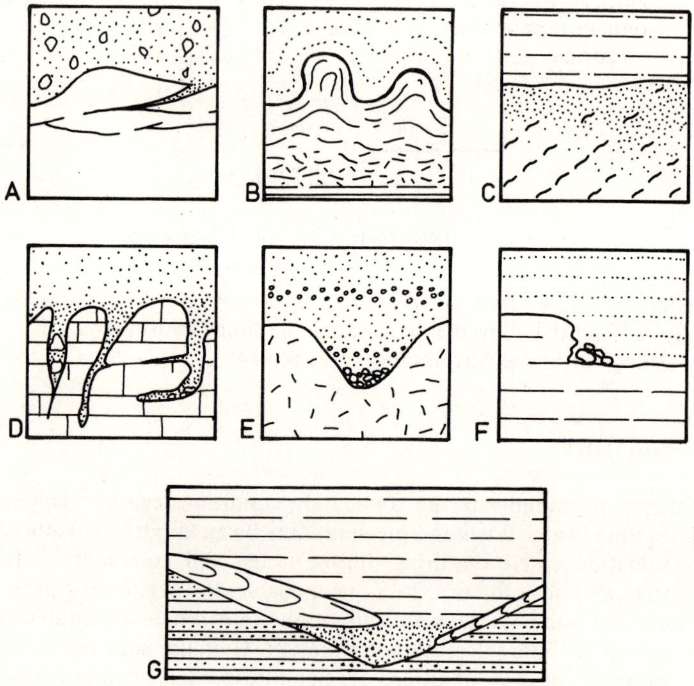

Figure 2.14 Differing rockhead situations. (A) till on rock with glacially-induced shears containing till infilling, (B) bulged periglacially disturbed rockhead, (C) weathered rock below rockhead, (D) solution-affected limestone, (E) buried channel with alluvial fill, (F) buried wind-eroded and undercut rock face, (G) buried channel with slope failures on flanks

which illustrates a different origin to the rockhead surface and, by implication, the superincumbent overlying soils. Each of these cases poses different technical questions as presented by the variation in rockhead geometry, the preferential concentration of specific soil types above rockhead, the degree of penetration of the overburden into the rock mass as a whole and the extent to which the upper part of the rock mass has been subject to deterioration in engineering properties. In conclusion, therefore, a consideration of the engineering aspects of rockhead presents a microcosm of the potential influence of geology on foundations.

References

1. Warkentin, B. and Bozozuk, M., 'Shrinking and swelling of two Canadian clays,' *Proc. 5th Int. Conf. Soil Mech. and Found. Eng.*, **1**, 851 (1961).
2. Coates, F.W. and Taylor, R.S., 'Hartlepool Power Station: major civil engineering features,' *Proc. ICE.*, **60**, 95 (1976).
3. Hobbs, N.B., 'Factors affecting the prediction of settlement of structures on rocks with particular reference to Chalk and Trias,' *article in Settlement of Structures*, British Geotechnical Society, Pentech Press, 579 (1975).
4. Malone, A.W., 'Elastic wave measurement in rock engineering,' *PhD Thesis,* University of London.
5. Bieniawski, Z.T., 'Engineering classification of jointed rock masses,' *Trans. S. Afr. Inst. Civ. Engrs.*, **15**, 355 (1973).
6. Phillips, L. and Knill, J.L., 'Foundation construction at the Bristol Royal Infirmary – Phase 1,' *Quart. J. Engg. Geol.*, **6**, 207 (1973).
7. Bell, F.G. (Editor), 'Ground conditions in mining areas', *article in Methods of treatment of unstable ground*, Newnes-Butterworths, London, 112 (1975).
8. Knill, J.L., 'Rock conditions in the Tyne Tunnels, north eastern England', *Bull. Ass. Eng. Geol.*, 10, 1 (1973).
9. Polo-Chiapolini, C.I. Characteristiques geomecaniques des roches du bassin houiller de Liège (Belgique),' *Mem. CERES* new series, No. 47 (1974).
10. Jennings, J.E., Brinto, A.B.A., Louw, A. and Gowan, G.D., 'Sinkholes and subsidences in the Transvaal Dolomite of South Africa, *Proc. 6th Int. Conf. Soil Mech. Found. Eng*, Montreal, **1**, 51–54 (1965).
11. Poland, J.F. and Davis, G.H., 'Land subsidence due to withdrawal of fluids,' *Reviews in Engineering Geology, Am. Geol. Soc.*, 190--269 (1963).
12. Bell, F.G., 'Salt and subsidence in Cheshire, England', *Engng Geol*, **9**, 237–247 (1975).
13. Kresse, F.C., Baldwin Hills Reservoir Failure, *article in Engineering Geology in Southern California*, Ed by Lung, R. and Proctor, R., *Spec. publ. Ass. Eng. Geol.*, (1968).

Chapter 3

Modern Models for Mechanical Behaviour of Soil

INTRODUCTION

In the field of foundation problems and of site conditions wide experience is all important. However the description of the character of the ground is also a problem of language. When we use the words 'overconsolidated clay' we have in mind a model of elastic and plastic compression of clay in Terzaghi's oedometer, in which we conceive of soil as a spring in compression in one direction only. However in a modern computation we need to know lateral stresses as well as vertical stresses in the ground, and the language of one-dimensional compression is

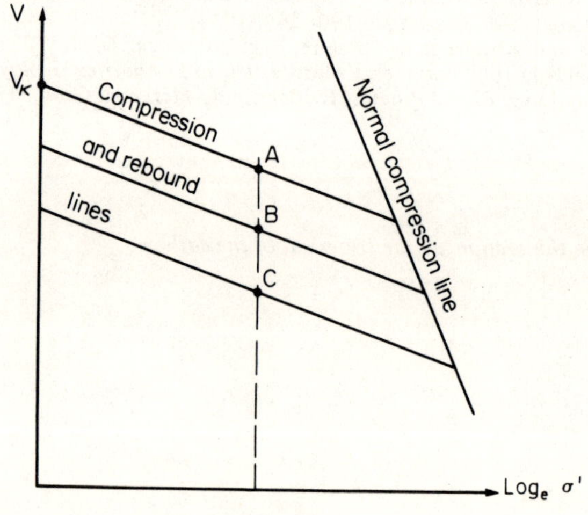

Figure 3.1

inadequate to convey all the information that will be needed. It is therefore worth spending a little time reviewing a modern model for the compression, yielding, and rupture of soil.

If we plot data of one-dimensional compression of clay in Figure 3.1, we see a family of parallel lines for compression and rebound on a graph of specific volume V against natural logarithm of one-dimensional effective compressive stress σ'. Each of these lines has an equation

$$V + \kappa \log_e \sigma' = \text{constant} = V\kappa \tag{3.1}$$

where $V\kappa$ is the value of the intercept of any specific line with the V axis. For example in Figure 3.1 the value of $V\kappa$ which is indicated corresponds to the line through the point A.

Considering the three points A, B and C in Figure 3.1, when the stress is the same the compressibility of the soil is the same

$$\frac{dV}{d\sigma'} = -\frac{\kappa}{\sigma'} \tag{3.2}$$

This is understandable if we look at Figure 3.2, where we see in (a) an open structure of particles most of which are carrying stress, and in (c) and (b) an

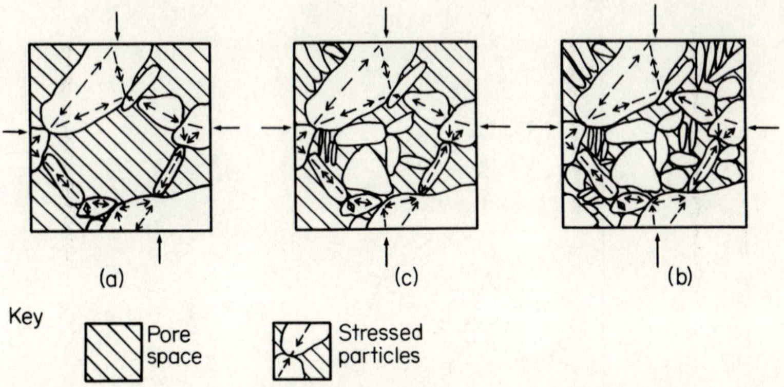

Figure 3.2

increasing proportion of lightly stressed or unstressed particles which do not participate in the load carrying structure. As we go from (a) to (c) to (b) we are filling in the voids of the soil structure without altering its elastic stiffness: equation (3.2) shows that soil stiffness changes in a major way with effective stress.

ANALYSIS OF SOIL

If we now impose shear stresses on the aggregations of particles represented by Figure 3.2(a), (b) and (c), and permit drainage of pore fluid, we may expect slippage of particles and volume changes. Particles in an open packing under a heavy pressure at (a) will compact during shear distortion. Particles in a dense packing under light pressure at (b) will dilate during shear distortion. Between these limits there will be a density of packing, let us say at (c) at which during shear distortion a succession of load carrying structures of stressed particles will form and collapse, with successive new structures being formed at about the same density of packing. In this shearing motion, a certain proportion of the particles will at one time form part of the load carrying structure, and at other times be relatively lightly stressed or unstressed and have the role of 'filler' particles filling voids. The notion of the critical state is that there exists one certain critical packing of particles or critical void ratio, at which continuous flow is possible at constant mean normal effective stress p'. If we write

$$p' = \frac{1}{3}(\sigma_1' + \sigma_2' + \sigma_3') \tag{3.3}$$

and if in Figure 3.3 we plot values of V against $\log_e p'$, then Schofield and Wroth[1] quote experimental evidence that the ultimate state of any soil (Γ) sample during continuous remoulding and shear flow will lie on a 'critical state' line with equation

$$V + \lambda \log_e p' = \Gamma$$

shown in Figure 3.3. We can now distinguish between two different types of overconsolidated state of soil. There are states for which the combinations of

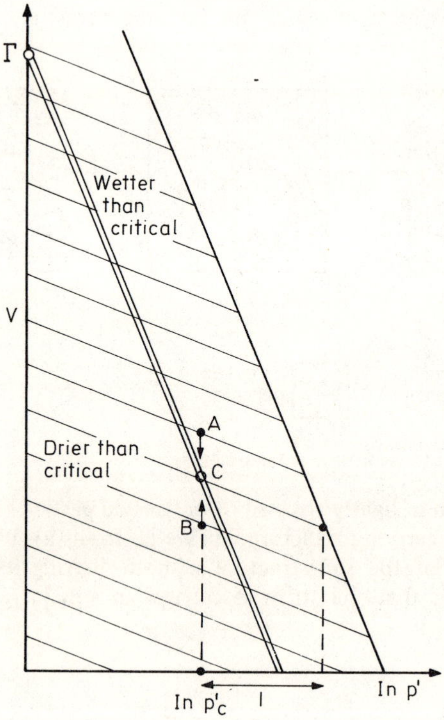

Figure 3.3

specific volume V and mean normal effective stress p' lie further away from the origin than the line of critical states, so that

$$V + \lambda \log_e p' > \Gamma \tag{3.5}$$

and these states have been called 'wetter than critical'. There are also states of specific volume V and mean normal effective stress p' such that

$$V + \lambda \log_e p' > \Gamma \tag{3.6}$$

and these states have been called 'drier than critical'.

In Figure 3.3 the arrows from A and from B indicate paths of drained tests at constant mean normal effective stress p'. We could equally study the paths of tests conducted without drainage, that is to say at constant specific volume V, and these would be indicated by arrows approaching the critical state line with changing values of lnp'. One of the successes of Critical State Soil Mechanics was the prediction of the successive yielding states along the path by which undrained tests approached critical state from states wetter than critical.

The introduction of the mean normal effective stress p' paid due regard to all three principal stress components and got away from the problem of regarding any one component as all important: p' is what can be described mathematically as an 'invariant', meaning that, and whatever axes are taken for definition of stress components, the mean of the three normal stresses will remain the same, invariant with respect to rotation of axes of reference. There was then a problem of defining a parameter which measures in a general way the level of shear stress, for which it proved to be mathematically consistent to define a scalar measure of shear stress intensity

$$q^* = \sqrt{\left(\frac{(\sigma_2 - \sigma_3)^2 + (\sigma_3' - \sigma_1')^2 + (\sigma_1' - \sigma_2')^2}{2}\right)} \tag{3.7}$$

This looks less formidable in the triaxial test conditions where $\sigma_2' = \sigma_3'$, in which case equation (3.7) reduces to

$$q = (\sigma_1' - \sigma_3') \tag{3.8}$$

which is then defined as the 'deviator stress'. The two parameters p' and q prove to be useful in descriptions of the yielding of soil in states wetter than critical, before critical states are reached.

It is perhaps worth emphasising at this point that whenever the parameters p' and q are used it is implicit in their use that the soil is yielding as a plastic-continuum, or is deforming as an elastic continuum, and that once a rupture surface passes through a point there may be different principal stresses in the soil parallel to the rupture surface to either side of the rupture surface, so the state of stress there cannot be described by these continuum parameters. At a point on the rupture surface what matters are the components of effective stress across that rupture surface.

CAM CLAY MODEL

Continuing with the analysis of the yielding of the continuum prior to the formation of any rupture surface, we come to the original model of yielding of clay which was called Cam Clay. This model can be regarded as deriving from the work of D. W. Taylor[2], who studied the yielding of sand in shear boxes, Figure 3.4(a), and observed the way that the peak value of shear stress τ corresponded to the maximum rate of dilatation. In these tests the shear displacement x is increased steadily and the shear stress τ that is required is observed. The normal stress σ' is held constant and the dilatation y is observed.

In Figure 3.4(b), the curve B shows the peak stress from a dense test of sand, and Figure 3.4(c) shows that this peak coincides with the maximum rate of dilatation on the curve B for this dense test. Taylor showed that these data of failure can be expressed in the form of a work equation

$$\tau dx = \mu\sigma' dx + \sigma' dy \tag{3.9}$$

in which we see on the left hand side the product τdx being the rate work is put into shear displacement of the shear box (which is equal to the product of a coefficient of friction μ times the normal stress σ' times the movement of

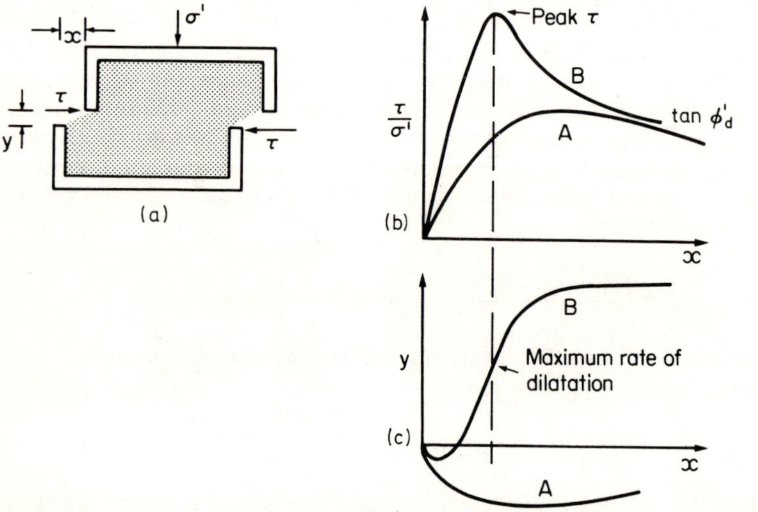

Figure 3.4

shear displacement) and, on the right-hand side the rate at which work is done against friction, plus the product $\sigma' dy$ (which is the rate at which work is done in pushing back the normal stress σ' on the boundary which tries to prevent dilatation for dense soil). We can rewrite Taylor's equation as

$$\frac{\tau}{\sigma'} = \mu + \frac{dy}{dx} \tag{3.10}$$

in which form we see that the stress ratio at failure as the sum of a coefficient of friction and a rate of dilatation. If we replace the stress ratio τ/σ' by the letter R and the dilatation rate dy/dx by the letter D, and consider states before and after failure, then we can plot equation (3.10) as a line in Figure 3.5. The line as drawn is based on some simple notions of work, and it is possible to progress beyond these simple notions to direct experimental measurements in the manner of Rowe[3] who found experimental lines of slope slightly different from that shown in Figure 3.5, which leads to a revision of equation (3.10), although we will not concern ourselves with that here.

Since soil in a critical state neither dilates nor contracts it is appropriate to replace equation (3.10) by a new equation

$$\frac{q}{p'} = M \tag{3.11}$$

for the yielding of soil in critical states. Soil in states denser than critical (Figure 3.6(b)) will tend to reach a peak and then rupture on a distinct failure plane. Soil in states wetter than critical, Figure 3.6(a), will tend to yield progressively through the whole volume of a sample as it approaches critical states. If we replace the dilatancy term in equation (3.10) by a generalised expression for the

Figure 3.5

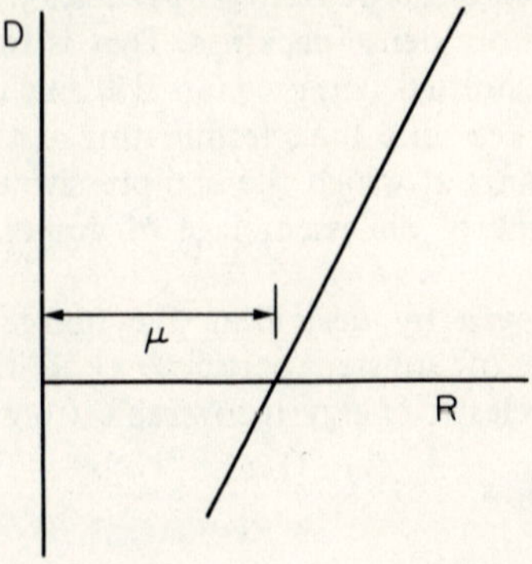

Figure 3.6

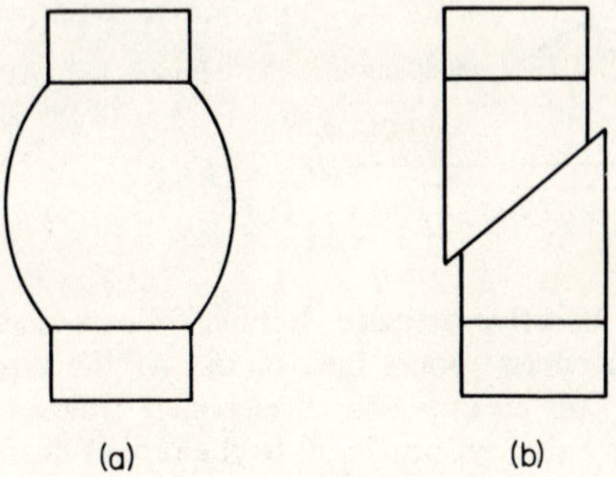

shift of swelling line κ with increase of generalised distortion $d\epsilon$, then an expression for the work done in yielding of an isotropic granular continuum is

$$\frac{q}{p'} = M - \frac{dV\kappa}{d\epsilon} \tag{3.12}$$

which is a generalised expression analagous to equation (3.10). With a further assumption that the yielding of this ideal soil obeys the associated flow rule of the theory of plasticity, Schofield and Wroth derived an expression for the yield locus of this ideal soil, replacing the term for plastic volume change $(dV\kappa)/d\epsilon$ by the normal to the yield locus dq/dp', giving

$$\frac{q}{p'} = M + \frac{dq}{dp'} \tag{3.13}$$

which, when integrated, gives

$$\frac{q}{Mp'} = 1 - \log_e(p'/p'_c) \tag{3.14}$$

which is the curve sketched in Figure 3.7. At point C in Figure 3.7, the value of q equals Mp'_c and the strength at the critical state is seen to be the maximum strength at yield for soil on this particular swelling and rebound curve. As mean normal effective stress p' increases in the region A in Figure 3.7, we see that the soil is less and less capable of sustaining a shear stress level q without yielding. The yield curve ultimately reaches the p' axis at a point $p' = 2718\ p'_c$. The point is most significant because it predicts that a pressure will be reached at which this assemblage of granular particles will spontaneously shear and yield and collapse to more dense packings. That is to say the model predicts that the phenomenon of normal compression will occur in Figure 3.1 with successive compression and rebound lines terminating at the normal compression line at a succession of points at which the compressive effective stress will cause plastic collapse of volume of an assemblage of granular particles such as is shown in Figure 3.2.

The model began by describing the non-linear compressibility of a load-bearing structure of stressed particles, equation (3.2). From this equation it follows that the elastic energy recoverable from the stressed structure increases

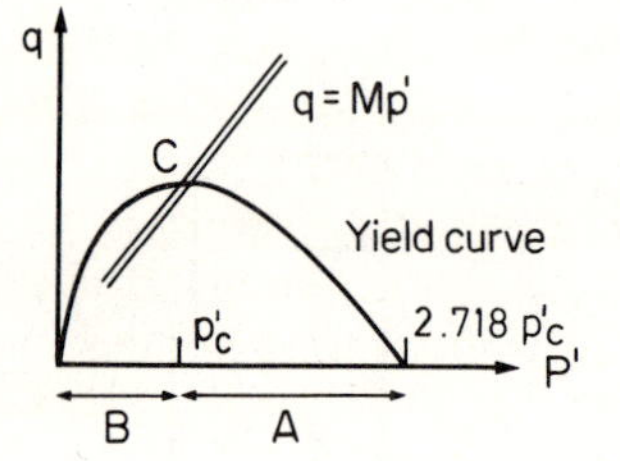

Figure 3.7 (After Schofield and Wroth)

linearly with effective pressure. It then follows that we can simply calculate the rate at which energy pours into or out of the elastically stressed structure of soil particles, for example as a specimen follows an undrained test path in Figure 3.3, as being proportional to the rate of change of mean normal effective stress p'. The model then incorporates a second statement about the general character of frictional dissipation of energy when slippages occur within a load-bearing structure of soil particles, and successive stressed structures deform and break down as slippages occur here and there within the structure of stressed particles. The simple statement that is made is that the rate at which work is dissipated in frictional yielding is the product of a friction constant M times the mean normal effective stress p' times the rate at which generalised shear distortion ϵ is occurring. This statement is embodied in equation (3.12). A third statement is then introduced that the yielding of this ideal soil shall obey the associated flow rule of the theory of plasticity. This statement is embodied in equation (3.13). This equation can be integrated (equation (3.14)) giving the yield curve plotted in Figure 3.7.

The process of integration requires introduction of a constant. The pressure p'_c is seen in Figure 3.3 to define the point at which one particular compression and rebound line intersects the line of critical states. If we were concerned with the yielding of soil in any other density of packing represented by another parallel compression and rebound line, we would simply find the mean normal effective pressure at which that line intersected the lines of critical states and use that instead of p'_c.

The point at which the yield curve reaches the p' axis in Figure 3.7, can be found simply by putting $q = 0$ in equation (3.14). This shows that $ln(p'/p'_c) = 1$ will define the maximum range of elastic states wetter than critical. Hence in

Figure 3.3, the spacing between the critical state line and the normal compression line is such that they intersect each swelling and rebound line at points whose abscissa are unit distance apart on the lnp' axis. If the line of critical state equation (3.4) has slope λ then it follows that the line of normal plastic compression is parallel and has the same slope. If the displacement between these lines in the V direction is calculated this is found to be simply $(\lambda - \kappa)$. Thus the model makes a prediction that the change of water content in drained compression with shear distortion at constant mean normal effective stress p', is independent of p' and can be calculated from constants λ and κ. In this respect the model corresponds closely to the data of the best triaxial tests.

The close fit of the model to these data, and the ability of the model to take account of all components of the tensors of stress and stress increment and strain increment, have made the model attractive to those concerned with finite element calculations. The model cannot be used directly in its simple form for these computations, without incorporating some additional features. A variety of additional features have been introduced by various workers in this field, and so a variety of Cam Clay models have come into existence. We need not consider them all here but it is important to note that each model incorporates the line of critical states. Whichever model is considered, continuous shearing will ultimately bring an ideal material into a critical state. There may be important differences in the predictions of behaviour during yielding, but all predictions reach common ground when the critical states are reached. This has consequences in what has been called 'critical state design'.

CRITICAL STATE DESIGN

Critical state design in soil mechanics is analagous to plastic design in steel frame structures. When a steel frame yields at a sufficient number of points the heavily loaded structures ceases to be a rigid frame, and degenerates into a mechanism of rigid bars connected by stiff plastic hinges. During plastic collapse, the work done by the moving loads is equal to the work dissipated at the yielding plastic hinges in the steel frame structure. Similarly when a body of soil yield, because either its own self weight or the external loads are too great for its strength, a failure mechanism will develop within the body of soil. Although the geometric principles of failure mechanisms of soil continua are less well understood than for steel frame structures, certain features of the mechanisms are well understood. In regions where the effective stresses are relatively high the soil will yield and flow as a plastic continuum, with positive pore water pressures in the interior of undrained regions, and with consolidation as water drains out of these regions. In contrast, regions with low mean normal effective stress are liable to exhibit tensile cracking. Relatively low mean normal effective stress may occur near unloaded boundaries of soil bodies in which there are not high stresses parallel to the boundary, but they may also occur at depth if high water pressures are present, for example, in hydraulic fracture of the core of an earth dam. In describing these regions we have used the words 'relatively' highly stressed or 'relatively' lightly stressed and this language can be given precise meaning in relation to Figure 3.3. Soil is relatively heavily stressed when it is in states wetter than critical. We do not speak only of a relative density, or of a relative pressure, but of the combination of density and pressure that is involved in the equation

$$V + \lambda \ln p' = V_\lambda \tag{3.15}$$

This is not really a new use of language; when the blow counts of a standard penetration test are used to define the relative density of soil they are corrected for the effective pressure in this way. Equally when the risk of hydraulic fracture is assessed it is appreciated that use of very densely compacted soil in the core of an earth dam will introduce more risk because relatively higher stresses are needed to bring that soil into ductile critical states. Between these regions of highly stressed plastically yielding soil, and lightly stressed regions of tensile cracking or hydraulic fracture, there lies the region of states drier than critical in which there occur the rupture surfaces that are classically associated with the name of Coulomb. Critical State Soil Mechanics has some simple statements to make about the region of ruptures.

In Figure 3.6(b) we see a specimen of soil ruptured into two halves. The existence of a little friction at each end cap will slightly reduce the stress ratio at the ends below its value near the middle of the specimen. The soil near the centre of the specimen will reach the peak stress ratio, illustrated for example in Figure 3.4 (b), and as a rupture propagates across the specimen, the stress ratio that can be sustained across the rupture plane thereafter falls. The fall in stress ratio on the rupture surface also reduces the stress ratio at the ends, thereby unloading them and ensuring that further strain leads to further displacement on the rupture zone while the soil near the ends remains unruptured and intact. This is the well-known 'strain softening' behaviour which localises displacements within rupture zones, and contrasts with the strain hardening behaviour illustrated in Figure 3.6 (a) where deformation spreads throughout a plastically distorting specimen.

Where a rupture surface is very thin and displacements across the surface become very large, there is a marked orientation of plate-like particles parallel to the rupture zone orientation, and stress ratios fall from peak values to residual values, often associated with quite low angles of friction. However, Figure 3.4 (b) also indicates the way in which the strength of loose and of dense sand approaches a common value near the peak strength of loose or disturbed specimens, before the long decline to residual values begins with continued displacements across the rupture surface. If we identify the peak strength of loose or disturbed soil specimens in drained tests by a drained friction angle ϕ'_d then we can approximately calculate (Schofield and Wroth[1]):

$$\phi'_d = \arc\sin(3M/(6+M)) \simeq 25\,M^\circ \tag{3.16}$$

This calculation implies that soil in the thin rupture zone is at or near critical states, as is appropriate in calculations in which a body of stiff or stiff fissured clay becomes ruptured for the first time. This principle should not be applied to calculate the stress ratio required to reactivate an existing rupture surface on which large displacements have occurred in the past. Such caution applies particularly to thin seams of clay between horizontally bedded rocks which may have in the past been displaced parallel to the bedding planes. Where a large body of stiff clay becomes distorted and shatters into a rubble of inumerable small blocks sliding upon each other, we can envisage a softening in the vicinity of each rupture surface under the effective stress that acts in the soil at the depth being considered.

The softening soil follows a drained path to the critical states in the manner illustrated by the arrow from B in Figure 3.3. Whatever may be the peak strength in region B in Figure 3.7, we cannot expect that the peak strength will be available everywhere at the same time on all rupture surfaces. We must anticipate that progressive failure will cause stress ratios first on one surface, and then on another, to fall from peak value to critical state values in the manner indicated in Figure 3.4b. This is consistent with the practice of calculating the ultimate pressure of stiff fissured clay against retaining walls on the basis of soil properties $\phi = \phi'_d$ only. It will seem strange to some to neglect cohesion even though the soil in question is evidently a clay, but the language of Critical State Soil Mechanics allows us to uncouple the familiar parameters c and ϕ respectively from their classical association with clay and with sand.

The site investigation of difficult ground often involves retrieval of seriously disturbed specimens, from which it is possible only to establish the liquid and plastic limits of the soil and its *in situ* moisture content at various depths. Critical State Soil Mechanics provides a rational framework within which such data can be used as the basis of calculations. These calculations can become quite sophisticated, for example, in the field of offshore soil mechanics (George and Wood[4]). When such calculations are used to interpret the data of scattered strengths of specimens of stiff fissured clay retrieved through the depth of a layer, they will fit the lower boundary of the zone of scatter. The temptation to rely upon the peak strength of intact blocks may be great, but if the lower bound calculation is adopted a smaller factor of safety may be adopted. It is only appropriate to design on the basis of peak strength of intact soil in circumstances where it is known from a very full site investigation that there has been no damage to the mass of ground under consideration, or where a body of soil such as an embankment has been constructed and it is known that there have been no ruptures since construction.

CONCLUSION

A final comment is relevant on the behaviour of thick layers of soil in Earth's gravity. In Figure 3.7 it can be seen that there is a region B which corresponds to the upper part of a thick layer of elastically compressed soil, and a region A which corresponds to a lower part of a thick layer of elastically compressed soil. Below A the soil will experience plastic compression. Within the depth B we will see the behaviour of the soil as a crust of essentially rigid blocks, forming a surface carpet over the layer. At greater depths the soil will deform plastically. Certain engineering problems are best solved by relying on the peak strength of intact blocks of this crust, placing the foundations for light structures above the critical depth of the crust. We see from Figure 3.3 that where the soil is loose and the specific volume is large, the crust becomes thin, and foundation engineering then may involve contained regions of yielding below a structure. Although yielding may involve transient positive pore pressures, in the long term the strength will increase because of this yielding under high mean normal effective stress. Thus for example the solid piles driven into such a layer will in time have strengths set up to values above those that can be achieved for bored piles.

Much of the language of current site investigation implies that the tendency of soil to crack, or to rupture, or to yield, requires the use of different names for soil when it is found on site in a lightly stressed and alternatively in a heavily stressed location. Change of stress conditions during construction may well carry the soil from one state to another, which has to be covered in present site investigation reports by discussions of the various overconsolidation ratios of samples at various depths. If new models of the yielding of soil eventually allow a new precision of language in describing the states of soil under changing stress, perhaps this will contribute to solution of some problems of what at present has to be called 'difficult ground'.

References

1. Schofield, A.N. and Wroth, C.P., *Critical state soil mechanics*, McGraw Hill, London (1968)
2. Taylor, D.W., *Fundamentals of soil mechanics*, Wiley, New York (1948)
3. Rowe, P.W., 'The relation between the shear strength of sands in triaxial compression, plain strain and direct shear,' *Geotechnique,* **19**, 75–86 (1969)
4. George, P. and Wood, D.M. (ed), *Offshore soil mechanics*, Lloyds Register of Shipping and Cambridge University Engineering Department (1976)

Chapter 4

Computer Predictions in Difficult Soil Conditions

INTRODUCTION

One of the main areas of University research in Civil Engineering is the use of computers for the solution of problems in design and construction. In the UK, very substantial sums of money have been devoted to capital and recurrent expenditure on computers. For example, the University of Manchester Regional Computer Centre houses the greatest concentration of computing power outside the USA at a capital cost in excess of £10 million. If a purpose of research is to generate changes in established practice, the question arises whether any changes are detectable in such an applied science as foundation engineering.

That changes have occurred in many fields of engineering due to the introduction of computers can hardly be denied. Indeed computers were developed mainly to assist in the design of projects which would not have been possible without them. This form of development was particaulrly striking in the UK where engineering companies built the first machines (Ferranti Mercury, English Electric Deuce and so on), to assist in the design of early jet aircraft.

The area of civil engineering most affected by the advent of computers is probably structural analysis and design, as a direct spin-off from airframe analysis and design. Simple skeletal plane frames were first analysed in the late 1940s, and at Manchester, Livesley[1] wrote programs for the Ferranti Mark I which enabled 10-joint frames (1952) and subsequently 50-joint frames (1954) to be analysed. A measure of progress in this field is that space structures with tens of thousands of joints are now commonly analysed.

Foundation engineering has been understandably slower in exploiting computers, because the problems involved in foundation design are clearly more complex than those involved in dealing with frames. Obvious additional complexities are:

1. Foundations consist of essentially continuous bodies of soil or rock which present greater analytical problems than do skeletal structures.
2. These continuous bodies are often inhomogenous and may be interlaced with joints and fissures.
3. They are composed of material whose deformations respond non-linearly to stress changes and may be large enough relative to the scale of the foundation to cause geometrical nonlinearity as well.

4. Soil (and to a lesser extent rock) can behave in a way which is dominated not by the total external loads which are placed on it, but rather by the way in which these loads are distributed between the solid and fluid phases of the material (i.e. the effective and pore pressures).
5. Permanent deformations, with plastic flow, are quite common under working loads.

Analysis to take account of the first two of these complications became a possibility with the invention of matrix methods of structural analysis and in particular of the finite element concept, arguably first introduced by Turner *et al*[2]. This method is quite widely used for foundations which can be assumed to behave linearly and with small strains. A typical example of its use in practice is given by Hooper[3]. Of course such a foundation would not normally qualify for the description 'difficult'.

By 'difficult' ground, the writer chooses to imply relatively soft ground in which movements cannot simply be computed by the assumption of linear elasticity, and in which substantial excess pore water pressures can be developed under load. This definition is quite subjective and differs from that used in other chapters. Concern is also limited to loaded ground, rather than to excavations prior to founding.

Although research is continuing on rather complicated computer models of soil behaviour the present purpose is to show how established models can be used in foundation work. The computer techniques described are well enough developed to be available as standard packages on commercial machines, although being an order of magnitude more expensive than the simple linear analyses.

DESIGN AS YOU GO

If new analytical techniques based on computers are proposed it is as well to examine the nature of the current methods which they seek to replace. A prominent feature of methods of design in soil mechanics and foundation engineering is the philosophy of 'designing as you go'[4]. Simplified calculation procedures based on very idealised ground properties are sufficient (and indeed the site need not be too rigorously investigated beforehand). All of this apparent laxity is permissible because, as construction of the project proceeds, careful monitoring on site will show to what extent the design assumptions are being violated. If such violation becomes excessive, the design is modified, for example, to involve slowing down the rate of construction to allow time for excess pore water pressures to dissipate, or indeed to involve a complete change of the fundamental design concept.

This basic philosophy can be faulted on several grounds. For example, it can lead to conservative foundation designs because the inadequacy of the site investigation leads to the installation of artificial drainage provisions (sand, drains, etc) where these are not really required. They are subsequently justified on the grounds of 'better safe than sorry' but are none the less expensive. More critically, if the maximum load on a foundation is of a transient nature and does not occur till after completion of the works how can the design be modified?

Typical transient loads arise from wind, wave and earthquake forces. In these cases it is necessary to make a genuine prediction of the likely behaviour of the structure under working load[5]. To do this, a model of the foundation must be made.

SITE MODELLING

It cannot be suggested that all foundation projects or even all foundation projects in difficult ground will benefit from computer analysis and predictions. As a rough guide, works could be classified under three main headings following the suggestions of Rowe[6].

Class A refers to major projects such as large dams, high embankments, deep cuttings, docks, high retaining walls and heavy or deep foundations.
Class B refers to more limited projects of similar type where difficult ground occurs.
Class C refers to small building work comprising housing estates, small factories, three- and four-storey buildings, sewers, pedestrian subways and highways at or close to ground level, or repetitive work in known areas.

It is to foundations for Class A projects that it is suggested computer predictions will be applicable. The site must first be probed by a site investigation of a quality sufficient to establish an adequate *model* of what the ground looks like.

This model could be a *physical* one (e.g. constructed in a centrifuge) or a *numerical* one constructed in a computer. The relative merits of physical and numerical modelling have been discussed elsewhere[7] and present concern is with numerical modelling.

The alternative to models is a field test, usually ruled out because the only appropriate scale in foundation engineering is usually prototype scale.

To make the model, information is needed about geometry and about material properties, the quality of which depends on the site investigation techniques employed. For high quality work, sampling or *in situ* testing should reveal a *continuous* picture of the strata being penetrated. An example of a length of continuous 50 mm diameter split and partially dried core is shown in Figure 4.1. Such samples need not be of high mechanical quality. If sample recovery is poor as it can be in the case of hetergeneous sandy deposits, or site access is difficult as in offshore work, the Dutch cone provides a continuous picture of penetration resistance with depth typified by Figure 4.2.

On the basis of wide experience, Rowe[6] has advanced the following opinion: 'Given an adequate number of initial small diameter borings, the majority of sites I have examined in Britain could be represented for all practical purposes by a series of representative strata with properties based on relevant representative samples and tests.' In practice the number of these representative strata is small, typically four or six.

In the most general terms, the material properties required from a laboratory or *in situ* test programme are the *stiffness* and *strength* of the ground together with its *drainage* properties. *In situ* tests, at large enough scale, are an attractive if expensive means of arriving at the appropriate data. They exclude the problems associated with sample extraction (but often not with stress relief in the ground). They can be difficult to conduct at locations far removed from ground surface (an exception would be pressuremeter tests) and can be totally irrelevant if the property being measured is strongly dependent upon stress level (stiffness of a sand or drainage rate of a soft alluvial deposit for example).

Laboratory tests should be conducted on specimens which are large enough to contain sufficient 'fabric' (fissures, joints, bedding, and other non-uniformities) to be sufficiently representative of the foundation under consideration. At the present, drill holes some 260 mm in diameter appear to be the maximum practically obtainable and provision of kentledge for deep sampling in stiff materials can be expensive. Nevertheless, large samples have been extracted from depths of the order of 100 m in stiff subsoils.

The effects of sample size and laboratory testing technique on the vital soil parameters are shown in Figures 4.3 and 4.4. The former shows that undrained strength C_u of fissured, overconsolidated clays can be overestimated by testing

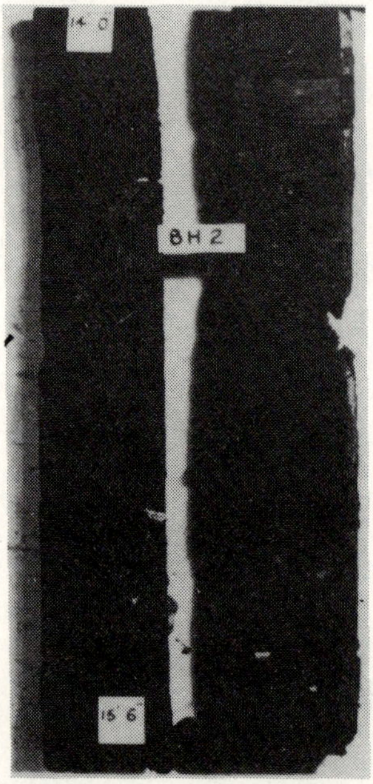

Figure 4.1 Split and partially dried continuous cores

specimens which are too small, or under non-uniform stress conditions, or at too fast a rate. For similar reasons the stiffness of fissured ground (defined by a modulus E) will tend to be overestimated by tests on little samples. Figure 4.4 shows that (a) small oedometer tests can lead to drainage rates c_v which are in error by at least two orders of magnitude and that (b) *in situ* tests are of no help because they cannot be conducted at the right ranges of effective stress prior to construction.

In summary, the adequate modelling of a site requires a site investigation of high quality. Continuous or closely consecutive probing is essential to build up a picture of site geometry. In order to make good predictions, non-standard *in situ* and laboratory testing is necessary. All of this represents a significant departure from 'design as you go' methods, but of course the computer predictions should be checked, and the design modified if events in the course of construction prove to be unexpected. It is not anticipated that the development of predictive

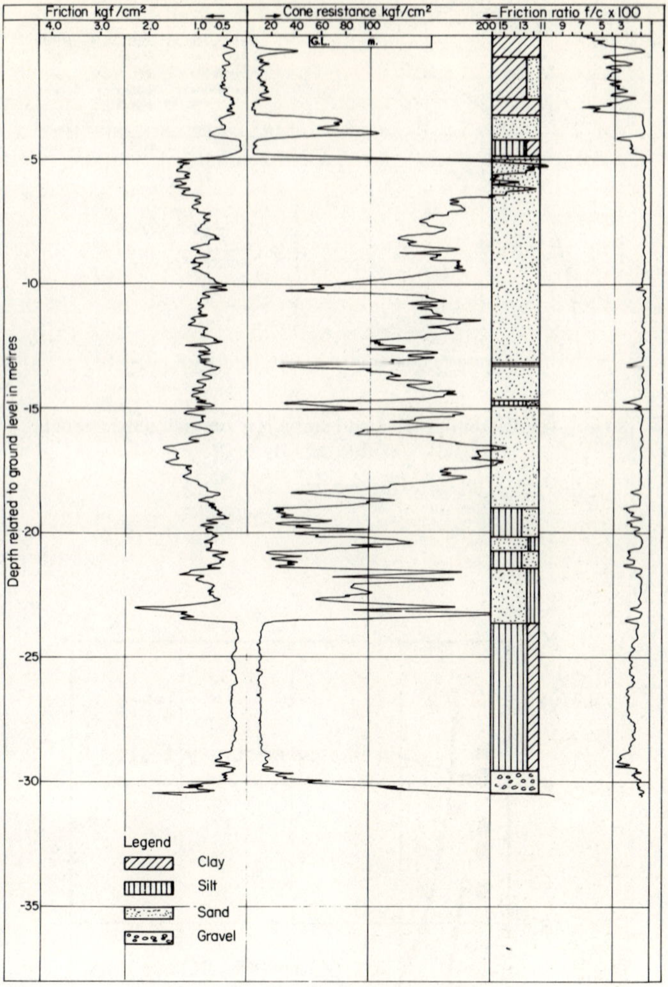

Figure 4.2 Cone penetrometer traces

techniques will have much influence on the need for provision of field instrumentation. However, a better interpretation of the field results should be possible.

PREDICTION TECHNIQUE I

The first prediction technique to be considered, involving the use of a computer, is concerned with the generation and dissipation of excess porewater pressures in foundation soils. The variable factors in the analysis are the rate at which load is applied, the stiffness of the structure applying the load relative to the stiffness of the ground, and the manner in which the properties of the ground (its stiffness and permeability) change as consolidation proceeds. Foundation conditions of this kind which are likely to be 'difficult' would tend to involve softer soils, for

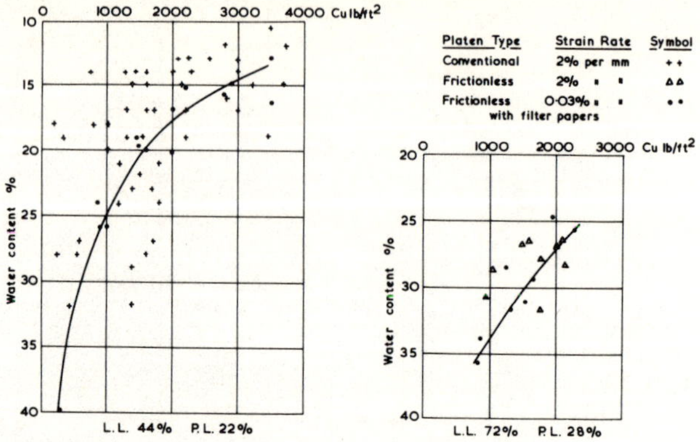

Figure 4.3 Effect of platen and strain rate on undrained strength of fissured clays

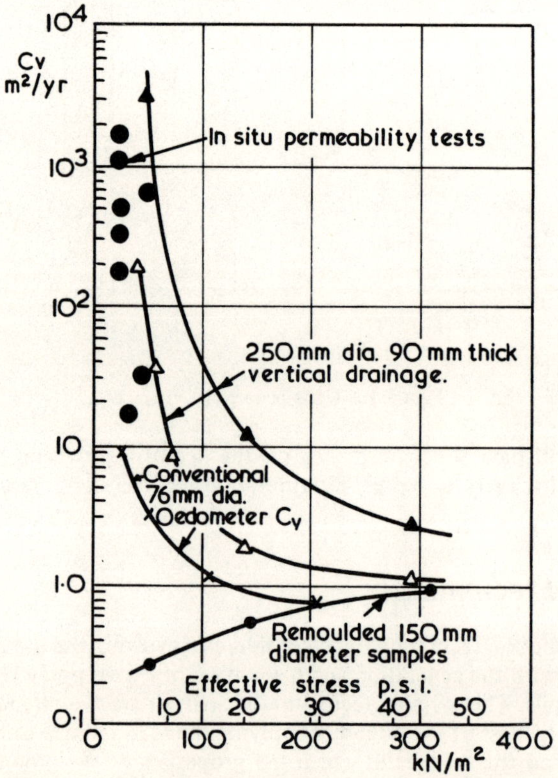

Figure 4.4 Effect of sample size on drainage properties

example, alluvial or estuarine deposits. When, for example, embankments are being constructed on such soils, failure of the foundation will be imminent if the excess porewater pressures within it attain a head of water approximately equal to the height of the embankment.

The theoretical analysis of this type of problem is due to Biot, and considers the soil to consist of a porous elastic solid, whose pores are filled with an incompressible fluid (water). The governing equations and method of solution using the computer are described by Smith and Hobbs[8]. The input parameters required for such a computer program are:

(i) The variation of load on the foundation as a function of time.
(ii) The geometry of the soil horizons.
(iii) The stiffness of the load relative to that of the soil.
(iv) The variation of the soil parameters of permeability k (or coefficient of consolidation c_v) and stiffness E', ν' (or coefficient of compressibility m_v) with changing effective stress (see Figure 4.4). This data should preferably be obtained from 'large', say 250 mm, diameter oedometer or triaxial cell tests.

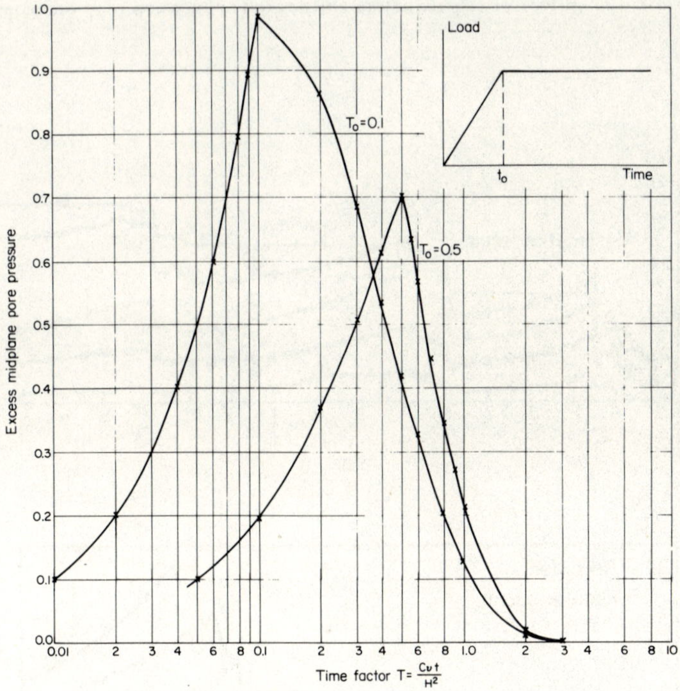

Figure 4.5 Effect of rate of loading on foundation pore pressures

The importance of input parameter (i), the rate of loading, can be judged from Figure 4.5 which illustrates the classical one-dimensional consolidation problem for a clay leyer with the innovation that the load increases linearly from time 0 to time t_o and thereafter remains constant. This differs from the standard solutions in which the full value of the load is instantaneously applied at time 0. It can be seen from the diagram that if the time required for full dissipation of

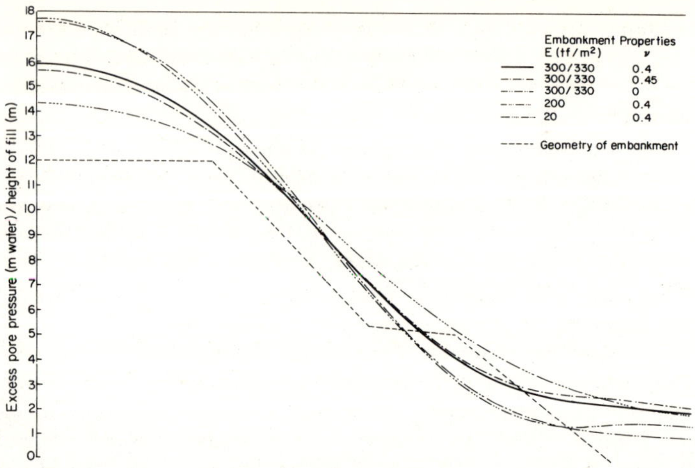

Figure 4.6 Effect of superstructure stiffness on foundation pore pressures

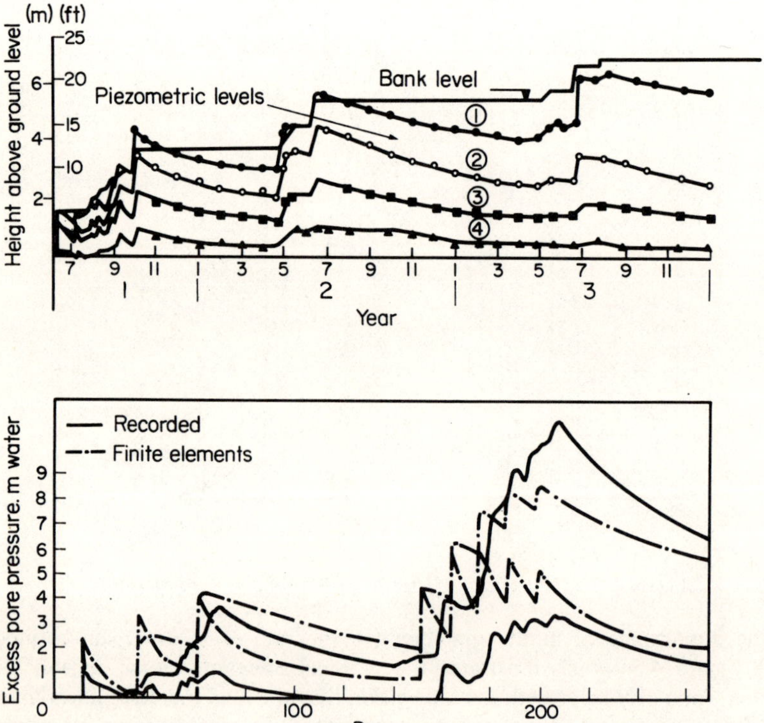

Figure 4.7 Effect of permeability changes on foundation pore pressures

excess midplane pore pressure in the layer is, say, four months, then if the load is applied gradually over a two-week period, the maximum excess midplane pore pressure will be 30% less than if the load were applied 'instantaneously', i.e. in four days or less.

The effect of superstructure stiffness relative to that of the foundation is illustrated in Figure 4.6, for the case of an elastic solid embankment bearing on a compressible subsoil. If the embankment is very stiff (high E') or has a low Poisson's ratio ν', the excess porewater pressures generated in the foundation are higher under the centre, and lower under the edges, than would be the case for more flexible superstructures.

Finally, Figure 4.7 shows the effects of changing foundation material properties (k and m_v) with changing effective stress. These effects are somewhat complicated, because at higher loads, the excess pore pressures and their gradients are higher and therefore consolidation tends to occur more quickly immediately on application of a fresh load. However, the trend of the dissipation curves is clearly towards a flatter slope approaching the end of each phase of consolidation under a newly imposed load. This reflects the inherent tendency for a decrease of permeability in soft foundation soils with increasing effective stress.

PREDICTION TECHNIQUE II

The first prediction technique made use of the power of computers in dealing with some aspects of soil foundations as they really are, namely, two-phase, inhomogeneous, stress dependent solids. However, a very important aspect of soil performance was not treated, namely, the tendency for many soils to sustain permanent (plastic) deformations under load. In some cases the load in question is 'working load' and limited plastic flow is permitted to occur because it would be ultra-conservative to prevent it, while in other cases the placing of the foun-

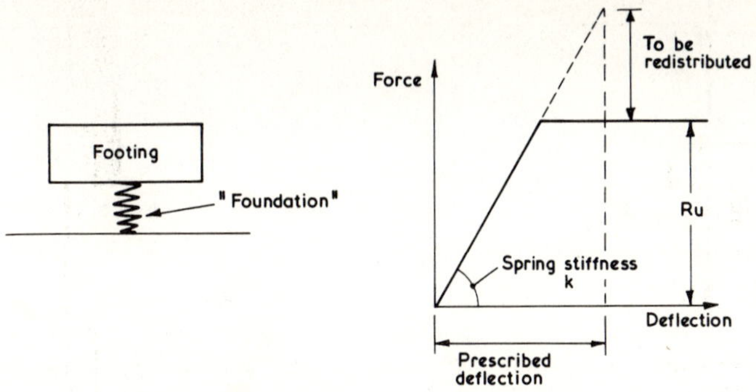

Figure 4.8 One degree of freedom foundation

dation *necessitates* failing the ground beforehand, as in the case of driven piles or gravity caissons for example. The second prediction technique described is concerned with this aspect of foundation analysis.

The principles of elasto-plastic analysis may best be described by considering the trivial one degree of freedom foundation system shown in Figure 4.8. The foundation is idealised by a single spring which has the force-deformation properties illustrated in the diagram. This allows the foundation to deform elastically under load until a limiting force Ru is reached. Therefore, unlimited deformation (total collapse) ensues. Such a system could only be analysed after Ru has been attained by prescribing some displacement and analysing for the force in the spring. As shown, this could be done by assuming the spring remains elastic, with its original stiffness k, but that the elastic load in excess of Ru is redistributed as an extra internal force. Of course it is this extra internal force that causes the extra (plastic) movement in the 'spring'.

A real foundation consists of an infinite number of degrees of freedom with nonlinear force-deformation relationships which are in general stress dependent. Ultimately, we shall use finite element models to replace the infinity of freedoms, but the principles involved may be clarified by studying systems such as that in Figure 4.9. This shows a pile, together with its cap, cushion and driving hammer, represented by a series of rectangular elements restrained from any

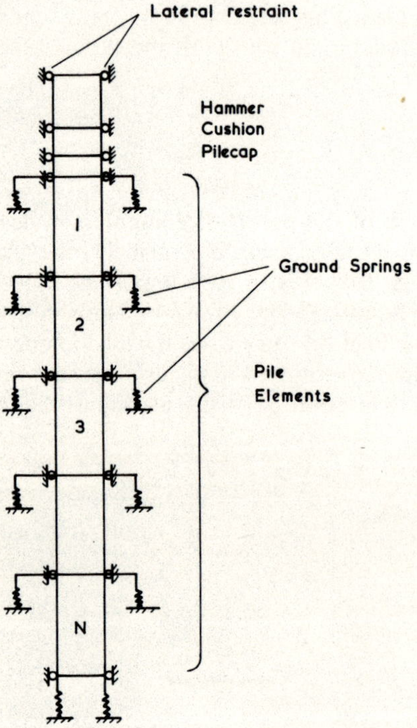

Figure 4.9 Idealisation of pile and ground

lateral movement. The ground resistance is represented by a series of springs, attached to the nodes of the finite element mesh, and having the force-deflection characteristics shown in Figure 4.10. Note that it is assumed that the ground may be subjected to compressive *or* tensile forces, but there is assumed to be no Bauschinger effect and R_u is a constant. This can really only be a tenable assumption for saturated, monotonically loaded clay in the undrained state, since in general soil resistance is strongly dependent on *effective* mean normal pressure (as analysed in prediction technique *I*) and on alternating loads. Limiting resistance is assumed to be attained at a displacement Q, called the 'quake'.

Any problem characterised by load-displacement relationships of the type shown will in general be strongly nonlinear, and will have to be analysed by step-by-step application of loads with iterative redistribution of load excesses. Physically, the load diffuses down the pile and is partially carried by the uppermost soil springs. If their ultimate resistance is exceeded, this excess load is passed

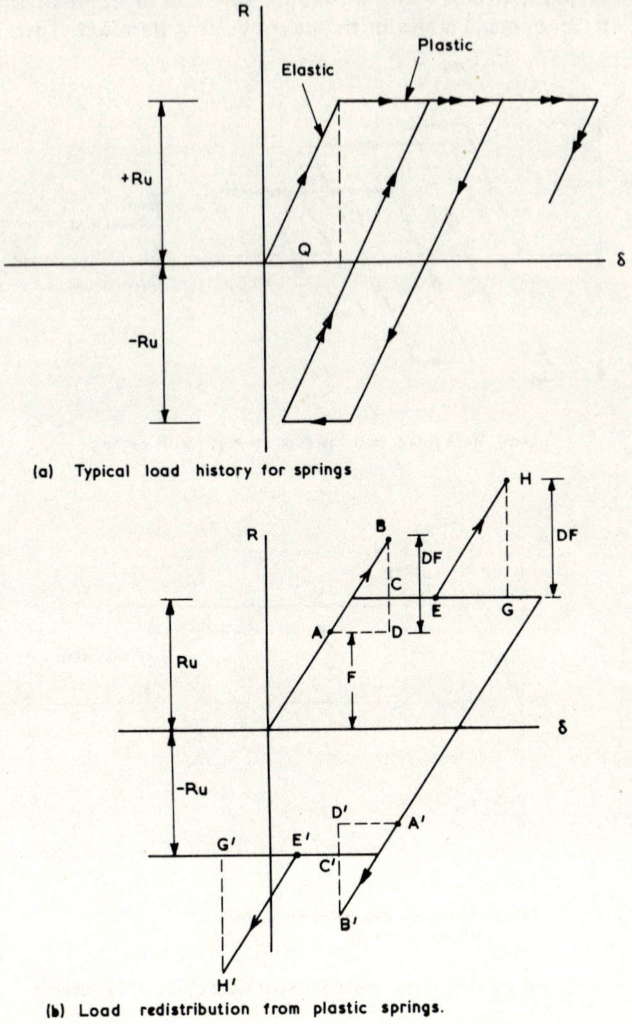

Figure 4.10 Soil spring behaviour (no weakening)

down to the next pair of springs below and so on. The pile can only be pushed into the ground when all the springs have been loaded to their ultimate resistances.

This physical redistribution of load is precisely mirrored in the computation technique used for analysing such problems. The load redistribution from overloaded springs is a simple example of an 'initial stress' computation. The current force in a spring is denoted by F (Figure 4.10) and the change due to the next increment of total load by DF. If the latter is such as to reduce the absolute value of F or if abs $(F + DF)$ is less than abs (R_u) then elastic behaviour results. On the other hand, if the current state is at A or E and DF increases the absolute value of $F + DF$ beyond R_u, then plastic yielding occurs. In the former case the excess load to be redistributed to other springs is BC and in the latter HG. In the computations, these excess loads are simply added in as internal, self-equilibrating loads at the element nodes in the next cycle of iteration. Their effect is to

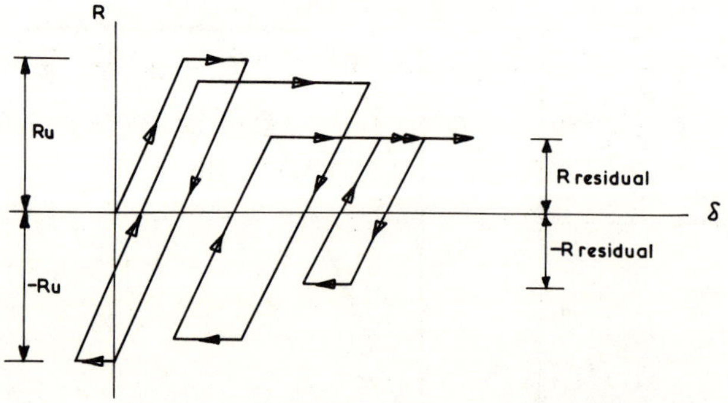

(a) Weakening to a residual strength with cycling

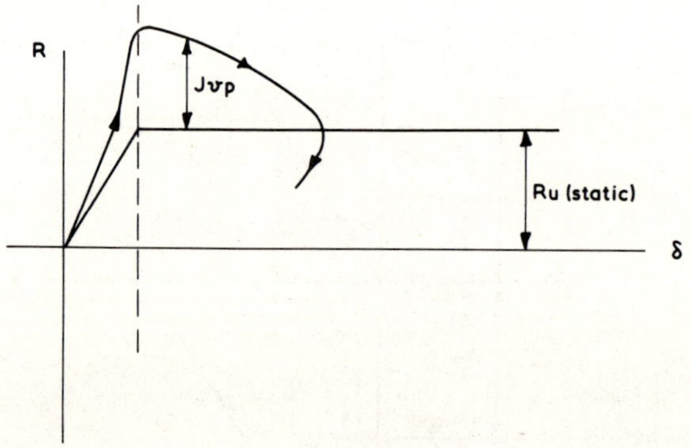

(b) Additional soil resistance due to viscous damping.

Figure 4.11 Soil spring behaviour for weakening and damping

cause increased deformations in the springs while limiting the spring forces. When a substantial number of springs has yielded, the iteration process for each load increment can take twenty iterations or more to reach convergence, and in the limit, the computation will not converge at all when the pile 'fails'. In practice it is usual to set a limit on the number of iterations allowed, say thirty, and to consider failure to have occurred if this limit is exceeded.

In general, soil resistance-deformation relationships are more complicated than those shown in Figure 4.10. Two of the important differences are shown in Figure 4.11. The first is a progressive weakening to a residual strength as a function of number of repetitions of cyclic load or of cyclic displacement amplitude. This can readily be incorporated in the present type of calculation.

The second effect is thought to be present during pile driving when high velocities of penetration are attained. Soil resistance is assumed to increase in proportion to the velocity of the pile relative to the soil. In any event, the 'initial stress' load distribution procedure can readily be made to ensure that any desired level of force is not exceeded in the soil springs.

STATIC AXIALLY LOADED PILE ANALYSIS INCLUDING CYCLIC LOADS

The above procedure has been applied in the analysis of a field test pile described by McCammon and Golder[9]. The same load test has recently been analysed by the alternative 'beam column' process[10], and pile and soil data are shown in Figure 4.12. Side quake was taken to be 0.026 ft and end quake to be 0.1 ft. In total, 85 pile elements were used. The computed load-settlement curve to 'failure' at about 850 kips load is shown in Figure 4.13(a) and is in excellent agreement with the measured behaviour.

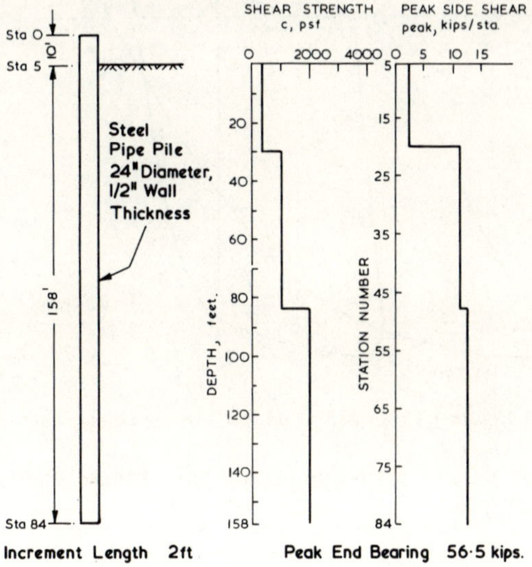

Figure 4.12 Details of test pile

A second calculation of interest especially in an offshore structures context results in the behaviour shown in Figure 4.13(b). Load is assumed to be cycled between ± 850 kips with the further assumption that the end resistance in tension equals that in compression. There is marked hysteresis, although such severe loads are inconceivable in a properly designed pile. Reversals of load also produce interesting stress reversals in the piling. If the pile is loaded to 850 kips, unloaded, and then subjected to a tension of 150 kips, the stress distribution in the pile is as shown in Figure 4.14.

A more usual offshore loading condition is compressive loading followed by tensile loading to about one half of maximum compressive load. When the above test pile is subjected to cyclic loading of +700 kips and –400 kips, the cap displacement with cycling follows the path shown in Figure 4.15. Clearly there is permanent compressive movement, although at lower loads, and without soil weakening, essentially elastic response can occur.

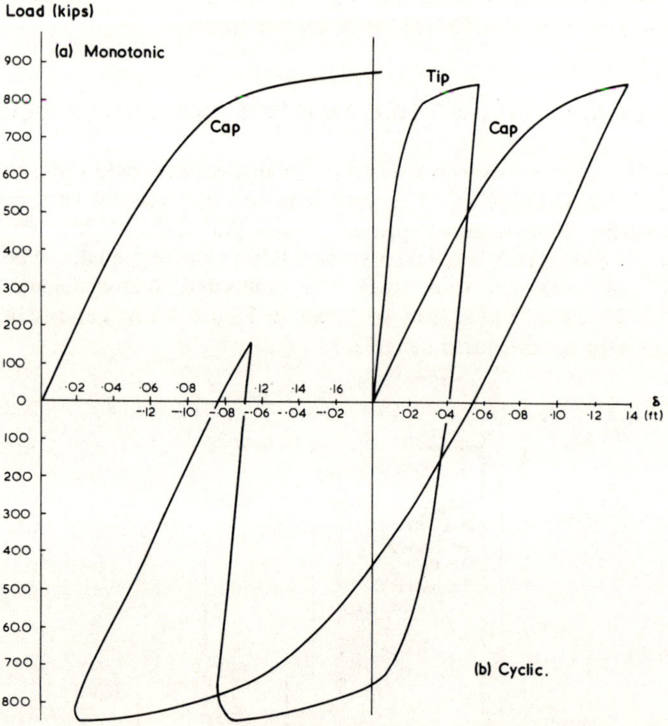

Figure 4.13 Load test on McCammon-Golder pile

The foundation parameters required as input to these analyses are simply *Ru* and *Q* for each soil spring.

PILE DRIVEABILITY

The idea of treating soil as an elasto-plastic material in the above way is not new. It was developed by E.A.L. Smith in the early 1950's and definitively published in 1960[11] in connection with pile driveability. Smith wrote computer programs to analyse dynamic wave propagation during pile driving by the one dimensional wave equation. His work is not widely used in the UK but has received more attention in the USA.

The pressing need to predict accurately the driveability of piles at offshore locations has led to awakened interest in the method and it is to be hoped that its use will expand in onshore work. Finite element idealisations of pile and ground allow extension of Smith's treatment to more than one dimension.

EXTENSIONS TO SOLIDS BY FINITE ELEMENTS

Generalising from one dimensional springs to three dimensional solids we merely replace lineal force by a generalised measure of stress, $\bar{\sigma}$, for example, the von Mises equivalent stress, and lineal displacement by a generalised measure of strain, $\bar{\epsilon}$, again, for example, the von Mises measure[12]. Foundations composed of saturated, undrained clay can be approximated by a simple von Mises material

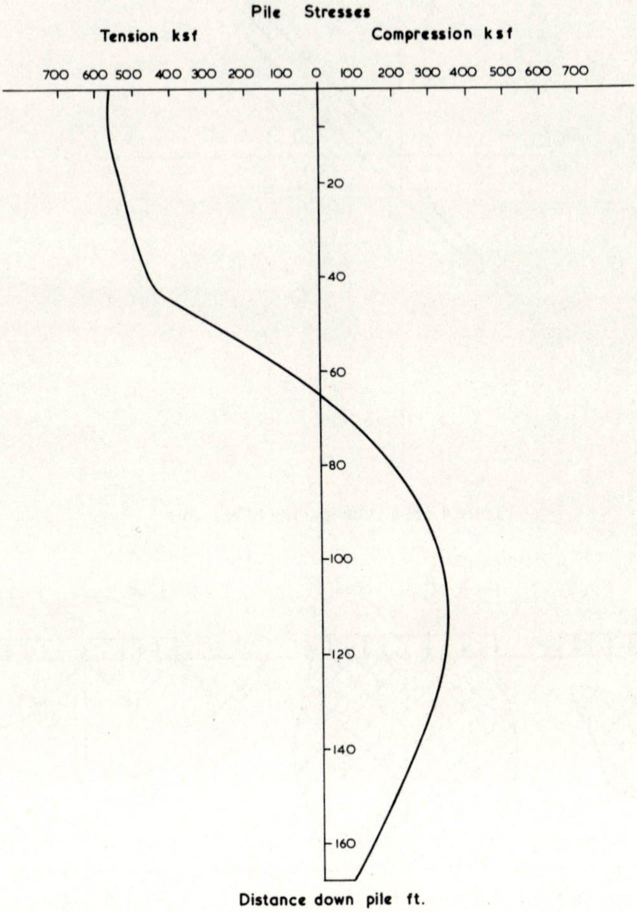

Figure 4.14 Pile stresses during cycling

with constant stiffness, E/C_u, Poisson's ratio about one half, and strength, C_u, as shown in Figure 4.16. This also shows finite element computations of the spread of yield under a vertically loaded footing at 'collapse' (defined by failure of the computation to achieve a convergence criterion as described previously) and at a factor of safety of 1.7. Close agreement is obtained between the computed collapse load and the Prandt or Hill values usually used in foundation design. Contained plastic deformation does not impair the performance of the foundation for non-cyclic loads.

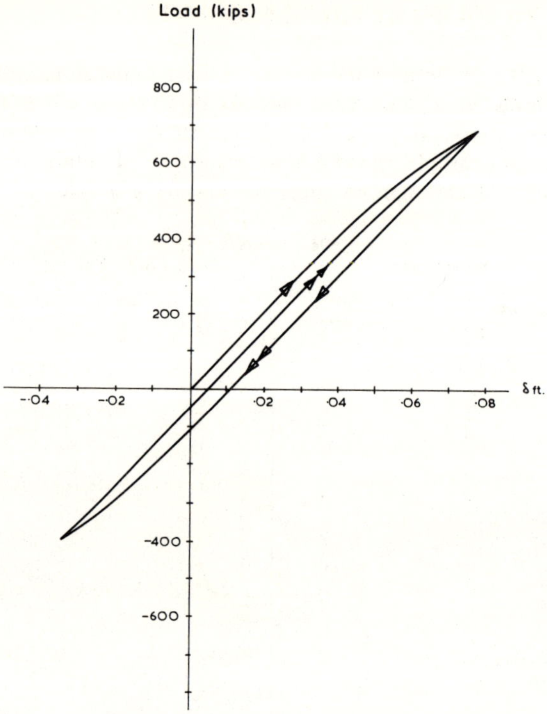

Figure 4.15 Cyclic loading of test pile

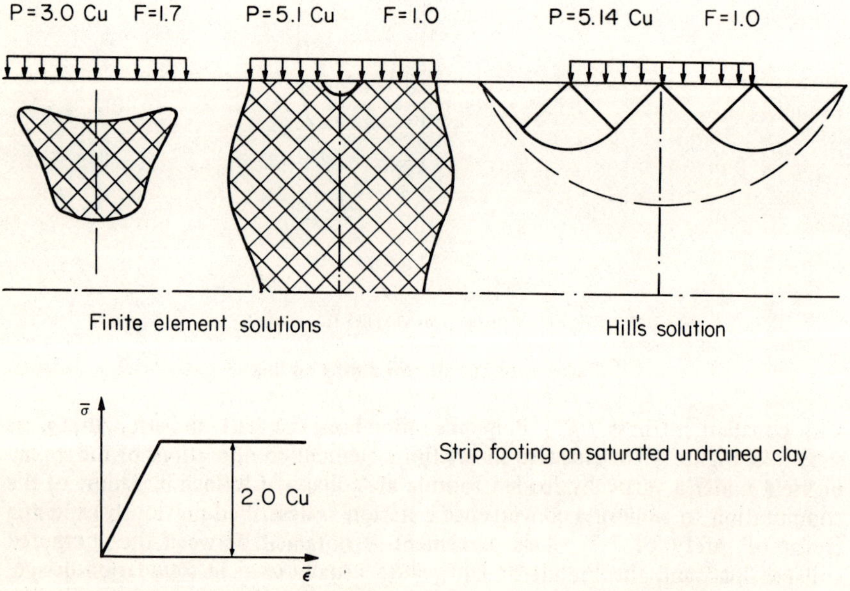

Figure 4.16 Strip footing on saturated undrained clay

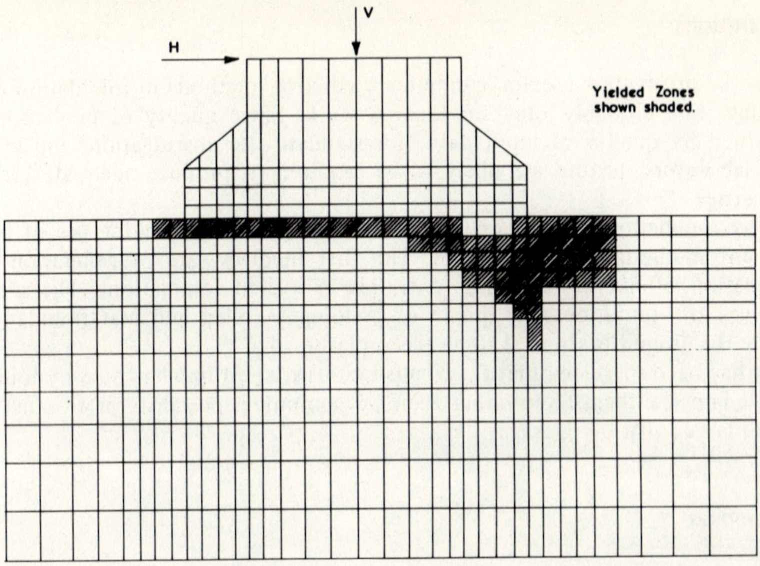

Figure 4.17 Failure of foundation subject to inclined loading

The very different yielding pattern underneath a foundation subjected to large horizontal, as well as vertical, load is shown in Figure 4.17[13].

The input parameters required for these analyses are merely undrained strength, C_u, and undrained stiffness, E, of the foundation material.

COMPUTER USAGE POLICY

Since there are many ways of gaining access to computers, the foundation engineer may wonder which approach would be best for him. In high quality work, where the use of a computer is justified, the writer has no doubt that the engineer himself should be responsible for programming his own problems. He should not be reduced to the status of a clerk, entering numbers on a coding form. Despite the apparent personnel cost to an organisation of this approach, it has the great benefit that the set-up costs involved in running commercial packages and the costs of interpreting abortive or defective runs are avoided. For many problems, the set-up charge (a fixed charge for accessing a large package) exceeds the cost of the computation itself.

Engineer-programmers can be greatly assisted by the adoption of modular or building block program structures. Unfortunately the architecture of many of the most powerful USA designed computers does not lend itself to these structures. Similarly on the software side most engineers' computer programs are written in FORTRAN (because of its popularity in the USA) and this does not lend itself to a modular technique. In the long term, the writer believes that programming languages will become more ALGOL-like and less FORTRAN-like, although holding no particular brief for ALGOL 60. At any rate, given all the other advances which have occurred, it seems absurd that major software is twenty years old in such a rapidly advancing field.

Conclusions

There is an obvious use for computer prediction methods in foundation engineering, but probably only on Class A work. Since quality of predictions is governed by quality of input data, non-standard site investigations and *in situ* and laboratory testing are likely to be required to produce adequate ground properties.

Two prediction techniques have been described, both making use of finite element methods of computation. The first involves rate of generation and dissipation of excess porewater pressures in loaded foundations. The second involves the initiation and spread of yielding in piled and mat foundations, where the ground is assumed to be elasto-plastic.

It has been suggested that these calculations are best implemented by foundation engineers themselves rather than by computer specialists or through the use of large software packages.

References

1. Livesley, R.K., 'The Place of Digital Computers in Civil Engineering,' *Proc. Inst. Civ. Engrs.*, **15**, 15 (1960).
2. Turner, M.J., Clough, R.W., Martin, H.C. and Topp, L.J., 'Stiffness and Deflection Analysis of Complex Structures,' *J. Aeronautical Sciences*, **23**, 16, 805 (1956).
3. Hooper, J.A., 'Observations on the Behaviour of a Piled-Raft Foundation on London Clay,' *Proc. Inst. Civ. Engrs.*, **55**, Part 2, 855 (1973).
4. Peck, R.B., 'Advantages and Limitations of the Observational Method in Applied Soil Mechanics,' *Geotechnique*, **19**, 2, 171 (1969).
5. Lambe, T.W., 'Predictions in Soil Engineering,' *Geotechnique*, **23**, 2, 151 (1973).
6. Rowe, P.W., 'The Relevance of Soil Fabric to Site Investigation Practice,' *Geotechnique*, **22**, 2, 195 (1972).
7. Smith, I.M., 'Numerical and Physical Modelling,' Chapter 17 in *Numerical Methods in Geotechnical Engineering*, McGraw Hill, New York.
8. Smith, I.M. and Hobbs, R., 'Biot Analysis of Consolidation beneath Embankments,' *Geotechnique*, **26**, 1, 149 (1976).
9. McCammon, N.R. and Golder, H.Q., 'Some Loading Tests on Long Pipe Piles,' *Geotechnique*, **20**, 2, 171 (1970).
10. Meyer, P.L., Holmquist, D.V. and Matlock, H., 'Computer Predictions for Axially Loaded Piles with Nonlinear Supports,' *O.T.C. Paper 2186*, Houston (1975).
11. Smith, E.A.L., 'Pile Driving Analysis by the Wave Equation,' *Proc. A.S.C.E.*, **86**, SM4 (1960).
12. Smith, I.M. and Hobbs, R., 'Finite Element Analysis of Centrifuged and Built-Up Slopes,' *Geotechnique*, **24**, 4, 531 (1974).
13. Smith, I.M., 'Aspects of the Analysis of Gravity Offshore Structures,' *Numerical Methods in Geomechanics*, v. 2, p. 957, Blacksburg, Virginia (1976).

Chapter 5

Foundations on Rock

Foundation engineering is normally considered a major application of soil mechanics in engineering practice, but a minor application of rock mechanics. It is easy to see why when most of the Biblical metaphors for firm foundations include rock. Yet today there are many structures of sufficient weight to cause significant deformation leading possibly to failure in the typically weathered, weathering and open-fissured rocks which make up most of the near surface rocks on which structural foundations are based.

It is important at this point to stress that foundation design must take into account both bearing capacity and settlement, and to state the two basic rules for foundation design:

(a) The bearing stress of a structure must be less than the *bearing capacity* of the ground material.
(b) The settlement of the structure caused by deformation of the ground material must be less than the *allowable settlement* for the structure.

The *allowable settlement* of course depends on the sensitivity of the structure. Sowers[1] and Bjerrum[2] among others, quote limiting differential settlements in terms of distortion (in practice, the ratio between maximum settlement and foundation length) of 1% for low rise brick and panel walls down to 0.001% for delicate tracking equipment such as a radio telescope. Thus the allowable bearing capacity is the bearing stress at which allowable settlement occurs. The ultimate bearing capacity, at which the foundation rock will fail, although worth considering, has less engineering significance.

BEARING CAPACITY

The simplest approach to bearing capacity, based on previous experience, is that used in various codes of practice. British Standard CP 2004[3] quoted in Table 5.1 indicates safe bearing pressures for some typical unweathered rock types.

Peck *et al*[4] summarise similar United States Building codes (Table 5.2) although they prefer the New York City code which quotes safe bearing pressures for specific geologic formations on the basis of lithology and condition. In the latter case a hard sound rock is defined as one which 'rings when struck with a pick or bar; does not disintegrate after exposure to air or water; breaks

Table 5.1 SAFE BEARING PRESSURES – BRITISH PRACTICE, CP 2004

Igneous and gneissic rocks in sound condition	10 MN/m^2
Massively bedded limestones and hard sandstones	4 MN/m^2
Schists and slates	3 MN/m^2
Hard shales, mudstones and soft sandstones	2 MN/m^2
Clay shales	1 MN/m^2
Hard solid chalk	0.6 MN/m^2
Thinly bedded limestones and sandstones; heavily shattered rocks	To be assessed after inspection

with a sharp fracture; cracks are unweathered and less than 10 mm wide, and no closer than 0.9 m apart; core recovery with a double tube diamond core barrel is generally 86% or greater for each 1.5 m run.'

The emphasis on unconfined compressive strength (S_C) of the Uniform Building Code is essentially the conventional approach to design in massive, unweathered rock. If a foundation is to be constructed on such rock, then an unconfined compressive strength measured using a standard laboratory test on an unconfined sample (ASTM[5]) or estimated from a field point load or Schmidt hammer test (see Rankilor[6]) can be used to determine a safe bearing pressure for a rock type.

Table 5.2 SAFE BEARING PRESSURES – UNITED STATES PRACTICE (AFTER PECK *et al*[4], 1974)

Rock type	*Bearing pressure* MN/m^2			
	BOCA	*National*	*Uniform*	*Los Angelos*
Massive crystalline bedrock, including granite, diorite, gneiss, basalt, hard limestone and dolerite	10	10	0.2 S_C*	1
Foliated rocks such as schist or slate in sound condition	4	4	0.2 S_C	0.4
Bedded limestone in sound condition	4	1.5	0.2 S_C	0.4
Sedimentary rocks including hard shales and sandstones	2.5	1.5	0.2 S_C	0.3
Soft or broken bedrock (excluding shale) and soft limestone	1	–	0.2 S_C	–
Soft shale	0.4	–	0.2 S_C	–

S_C = Unconfined compressive strength

The relationship between compressive strength and deformation modulus for rocks established by various workers (see Judd and Huber[7]) can also be used to estimate settlement in *intact* rocks if a stress-strain relationship can be estimated. A typical example of relationships for groups of rocks put together by Deere and Miller[8] is illustrated in Figure 5.1.

Once this basic information is known, together with data on foundation loading, then stability analyses for the foundation can be carried out using either simple two-dimensional limit-equilibrium approaches or two or three dimensional finite element modelling techniques. But of course rocks, particularly near surface rocks, are not continuous intact masses. They are discontinuous, fissured and weathered and their properties change with depth. In particular the presence

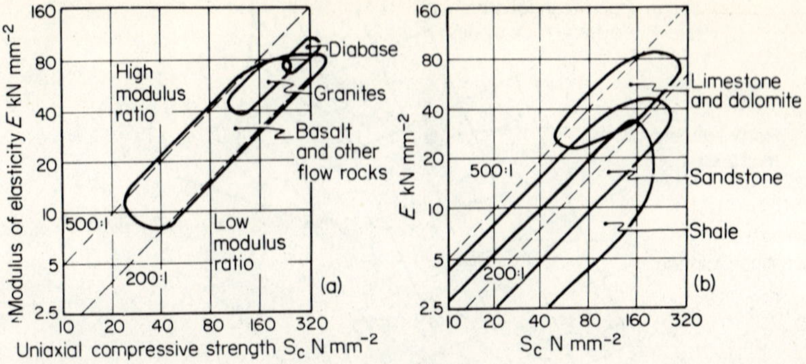

Figure 5.1 Modulus ratios for typical lithological rock groups. (after Deere and Miller 1966)

of discontinuities can have a disproportionate effect on bearing capacity and settlement.

EFFECTS OF DISCONTINUITIES ON BEARING CAPACITY

It is simplest to consider initially the basic stability mechanisms of a jointed rock. The probability of failure of a foundation, in rock of sufficiently high bearing capacity to resist intrinsic failure, through sliding along joints will probably be limited to very heavily loaded foundations such as dams. There is a limited literature on rock mass foundation stability, with particular emphasis on dam foundations, mainly in the Proceedings of the 2nd International Congress of the IAEG[9] and of the 3rd Congress of the ISRM[10] although the approach in both cases, as befits an area of research involving geology, is rhetorical rather than computational.

John[11] in his usual succinct way points out the basic importance of the geological structure in determining the stability of rock foundations. Figure 5.2 illustrates the essential difference between the overall system and its structural and fabric elements. In order to determine the stability of the overall system it is necessary to consider the possibility of sliding along major or minor joint sets under the loading conditions imposed by the foundations.

This involves mapping the complete pattern of discontinuities in the rock over the whole of the foundation area. These can then be plotted on an equal area stereographic projection as poles to planes to isolate the major joint orientation concentrations. The method of joint surveying and also of stereographic representation is outlined in chapter 16 and by Attewell and Farmer[12] and Hoek and Bray[13]. Londe[14] suggested that the discontinuities should also be classified in distinct populations as micro-fissures, fissures, joints, shear planes and faults and that the spacings and characteristics of each should be noted and recorded as histograms (Figure 5.3).

A particular example of a dam foundation analysis which illustrates the particular approach to foundation stability is that of Henkel *et al*[15] who carried out a design study for the Monar Dam foundation (Figure 5.4). This dam is founded upon heavily jointed but relatively unweathered psammitic granulites of

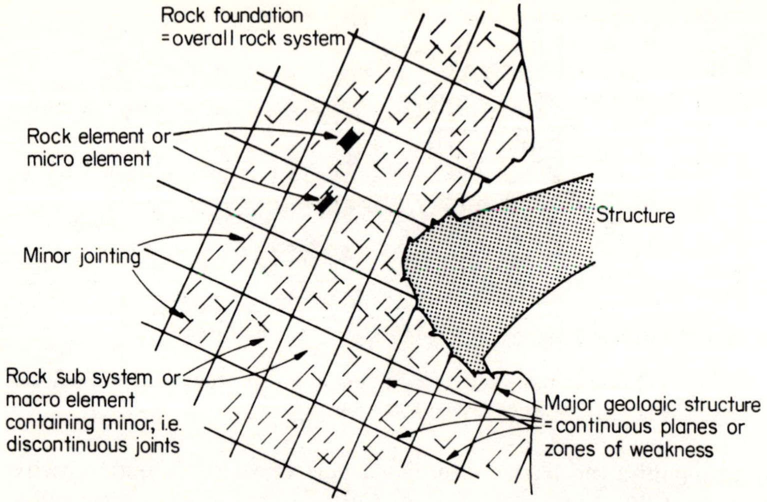

Figure 5.2 Structural elements of massive rock (after John 1974)

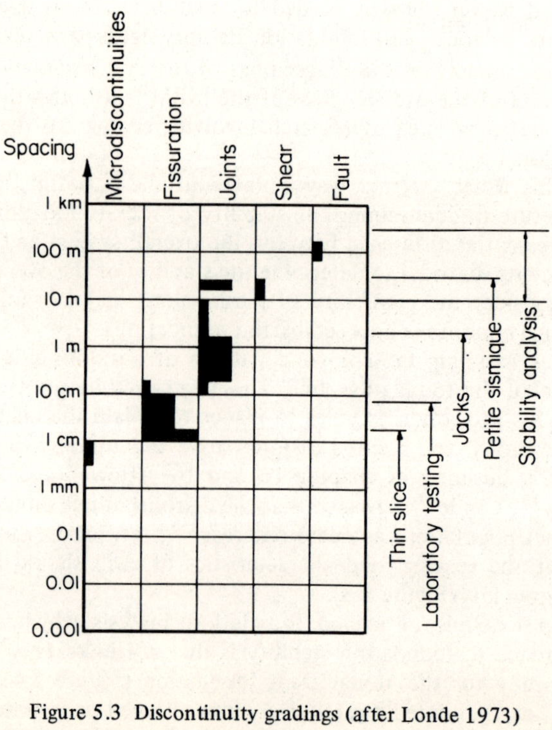

Figure 5.3 Discontinuity gradings (after Londe 1973)

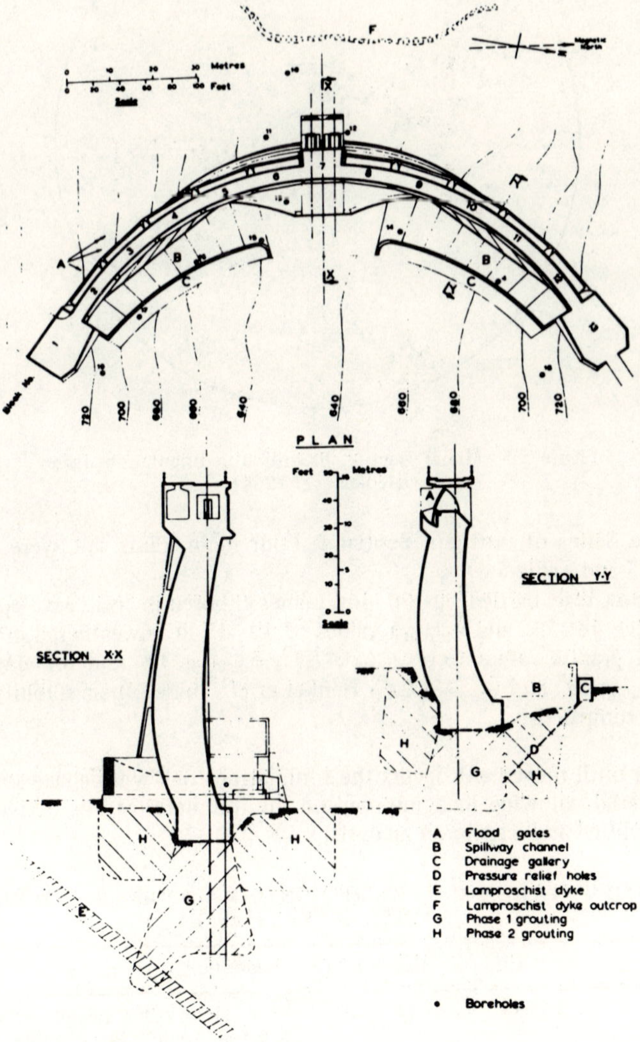

Figure 5.4 Monar Dam – plan and section (after Henkel *et al* 1964)

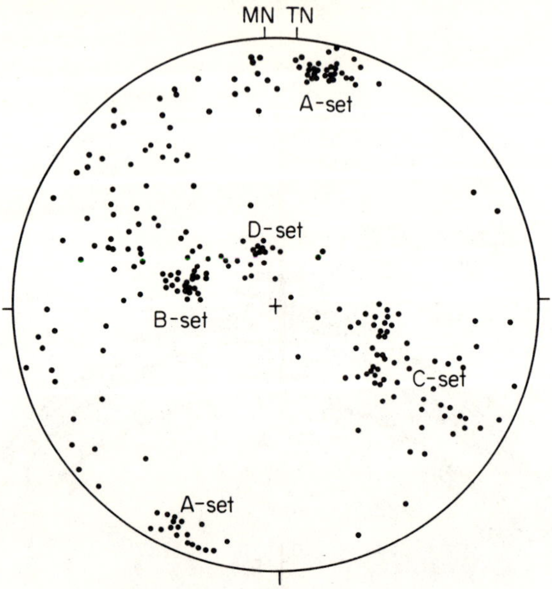

Figure 5.5 Monar Dam — discontinuity orientations (after Henkel *et al* 1964)

the Moine Series of northern Scotland. Four main joints sets were identified (Figure 5.5 and Table 5.3).

Shear box tests carried out on joint planes exposed in cores gave ϕ_{peak} values in the range 48–55° and $\phi_{residual}$ values of 40–45 in unweathered granite. For weathered granite values were ϕ_{peak} 47.5°, $\phi_{residual}$ 39° and for clay covered granite ϕ_{peak} 25° and $\phi_{residual}$ 22°. Henkel *et al*[15] based their stability analysis on the presumptions:

(a) that until movement occurs the foundation rock will behave as an elastic material, allowing load imposed on the foundation to be derived from a simplified arch-cantilever analysis;

Table 5.3 ORIENTATION AND DESCRIPTION OF MAIN JOINT SETS MONAR DAM (AFTER HENKEL *et al*[15])

Set	*Dip*	*Mag bearing*	*Description*
A	83°	14°	Strike along valley. Length ≅ 6 m, spacing ≅ 0.4 m. Smooth surfaces filled with debris near surface
B	31°	284°	Dip upstream. Well developed both banks. Length ≅ 9 m, spacing ≅ 1 m. Irregular surfaces; no infilling
C	40°	110°	Dip downstream. Well developed on north bank
D	20°	345°	Gentle dip upstream. Minor jointing only exposed on south bank
Lamproschist Dyke	38½°	86°	1.8 m thick, 18 m below foundation

(b) that the possibility of sliding is limited to those families of joints which intersect the ground surface on the side of the foundation subject to major directional thrust (i.e the downstream side of the dam) and pass below the foundation.

Together these make a sound statement on which to base general rock foundation stability analyses. In the specific case of the Monar Dam it meant that the danger was confined to the B joints, bounded by A joints which were assumed vertical for the purposes of analysis.

The forces applied to a wedge of rock by foundation elements, and their respective orientations are illustrated in Figure 5.6. The forces exerted by the

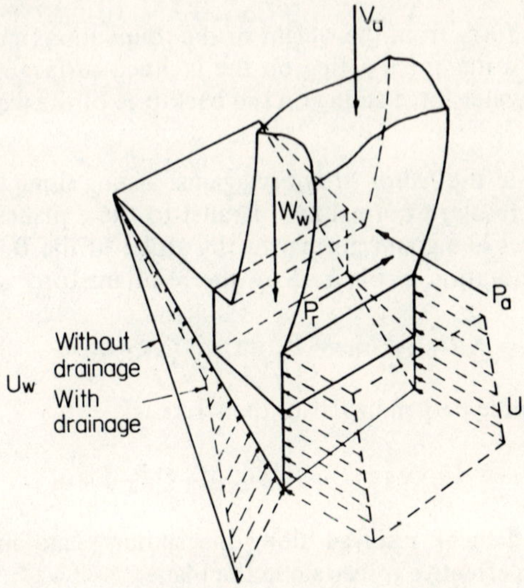

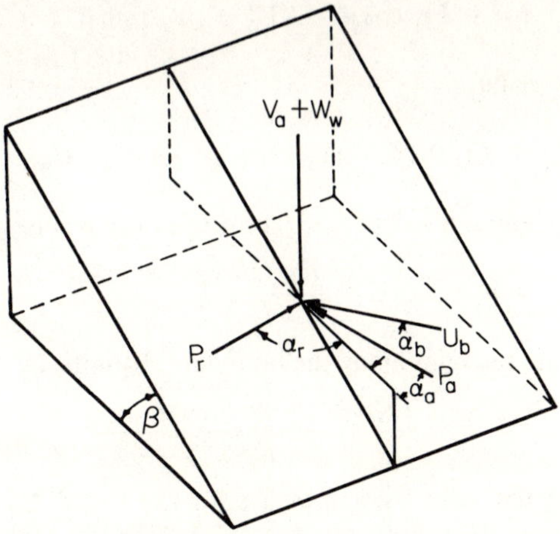

Figure 5.6 (a) Forces applied to rock wedge; (b) Orientation of forces applied to rock wedge (after Henkel *et al* 1964)

dam may be resolved in three orthogonal directions from computed horizontal and vertical thrusts, shear forces and bending moments as:

V_a, a vertical force from the weight of the dam,
P_a, a horizontal force parallel to the dam axis, and
P_r, a horizontal force perpendicular to the dam axis.

The potential rock wedge is assumed bounded by a B joint passing through the lowest point of the foundation concrete and the forces acting on the rock wedge may be represented as:

W_w the vertical force from the weight of the foundations and rock wedge,
U_w the fissure water force acting on the inclined surface of the wedge, and
U_b the fissure water force acting on the back face of the wedge.

In order to calculate the *factor of safety* against sliding along the B joint planes, all forces must be resolved normal and parallel to these planes. The first step is to resolve the forces along and normal to the strike of the B joint planes. Then using the angular notation in Figure 5.6b the resultant force along the strike is:

$$F_A = U_b \sin \alpha_b + P_a \sin \alpha_a - P_r \sin \alpha_r \tag{5.1}$$

and the resultant force perpendicular to the strike is:

$$F_D = U_b \cos \alpha_b + P_a \cos \alpha_a + P_r \cos \alpha_r \tag{5.2}$$

These forces can then be resolved along the sliding plane and normal to the sliding plane, giving effective forces along the plane:

$$F_S = F_D \cos \beta - (V_a + W_w) \sin \beta \tag{5.3}$$

and normal to the plane

$$F_N = F_D \sin \beta + (V_a + W_w) \cos \beta - U_w \tag{5.4}$$

Then if there is no lateral restraint, the factor of safety will be given by

$$FoS = \frac{F_N \tan \phi_B}{F_s} \tag{5.5}$$

and if there is lateral restraint along the boundary A joints, by:

$$FoS = \frac{F_N \tan \phi_B + F_A \tan \phi_A}{F_s} \tag{5.6}$$

where ϕ_B, ϕ_A are the peak friction angles for the B and A joints respectively.

Henkel *et al*, using $\phi_B = 50°$, and $\phi_A = 25°$, calculated safety factors against sliding of about 3 for various sections of the dam, orientated at different directions to the wedge (due to the dam curvature). Expansion of the method and more detailed analysis of the Monar Dam is included in Attewell and Farmer[12].

SHEAR RESISTANCE OF FISSURES IN ROCKS (see also Chapter 16)

The shear resistance of fissures in rocks – essential for stability calculations – can be determined by small or large scale *in-situ* or laboratory experiment. Attewell and Farmer[12] describe in detail *in-situ* experiments. Probably the best known laboratory tool is the Robertson Research joint shear box described by Hoek and Bray[13] and designed to determine the shear resistance along discontinuities exposed in 50 to 100 mm cores.

There remains, however, a need for a simple empirical method for determining joint shear strength based on easily obtained field observations, and Barton's[16,17] approach has much to recommend it. He shows that peak (as distinct from residual) shear resistance along a discontinuity is affected by the degree of surface roughness, rock strength, degree of weathering, mineralogy and the presence or absence of water. It will of course also be affected by the presence of infill material.

Barton, as the result of a comprehensive investigation, shows that the angle of surface friction in effective stress terms along a discontinuity can be represented by an empirical equation of the form:

$$\phi' = \phi_b + (JRC) \log_{10} \left(\frac{JCS}{\sigma'_n} \right) \tag{5.7}$$

where σ'_n is the effective normal stress on the discontinuity plane;
ϕ_b is the residual angle of friction;
JRC is the joint roughness coefficient, and
JCS is the effective joint wall compressive strength.

The latter three parameters are defined by experimental technique. The *residual* or basic friction angle is obtained from residual shear tests on flat sand blasted or sawn surfaces. Its value ranges from 25° to 35° in most rocks although less in those containing large amounts of sheet silicates. The *joint roughness coefficient* has values ranging from zero in smooth rock ($\phi = \phi_b$) to 20 in rough rock. It can of course be computed from a shear test when:

$$JRC = \frac{\tan^{-1}(\tau/\sigma'_n) - \phi_b}{\log_{10}(JCS/\sigma'_n)} \tag{5.8}$$

although there is a touch of circular logic about the idea.

The *joint compressive strength* can be determined directly by Schmidt hammer tests or indirectly from point loading tests on cores and correlation of the resultant strength index (see Broch and Franklin[18]) with laboratory 'compressive strength' values. An allowance must then be made for weathering which may be as much as 80% strength loss in extreme cases. The concept of compressive strength is not a good one, but in this case it gives a rough indication of the resistance of asperities to shearing at the interface. Cohesion at the interfaces can usually be ignored in weathered rocks.

FOUNDATION SETTLEMENT

Whilst foundation failure when bearing pressures exceed bearing capacity must be guarded against, a more important aspect of foundation design is settlement.

The simplest starting point is the classical equation for surface settlement s below a uniformly loaded circular area of radius r on an elastic half-space:

$$s = q_s \frac{S}{E} I_p \qquad (5.9)$$

where

q_s is the uniform load per unit area;
E is the modulus of elasticity of the material under the foundation;
I_p is an influence factor depending on Poisson's ratio (ν) and the distance of the surface point from the centre of the loaded area; and equal to $2(1 - \nu^2)$ when $r = 0$

Where the foundation area is not circular, not uniformly loaded and, typical of most earth materials, E and ν vary with depth, then the area may be divided into small areas which may be assumed subject to a point load at their centre. If the material is assumed elastic then the vertical $\Delta\sigma_x$, $\Delta\sigma_y$ stress increments at any vertical depth and horizontal displacement below the point load can be computed using the Boussinesq equations. By superposition – possibly using influence charts – an incremental stress profile can be built up below any point in the foundation. The method is explained in most Soil Mechanics texts e.g. Peck *et al*[4] and Capper and Cassie[15] are recommended. Then the settlement beneath a particular point on the foundation will be given by:

$$\Sigma \frac{I}{E_z} \left[\Delta\sigma_z - \nu_z (\Delta\sigma_x + \Delta\sigma_y) \delta z \right] \qquad (5.10)$$

where

E_z is the deformation modulus
ν_z is the lateral/vertical strain ratio over the depth increment δ_z.

The terms *modulus of elasticity* and *Poisson's ratio* are deliberately excluded because we are talking about a non-elastic real material. Nevertheless the measurement and interpretation of these rock properties are obviously critical to the estimation of settlement, beneath a foundation. Indeed provided we can assess accurately the deformation modulus or deformability of a rock mass, then we can forget about other factors such as foundation rigidity, embedment, non-linearity, anisotropy and roughness of the foundation base which might be important in the case of a foundation on a perfectly elastic rock.

DEFORMATION OF MASSIVE FISSURED ROCK

The influence and effect of discontinuities on deformation modulus has been recognised for some time. Deere[20] and Deere *et al*[21] developed the concept of Rock Quality Designation or RQD as a simple way of classifying the rock in terms of discontinuity intensity. RQD is essentially the ratio, expressed in percentage terms, between the length of sound core recovered in lengths greater than 100 mm and the total length of coring run. It is not a perfect indicator of joint spacing since the quality of core recovery is obviously a major factor in determining RQD and care has to be taken to distinguish between natural fissures and those caused by drilling.

RQD can however be related quite closely to seismic velocity and mass/intact rock velocity and modulus ratios. Table 5.4 based on Deere *et al*[22] and Coon and Merritt[23] gives typical values associated with RQD ranges.

Table 5.4 RQD, VELOCITY RATIOS AND ROCK MASS FACTORS

RQD, %	*Rock description*	*Fracture frequency per m*	*Velocity ratio* V_f^2/V_e^2	*Rock Mass factor* E_d/E_{t50}
0–25	very poor	> 15	0–0.2	> 0.2
25 50	poor	8–15	0.2–0.4	> 0.2
50 75	fair	5–8	0.4 0.6	0.2–0.5
75–90	good	1–5	0.6–0.8	0.5–0.8
90–100	excellent	> 1	0.8–1.0	0.8–1.0

In Table 5.4 V_f is the field seismic velocity, V_e, the laboratory velocity, E_d the deformation modulus computed from V_f and E_{t50} the laboratory tangent modulus at 50% compressive strength. The rock mass factor, or as Coon and Merritt call it the *deformability* of a rock mass has been developed into quite an important concept by Hobbs[24, 25]. Hobbes defines the *rock mass factor*, as the ratio of the deformation modulus of a rock mass in any readily identifiable lithological and structural component to that of the deformation modulus of the intact rock comprising the component. The magnitude of *j* depends, as Hobbs points out, on the method of measuring the rock deformation modulus. He

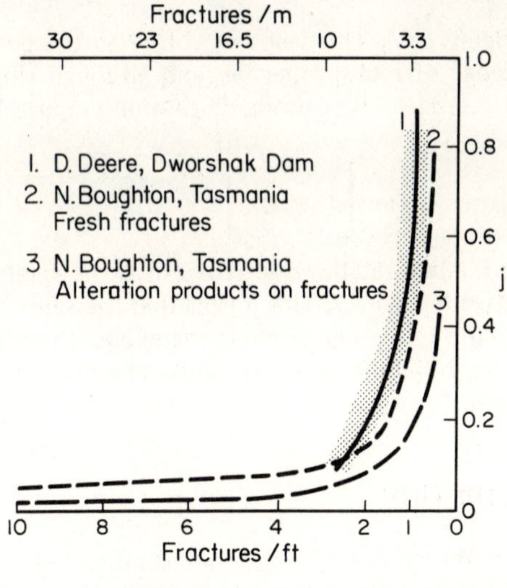

Figure 5.7 Relationship between rock mass factor and fracture frequency (after Hobbs 1975)

suggests, on the basis of previous work, that a plate loading test, scaled to the fracture spacing and with strain measured through a deep central hole in the plate, is the preferred method.

Figure 5.7, based on the work of Deere *et al*[21], Boughton[26] and Hobbs[25] shows how the *j* value can be related to fracture frequency for various rocks, although the scatter inherent in the results implies that fissure width is also an important factor in rock mass deformation. It is interesting in this context to remember Snow's[27] assessment of fissure frequencies and widths during permeability measurements, and his conclusion that, irrespective of rock type, fissure spacings increase and fissure widths decrease significantly over 100 m depth in

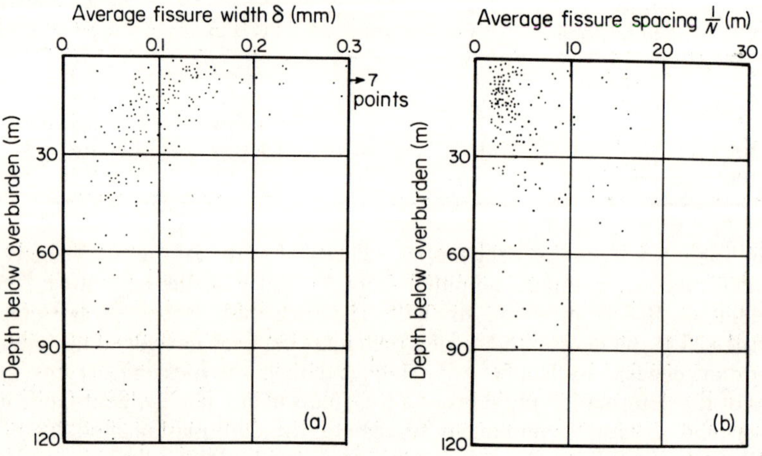

Figure 5.8 Fissure spacings and widths over 100 m depth in various dam site rocks (after Snow 1968)

the Earth's crust (Figure 5.8). This leads of course to the possibility of computing j from a knowledge of rock properties, and although Hobbs[25] suggests one or two slightly exotic models, it is probable that an empirical expression of the form:

$$j = \left[1 + Kn\right]^{-1} \tag{5.11}$$

will provide an approximate relationship where n is the fracture frequency and K is related to the fissure porosity. This means that the value of j will increase in virtually all cases from a minimum value at the exposed weathered rock surface (values in Table 5.4 are minimum values) to unity at a depth approaching 100 m.

SETTLEMENT PREDICTION

The basic methods of settlement prediction are summarised in equations 5.9 and 5.10, and not withstanding the indelicacy of using fissure porosity controlled deformation modulus in a standard solution based on an elastic half space, they probably represent the best methods available. Certainly elastic solutions can be used for settlement prediction on rock foundations with rather more confidence than in soil foundations, if only because the magnitude and even the likelihood of primary and particularly secondary consolidation will be much reduced. The problem therefore comes down to predicting the rock mass deformation modulus or the rock mass factor accurately.

The simplest method is to carry out a plate test and scale up the settlement. Thus if the plate and foundation are loaded to the same bearing pressure:

$$\frac{\delta_f}{\delta_p} = \left(\frac{R_f}{R_p}\right)^a \qquad (5.12)$$

where

δ is the settlement;

R is the radius;

f, p refer to foundation and plate respectively, and

a depends on the plate dimensions but is approximately unity.

Otherwise plate tests may be used at depth above the water table to estimate the deformation modulus profile. Hobbs[25] gives examples of this for the Chalk and Triassic rocks.

As a useful rule of thumb, j for the surface strata may be predicted from Table 5.4 and assumed to rise to $j = 1$ at a depth of 100 m. Provided always that a detailed geological description of relevant cores does not indicate an anomalous situation.

References

1. Sowers, G.F., 'Shallow foundations' *in Foundation Engineering*, ed. G.A. Leonards, McGraw-Hill, New York, pp 525 onwards (1962)
2. Bjerrum, L., Contribution to discussion *European Conference on Soil Mechanics Foundation Engineering*, Wiesbaden, 125–126 (1963)
3. British Standards Institution, Code of Practice 2004; 1972 *Foundations*
4. Peck, R.B., Hanson, W.E. and Thornburn, T.H., *Foundation engineering*, 2nd Ed, Wiley, New York (1974)
5. American Society for Testing and Materials, *Standard method of test for unconfined compressive strength of rock core specimens*, Designation D2938 (1975)
6. Rankilor, P.R., 'A suggested field system of logging rock cores for engineering purposes,' *Bull. Assoc. Eng. Geol.*, **11**, 247–258 (1974)
7. Judd, W.R. and Huber, C., 'Correlation of rock properties by statistical methods,' *Int. Symp. Mining Research*, **2**, 621–648, Pergamon (1962)
8. Deere, D.U. and Miller, R.P., *Engineering classification and index properties for intact rock*, Air Force Weapons Lab. Tech. Report AFWL-TR-G5-116, Kirtland, New Mexico (1966)
9. International Association of Engineering Geology, *Proceedings of the Second International Congress*, Sao Paulo, Brazil (1974)
10. International Society for Rock Mechanics, *Advances in rock mechanics*, Proceedings of the 3rd Congress, Denver, Colorado (1974)
11. John, K.W., 'Geologists and civil engineers in the design of rock foundations of dams,' *Proc. 2nd Int. Cong. Int. Assoc. Eng. Geol.*, Sao Paulo, paper VI-PC-3 (1974)
12. Attewell, P.B. and Farmer, I.W., *Principles of engineering geology*, Chapman and Hall, London (1976)
13. Hoek, E. and Bray, J.W., *Rock slope engineering*, Inst. of Min. Met., London (1974)
14. Londe, P., 'The role of rock mechanics in the reconaissance of rock foundations,' *Quart. J. Eng. Geol.*, **6**, 58–74 (1973)
15. Henkel, D.J., Knill, J.L., Lloyd, D.G. and Skempton, A.W., 'Stability of the foundations of the Monar Dam,' *Proc. 8th Int. Cong. on Large Dams*, **1**, pp 425–441 (1964)
16. Barton, N., 'Review of a new shear strength criterion for rock joints,' *Eng. Geol.*, **7**, 287–331 (1973)

17. Barton, N., 'Estimating the shear strength of rock joints,' *Proc. 3rd Int. Soc. Rock Mech. Conf.* Vol II A, 219–220 (1974)
18. Broch, E. and Franklin, J.A., 'The point load strength test', *Int. J. Rock Mech. Min. Sci.*, 9, 669–697 (1972)
19. Capper, P.L. and Cassie, W.F., *The mechanics of engineering soils*, 5th Ed, Chapman and Hall, London (1969)
20. Deere, D.U., 'The technical description of rock cores for engineering purposes,' *Rock Mech. Eng. Geol.*, **11**, 17–22 (1964)
21. Deere, D.U., Hendron, A.J., Patton, F.D. and Cording, E.J., 'Design of surface and near-surface construction in rock,' *Proc. 8th Symp. Rock Mechs. Minnesota*, Amer. Inst. Min. Eng. 237–303 (1966)
22. Deere, D.U., Merritt, A.H. and Coon, R.F., *Engineering classification of in-situ rock* NTIS U.S. Dept. of Commerce, Springfield, Va (1969)
23. Coon, R.F. and Merritt, A.H., 'Predicting *in-situ* modulus of deformation using rock quality indexes,' *In-situ testing in rock*. ASTM STP pp 154–173 (1970)
24. Hobbs, H.B., 'Effects of non-linearity on the prediction of settlements of foundations on rock,' *Quart. J. Eng. Geol.*, **6**, 153–168 (1973)
25. Hobbs, N.B., 'The prediction of settlement of structures on rock,' *Conference on settlement of structures*, Cambridge, Pentech Press, London, 579–610 (1975)
26. Boughton, N.O., 'Correlation of measured foundation modulus with *in-situ* rock properties,' *Proc. Int. Symp. Rock Mech*, Madrid (1968)
27. Snow, D.T., 'Rock fracture spacings, openings and porosities,' *J. Soil Mech. Found. Div.* Amer. Soc. Civ. Eng., **94**, 73–91 (1968)

Chapter 6

Properties of Mining Wastes with Respect to Foundations

INTRODUCTION

The most common use of mining wastes in civil engineering practice is as a fill material for roadworks, industrial estate and housing developments. During the past decade many spoil heap reclamation schemes in derelict mining areas in Britain have specifically included some form of building development as well. Like motorway construction, the methods adopted for sites restored for building purposes follow Section 600 of the Ministry of Transport's (now DOE), 1969 *Specification for Road and Bridge Works*[1]. Layering and compaction enables sites to be built on shortly after completion, thus avoiding an excessive settlement period.

Table 6.1 demonstrates why this chapter should be devoted to coarse colliery discard in Britain. Out of the *total* waste production of about 163 million tonnes per annum, coal-mining wastes from underground extraction represent nearly 36%. These wastes are responsible for about 30% of the derelict land in Britain. The annual production from other mining activities is only some 3 million tonnes. This latter figure should be compared with the USA, where in 1974 the estimated quantity of waste from the metals mining industry alone was around 793 million tonnes (23.3% overburden, 47.2% waste rock, 29.5% tailings; weight ratio, waste:ore = 1.86). An estimate of coarse colliery discard produced every year in the USA is 52.6 million tonnes, with fine discard running at between 13 and 26 million tonnes*. In other words, the balance in mining wastes and quantities is completely different from that of Britain.

Chemical analyses of tailings (not including coal-rich slurries which may be reclaimed as low-grade fuel, see Table 6.1) produced by the coal preparation process show that organic carbon (coaly material) accounts for an average of about 38% by weight of the total constituents. The reported specific gravities given in Table 6.2 also highlight this fact since the specific gravity of coaly materials is generally about 1.3 to 1.5. Moreover, these fine-grained, low density materials are usually difficult to handle because of the high moisture contents commonly measured in colliery lagoon samples. Nevertheless, there are a few reports in the British and North American technical literature of the limited use on tailings, mixed with coarse discard to improve grading characteristics, for use as roadwork fill.

*A recent survey suggests that coal-mine wastes may reach 120 M tonnes per annum.

Table 6.1 QUANTITIES OF MAJOR WASTES IN BRITAIN

Material/origin	*Approx. annual production* M/tonnes/year		*Toral quantity available* M tonnes	*Approx. annual use* M tonnes	*Disposal methods*	*Main uses*
1. Coal Mining						
Colliery shale (coarse discard)	c.55	*58*	3000	8	Coarse discard mainly to spoil heaps on land, some disposal to old quarries etc & about 10% disposed of at sea. Fine discard to lagoons, infilling shafts; also pressing or other treatments for mixing with coarse discard for spoil heap disposal.	Coarse discard mainly as fill. Other uses include light-weight aggregate, brick manufacture. Slurry re-claimed for power station use. Some coal reclaimed from coarse discard down to 10% by weight content.
Slurry & tailings	c.3					
2. Mining other than coal		*3*	Not known	Not known		
e.g. (a) Fluorspar, barytes, galena mining		0.23			Spoil heaps, tailings lagoons	Minor use as road metal, concrete aggregare
(b) Tin mining (tailings from both)		0.46			Tailings, lagoons or discharged into sea	Minor use as concrete aggregate
3. China clay quarrying						
Sand		*10*	125		Sand and overburden to spoil heaps, micaceous residue to lagoons and disused pits	Sand used as fill, as fine aggregate in concrete, manu-facture of blocks & bricks
Overburden / Micaceous residue		*12*	155	1		
4. Other quarrying		*27*	Not known	Not known	Some utilisation, remainder tipped	Aggregates for road con-struction & concrete.
e.g. slate waste		1.2	300–500	0.03		Fill, inert filler, granules, expanded aggregate & filter medium

5. Power Stations						
Pulverised fuel ash	c.7.5	12	Not known	6	Lagoons, infilling old clay pits etc; some disposal at sea	As fill, manufacture of cement, concrete blocks, lightweight aggregate, bricks
Furnace bottom ash	c.2.5					
Furnace clinker	c.2.0			2	All used	Concrete blocks
6. Domestic & trade refuse		*18* (of which 0.7 is incinerated)	Not known	Not known	Mostly dumped	As fill & used for covering refuse tips
7. Industrial waste		*23*				
e.g. Blast furnace slag, steel-making slag, zinc-lead slag, copper slag, tin slag	c.14		Not known	Nearly all utilised	Minor tipping	e.g. roadstone
Total		*163*				
Imported fill for major roadworks, 1969		26				
Natural aggregates used in road construction, 1968		59				

*Data taken mainly from: Gutt *et al*[2], Sherwood[3] and Royal Commission on Environmental Pollution[4]

It is eminently clear however, that of the mining wastes in Britain it is primarily *coarse* discard from underground coal extraction that has the greatest utilization potential. Research into fresh discard materials, old spoil heaps, and the new generation of layered and compacted spoil heaps and lagoon embankments has elucidated most of the properties, and characterised the behaviour of these materials. It should be borne in mind that the new generations of spoil heaps and lagoon embankments are major structures which are being constructed according to both well established and modified civil engineering practices.

Table 6.2 COLLIERY TAILINGS

	Britain (Ranges to date)	*USA (Appalachia)*‡
Specific gravity	1.53–2.41	1.34–1.66
Moisture content %	11*–64†	23–65
Dry density, Mg/m^3	0.77–1.90	0.75–1.17

*Generally in excess of 20%
†*Note.* Since tailings are initially a suspension the upper limits of moisture content and lower limits of density should be treated with caution. The values are taken from lagoons with a sufficiently strong surface crust for sampling purposes.
‡from Holubec[5]

With respect to earthworks construction and the founding of structures on or within these materials it is necessary to assess their suitability in terms of the following factors:

(a) Lithology (rock types), chemical and mineralogical characteristics.
(b) Engineering properties and their relationship with (a).
(c) Short and long-term behaviour with respect to (a) and (b).
(d) Current construction specifications.
(e) Potential hazards and environmental pollution.

ORIGIN AND COMPOSITION OF COLLIERY WASTES

The 'coarsening upwards' grain size sequence between two coal seams, which is defined by an ideal cyclothem is: coal-mudstone/shale-siltstone-sandstone-seatearth-coal. Although certain members may be absent, fresh discards are usually dominated by the argillaceous rocks immediately associated with coal seams. Mudstone/shale, siltstone, minor sandstone quantities and seatearths (which can range from highly plastic fireclay through carbonaceous mudstone to siliceous sandstone) are the principal rock types found in colliery spoil heaps.

The coal preparation process produces varying degrees of slaking, and in any case the rock materials are usually pre-crushed to a top size of between 76 mm and 203 mm. Waste rock materials from underground roadway drivages, however, often by-pass the preparation process. These materials are usually recognised in spoil heaps from the higher proportion of stronger rock types present and the wider range of grain sizes, including cobbles and boulders.

Spoil heaps containing a high seatearth content (>50%) are commonly fine-grained, and clayey in appearance. They tend to dry out to a pale grey colour after washing which compares with the darker greys of heaps containing a higher proportion of roof materials. There is normally a pronounced difference in the breakdown of the two types of materials (see Figure 6.1). Rock fragments in coarse discards are plate-shaped as illustrated in Figure 6.2, although this may not be so pronounced in seatearth-rich materials (e.g. at Gedling Colliery).

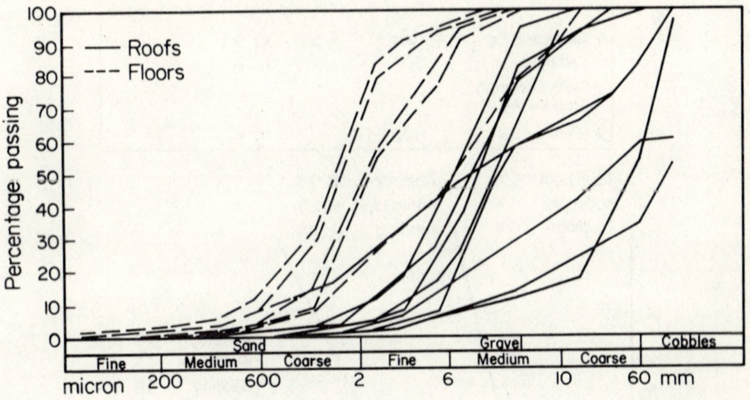

Figure 6.1 Showing degradation of roof and floor rocks over a 12-month period (from NCB[6])

Semi-quantitative X-ray analysis and organic carbon assessments are a useful guide to composition, as well as being an aid in predicting the engineering behaviour of a particular spoil. It must be remembered, however, that in old spoil heaps the type of waste may have changed over the years as new seams have been worked, or the mineralogy may have altered because of combustion. Adequate sampling from boreholes, or from trenches which are sited so as to sample the historical tipping sequence is therefore necessary.

The broad geographical distribution of clay minerals in colliery discards is shown in Figure 6.3. It will be observed that kaolinite, an inert clay mineral, tends to be dominant in the northern coalfields, whilst expandable mixed-layer clay is quite common in the Yorkshire coalfield. The relevance of the clay mineral balance is important with respect to physical breakdown which will be examined in the later section dealing with physical disintegration of shaly rocks. The averaged mineralogical analyses given in Table 6.3 emphasise a number of important features, including the constraints imposed by geographical variations. Illite is on average the most common clay mineral and in spoil heaps with a sampling bias towards south Wales and Yorkshire (column a), mixed-layer clay is the second most important type. Two Scottish samples included in Collins[9] detailed mineral survey, (column b), include considerable quantities of kaolinite and consequently the clay mineral balance is different. The eight specimens analysed by Collins are from the 1 to 3 m zone of the requisite spoil heaps and a minor degree of chemical weathering is suggested by the sulphate mineral quantities present and possibly by the low level of mixed-layer clay in one or two samples.

A noteworthy feature of both sets of analyses is the content of carbonaceous (coaly) material. Analysis of rocks found in discards implies that on average inherent organic matter is unlikely to account for more than 2.5% by weight.

The remainder of the carbonaceous matter is largely coal waste, which has an important effect on both the mechanical properties of discards and on their longer-term chemical and thermal behaviour. In a number of very old tips the average coal content was found to be exceedingly high, the range for 3 spoil heaps in Co. Durham, for example, being 19 to 22%.

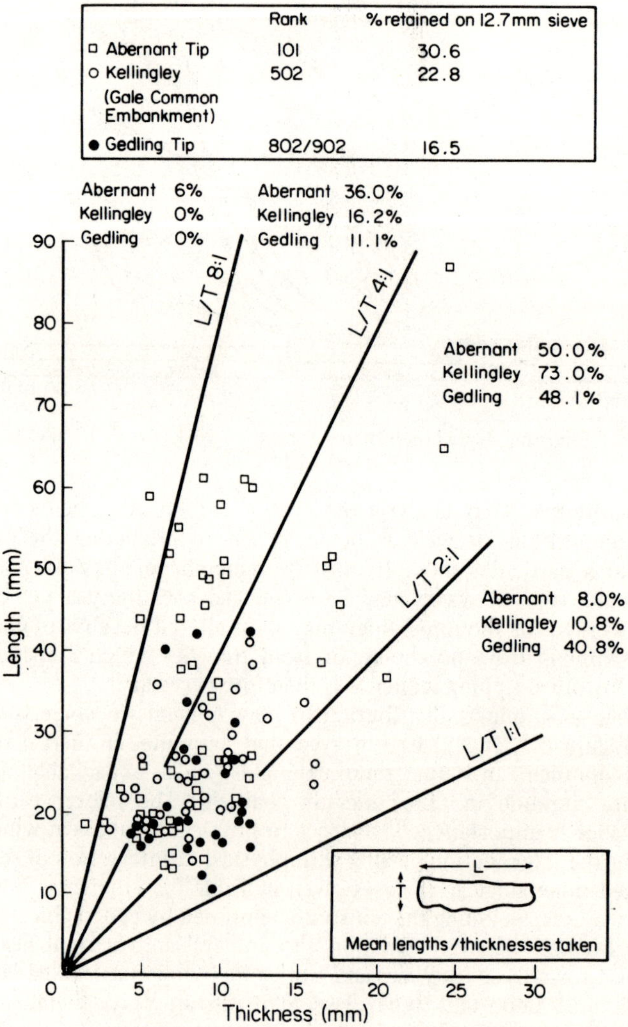

Figure 6.2 Particle shape of coarse discard fragments. Comparison of materials associated with coals of differing ranks

Statistical comparisons of means by Student's t test implies that at the 95% confidence level the quartz content is higher and the clay content lower in spoils from South Wales. This suggests that in general the fragmental 'shale' components could therefore be expected to be a little stronger in the latter materials.

Routine X-ray analyses (with confirmatory major element geochemical

Table 6.3 COLLIERY SPOIL HEAPS QUANTITATIVE MINERALOGICAL ANALYSIS (averaged to 0.5%, wt.%)

	*74 specimens** *15 spoil heaps*	*8 specimens†* *8 spoil heaps*
Quartz	17.5	19.5
Illite	31.5	24.0
Mixed-layer clay	26.0	12.0
Kaolinite	10.5	16.0
Chlorite	0.5	4.0
Allophane (assigned from amorphous Si & Al)	N.D.	1.5
Carbonates‡	1.0	4.0
Feldspars	Tr	0.5
Sulphates	Tr	1.0★
Pyrite	Tr	≏0.5
Carbonaceous (coaly) matter	13.0	16.0

*From Taylor[8] : Constituents including Trace sulphates, pyrite, feldspar, rutile, apatite and Fe^{3+} minerals = 2%

†From Collins[9] : trace Limonite, haematite, ? magnetite and ? anatase identified.

‡Siderite, dominant carbonate.

★Jarosite, alunite, trace gypsum.

N.D. Not determined

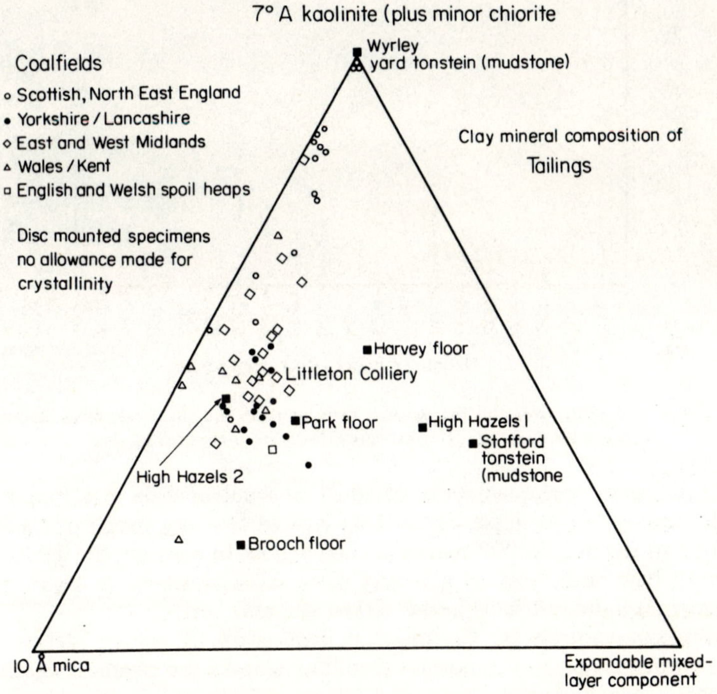

Figure 6.3 Showing clay mineral variation in British coalfields[7]. Also included are other rock types referred to in the text

analyses in some cases), provide an insight into the nature of these materials. From an engineering aspect, sulphate analyses are also necessary if concrete structures or cemented materials are involved. The organic carbon (coal) content can be determined approximately in a refractory oven on samples which have been used for moisture content determinations. The temperature should be raised very slowly until constant weight is attained at 375°C.

PHYSICAL DISINTEGRATION OF SHALY ROCKS

It was recognized over twenty years ago that the coalification process, which is a function of depth of burial, and hence expulsion of volatiles consequent upon temperature rise, does appear to have some effect on the associated rocks as well. The concept of an incipient 'shale rank' emerged.

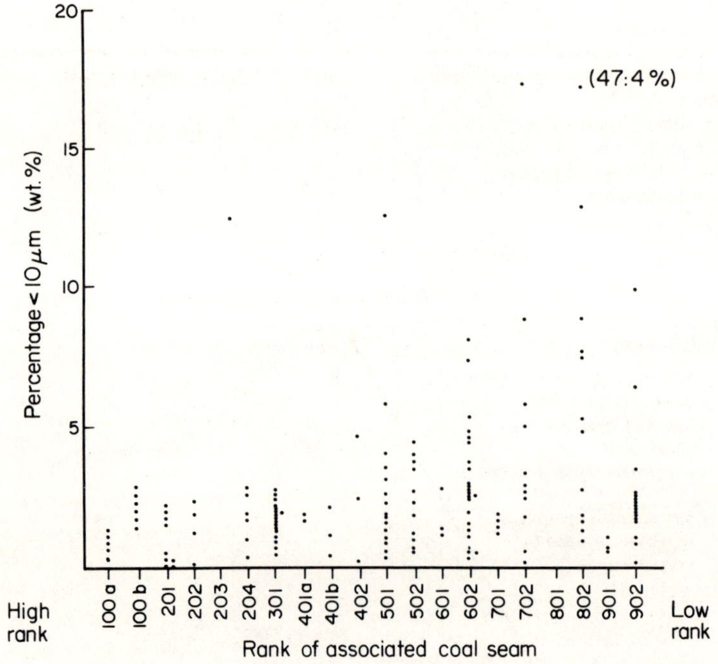

Figure 6.4 End-over-end breakdown tests of roof and floor measures associated with coals of differing ranks. Data from Berkovitch *et al*[10]

End-over-end breakdown tests designed to simulate shale breakdown during coal processing (for example Figure 6.4) showed that the spread of breakdown increased in the 500 to 900 coal rank code region. In contrast, the debris associated with high rank coals are generally more stable in water. In Figure 6.4 the breakdown is more restricted in the 100 to 402 rank region.

The major controls on the physical breakdown of colliery debris due to water, (which is far more important than low temperature chemical weathering), have been recognised as:

(a) The incidence of sedimentary structures and discontinuities (planes of weakness), such as bedding planes, laminations, joints and slickensides.

(b) Slaking or compressed air breakage, which is a function of cycles of desiccation and saturation. This starts at the washery and ceases only when the material is buried within an embankment.

(c) Interlayer expansion of expandable 10 Å mixed-layer mica-montmorillonite clay minerals, which appears to be important when exchangeable sodium (Na) is high.

Sedimentary structures and discontinuities are familiar to most ground engineering practitioners and the development of high suctions in shales can be demonstrated by conventional soil suction methods. Individual thin slivers of shale and mudstone were used for each measurement in the suction plate and pressure membrane experiments demonstrated in Figure 6.5 and 6.6.

The capillary model which represents the pore space within a soil or rock as a series of capillaries of tortuous configuration is a simple, reasonably valid approach on which to base suction pressure ideas. The term suction implies the existence of a pressure deficiency or moisture tension in some part of the water in the voids or capillaries.

On the pF scale, the logarithm to base ten of the suction expressed in centimetres of water is equivalent to the pF value. Hence 10 cm of water equals pF 1 and 1000 cm equals pF 3 (approximately the negative equivalent of atmospheric pressure). The specimens in equilibrium with free water have a pF of almost zero, and oven-dried material a pF of almost 7.

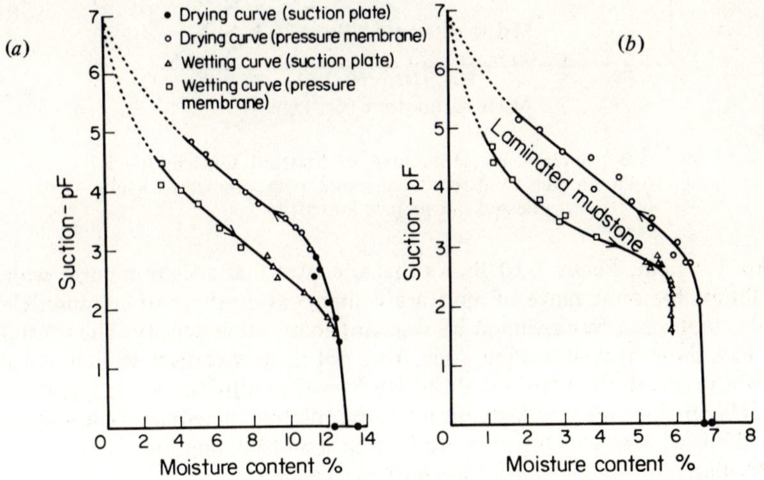

Figure 6.5 Suction pressure curves for (*a*) fissile shale (High Hazels) and (*b*) mudstone (Tinsley)

Over a large part of their suction curves the typical laminated mudstone and the fissile shale (Figure 6.5) have suction characteristics which are more akin to non-shrinkable, incompressible materials than to shrinkable materials. The sigmoidal shape of the wetting and drying curves can be accounted for by the fact that in general a porous body will not contain pores of uniform size and shape that can be emptied at the same suction. Also, the greater disparity in size between capillaries and voids the more marked will be the hysteresis between the wetting and drying curves.

The vertical parts of the drying curves indicate that considerable suctions can be applied to the pore water without a change in moisture content. Under *in situ* conditions it may therefore be expected that rocks of this type should be close to full saturation.

The saturation moisture content of the High Hazels 2 shale is not absolute in terms of a perfectly rigid skeleton and even by making the specimens small, slaking was only reduced and not eliminated. Similarly, the non-coalescence of wetting and drying curves (Tinsley mudstone) is a function of entrapped air.

Comparison of the above rocks with an extreme type of thin, Coal Measures mudstone band with a very high expandable mixed-layer clay content (the

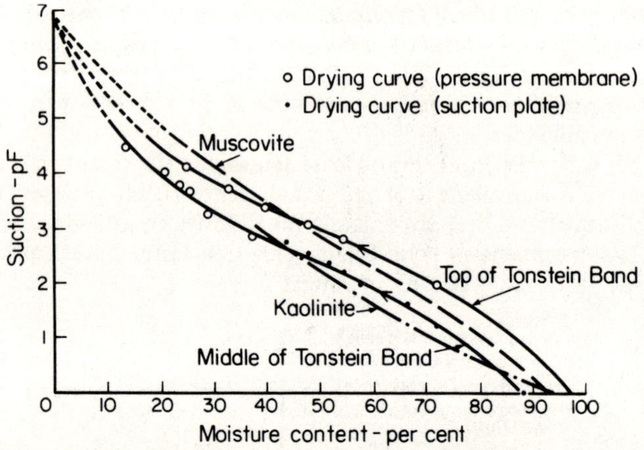

Figure 6.6 Suction pressure curve of Stafford tonstein (mica-montmorillonite-rich mudstone) compared with compacted kaolinite and fine-grained muscovite

Stafford Tonstein, Figure 6.6) shows that the latter has a suction curve which spans almost the same range of moisture contents as do those of common clay minerals compacted to maximum BS standard compaction density. The tonstein curve, like those of compressible soils, does not show a vertical section and its shape is a result of air entry and shrinkage over the entire suction range. Fortunately, clay-rocks with very high mica-montmorillonite mixed-layer contents are rare in the Coal Measures but they do help in understanding the breakdown of other argillaceous types which are susceptible to water.

With more typical shales and mudstones such as those considered in Figure 6.5, it can be concluded that with extreme desiccation (high suctions) the bulk of the voids will be filled with air, which, on rapid immersion in water, is pressurised by the capillary pressures developed in the outer pores. Failure of the mineral skeleton along the weakest plane ensues (e.g. along a bedding plane or lamination) and an increased surface area is then exposed to a further cycle. Work by Van Eeckhout[11] indicates that a tensile failure mechanism is operative. The reduced breakdown obtained by de-airing specimens in slaking tests has provided supporting evidence on the significance of compressed air breakage. Normal shales and mudstones can consequently be reduced to sand and gravel sizes within months or even weeks of excavation.

The breakdown of types which are *extremely* susceptible to water – Stafford tonstein, Brooch and Park seatearths, High Hazels 1, roof shale and Harvey seat-earth (Figure 6.3) – is more explicable in terms of *intra*-particle expansion of the mixed-layer clay content, especially when sodium (Na) is the dominant exchangeable cation. The randomly interstratified mixed-layer clay content alone does not appear to necessarily imply increased breakdown due to *intra*-particle expansion. Some strong, South Wales spoil heap materials contain quite high quantities of the mineral and the average content shown in Table 6.3, column (a) is quite high. Evidence to date infers that inherent organic matter may inhibit breakdown by *intra*particle expansion, perhaps more particularly in seatearths.

Table 6.4 EXCHANGEABLE CATIONS (m-equiv./100 g)

	Na	*K*	*Ca*	*Mg*
I Reference clay minerals				
1. Montmorillonite (Wyoming)* (API 25b)	54	–	11	15
2. Illite (Fithian)* (API 35)	–	–	17	3
3. Kaolinite (Bath, S. Carolina)* (API 5)	–	–	0.5	0.4
II Carboniferous shales/mudstones/seatearths				
1. Stafford tonstein†	14.2	4.9	6.1	2.5
2. Wyrley Yard tonstein	–	0.2	0.6	0.5
3. Ryder shale‡	15	5	–	–
4. Park seatearth §	12.0	1.8	1.0	2.2
5. Brooch seatearth §	10.1	2.3	1.8	2.4
6. High Hazels I shale	9.6	2.2	3.5	2.8
7. Harvey seatearth	5.3	2.8	3.9	1.2
8. Illitic marine shale†	1.6	1.1	3.6	12.9
III Spoil heap materials				
1. Littleton tip (toe of slip)	1.5	2.5	7.9	2.6
2. Littleton tip (failure plane)	4.4	2.8	6.3	4.0
3. Askern tip §	9.6	1.9	2.4	3.1
4. Askern tip §	8.8	1.7	2.5	3.1
5. *Average for 74 spoil heap specimens* §	2.1	1.5	7.3	4.8

Data from:
*Carroll and Starkey[12]
†Taylor and Spears[7]
‡Badger *et al*[13] (values may be too high: dil. HCl used)
§Taylor[8]

Cation exchange values given in Table 6.4 demonstrate the major differences in the apportionment of exchangeable cations in reference to clays, with expansive Na-montmorillonite at one extreme and kaolinite at the other. Apart from the kaolinitic Wyrley Yard tonstein and the illinitic marine shale, which are included for comparison (low exchangeable Na), the specimens listed under Carboniferous shales, mudstones and seatearths are ones which are unstable in water, or have proved to be troublesome underground.

Measurement of uplift pressures on a number of intact, air-dried specimens in a small 25 mm diameter oedometer which are illustrated in Figure 6.7 gives some indication of the effect of exchangeable Na in the behaviour of expandable mica-montmorillonite mixed-layer clay.

The evidence from spoil heap samples to date suggests that rapidly degradable materials with enhanced exchangeable Na levels (including Askern, Table 6.4) are restricted. The average value for 74 spoil heap specimens is low and it is possible that once on a spoil heap or within an embankment, *intra*particle expansion may be restricted since Na in the interlayer structural sites could be subject to exchange by the readily available divalent cations from sulphates and carbonates which help preclude expansion. Further research is required in this area.

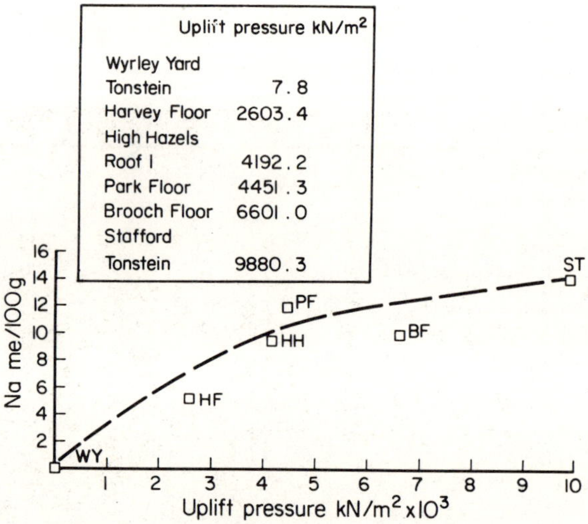

Figure 6.7 Uplift pressures measured on air-dried roof and floor rocks containing variable amounts of mixed-layer clay (see Figure 6.3). Exchangeable sodium (Na) in m-equivalents/ 100 g shown on abscissa

In any case the proportion of stronger rock types will have an 'overprinting' effect on the overall degree of breakdown of particular discards. This means that the composition of materials from disposal points or a spoil heap should be considered on an individual basis. The breakdown of the 7 ranked samples over a 10-month period shown in Figure 6.8 demonstrates this point. Samples 5 to 7 (Ranks 700, 802, 902) do show the greatest degree of breakdown, although the high organic carbon value (25%) of sample 7 has resulted in a surprising amount of large coaly and carbonaceous material remaining at the end of the test period. In all samples, coal, coaly shale, ironstone nodules and sandstone proved resistant to the weather.

WEATHERING AND DEGRADATION

In addition to the breakdown processes already discussed, further physical degradation during transportation, tipping and spreading will occur, particularly with degradable discards like Askern shown in Figure 6.10.

When tipped in a *loose* state long-term chemical and mineralogical changes begin, the consequences of which need not result in ground engineering problems, so long as they are appreciated.

The minerals that are most prone to weathering are non-detrital in origin, namely pyrite and carbonates, i.e. calcite, ankerite (a mixed CaMgFe carbonate) and to a lesser extent, ferrous carbonate (siderite). It is the oxidation of sulphide minerals, principally pyrite (FeS_2), that is the common denominator of acidity and sulphate genesis. Although pyrite is on average a very minor component of coarse discard (see Table 6.3) it can be concentrated in coal, some organic roof shales and in coaly seatearth materials. Finely disseminated pyrite in concentrations which are not detectable by X-ray analysis, can still be instrumental in chemical weathering.

Figure 6.8 Comparison of 7 'ranked' fresh discard samples after exposure to atmosphere for 10 month time interval. Samples: 1 = rank 101 spoil, 2 = rank 301, 3 = rank 502, 4 = rank 600, 5 = rank 700, 6 = rank 802, 7 = rank 902 (from Ratsey[30])

The oxidation of pyrite and the consequent formation of sulphates is still a matter of debate. There is good evidence that in warm, oxidising environments bacteria are involved as a catalyst, or possibly more directly in bacterial oxidation[14]. The highly exothermic reaction is a multi-stage one and is probably as follows[15]:

$$\text{Stage 1: } 2FeS_2 + 7O_2 + 8H_2O \rightarrow 2FeSO_4 \text{ (ferrous sulphate)} + 2H_2SO_4 \text{ (sulphuric acid)}$$

$$\text{Stage 2: (Bacteria required); } 4FeSO_4 + O_2 + 2H_2SO_4 \rightarrow 2Fe_2(SO_4)_3 \text{ (ferric sulphate)} + 2H_2O$$

$$\text{Stage 3: (oxidation by ferric sulphate); } 7Fe_2(SO_4)_3 + FeS_2 + 8H_2O \rightarrow 15FeSO_4 + 8H_2SO_4$$

The reaction of sulphuric acid with carbonates and ions in the clay minerals will lead to the production of secondary sulphates, the other major weathering products being hydrated iron oxides, usually in an amorphous state. Loose tipped fresh discard, which is commonly alkaline or neutral can become progressively acidic; drainage water can consequently contain several hundred parts per thousand of sulphates.

The evidence from old, initially loose, unburnt spoil heaps, however, implies that intense weathering is limited to depths of about 1 m. At a lower level of activity chemical weathering does not seem to extend to depths in excess of 3 m. For example, the investigation at Yorkshire Main colliery showed that the sulphate content only increased significantly in the top 3 m of the 50-year-old tip (Figure 6.9). The survival of pyrite without any obvious or systematic

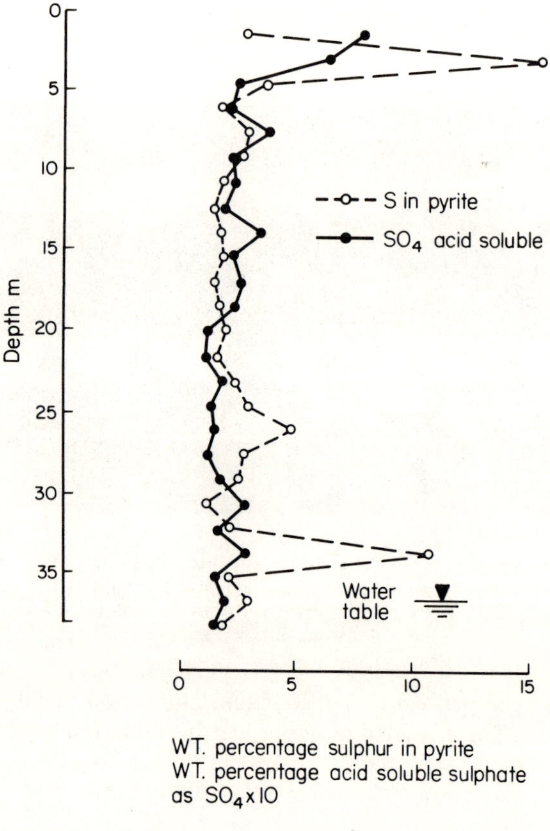

Figure 6.9 Profile showing acid soluble sulphate (as SO_4) and pyrite (as %.S) through Yorkshire Main Tip (from Spears *et al*[16])

decrease in its amount with depth (i.e. age) is yet another indication that weathering of even loose tipped materials is generally restricted. Grain size distributions have again shown no systematic variations with depth that would support progressive degradation with time. In addition to superficial chemical weathering, continuous seepages in loosely placed discard may display a narrow zone immediately above the phreatic surface in which chemical changes have taken place.

It is pertinent to record, however, that the effective shear strength of materials is not necessarily affected by observed chemical weathering. From the wide range of materials shown in Table 6.5 it can be concluded that any decrease in the shear strength of layered and compacted fresh discard is small (e.g.

Table 6.5 SHEAR STRENGTH COMPARISONS WITH RESPECT TO SUPERFICIAL WEATHERING

Location	*Sample*	*Shear strength** ϕ'_e degrees
1. Gedling spoil heap (Notts)	(a) Surface of tip, 6 months old	22.0†
	(b) 3 m deep, 3–4 months old	28.0
	(c) Fresh material from washery	30.0
2. Kellingley spoil (Yorks)	(a) Uncompacted, 2.2 m deep subject to fluctuating seepage, 8 years old	28.5
	(b) Uncompacted, below water table, 2.5 m deep, 8 years old	34.5
	(c) Embankment BS compaction density	35.0
	(d) Stockpile, fresh material	36.0
3. Littleton spoil heap (Staffs)	(a) Buried spoil, re-excavated ⩾ 20 years old	27.0
	(b) 0.2–0.5 m deep, c. 20 years old	27.5
4. Oakdale spoil heap (S. Wales)	(a) Surface of tip, 6 months old	34.0
	(b) Fresh material from washery	33.0
5. Yorkshire main spoil heap	(a) 0.76–1.22 m deep, c. 12 years old	33.5
	(b) 3.81–4.27 m deep	34.0
	(c) 6.86–7.32 m deep	34.5
	(d) Fabricated samples from fresh parental roof & floor materials from underground	33.5–34.5

*ϕ'_e arbitrarily defined by McKechnie Thomson and Rodin[17] (useful for comparing shear strengths, given linear and curved Mohr envelopes).
†A small cohesion intercept may be realistic.
ϕ'_e slightly underestimates the shear strength of this material.

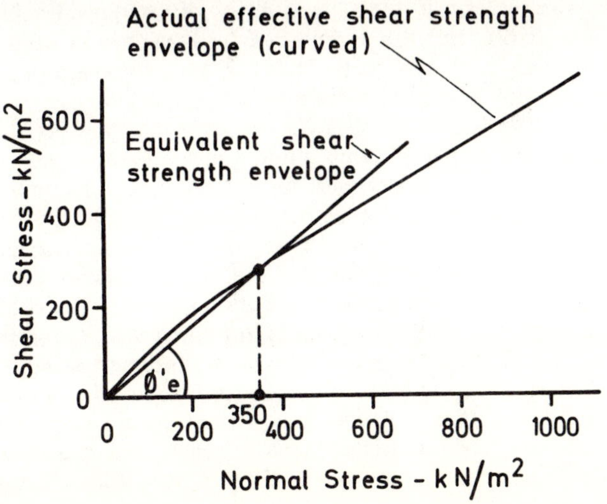

Kellingley, Oakdale and the 3 m depth Gedling sample). The near-surface weathering effects in the old Yorkshire Main and Littleton spoil heaps are minimal and it is only the modern surface material from Gedling and the Kellingley discard from just above the seepage level which show an appreciable fall-off in shear strength (21%). Physical degradation is quite appreciable in the seat-earth-rich spoil from Gedling (Figure 6.10) and the decrease in strength is almost certainly related to surface comminution during compaction. The highly weathered Kellingley material with the lowest shear strength was subject to fluctuating seepage and it is of interest that even uncompacted material from below the phreatic surface has a much higher ϕ'_e value.

Because of possible pollution or sulphate attack when cement or concrete is involved it may be wise to exclude spoil from the outer metre of a spoil heap for underfloor fill or embankment construction purposes. The upper parts of seepage horizons should also be treated with caution.

SPONTANEOUS IGNITION

In 1967 about 15% of the 2000 spoil heaps owned by the NCB alone were classified as burnt out and more than half as burning. Spontaneous heating in loosely tipped spoils is in essence an atmospheric oxidation (exothermic) process of coal. Most authorities now agree that the exothermic oxidation of pyrite is not a prime cause of self heating, although it will increase the chances of combustion occurring, especially when present in a finely divided form.

The permeability to air is one of the controlling factors as to whether oxidation will be sustained and the temperature continue to rise; other factors are the grain size of the discard and the specific surface area of the coaly materials present. There is a tendency in any case for coaly material to be concentrated in the smaller grain sizes. Intermediate grain sizes and voids enhance the heating process.

Many spoil heaps are in the 'partly burnt' category and this is important with respect to utilisation. The effects of heating can be dramatic in altering the chemical and mineralogical balance. A 10°C rise in temperature will involve a two or three-fold increase in the rate of chemical reactions. Kaolinite decomposes at around 550°C and this is followed by breakdown of the 10 Å illite/mixed-layer minerals at about 700°C. These decomposition temperatures could be much lower in spoil heaps over more protected time intervals.

The point about red or burnt shales therefore is that they are not all necessarily of the same mineralogical composition. Generally, the higher the temperature attained the more chemically inert the shale becomes. [Note that temperatures approaching 1000°C have been measured in burning tips (e.g. Brandon Tip, Co. Durham).] Burnt shales, which are commonly much stronger materials than their unburnt equivalents, have been used for construction purposes over the years and many of the best deposits have been used up. They conform to the Ministry of Transport's 1969 *Specification for Road and Bridge Works*[1] as a 'well-burnt non-plastic shale'. Dependent on them fulfilling grading requirements they may be used as a sub-base course in road pavements. Moreover, when stabilised with cement they may satisfy road base requirements under Clause 805 for 'soil cement', and also Clause 806 as a 'cement bound granular material' (see Sherwood[18] for details). Two drawbacks of burnt shales, however, are that

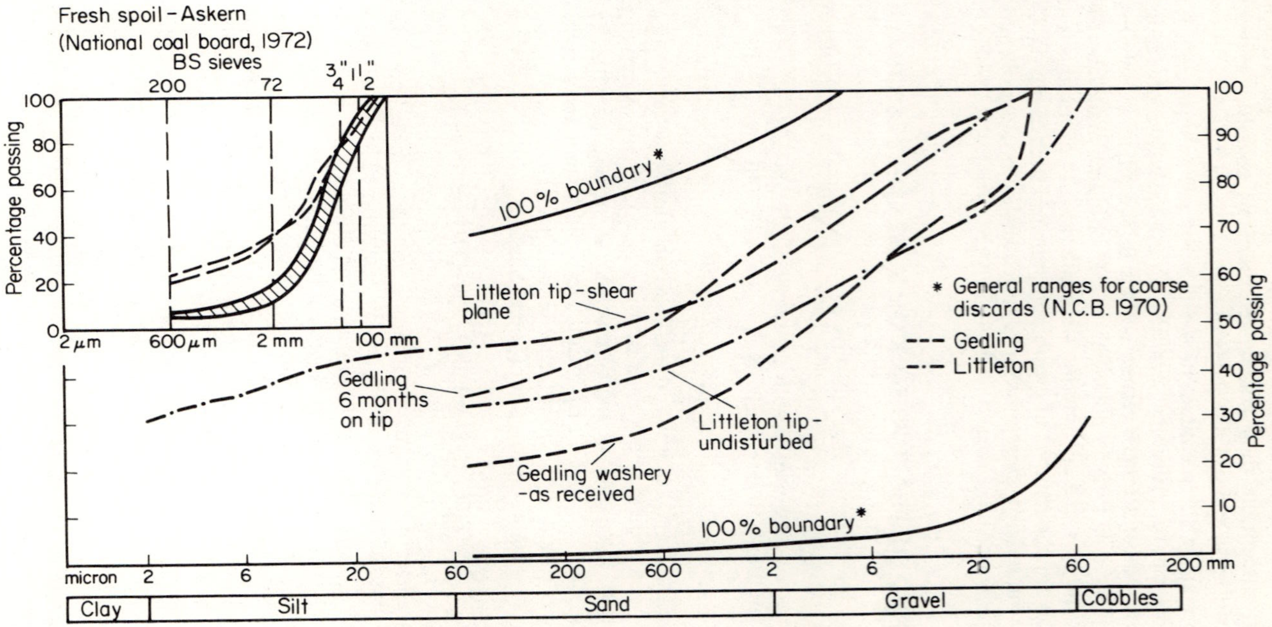

Figure 6.10 Particle distribution curves: (a) breakdown after 6 months exposure (Gedling Spoil heap), (b) *in situ* shear plane material compared with associated coarse discard (Littleton spoil heap), (c) mechanical breakdown after transportation and spreading (Askern: from NCB[6])

they are commonly frost susceptible and consequently may require up to 5% cement additive to render them suitable as sub-base materials; sulphate contents are very often higher than those of unburnt shales. Although some unburnt shales are frost susceptible their grading requirements usually conform to the requirements of Clause 805 of the Specification.

There is normally a sharp contact between burnt and unburnt shale in partly burnt heaps. This can be misleading because high sulphates, hydrated iron oxides (and often lower shear strength characteristics in the adjacent unburnt spoil) may considerably reduce their potential as construction materials. Chemical degradation of the clay and silt fraction under elevated temperatures is believed to be responsible for the removal of even stable minerals such as quartz. Unburnt or partly burnt material in close proximity to red shales should consequently be thoroughly investigated.

SULPHATES AND SULPHATE ATTACK ON CEMENTED MATERIALS

Secondary sulphate minerals in coarse discard can originate from two sources:

(a) As a component associated with the waste coal from underground.
(b) Formed within a waste heap or embankment fill, given the right conditions.

Gypsum ($CaSO_4.2H_2O$) may be found in association with carbonates in the cleat (joints) of coal. There is evidence in some unburnt heaps (even of average coal content) that this common sulphate species is introduced with the waste. The reaction of sulphuric acid, originating from pyrite oxidation, with carbonates in the waste and ions within clay minerals can produce innumerable soluble and partly soluble sulphates in old spoil heaps, e.g.

$$H_2SO_4 + CaCO_3 + 2H_2O \rightarrow CaSO_4.2H_2O \text{ (Gypsum)} + H_2CO_3$$

Calcium, magnesium, potassium, sodium and at a lower concentration, aluminium, are the sulphates most frequently encountered in water extractions. The position of jarosite which is found as a yellow-coloured veneer on shale particles is of interest. X-ray work suggests that it is usually the potassium variety ($KFe_3(SO_4)(OH)_6$) that is formed when sulphuric acid reacts with potassium from the micaceous minerals. Strictly speaking jarosite of good crystallinity is insoluble in water but it appears to be very mobile in spoil heaps. It could be that in a cryptocrystalline state it is soluble, particularly under acidic conditions in the micro-environment of shale particle surfaces. Shamburger *et al.*[19] also refer to it as a soluble mineral in their work on compacted shale embankments.

Although sparingly soluble (1.2 g SO_3/litre), calcium sulphate reacts with the tricalcium aluminate in Ordinary Portland Cement to produce ettringite, which, because of its greater volume causes the cement to break down. Whilst potassium and sodium sulphates can penetrate deeply into concrete to form further calcium sulphate from calcium hydroxide, it is magnesium sulphate which is particularly aggressive since it can react with all the four main constituents of cement. It should also be mentioned that sulphuric acid from pyrite oxidation can itself attack the surface of concrete, dissolving out of the cement gel and paste.

Sulphate attack of concrete floors of buildings laid on compacted burnt shale has resulted in serious problems[20,21]; the expansion and cracking of cement stabilized burnt shale due to sulphates has also been investigated[22].

RECOMMENDED SULPHATE LEVELS

Acceptable sulphate levels are to a large extent determined by:

(a) The proximity of the spoil materials to concrete or other cemented materials, and
(b) the solubility of the sulphate minerals in the spoil.

So long as the material is used as bulk fill that is not associated with cemented materials no restrictions on sulphate content need to be applied. Attention should be paid, however, to groundwater pollution effects due to acidity and contained sulphates.

The TRRL maximim limit placed on colliery shale used within 0.5 m of bridge abutments or roads containing cemented horizons is 2 g/litre (as SO_3), based on the 1:1 shale-water extract method (BS 1377: 1975[23]). Recommendations are given in BRS Digest 174[24] with respect to precautions which can be taken against elevated sulphate levels. With conventional house foundations these can be protected by adhesive plastics when pH values are somewhat lower than 6 (see Chapter 18).

Cement stabilisation of coarse discard raises the question of the continuous leaching of sparingly soluble sulphates; total sulphate contents must therefore be determined in this case. For stabilised burnt shales the Ministry of Transport's Specification[1] states that acid-soluble sulphates should not exceed 1% (Clause 805). Recent work shows that unburnt shales can also be successfully stabilised and can therefore probably meet the Specification as a soil cement road base material. Total sulphate limits have not yet been established and although they may be lower for unburnt shales than for burnt shales further investigations are necessary in this area. However, it has already been mentioned that materials from *unburnt* heaps (and disposal points) usually have substantially lower sulphate contents than burnt materials.

Pertinent to compacted burnt shale when used as an under-floor fill for houses, the Building Research Establishment recommends that the total sulphate content should not exceed 0.5%. Furthermore, a polythene or building paper membrane placed on top of the shale hardcore and linked to the damp proof course is also advocated.

MECHANICAL PROPERTIES

The engineering properties of coal mine discards have been investigated in detail during the last ten years (McKechnie Thomson and Rodin[17]; National Coal Board[6]). Shales have been used successfully for many decades as a compacted fill for the construction of dams and embankments, both in this country and abroad. It is clear from the earlier discussion that the only major difference between conventional shale fill and coarse discard is the coal component which is invariably present in the latter. Because of the shallower depths involved (i.e. reduced stress relief), disintegration due to sedimentary structures tends to be less for borrow pit materials than for fresh discards extracted from some hundreds of metres underground.

Over 7.5 million tonnes of unburnt coarse discard have been utilised in increasing amounts over the last few years for major roadworks and experience has enabled the 'susceptibility to combustion' restriction on its use to be withdrawn. Compaction in accordance with Clause 609 of the Ministry of Transport's Specification[1] relating to earthworks has shown that previous fears about spontaneous heating were based on the behaviour of loose tipped materials.

An embankment compacted to roadworks standards using spoil susceptible to combustion was constructed at Cortonwood, Yorkshire and instrumented with 180 thermocouples. Samples of spoil had a calorific value in the range 5699 to 8723 kJ/kg which compares with about 32564 kJ/kg for coal. Over a period of five years the only changes in temperature were ones attributable to changes in ambient temperature and a trench cut across the top of the embankment did not produce any increase in reaction; no indications of overheating were detected. The first major use of coarse discard for an Expressway embankment in the United States was similarly instrumented. Butler[25] reports that over the first six months the temperatures measured were normal and well within the predicted range.

Compaction requirements for colliery shale when used as a fill material are given in Table 6.6, on the basis that these materials will normally fall within the 'well-graded granular and dry cohesive soils' definition of Clause 601[1]. A number of field tests carried out on coarse coal mine discard in the USA have shown that good compaction is achieved with heavy vibratory compactors.

Reduction of air voids not only precludes spontaneous heating but generally increases the shear strength of the material. It also reduces seepage and hence possible pollution. Variation in particle strength, however, will greatly affect the densities achieved. Moreover, density is a function of specific gravity and the content of coaly materials must be considered when comparing compaction densities. The compaction test results for very coaly (low specific gravity) discards from Pennsylvania shown in Figure 6.11 emphasise these points. The improvement in compaction characteristics is demonstrated by the three cycle repetitive tests involving continual breakdown of the anthracitic spoil. The increase in density, which is of lower magnitude, was more readily achieved with less severe breakdown in the case of the bituminous spoils (2-cycle tests).

Bishop[26] drew attention to the sensitivity of loose spoil to placement moisture content in terms of density and subsequent volume decrease on saturation. Recent work at Durham University suggests that materials with a significant fines content are more sensitive than stronger materials with fewer fines. This may be due to a metastable structure being maintained by the cohesion of the fines, until the placement moisture content attains an 'optimum' at which a minimum placement density is achieved.

It is important from both a compaction point of view and the subsequent moisture distribution to compact the materials as near to optimum conditions as is possible. The moisture content histograms published by the NCB[27], imply that

Table 6.6 COMPACTION REQUIREMENTS OF COLLIERY SHALE WHEN USED AS A FILL MATERIAL (After Sherwood[18])

Type of compaction plant	*Category*	*Max depth of compacted layer* (mm)	*Minimum no of passes*
Smooth-wheeled roller	*Mass per metre width of roll:*		
	More than 2100 kg but not more than 2700 kg	125	10
	More than 2700 kg but not more than 5400 kg	125	8
	More than 5400 kg	150	8
Grid roller	*Mass per metre width of roll:*		
	More than 2700 kg but not more than 5400 kg	Unsuitable	
	More than 5400 kg but not more than 8000 kg	125	12
	More than 8000 kg	150	12
Tamping roller	*Mass per metre width of roll:*		
	More than 4000 kg	150	12
Pneumatic-tyred roller	*Mass per wheel:*		
	More than 1000 kg but not more than 1500 kg	Unsuitable	
	More than 1500 kg but not more than 2000 kg	Unsuitable	
	More than 2000 kg but not more than 2500 kg	125	12
	More than 2500 kg but not more than 4000 kg	125	10
	More than 4000 kg but not more than 6000 kg	125	10
	More than 6000 kg but not more than 8000 kg	150	8
	More than 8000 kg but not more than 12000 kg	150	8
	More than 12000 kg	175	6
Vibrating roller	*Mass per metre width of vibrating roll:*		
	More than 270 kg but not more than 450 kg	75	16
	More than 450 kg but not more than 710 kg	75	12
	More than 710 kg but not more than 1300 kg	125	12
	More than 1300 kg but not more than 1800 kg	150	8
	More than 1800 kg but not more than 2300 kg	150	4
	More than 2300 kg but not more than 2900 kg	175	4
	More than 2900 kg but not more than 3600 kg	200	4
	More than 3600 kg but not more than 4300 kg	225	4
	More than 4300 kg but not more than 5000 kg	250	4
	More than 5000 kg	275	4
Vibrating-plate compactor	*Mass per unit area of base plate:*		
	More than 880 kg/m² but not more than 1100 kg/m²	Unsuitable	
	More than 1100 kg/m² but not more than 1200 kg/m²	75	10
	More than 1200 kg/m² but not more than 1400 kg/m²	75	6
	More than 1400 kg/m² but not more than 1800 kg/m²	125	6
	More than 1800 kg/m² but not more than 2100 kg/m²	150	5
	More than 2100 kg/m²	200	5
Vibro-tamper	*Mass:*		
	More than 50 kg but not more than 65 kg	100	3
	More than 65 kg but not more than 75 kg	125	3
	More than 75 kg	150	3
Power rammer	*Mass:*		
	100 kg	150	6
	More than 500 kg	275	12
Dropping-weight compactor	Mass of rammer more than 500 kg		
	Height of drop:		
	More than 1 m but not more than 2 m	600	8
	More than 2 m	600	4

materials in colliery spoil heaps are commonly close to optimum moisture content (BS 2.5 kg rammer).

The shear strength range of spoil heap materials ($\phi' = 25.5°$ to just over 40° for all coalfields) does not reflect the rank of the coals concerned. Using a linear regression K_f-line fit to grouped test data for 147 specimens from Welsh and English spoil heaps a variance ratio test (F-test) has shown that there is a statistical difference between the West Wales (anthracitic) spoils and the English

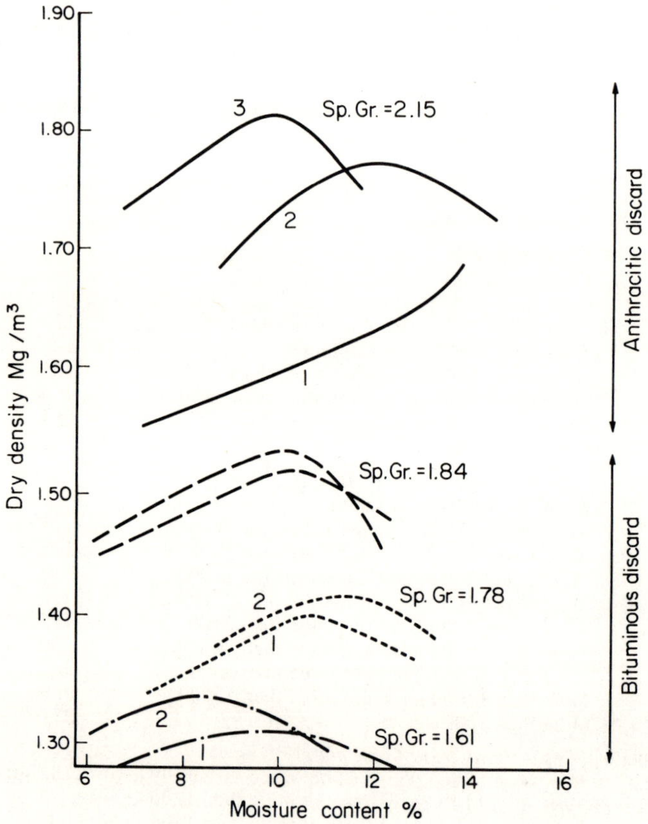

Figure 6.11 Compaction curves for coaly spoils (bituminous and anthracitic) from Pennsylvania (Data from Butler[25])

materials. In practical terms, however, this difference is very small – $\phi' = 36.5°$ (West Wales), 34.3° (England). The absence of a *significant* rank relationship is believed to be due to two major factors:

(a) over-printing effects of rock types of differing inherent strength;
(b) the effect of carbonaceous (coaly) materials.

Recent work[28] has shown that organic carbon (coal) content has a significant effect on shear strength, such that peak strength increases with increasing coal content. In this respect coal is enigmatic since it is responsible for many of the problems relating to coal mine discards (e.g. spontaneous heating, weathering and possible pollution via contained pyrite), whilst at the same time it enhances the effective angle of shearing resistance, ϕ'.

Drained triaxial strengths determined on 100 mm diameter specimens obtained from standard 'undisturbed' driven samples have been found to accord reasonably well with large-scale *in situ* double-box shear results[17]. Similarly, 0.3 m diameter samples have also shown that standard U100 samples are reliable for shear strength determinations[26]. Because the drained shear strength is not significantly different for loose or dense spoil samples the large shear-box (0.3 m by 0.3 m) has also proved to be a versatile testing machine in the normal effective pressure range 0 to 500 kN/m^2. Large shear-boxes with reversing facilities can be used to advantage since the post-peak shear strength behaviour can be investigated with a single sample using the step-loading principle.

With very few exceptions the effective peak shear strength envelope passes through the zero intercept so that cohesion, $c' = 0$. In other words, under drained conditions discards behave essentially as granular materials. Envelopes tend to be linear in the case of the stronger burnt and unburnt shales, but with more plastic materials curvature can be very marked indeed, because of particle breakdown as the normal pressure increases. This feature can be important in slope stability analyses of embankments and a number of curve fitting methods to triaxial test data are in use.

A general equation of the following form is commonly a satisfactory approximation:

$$\tau = m(\sigma')^z$$

where τ = shear stress,
σ' = effective stress normal to the plane of failure
m and z are constants,

i.e. $\log \tau = z \log \sigma' + \log m$

A simple technique is to determine failure stress points on the individual Mohr's circles and then to fit the shear strength envelope using a reduced major axis regression line to a log-log transform of the points (see Figure 6.12). Slope stability programs can then be modified to accommodate the relationship.

Particle breakdown and mineral re-orientation during the shearing process can lead to quite dramatic decreases in shear strength. Test results obtained by Ratsey[30] which are given in Table 6.7 show that for total displacements of only about 2.5 m (step loading) the shear strengths of certain low rank spoils (Askern, Birch Coppice) show a drop of 37% and 56% respectively. The shear strength determined on a failure surface exposed within the Littleton spoil heap was only 20.5°.

The particle size distribution of the Littleton shear plane is compared with undisturbed spoil from this spoil heap in Figure 6.10. A comparison between the field shear plane with one obtained in a laboratory test is also illustrated in Figure 6.13.

Inspection of residual shear strength data shows that very large shear displacements indeed may be required for residual strengths to be achieved in the low normal pressure range, e.g. Tip 2, Aberfan – over 1/6 mile at 100 kN/m^2

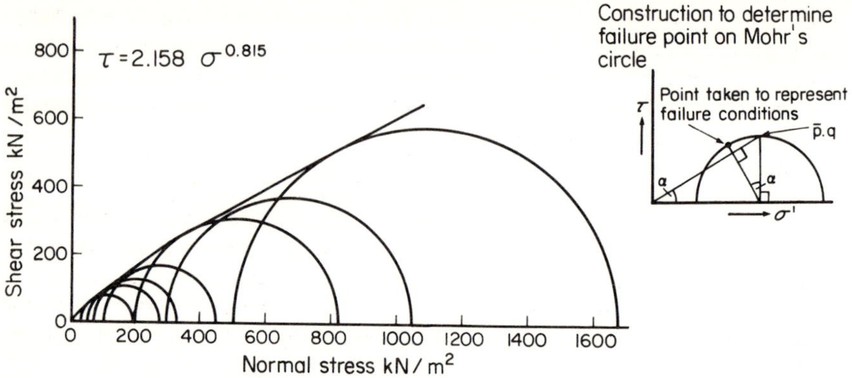

Figure 6.12 Taken from computer print-out of curved Mohr envelope fitted to test data obtained by McWilliam[29]. Inset shows construction for determining failure points used in this curve fitting technique. Logarithmic transformation of points facilitates statistical fitting of curved envelope

Figure 6.13 Showing section of shear plane exposed in Littleton tip in 1970 and similar material which has attained residual shear strength in 0.3 m × 0.3 m reversing shear-box. Shear plane specimen 0.23 m long

(Bishop[26]). At higher normal pressures (e.g. 400 to 1000 kN/m^2) the residual shear strengths are in keeping with expected values for the mineral components of the shale fragments[6]. This point is also demonstrated in Table 6.7 where the Bersham residual shear strength was found to be 13.7° on *fully disaggregated* material (c.f. 29.5° to 32°) at low normal pressures.

Table 6.7 SHEAR-BOX TESTS ON RANKED WASHERY DISCARDS (RATSEY[30])

(a) 0.3 × 0.3 m shear-box (–38 mm material)

Rank	*Colliery Washery*	*Area*	*Peak* ϕ' *degrees*	*Normal stress range* (kN/m^2)	*Disp-lacement* m	*Apparent residual* ϕ' *degrees*
101	Cynheidre	West Wales	37	[illegible]0.3–240.0	2.1	28
301	Morrison Busty	North Durham	42	8[illegible].3–267.1	1.35	31
502	Bersham	North Western (N. Wales)	40	80.[illegible]–347.1	1.9	29.5*
600	Easton	Scottish South	39	80.3–240.3	1.65	35
700	Askern	Doncaster, Yorks	35	53.6–320.4	2.6	22
802	Ollerton	N. Nottinghamshire	37	80.3–320.4	2.5	26
902	Birch Coppice	S. Midlands	35	53.6–320.4	2.5	15.5

(b) 60 × 60 mm shear-box (material 0.6–1.2 mm)

Rank	*Colliery washery*	*Normal stress range* kN/m^2	*Displacement* mm	*Apparent residual* ϕ' *degrees*
101	Cynheidre	61.3–200.1	820	28.5
310	Morrison Busty	18.2–188.7	540	36
502	Bersham	35.6–160.5	780	32*
600	Easton	87.8–133.7	600	35
700	Askern	110.1–302.5	820	17.5
802	Ollerton	17.0–285.4	760	20
902	Birch Coppice	133.7–285.4	720	15

*ϕ'_r = 13.7° on fully disaggregated material

With weak materials in particular, care should be exercised during compaction. Excessive crushing can produce a skin of highly degraded material of low shear strength which must then be scarified. Unless broken up such material could be incorporated as a potential shear plane, which, with the onset of failure, may be subject to a large reduction in shear strength.

HEAVE PHENOMENA

Reports of heave phenomena (other than direct sulphate attack on concrete) have been reported in the press and in the technical literature (e.g. Building Failure Sheet 14[21]). As opposed to the sulphate attack situation, swelling of compacted shale used for underfloor infill has been recognised by the fact that the fill is still in contact with the lifted floor slab. Factors influencing the heave of colliery discards are currently being investigated by the National Coal Board's Minestone Executive and Coal Research Establishment, together with a number of research workers in British Universities. In attempting to establish possible mechanisms it is germane to consider both mineralogical and physical processes in what is obviously an important research area.

Winmill[31] recorded the disintegration of coal as a consequence of volume increase of contained pyrite during oxidation. Considerable heave of reinforced floor slabs has been investigated in Canada in recent years. Biogenic oxidation of finely disseminated pyrite, with the consequent formation of gypsum and jarosite is believed to be responsible, under warm, relatively dry conditions with a plentiful air supply. The transformation of carbonates to sulphates can theoretically involve a considerable volume increase (e.g. calcite to gypsum, around 99%). There is a possibility that pressures up to 500 kN/m^2 could obtain in such processes.

This type of response has taken place in *in situ* shales and the larger available void spaces and air voids (5 to 10%) in compacted colliery discards might reduce its importance in the present context. Nevertheless, gypsum formation within bedding laminae of shale fragments on spoil tips cannot be ignored. The sealing of fresh surfaces or submergence of formation rocks in water have been used to preclude oxidation in the cases reported.

Physical swelling of clay minerals, and more particularly capillary effects, are a major cause for concern. Due regard should be paid to the moisture content on compaction, and desiccation of compacted layers should be prevented. Desiccation of large size materials from washery bunkers can be expected, particularly when run-of-mine materials, which have not been through a washery are included. The tests conducted on the Balderhead Dam shale fill[32] are of relevance to this problem. It was found that mudstone stored under water did not develop polygonal discontinuities recognised in desiccated specimens.

Unless desiccation and concomitant suction pressures are controlled, water will be entrained into the cores of the larger inclusions from both the fill matrix and adjacent soils and rocks. A longer-term heave is possible because hydration and equilibrium are governed by the greatly differing permeabilities of rock fragments and matrix. The consequent distribution of water under small pressure gradients may involve a considerable time element. The possibility that domestic heating and industrial processes could cause an increase in temperature in underfloor shale infill should be considered at the design stage. This may be important both in terms of sulphate production and induced desiccation.

Gibson[33] in referring to the expansion of shales in cement stabilization work advocates that the required water should be added to accommodate expansion before the cement is added.

Empirical tests are being carried out by the National Coal Board in their own work as a *guide* to the likely behaviour of particular compacted discards. Duplicate samples are compacted to the appropriate level at the 'as received' moisture content and after air drying. The specimens are then allowed to take up water under free swell conditions. The test is the repeated on two further samples under surcharge. A small surcharge of around 27 kN/m^2 generally results in the compression of the initially wetter sample of unburnt discard, whilst the heave of air dried material on flooding is reduced dramatically.

From a utilisation aspect it is important that research is continued into the causes of heave and that case histories are investigated in detail when they arise.

GENERAL

The skin which can form on areas of old spoil can give a false sense of security. Similarly, the fact that a 90% BS compaction density is often achieved naturally in old, loose tips should not be regarded as an indication that such areas are suitable for construction without the appropriate measures being taken. The likely presence of large voids and all types of mine rubbish means that large settlements should be anticipated, particularly if water is subsequently entrained into the material. In other words, from a foundation design point of view loose tipped mining wastes should not be treated as anything other than filled ground. Foundations should consequently be designed accordingly. As far as the writer is aware dynamic compaction has currently been used very infrequently as a ground improvement method in these circumstances. The results obtained at the Methil Colliery site[34] suggest that discards respond very favourably to the technique.

With respect to the choice between fresh discard from colliery bunkers or spoil heap materials as a source of fill, it is the author's opinion that on balance spoil heaps do have advantages – a definable quantity of material, a chance to establish the best and worst materials, generally a better grading, and the probability of a moisture content which is close to optimum. Sections of old spoil heaps may also contain sound, burnt shale which is now at a premium. Unburnt spoil heap materials have in fact been through a slaking and weathering test which cannot be surpassed in the laboratory. Like all degradable materials, however, good quality control measures, together with a laboratory evaluation of composition and engineering properties are a necessary background for the assessment of their performance in the field.

ACKNOWLEDGEMENTS

Many of the points discussed are a consequence of research funded by the National Coal Board during the past nine years. The writer would like to express his appreciation, and would qualify any views made in the text as being those of the author and not necessarily those of the Board. Much of the work carried out by past research students at Durham University is referenced accordingly. Other past and present research students whose help is gratefully acknowledged are: Messrs. A.E. Cobb, K.D. Philpott, G.B. Simpson and R.G. Hardy (Senior Experimental Officer). Warm thanks for advice and information are also due to: Messrs. J.T. Finey (Wimpey Laboratories Ltd)., D.A. Tanfield (N.C.B. Minestone Executive) and R.R. Backer (U.S. Bureau of Mines).

References

1. *Specification for road and bridge works*, Ministry of Transport, H.M.S.O., London, 1955 (1969)
2. Gutt, W., Nixon, P.J., Smith, M.A., Harrison, W.W. and Russell, A.D., 'A survey of the locations, disposal and protective uses of the major industrial by-products and waste materials,' *Building Research Establishment Current Paper CP/74*, Building Research Establishment, Garston (1974)

3. Sherwood, P.T., 'The use of waste and low-grade materials in road construction: 1 Guide to Materials Available', *TRRL Laboratory Report 647*, Transport and Road Research Laboratory, Crowthorne, Berks (1974)
4. Royal Commission on Environmental Pollution (4th Report), *Pollution Control Progress and Problems*, Command No. 5786, HMSO, London (1974)
5. Holubec, I., 'Geotechnical aspects of coal waste embankments,' *Can. Geotech. J.*, **13**, No. 1, 127–39 (1976)
6. National Coal Board, *Review of research on properties of spoil tip materials*, National Coal Board, London, 100 (1972)
7. Taylor, R.K. and Spears, D.A., 'The breakdown of British coal measure rocks', *Int. J. Rock Mech. Min. Sci.*, 7, 481–501 (1970)
8. Taylor, R.K., 'English and Welsh spoil heaps – mineralogical and mechanical interrelationships', *Engineering Geology*, **9**, 39–52 (1975)
9. Collins, R.J., 'A method for measuring the mineralogical variation of spoils from British collieries', *Clay Minerals*, **11**, 31–50 (1976)
10. Berkovitch, I., Manackerman, M. and Potter, N.M., 'The shale breakdown problem in coal washing. Part I: Assessing the breakdown of shales in water', *J. Inst. Fuel*, **32**, 579–589 (1959)
11. Van Eeckhout, E.M., 'The mechanism of strength reduction due to moisture in coal mine shales', *Inst. J. Rock Mech. Min. Sci. & Geomech. Abstra.*, **13**, 61–67 (1976)
12. Carroll, D. and Starkey, H.C., 'Effect of sea-water on clay minerals,' *7th National Conference on Clays and Clay Minerals*, Pergamon Press, London 80–101 (1958)
13. Badger, C.W., Cummings, A.D., and Whitmore, R.L., 'The disintegration of shales in water,' *J. Inst. Fuel*, **29**, 417–423 (1956)
14. Quigley, R.M., Zajic, J.E., McKyes, E. and Young, R.N., 'Biochemical alteration and heave of black shale; Detailed observations and interpretations,' *Can. J. Earth Sci.*, **10**, 1005–1015 (1973)
15. Penner, E., Eden, W.J. and Gillott, J.E., 'Floor heave due to biochemical weathering of shale', *Proc. 8th Internat. Conf. on Soil Mechanics and Foundation Engineering*, Moscow, **2 (2)**, Session 4, 151–158 (1973)
16. Spears, D.A., Taylor, R.K. and Till, R., 'A mineralogical investigation of a spoil heap at Yorkshire Main Colliery', *Q. Jl Engng Geol.*, **3**, No. 4, 239–252 (1971)
17. McKechnie Thomson, G. and Rodin, S., 'Colliery spoil tips – after Aberfan', *Proc. Instn C.E.*, Paper 7522, 59 (1972)
18. Sherwood, P.T., 'The use of waste and low-grade materials in road construction: 2 Colliery shale', *T.R.R.L. Laboratory Report 649*, Transport and Road Research Laboratory, Crowthorne, Berks (1975)
19. Shamburger, J.H., Patrick, D.M. and Lutton, R.J., 'Design and construction of compacted shale embankments' Vol. 1, Survey of problem areas and current practices', *Report No. FHWA-RD-75-61*, US Army Engineer Waterways Experiment Station, Vicksburg (1975)
20. Eldridge, H.J., 'Concrete floors on hardcore', *The Municipal Journal*, 2893–2894, 2896 (11 Sept. 1964)
21. Building Failure Sheet 14, 'Solid floor on a shale fill', *Building*, 102 (14 April, 1972)
22. Sherwood, P.T. and Ryley, M.D., 'The effect of sulphates in colliery shale on its use for roadmaking', *R.R.L. Report LR 324*, Road Research Laboratory, Crowthorne, Berks (1970)
23. British Standards Institution, *BS 1377:1975. Methods of testing soils for civil engineering purposes*, British Standard Institution, London (1975)
24. Building Research Establishment, *Concrete in sulphate-bearing soils and groundwaters*, Building Research Establishment Digest 174, HMSO (1975)
25. Butler, P.E., 'Utilisation of coal mine refuse in the construction of highway embankments', *Proc. 1st Symposium on Mine and Preparation Plant Disposal*, National Coal Association, Louisville, Kentucky, 238–255 (1974)
26. Bishop, A.W., 'The stability of tips and soil heaps', *Q. Jl. Engng Geol.*, 6, No. 3 & 4, 355–376 (1973)
27. National Coal Board Technical Handbook, *Spoil heaps and lagoons*, 2nd Draft, National Coal Board, London, 100 (1970)
28. Taylor, R.K., 'Influence of coal on the peak shear strength of colliery shales', *Geotechnique*, 24, No. 4, 683–688 (1974)

29. McWilliam, D.J., 'Compositional and shear strength characteristics of a spoil heap at Littleton Colliery, Staffordshire, M.Sc. Thesis, University of Durham (1975)
30. Ratsey, J., Shear strength characteristics of certain colliery discards with respect to rank, M.Sc. Thesis, University of Durham (1973)
31. Winmill, T.F., 'Atmospheric oxidation of iron pyrites', Pt IV, *Trans. Instn Min. Engrs.*, 51, 500 (1915–1916)
32. Kennard, M.F., Knill, J.L. and Vaughan, P.R., 'The geotechnical properties and behaviour of carboniferous shale at Balderhead Dam', *Q. Jl. Engng Geol.*, *1*, No. 1, 3–24 (1967)
33. Gibson, J., 'Useful materials from colliery spoil', *Mine and Quarry*, 24, 29–30, 33 (August 1975)
34. West, J.M. and Slocombe, B.C., 'Dynamic consolidation as an alternative foundation', *Ground Engineering*, *6*, No. 6, 52–54 (1973)

Chapter 7

Ground Water and Foundations

GENERAL PRINCIPLES

The strength of concrete may partly be attributed to the bond afforded by the cement between the particles of sand and gravel. Without cement, however, sand and gravel may form a strong material when there is some force to hold the grains together. A rubber football bladder filled with sand becomes very rigid under only atmospheric pressure when it is connected to a vacuum pump. If it is put in a pressure vessel so that a greater confining pressure can be applied, it can be made as strong as concrete. If a very large pressure can be applied, the limit to the strength of the sand mass is controlled by the strength of the grains themselves and if they are quartz this can be high. Before the ultimate strength is reached, the grains will begin to crush at their points of contact.

The seriousness of this problem depends on the size, the distribution of sizes (the grading) and the shape of the grains as well as their strength. Under a given confining pressure the force transmitted through each grain increases in proportion to the square of its diameter. Thus a rockfill, composed only of quarried gneiss with particles of at least 200 mm diameter, will suffer extensive crushing under a height of embankment that would cause no distress to a sand with particles of 1 mm diameter, each one of which would support a load only 1/40 000th of that carried by the rockfill particles.

It is for this reason that rockfill must be correctly graded so that the larger pieces are supported by sufficient fines to avoid high contact stresses. Bricks cannot be used to build a high wall without mortar because they would break and crush. The mortar provides a bed for each brick, the particles of sand, silt, lime and cement distribute the contact force safely over the whole face of the brick. The cement in concrete, as well as forming a bond, also provides a bed for the particles of sand and gravel so avoiding contact crushing. Material made up of smaller particles, such as silt and clay, suffers no contact crushing because the forces supported by each grain are small compared with its strength.

A confining pressure applied to an assemblage of particles such as sand, silt or clay causes a reduction of volume through deformation and rearrangement of the particles. It is mostly a one-way process, the soil does not spring back very much on release of pressure. When shearing stresses are applied to the soil they may cause it to expand or contract, depending on its initial state. In general, sands and gravels expand (or dilate) unless they are very loose, and wind blown silts (loess) and many clays suffer a reduction of volume on shearing.

What happens when the spaces between the particles (the pore space) are full of water? If the sand in our football bladder is saturated, an applied confining pressure merely tries to compress the water (and the surrounded grains) without increasing the force holding the grains together, so there is no increase of strength. If some shearing stresses are also applied and if the sand is dense enough, there may be some dilation, which will be opposed by the confining pressure so that some strength develops.

However, if loose silt was used instead of sand, any shear stresses would induce collapse and any strength would be lost, producing a material with the properties of a liquid. Some materials in tailings lagoons and some natural soils such as the Canadian and Norwegian quickclays exhibit these properties and can develop serious flow slides when disturbed.

EFFECTIVE STRESS THEORY

All soils have a strength proportional to the stress holding the particles together. This stress affects both the strength and the compression of the soil and it is equal to the confining pressure minus the pressure in the pore space. Because the pressure in the pore space was not effective in altering the soil but merely opposed the applied pressure, Terzaghi called it a neutral pressure when he formulated the effective stress theory, but it is now referred to as the pore pressure and the well-known theory is expressed as:

$$\tau = c' + (\sigma - u) \tan \phi'$$

where

τ, the shear strength of the soil;
σ, the confining pressure;
u, the pore pressure;
ϕ', the angle of shearing resistance, and
c', the effective cohesion of the soil.

Values of the soil parameters c' and ϕ', which can be assessed from a variety of laboratory and *in-situ* tests, are affected by soil type, geological history, existing stress ratio, density and many other factors. All soils, however, exhibit some strength, which in general can be improved by minimising pore pressures. Most of the problems encountered in foundation engineering practice are rooted in undesirable pore pressure and uncontrolled pore water.

Effective stress, and hence the strength, in a soil can be increased by increasing the confining pressure (e.g. adding a surcharge) and giving time and drainage paths for the excess pore water to escape. In a saturated soil, the volume of water that has to escape is equal to the settlement that will be caused by the additional load. The rate of drainage is proportional to the square of the length of the drainage path and inversely proportional to the permeability of the soil. While sands and gravel, with permeabilities of about 0.1 mm/sec may allow complete drainage during the time of construction, clays with permeabilities of 1×10^{-7} mm/sec will require a million times as long to drain, so that slow consolidation will continue for a long time after construction is complete.

Clays are usually much more compressible than sands so that a greater volume of water has to escape from them for a given increase in confining pressure.

Settlement can be caused by lowering the water level in a saturated soil because the resulting reduction of pore pressure increases the effective stress in the soil, causing compression. In the special case of a very loose soil, or more particularly a loose backfill, a rise of water level in the soil can also cause settlement. This is because the strength of the soil at the points of contact is reduced on wetting, allowing the soil skeleton to collapse.

The following six case histories illustrate these general principles.

PAD FOUNDATIONS ON GOOD GROUND

A multi-storey steel frame building was required on a site where there was a layer of sand and gravel over chalk. There were many other successful buildings in the area and the site investigation confirmed that the sand and gravel was fairly dense and 5 m thick. Below it, the chalk surface was weathered for 1 or 2 m before it became hard and fissured.

Each column would impose a load of 4000 kN on the foundation and this was spread through a reinforced concrete pad, 4 m × 4 m and 1 m thick. The water table was 2 m below the surface and presented no problem when the pad footings were constructed. At one end of the building an underpass had to be built which would require a trench 6 m deep and about 10 m wide, as shown in Figure 7.1. Two lines of steel sheet piles were driven to form the sides of the

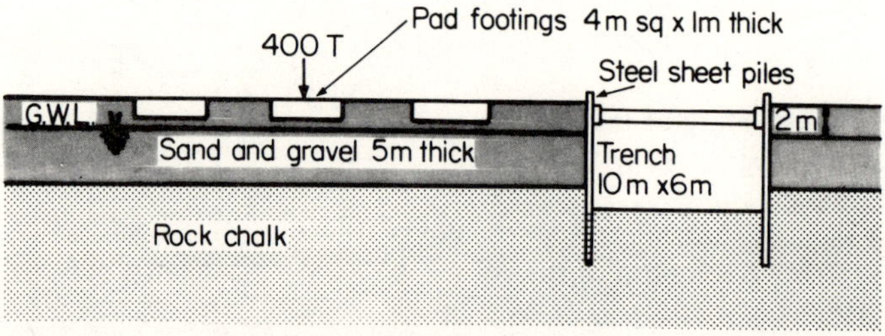

Figure 7.1 Section of foundations for the steel framed building, showing trench for underpass

trench, taken about 3 m below the design bottom for the trench, i.e. about 4 m into chalk. Excavation of the gravel soon reached the water table and a sump was formed for a pump to keep the excavation dry. The water flow increased as the excavation and the level of water in it, went down.

When chalk was reached, the excavated surface could not be kept dry, and chalk was soon worked up into a slurry. As formation level was approached, it was clear that it was going to be impossible to clean out this layer of slurry so that concrete could be placed on firm chalk. In a very short time, word got round that the sand and gravel was underlain by liquid chalk everywhere and that pad foundations would prove useless; piles should have been used. It was soon decided to make extra holes to investigate this chalk and carry out standard penetrations tests in it. Cased holes were driven through the gravel and pumped nearly dry before the SPTs were made. Near the chalk surface they gave very low values, almost confirming the liquid chalk suggestion, and the design for a piled foundation had advanced enough to show that it was going to be very expensive before further opinions were sought and it was realised that the whole situation had been caused by the upward flow of water through this chalk which had maintained the pore pressure sufficiently to make the effective stress zero when the restraining weight of the gravel layer had been removed. This caused the weathered, paste-like chalk to lose all strength and form boils of a type shown in Figure 7.2, round each outlet of the pore water. A similar thing occurred in the new boreholes, where the strength of chalk had been greatly reduced by the pressure of water rising into the pumped-out holes while the SPT's were being made.

Figure 7.2 Boils formed by pore water escaping into base of excavation

The solution to the problem was to keep the trench excavation completely full of water by pumping as the final excavation was made, then place the concrete underwater by tremi-pipe. To demonstrate that the pad foundations would be satisfactory, one of them was test loaded, as indicated in Figure 7.3. A cased hole was drilled through the centre of the pad and extended into the chalk for a settlement gauge. This consisted of a central rod attached to a pointed shoe and surrounded by a steel tube.

The shoe was driven down into the unweathered chalk, but the tube was only driven a short distance into the weathered chalk, so that the rod would indicate settlements in relation to the rock chalk and the tube would show the settlements caused by the softer, weathered chalk. Settlement points were attached to the corners of the pad and it was then loaded with kentledge which would produce a downward force of 4000 kN under the action of gravity. Observations were made by precise surveyors level and special staff, giving an accuracy of ± 0.1 mm. Under the full load, the pad only settled 3 mm and the base of the gravel (as indicated by the tube) 1 mm.

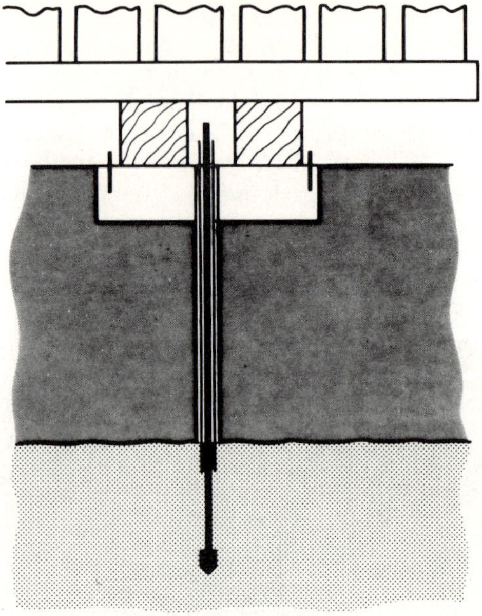

Figure 7.3 Section of test load on pad foundation showing settlement gauge

This case history illustrates a difficulty that can arise when the behaviour of water pressure in reducing the strength of the soil has not been clearly understood.

OIL TANKS ON POOR GROUND

When a soil is too weak to support the load from a proposed foundation, its strength may be increased by consolidating it under the weight of a temporary surcharge. The time that the surcharge has to be left in place depends on the compressibility and permeability of the soil and the lengths of the drainage paths. These may be reduced by the use of sand drains or special wicks inserted into the ground to tap any natural sand or silt layers that may separate the less permeable soil.

Oil tanks are fairly flexible structures and it is sometimes possible to use the weight of a tank and its test load to consolidate and so strengthen the soil beneath so that it becomes strong enough to support the weight of the full tank by the time it is full. This technique requires some skill to avoid failure or damaging distortions of the tank. If the tank is filled too rapidly, a rotational slip may be induced under the edge of the tank because the soil has not become strong enough to resist the applied forces. A tendency to edge failure can be expected to cause the greatest floor settlement to occur not at the centre, as might be expected from elastic theory, but at a position about ¼ radius from the edge of the tank. This tendency can be resisted by a temporary surcharge in the form of a bund built around the edge of the tank.

The technique of consolidating the soil by the weight of the tank and its test load was used at Teesmouth, where the softness of the soil was forcefully demonstrated when a crawler crane being used to erect an electricity pylon, broke through the thin dried crust and sank. It was only rescued by dismantling it and removing it in sections. The first four tanks of the new farm were placed on piles founded in underlying hard marl.

A site investigation showed that the soft soil was 4 m thick overlying sand resting on the marl. The *in-situ* shear strength, measured by a vane and shown in Figure 7.4, was less than 20 kN/m^2 corresponding to an ultimate bearing

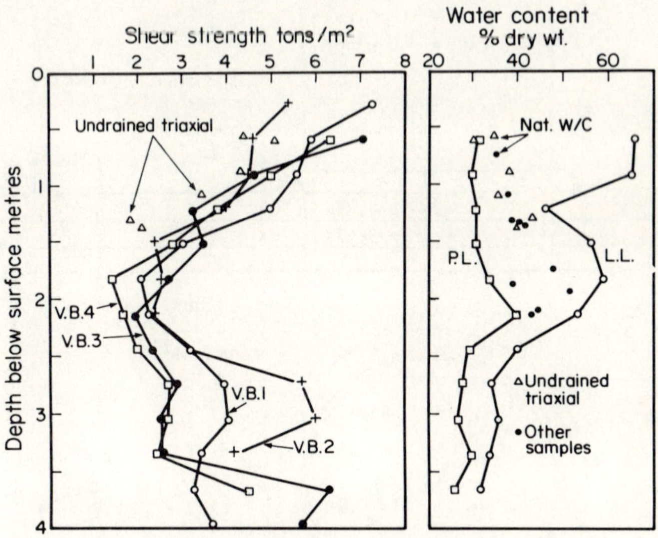

Figure 7.4 Shear strength of the soft soil measured by *in-situ* vane and undrained triaxial tests on undisturbed samples

capacity of less than 120 kN/m^2. The proposed tanks were to be about 15 m high and so would impose a bearing pressure of about 150 kN/m^2. A stability analysis indicated that for stability, the pore pressure must not rise to more than 60% of the applied load and the results from oedometer tests made on U4 samples gave a time for filling of three months to achieve this degree of pore pressure dissipation.

A small tank 14.6 m high and 13.7 m diameter was used to try out the method (see Penman and Watson[11]). Piezometers and settlement gauges were installed in the ground before the tank was built (Figure 7.5). Hand-made boreholes for these instruments revealed two sand layers which divided the soft soil into three layers of almost equal thickness, which could reduce the drainage path to a third and the time for the necessary drainage from 3 months to 10 days. To ensure that these layers could drain freely, a curtain of sand drains was placed through the soft soil on the line for the periphery of the tank. Estimates of settlement were of the order of 0.6 m, so a base pad of compacted rockfill was made 1.2 m thick at the edge and 1.4 m thick at the centre to keep the tank at an operational level after settlement and to allow for the expected greater central settlement.

When tank construction was complete, temporary connections were made to a large discharge pipe, in case it became necessary to rapidly reduce the load in the tank, and to a ring main through fire hoses to supply water to the tank.

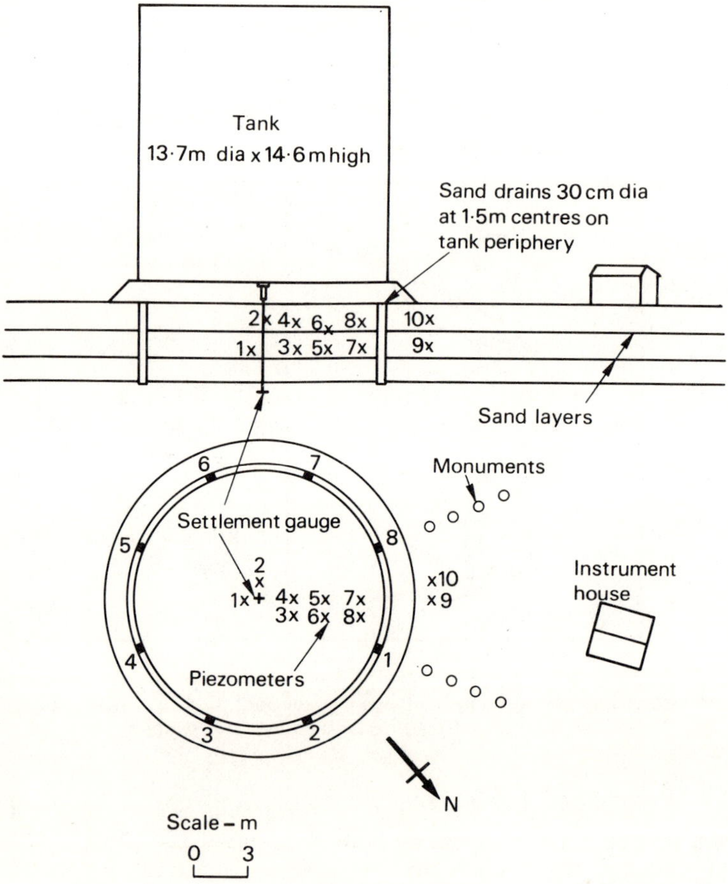

Figure 7.5 Section through small tank showing position of instruments and sand drain

Shelters were constructed for three precise surveyors levels which were positioned so that the heights of the eight brackets on the edge of the tank could be measured rapidly. The remote reading settlement gauges and the piezometers were connected to a small instrument house and from the time when water was first admitted to the tank, readings were taken at every half hour.

It was soon apparent that the supply of water could fill the tank in about two days and it was decided to continue filling at that rate until the pore pressures became dangerously high or the settlement became unacceptable.

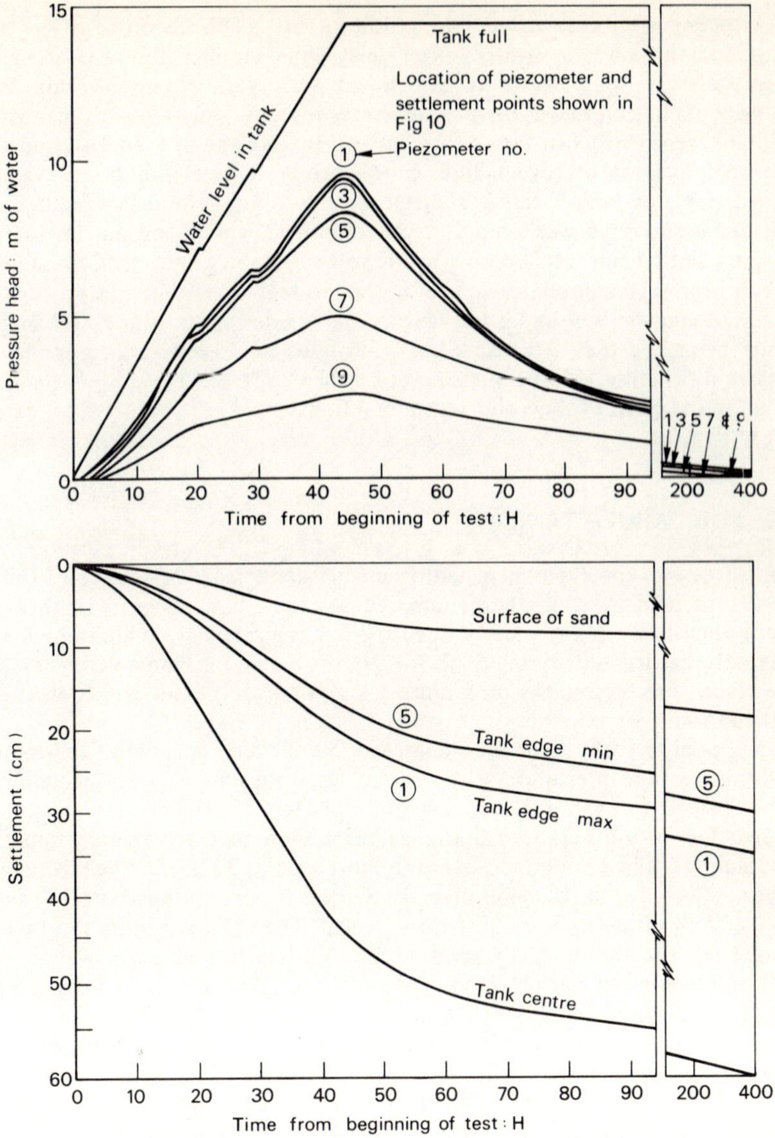

Figure 7.6 Pore pressures and settlements measured during initial filling of the small tank

After 20 hours, the water supply was cut off unexpectedly and it was found that the ring main was always isolated on that day of the week for fire testing. It took more than half an hour to have the supply returned, but the stop proved useful in showing that as soon as loading ceased, dissipation caused the pore pressure to begin falling immediately without any overrun. This was very reassuring and gave confidence in the control of the experiment. As filling continued, pore pressures began to rise again, as shown by the plotted results in Figure 7.6.

A second stop occurred in the middle of the night when one of the hoses burst and there was a slightly longer delay while another hose was fitted. The poie pressures again began to fall immediately, without any overrun. When filling could be continued, the pore pressures rose, but only the piezometers near the tank centre reached the danger limit of 60% of the applied load and there was no indication of edge failure. Consequently the fast filling rate was continued until the tank became full after only 44 hours. The tank was then kept full for about three weeks until it was emptied, painted and put into service.

Tank settlements, attributable to the soft soil, were greater than those due to one-dimensional consolidation and it was evident that some plastic flow had occurred allowing soil under the tank to move sideways into unloaded areas. It could be argued that if the tank had been filled more slowly, thereby allowing greater dissipation of pore pressure and hence a greater increase of strength at any given stage of loading, that less lateral flow would have occurred so reducing the settlements.

USE FOR LARGER TANKS

The success of the experiment with the small tank caused this type of foundation to be used for other, larger tanks on the same farm. Because of their size, the normal flow of test water limited the loading rate and, in addition, loading was only carried out during 8 working hours a day, excluding weekends, with the result that these tanks took quite a long time to fill and as a consequence, settled by smaller amounts.

A special test was made on one tank of 39 m diameter and 16.5 m height. In addition to pore pressure and settlement instrumentation, it had inclinometer tubes placed through the soft soil just outside its periphery to measure any lateral flow of the soil. Special arrangements were made for a water supply and the tank was filled continuously in the short time of 99 hours. The inclinometer tubes deflected to a parabolic shape showing a maximum horizontal movement of 0.107 m at the mid-depth of the soft soil. The volume of soft soil that had moved from under the tank accounted for a settlement that agreed well with the difference between the observed settlement and that calculated from the one-dimensional compressibility of the soil.

Once a soil has been compressed to improve its strength, it does not readily spring back on release of load, so that a tank made stable under consolidating action of its test load may subsequently be emptied and refilled without any serious loss of soil strength. The rebound of a tank on unloading part way through the consolidation process is illustrated in Figure 7.7. The centre came up about 0.03 m on unloading and only a further 0.01 m during the next three days. At the time of unloading it had already settled 0.31 m and on reloading continued to settle on a parallel curve to a final settlement of 0.4 m before the tank was emptied for painting and put into service.

EFFECTS OF LOWERING THE GROUND WATER LEVEL – MEXICO CITY

The effect of lowering the ground water level is to reduce the ambient pore pressure and so increase the effective stress in the soil. This increase of effective stress causes consolidation to a degree dependent on the compressibility of the soil. For every 1 m that the ground water level is lowered, an increased vertical effective stress is imposed on the soil equivalent to a load of 1 tonne/m^2 over the whole area.

In a well-known paper to the 2nd International Conference on Soil Mechanics and Foundation Engineering, Carrillo[2] revealed that there were areas of Mexico City settling at a rate of 1 mm/day. Using Terzaghi's consolidation theory, he

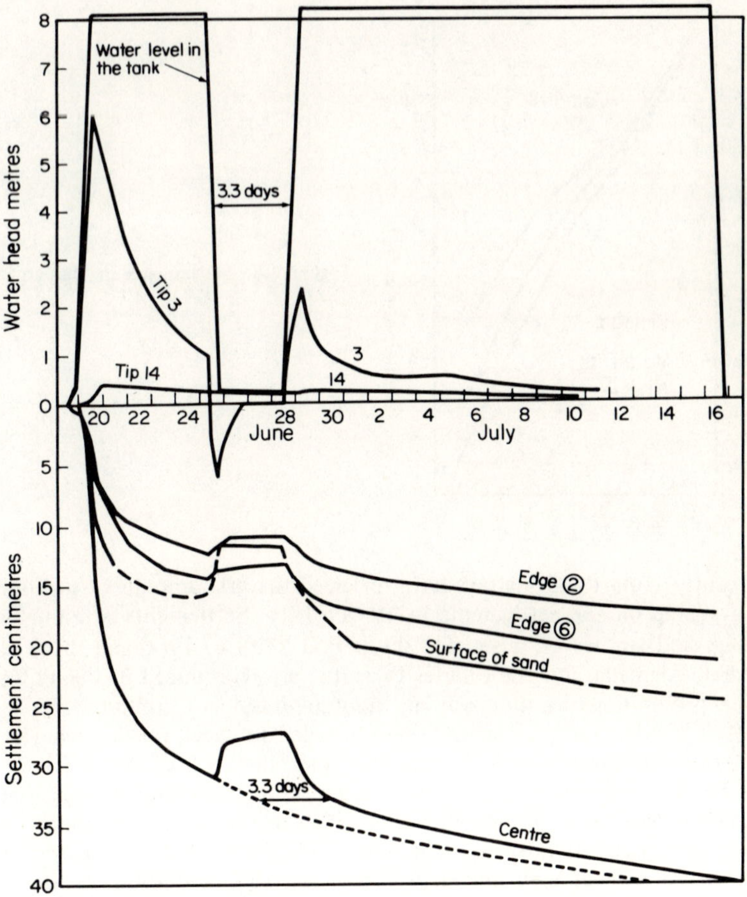

Figure 7.7 Rebound of a tank foundation on unloading

showed that for a clay containing sand aquifers, a uniform fall in the hydrostatic pressure in the aquifers due to pumping would cause an accelerating settlement.

He said that there were 2200 registered wells (and many not registered) extracting in 1944 at a rate of 7 m^3/sec. The wells exploit several aquifers, confined by very soft clay of volcanic origin, ranging in depth from 30 to 600 m. Drop in the static head of the wells ranges from 0.4 to 0.25 m/year. This was causing an accelerating settlement of the central part of the city, and settlements of the cathedral, the Alameda (Central Park) and the statue of Charles IV during the period 1939 to 1947 are shown in Figure 7.8. The settlement has serious consequences for the drainage of the city and for the buildings. Buildings on piles supported by a lower hard stratum, appear to be rising from the surrounding ground and negative skin friction causes heavy overloading on the piles, leading to some sudden differential settlements. Unpiled buildings adjoining those that are piled suffer due to the ground deformation caused by the supporting action of the piles, and ground tensions caused by the central depression are transmitted to buildings, causing cracking.

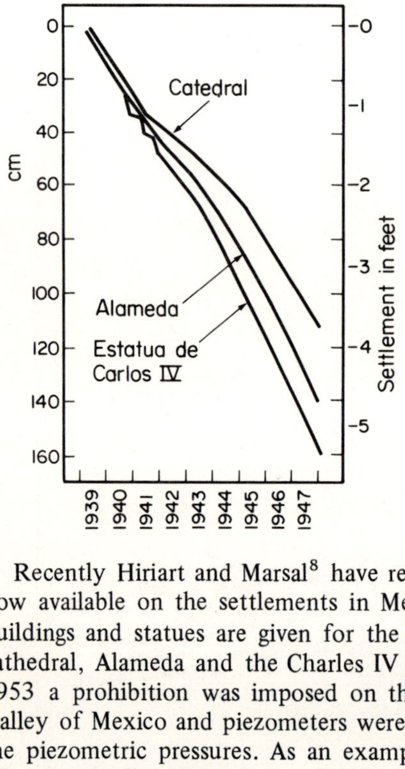

Figure 7.8 Settlements in Mexico City 1939–47

Recently Hiriart and Marsal[8] have reviewed the vast amount of information now available on the settlements in Mexico City. Settlements of a number of buildings and statues are given for the period 1898 to 1966 and those of the cathedral, Alameda and the Charles IV statue are reproduced in Figure 7.9. In 1953 a prohibition was imposed on the exploitation of ground water in the Valley of Mexico and piezometers were installed to check on the behaviour of the piezometric pressures. As an example, piezometers installed at 56, 74 and 95 m depth all registered virtually the same pore pressure and showed that the piezometric level was 30 m below the old water table, corresponding to an increase in vertical effective stress of almost 300 kN/m^2. Continuing draw-down was practically lineal with time at 4 m between 1953 and 1966, i.e. 0.3 m/yr.

An estimate of the water extraction rate since 1932 showed a general increase from 3 m^3/sec to about 9 m^3/sec in 1953. Since that time there has been a slow decline and, under the prohibition order, no more wells have been sunk. Rates of settlement of Alameda are given below:

1900–1938	0.047 m/yr
1938–1948	0.156 m/yr
1948–1956	0.350 m/yr
1956–1966	0.082 m/yr

Although serious damage has been caused to buildings, the works most affected by the subsidence are the sewers. The system, which once worked by

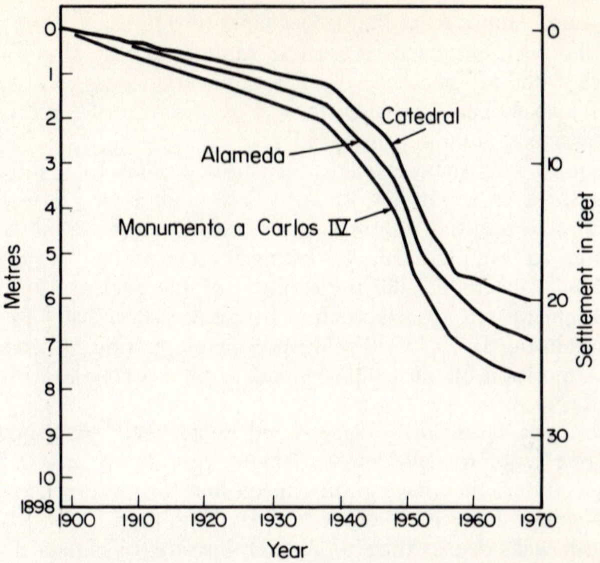

Figure 7.9 Settlements in Mexico City 1898–1968

gravity, now requires pumps. Flooding is the most serious threat facing the city and to overcome it the authorities are now driving a tunnel 6.9 m diameter and 48 km long to dispose of run-off and sewage from the urban area by gravity, using collectors at greater depth than the present ones.

EFFECT OF LOWERING THE GROUND WATER LEVEL – VENICE

This is a similar story to that of Mexico City except that the soil is much less compressible than that found in the Valley of Mexico and the ground water level has only been lowered about 10 m instead of 30 m. This is fortunate for Venice for otherwise it now might well be below sea level.

The Quaternary deposits at the head of the Adriatic consist of layers of sand, silty sand and silt, and some very silty clay. The low islands in the Venice lagoon, safe centres for early peoples, were particularly attractive because shallow wells tapped sand aquifers that yielded a ready supply of fresh water under artesian pressure that would allow a resident population to successfully withstand a prolonged siege. This feature, combined with the excellent harbour afforded by the lagoon accounts for the importance of Venice in the Middle Ages.

Deposits from the rivers in the lagoon threatened to make it too shallow for shipping, so between the 12th and 18th centuries the main rivers were diverted into extensive canal systems around the lagoon periphery. Before 1890 all fresh water used in Venice came from wells in the shallower aquifers, but after that date it was supplied almost entirely by well fields on the main land about 25 km to the north west. Unemployment in the area led the government to encourage industry to move near Venice. Factories built on the mainland obtained a water supply from the deeper aquifers and since 1930 progressively more water has been extracted from these aquifers that lie at depths of 200 to 300 m. In 1858 the artesian head in shallow Venetian wells was typically about 4 m above sea level. This had fallen to 3 m by the turn of the century and by 1970 was down to about 6 m below sea level.

This fall of 10 m in the piezometric level is equivalent to an increase in vertical effective stress of about 100 kN/m^2. This increase has caused compressive strain in the soil resulting in surface settlement which, combined with the eustatic rise in sea level, has caused flooding at an increasing frequency resulting in international concern for the preservation of the ancient city. In 1970 the Italian Government set up a laboratory in Venice specifically to study these problems and during 1971, a 950 m deep borehole was put down to determine the exact composition of the sediments and to obtain samples with a minimum of disturbance.

The project was described by Ricceri and Butterfield[14] who showed that the rates of fall of piezometric level and settlement were closely linked. From a large mass of observations, they have given a mean line for piezometric levels with a slope of 60 mm/yr between 1900 and 1935. This then increased to a rate of 260 mm/yr for wells deeper than 100 m and this rate has remained steady up to 1970. Settlement from 1900 to 1935 was occurring at a rate of 1 mm/yr, but after 1935 it increased to 4 mm/yr and since about 1952 it has become 6 mm/yr. The increase in rate of settlement in 1935 could be attributed to the increased rate of extraction, but the mean fall in piezometric level since 1935 has remained almost constant. The increase in rate of settlement evident since 1952 may be due to the effective stress reaching a value greater than any previous stress that has acted on the soil, i.e. the pressure-settlement relationship may now lie on the virgin compression curve for the soil.

SOLUTIONS TO THE PROBLEM

Various schemes have been proposed for arresting settlement or avoiding flooding. One was to build a diaphragm wall right round the city to prevent further extraction of water from the aquifers under the city, but it would appear to be cheaper to supply water to the factories from reservoirs in the mountains and prohibit further extraction of water from the ground. Unfortunately even such drastic action could do no more than halt the settlement. The reduction of vertical stress caused by raising the piezometric levels to their original position would produce negligible expansion in the sand, silt and silty clay sediments. Flooding could be avoided by a sea defence embankment built to surround the city, like the flood banks that prevent the Thames from covering Canvey Island. Alternatively, the water level in the lagoon could be held down during periods of high water in the Adriatic by fitting floodgates to the entrances to the lagoon, a concept similar to that of the Thames barrage. This seems to be the solution favoured and estimates are being invited for the construction of these floodgates.

The relative compressibilities of the soils at Venice and Mexico are indicated by the fact that a fall of piezometric level of 30 m at Mexico has produced a settlement of about 7 m, while under Venice a fall of 10 m has caused settlements of only about 0.2 m.

EFFECT OF RAISING THE GROUND WATER LEVEL – RECLAIMED LAND

Opencast methods for winning coal, iron ore and other minerals leave loosely dumped backfill. The economics of the mining process prevents the backfill from being placed and compacted in thin layers, a technique which could make the backfill as strong and no more compressible than the original ground.

Often large areas of ex-opencast mined land are left near towns whose expansion requires building construction on the land. It is widely thought that if the replaced soil is left long enough, all settlement will be complete and it will be safe to build on it. In fact, the settlement that occurs under the self-weight of the fill depends on several factors, but particularly on the behaviour of the water level in the fill. While mining is in progress, the water level is kept below the bottom of the working by pumping or drainage, but when it is complete that water level is often allowed to rise and it causes collapse settlement in the fill.

FIELD MEASUREMENTS

Measurements are currently being made by the Building Research Station as part of a joint experiment with the National Coal Board to observe this effect. Magnet extensometers have been placed in boreholes passing right through the backfill into virgin ground at a site where mining had just stopped and where the drainage pump that was keeping the water level at the base of the working would soon be removed. The extensometer, which has been described by Marsland and Quarterman[9], consists essentially of a number of ring magnets placed at intervals throughout the depth of the borehole as shown in Figure 7.10. The distances between them can be measured by a steel tape supporting a sensing unit that is activated by the ring magnets. The tape drum is mounted in a small frame that is securely fixed over each borehole and measurements made where the tape slides past a vernier on the frame. Reproducible accuracies of ± 1 mm are found in boreholes 40 m deep.

Movement of the magnets in the fill can be measured in relation to the magnets attached to virgin ground near the bottom of the borehole, on the assumption that they do not move, or from the surface by measuring the level of each borehole mouth by precise surveying methods.

The magnet extensometers were fitted with a cylindrical porous stone so that with the inner vertical tube they formed standpipe piezometers to measure the level of the water table in the fill. Some initial results from this experiment have been given by Charles[3]. He described one of the extensometers, passing through 63 m depth of backfill into bedrock and showed the vertical compression that occurred between magnets 4 and 5 and between 5 and 6 as reproduced in Figure 7.11. During the first half of 1974, while the water table was below magnet 4, the fill showed little change in vertical strain, but as the water level rose from 4 to 6, the fill between 4 and 5 showed slight expansion, while that between 5 and 6 indicated a collapse settlement of more than 1%. Charles explained the difference in behaviour at the two levels by the fact that although loose unsaturated fills may be liable to settlement on saturation, a parent shale material may itself swell. He suggested that if these movements, caused by a rising water table occurred at a restored opencast site subsequent to building on the site, severe settlement problems could result.

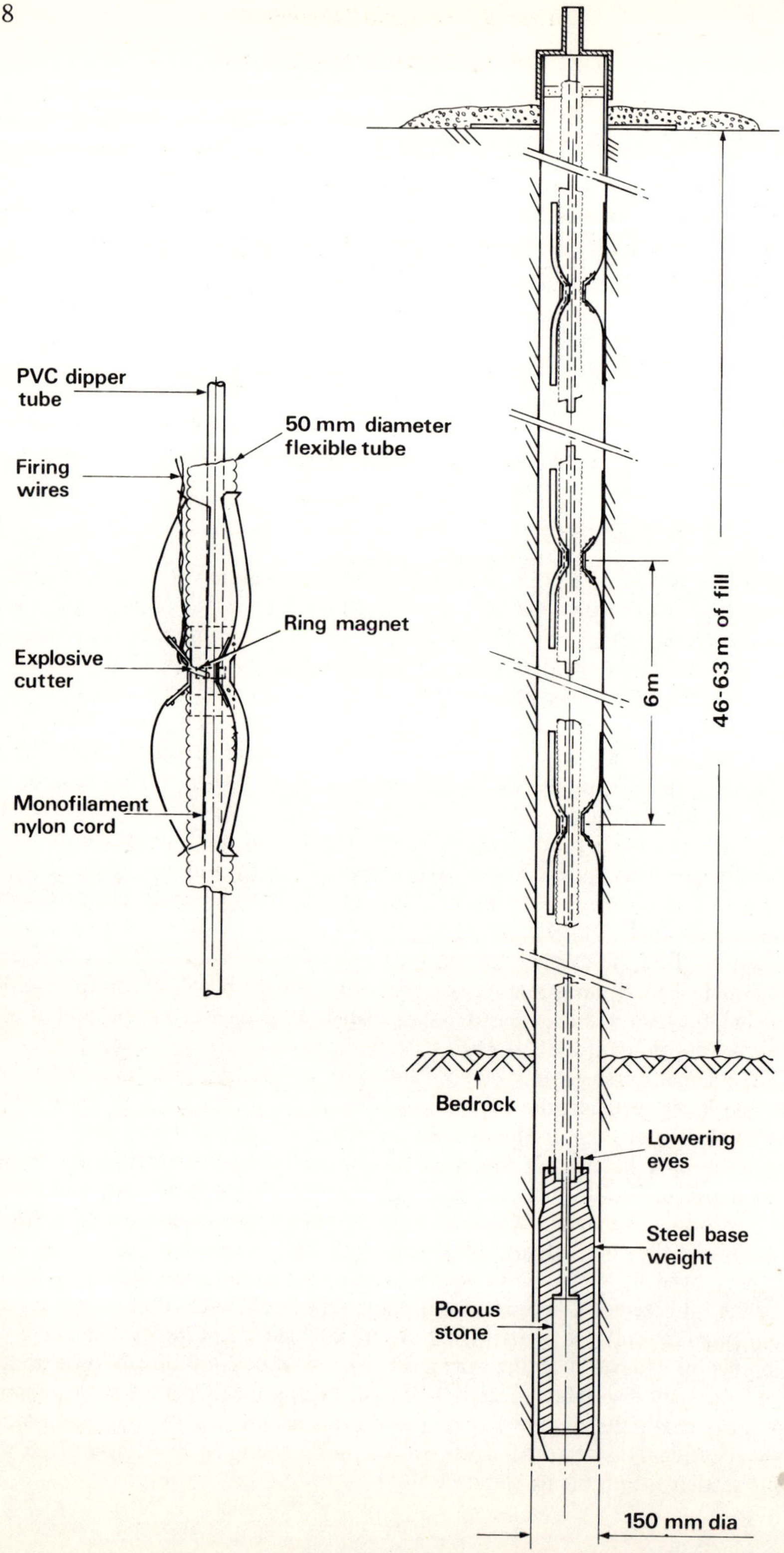

Figure 7.10 Magnet gauge with piezometer tip in bottom weight

COMPARISON OF GROUND TREATMENT METHODS

Damage to houses built on ex-opencast mines has been reported by Guest[6]. The backfill was 9 m deep and a housing estate of 350 dwellings was under construction. A differential settlement of 200 mm over a distance of less than 45 m caused irrepairable damage to one block of 8 houses. The settlement was associated with drain trenches that had been opened nearby and attracted surface run-off so providing a concentration of water in the backfill.

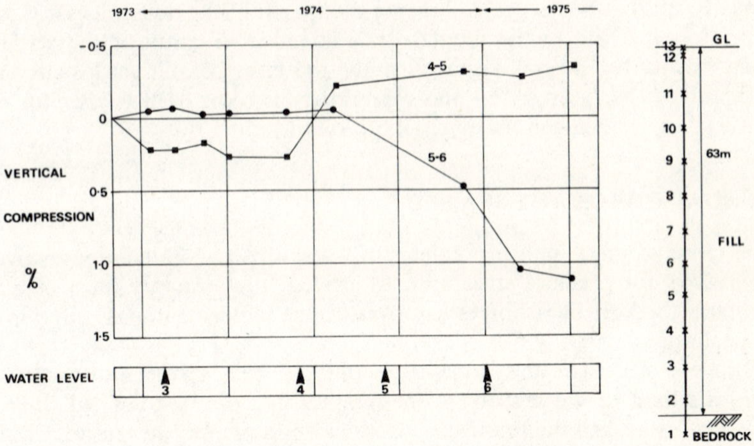

Figure 7.11 Effect of rising water table on vertical compression of backfill

A joint experiment between the Building Research Station and Corby Development Corporation is currently being carried out to compare three methods of ground treatment on the loose backfill left by opencast ironstone workings, to make it suitable for traditional two-storey dwellings. At the site the ironstone had been overlain by oolitic limestone and boulder clay of almost equal ticknesses. The ironstone was bared by a large walking dragline working from the surface of the limestone. The advancing faces of the clay and limestone were stepped back so that after the ironstone seam had been excavated by blasting and face shovels, the excavated overburden could be dumped back with the limestone at the bottom overlain by the clay, so following, as closely as practicable, the original sequence.

An area of land, where the backfill was 24 m thick, was divided into four squares each 50 m × 50 m and five magnet extensometers were placed in each of the areas to be treated. The three methods of ground treatment consisted of application of a 9 m high surcharge, dynamic compaction and inundation. Since this chapter deals with the problem of ground water and foundation, reference will only be made to the inundation part of the experiment. Preliminary results have been given by Charles[4] who explained that water was introduced into the ground from trenches about 1 m deep at 10 m centres dug across the site then filled and kept full of water. Over a period of 3 months, surface settlements of up to 140 mm have been measured and the magnet extensometers have shown that the main compression of the fill occurred between 2 and 5 m depth.

Since treatment was carried out, traditional two-storey brick dwellings in blocks of one, two and three houses have been built using traditional strip footings. Special sockets of a type described by Cheney[5] were built into the walls just below damp-proof course at the corners of each house so that settlement measurements can be made with a precision staff put on a plug fitted to the sockets. House construction was completed during the spring of 1976 and at an Open Day held in June it was revealed that the average settlement of the fill due to inundation at a depth of 4 m was 40 mm, while at a depth of 10 m it was less than 10 mm. Settlements of the houses during construction were small (about 10 mm) as was expected because the weight of the house does not cause a large stress increase in the ground. It is intended to continue the settlement measurements for several years to study the effect of fill settlement on the behaviour of the houses, and photogrammetry is being used to monitor movement and any cracking in the external brickwork.

FOUNDATIONS ON THE SEA BED

An extreme example of the problems of water affecting foundations is given by foundations for offshore structures. In general, the structure must be able to withstand the horizontal forces and overturning moments imposed by wave and wind action.

Offshore structures are often supported by piles driven into the sea bed. There is a limit to the severity of the weather that can be tolerated while piles are being driven and the foundation work can become very expensive because of the long periods of standing time caused by adverse weather. Installation time can be reduced considerably by using gravity type foundations. These, as with the foundations of most structures on land, rely on the weight of the structure to hold it to the ground with sufficient force to resist the horizontal forces and overturning moments imposed by environmental conditions. The horizontal forces imposed by waves on an offshore structure are much greater than those caused by wind on a land-based structure so that in general the foundation width must be much greater for an offshore structure. It is usual to reduce the size of the part of the structure affected by the waves to a minimum.

The contact pressure under a gravity base must be enough to remain positive over the whole contact area under the maximum overturning moment. The contact pressure and surface area must also be large enough to prevent horizontal sliding. The size and shape of the foundation depends on the nature of the soil forming the sea bed. A large area would be required if it is to be supported by a soft clay, whereas a dense sand may require special scour protection to avoid undermining at the foundation edges. The fact that it is submerged reduces both the effective weight of the structure and the effective stress and hence the strength of the soil. The strength of a sand may be almost halved when it is submerged, but it is, of course, reduced no more by a great depth of water over it than it is with water just above ground level.

EXAMPLES OF GRAVITY FOUNDATIONS

Gravity foundations were chosen for the submarine defence towers built during the Great War of 1914–18. The plan was to form a submarine barrier across the

English Channel with a number of towers connected by nets. The steel towers had cellular reinforced concrete bases of such a size that the whole structure would float. At the chosen sites, the bases were to be flooded to sink the towers to rest on the sea bed. Construction of two towers was begun at Shoreham, but only one was completed before the end of the War. It was used to replace the lightship marking the shoals just north east of the Isle of Wight and is now the well-known Nab Tower, a navigational aid in the shipping lane from Southampton.

A similar principle was used for the construction of forts to carry anti-aircraft guns during the Second World War of 1939–45. Each fort had two towers supporting a deck unit from a gravity foundation shaped as a flat-bottomed barge. Six forts were used to guard the entrance of the Thames and Harwich. They were towed to site, sunk in only 3 minutes and were immediately operational. A description of the construction and placing of these forts has been given by Posford[13].

Several lighthouses in the Baltic have been constructed with this type of foundation and both the Kish Bank lighthouse in the Irish Sea and the Royal Sovereign in the English Channel used this principle. The construction and placement of the Royal Sovereign have been described by Antonakis[1]. The advantage of gravity over piled foundations in saving installation time has been recognised for offshore oil production platforms in the North Sea. Four reinforced concrete Condeep platforms have already been built in Stavanger Fjord and two of these are on site in the Brent oil field.

EXPERIMENTAL TOWER TO MEASURE WAVE FORCES

Recently an experimental tower has been placed in about 9 m depth of water off the south coast of England. Its prime purpose was to measure wave forces, but it has also been used to study the behaviour of a simple gravity type foundation. The tower has been described by Wheatley[15] and is shown by Figure 7.12. The small experimental tower is about 3 m diameter and 14 m high. All the forces imposed on it by the waves will be measured and this gives a unique opportunity to study the behaviour of a gravity base subjected to known horizontal forces and overturning moments.

Good civil engineering practice demands a positive effective contact pressure between the base slab and sea bed under the worst storm conditions, but because the tower would not be manned during severe storms it was decided to provide a factor of safety of only 1.5 against the expected normal storms to produce larger and more easily measurable contact pressures. The steel tower was almost bouyant in 9 m depth of water so that the reinforced concrete base pad had to provide both sufficient weight and a large enough diameter to resist the expected horizontal forces. Positive contact over the whole base would require a diameter of 15 m, but by limiting the design requirement to a factor of safety of 1.5 against the normal storm, the diameter was reduced to 10.5 m. This would allow the factor of safety to fall to 1 under the expected 100-year storm.

Design was based on the methods given by Meyerhof[10] and Brinch Hansen[7] for a circular pad on sand subject to inclined loads. The potential bearing capacity of dry sand is approximately halved when it is saturated, with a water table at, or above, the surface of the sand. As a horizontal force acting on the upper part of the tower increases, thus widening the angle to the vertical and increasing the eccentricity of the resultant force applied to the base, so the area of positive effective contact between the base and the sand decreases until a bearing capacity failure occurs. The forces expected to act on the tower during a storm with 5.5 m waves and the calculated areas of effective contact are shown in Figure 7.14.

Figure 7.12 The experimental tower being placed on a barge for transport to site

Eleven earth pressure cells were cast into the concrete base slab during construction so as to be flush with the underside. The surface of the sand at the site for the tower was rippled, with sand waves about 0.02 m high at a spacing of about 0.15 m, as commonly seen on the beach at low tide. A long straight-edge placed on the sand by divers showed that the surface was almost completely level, and when the tower was lowered to position by crane, it stood vertical to within 13 minutes of arc. Few of the earth pressure cells showed any appreciable contact pressure and it was assumed that most of them lay between the sand ripples. It was expected that as stress variations occurred under wave action, the base pad would slowly bed-in until a uniform contact pressure developed.

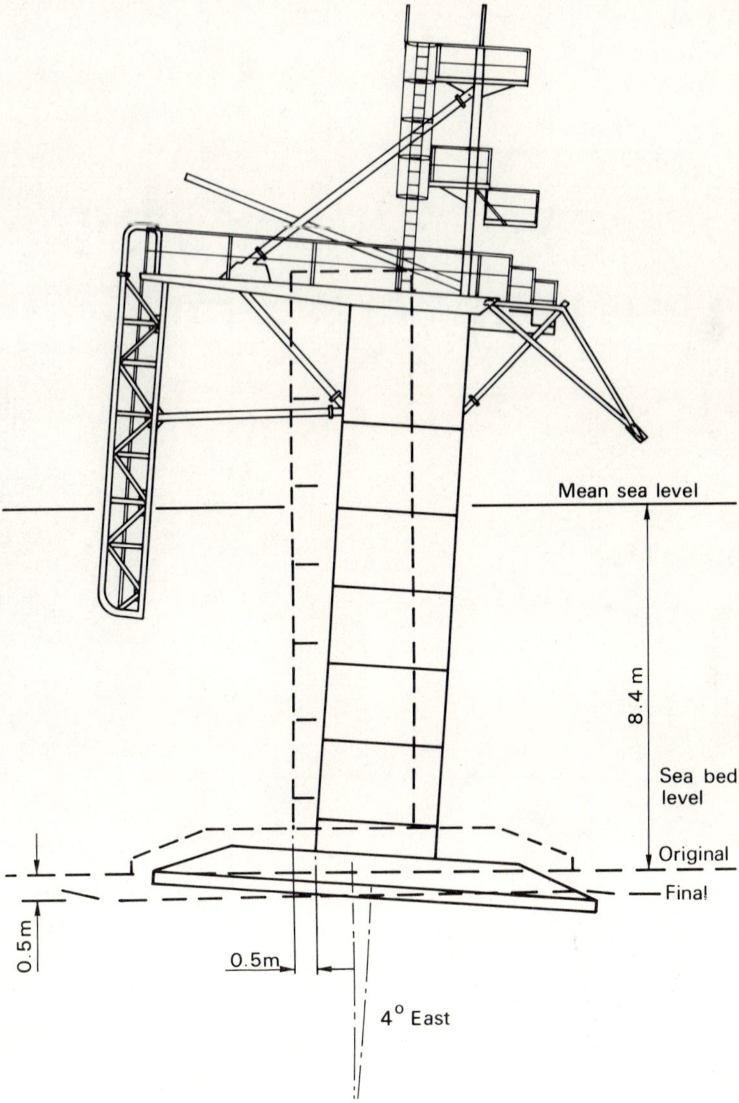

Figure 7.13 Foundation of experimental tower after failure

Only thirteen days after the tower was placed, it was struck by a very severe storm and afterwards was found to have a tilt of nearly 4° in the direction of the storm and to have settled almost 0.5 m. Divers found that the down-storm edge of the base had been pressed down into the sand and had pushed up a mound of sand in front of it. This was consistent with a rotational slip in the sand under a

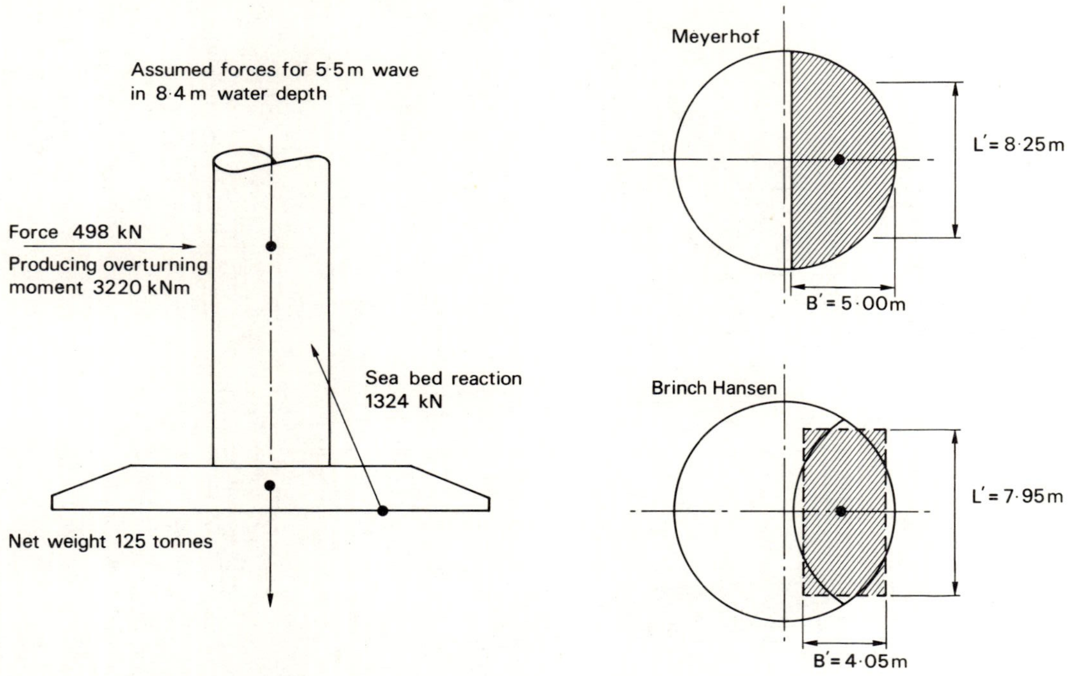

Figure 7.14 Assumed forces and contact areas

'footprint' area of the base slab and agreed with the failure conditions postulated by both Meyerhof and Brinch Hansen. The failure has been described by Penman and Gallagher[12] and the attitude of the tower is shown in Figure 7,13.

After failure, effective contact pressure was found only under the centre of the base and its measured value corresponded with the ultimate bearing capacity of the sand. Together with the fact that the tower was rocking over a magnitude of about ¾°, this indicated that the base was supported on a central mound of sand whose size was governed by the bearing capacity of the sand.

Rotational failure under the edge of a foundation slab can be resisted by a surcharge placed just beyond the edge of the foundation. To correct the tilt, the tower was raised with bouyancy tanks and moved a few metres to another part of the flat site, and to improve the factor of safety a surcharge of gravel was placed around the base pad.

References

1. Antonakis, C.J., 'A problem of designing and building for a structure at sea,' *Proc. Inst. C.E.*, Part 1, **52**, pp 95–126 (1972)
2. Carrillo, N., 'Influence of artesian wells on the sinking of Mexico City' *Proc 2nd Int Conf SM & FE*, 7, pp 156–159 (1948)
3. Charles, J.A., 'Contribution to discussion. Ground treatment by deep compaction.' *Inst C.E.*, pp 149–150 (1975)
4. Charles, J.A., 'Contribution to discussion. Ground treatment by deep compaction,' *Inst C.E.*, pp 96–98 (1975)
5. Cheney, J.E., *Techniques and equipment using the surveyor's level for accurate measurement of building movement.* Proc. BGS Conf Field Instrumentation in Geotechnical Engineering, Butterworths, pp 85–99 (1974)
6. Guest, J.E., *Contribution to discussion.* Proc, BGS Conf Settlement of Structures, Pentech Press, London, p. 662 (1974)
7. Hansen, Brinch, *A revised and extended formula for bearing capacity*. The Danish Geotech Inst, Bulletin No 28 (1970)
8. Hiriart, F. and Marsal, R.J., *The subsidence of Mexico City*, Nabor Carrillo commemorative volume. Mexican Geotechnical Soc, pp 109–147 (1969)
9. Marsland, A. and Quarterman, R., 'Further developments of multipoint magnetic extensometer for use in highly compressible ground,' *Geotechnique*, **24**, No 3, pp 429–433 (1974)
10. Meyerhof, G.G., 'The bearing capacity of foundations under eccentric and inclined loads,' *Proc 3rd Int Conf SM & FE, Zurich*, **1**, pp 440–445 (1953)
11. Penman, A.D.M. and Watson, G.H., 'The improvement of a tank foundation by the weight of its own test load,' *Proc 6th Int Conf SM & FE, Montreal*, **2**, pp 169–173 (1965)
12. Penman, A.D.M. and Gallagher, K.A., 'Preliminary study of a gravity foundation failure,' *Ground Engineering*, **8**, No 4, pp 15–20 and 44 (1976)
13. Posford, J.A., 'The construction of Britain's sea forts. The Civil Engineer in War,' *Proc Inst C.E. Symp*, **3**, pp 132–163 (1948)
14. Ricceri, G. and Butterfield, R., 'An analysis of compressibility data from a deep borehole in Venice,' *Geotechnique*, **24**, No 2, pp 175–192 (1974)
15. Wheatley, J.H.W., 'British offshore research facility,' *Trans Royal Inst Naval Arch*, Paper No 8, Spring Meeting (1976). (Also NPL report, Mar Sci R248.)

Chapter 8

Site Investigation in Difficult Ground

INTRODUCTION

In the British Isles, site investigation techniques have tended to become more or less standard, and are fully described in codes of practice[1] and specifications[2]. Soils are drilled by cable percussion rigs with a minimum hole size of about 150 mm diameter. 'Undisturbed' samples are taken with the 100 mm diameter general purpose sampler and standard penetration tests[2] are carried out in order to obtain an empirical indication of the consistency of the ground. Rocks are drilled by rotary coring using N and H size (approximately 55 mm and 75 mm diameter core sizes) double-tube barrels. In addition, shallow trial pits (maximum depth about 4 m can be dug in soil or weak rock using a light tractor-mounted hydraulic excavators.

'Difficult' ground has been interpreted as those conditions in soils or rocks where these standard techniques of site investigation fail to give reliable data on the real character and properties of the ground.

THE OBJECTIVES OF A SITE INVESTIGATION

The basic objectives of any site investigation are the determination of the following:

(a) The character of the ground – the identification of the various soil and rock types and their structure (or fabric), together with the determination of the distribution and orientation of the discontinuities (bedding planes, joints, fissures etc) which tend to influence the behaviour of the general mass of the ground.
(b) The character of the ground water – the measurement of the piezometric pressures and the location of the more permeable or water bearing zones.
(c) The engineering properties of the ground – measurement by laboratory and field testing as relevant to the project under investigation.

The detailed planning of a site investigation will depend on the engineering requirements of the particular project but the general plan should cover the three objectives in a logical order so that the general conditions are established (objectives 'a' and 'b') before proceeding with the programme of testing required for objective C. 'Difficult' ground can cause problems in the correct evaluation of one or all of these objectives.

DIFFICULTIES IN DETERMINING THE CHARACTERISTICS OF THE GROUND

The characteristics of the ground will be determined largely by an examination of samples obtained from boreholes or from shallow trial pits. If it is assumed that a sufficient number of boreholes can be sunk, difficulties may still arise because either the samples are not of sufficient quality or the samples, irrespective of their quality, are not sufficiently representative of the general characteristics of the mass of the ground.

(a) Samples of insufficient quality can be obtained after, even though the methods of boring which are generally used in site investigations have been developed to give the best possible indication of the character of the ground. This is because serious difficulties can be encountered in certain ground conditions. Typical conditions and some alternative expedients are listed below:

(i) Non-cohesive soils beneath the water table

When disturbed samples are obtained from the borehole tools or excavator bucket, it often happens that the fine fraction of the soil sample has been washed away. This frequently occurs with samples of sands and gravels obtained by the borehole shell. The best expedient, if the soil is suitable, is to take tube samples.

The 100 mm general purpose sampler[1] is suitable for soils containing particles up to about fine gravel size and when the sample barrel is extended by screw couplings to a length of about 1.5 m and is fitted with a latch type sample retainer, it will generally recover satisfactory samples of completely cohesionless sand. The standard penetration test sampler[2] will recover samples of sand but often, the short length of the sample, (about 100 to 150 mm) makes it difficult to sample below the disturbed material in the bottom of the borehole.

An alternative expedient, for use in coarse grained soils which are unsuitable for the use of a tube sampler, is the borehole permeability test[1]. The measured permeability is a good indicator of the particle size of the fine soil constituting the matrix of the coarser material.

The final expedient is a visual inspection of the soil in a trial excavation, but if the ground lies below the water table, dewatering may be required in order to form a stable, and reasonable dry excavation.

(ii) Samples which are too badly disturbed to be reliably identified

For example, when a cable percussion rig is being used it is necessary to use a heavy chisel in order to break up hard ground, and then shell out the chisel debris, the disturbed samples recovered in the shell cannot be reliably identified. If the ground is also unsuitable for recovering tube samples, serious difficulties can occur. This situation often occurs in soils containing boulders and in rock, and can lead to a failure to correctly identify rock-head when overlain by boulders. The ideal expedient is a visual examination of the ground in a trial pit, but if this is not practicable, rotary core drilling can be tried. This will generally recover identifiable samples of the bed rock and may recover samples of the boulder deposit if the boulders are set in a clay matrix. It may be necessary to use the larger sizes of core barrels, see (iii) below.

(iii) Rotary cored boreholes where the sample recovery is inadequate

With many rock conditions it is not possible to obtain adequate core recovery using the standard core barrels. Typical examples are weak friable rock, jointed rock, or thin layers of weak rock within a harder rock mass. Often, in engineering projects, the occurrence of thin layers of weak material is a factor of critical importance and yet, even if the recovery of solid core were a normally acceptable 90% of the length drilled, these critical layers might be missed completely. In weak rocks (e.g. chalk) in dry conditions, visual inspection in a shaft sunk by a bored pile auger is the best expedient[3]. In other circumstances, it may be possible to obtain better samples using larger diameter rotary core barrels (up to 110 mm diameter core size) and possibly, in addition, fitted with a third removable liner so that the core may be stored and examined without having suffered the disturbance of being extruded into a core box. In some conditions of weak rock, Denison type samplers[4] may give useful results. These have an outer rotary barrel within which a sample tube with a sharp cutting edge is driven into the ground ahead of the rotary cutter. Denison type samplers may also have a third, removable liner for containing the sample. Indirect methods can give some indication of the character of the ground. Borehole permeability tests using packers[1] will give an indication of the presence of open fissures and in some conditions, down-the-hole geophysics techniques[5] may give useful results.

(b) Samples not representative of the ground mass

The properties of the ground mass depend both on the properties of the discontinuities and on the properties of the solid material contained between the discontinuities. Often, the linear scale of the distribution of these discontinuities is such that little or no information about the ground mass can be obtained by an examination of the relatively small scale borehole samples. This difficulty occurs frequently in rock investigations and can also occur in soil investigations, particularly when permeability is important. Possible expedients are:

(i) a visual examination of the ground in large exposures where the geometry and character of the discontinuities can be studied.

(ii) taking the maximum practical size of borehole sample, e.g. a 100 mm diameter soil sample is of much greater value than the alternative 37 mm size.

DIFFICULTIES IN DETERMINING THE CHARACTER OF THE GROUND-WATER

Generally ground water conditions cannot be determined with any useful accuracy from observations of borehole water levels while the borehole is being sunk. The addition of water to facilitate boring or the use of casing to exclude ground water may actually conceal the true ground water conditions. The correct expedient is to set piezometers between grout seals at critical positions which have been identified in the preliminary boreholes.

In general, each borehole should contain only one piezometer set at the bottom of the borehole. If otherwise, it is necessary to set a grout seal below the piezometer which is time-consuming and also difficult to do satisfactorily. Often, the best practice is to drill special boreholes, one for each piezometer. Once installed, piezometers can be observed over an extended period so that the true piezometric pressures can be determined.

DIFFICULTIES IN DETERMINING THE PROPERTIES OF THE GROUND

The properties of the ground will be determined by laboratory tests on 'undisturbed' samples taken with the 100 mm general purpose sampler[1], on disturbed samples obtained from the borehole tools, and by standard penetration tests[2]. In ground conditions where tests on 'undisturbed' tube samples are likely to give unreliable results, standard penetration tests[2] are used to give an indication of the consistency of the ground. Using these standard procedures, difficulties can occur through three main causes, the samples not being of sufficient quality, the standard penetration tests giving erroneous results, and neither the tube samples, nor the standard penetration tests, being sufficiently representative of the properties of the ground mass in its entirety.

(a) Samples not being of sufficient quality for laboratory testing. A classification of sample quality, which is frequently used[6,7] is given in Table 8.1.

Table 8.1 CLASSIFICATION OF SAMPLE QUALITY

Quality	*Properties that can be reliably determined by laboratory tests*
Class 1	Classification, moisture content, density, strength, deformation and consolidation characteristics
Class 2	Classification, moisture content and density
Class 3	Classification and moisture content
Class 4	Classification
Class 5	None

It is important that a site investigation is planned having due regard to the sample quality which will be given by the various techniques. A number of frequently encountered difficulties of sample quality are listed below.

(i) Sands

Tube samples are normally not better than Class 4 below water level or Class

3 if above. Disturbed samples from below water level are likely to be only water level are likely to be only Class 5.

(ii) Coarse grained non-cohesive soils (e.g. gravels)

Tube sampling is generally impracticable. Disturbed samples from below water table are likely to be only Class 5.

(iii) Loose silts and soft clays

The 100 mm tube sampler often causes some disturbance, particularly with sensitive soils, and samples are generally Class 2 quality.

(iv) Stiff clays and weak rock (e.g. chalk)

The 100 mm tubes sampler causes varying degrees of disturbance through shattering the brittle material. Often samples are only Class 2 or 3 and the strength is underestimated.

(v) Cohesive soils containing hard granular material (e.g. till)

With the sampler, there is often serious disturbance caused by the granular material, sample quality may be as low as Class 4.

(vi) Rock

Rotary core samples are generally only suitable for tests on intact material; the moisture content may have been altered by the flushing process and the sample quality may vary considerably.

(vii) Laboratory tests not appropriate to the soil type

The unconsolidated undrained triaxial test is used for measuring the total stress shear strength parameter (cohesion) which is required for the evaluation of the allowable bearing capacity of foundations. This test often gives unrepresentative results in cohesive soils containing appreciable proportions of silt and fine sand and leads to an underestimation of the bearing capacity.

There are a number of expedients which can be used in various ground conditions when it is suspected that the samples are of inadequate quality:

(i) Standard penetration test[2]. This gives an indication of the strength and density; it can be a useful check on the results obtained by a laboratory testing and is suitable for most ground conditions.
(ii) Static penetration test (Dutch cone)[1]. This gives an indication of both strength and stratification and is suitable for soils up to medium strength, (firm clays or medium density sands).

(iii) Vane test[2]. Measures the *in situ* strength of soft to firm clays and is particularly suitable for sensitive clays.
(iv) Thin-wall tube samplers, including piston samplers[1]. Used in silt and soft clay, samples are generally Class 1 quality.
(v) *In situ* Borehole tests. A number of *in situ* tests are frequently used[1], e.g. plate test, permeability test.

(b) The Standard penetration test giving erroneous results even when the specified procedure is strictly followed[2]. Large errors can arise due to the ground having been disturbed while the borehole is being sunk. When boring in granular soils below the water table, the soil can be loosened by the action of the shell and in some circumstances might 'blow' many metres up the borehole. This effect is most prevalent in uniformly graded sands and causes the standard penetration test to underestimate the consistency of the undisturbed soil. The effect can to some extent be reduced by keeping the borehole full of water but it is doubtful if this expedient will completely eliminate the problem. When consistency measurements are important to the project (e.g. in designing driven piles) and there is a possibility that the sand might have been loosened during boring, the consistency should be investigated using the static, Dutch cone, equipment[1].

(c) The samples and borehole tests are not sufficiently representative of the ground mass in its entirety. This difficulty occurs to varying degrees in almost all types of ground and is caused by the ground properties being influenced by discontinuities (e.g. bedding planes or joints) which occur on a scale larger than that of the sample dimensions. The problem can be recognised by examining exposures of the ground where the pattern and character of the discontinuities can be studied and their influence on the properties of the ground can be assessed. There are three main expedients:

(i) Taking the largest size of sample that is practical. Tests on samples 100 mm diameter in general give more reprsentative results than tests on 37 mm diameter samples. Samples up to 250 mm diameter are often used for studies of consolidation.
(ii) Taking standard 100 mm diameter samples at critical locations and at special orientations in order to obtain samples of each component of the ground structure.
(iii) Carrying out large scale field tests in which the scale of the test is selected so that it is compatible with the pattern of the discontinuities e.g. plate load test, triaxial embankment, pumping test.

CONCLUSIONS

The standard techniques used in routine site investigations have been developed over many years. In general they are simple to carry out, reasonably fast, and give good results. However, in certain ground conditions, designated as 'difficult', they can give inadequate or misleading results and alternative expedients have been suggested.

On sites where there is no previous experience of the ground conditions, it is suggested that the investigation should be carried out in two stages. The first stage would employ standard techniques and would investigate the character of the ground and of the ground water; at the same time any 'difficult' ground conditions would be recognised. The second stage would cover the detailed programme of sampling and testing to determine the engineering properties of the ground and would employ any special techniques which were required for difficult ground.

References

1. CP 2001: 1976 (revision): *Draft standard code of practice for site investigations,* British Standards Institution
2. BS 1377: 1975: *Methods of test for soils for civil engineering purposes,* British Standards Institution
3. Ward, W.H., Burland, J.B. and Gallois, R.W., 'Geotechnical assessment of a site at Munford, Norfolk, for a large proton accelerator', *Geotechnique,* **18**, 399–431 (1968)
4. Terzaghi, K. and Peck, R.B., *Soil mechanics in engineering practice.* 2nd Edit. Wiley International (1967)
5. Kennett, P., 'Geophysical borehole logs as an aid to ground engineering'. *Ground Engineering* (Sept 1971)
6. Idel, K.H., Muhs, H. and Van Soos, P., 'Proposals for quality classes in soil sampling and the importance of boring methods and sampling equipment', *Speciality session No. 1., Proc. 7th Int. Conf. Soil Mech.* (1969)
7. Rowe, P.W., 'The relevance of soil fabric to site investigation process', *Geotechnique,* **22**, 193 (1972)

Acknowledgement: The Author is endebted to Messrs Foundation Engineering Ltd. for permission to publish this paper.

Chapter 9

In Situ Testing and Geophysical Surveying

INTRODUCTION

Field testing and less frequently geophysical surveying, form an integral part of a major site investigation. In recent years field testing has been given increasing attention and notable advances have been made. The basic reason for this is that field tests provide, for design purposes, a more realistic appraisal of the geotechnical character of the site conditions than can be obtained from laboratory tests on samples. This is because small samples cannot give a truly accurate representation of the soil or rock mass from which they were obtained. For example, the engineering behaviour and hydrogeological properties of rocks depend, frequently to a large degree, on the nature and geometry of the discontinuities, but the influence of the latter cannot be assessed from small samples tested in the laboratory. What is more sampling invariably involves some amount of disturbance (see chapter 8), the greater this is then the less representative are the results of certain tests. Results gathered in the field therefore provide a desirable check against those obtained in the laboratory. Thus, although the costs of *in situ* investigations may be higher than those incurred by laboratory testing the dividends are worthwhile since they afford a means of checking and improving the available design methods.

PENETROMETERS

According to Sanglerat[1] penetrometers evolved from the need to acquire data on subsurface soils which could not readily be obtained by other means. Basically a penetrometer consists of a conical point attached to a drive rod which is forced into the ground either by hammer blows or by jacking. Hence two types of penetrometer tests are recognised, the dynamic and the static. Both methods measure the resistance to penetration offered by the soil at any particular depth. Penetration of the cone forces the soil aside, creating a complex shear failure and thus provides an indirect measure of the *in situ* shear strength of the soil.

STANDARD PENETRATION TEST (SPT)

The most widely used dynamic method is the standard penetration test (SPT). This empirical test, done with the aid of a cable and tool rig, consists of driving a split-spoon sampler into the soil at the base of a borehole. Drivage is accomplished by a trip hammer and the relative density or consistency of a soil may be determined from the blow-count or *N* value (Table 9.1). The standard penetration test can also be used in soft rocks and in the weathered zones of harder rocks. In deep boreholes the standard penetration test suffers the disadvantage that the load must always be applied at the top of the rods so that some of the energy from the blows is dissipated. The energy loss becomes more significant with increasing depth so that the test results become more suspect.

An excessively high value of the penetration resistance is obtained if the sample is compressed during drivage. Moreover test data are of no value if the soil to be tested has been disturbed. This is especially the case if heavy driving is required to advance the casing. Disturbance is often created when boring is done by shelling, this principally being due to the suction created as the shell is withdrawn (see Rodin[5]). In such instances it is recommended that the soil to a depth of a further 0.5 m be carefully removed by a smaller diameter shell before the test is carried out. Disturbance is also developed when excess hydrostatic pressure exists at the bottom of the hole, causing a rapid inward flow of water. If sand is subject to liquefaction then, after the initial hammer blows, it starts to flow and partly fills the borehole. If there is any doubt or if the blow count is

Table 9.1 RELATIVE DENSITY AND CONSISTENCY OF SOIL
(After Terzaghi and Peck[2], and Sanglerat[1])

A. Relative density of sand and SPT values, and relationship to static cone resistance and angle of internal friction

SPT (N)	*Relative density*	*Description of compactness*	*Static cone* resistance* (q_c)	*Angle of internal friction* (ϕ)
Under 5	0.2	Very loose	Under 2.0	Under 30°
5–10	0.2–0.4	Loose	2.0–4.0	30°–35°
11–30	0.4–0.6	Medium dense	4.0–12	35°–40°
31–50	0.6–0.8	Dense	12–20	40°–45°
Over 50	0.8–1	Very dense	Over 20	Over 45°

*These values are approximate. Rodin *ét al*[3] demonstrated that there is no unique relationship between STP and static cone resistance values, it should be assessed at each site.

B. N-values, consistency and unconfined compressive strength of cohesive soils (not always reliable)

N	*Consistency*	*Unconfined compressive strength (CP2004:1972)*[4] N/m²
Under 2	Very soft	Under 20
2–4	Soft	20–40
5–8	Firm	40–75
9–15	Stiff	75–150
16–30	Very stiff	150–300
Over 30	Hard	Over 300

less than ten, the test should be repeated, driving the sampler an additional 300 mm.

In gravel deposits the value of N, if accurately determined, is of great value, but care must be taken to determine whether a large gravel size may have influenced the results. Usually in the case of gravel, only the lowest values of N are taken into account. The lowest values of the angle of internal friction in Table 9.1 are conservative estimates for uniform, clean sand and they should be reduced by at least 5° for clayey sand. These values, as well as the upper values of the angle of internal friction which apply to well graded sand, may be increased by 5° for gravelly sand. The position of the water table, the depth of the foundation and the grain size of the material tested must be taken into account in analysing the results of the tests.

Terzaghi and Peck[2] suggested that for loose, very fine or silty submerged sand, positive pore water pressures might develop in the soil during the standard

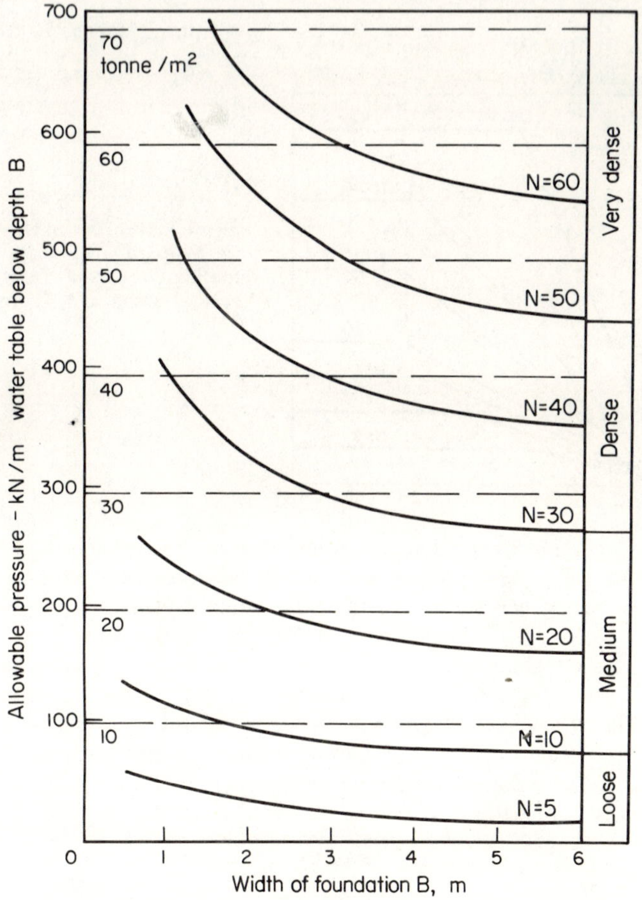

Figure 9.1(a) Chart for estimating the allowable bearing pressure for 25 mm settlement on sand on the basis of the results of the standard penetration test. (After Terzaghi and Peck[2])

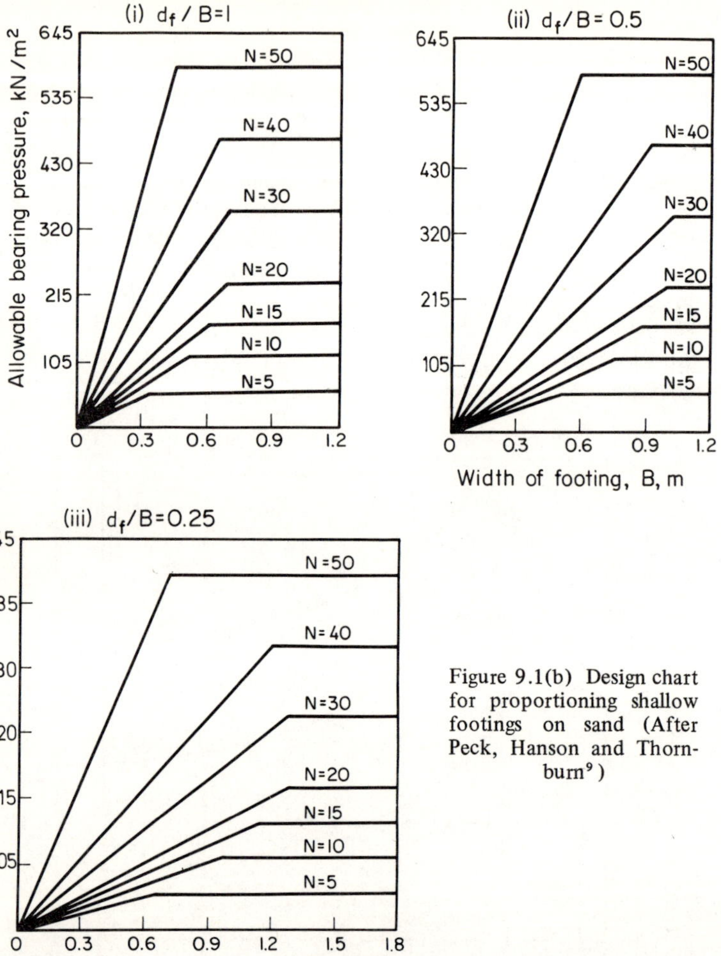

Figure 9.1(b) Design chart for proportioning shallow footings on sand (After Peck, Hanson and Thornburn[9])

penetration test, due to the dynamic application of the load and the low permeability of the soil. These positive pore water pressures would reduce shearing resistance opposing penetration so that the standard penetration value of such soils would decrease with increasing submergence. On the other hand, in dense, very fine or silty submerged sand, the penetration test might produce negative pore water pressures which would increase the resistance to penetration and hence increase the N value. Consequently Terzaghi and Peck suggested that for submerged very fine or silty sand with a standard penetration value N' greater than 15, the relative density would be nearly equal to that of a dry sand with a standard penetration value N where:

$$N = 15 + ½(N' - 15)$$

Terzaghi and Peck[2] showed that the number of blows could be related to the allowable bearing pressure for various footings widths (Figure 9.1a), but subsequent observation on structures has indicated that their method was unduly conservative (see Sutherland[6]). These curves represent the relation between the widest width of the footings (*B*) and the soil pressure required to produce 25 mm of settlement. In other words, if *B* is the width of the largest footing supporting the structure and all of the footings are proportioned in accordance with the allowable bearing pressure corresponding to *B* then the maximum settlement of the foundation should not exceed 25 mm or the differential settlement 19 mm. The *N* value used with these curves should be the average of a series of tests performed between the base of the footings and a further depth (equal to *B*), one test being carried out every 0.75 m. These allowable bearing pressures relate to a non-cohesive soil in which the water table is located at a depth of at least twice the foundation width. If the water table is at or near foundation level, being very close to the surface, then twice the amount of

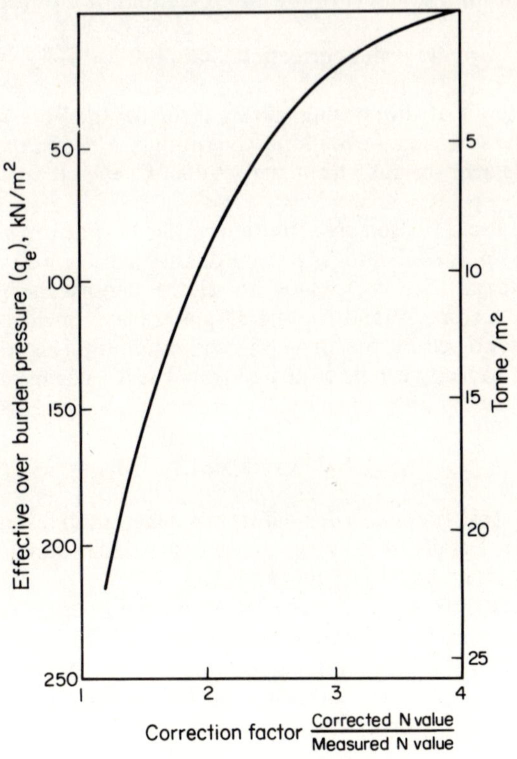

Figure 9.2 Correction factors for standard penetration test (After Gibbs and Holtz[7])

settlement can occur. If this is to be avoided the load should be reduced accordingly. The curves were based on the behaviour of shallow footings (i.e. depth $\lesseqgtr$ width), consequently corrections have to be applied for overburden pressure. Corrections were proposed by Gibbs and Holtz[7] (Figure 9.2) and have been subsequently modified by Tomlinson[8] and by Peck, Hanson and Thornburn[9] (Figure 9.1b).

Meyerhof[10] recommended that the allowable bearing pressure obtained from the Terzaghi and Peck curves should be increased by 50% and that the presence of the water table should be ignored since its presence is recorded by the blow-count. Subsequently Meyerhof[11] suggested that the amount of settlement (S) likely to take place in sand could be estimated from the following expression:

$$S = P\sqrt{\left(\frac{B}{2N}\right)}$$

where

P is the net foundation pressure in tsf;
B is the width of footing in inches, and
N is the average value within a depth equal to B.

The original curves have been modified by Peck, Hanson and Thornburn[9] (Figure 9.16) and they gave the following correction for the ground water table:

$$\text{Ground water correction factor} = 0.5 + \frac{d_w}{d + B}$$

where d is the depth of the footing and d_w is the depth of the water table from the surface of the surcharge surrounding the footing. A review of the methods of calculating settlement in sands from the results of penetration tests is given by Jordan[12].

According to Peck, Hanson and Thornburn[9] the bearing capacity of a footing on non-cohesive soil is related to its relative density and the width of the footing. The angle of internal friction is related to relative density and therefore to the bearing capacity factors N_γ and N_q and all are at least roughly correlated with the N values determined by the standard penetration test (Figure 9.3). Thus the ultimate bearing capacity can be evaluated from the N value by using the following expression, the bearing capacity factors being obtained from Figure 9.3:

$$q_d = \tfrac{1}{2}B\gamma N_\gamma + \gamma d_f(N - 1).$$

The angle of internal friction as derived from a penetration test can be made use of in pile design. Details of how the N value is used in foundation design are provided by Sanglerat[1] and Tomlinson[8].

Wakeling[13] has shown that the results of standard penetration tests can be correlated in a general way with the classification of chalk proposed by Ward, Burland and Gallois[14]. Like other authors he noted the importance of the position of the water table, pointing out that usually low SPT values are obtained from submerged chalk.

THE DUTCH CONE PENETROMETER

In recent years there has been a notable increase in the use of static cone penetrometers (Figure 9.4) in soil investigations. This, according to Mitchell and Durgunoglu[15], is due to the fact that they are, firstly, relatively quick and economical to perform and, secondly, provide reliable information on *in situ* soil characteristics.

The most widely employed static method involves the Dutch cone penetrometer. This is especially suited to soft clays and loose sands which are particularly difficult to sample. The Dutch cone penetrometer test is performed by advancing a cone, with a base of 1000 mm^2, at a rate of 20 mm/s, into the soil and measuring the resistance offered against the cone. The cone resistance is recorded against depth thereby allowing a profile of the variation in *in situ* density and consistency to be derived (Figure 4.2).

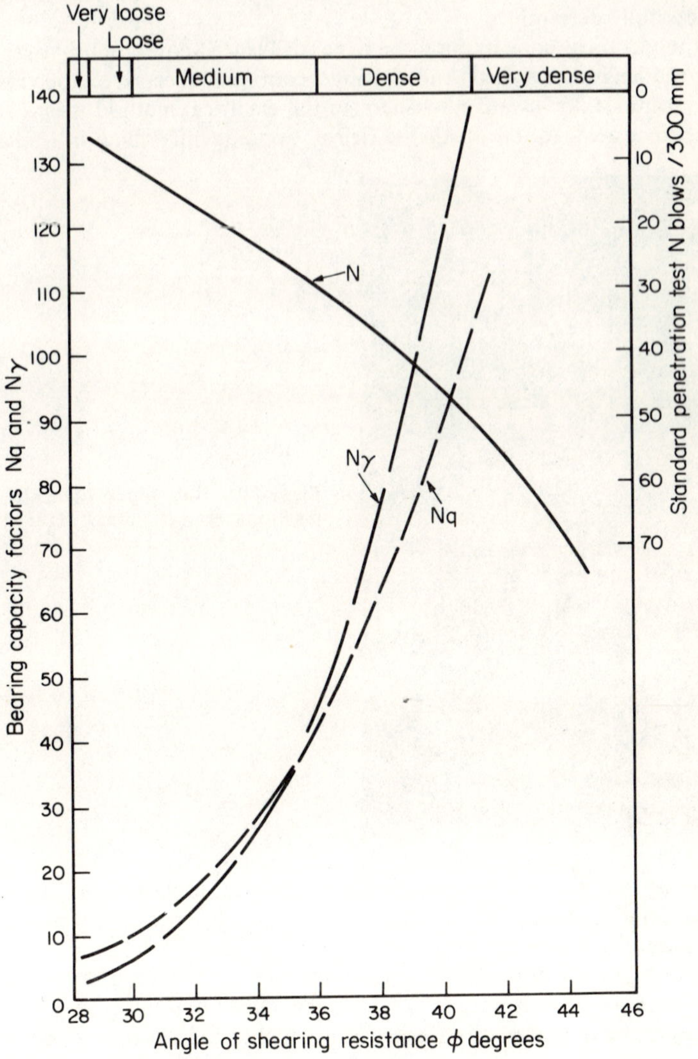

Figure 9.3 Relationship between the angle of shearing resistance, bearing capacity factors and the N values from the standard penetration test (After Peck, Hanson and Thornburn[9])

With a mechanical penetrometer readings are generally taken at intervals of 200 mm. The cone and surrounding tubes are sunk to the required depth and the cone is then intruded a further 70 mm into the soil, the cone resistance then being recorded. A sleeve, located just above the cone, measures the side friction at the same time as the cone resistance. Because the sounding tubes are at rest when a reading is taken, only small frictional forces are developed along the shaft. This means that the data obtained by discontinuous mechanical soundings are not much influenced by such forces and depend mainly on the properties of the soil at the depth where measurements are taken. As a result with a mechanical penetrometer, the friction ratio, the ratio of sleeve to cone resistance, cannot be successfully determined.

In the electrical cone method the cone is advanced continuously over a metre depth, the length of a single tube, giving a continuous record of cone resistance. The continuous downward movement in the electrical method means that frictional forces are exerted along the sleeve. Consequently there is a reliable and

Figure 9.4 Ten tonne hydraulic static penetrometer (Courtesy Cementation Ltd)

continuous record of sleeve resistance and hence friction ratio. The marked difference in the values of friction ratio obtained by mechanical and electrical penetrometer methods has been pointed out by Joustra[16]. Heijnen[17] maintained that these frictional forces gave rise to an increase in the level of effective stresses around the cone, the magnitude of which depended upon soil type. For example, it is substantial in sandy soils. Consequently higher cone resistances are obtained with an electrical cone than with the discontinuous mechanical method. Indeed according to Heijnen practical experience has shown differences ranging from 20 to 100%.

Several attempts have been made to refine the cone penetrometer method. For instance, Mitchell and Durgunoglu[15] carried out model tests with static cone penetrometers in dense cohesionless and low cohesion soils in an attempt to define the failure mechanism more fully. This they did in terms of soil friction angle, cohesion and density on the one hand and cone apex angle, penetrometer roughness and relative depths on the other. For example, they suggested that the ultimate resistance (q_c) could be obtained by using the following expression:

$$q_c = cN_c \epsilon_c + D\gamma N_\gamma \epsilon_\gamma$$

where c is the unit cohesion, D is the penetrometer diameter, γ is the unit weight of the soil, N_c, N_γ are bearing capacity factors and ϵ_c, ϵ_γ are shape factors.

The relationship between the N values of the SPT test and those of static cone resistance (q_c) is expressed in the ratio $n = q_c/N$. This ratio varies according to soil type (see Sanglerat[1]) as follows:

Soil type	q_c/N
Silts, sandy silts, and slightly cohesive silt-sand mixtures	2.0
Clean, fine to medium sands and slightly silty sands	3–4
Coarse sands and sands with little gravel	5–6
Sandy gravels and gravel	8–10

Terzaghi[18] related core resistance (q_c) to undrained shear strength by the following expression:

$$q_c = N_k\tau + \gamma.z$$

where N_k is the cone factor, τ is the undrained shear strength, γ is the total unit weight of the soil and z is the depth of penetration. Although subsequent correlations between these two factors have often shown a wide scatter of results, investigations continue to be made in an attempt to determine the relationship between cone resistance and shear strength. One of the most recent has been by Lunde *et al*[19] who made correlations between the cone resistance and shear vane strength of some Scandinavian soft to medium stiff clays. They showed that for marine clays the N_k factor averaged 17 whilst for brackish-marine clay the range was 11 to 13. Nevertheless they found themselves obliged to agree with Schmertmann[20] that the correlation between cone resistance and undrained shear strength should always be assessed locally.

Meyerhof, quoted in Jordan[12], proposed that the static cone resistance for dry, non-cohesive soils could be used to estimate the allowable bearing pressure required to ensure that footings did not settle more than 25 mm. For footings up to 1.2 m wide the relationship is:

$$\text{Allowable bearing pressure } (q_a) = 3.6\, q_c \text{ kN/m}^2$$

whilst if the footings exceed this width it is:

$$q_a = 2.1 q_c \text{ kN/m}^2$$

These values of allowable bearing pressure should be halved if the soil in the stressed zone is beneath the water table. Meyerhof suggested that a conservative estimate of settlement could be made from the following expression:

$$S = \frac{PB}{2q_c}$$

The settlement of shallow foundations in non-cohesive soils can be estimated from the results of static penetration tests by using Buisman's formula (see Buisman[22], and also De Beer and Martens[23]) which is as follows:

$$\text{Constant of compressibility } (C) = 1.5\frac{q_c}{p_o}$$

where p_o is the overburden pressure at the depth where the static cone resistance is tested.

After correlating the values of static penetrometer tests with the compressibility characteristics of cohesive soils Gielly, Lareal and Sanglerat[24] concluded that the resistance values could also be used to derive the compressibility constant. They suggested that with the resistance values and those of water content, the compression index (C_c) could be approximated from Table 9.2. The value of compression index, together with that of void ratio could then be substituted in the following expression to estimate the compressibility constant:

$$C = 2.3\,\frac{1+e}{C_c}$$

These three authors stated that the method they outlined could only be used in the soils where the variation in resistance values was low. Although it needed to be refined by further research it could be used for providing a quick estimate of the range of settlements.

Table 9.2 RELATIONSHIP OF STATIC CONE RESISTANCE AND WATER CONTENT TO COMPRESSION INDEX

Static cone resistance (kN/m²)	*Water content* (%)	*Compression index*
Over 1200	Under 30	Under 0.2
	Under 40	Under 0.3
600–1200	40–100	0.3–0.7
	100–130	0.7–1
Under 600	Over 130	Over 1

Once C has been obtained, for either type of soil, the amount of settlement (S) can be calculated from the following expression:

$$S = \frac{H}{C}\log_e\frac{p_o + p_v}{p_o}$$

where H is thickness of layer in metres and p_v is the vertical stress in kN/m² induced at the centre of the layer by the net foundation pressure. If the stressed beneath a foundation consists of several layers of soil then the settlement is calculated for each layer and summed to give the total expected settlement below the foundation.

The values of cone resistance also provide a rapid appreciation of the ultimate loads which can be carried by piles. The ultimate load bearing resistance of a pile being taken as the average resistance value over a depth equal to three times the pile diameter above, and one pile diameter below the position of the end of the pile. The skin fraction on the pile shaft can also be derived from the cone resistance value (see Meyerhof[21]).

THE SHEAR VANE TEST

Vane tests are often used to measure the *in situ* undrained shear strength of very soft to firm clays. Carlson[25] and Skempton[26] were among the first workers to perform *in situ* shear vane tests and both suggested that the soil strength determined by these tests was often higher than that obtained by unconfined compression tests in the laboratory and agreed better with the computed average shear strength along slip surfaces in slides. However, the claim has been disputed recently by La Rochelle, Roy and Tavenas[27] who concluded that the vane test seemed to grossly underestimate the *in situ* shear strength of clays when compared with the results of compression tests on good quality samples. They attributed this underestimate to the anisotropy of clay deposits and the disturbance they undergo when the vane is intruded. Previously Aas[28] had also suggested that the discrepancy between the results of laboratory tests on clay and those obtained by vane measurements in the field was probably due to the anisotropy of its shear strength in that the undrained shear strength is dependent on the initial effective normal stress and the shear stress in the failure plane and that *in situ* these stresses might vary in different directions. This would certainly apply in fissured clays.

In its simplest form the shear vane apparatus consists of four blades arranged in cruciform fashion and attached to the end of a rod (Figure 9.5). The height of the vane is usually twice its width and for soils with a shear strength of up to 50 kN/m^2 a 150 mm $\times$ 75 mm vane is suitable. For soils with shear strengths exceeding this value a smaller vane, 100 mm $\times$ 50 mm, is satisfactory.

To eliminate the effects of friction of the soil on the vane rods during the test, all rotating parts, other than the vane, are enclosed in guide tubes. The vane is housed in a protective shoe when tests are carried out by direct penetration from the surface. Tests can also be done down a borehole, at different depths, as it is advanced. The vane and rod are pushed into the soil from the surface or the base of a borehole to a point 0.5 m above the required depth (see BS 1377[29]). Then the vane is pushed out of the protective shoe and advanced to the test position. It is then rotated at a rate of 6–12° per minute, and the maximum torque required for rotation is recorded.

It is assumed that the soil fails along the surface of a cylinder, the diameter and height of which are equal to those of the vane. In addition, it is assumed that the shear stresses are fully mobilised and uniformly distributed across the entire failure surface although not necessarily in equal magnitude for horizontal and vertical surfaces. After failure a second torque reading may be made to assess the remoulded strength of the soil, and so the sensitivity of the clay. The vane is first rotated about a dozen times to remould the clay and then after 10 minutes the test is carried out. The shearing resistance (τ) is obtained from the following expression:

$$\tau = \frac{M}{\pi\left(\frac{D^2 h}{2} + \frac{D^3}{6}\right)}$$

where

D is the diameter of vane;
h is the height of vane, and
M is the maximum torque.

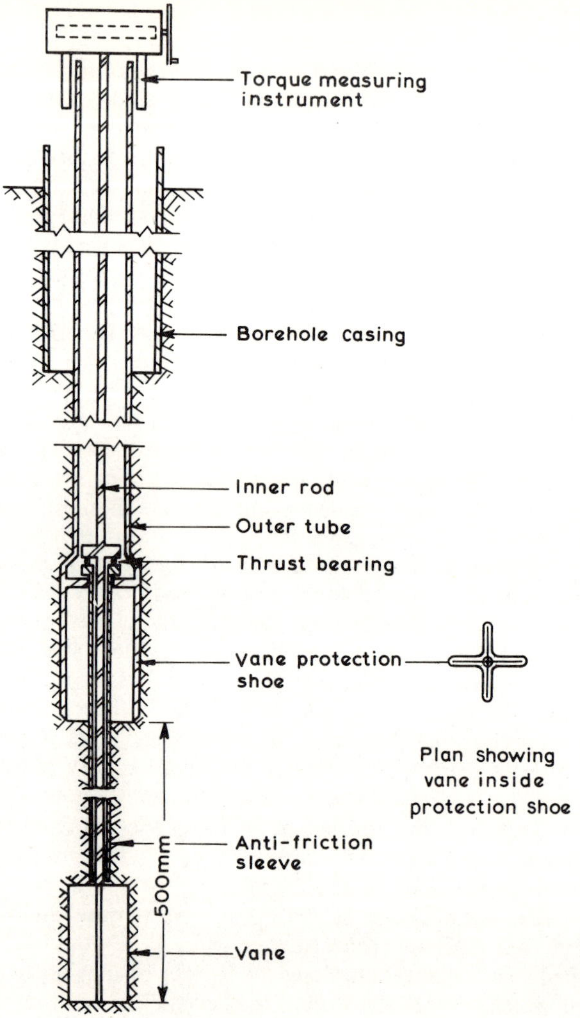

Figure 9.5 Shear vane apparatus

As the height-width ratio of the vane is 2:1 the above expression can be simplified as follows:

$$\tau = \frac{M}{3.66D^3}$$

Tests in clays with a high organic content or with pockets of sand or silt are likely to produce erratic results. The results should therefore be related to borehole evidence.

Wiesel[30] carried out tests to assess the influence of vane shape and rotation rate on the measured shear strength of soft normally consolidated clay. His tests revealed that the undrained shear strength was lower in the horizontal than vertical plane, and that the relationship between the shear strength and rotation rate followed a power function. For further details on shear vane testing, see Flaate[31].

PLATE BEARING TESTS

The basic purpose of the plate bearing test is that a loaded plate gives rise to settlement which is a function of the effective modulus of deformation of the ground beneath it. It is generally assumed that the load is uniformly distributed, and total load and average ground displacement are measured. This test is particularly useful where soils are difficult to sample as, for example, fills.

The size of the plate is an important factor, it usually varying in diameter from 300 mm to a metre for soils testing. However, except in the case of narrow footings, the zone affected by the loaded plate is much smaller than that which eventually will be affected by the structural load. In such a situation a weaker horizon at depth or a high water table in non-cohesive soils may not be discerned by a plate bearing test but they can have a significant influence on the performance of the ground conditions once the structure is erected. Consequently to avoid misinterpretation of the results of plate bearing tests they should be referred to borehole evidence.

Ideally plate bearing tests should be carried out, not just at foundation level, but at various depths within the zone of influence of the foundation structure. In particular in a heterogeneous sequence of soils care must be taken to choose test levels which provide an overall picture of the various types of soil involved within the zone which is to be stressed by the structural load. The screw plate is a recent development of the plate bearing test. It is rotated into the ground to a series of chosen depths, at each of which a loading test is performed. Moreover no excavation is required and tests can be done below water table.

The plate should be properly bedded and the test carried out on undisturbed material so that reliable results may be obtained. Dial gauges for measuring displacement must be provided with a firm measuring base, usually in the form of a steel bridge spanning the test area and provided with a stable anchorage about three or four plate diameters away. The load is usually applied by jacking against kentledge, it being applied in increments, successive increments being added after settlement has ceased. Each incremental loading may amount to one fifth of the proposed bearing pressure or they may be added in steps of 25 to 50 kN/m^2 (these are smaller in soft soils, i.e., where the settlement under the first increment of 25 kN/m^2 is greater than 0.002 D, where D = diameter of plate). Generally the ground is subjected to cyclic loading and unloading. The test is continued up to 2 or 3 times the proposed load of the structure, or in clays until settlement equal to 10 to 20% of the plate dimension is reached. It can be assumed that the stage of primary consolidation in clays has been completed when, after placement of the final increment, the rate of settlement becomes less than 0.1 mm in 2 hours. Consequently the ultimate bearing capacity at which settlement continues without increasing the load is rarely reached. Although it is generally only necessary to show that a foundation is capable of carrying a given load rather than determining the point at which it will fail, Tomlinson[9] maintained that the test should be taken to failure whenever economically possible. Large plate bearing tests are, however, expensive and may need to be carried out over an appreciable period of time. The material beneath the test plate should be examined to a depth of 1 m after the test.

Marsland carried out a series of plate loading tests in the London Clay. In 1972[32] he described tests made in two boreholes of 0.3 m diameter and carried out at intervals of 1.5 m to a depth of 15 m below the top of the clay. The plates had a diameter of 292 mm and the boreholes were either partially or completely lined. The load was applied at a constant rate of penetration of 2.5 mm/min and the bearing pressure value was taken at a penetration of 15% of the plate diameter. He showed that the undrained shear strength of the London Clay varied between 98 and 200 kN/m^2.

In a comparison of the results of large scale plate bearing tests in the London Clay with those obtained from static cone penetration tests Marsland[33] noted that the bearing pressures measured in cone tests were 2.4 to 2.65 times those measured in load tests. This was largely due to the difference in size between cone and plate, the former being too small to take account of sufficient fissures to be representative of the soil mass. Marsland[34] had previously shown that plate load tests carried out on fissured clay gave deformation moduli that were 1.8 to 4.8 times higher than those determined by triaxial tests.

Problems may be experienced in soils subject to stress relief leading to heave. The test, in such cases, should be carried out as quickly as possible. In some instances it may provide a conservative estimate of Young's modulus.

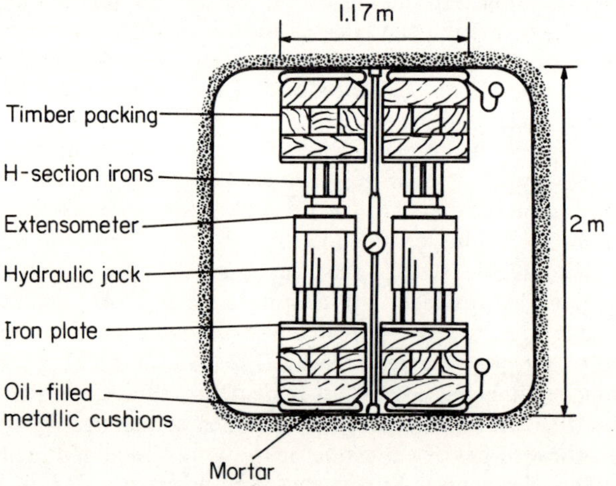

Figure 9.6 Large scale plate loading apparatus (After Rocha, M., Serafim, J.L., Da Silveira, A.F. and Neto, J.M.R. Deformability of foundation rocks, Proc. 5th Int. Cong. on Large Dams, Paris, 3, 531–559)

Rock masses are heterogeneous, discontinuous assemblages of material and as a consequence laboratory tests give overestimates of their stiffness as compared with the results obtained from *in situ* field tests. Overestimation of stiffness by factors of 5 to 15 is quite common according to Stagg[35]. Hence plate bearing tests are carried out on rock masses in an attempt to assess the influence of discontinuities, and the loaded area must be greater than the distance between discontinuities. Even so, as far as large structures such as concrete dams are concerned, Rocha[36] maintained that the heterogeneity and joint systems within rock masses require that tests should really be carried out on volumes so large that the cost and time involved would be prohibitive. Hence the results of *in situ* tests made on comparatively small volumes are often controversial. In the case of concrete dams, loading of the order of several thousand kilonewtons is required to obtain measurable deformation of representative areas (see Rice[37]), but for economic reasons the area of rock loaded rarely exceeds 1 m^2 (Figure 9.6). Thus the zone of influence usually does not extend beyond a depth of 2 m.

Because of the difficulty in providing a satisfactory reaction point for jacks to bear against, tests are usually carried out in specially excavated galleries which provide a sufficiently strong reaction point for the loading jacks to bear against (see Serafim[38]). Unfortunately as a gallery is excavated, the rock to be tested is disturbed by the removal of other rock and the modification of the state of stress in the rock mass, indeed the stress relief may give rise to fractures which may alter deformability significantly. Care must therefore be taken with excavation.

In addition to the measurement of plate displacement Stagg[35] suggested that it was advisable to measure the normal surface displacement at several points outside the loaded area since the latter provided a check against errors and allowed a better assessment of deformation properties at depth. This is because displacements outside the loaded area are influenced to a much greater relative extent by the behaviour of rock well below that in immediate contact with the plate. He suggested that something like 80% of plate displacement is attributable to that rock within a depth equal to double the plate diameter.

The mass deformation modulus (E_m) can be obtained from the following expression:

$$E_m = \frac{P(1-\nu)^2}{2\delta D}$$

where

P is the load on the plate;
ν is Poisson's ratio;
δ is the plate displacement, and
D diameter of plate (see Coates and Gyenge[39]).

As in loading soils, so rocks are also loaded and unloaded in cycles, allowing hysteresis curves to be plotted. Such cycling provides an indication of the amount of irrecoverable strain. The rock usually stiffens appreciably after the first few cycles, due to joint closure. Usually the secant modulus of deformation is determined from the second loading curve because of the tendency for the rock to behave in a more elastic manner as it stiffens. Short-term creep measurements may be made at any stage by observing displacement at constant load. The largest plate bearing test done in Britain was carried out at a site at Mundford, Norfolk, on the Middle Chalk (see Ward, Burland and Gattois[14]). This site had been offered by the government for a giant proton accelerator for CERN, thus an unusually accurate assessment of ground conditions was required. A water tank, 18.3 m in diameter and weighing 4500 tonnes, was used to apply the load, which averaged 180 kN/m^2. At the same time a series of smaller scale plate loading tests were performed on the site, with loadings of up to 1.67 MN/m^2.

Deflections were measured at various depths in shafts sunk to just above the water table, a depth of 16.3 m, below and around the tank. Steel tubes were extended from the base of each shaft so that deflections at greater depths could be monitored. A short-term loading cycle was carried out first to investigate the immediate ground behaviour. The linearity and recoverability of the observed displacements and strains suggested that the ground beneath the tank behaved in a reasonably elastic manner for this range of applied stress. This was followed by a long term tank test, the purpose of which was to allow a qualitative assessment to be made of the creep properties of chalk. It showed that creep in the two weakest grades of chalk was appreciable.

The geotechnical work at Mundford demonstrated the importance of gaining a detailed picture of the *in situ* structure of the rock. Ward *et al* were able to correlate the creep properties and values of Young's modulus with their visual classification of the Chalk (Table 1.13). Furthermore Burland and Lord[40] showed that there was also a reasonable relationship between the E values obtained in the field and those derived in the laboratory.

A long term plate bearing test of many months duration was carried out on the Bunter Sandstone by Moore[41]. Contact stresses of up to 5.6 MN/m^2 were provided by 360 tonnes kentledge on a plate 0.91 m diameter. As far as footings are concerned this afforded the opportunity of extending stress levels beyond

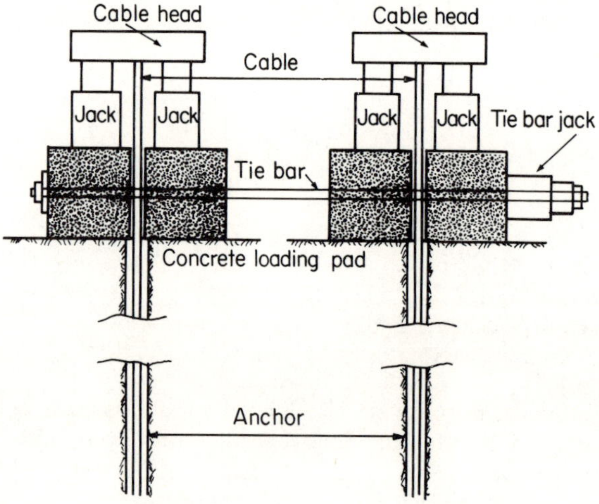

Figure 9.7 The 'double-cable' test (After Zienkiewicz and Stagg[42])

the limits currently used in the construction industry. Creep accounted, on average, for 25% of the total settlement (4 mm) and Young's modulus appeared to increase with depth.

The Freyssinet jack can be regarded as a variation of the plate bearing test. This is placed in a narrow slit in the rock mass and then grouted into position so that each face is in uniform contact with the rock. Pressure is then applied to the jack.

CABLE TESTS

As pointed out above one of the problems in using a plate bearing test in a rock mass is the difficulty in obtaining a suitable reaction point. Hence tests are usually carried out in galleries, which are very costly to excavate. However, the problem of a suitable reaction point is overcome in a cable test where the reaction support is provided by a cable anchored in a borehole at a suitably large distance from the surface loading pad. Zienkiewicz and Stagg[42] recommended a minimum depth for anchorage of between six and eight times the bearing pad diameter. They maintained that the loading area could be made large enough, in relation to the borehole cutting through it, to justify neglecting the influence of the borehole in subsequent analysis.

A greater area of the rock mass can be tested with two cables. Loads of up to 10 MN can be applied by a single cable and several cables can be used where necessary. With two adjacent cables (Figure 9.7) loads tangential to the surface can be applied and information can be obtained about the variation of elastic moduli with direction of load. If it is assumed that the rock mass is isotropic, admittedly this is rarely, if ever, the case, then the vertical displacement (w) may be related to the applied load by the Boussinesq equation:

$$w = \frac{P(1 - \nu^2)}{E_r}$$

where E_r is the distance from a point normal load, P

The advantages of the cable test outlined by Zienkiewicz and Stagg[42] are as follows:

(a) It is possible to assess the modulus of deformation in the direction of loading and at the same position where these loads will occur.
(b) A large area of a rock mass can be tested with two cables and tests can be carried out at various levels to determine the variation in moduli with depth.
(c) *In situ* shear tests can also be performed with two cables.
(d) The test is not sensitive to minor inaccuracies in the borehole so that percussion drilling can be employed.
(e) The holes drilled for site exploration can be used for cable testing thereby effecting savings in cost.

PRESSUREMETERS AND DILATOMETERS

The Menard pressuremeter (Figure 9.8) is used to determine the *in situ* strength and deformation modulus of soils and weak rocks. This pressuremeter consists essentially of a probe which is placed in a borehole at the appropriate depth and then expanded. Where possible the test is carried out in an unlined hole but if necessary a special slotted casing is used to provide support. The probe consists of a cylindrical metal body over which are fitted three cylinders. A rubber membrane covers the cylinders and is clamped between them to give three independent cells. The cells are inflated with water and a common gas pressure is applied by a volumeter located at the surface, thus a radial stress is applied to the soil. The deformations produced by the central cell are indicated on the volumeter.

Pressure-deformation relationships are established by observing the volume of water injected into the probe during incremental pressure increases. On the pressure-volume curve (Figure 9.9) there is an initial lead in phase where the earth pressure (P_r) condition is being restored. This is followed by an assumed

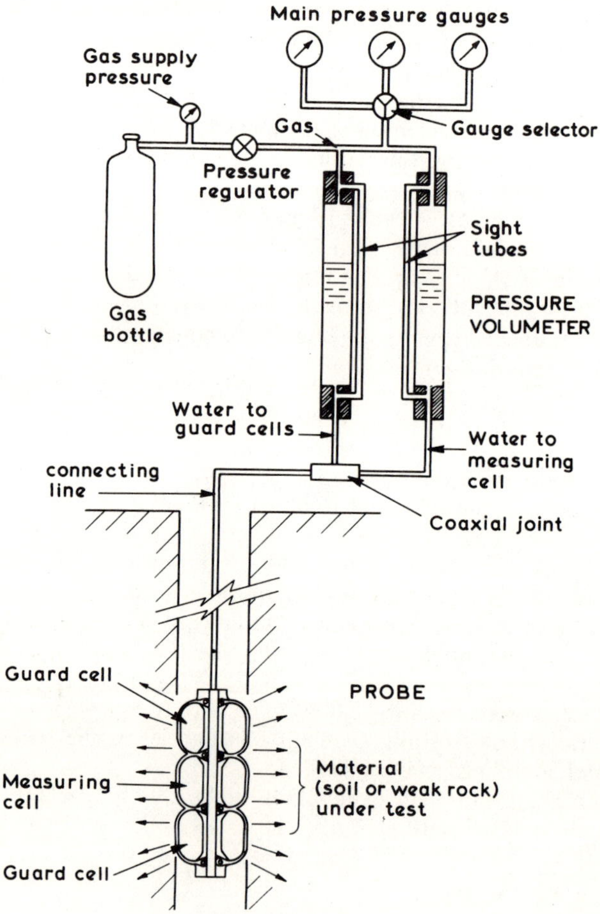

Figure 9.8 The Menard pressuremeter

elastic phase where the response is linear and from which the modulus of deformation (E_p) is derived. It is customary to unload the ground towards the end of the elastic phase and to repeat the test. This results in a flatter line giving the modulus E^+. Broadly speaking the ratio of E_p to E^+ in soils is a function of the soil type and in rocks of the degree of fissuring (in perfectly elastic materials the ratio would be unity). At higher pressures the plastic phase is entered and the volume increases rapidly until the limit pressure (P_L) is reached where the curve becomes asymptotic to the vertical. This provides an indication of the ultimate bearing capacity.

Four volume readings are made at each incremental pressure at time intervals of 15, 30 and 60 and 120 sec after the pressure has stabilised. The plot of the volume change in this period against pressure is known as the creep curve and closely follows the three phases of the pressure-volume curve. During the elastic phase the incremental time-volume change is constant and the flow of ground

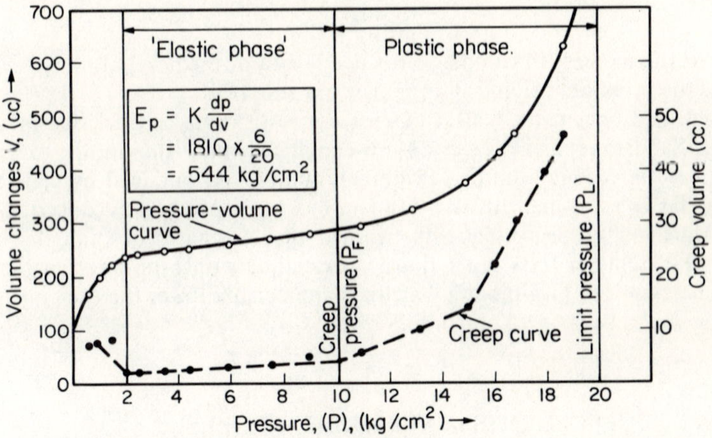

Figure 9.9 Pressuremeter test curves for silty mudstone of Coal Measures age (After Meigh and Greenland[46])

only occurs when the creep pressure P_f is exceeded. The creep pressure is that at which the curve makes a definite upward turn, which generally agrees with the upper limit of the linear phase of the pressure-volume curve. Hobbs and Dixon[43] found that the limit pressure was generally double the creep pressure.

Gibson and Anderson[44] used elastic theory to develop the following expression to obtain the elastic modulus:

$$E_p = \frac{2\Delta P}{\Delta P}(1 + \nu) V_c$$

where

$\Delta P/\Delta V$ is the gradient of the pressure-volume curve;

ν is Poisson's ratio, and

V_c is the initial volume of the measuring cell.

In their interpretation of pressuremeter measurements Gibson and Anderson assumed that the clay under test was saturated, isotropic and incompressible, and that it was an ideal elastic-plastic material.

Ladanyi[45] extended their interpretation to derive a complete undrained stress-strain curve for a clay allowing for non-linear stress-strain behaviour. He argued that the mobilised shear strength at any stage of the test can be related to the undrained shear strain under undrained plane strain conditions. The whole undrained stress-strain behaviour can be estimated *in situ* up to relatively large strains. For overconsolidated clays Ladanyi proposed that the deformation modulus can be determined from the initial straight portion of the adjusted or 'true' pressuremeter curve (Figure 9.10) using

$$E_p = \frac{3\Delta P_i}{\Delta\left(\frac{\Delta V}{V}\right)}$$

where

ΔV is the increase in volume of the test length of the borehole;

V is the borehole volume at any other pressure (P_i).

Meigh and Greenland[46] after extensive investigations in Coal Measures rocks, Bunter Sandstone and Keuper Marl concluded that the results of Menard pressuremeter testing accorded favourably with those obtained by plate bearing tests as far as assessment of the ultimate bearing capacity was concerned. However, Marsland[47] has subsequently warned that estimates of strength obtained from pressuremeter tests can seriously overestimate operational strengths and at the same time can give moduli which are appreciably lower than the operational values.

RECENT DEVELOPMENTS IN PRESSUREMETERS

Several modified versions of pressuremeters have been developed recently, for example, Wroth and Hughes[48], designed an instrument which had a special cutting head so that it could be drilled into soft ground to form a cylindrical cavity of exact dimensions and thereby create a minimum of disturbance. The instrument measures the lateral stress, undrained stress-strain properties and the peak stress ratio of soft clays *in situ*. It has also been used with success in sands.

McKinley and Anderson[49] also developed a new type of pressuremeter to use in till. There are no guard cells on this model and it is inflated by air. Basically the probe consists of a central expansion monitoring device over which the membrane is fitted. This type of pressuremeter enables soils to be tested at a constant rate of radial expansion. The rate of expansion, can, however, be varied, programmed to allow cyclic loading or operated manually. The test to failure takes about 20 to 30 minutes. It can be used in 150 mm boreholes to give closely spaced determinations of undrained shear strength and deformation at constant rate.

DILATOMETERS

As pointed out previously, large scale loading tests in rock masses suffer two serious disadvantages, namely, no matter how carefully a gallery is excavated the rock mass in question is inevitably disturbed, and since, for economic reasons, the loaded area needs to be limited the representativeness of the results is adversely affected. Dilatometers, used to measure the deformability of rocks in diamond drilled holes, overcome these two weaknesses since the rock remains almost undisturbed and several tests can be done down the hole. According to

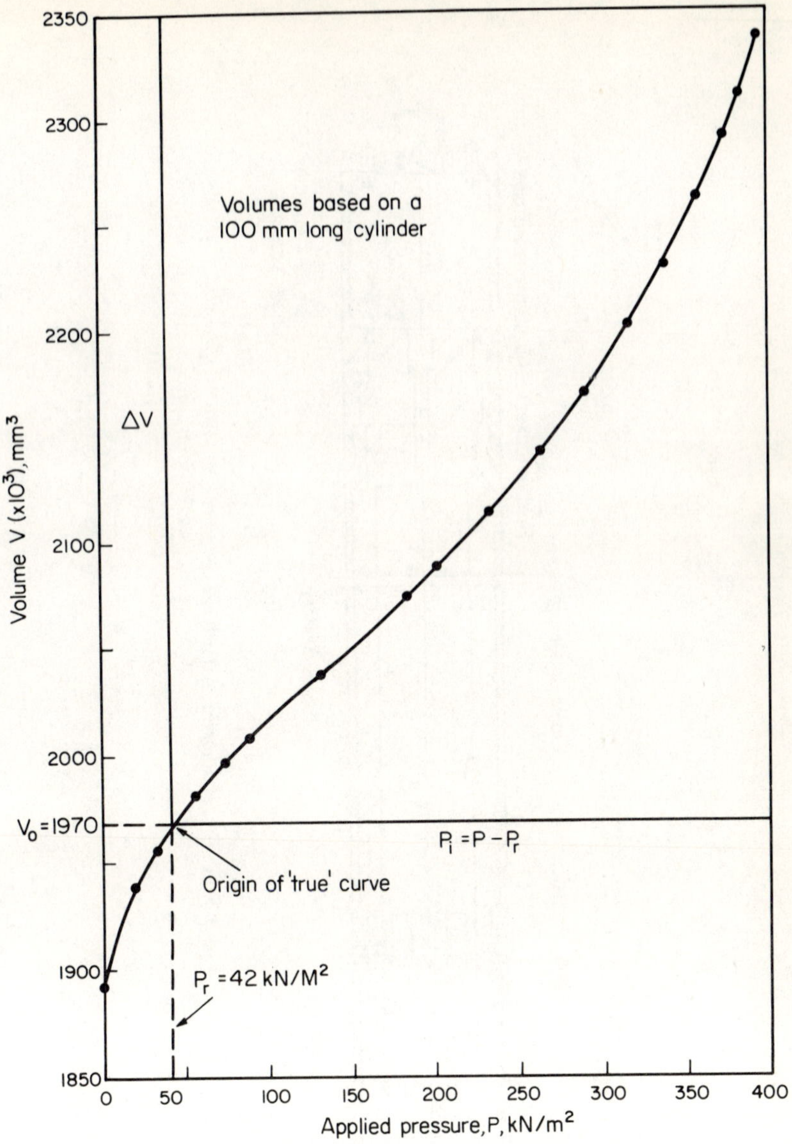

Figure 9.10 'True' pressuremeter curve (After McKinlay and Anderson[49])

Rocha *et al*[52] such testing allows an assessment of the distribution of deformability in a rock mass as well as considering the effects of anisotropy. They, in fact, suggested that the test results could be considered as a quality index. Dilatometer tests can also be carried out beneath the water table, at appreciable and variable depths and at higher pressures than loading tests. Like the plate bearing test, cyclic loading and unloading can be carried out.

Dilatometers may be up to or just over 1.5 m in length and vary from 150 to 300 mm in diameter (Figure 9.11). They can impose loads on the borehole walls of up to 15 MN/m^2. Diametral strains can be determined either directly

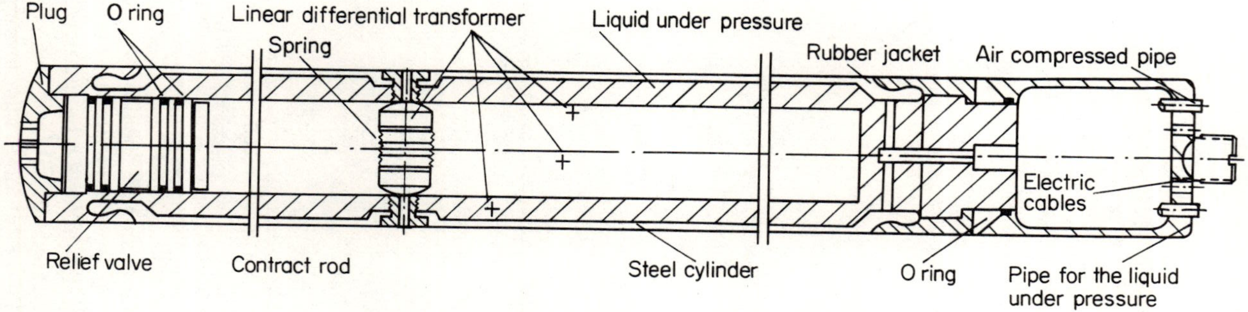

Figure 9.11 The dilatometer (After Rocha *et al*[52])

along two perpendicular diameters or by measuring the volume of liquid pumped into the instrument. The latter provides a less accurate measurement and tends only to be used when the rock is very deformable. According to Rocha *et al*[52] the modulus of deformation (E) can be obtained from the following expression:

$$E = 2\frac{1+\nu}{\Delta\delta}rP$$

where
r is the radius of the hole;
P is the pressure applied, and
$\Delta\delta$ is the deformation of the diameter of the hole.

If, however, the rock is fissured they suggested that the following expression should be used:

$$E = \frac{2rP}{\Delta\delta}\left(\log\sqrt{\left(\frac{P}{\sigma_t}\right)} + 1 + \nu\right)$$

where σ_t is the tensile strength of the rock. They also suggested that the value of diametral formation for fissured rocks should be taken from the second loading curve.

HYDROSTATIC PRESSURE CHAMBERS

Hydrostatic pressure chambers are used to measure the reaction of a rock mass to stress over large areas, giving values of Young's modulus, elastic recovery, inelastic deformation and creep. The results are used to evaluate the behaviour of dam foundations and related strain distribution in the structure. Hydrostatic chambers cover a much larger surface area than other test methods and so provide better results of mass behaviour. However, because of their cost these tests are used sparingly. These chambers may exceed 5 m in length and 2 m in diameter (see Dodds[50] and Deere *et al*[51]). They are lined with reinforced concrete which is divided into a number of independent sections, and reinforced concrete plugs close both ends. A rubber lining prevents water loss and in sound rock water pressures may reach 5 MN/m^2. Radial deformations are measured by special extensometers. The modulus of radial deformation (E_r) is obtained from the following expression:

$$E_r = \frac{r_i^2\, p_w}{r\delta_r}$$

where
r is the radius of deformation;
p_w is the water pressure;
r_i is the internal radius of the pressurized test section, and
σ_r is the deformation of the rock behind the skin of the chamber at radius r.

IN SITU SHEAR STRENGTH – SHEAR TESTING

Because the shear strength of rock masses is a fundamental property it must be carefully investigated for design purposes when large structures such as dams are

Table 9.3 SHEARING STRENGTH OF SOME ROCKS TESTED *IN SITU (after Rocha*[52]*)*

Rock type	*Quality index**	*Cohesion* kN/m^2	*Angle of friction*
Granite	3	1275	62°
Shales I	0.8–2.0	196–686	55°–69°
Shales II	1.0–1.3	98–392	56°–63°
Sandstone	–	196	53°–60°

*Ratio of the weight of water absorbed by the rock dried at 105°C to its dry weight. This provides a measure of the degree of alteration that a rock has undergone.
I Normal to lamination
II Parallel to lamination

concerned. The shear characteristics in a rock mass, particularly along discontinuities, can be tested *in situ* by shear tests performed on large surfaces of rock. The shear strength of anisotropic rocks must be determined along various planes and directions in order to obtain complete information for design. Tests performed on the same rock mass at different normal loadings provide an estimate of its angle of internal friction and the cohesion (see Table 9.3).

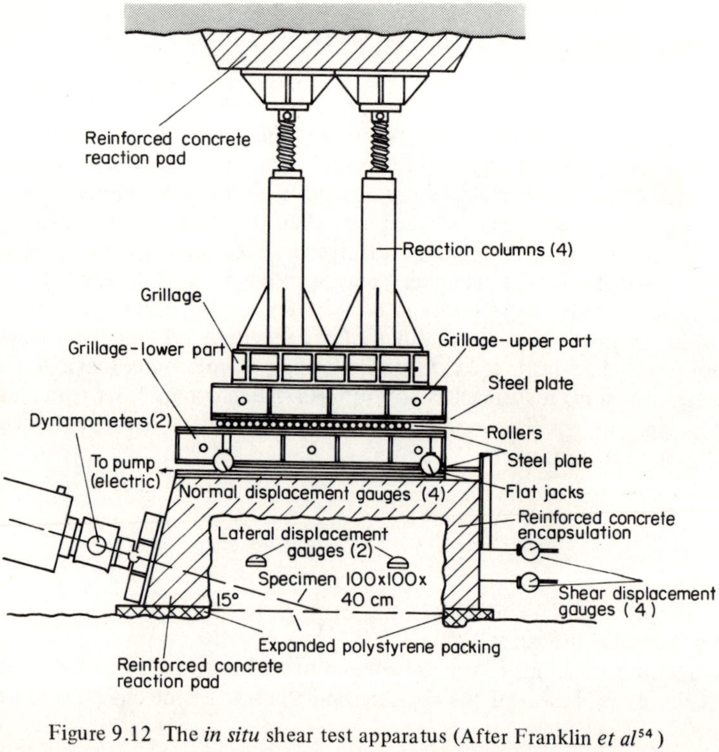

Figure 9.12 The *in situ* shear test apparatus (After Franklin *et al*[54])

Large scale shear tests are made inside galleries, where reaction points for the jacks are readily available. The test sample is excavated and then encapsulated (Figure 9.12) to hold it intact during shearing and to provide a flat, correctly inclined bearing surface. Usually the sample block of rock is sheared from the rock mass whilst a horizontal jack exerts a vertical load. The International Society for Rock Mechanics (ISRM) have proposed in their draft standard on *Suggested Methods for Determining Shear Strength* that samples 700 × 700 mm should be tested. There is considerable evidence to show that in many circumstances the measured shear strength may be independent of the size of specimen, although the displacements at which this strength is mobilised are likely to be greater with larger specimens. In the ISRM method there is an upper limit of 0.2 mm/min placed upon the rate of shearing displacement before reaching peak strength which generally affords enough time for the dissipation of pore pressure for the test to be regarded as drained.

The principal problem in testing small specimens is the limited amount of shear displacement achievable (generally about 10% of sample length). The available displacement may be utilised either at a constant normal stress, with several increments of normal stress as in the multi-stage test or by a method involving cycling of normal stress (Figure 9.13). It may be impossible to reach a residual strength value if the specimen is too small.

By contrast, tests on samples 5 to 10 m long may yield an unrealistic shear displacement gradient along the base of the block. It may be of advantage to test a rectangular rather than a square block, it being longer in the shear path direction. This allows a greater shear displacement for the same contact area so that there is a greater likelihood of reaching a residual strength.

During the test the normal stress remains constant whilst the inclined force is applied. Normal and tangential deformations are recorded as the inclined load is increased. Serafim[38] noted that at a certain inclined load there was a volume increase in the rock being sheared, likening it to dilatancy in a plastic body when the yield strength is reached. The disruption of the structure of a rock mass when subjected to shear sometimes begins long before the maximum tangential load is reached. The volume increase of the sheared mass can be regarded as a criterion for defining shear strength.

The value of the shear test in very jointed and heterogeneous rock masses is, according to Lane[53], severely limited both because of the difficulty in isolating undisturbed test blocks and because the results cannot be translated to the scale of conditions of the actual structure. Laminated rocks such as shales are usually difficult to prepare, being easily disturbed. Disturbance can also be brought about by the relief of residual stress.

In a recent review of *in situ* shear testing Franklin, Manailoglou and Sherwood[54] pointed out that although only one peak strength can be determined per block sample it was possible to evaluate a number of residual strengths at different normal loads using the same specimen block. The number of determinations required is influenced by the incidence of discontinuities. As it is only occasionally that stability problems can be resolved to sliding on a unique and predetermined plane, several determinations may be required to assess the variation in strength within a given family of discontinuities. At times, however, it may only be necessary to assess the behaviour of the weakest surface within a critical family.

Krsmanovic and Popovic[55] performed direct shear tests on limestone in a gallery. The test blocks were 2.8 m × 1.8 m, the shearing surface approximating to 5 m^2. Maximum normal pressures of up to 2.5 MN/m^2 were applied and the hydraulic jacks exerted a maximum shear of 30 MN. Deformations of the

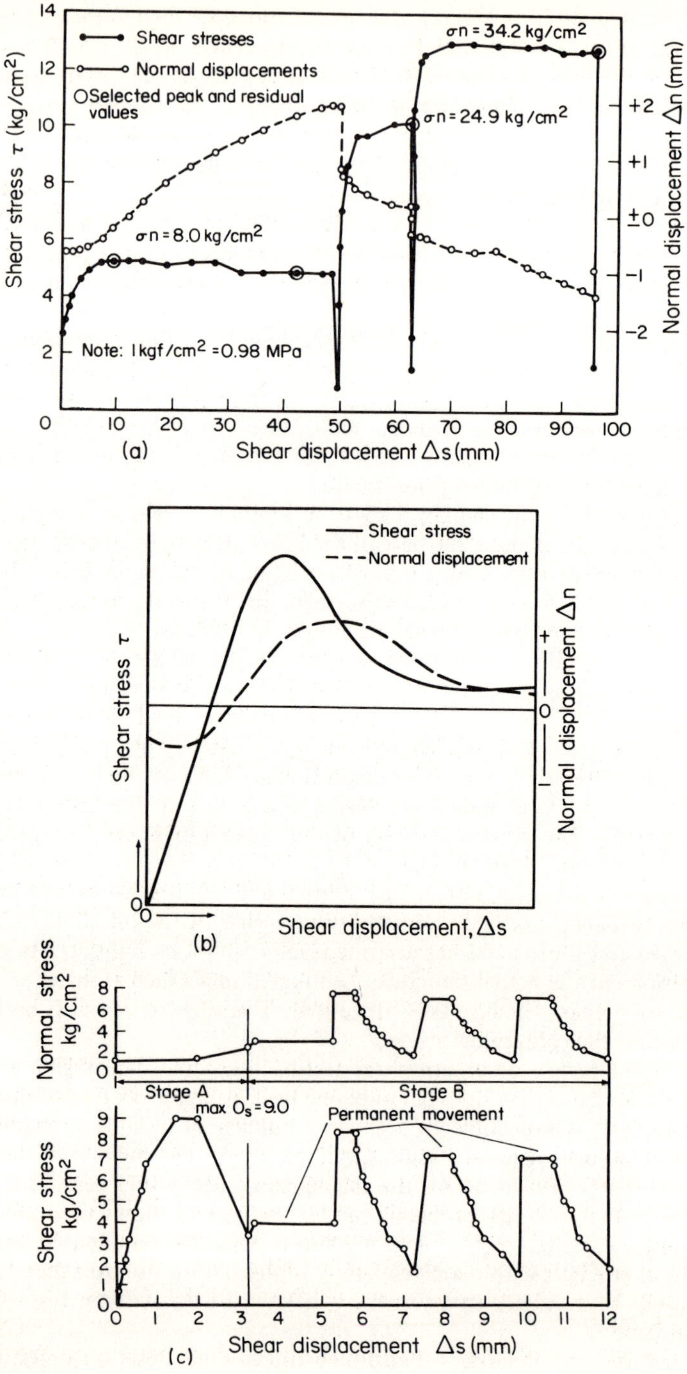

Figure 9.13 Shear test displacement graphs (From Franklin *et al*[54])

blocks were irregular, this being due to the presence of discontinuities. They found that the shear strength was highly influenced by the roughness of the surfaces of the discontinuities. The *in situ* shear tests carried out on mudstone bands in massive quartzite at the Muda dam site have been described by James[56].

TRIAXIAL TESTING

The first attempt to perform *in situ* triaxial tests with hydraulic jacks and flat jacks on a rock cube was made for the Kurobe IV dam project in Japan (see John[57]), although the idea had been conceived by Muller[58] in 1943. A more recent testing programme at Ghiona in Greece has been described by Voort and Logters[59]. A test chamber was excavated in the rock mass and three 1 m^3 samples were prepared at its base, thereby ensuring saturation during the tests. Four slots were cut in each block of rock (Figure 9.14b) into which the flat jacks were fitted and sealed with cement grout. A 1 m^2 steel loading platen with a central 50 mm hole and seatings for four hydraulic jacks (each capable of exerting a load of 100 metric tons) was placed on top of the rock sample block. Rigid steel extension pieces were installed between these jacks and a concrete abutment attached to the roof.

A three-point rod extensometer was inserted into a central borehole within the rock cube in order to determine the vertical displacement of the specimen by transducers at three different depths. Four 50 mm diameter holes were drilled outside the flat jack positions to a point near the centre where an anchor was fastened with grout. This anchor was connected to a reference frame above the sample by a thin steel tube thus forming a pendulum. Horizontal displacements of the steel tubes were measured by transducers.

Because the rock sample could be loaded with different loads in three orthogonal directions it was possible to produce a great number of states of stress in the centre of the specimen. The extensometer readings at different depths inside the sample proved that the rate of compression decreased with growing distance from the loading plate. Strength and deformational behaviour varied with direction of loading and this was attributable to the anisotropy of the rock mass.

ASSESSMENT OF *IN SITU* PERMEABILITY AND PORE WATER PRESSURES

Accurate recording of ground water conditions is important, particularly if excavation level extends beneath the water table. Not only should the water levels be observed in boreholes, say twice daily, but at least one stand-pipe should be installed for long term observation. Piezometers may be installed in boreholes and *in situ* permeability tests carried out. In some instances the ground water may contain substances in great enough quantity to adversely affect concrete, the sulphate content and pH value being of particular interest. A chemical analysis of the ground water is then required to assess the need for special precautions.

The position of the water table is readily determined in homogeneous relatively permeable ground, the hydrostatic head increasing at a linear rate below

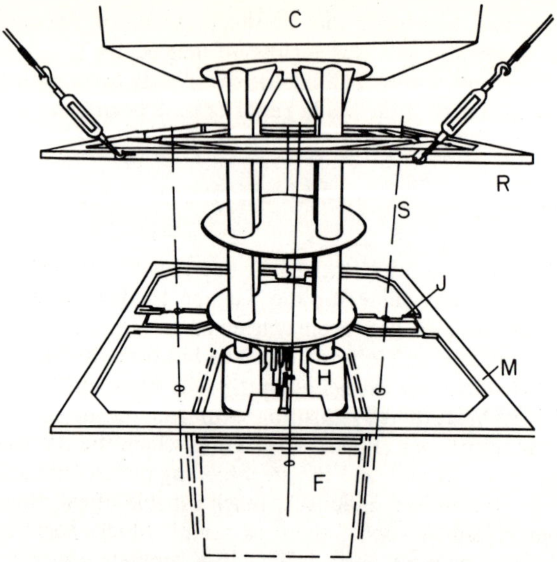

Figure 9.14(a) *In situ* triaxial test apparatus. Loading and measuring equipment arranged for the test.
F. Flat jack
H. Hydraulic jack
C. Counterweight
M. Measurement frame
R. Reference frame
S. Rod pendulum
J. Inductive transducer

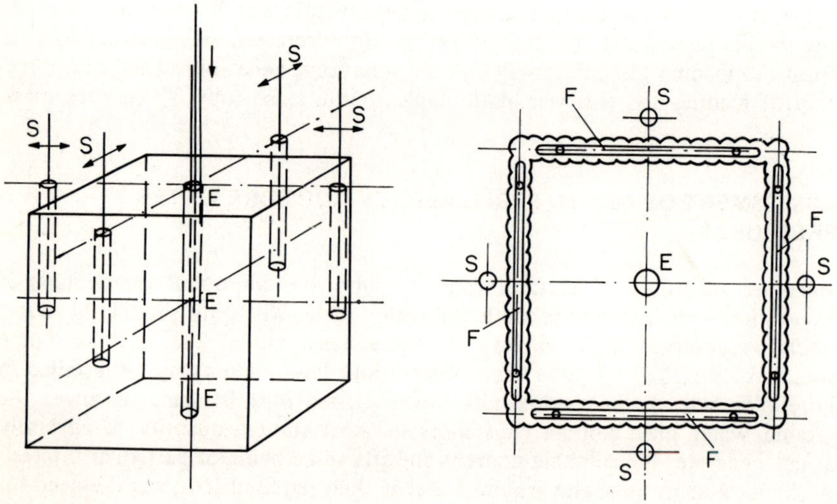

Figure 9.14(b) Orientation of boreholes for extensometer (E), rod pendulums (S) and flat jacks (F). (After Voort and Logters[59])

this level. A stand-pipe consisting of a perforated tube placed in a borehole and surrounded by gravel to near the surface, with a seal of 1 m at the top to keep out surface drainage, records the upper free water level. However, differences in the character of soil or rock types mean varying permeabilities which in turn mean variable water pressures, at different depths.

An initial assessment of the magnitude and the variability of the *in situ* coefficient of permeability can be obtained from tests carried out in boreholes as the hole is advanced. By artificially raising the level of water in the borehole (falling head test) above that in the surrounding ground the flow rate from the borehole can be measured. However, in very permeable soils it may not be possible to raise the level of water in the borehole. Conversely the water level in the borehole can be artificially depressed (rising head test) so allowing the rate of water flow into the borehole to be assessed. Whenever possible a rising and a falling head test should be carried out at each required level and the results averaged. Once a steady state has been attained the permeability (k) can be determined by measuring the rate of flow of water (q) under a constant applied change in head (H) by using the following expression of Hvorslev[60]:

$$k = \frac{q}{F.H}$$

where F is a shape factor. For example, F in the case of a lined borehole with a constant diameter is 2.75 D (D = diameter).

Other shape factors are given by Hvorslev. In a rising or a falling head test in which the piezometric head varies with time the permeability is determined from the expression:

$$k = \frac{A}{F(t_2 - t_1)} \times \log_e\left(\frac{H_1}{H_2}\right)$$

where

H_1 and H_2 are the piezometric heads at times t_1 and t_2 respectively;

A is the inner cross sectional area of the casing in the borehole.

The test procedure involves observing the water level in the casing at given times, then a graph of water level against time is constructed (Figure 9.15).

In low permeability, cohesive soils it is preferable to carry out the test over an unlined section of borehole. However, in soft cohesive soils, because the walls of the borehole may collapse, a gravel filter needs to be installed over the test length. Casing can be used when testing in sands and gravels and the borehole must be kept topped up with water to prevent piping.

The permeability of an individual bed of rock can be determined by a water injection or packer test carried out in a drill hole. This is done by sealing off a length of uncased hole with packers and injecting water under pressure into the test section. Usually because it is more convenient, these permeability tests are carried out after a hole has been drilled. Two packers are used to seal off selected test lengths and the tests are performed from the base of the hole upwards. The hole must be flushed to remove sediment prior to a test being performed. Water is generally pumped into the test section at steady pressures for periods of 15

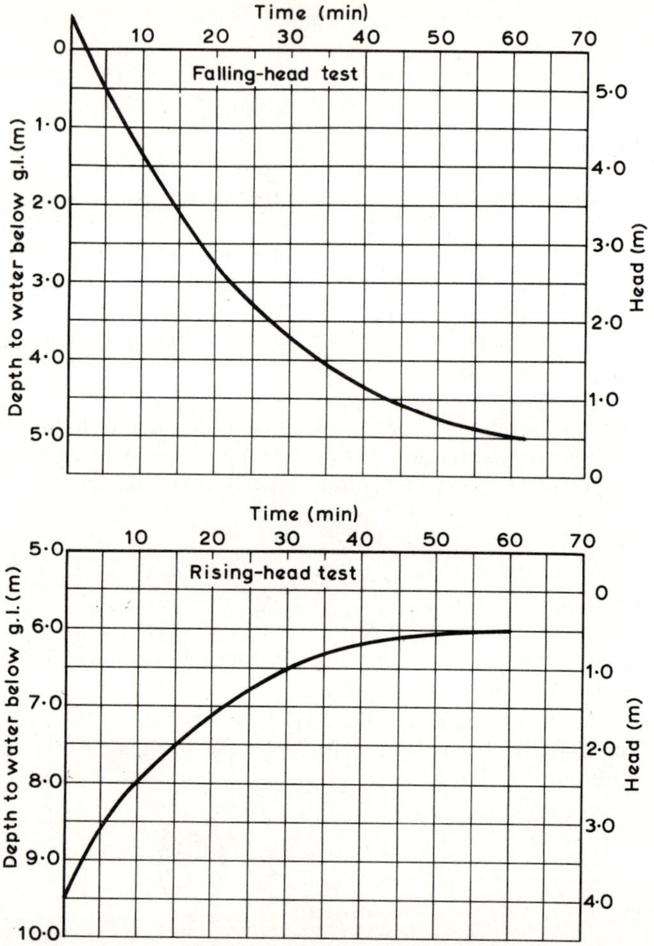

Figure 9.15 Rising and falling head permeability tests

min, readings of water absorption being taken every 5 min. The test usually consists of five cycles at successive pressures of 6, 12, 18, 12 and 6 kN/m^2 for every metre depth of packer below the surface (see Dixon and Clarke[61]). The permeability from packer tests can be derived from

$$k = \frac{q}{C_s H r}$$

where

q is the steady flow rate under an effective applied head (H);
C_s is a constant depending upon the length and diameter of the test section;
r is the radius of the drill hole.

Piezometers are used to measure water pressures. They are sealed into each zone where it is required to determine the water pressure. The installation of a piezometer must be carefully carried out otherwise its performance and measurements may be seriously affected. If more than one piezometer is installed in one borehole the likelihood of failure increases. After installation the ability of a piezometer to function should be checked by topping up and it should be observed daily until an apparent equilibrium is attained. Thereafter the intervals between observations can be lengthened to a week or a month but ideally should continue for about a year. A further account of the measurement of ground water pressures by piezometers is given in chapter 16. The direction of water flow can be determined by monitoring with either dyes or radioactive tracers (see Mather *et al*[62]) which are introduced into the ground water via boreholes.

NOTES ON FIELD INSTRUMENTATION

When some degree of risk is involved in construction then some type of field instrumentation may be required in order to provide a continual check on the stability of the structure during its life-span. Furthermore field observations of both the magnitude and rate of subsurface ground movements are needed in connection with deep excavations, natural slope stability and earth and rockfill dam construction. Such instrumentation needs to assess the pore water pressure (see previous section and chapters 6 and 16), deformation, stress and strain in the ground (see chapter 10). Field monitoring methods both at ground surface and underground make use of a vast range of instruments (see Dunnicliff[63]; Gould and Dunnicliff[64]; Hanna[65]; Green[66]).

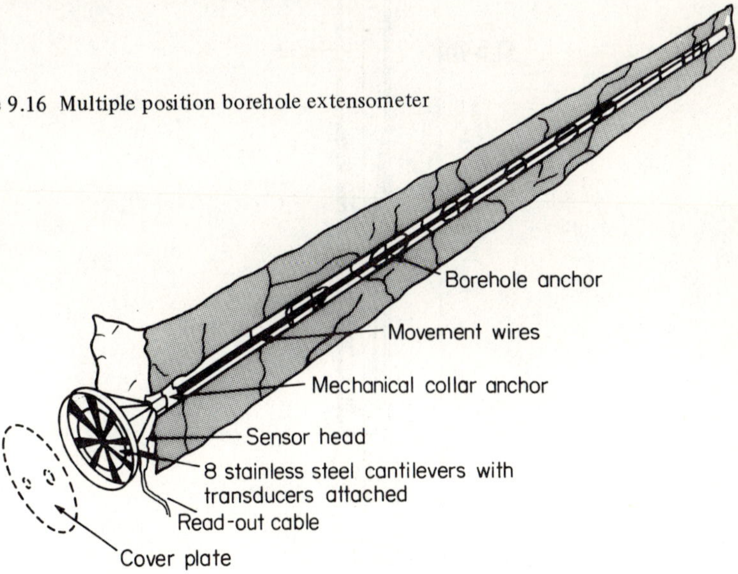

Figure 9.16 Multiple position borehole extensometer

Surface deformation either in the form of settlements or horizontal movements can be assessed by precise surveying methods (see Ashkenazi[67]; Liu, Aldrich and Dugan[68]), the use of EDM or laser equipment providing particularly accurate results (see Penman and Charles[69]). Morton and Au[70] described the results of several years observations of settlement on eight structures in London. In order to record the settlement reference marks were positioned at ground or basement level on the structures concerned and the readings were taken using

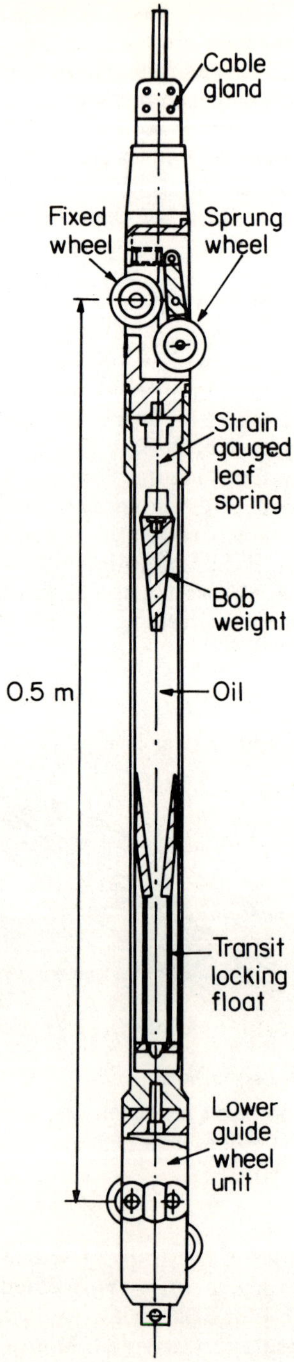

Figure 9.17a An inclinometer

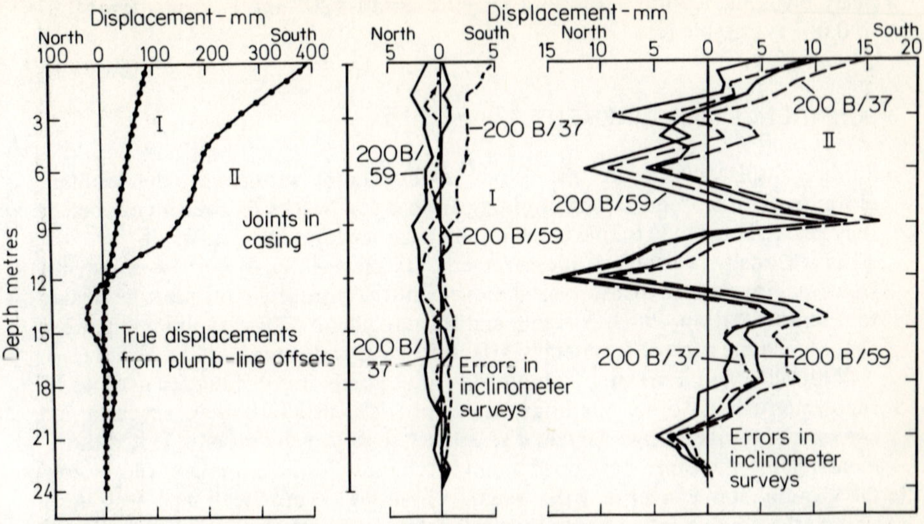

Figure 9.17b Slope indicator. Actual and measured displacements (After Green[3]). Green found that errors were not cumulative, but were self-cancelling. They were similar for 'up' and 'down' readings and greatest at the joints in the casing. Because they were generally small they were acceptable in most field situations

precise surveying methods. The observations were related to bench marks located between 50 and 100 m distant from the structure. Vertical movements can also be determined by settlement tubes or by water-level or mercury filled gauges. According to Dixon and Clarke[61] the most convenient method of relating settlement to a central point is by using a pneumatically operated settlement pot, the change in level between the required point and the central datum being indicated by a pressure change in a mercury column. However, unless the datum can be related to a bench mark outside the area of influence by precise surveying, then this system only provides an indication of relative settlement.

Borehole extensometers are used to measure the vertical displacement of the ground at different depths (Figure 9.16). In such cases, a series of anchor points at various depths in a borehole are connected to the surface using high tensile steel wires. The displacement of each anchor point relative to the surface can be monitored continuously. Burland, Moore and Smith[71] described a precise borehole extensometer which consists of circular magnets embedded in the ground, which act as markers, and reed switch sensors moving in a central access tube which locate the positions of the magnets. Measurements of surface movements, only provide a partial picture of ground behaviour as affected by a particular structure. For a complete picture they should be combined with observations of subsurface movement.

An inclinometer (Figure 9.17) is used to measure horizontal movements below ground and relies on the measurement of the angle a pendulum makes with the vertical at given positions in a specially cased borehole (see Cornforth[72]). The gravity operated pendulum transmits electrical signals to the recorder and a vertical profile is thereby obtained. Sets of readings over a period of time enable both the magnitude and rate of horizontal movement to be determined. For many civil engineering purposes a range of about ±20° and a sensitivity of 0.01 to 0.05° is suitable (see Green[73]).

SUBSIDENCE MEASUREMENT

Surface displacements are also produced by mining subsidence, the resultant movements of ground taking anything from a few weeks to over a year before they cease. This tends to make measurement procedures more difficult.

Traditionally subsidence measurement has been done by precise surveying. However, in order to measure total movement the measurements must be related to a remote datum which is stable and this provides a difficulty in mining subsidence since the surface is generally affected over a wide area.

Whittaker and Forrester[74] contended that surveying techniques need to be supplemented by other methods which provide, in addition to accurate and consistent results, a high degree of sensitivity to detect the onset of initial movements. They therefore developed a rapid technique for measurement of horizontal ground strain over a base length of about 3 m. Steel tape was used, observations being taken between surface stations (vertical steel tubes concreted into the ground, the tops of which were machined to accept special registration plugs designed to carry the tape suspension, tensioning and measuring components). The station had a protective cover. They also showed that tilt could be measured directly by mounting a sensing head (a BAC electrolevel) on special field stations which could be installed as required. Littlejohn[75] described the measurement of horizontal strains, differential settlements and changes in slope, monitored at stations at 1.22 m intervals in brick walls subjected to mining subsidence. Measurements were taken each month whilst the site was affected by the extraction of coal and deformations were related to the position of the advancing face.

MEASUREMENT OF STRESS AND CONTACT PRESSURES

The measurement of stress, contact pressures and stress change may be made in two ways. Strain may be measured and then converted to stress or stress may be measured directly by an earth pressure cell such as the Glotzl cell (see Haws *et al*[76]). This is a hydraulic cell which has a high stiffness at constant temperature and is used for measuring contact pressures. This flat diaphragm cell is usually located at the soil-foundation contact. An earth cell must be placed in position in such a way as to minimise disturbance of the stress and strain distribution.

Irwin[77] described an earth pressure cell which was developed at the Transport and Road Research Laboratory for measuring static or slowly changing total stresses in earth masses.

Another earth pressure cell was developed by Thomas and Ward[78] at the Building Research Station. Ideally an earth pressure cell should have the same elastic properties as the surrounding soil. This, of course, cannot be attained and in order to minimise the magnitude of error, i.e. the cell action factor, the ratio of the thockness to the diameter of the cell should not exceed 0.2 and the ratio of the diameter to the deflection of the diaphragm must be 2000 or greater.

STRAIN MEASUREMENTS

Most strain measurements in soils and soft rocks are deformation measurements which are interpreted in terms of strain. Strain is sometimes measured directly in hard rocks. This can be done by mounting strain gauges onto the ends of an insertion in a borehole, so monitoring deformation of the borehole.

A strain cell has been described by Brown[79]. He noted that such a device was required to move with the soil without, however, causing it to be reinforced. In order to record strain it was necessary to monitor the relative movements of two fixed points at either end of a gauge length. He went on to state that strain cells with a positive connection between their end plates had difficulty in measuring small strains. This he suggested could be overcome by substituting separated strain cells at either end of the gauge length.

SEISMIC SURVEYING

Geophysical methods are used to determine the geological sequence and structure of subsurface rocks by the measurement of certain physical properties or forces. Thus a geophysical exploration may be included in a site investigation in order to provide subsurface information over a large area at a reasonable price. The information obtained may aid the location of test holes in critical areas and prevent unnecessary repetitive drilling in uniform ground. Not only can a geophysical survey help locate the position of boreholes but it also complements them in that boreholes provide information about the strata where they are sunk but tell nothing about the ground between whereas geophysical methods can detect variations in subsurface conditions between boreholes.

In a geophysical survey measurements of the variations in certain physical properties are usually taken in a traverse across the surface, although they may be made in order to log a borehole. Anomalies in the physical properties measured generally reflect anomalies in the geological conditions. The ease of recognising and interpretating these anomalies depends on the contrast in physical properties which in turn influences the choice of the method employed. The choice of method used for a particular survey may not be difficult to make. The character and situation of the site have to be taken into account, especially in built up areas, which may be unsuitable for one or other of the geophysical methods either because of the presence of old buildings or services on the site or interference from some source. When dealing with layered rocks, provided their geological structure is not too complex, seismic methods have a distinct advantage in that they give far more detailed information than any other method. On the other hand for small scale work where the structures are simple, electrical methods may be preferred. On occasions more than one method may be used to attack the same problem.

Generally speaking observations should be close enough for correlation between them to be obvious, so enabling interpretation to be carried out without ambiguity. Nevertheless it must be admitted that an accurate and unambiguous interpretation of geophysical data is only possible where the subsurface structure is simple and even then there is no guarantee that this will be achieved (see Griffiths and King[80]). For instance, it may seem a simple problem to determine depth to bedrock below unconsolidated deposits, yet it may prove impossible to find this interface because of the inhomogeneous nature of the overburden, for even slight variations in lithology, grain size and water content can upset interpretation. No overburden is ever truly uniform but all interpretation techniques allow for some latitude in this respect and will work if the departures from homogeneity are not too great. Boreholes to provide correlation for the geophysical measurements and facilitate their interpretation are an essential part of any geophysical survey.

The sudden release of energy from the detonation of an explosive charge in the ground or by mechanical pounding of the surface generates seismic shock waves which radiate out in hemi-spherical wave fronts from the point of impact. The waves generated are compressional (*P*), dilational shear (*S*) and surface waves.

The velocities of the *P* and *S* waves are given by:

$$V_p = \sqrt{\left(\frac{K + \frac{4}{3}G}{\rho}\right)} \quad \text{or} \quad V_p = \frac{E(1-\nu)}{\rho(1+\nu)(1-2\nu)}$$

$$V_s = \sqrt{\left(\frac{G}{\rho}\right)} \quad \text{or} \quad V_s = \frac{E}{2\rho(1+\nu)}$$

where K is the bulk modulus or compressibility, G the shear modulus or rigidity and ρ is the density.

These velocities generally increase with depth below the surface since the elastic moduli increase more rapidly with depth than density. The compressional waves travel faster and are more easily recorded than the shear waves. They have therefore been used almost exclusively is seismic exploration. In general the velocities of shock waves in crystalline rocks are high to very high (Table 9.4).

Table 9.4 VELOCITIES OF COMPRESSIONAL WAVES OF SOME COMMON ROCKS

Igneous rocks	V_p (km/sec)	*Sedimentary rocks*	V_p (km/sec)
Basalt	5.1–6.4	Gypsum	2.0–3.5
Dolerite	5.8–6.6	Limestone	2.8–7.0
Gabbro	6.5–6.7	Sandstone	1.4–4.4
Granite	5.5–6.1	Shale	2.1–4.4
Metamorphic rocks		*Unconsolidated deposits*	
Gneiss	3.7–7.0	Alluvium	0.3–0.6
Marble	3.7–6.9	Sands and gravels	0.3–1.8
Quartzite	5.6–6.1	Clay (wet)	1.5–2.0
Schist	3.5–5.7	Clay (sandy)	2.0–2.4
Slate	3.5–5.4		

Velocities in sedimentary rocks increase concomitantly with consolidation and decrease in pore fluids, and with increase in the degree of cementation and diagenesis. Unconsolidated sedimentary accumulations have maximum velocities varying as a function of the mineralogy, the volume of voids, either air filled or water filled, and grain size. Most rocks and unconsolidated deposits are anisotropic.

When seismic waves pass from one layer to another some energy is reflected back towards the surface whilst the remainder is refracted. Thus two methods of seismic surveying can be distinguished, that is, seismic reflection and seismic refraction. Measurement of the time taken from the generation of the shock waves until they are recorded by the detector array forms the basis of the two methods.

SEISMIC REFRACTION

In the seismic refraction method one ray approaches the interface between two rock types at a critical angle which means that if the ray is passing from a low (V_0) to a high velocity (V_1) layer it will be refracted along the upper boundary of the latter layer. After refraction the pulse travels along the interface with

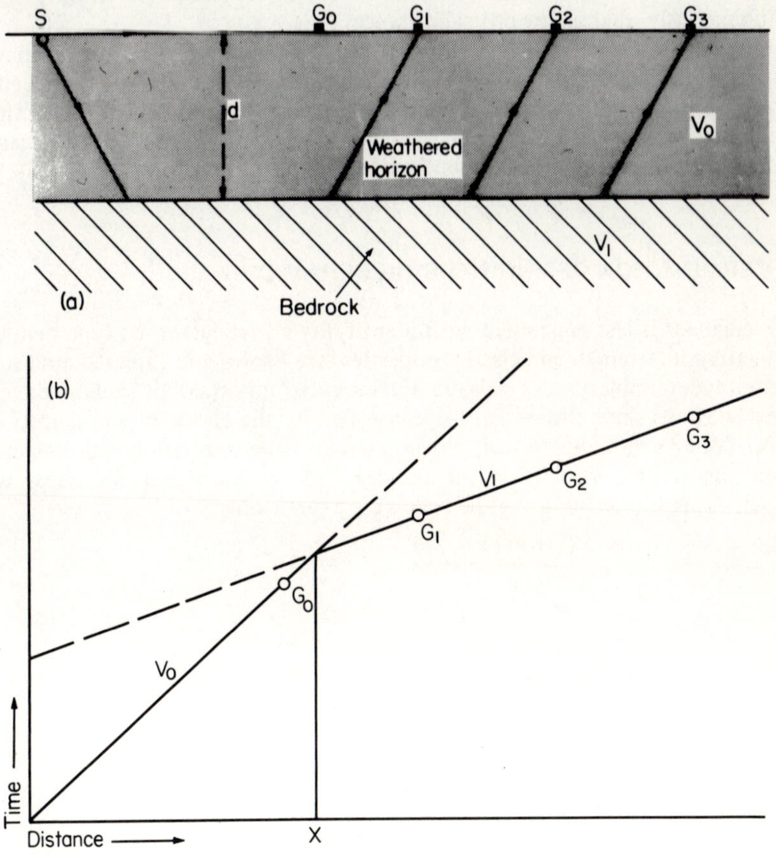

Figure 9.18 Seismic refraction. (a) Geophone spread. S. Origin of shock wave; G. Geophone; d. Depth to bedrock; (b) Time-distance graph. V_0. Velocity of shock waves through the weathered horizon (low velocity); V_1. Velocity of shock waves through the bedrock (high velocity)

velocity V_1. The material at the boundary is subjected to oscillating stress from below. This generates new disturbances along the boundary which travel upwards through the low velocity rock and eventually reach the surface.

At short distances from the point where the shock waves are generated the geophones record direct waves whilst at a critical distance both the direct and refracted waves arrive at the same time. Beyond this, because the rays refracted along the high velocity layer travel faster than those through the low velocity layer above, they reach the geophones first. In refraction work the object is to develop a time-distance graph which involves plotting arrival times against geophone spacing (Figure 9.18). In the simple case of refraction by a single high velocity layer at depth, the depth (d) to the high velocity layer can be obtained from the graph by using the expression:

$$d = \frac{x}{2}\sqrt{\left(\frac{V_1 - V_0}{V_1 + V_0}\right)}$$

where x is the distance to the point where the direct and refracted waves arrive simultaneously, that is, the critical distance.

On many sites where it is required to determine depth to bedrock it may be sufficient to record from two shot point distances at each end of the receiving spread. By traversing in both directions the angle of dip can be determined. According to Wantland[81] an average accuracy of approximately 5–10% can be realized by seismuc methods to depths of approximately 33 m.

DETERMINATION OF ENGINEERING BEHAVIOUR

The engineer is less concerned with identifying a particular rock type than with estimating its strength and elastic properties (see Brown and Robertshaw[82]). The accurate determination of seismic velocities is important in such engineering considerations since the velocity is controlled by the elasticity and density of a rock. Young's modulus (E) and Poisson's ratio (ν) consequently can be derived from the seismic velocity if the density and compressional and shear wave velocities are known by using the following expressions:

$$E = \rho V_\rho^{\,2} \frac{(1+\nu)(1-2\nu)}{(1-\nu)}$$

or

$$E = 2V_s^{\,2}\,\rho\,(1+\nu)$$

or

$$E = \frac{V_s^{\,2}\,\rho\,3(V_p/V_s)^2 - 4}{(V_p/V_s)^2 - 1}$$

$$\nu = \frac{\tfrac{1}{2}(V_p/V_s)^2 - 1}{(V_p/V_s)^2 - 1}$$

These dynamic moduli correspond to the initial tangent moduli of the stress-strain curve for an instantaneously applied load and are usually higher than those obtained in static tests.

As pointed out above, it is usual only to determine the compressional wave velocities. Consequently the normal difficulty involved in assessing the elastic parameters is the measurement of the shear wave velocity (see crosshole seismic testing below). Since shear waves travel more slowly than compression waves the first shear wave arrivals from small explosive charges cannot be identified because they are indistinguishable from later arriving compressional waves which have taken longer travel paths. Shear waves can only be identified if the energy source generates shear waves with larger amplitudes than the compressive waves. Measurements of shear and compressional wave velocities by Evison[83], indicated that as a rock mass becomes less compact the compressional velocity decreases and the *in situ* value of Poisson's ratio increases. Onodera[84] reported values of Poisson's ratio for hard to moderate *in situ* rock ranging from 0.27 to 0.35. He further observed that the less competent rock masses in which problems developed appeared to have higher values of Poisson's ratio. The available data suggests that for fairly competent rock masses, the compressional wave velocity and an assumed value of Poisson's ratio ranging from 0.27 to 0.35 may be used to evaluate a dynamic Young's modulus for engineering purposes. For less competent rock masses it is advisable to obtain shear wave velocities since an underestimate of Poisson's ratio could result in a serious overestimate of Young's modulus.

The frequency and nature of discontinuities within the area for a proposed structure are significant factors which determine the static deformability of the rock mass involved. A highly fractured or weathered rock will exhibit a lower compressional velocity than a sound rock mass. The effect of discontinuities within a rock mass can be estimated by comparing the *in situ* compressional wave velocity (V_{pf}) with the laboratory sonic velocity (V_{pl}) of an intact core obtained from the same rock mass. The difference between these two velocities reflects the influence of discontinuities in the rock mass. Hence the velocity ratio V_{pf}/V_{pl}, has been proposed as a rock quality index. For high quality, massive rock with only a few tight joints the velocity ratio approaches unity whilst lower and lower values below unity are recorded as rocks become more jointed and fractured. Deere[85] has shown that the RQD can be correlated with the velocity ratio (Table 1.7). Similarly Knill[86] demonstrated that there is a near linear relationship between fracture index and velocity ratio. This fact could be used to assess grout take.

The mass factor can also be related to these properties (see Hobbs[87]). Grainger *et al*[88] showed that it was possible to correlate the engineering grades of chalk established by Ward *et al*[14] at Mundford with seismic velocities. They found that the seismic velocity increased with depth in clearly defined steps which broadly corresponded with the grades of chalk. As the grade of chalk is very much influenced by fracturing they were also able to establish a calibration of seismic velocity with the degree of fracturing.

CROSSHOLE SEISMIC TESTING

The technique of crosshole seismic testing has been used on sites where the structures to be erected had a high stability requirement (see Stokoe and Woods[89]; Stokoe and Richart[90]; Ballard[91]). In such cases an accurate assessment of the elastic moduli of the rocks concerned is necessary. The aforementioned technique allows the determination of the compressional and shear wave velocities. In most cases the crosshole technique is used in conjunction with surface refraction seismic surveys. This method permits the determination of horizontal velocities and layering as well as detecting low velocity zones if they are thick with respect to the source and receiver spacing.

Crosshole surveys are usually carried out using two or more boreholes into which are placed the vibratory seismic source and transducers at known elevations. Borehole spacing and source elevations depend upon the site conditions. If little is known about the subsurface velocity characteristics at the site then the borehole spacing should ideally be less than 6.1 m. The seismic source and receivers are relocated after the determination of the *S* and *P* wave velocities.

The velocity scanning technique is also a borehole to borehole or borehole to surface method. McCann *et al*[92] used a fairly high power sparker, enclosed in a liquid filled cylinder in one borehole, and sensitive hydrophone receivers in the others. Although the boreholes have to be filled with fluid the casing does not affect the results. The method has been used between boreholes as much as 100 m apart. This technique can aid the detection of subsurface cavities, open fractures, zones of weakness and other discontinuities. Fault planes can be investigated by varying the positions of transmitter and receiver on either side of the fault.

RESISTIVITY SURVEYING

Since most of the principal rock forming minerals, as well as many others, are practically insulators, the resistivity of rocks and soils is determined by the amount of conducting mineral constituents and the content of mineralized water in the pores. The latter condition is by far the most important factor and, in fact, most rocks and soils conduct an electric current only because they contain water. Weathering increases the porosity of a rock mass and therefore its conductivity.

The widely differing resistivity of the various types of impregnating water alone can cause variations in the resistivity of rock formations ranging from a few tenths of an ohm metre to hundreds of ohm metres, the more saline the water the lower the resistivity. Thus, since the resistivity is not a characteristic

Table 9.5 RESISTIVITY VALUES OF SOME COMMON ROCK TYPES

Rock type	*Resistivity* (ohm-m)
Topsoil	5–50
Peat and clay	8–50
Clay, sand and gravel mixtures	40–250
Saturated sand and gravel	40–100
Moist to dry sand and gravel	100–3000
Mudstones, marls and shales	8–100
Sandstones and limestones	100–1000
Crystalline rocks	200–10000

parameter for the rock mass itself, it is not possible to identify it solely by means of the resistivity value. In general, however, sediments are better conductors than igneous rocks and among the latter the basic types are better conductors than than the acid types (Table 9.5). Clayey material has a higher conductivity than material of sandy type, while compacted material is a better conductor than unconsolidated material. It is accordingly possible to distinguish between major groups.

In the resistivity method an electric current is introduced into the ground usually by means of two current electrodes and the potential difference between two potential electrodes is measured. The ohms value is converted to apparent resistivity by use of a factor that depends on the particular electrode configuration in use. The electrodes are normally arranged along a straight line, the potential electrodes being placed inside the current electrodes and both are symmetrically disposed with respect to the centre of the configuration. The most frequently used symmetric configurations are those introduced by Wenner and by Schlumberger. The expressions used to compute the apparent resistivity (ρ_a) for the Wenner and Schlumberger configurations are:

Wenner: $$\rho_a = 2\pi aR$$

Schlumberger: $$\rho_a = \frac{\pi(L^2 - l^2)}{2l} \times R$$

where

a, *L* and *l* are explained in Figure 9.19;

R is the resistivity reading.

In the Wenner configuration the distances between all four electrodes are equal. The spacings can be progressively increased, keeping the centre of the array fixed or the whole array, with fixed spacings, can be shifted along a given

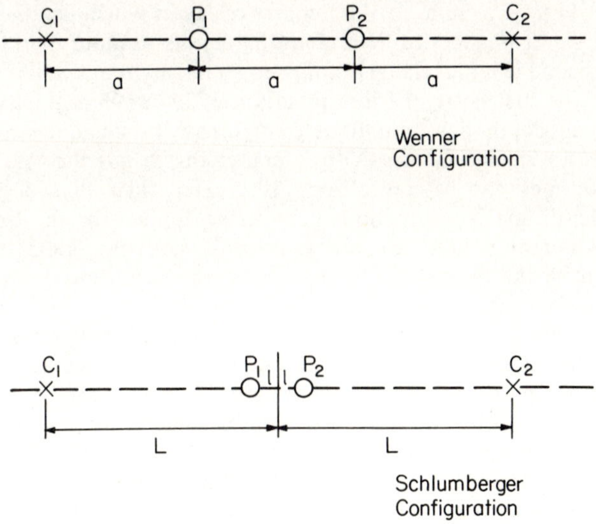

Figure 9.19 Wenner and Schlumberger configurations

line. In the Schlumberger arrangement the potential electrodes maintain a constant separation about the station whilst the current electrodes are moved outwards after each reading.

The depth of current penetration increases with increasing electrode distance.

Indeed 70% of the total current passes above a depth about equal to the electrode separation. Analysis of the variation in the value of apparent resistivity with respect to electrode separation, which is taken to approximate depth of penetration, enables inferences to be drawn about the subsurface formations.

Horizontal profiling is used to determine the variations in the apparent resistivity in a horizontal direction at a preselected depth. For this purpose an electrode configuration, with suitable interelectrode spacing, is moved as an entity along straight traverses, resistivity determinations being made at stations at regular intervals. The data of a constant separation survey may be used to construct a contour map of lines of equal resistivity. These maps are often extremely useful in locating areas of anomalous resistivity such as gravel pockets and the trend of buried channels.

Electrical sounding furnishes information on the vertical succession of different conducting zones and their individual thicknesses and resistivities. This method is particularly valuable for investigations on horizontally or nearly horizontally stratified ground. In electrical sounding, the midpoint of the electrode configuration is fixed at the observation station while the length of the configuration is gradually increased. As a result the current penetrates deeper and deeper, the apparent resistivity being measured each time the current electrodes are moved outwards. The readings therefore become more and more affected by the resistivity conditions at increasing depths.

The Schlumberger configuration is preferable to the Wenner configuration for depth sounding because the field procedure is quicker and simpler, and master curves are more readily available to analyse the results (see Tagg[93]). According to Enslin[94] there is a limit to the number of layers which can be interpreted. Generally it is not possible to determine the depths to more than three or four layers. If a second layer is relatively thin and its resistivity much larger or smaller than that of the first layer, the interpretation of its lower contact will be inaccurate. For all depth determinations from resistivity soundings it is assumed that there is no change in resistivity laterally; this is not the case in practice. Indeed sometimes the lateral change is greater than that occurring with increasing depth and so corrections have to be applied for the lateral effects when depth determinations are made. In some cases the lateral effect is the major feature of the resistivity sounding curve and depth interpretation can be very innaccurate.

Ground resistivities are never uniform so that a single depth determination cannot be relied upon. Accordingly readings may be repeated at the same spot but with the spread in a different azimuth and a few more sets of measurements are usually taken nearby. When the position of a steeply dipping interface like a fault or igneous contact has been found, if there is a notable difference in resistivity across the boundary, then its position can be determined by making a single constant separation traverse across the presumed locality of the interface. Where bedrock is covered by overburden then a wide enough separation must be chosen to ensure that a reasonable proportion of the current flows through the bedrock otherwise any lateral variation in the thickness of the overburden or the resistivity, will affect the measurements and if large enough could conceal an anomaly.

Geophysical logging of boreholes can yield considerable supplementary details and aid interpretation of surface geophysical surveys. The results can help identification of individual horizons and correlation between boreholes, as well as providing information relating to rock properties. Logging can be done by electrical, sonic or radioactive methods (Table 9.6).

COMPARISON OF SEISMIC RESISTIVITY METHODS

Both seismic refraction and resistivity are suited to shallow investigations of the type commonly required in civil engineering. Resistivity surveys are more economic than seismic refraction surveys but they can give rather large errors in depth sounding (± 20%). This compares with an error of ± 10% in seismic refraction work so that in some instances it may well be worth the extra cost for a more accurate survey.

In a comparison of seismic with resistivity methods Moore[95] remarked that seismic refraction will normally have a good application to foundation studies for dams, buildings and bridges, and can be used to good advantage in determining the character of rock present in proposed excavations and shallow tunnels. By contrast seismic refraction proves of little value in locating construction materials such as sands and gravels. It may also be rather ineffective in detailed studies of landslide areas. This method cannot generally detect the presence of an inversion, that is, a low velocity layer beneath a high velocity layer.

The resistivity method may also be used in foundation studies when proper calibration of the test is made over known materials. It has been used in slope design problems and to provide information relating to landslides with particular regard to the position of the slip surface. Resistivity has been used successfully to locate construction materials (see Vann[96]).

MAGNETIC AND GRAVITY METHODS

The magnetic and gravity methods are only occasionally used in the civil engineering industry. However, the proton magnetometer has been used with some success in the location of old mine shafts (see Cox and Helliwell[98]) the shafts producing magnetic anomalies on isomagnetic contour maps. Interpretation of magnetic anomalies is usually qualitative and depth determinations are the exception rather than the rule. However, surveys may not be conclusive in areas subject to external magnetic influences such as power cables and buildings. The proton magnetometer can also, in some circumstances, indicate local areas of fill material and can therefore be used to supplement borehole evidence. The effectiveness of the magnetometer for this purpose depends on there being a magnetic susceptibility contrast between the fill and the adjacent undisturbed material.

Like the magnetic survey, the basic aim of a gravity survey is to produce a gravity map, the measured variations in acceleration due to gravity being used to construct the isogals. Gravity methods are mainly used in regional reconnaissance surveys to reveal anomalies which may be subsequently investigated by other methods. Since the gravitational effects of geological bodies are proportional to the contrast in density between them and their surroundings, gravity methods are suitable for the location of structures in stratified formations. Eaton, Martin and Murphy[99] reviewed the application of gravity measurements to some problems associated with civil engineering such as reconnaissance mapping of

Table 9.6 ENGINEERING USES OF BOREHOLE GEOPHYSICAL LOGS
(After Institution of Civil Engineers[97])

Type of log	*Parameters measured*	*Application*
1. Electrical logs	Continuous records of:	
(a) single point resistivity log	Apparent resistivity between single electrode in borehole and ground surface	Lithostratigraphical correlation
(b) Normal resistivity log	Apparent resistivity of borehole wall measured by	As for 1(a) and rock quality/ porosity/ groundwater salinity
(c) Induction (resistivity) log	Focused beyond borehole	
(d) Laterolog (resistivity) log	invaded zone	
(e) Microlaterolog (resistivity)	Resistivity of invaded zone	Estimation of formation permeability
(f) Spontaneous potential (SP) log	Natural potential differences in borehole wall	In conjunction with resistivity, lithological identification and correlation
2. Sonic logs		
(a) Sonic or continuous velolog	Continuous record of (*P* wave) velocity in the borehole wall	Determination of rock quality, degree of fracturing, porosity
(b) 3-D sonic log	Continuous record of both *P* and *S* wave velocity in the borehole wall	As for 2(a) and determination of dynamic moduli in homogeneous material (with density log)
(c) Inter-borehole sonic log (see above, McCann *et al*[92])	Record of apparent velocity between adjacent boreholes at discrete depths	Location of cavities and determination of rock quality between boreholes
3. Radioactive logs		
(a) Gamma log	Rapid sampling with continuous record of natural gamma ray emission	Estimation of shale/clay content of rock
(b) Gamma-gamma or Formation density log	Gamma ray intensity after bombardment by gamma ray source	Determination of electron density and hence rock density
(c) Neutron log	Quantity of hydrogen in rock around borehole i.e. water and hydrocarbons	Determination of porosity and shale content

suballuvial bedrock topography and the *in situ* measurement of aquifer porosity. Interpretation of data again tends to produce a qualitative answer.

References

1. Sanglerat, G., *The penetrometer and soil exploration*, Elsevier, Amsterdam (1972)
2. Terzaghi, K. and Peck, R.B., *Soil mechanics in engineering practice*, Wiley, New York (1948, 1967)
3. Rodin, S., Corbett, B.O., Sherwood, D.E. and Thorburn, S., 'Penetration testing in the United Kingdom,' *Proc. European Symp. on Penetration Testing*, Stockholm, 2.2, (1974)

4. CP 2004, *Code of Practice on foundations*, British Standards Institution, London (1972)
5. Rodin, S., 'Experiences with penetrometers with particular reference to the standard penetration test,' *Proc. 5th Conf. Soil Mech. Found. Eng.*, Paris, **1**, 517–521 (1961)
6. Sutherland, H.B., *Granular materials (article in Settlement of Structures)* Pentech Press, London, 473–499 (1975)
7. Gibbs, H.J. and Holtz, W.G., 'Research on determining the density of sands by spoon penetration testing,' *Proc. 4th Int. Conf. Soil Mech. Found. Eng.*, London, **1**, 35–39 (1957)
8. Tomlinson, M.J., *Foundation design and construction*, Pitman, London (1969; 1975)
9. Peck, R.B., Hanson, W.E. and Thornburn, T.H., *Foundation engineering*, Wiley, New York (1974)
10. Meyerhof, G.G., 'Shallow foundations,' *ASCE (Div. Soil Mech. Found. Eng.)*, **91**, 21–31 (1965)
11. Meyerhof, G.G., 'State-of-the-art of penetration testing in countries outside Europe,' *Proc. European Symp. Penetration Testing*, Stockholm, **2** (1974)
12. Jordan, E.E., 'Settlement in sand – methods of calculating and factors affecting Ground Engineering,' **10**, 31–37 (Jan. 1977)
13. Wakeling, T.R.M., 'A comparison of the results of standard site investigation methods against the results of a detailed geotechnical investigation on Middle Chalk at Mundford, Norfolk,' *(article in In-situ investigation in Soils and Rocks) British Geotechnical Society*, London, 17–22 (1900)
14. Ward, W.H., Burland, J.B. and Gallois, R.W., 'Geotechnical assessment of a site at Mundford, Norfolk for a large proton accelerator,' *Geotechnique*, **18**, 399–341 (1968)
15. Mitchell, J.K. and Durgunoglu, H.T., '*In situ* strength by static cone penetration test,' *Proc. 8th Int. Conf. Soil Mech. Found. Eng.*, Moscow, **1**, 279–286 (1973)
16. Joustra, I.R.K., 'New developments of the Dutch cone penetration test,' *Proc. 8th Int. Conf. Soil Mech. Found. Eng.*, Moscow, **1**, 199–201 (1973)
17. Heijnen, W.J., 'The Dutch cone test: a study of the shape of the electrical cone,' *Proc. 8th Int. Conf. Soil Mech. Found. Eng.*, **1**, 181–184 (1973)
18. Terzaghi, K., *Theoretical soil mechanics*, Wiley, New York (1943)
19. Lunde, T., Eide, O. and De Ruiter, J., 'Correlations between cone resistance and vane shear strength in some Scandinavian soft to medium stiff clays,' *Canadian Geotechnical Jour.*, **13**, 430–441 (1976)
20. Schmertmann, J.H., 'Measurement of *in situ* shear strength. State of art report,' *Proc. ASCE Conf. in situ Measurements of Soil Properties*, Raliegh, **2**, 57–138 (1975)
21. Meyerhof, G.G., 'Penetration tests and bearing capacity of cohesionless soils'. *ASCE (Div. Soil Mech. Found. Eng.)*, Paper No. 866, **82**, (1956)
22. Buisman, A.S.K., *Grondmechanica*, Waltman, Delft (1940)
23. De Beer, E. and Martens, A., 'Method of computation of an upper limit for the influence of the heterogeneity of sand layers in the settlement of bridges,' *Proc. 4th Int. Conf. Soil Mech. Found. Eng.*, London, **1**, 275–282 (1957)
24. Gielly, J., Lareal, P. and Sanglerat, G., 'Correlation between *in situ* penetrometer tests and the compressibility characteristics of soils,' *(article in In-situ investigations in Soils and Rocks) British Geotechnical Society*, London, 167–172 (1970)
25. Carlson, L., 'Determination *in situ* of the shear strength of undisturbed clay by means of a rotating auger,' *Proc. 2nd Int. Conf. Soil Mech. Found. Eng.*, Stockholm, **1**, 265 (1948)
26. Skempton, A.W., 'Vane tests in the alluvial plain of the River Forth near Grangemouth,' *Geotechnique*, **1**, 111 (1948)
27. La Rochelle, P., Roy, M. and Tavenas, F., 'Field measurements of cohesion in Champlain clays,' *Proc. 8th Int. Conf. Soil Mech. Found. Eng.*, Moscow, **1.1**, 229–236 (1973)
28. Aas, G., 'A study of the effect of vane shape and rate of strain on the measured values of *in situ* shear strength of clays,' *Proc. 6th Int. Conf. Soil Mech. Found. Eng.*, Paris, **1**, 141–145 (1965)
29. BS 1377: *Methods of test for soils for civil engineering purposes*, British Standards Institution
30. Wiesel, G.E., 'Some factors influencing *in situ* vane test results,' *Proc. 8th Int. Conf. Soil Mech. Found. Eng.*, Moscow, **1**, 475–479 (1973)
31. Flaate, K., 'Factors influencing the results of vane tests,' *Canadian Geotech. Jour.*, **3**, 18–31 (1966)

32. Marsland, A., 'Clays subjected to *in situ* plate tests,' *Ground Engineering*, **5**, 24–31 (Nov. 1972)
33. Marsland, A., 'Comparison of the results from static penetration tests and large *in situ* plate tests in London Clay,' *Proc. European Symp. Penetration Testing*, Stockholm, **2.2**, 245–252 (1974)
34. Marsland, A., 'Large *in situ* tests to measure the properties of stiff fissured clays,' *Proc. 1st Australian-New Zealand Conf. on Geomechanics*, Melbourne, **1**, 180–189 (1971)
35. Stagg, K.G., In situ tests on the rock mass *article in Rock Mechanics in Engineering Practice*, (Ed. by Stagg, K.G. and Zienkiewicz, O.C.), Wiley, London, 125–156 (1968)
36. Rocha, M., 'Mechanical behaviour of rock foundations' in concrete dams,' *8th Int. Cong. Large Dams*, Edinburgh, **1**, 785–831 (1964)
37. Rice, O.L., '*In situ* testing of foundations and abutments for dams,' *8th Int. Cong. Large Dams*, Edinburgh, **1**, 87–100 (1964)
38. Serafim, J.L., (Ed. by Judd, W.R.) 'Rock mechanics consideration in the design of concrete dams,' *article in State of Stress in the Earth's crust*, Elsevier, New York, 611–650 (1964)
39. Coates, D.F. and Gyenge, M., 'Plate load testing on rock for deformation and strength properties,' *(article in Testing Techniques for Rock Mechanics), ASTM STP 402*, 19–40 (1966)
40. Burland, J.B. and Lord, J.A., 'The load deformation behaviour of middle chalk at Mundford, Norfolk: a comparison between full-scale performance and *in situ* laboratory measurements,' *(article in In situ investigations in Soils and Rocks) British Geotechnical Society*, 13–15 (1970)
41. Moore, J.F.A., 'A long term plate test on Bunter sandstone,' *Proc. 3rd Int. Soc. Rock Mech. Cong.*, Denver, **2**, 724–732 (1974)
42. Zienkiewicz, O.C. and Stagg, K.G., 'The cable method of *in situ* testing,' *1st Int. Soc. Rock Mech. Cong.*, Lisbon, **3**, 667–672 (1966)
43. Hobbs, N.B. and Dixon, J.C., '*In situ* testing for bridge foundations in the Devonian marl,' *(article in In situ investigations in soils and rocks) British Geotechnical Society*, 31–38 (1970)
44. Gibson, R.E. and Anderson, W.F., '*In situ* measurement of soil properties with the pressuremeter, *Civ. Eng. Publ. Works Rev.*, **56**, 615–618 (1961)
45. Ladanyi, B., 'Evaluation of pressuremeter tests in sensitive clays,' *Canadian Geotechnical Jour.*, **9**, 313–319 (1972)
46. Meigh, A.C. and Greenland, S.W., '*In situ* testing of soft rocks,' *Proc. 6th Int. Conf. Soil Mech. Found. Eng.*, Paris, **1**, 73–76 (1965)
47. Marsland, A., 'The evaluation of the engineering design parameters for glacial clays,' *Q. Jour. Eng. Geol.*, **10**, 1–26 (1976)
48. Wroth, C.P. and Hughes, J.M.O., 'An instrument for the *in situ* measurement of the properties of soft clays,' *Proc. 8th Int. Conf. Soil Mech. Found. Eng.*, Moscow, **1**, 487–494 (1973)
49. McKinley, D.G. and Anderson, W.F., 'Determination of the modulus of deformation of a till using a pressuremeter,' *Ground Engineering*, **8**, 51–54 (Nov. 1975)
50. Dodds, R.K., 'Measurement and analysis of rock physical properties on the Dez Project, Iran,' *(article in Testing Techniques for Rock Mechanics) ASTM, Publ. No. 402*, Philadelphia (1966)
51. Deere, D.U., Peck, R.B., Monsees, J.E. and Schmidt, B., *Design of tunnel liners and support systems*, Final Report on Contract No. 3-0152 to Office of High Speed Ground Transportation, U.S. Dept. Transportation, Washington, D.C. (1969)
52. Rocha, M., Da Silveira, A., Grossman, N. and De Oliveira, E., 'Determination of the deformability of rock masses along boreholes,' *Proc. 1st Int. Soc. Rock Mech. Cong.*, Lisbon, **3**, 697–704 (1966)
53. Lane, R.G.T., 'Rock foundations: diagnosis of mechanical properties and treatment,' *Proc. 8th Int. Conf. Large Dams*, Edinburgh, **1**, 141–165 (1964)
54. Franklin, J.A., Manailoglou, J. and Sheerwood, D., 'Field determination of direct shear strength,' *Proc. 3rd Cong. Int. Soc. Rock Mech.*, Denver, **2**, 233–240 (1974)
55. Krsmanovic, D. and Popovic, M., 'Large scale field tests of the shear strength of limestone,' *Proc. 1st Cong., Int. Soc. Rock. Mech.*, Lisbon, **3**, 773–779 (1966)
56. James, P.M., '*In situ* shear tests at Muda Dam,' *(article in In Situ Investigations in Soils and Rocks) British Geotechnical Society*, 75–82 (1970)

57. John, K., 'Die Praxis der felsgrossversuche beschreiben am biespiel der arbeiten an der Kurobe-IV-Staumauer in Japan,' *Geologie und Bauwesen*, Jg, 27, H.3 (1961)
58. Muller, L., 'Wo steht die ingenieurgeologie?' *Geologie und Bauwesen*, Jg. 15. H.3 (1943)
59. Voort, H. and Logters, G., '*In situ* determination of the deformational behaviour of a cubical rock mass sample under triaxial load,' *Rock Mechanics*, **6**, 65–79 (1974)
60. Hvorslev, M.J., 'Time lag and soil permeability in groundwater observations,' *Bull. No. 36*, Waterways Experimental Station, Vicksburg (1951)
61. Dixon, J.C. and Clarke, K.B., (Ed. by Bell, F.G.) 'Field investigation techniques,' *(article in Site Investigations in Areas of Mining Subsidence)*, Newnes-Butterworths, London, 40–74 (1975)
62. Mather, J.D., Gray, D.A., Allen, R.A. and Smith, D.B., 'Ground water recharge in the lower greensand of the London Basin – results from tritium and carbon 14 determinations,' *Q. Jour. Engng. Geol.*, **6**, 141–152 (1975)
63. Dunnicliff, C.J., 'Equipment for field deformation measurements,' *Proc. 4th Panamerican Conf. Soil Mech. Found. Eng.*, **2**, 319–332 (1971)
64. Gould, J.P. and Dunnicliff, C.J., 'Accuracy of field deformation measurements,' *Proc. 4th Panamerican Conf. Soil. Mech. Found. Eng.*, **1**, 313–366 (1971)
65. Hanna, T.H., *Foundation instrumentation,* Trans. Tech Publications, Clausthal (1973)
66. Green, G.E., 'Equipment: general report,' *(article in Field Instrumentation in Geotechnical Engineering)* Butterworths, London, 517–526 (1974)
67. Ashkenazi, V., 'The measurement of spatial deformations by geodetic methods,' *(article in Field Instrumentation in Geotechnical Engineering)* Butterworths, London, 1–14 (1974)
68. Liu, T.K., Aldrich, H.P. and Dugan, J.P., 'Predicted and measured settlement of the Colonade Building, The First Church of Christ Scientist, Boston, Massachusetts *(article in Settlement of Structures)* Pentech Press, London, 106–109 (1975)
69. Penman, A.D.M. and Charles, J.A., *Constructional deformations in a rockfill dam*, Building Research Station, Watford, CP 19/72 (1972)
70. Morton, K. and Au, E., 'Settlement observations on eight structures in London,' *(article in Settlement of Structures)* Pentech Press, London, 183–203 (1975)
71. Burland, J.B., Moore, J.F.A. and Smith, P.D.K., 'A simple and precise borehole extensometer,' *Geotechnique*, **22**, 174–177 (1971)
72. Cornforth, D.H., 'Performance characteristics of the slope indicator series 200-B inclinometer,' *(article in Field Instrumentation in Geotechnical Engineering)* Butterworths, London, 126–135 (1974)
73. Green, G.E., 'Principles and performance of two inclinometers for measuring horizontal ground movements,' *(article in Field Instrumentation in Geotechnical Engineering)* Butterworths, London, 166–179 (1974)
74. Whittaker, B.N. and Forrester, D.J., 'Measurement of ground strain and tilt arising from mining subsidence,' *(article in Field Instrumentation in Geotechical Engineering)* Butterworths, London, 437–447 (1974)
75. Littlejohn, G.S., 'Observations of brick walls subjected to mining subsidence,' *(article in Settlement of Structures)* Pentech Press, London, 384–393 (1975)
76. Haws, E.T., Lippard, D.C., Tabb, R. and Burland, J.B., 'Foundation instrumentation for the National Westminster Bank Tower,' *(article in Field Instrumentation in Geotechnical Engineering)* Butterworths, London, 180–193 (1974)
77. Irwin, M.J., 'Instruments developed by the TRRL for studying the behaviour of earthworks,' *(article in Field Instrumentation in Geotechnical Engineering)* Butterworths, London, 194–206 (1974)
78. Thomas, H.S.H. and Ward, W.H., 'The design, construction and performance of a vibrating wire earth pressure cell,' *Geotechnique*, **19**, 39–51 (1969)
79. Brown, S.F., 'The measure of *in situ* stress and strain in soils,' *(article in Field Instrumentation in Geotechnical Engineering)* Butterworths, London, 38–51 (1974)
80. Griffiths, D.H. and King, R.F., *Applied geophysics for engineers and geologists*, Pergamon Press, Oxford (1969)
81. Wantland, D., 'Application of geophysical methods to problems in civil engineering,' *Canadian Inst. Min. Met.*, **46**, 134–142 (1953)
82. Brown, P.D. and Robertshaw, J., 'The *in situ* measurement of Young's modulus for rock by a dynamic method,' *Geotechnique*, **3**, 283–286 (1953)
83. Evison, F.F., 'The seismic determination of Young's modulus and Poisson's ratio for rocks *in situ*,' *Geotechnique*, **6**, (1956)

84. Onodera, T.F., 'Dynamic investigation of foundation rocks *in situ,' Proc. 5th Symp. Rock. Mech.*, Minnesota, Pergamon, New York, 517–533 (1963)
85. Deere, D.U., 'Geological considerations,' *(article in Rock Mechanics in Engineering Practice)* (Ed. by Stagg, K.G. and Zienkiewicz, O.C.) Wiley, London, 1–20 (1968)
86. Knill, J.L., 'The application of seismic methods in the prediction of grout take in rocks,' *(article in In Situ Investigations in Soils and Rocks) British Geotechnical Society*, 93–100 (1970)
87. Hobbs, N.B., 'Factors affecting the prediction of settlement of structures on rock: with particular reference to the Chalk and Trias,' *(article in Settlement of Structures)* Pentech Press, London, 579–610 (1975)
88. Grainger, P., McCann, D.M. and Gallois, R.W., 'The application of the seismic refraction technique to the study of fracturing of the middle chalk at Mundford, Norfolk,' *Geotechnique,* **23**, 219–232 (1973)
89. Stokoe, K.H. and Woods, R.D., '*In situ* shear wave velocity by cross hole method,' *ASCE (Div. Soil Mech. Found. Eng.)*, **98**, Paper 8904, 443–460 (1972)
90. Stokoe, K.H. and Richart, F.E., '*In situ* and laboratory shear wave velocities,' *Proc. 8th Int. Conf. Soil Mech. Found. Eng.*, Moscow, **1**, 403–409 (1973)
91. Ballard, R.F., 'Method for crosshole seismic testing,' *ASCE (Jour. Geo. Eng. Div.)*, 102, No. GT12, 1261–1273 (1976)
92. McCann, D.M., Grainger, P. and McCann, C., 'Inter-borehole acoustic measurements and their use in engineering geology,' *Geophys. Pros.*, **23**, 50–69 (1975)
93. Tagg, G.F., 'Interpretation of earth resistivity measurements,' *Trans. Am. Inst. Min. Met. Engr (Geophysical Prospecting)*, **138**, 399–407 (1940)
94. Enslin, J.F., 'Geophysics as an aid to foundation engineering,' *Trans. S. Af. Inst. Civ. Eng.*, **3**, 49–60 (1953)
95. Moore, R.W., 'Geophysics efficient in exploring the subsurface,' *ASCE (Div. Soil Mech. Found. Eng.)* SM3, **87**, 69–75 (1961)
96. Vann, G., 'Location and evaluation of sand and gravel deposits by geophysical methods and drilling,' *(article in Opencast Mining, Quarrying and Alluvial Mining) Inst. Min. Met.*, London, 3–19 (1968)
97. Institution of Civil Engineers, *Manual of Applied Geology for Engineering*, ICE, London, 129–130 (1976)
98. Cox, M.R. and Helliwell, E.G., 'Limitations on the use of magnetic methods for the detection of abandoned coal mine shafts,' *Chartered Surveyor: LHM Quarterly*, **4**, 17–22 (1976)
99. Eaton, G.P., Martin, N.W. and Murphy, M.A., 'Application of gravity measurements to some problems in engineering geology,' *Bull. Ass. Eng. Geol.*, **3**, 6–21 (1965)

Chapter 10

Foundation Instrumentation

INTRODUCTION

The pressure to maintain and enhance the environment is growing. One result is that major engineering works are being forced to use derelict land and difficult sites; sites which in the past would have been avoided are now being sought and used although their use imposes restrictions on the engineer and his design. The result is that settlements and strains are inevitable in a structure and have to be allowed for in the design and accommodated during construction. In addition, care may be required during construction to ensure the stability of the works.

As the use of sites with difficult foundation conditions grows, designs become more sophisticated and the need for foundation instrumentation increases. This is reflected in the instrumentation techniques and types which are now available to the engineer, and the increase in quality and quantity of technical publications devoted to the subject of field instrumentation.

Instrumentation is part of modern technology and is continually evolving to meet the requirements of the engineer. The most successful instruments appear to have been based upon developments originating from research institutes or universities which are subsequently developed for commercial use.

INSTRUMENTATION TYPES

There are many types of instruments available and improvements are frequent (see Hanna[1]). Because of the rate of innovation and development it is inappropriate to describe specific instruments or equipment, but rather to refer the reader to the current technical literature and the specialist equipment suppliers.

For the engineer working with difficult foundations the instrumentation with which he is mostly concerned can be collectively grouped into the classification given by Green[2] (Figure 10.1).

PORE WATER PRESSURES

Poor ground is frequently synonymous with ground water, high moisture contents and high water tables. Pore water pressure measurements are considered to

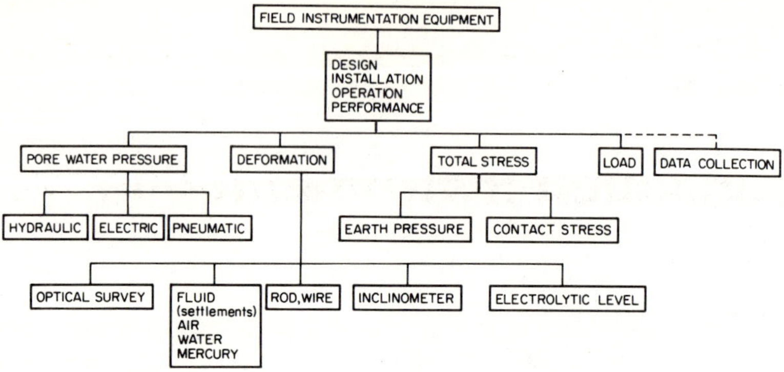

Figure 10.1 Classification of instrumentation (After Green[2])

be important by many engineers when dealing with difficult soils although this is not necessarily so in all countries, notably the Soviet Union.

The use of pore pressures to check the effectiveness of drainage systems and control analyses based upon effective stress criteria is an established field technique (see Bishop *et al*[3]). Details concerning the principles of pore pressure measurements and the possible errors have been reviewed by Vaughan[4].

DEFORMATION MEASUREMENTS

Because of the problem of uncertainty caused when instruments are inserted into soil, there has been a tendency towards ground strain measurements. This has occurred at a time of general advance in strain measuring instruments which now provide a high degree of accuracy.

Strain measurements are suitable when considering the geotechnical aspects of soil/structure interaction or when monitoring the construction of earth structures. However, the problem of assessing the forces applied to any structure associated with these strains is a function of the soil properties. In areas of difficult ground these properties are usually very complex.

TOTAL STRESS

Earth pressure measurements have long been acknowledged as being desirable, particularly to the engineer engaged in the design of earth retaining structures or bridge substructures. However, it must be recognised that reliable, accurate total stress measurements either within a soil mass or at a boundary are difficult to achieve.

It is argued that the presence of an earth pressure cell changes the properties of the soil. However, in practical circumstances 90% accuracy is all that is required and this can be achieved with modern instruments if used with care. Using the criteria of 90% accuracy the problem of cell action factor and the effect of environmental changes can be considered as a second order problem (see Jones[5]).

LOAD CELLS

Load measurements are easy and usually reliable. Measuring devices are normally based upon the use of vibrating wire gauges, resistance strain gauges, hydraulic load indicators or photoelastic load indicators. The latter require visual access but can be used to provide overall load patterns if used in the form of a coating.

DATA COLLECTION

The large scale use of instruments on sites poses a significant problem with regard to the volume of data collected and the need to rapidly assess the implications of the readings taken. As a result, the use of automatic data logging equipment is now common on large sites and the use of minicomputers is growing.

If minicomputers are used, care must be taken to ensure the reliability of the computer systems employed as frequently the scheme under consideration becomes reliant on these sophisticated tools. The environmental conditions with regard to computers on site must be carefully studied.

IDEAL INSTRUMENTS

It has already been stated that the number and types of gauges available for field instrumentation is very large. In general terms the ideal instrument requires the following attributes:

1. Simple principle of design and construction.
2. Proven durability in use.
3. Accuracy for short and long term measurements.
4. Ease of calibration and recalibration.
5. Non-sensitive to environmental changes.
6. Simple installation procedure.
7. Minimum interference due to installation.
8. A reliable output unit or comparator (if used).
9. Minimum overall cost.

The development of instruments is rapid and it may not be possible to establish all these attributes, i.e. the long term durability in use of the new instruments can only be established with time. In addition, instruments which in the past have proved to be unreliable or inaccurate are frequently withdrawn and replaced by a similar but improved instrument e.g. the bonded resistance strain gauge inclinometer described by Green[6].

The best practice relating to the use of field instruments results from discussions concerning the particular application with the instrument manufacturers and also with engineers who have practical experience of a particular problem and the instruments it is proposed to use.

INSTRUMENTS IN DIFFICULT GROUND

Where the emphasis is on foundations in difficult ground the majority of the instruments used fall under the following types: piezometers; inclinometers; settlement and movement gauges; earth pressure cells.

PRACTICAL ASPECTS

For all instrumentation projects the objectives of the instrumentation must be borne in mind at all times and the instrumentation developed accordingly. This point needs emphasising as the implications of it are frequently not understood.

If a proposed instrumentation project is not expensive it will be only a minor part of a contract. Unfortunately the treatment afforded the instrumentation is frequently reflected in this fact and insufficient attention is given to its importance. The contractor's men are frequently uninformed of the instrumentation and its importance and the whole objective of the project can be missed in the turmoil of general site conditions.

INSTALLATION

To ensure that instruments are installed correctly without mishap the following conditions should be adhered to:

(a) Explain the objectives of the instrumentation to the resident engineer and also the contractor's agent and obtain their co-operation.
(b) Obtain the co-operation of the professional site staff in whose area the instrumentation is to be used.
(c) Explain and seek the assistance of the contractor's men working on the structure or work. This is particularly important as the involvement of the men on site is the best guarantee against damage and abortive work.

PROTECTION

Instrumentation can represent a considerable investment and therefore needs protection. As most damage is accidental and results from lack of knowledge of the existence of the instrument the most obvious protection is the use of brightly coloured obstacles placed around the instrument. Damage and loss of some equipment is inevitable. Wrightman[7] suggests that 10% loss is to be expected and must be allowed for when designing the instrumentation programme.

Potential problems caused from the loss of critical instruments must be recognised from the outset and the necessary duplication of instruments provided. Where possible the instruments should be arranged to be complimentary or self-checking in character (see Wilkes[8]). From a practical aspect the following protection procedure is usual:

(a) Keep instrumentation neat and tidy – a professional attitude attracts respect.

(b) Wherever possible install instruments in consultation with the contractor's programme of work, bearing in mind the problems associated with trench digging in particular.
(c) Protect from vandals – if necessary use a night watchman (vandals are not logical and sometimes the precautions taken to detract vandals prove to be an attraction).

LONG TERM TESTS AND HIDDEN COSTS

If the instrumentation is to cover long term measurements this must be allowed for from the start. The correct instruments must be used and proper systems established to ensure that readings are taken correctly and the results processed in a logical manner. Frequently the latter implies costs well in excess of the apparent initial installation.

The measurements conducted on St. Paul's Cathedral since 1923 to record settlements and movements are an illustration of the care and dedication that is required in this respect (see Thomas and Fisher[9]).

RECORDING AND ANALYSIS

With major instrumentation projects preplanning is vital, particularly in the training of instrumentation engineers and instrumentation technicians. With the use of on-site computers to analyse results, several months work prior to installation of the instruments may be necessary in order to prepare the necessary computer programs and familiarise those doing the work with their operation.

From a practical point of view it has been found useful if the instrumentation engineer concerned with the project is familiar not only with the workings of the various computer systems but has been partly responsible for their development. He is therefore capable of making on site changes to the systems in use. Sites are frequently established in remote areas and self reliance in this respect can be important.

PURPOSE OF INSTRUMENTATION

The majority of soil instrumentation used in civil engineering falls under the following headings:

1. Research

These are usually long term studies to study the validity of a design assumption. Alternatively they are used to search for more efficient design methods or as a quantitive assessment of general behaviour. Examples of these are the instrumentation programmes done with the co-operation of the Department of the Environment (now the Department of Transportation) on M1, M18 and M62 to look at the design assumptions relating to earth retaining structures (see Jones[10]).

2. Check on specific design assumptions or to monitor construction

In monitoring construction the check is being run to ensure that construction is following the designer's intention. However, the construction is usually not dependent on the actual instrumentation results (see Chang *et al*[11]).

3. As an aid to construction

In these circumstances the instrumentation is an intricate part of the construction sequence. An example is the use of instruments to measure the design load required in rock anchors (see Basset[12]).

4. Control of the works

With this instrumentation, site work is partly or wholly dependent upon the readings obtained from the instrumentation and critical decisions concerning the next stage in construction are made, based upon the evidence that they give. With this construction technique there is an implied assumption that the behaviour of the structure during construction is understood and known and the instruments have been positioned accordingly to record this behaviour. This in turn demands a 'before the event' or Class A prediction to be made (see Lambe[13]). In difficult ground conditions this may be difficult to achieve.

In difficult ground conditions most instrumentation appears to fall under the categories of (2) and (4) i.e. monitoring and control of the works.

EXAMPLE ILLUSTRATING USE OF MODERN INSTRUMENTATION TECHNIQUES TO MONITOR CONTROL OF MAJOR WORKS IN AREA OF DIFFICULT GROUND

The preceding remarks have been general in nature. The points that they make can best be illustrated by reference to a practical example in which not only the instruments, but the instrumentation techniques, recording techniques, costs and associated design assumptions are all considered.

The South Humberside Motorway M180 crosses the river Trent near the village of West Butterwick. Investigations of the Trent site indicated a very soft alluvial subsoil including peat, overlying a brown medium grained sand. The laboratory tests from the soil survey indicated difficult design conditions for a 12 m high approach embankment to the river Trent and also that there were potential problems of instability during construction, together with post construction settlements. The site falls under the definition of difficult ground.

EMBANKMENT DESIGN AND CONSTRUCTION

A conventional embankment could not be designed but a suitable design using heavy earth bunds and a lightweight PFA infill embankment standing at a slope of 1 in 3 was shown to be possible. Design was based on slip circle considerations using total stress analysis (Figure 10.2).

The difficult site conditions indicated that the rate of construction of the embankment would be critical to the overall stability and precautions necessary to monitor behaviour during construction were taken. Sections of the embankment were instrumented and control of stability was monitored through the use

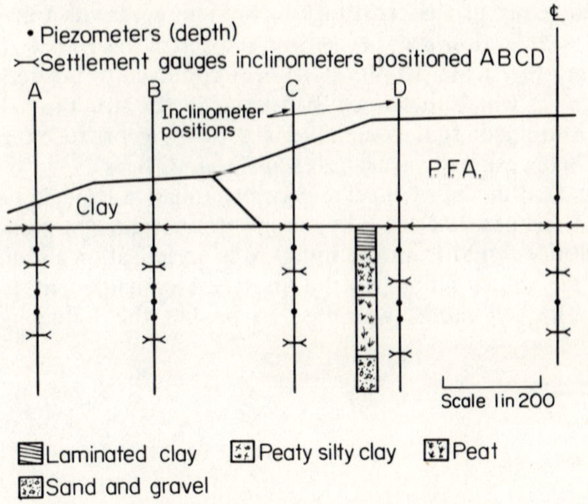

Figure 10.2 Total stress analysis

of pore water pressure measurements augmented by settlement and inclinometer readings of the strain behaviour of the embankment, i.e. the bulk of the instruments used fall under the category of piezometers, inclinometers and settlement movement gauges.

The use of conventional design methods does not permit accurate assessment of expected pore water pressures prior to construction and the use of linear elastic theory has been shown to be accurate only at the centre line of an embankment (see Schofield and Wroth[14]). In order to safely monitor the construction of a difficult embankment of this type prior knowledge of its behaviour during construction is required. To overcome this problem TRRL (Transport and Road Research Laboratory) have produced a method of control based upon the use of stability charts drawn up from initial pore pressure readings recorded in the early stages of construction, assuming a linear distribution of the major principal stresses beneath the embankment (see Symons[15], Symons and Murray[16], Murray and Symons[17]) The validity of this method is open to question (see Jones and Edwards[18]).

BEFORE THE EVENT PREDICTION OF BEHAVIOUR

The limitations of the conventional calculations used in the M180 embankment were recognised and a series of centrifuge tests based upon the proposed design were undertaken (see Bassett and Horner[19]).

The advantages of the centrifuge method is that it subjects the model of the prototype to the identical stress conditions which will occur during and after construction. The method is therefore capable of being used to provide a true before the event prediction of settlements, rate of settlement, vertical, horizontal and volumetric strains and pore pressures. From these results potential failure methods can be identified and control methods, based upon strain measurements made during construction, can be produced.

The development of the centrifuge technique represents a major advance in soil mechanics. It is, however, expensive and restrictive in use. The use of soil models on the other hand permits a different approach to the design of geotechnical structures in which the overall behaviour of the structure under consideration can be examined. It is complimentary to the centrifuge tests but has the advantage of being generally available to all design offices.

A mathematical model of the Trent embankment was made using non-linear finite element techniques using a hypobolic stress-strain soil model, based upon the soil properties found from the initial soils investigation and the materials to be used in the construction of the proposed embankment (see Jones and Edwards[20]). The soil model was used to predict the following construction details:

Pore water pressures

These were derived from the results of the increments of the principle stress (Figure 10.3). It has been shown by TRRL that a prediction of the expected

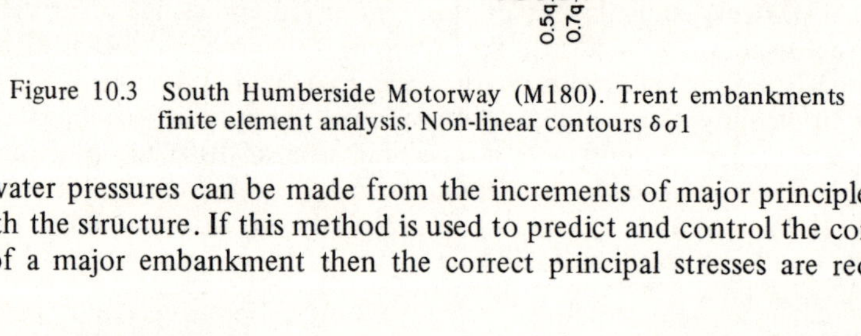

Figure 10.3 South Humberside Motorway (M180). Trent embankments finite element analysis. Non-linear contours $\delta\sigma 1$

pore water pressures can be made from the increments of major principle stress beneath the structure. If this method is used to predict and control the construction of a major embankment then the correct principal stresses are required.

Vertical settlement readings

These predict the vertical settlements taking place under the embankment structure at discreet stages of the construction (Figure 10.4).

Horizontal strains

The results of the horizontal displacements provide a prediction of the inclinometer readings which will occur during construction (Figure 10.5).

Effective stress analysis

The finite element analysis is normally based upon a total stress analysis. The analysis therefore ignores the effective gain in strength of materials with time.

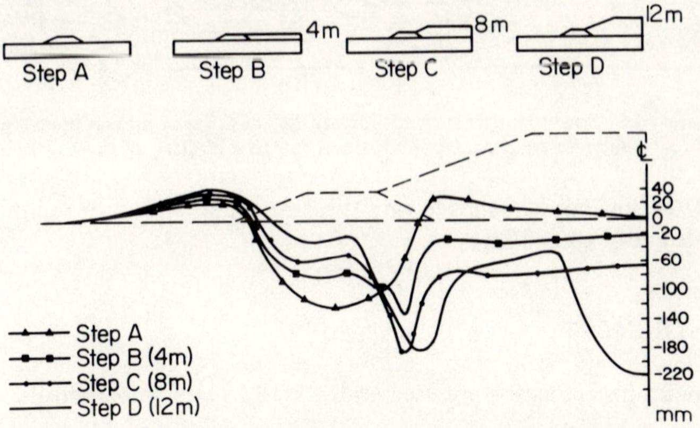

Figure 10.4 South Humberside Motorway (M180). Trent ambankments finite element analysis. Non-linear vertical displacements at G.L.

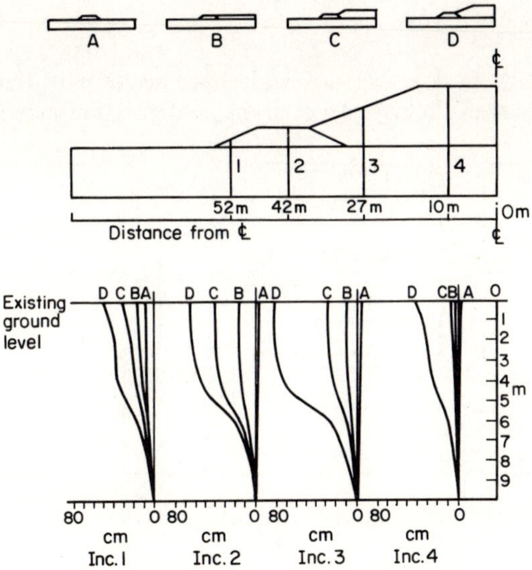

Figure 10.5 South Humberside Motorway (M180). Trent embankments finite element analysis. Positions and plots of inclinometers

Effective stress analysis can be conducted during construction of the works using the non-linear finite element method coupled with effective soil parameters taken from on site readings to assess the effective stability at discreet stages of construction. The advantage of the finite element method over other methods is that it is able to predict deformations related to the recalculated factor of

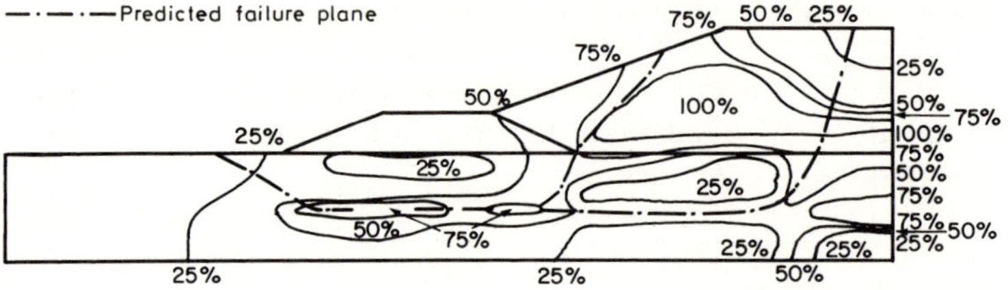

Figure 10.6 South Humberside Motorway (M180). Trent embankment finite element analysis. Non-linear percentage mobilised strength

safety. If the inclinometers agree with these deformations a true factor of safety has been found (Figure 10.6).

SITE INSTRUMENTS

The following instruments were used on the M180 Trent embankment:

Piezometers	192
Inclinometers	57
Profile gauges	4 (33 reading points each),
Settlement points	134

The instruments were positioned with three sections to the west and three sections to the east of the river Trent in the configuration shown in Figure 10.2.

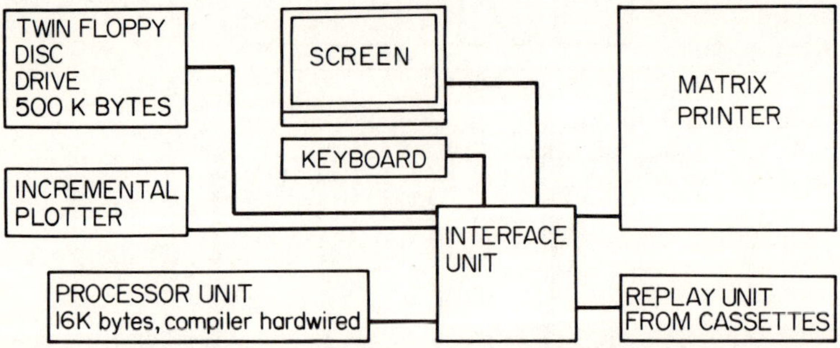

Figure 10.7 Computer configuration for instrumentation of the Trent embankment

Piezometers were read on a continuous 12 hour cycle from 5 gauge houses. Readings were recorded on paper tape and magnetic tape. The average depth of the inclinometers was 20 m and two readings were taken at each 0.5 m point.

Inclinometer readings were continuous and recordings were booked manually or recorded on magnetic tape. Complete sets of readings (225 number) were taken weekly and recorded on the computer.

Control of construction was produced through a series of cyclic total and effective stress analyses conducted daily on site during construction, coupled with a continuous plot of the strains within the embankment produced from the settlement gauges, profile gauges and the inclinometers. This was made possible by the use of a preprogrammed computer installed permanently on site, coupled with semi-automatic recording of instrumentation data and automatic plotting. All the instrumentation data was stored on the computer including the initial base readings. The computer programs required for the exercise were prepared by the engineer responsible for the instrumentation prior to the start of the contract. The configuration of the computer used for this project is shown in Figure 10.7.

The M180 is a multi-million pound motorway contract. The instrumentation of the Trent embankment cost approximately £70 000 (1976). The cost of the minicomputer was £20 000 (1976). One engineer and three technicians worked full time on the scheme.

ACKNOWLEDGEMENT

The author would like to acknowledge the help of the West Yorkshire Metropolitan County Council for their cooperation in the preparation of this chapter.

References

1. Hanna, T.H., *'Foundation Instrumentation*,. No. 3 of Volume 1 of Series on Rock and Soil Mechanics. *Trans Tech Publications*, Clausthal (1973)
2. Green, G.E., 'Equipment design, installation and performance'. *Symposium Field Instrumentation.* British Geotechnical Soc. Part II, Butterworths, London (1973)
3. Bishop, A.W., Vaughan, P.R. and Green, G.E., Report on Speciality Session 'Pore pressure measurements in the field and in the laboratory'. *Proc. 7th Int. Conf. Soil Mech and Found,* **3**, 427 (1969)
4. Vaughan, P.R., 'The measurement of pore pressures with piezometers'. *Symposium Field Instrumentation*, British Geotechnical Society, Part I, Butterworths, London (1973)
5. Jones, C.J.F.P., 'Earth retaining structures', Ph.D. Thesis, *University of Leeds* (1973)
6. Green, G.E., 'Principles and performance of two inclinometers for measuring horizontal ground movements', *Symposium Field Instrumentation*, British Geotechnical Society, Part I, Butterworths, London (1973)
7. Wrightman, J., 'Some organisation and benefit aspects of instrumentation'. *Symposium Field Instrumentation*, British Geotechnical Society, Part I, Butterworths, London (1973)
8. Wilkes, P.F., 'Instrumentation for the King's Lynn Southern Bypass', *Symposium Field Instrumentation*, British Geotechnical Society, Part I, Butterworths, London (1973)
9. Thomas, J.B. and Fisher, A.W., 'St. Paul's Cathedral – measurements of settlements and movements of the structure'. *Conference Settlement of Structures*, British Geotechnical Society, Part I, Pentech, London (1974)
10. Jones, C.J.F.P., 'Field measurements of earth pressures against motorway retaining walls and bridge abutments', *Symposium Field Instrumentation*, British Geotechnical Society, Part I, Butterworths, London (1973)

11. Chang, J.C., Forsyth, R.A. and Smith, T., 'Reinforced earth highway embankment – road 39', *Highway Focus,* **4**, 1, January (1972)
12. Bassett, R.H., Discussion on soil anchors, Conference Ground Engineering, *Institution of Civil Engineers*, 89–94, London (1970)
13. Lambe, T.W., 'Predictions in soil engineering', *Geotechnique,* **23**, 149–202 (1973)
14. Schofield, A.N. and Wroth, C.P., *Critical state soil mechanics*, McGraw-Hill, London (1968)
15. Symons, I.F., *Assessment and control of stability for road embankments constructed on soft subsoils*, Department of Environment, TRRL Report 711, Crowthorne (1976)
16. Symons, I.F. and Murray, R.T., *Embankments on soft foundations. Settlement and stability at over causeway bypass*, Department of Environment, TRRL Report 675, Crowthorne (1975)
17. Murray, R.T. and Symons, I.F., *Embankments on soft foundations. Settlement and stability study at Tickton in Yorkshire*, Department of Environment, TRRL Report 643, Crowthorne (1974)
18. Jones, C.J.F.P. and Edwards, L.W., Supplementary Report, *Symposium embankments on soft ground*, Transport and Road Research Laboratory, Crowthorne, April (1976)
19. Bassett, R.H. and Horner, J., Private communication (1975)
20. Jones, C.J.F.P. and Edwards, L.W., *Finite element analysis of Trent embankments on M180*. Engineering Development Report, West Yorkshire Metropolitan County Council, May (1976)

Chapter 11

Settlement and Foundations

INTRODUCTION

One of the most common applications of soil mechanics in civil engineering occurs when the foundations of a structure are being designed. The basic problem lies in the choice of an allowable bearing pressure which satisfies two main criteria:

(i) The allowable bearing pressure must provide an adequate factor of safety against rupture of the soil.

(ii) The resulting total and differential settlements must be kept within limits to prevent detrimental effects on the structure.

In granular soils and heavily over consolidated cohesive soils, settlement considerations usually provide the most critical criterion, whereas in normally consolidated and lightly over consolidated cohesive soil, the criterion of soil rupture generally predominates. For these latter materials, it is of course necessary for the foundation engineer to be able to estimate the settlements which will arise from a particular foundation design. In all cases, if the method of assessing settlement for a given pressure is conservative, then the footings or foundation will be over-designed, and it is therefore apparent that reliable accurate methods for estimating the settlement of foundations are essential for rational design.

The problems facing the engineer are complex. Not only must he be able to predict the magnitude and rate of settlement but also he must assess whether the completed structure will be capable of withstanding these settlements during its useful working life. In order to do this, the designer needs an understanding of five groups of topics:

(i) The basic causes under which settlement can arise;

(ii) The classification and description of the materials involved, and the measurement of their relevant properties;

(iii) Methods of calculating settlement;

(iv) Comparisons between the predictions of settlement and field observations;

(v) Observations of the damage caused to a structure by settlement of the foundations.

Considerable understanding of topics (i) to (iii) has developed since the scientific study of soil mechanics began, but the available information on topics (iv) and (v) remains unsatisfactory. It may be said, with some justification, that until an adequate body of case histories exists, the scientific study of settlement in any type of soil will be incomplete and poorly understood.

CAUSE OF SETTLEMENT

Settlement of foundations may occur from a number of causes, including:

(i) The influence of static loads, such as those imposed by the mass of a structure;

(ii) The influence of dynamic loads, such as those imposed by machinery;

(iii) The effects of vibrations, particularly in loose granular soils;

(iv) Changes in moisture content, for example, from seasonal fluctuations in the water table, from the absorption of water by vegetation, from the effects of artificial agencies such as beneath kilns and furnaces;

(v) Deterioration of the foundation caused by the corrosive action of agents in the soil acting on the foundation (see Bessey and Lea[1]).

(vi) Deterioration of the soil. The collapse of the soil structure of loess is a well-known example of this type of phenomenon (see Abelov and Askalonov[2]).

(vii) Mining subsidence (see chapter 12) and loss of support caused by nearby construction activities.

A treatment of all of the above causes of settlement is beyond the scope of this chapter. Attention will be focused on settlements caused by static loads which occur most widely in practice, although it must be appreciated that in particular situations, the effects of any of the remaining topics listed above may be of equal significance to that of the static loading. The various components of settlement will first be discussed, followed by a review of various methods used for predicting settlement. This will be followed by a comparison between predicted and observed settlements in granular and cohesive materials including a brief description of methods of monitoring the settlement of structures. Finally the question of the allowable settlements of a structure, i.e. the amount of settlement that a structure can tolerate, will be discussed.

BASIC COMPONENTS OF SETTLEMENT

Consider a small element of soil at some depth below ground level, as indicated in Figure 11.1. This element will be subjected to initial stresses which are primarily functions of the depth of the element beneath the soil surface, the bulk

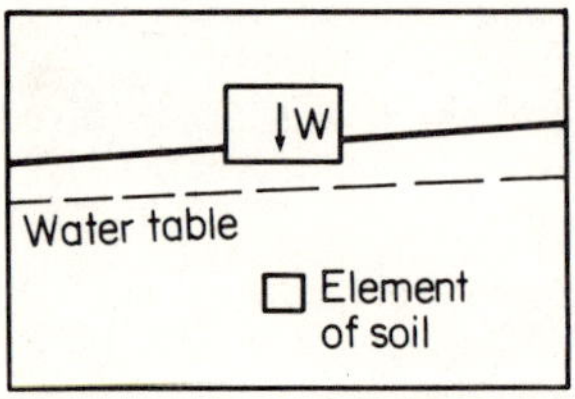

Figure 11.1 Typical soil element

properties of the soil, the position of the water table, and the past stress history of the soil (i.e. its degree of over consolidation).

When a building is constructed, the loads acting on the element will increase, and the behaviour of the element can be approximately represented by the behaviour of a relatively thin laboratory specimen of the soil when tested under one dimensional consolidation conditions. Two extreme types of behaviour are shown in Figure 11.2(a) and 11.2(b).

In Figure 11.2(a), it is considered that the loading is applied to the element under undrained conditions (A-B), inducing a change in pore water pressure in the soil. Since a condition of undrained loading is assumed, there will be no

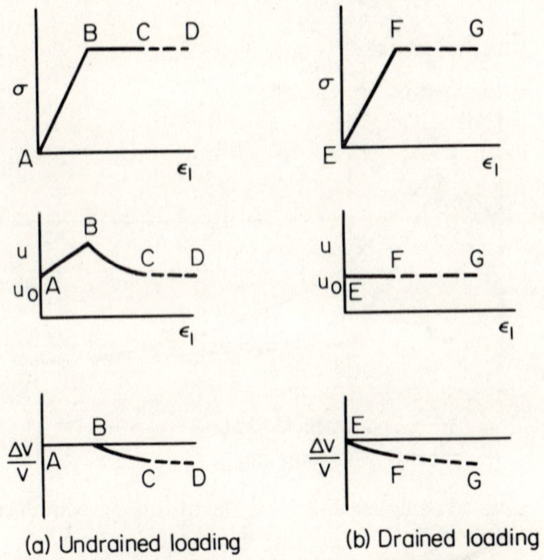

Figure 11.2 Behaviour of soil element under one-dimensional consolidation in laboratory

volume change in a saturated specimen. With time, the excess pore water pressures will dissipate, producing a change in volume under constant load (B-C) and this may be followed by long-term creep at constant effective stress.

In Figure 11.2(b), the loading on the soil is assumed to take place under drained conditions (EF), that is no excess pore water pressures are set up in the soil, and volume change does take place due to the drainage of water from the soil. This drained loading condition may be followed by long-term creep under constant effective stress (FG). This simplified picture strictly represents the special field loading condition in which strains can only occur under one-dimensional conditions, i.e. the loading of a relatively thin soil layer over a large area; under these conditions, no immediate settlement can occur in a saturated soil, and hence only settlement caused by drainage (i.e. consolidation) or creep need be considered.

The type of settlement caused by loading over fairly limited areas is shown in Figure 11.3 for the two simplified cases of undrained loading and drained loading. For the undrained loading case, during excavation heave will occur caused by elastic unloading of the soil and producing shear strain (1-2). During construction (i.e. loading under undrained conditions) an immediate settlement will be caused, again arising from shear deformations (2-3). Following loading under undrained conditions, and the development of excess pore water pressures, settlement will arise from consolidation of the soil produced by the dissipation of excess pore water pressures (3-4). This component of settlement will be time-dependent and its rate will be influenced by the coefficient of permeability of the soil. The consolidation settlement will arise mainly from volumetric deformations although shear deformations (or strains) may also be involved. Finally settlement caused by creep (4-5) may occur; these creep settlements (caused by secondary consolidation) occur after the dissipation of excess pore water pressure and are usually time-dependent. Both shear and volumetric deformations occur during creep.

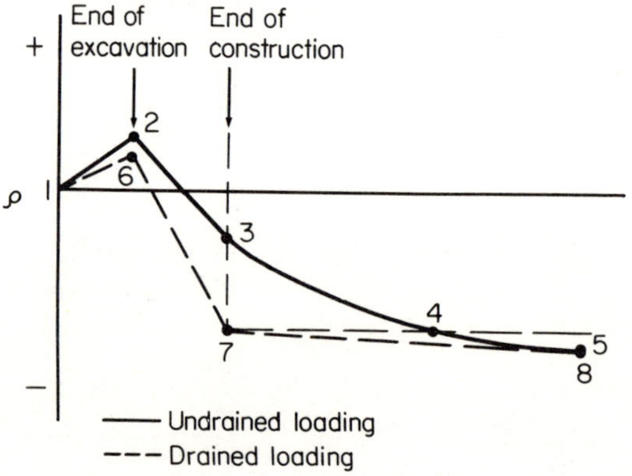

Figure 11.3 Settlement under field loading conditions

For the drained loading case illustrated in Figure 11.3, the process of excavation is accompanied by heave involving both shear and volumetric deformations (1-6). Similarly the construction period will involve settlements produced both by elastic deformation of the soil under the increase in loading, and settlement produced by consolidation (6-7). Finally some settlement produced by long-term creep may also occur (7-8).

METHODS OF SETTLEMENT PREDICTIONS IN CLAYS

In both the drained and undrained loading conditions considered above it is possible to identify three types of settlement:

(a) Immediate (or undrained) settlement, which occurs on application of the load;

(b) Consolidation (or drained) settlement, which arises from drainage caused by the dissipation of excess pore water pressures, and is therefore time-dependent;

(c) Creep settlement (or secondary consolidation) which occurs under constant effective stress following the dissipation of excess pore water pressures, and which is time-dependent.

The essential difference between undrained and drained loading occurs as a result of the time taken for drainage to take place in a particular soil, and is therefore related to the coefficient of permeability of the soil. In a soil possessing a high coefficient of permeability (e.g. sand or gravel), the loading condition can usually be represented by the drained case; whereas in soils which have a low coefficient of permeability, (e.g. clay) the loading conditions can generally be represented by the undrained case, It must, however, be emphasised that in this latter case, it is difficult in practice to isolate the individual components of settlement since they may occur simultaneously.

The total final settlement s of a foundation is therefore given by:

$$s = s_i + s_c + s_s \tag{11.1}$$

where s_i = immediate settlement
s_c = consolidation settlement, and
s_s = creep settlement.

Leaving aside the creep settlement, at any time t after the application of the load on the foundation, the settlement s_t is given by

$$s_t = s_i + U_s s_c \tag{11.2}$$

where U_s = degree of consolidation settlement.

Thus, apart from creep effects, a settlement analysis requires the estimation of s_i, s_c and U_s.

IMMEDIATE SETTLEMENT

The immediate settlements of foundations in clay are determined from procedures based on the mathematical theory of elasticity. The changes in stress on elements of the soil produced by changes in soil loading are obtained in terms of total stress. By the use of the theory of elasticity, the foundation engineer can obtain the displacements or settlements caused by loads covering areas of various geometrical shapes. The procedure necessitates the determination of the soil parameters Young's Modulus E and Poisson's Ratio ν.

It is important to realise that these parameters vary with time, and two extreme conditions can be distinguished:

(i) The immediate undrained case corresponding to the instant loading condition, with $\nu = 0.5$ for the idealised undrained case;

(ii) The final drained case related to changes in effective stress, and corresponding to the complete dissipation of excess pore water pressures.

The assumption of constant elastic soil parameters does not imply that real soil behaves as an ideal elastic solid (Davis and Poulos[3]), but similarities do exist between the behaviour of real soil and an ideal elastic solid, largely because of the relatively high factors of safety against overall failure which occur in many cases. The accuracy of the prediction of immediate settlement will depend on the accuracy and reliability of the values used for the elastic constants, which must be determined experimentally. The experimental conditions must simulate the range of stresses and type of deformation in the field if reliable results are to be obtained. The distributions of stress in an elastic solid caused by a variety of loading conditions are discussed by Poulos and Davis[4].

In a layered soil, the total final settlement may be obtained by summation of the vertical strains in each layer (see Poulos and Davis[4]).

$$\rho_t = \Sigma \frac{1}{E} \left\{ \sigma_t - \nu (\sigma_x + \sigma_y) \right\} \delta h \tag{11.3}$$

where ρ_t is the total final settlement,

E, ν are the values of the elastic parameters of the soil skeleton for the appropriate changes in stress in each layer,
σ_x, σ_y, σ_t are the increases in stress caused by the applied loading, calculated from elastic theory for the appropriate value of ν,
δh is the thickness of each layer or stratum.

If the soil profile is reasonably homogeneous then appropriate values of E and ν can be given to the whole thickness of the stratum, whence,

$$\rho_z = \frac{q.B.I}{E} \qquad (11.4)$$

where
q = foundation pressure;
B = some convenient dimesnion of the foundation (i.e. width);
I = an influence factor given by elastic theory.

Since a saturated clay layer in the short term behaves essentially as a loaded incompressible elastic layer, the immediate settlement can be calculated by using

$$\rho_i = \frac{q.B.I_u}{E_u} \qquad (11.5)$$

where I_u is an influence factor for $\nu = \nu_u = 0.5$ and E_u is the undrained Young's modulus of the soil.

Various solutions for I have been presented by Poulos and Davis[4] (1974) for deep and shallow foundations. A particularly useful chart is that given by Janby, Bjerrum and Kjaensli[5] shown in Figure 11.4, from which the average immediate settlement of uniformly loaded flexible loaded areas can be obtained in terms of μ_0 and μ_1. The immediate settlement is calculated by putting $I = \mu_0\mu_1$ in equation (11.5). Gibson[6] showed that the variation in modulus with depth can have a significant effect on the magnitude of the surface displacements.

The chart in Figure 11.4 can also be used to deal with the case where E varies with depth using the method described by Simons and Menzies.

The effects of heterogeneity and anisotropy have been considered by a number of workers, including Lambe[7], Gibson[6], Gibson and Sills[8], and Burland, Sills and Gibson[9]. It has been shown that the vertical settlement is very sensitive to the magnitude of the horizontal modulus E_h, and for the special case where $E_v = 0$ at the surface and a linear variation with depth, the maximum settlement occurs near the edge of the loaded area because of the existence of large lateral strains at the edge (Figure 11.5).

ESTIMATES OF UNDRAINED MODULUS E_u

It is clear from equation (11.5) that accurate predictions of the immediate settlement of foundations on clay will depend on the use of the appropriate values of the undrained modulus E_u. There are basically three ways in which an estimate of the undrained modulus may be obtained:

(i) From laboratory or *in situ* tests;
(ii) By published correlations between E_u and the undrained shear strength c_u;
(iii) By back analysis of the measured deformations of full-scale structures (e.g. retaining walls, tunnels).

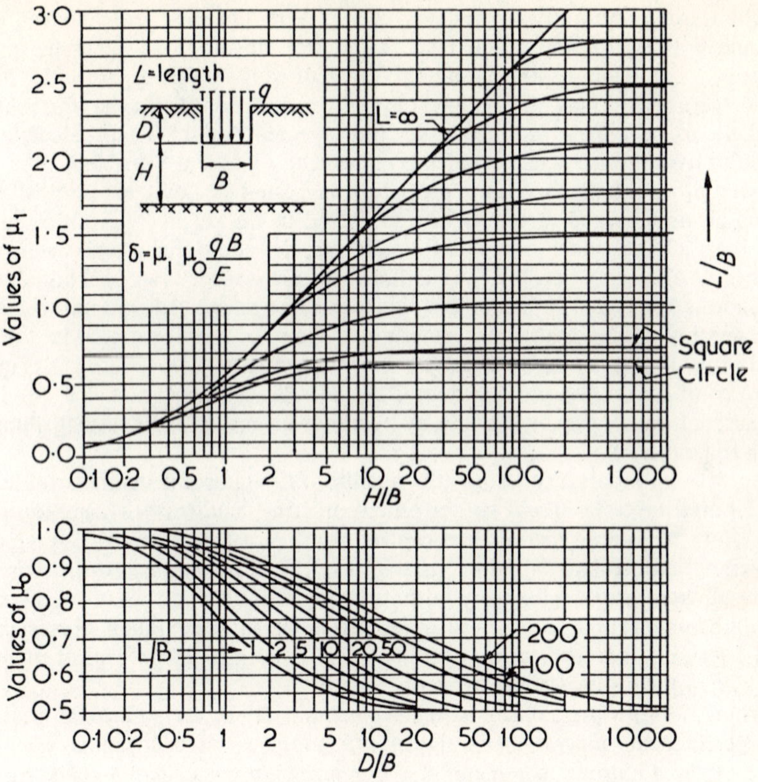

Figure 11.4 Average immediate settlement of uniformly loaded flexible areas (After Janby, Bjerrum and Kjaeonsli[5])

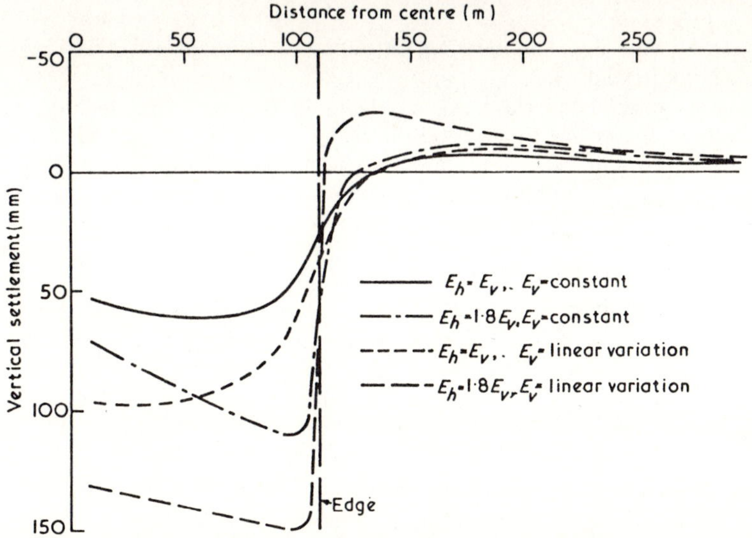

Figure 11.5 Influence of heterogeneity and anisotrophy on immediate settlement (after Rodrigues)

The earliest laboratory estimates of E_u were obtained from conventional undrained triaxial compression tests, where the modulus was calculated over a range of axial loading corresponding to one-half of the ultimate. Since the stress-strain relationship may not be linear even over this restricted range, this method is likely to be grossly unreliable. An improvement would be to calculate the modulus over the stress range which is relevant to a particular problem.

Even more satisfactory is the procedure suggested by Davis and Poulos[10,3], in which an undisturbed sample is first subjected to the original geostatic stresses, and then adding under undrained conditions, the additional stresses which are applied to the soil when the foundation loads are added. Tests on London Clay by Simons and Som[11] indicate that this technique gives elastic moduli which are considerably higher than those determined from conventional triaxial tests. A similar result was found by Marsland[12,13,14] from large plate loading tests in the bottom of boreholes in London Clay. Marsland found that the modulus measured from *in situ* tests could be nearly five times as high as that obtained from triaxial tests.

As with all laboratory testing, the quality of the sample is significant. Ideally, the laboratory tests should be performed on truly 'undisturbed' samples, since the effect of sample disturbance can be significant (see Simons[15], Raymond, Townsend and Lojkase[16]). More reliable results can be obtained if the sample is reconsolidated in the laboratory, although opinions differ on what stress system should be used. Thus Simons[15] suggests consolidation to the initial stress system, whilst Raymond *et al*[16] recommend isotropic consolidation to a stress of about one-half of the initial vertical stress.

Where anisotropic consolidation is recommended, an additional problem occurs, since the magnitude of the *in situ* lateral stress will depend upon the value of the additional parameter K_0'. For normally consolidated clays, K_0' can be obtained with reasonable accuracy from the equation

$$K_0' = 0.95 - \sin\phi' \tag{11.6}$$

where ϕ' is the angle of shearing resistance in terms of effective stress (c.f. Brooker and Ireland[17]).

In over consolidated clays, K_0' is related to the past stress history, and in particular to the degree of over-consolidation. A number of methods for determining K_0' of over consolidated clays have been suggested, including:

(a) The use of laboratory pore pressure measurements and the theoretical relationship between pore pressure coefficient A and K_0' (see Skempton[18], Blight[19]). The usefulness of this method may be affected by sample disturbance.

(b) The use of the triaxial 'Kink' test, suggested by Poulos and Davis[20]. In this method, an undisturbed sample is subjected to a triaxial loading condition in which the vertical stress is equal to the *in situ* vertical effective overburden stress, and a horizontal stress of about 0.4 times this value. The vertical stress is maintained constant whilst the horizontal stress is increased in a number of small steps, and volume change is allowed to take place under each increment of horizontal stress. From a plot of volume change against horizontal effective stress it is possible to estimate the *in situ* horizontal stress, since a 'Kink' appears in the plot at the corresponding value. The 'Kink' is pronounced in the case of normally consolidated soils, but less clearly defined in the case of over-consolidated soils.

(c) Bjerrum and Anderson[21] have used the techniques of 'hydraulic fracturing' in fine-grained soils where the K_0' value is less than 1. The method involves the formation of a vertical crack in the soil around a piezometer tip, and the measurement of the water pressure at which the crack closes is used to calculate the value of K_0'.

(d) The use of laboratory oedometer tests to estimate the preconsolidation pressure and hence the value of the over consolidation ratio. Then by the use of correlations between over consolidation ratio, and K_0' (Ladd[22]), the effective lateral stress can be calculated.

(e) The use of the Camkometer (Wroth and Hughes[23]). This sophisticated self-boring pressuremeter has been designed to minimise the disturbance of the soil and can be used to estimate K_0'. In soft clays, the Camkometer has given results which are in agreement with the results obtained from hydraulic fracturing, but no results are available for stiff clays.

As well as the type of consolidation in the laboratory measurement of E_u, a number of other factors may well be significant, including the stress level, consolidation period and rate of strain.

In view of the considerable difficulties in obtaining high quality undisturbed samples and the other problems associated with a laboratory determination of E_u, correlations between the undrained modulus E_u and the undrained shear strength have been suggested by a number of workers. For example, Bjerrum[24] quotes a range of 250–500 for the ratio E_u/c_u, D'Appolonia *et al*[25] suggest a range from 1000 for inorganic clays to 120 for highly organic clays, Bjerrum[26] suggesting 500–1500. Simons[27] summarised the values of E_u/c_u for a number of structures on normally and lightly over consolidated clays and found that the ratio ranged from 40–3000.

The use of field observations and back analysis to estimate values of the undrained modulus has been used on a number of occasions. For example, from the deflections of a retaining wall (Cole and Burland[28]) and from the deflection of tunnels below an excavation (Ward[29]).

A comparison between the values of the undrained modulus obtained from back-analysis and those obtained from undrained cylindrical compression tests was performed by Atkinson[30], who found that the values obtained from back analysis were about five times the values from laboratory tests. Atkinson considered that part of this difference could be explained by allowing for anisotropy in the back-analyses, but at least half of the difference is probably caused by sample disturbance, stress path effects and difficulties associated with the accurate measurement of strains during laboratory tests.

Before leaving the topic of immediate settlement of clays, it is of interest to note that Simons and Som[11] analysed a number of case histories in normally-consolidated and over consolidated clays and found the following ranges for the ratio s_i/s, (where s_i = immediate settlement, and s = total settlement);

Over consolidated clay	s_i/s range 0.315–0.735 average 0.575.
Normally consolidated clay	s_i/s range 0.077–0.212 average 0.156.

Thus, the immediate settlement accounts for a much higher percentage of the total settlement in the case of over consolidated clays.

CONSOLIDATION SETTLEMENT

The theory of one-dimensional consolidation was originally developed by Terzaghi[31] and has formed the basis for many of the predictions of magnitude and rate of consolidation settlement. Basically the method involves the summation of vertical strains beneath the loaded area, using soil properties determined from the oedometer apparatus. The resulting settlement (here termed s_{oed}) is calculated from:

$$s_{oed} = \Sigma m_v \sigma_t \delta_h \tag{11.7}$$

where
m_v = coefficient of volume change for the appropriate stress range and increment;
σ_t = increase in vertical stress at the centre of a layer due to the applied load (σ_t calculated from elastic theory); and
δh = thickness of each layer considered.

The basic one-dimensional method assumed that $s_{oed} = s_c$. The degree of settlement U_s is calculated from one-dimensional consolidation theory, as developed by Terzaghi[31].

A number of refinements have been suggested to the original Terzaghi procedure. As mentioned previously, high standards of sampling and testing must be employed if reliable settlement predictions are to be obtained. The influence of sample disturbance has been recognised for many years (see Terzaghi[32]). Procedures have been suggested for the correction of laboratory stress-strain curves in view of sample disturbance.

The effects of sample disturbance depend to a large degree on the past stress history of the soil; thus for normally consolidated clays, sample disturbance will lead to under-estimates of settlement, whilst the converse is true in the case of *over*-consolidated soils. Probably the most widely used method for correcting the results of laboratory oedometer tests in order to allow for sample disturbance is that suggested by Schmertmann[33]. A significant development to the basic Terzaghi method was proposed by Skempton and Bjerrum[34]. In the oedometer test, because the sample is laterally confined, the induced excess pore water pressure will always equal the applied increment of vertical stress, irrespective of the value of the pore-pressure parameter A (Simons and Menzies[36]). In practice, however, an element of soil beneath a loaded area will undergo lateral deformation and the induced pore-water pressure will be a function of the A-value of the soil, and generally will be less than the applied increment of vertical stress. Skempton and Bjerrum proposed the use of a factor to correct the results of computations based on the results of one-dimensional consolidation tests, i.e.

$$s_c = \mu s_{oed} \tag{11.8}$$

The factor μ is a function of the geometry of the loaded area and the value of A (Figure 11.6).

Although the Skempton and Bjerrum[34] approach represents a distinct improvement over the classical Terzaghi method, it is still largely based on data obtained from one-dimensional consolidation tests and the influence of lateral stresses on deformation properties has not been taken into account. A logical development would be to subject the laboratory specimens to the same stress changes as would occur in the field, and to obtain the various soil parameters for the correct changes in stress. This is the concept of stress path testing, and it represents a significant advance in settlement analysis. The method is described by

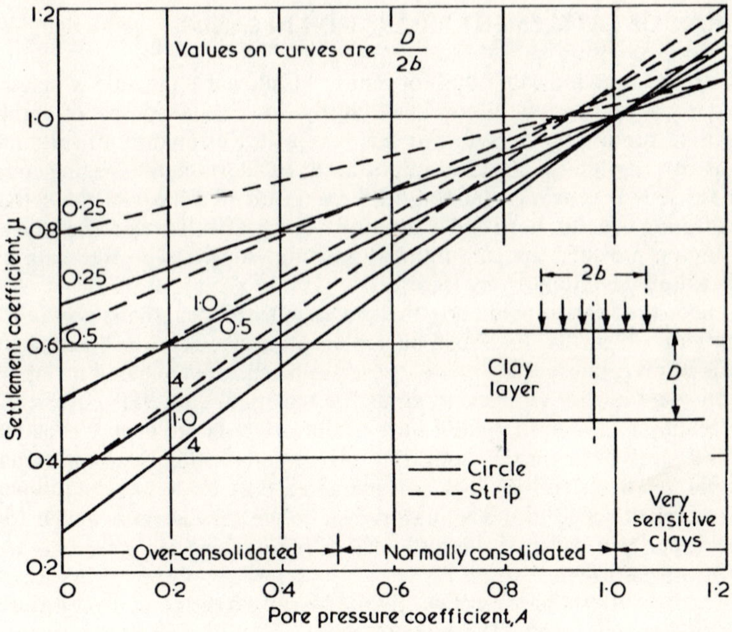

Figure 11.6 Correction factor for pore pressures set up under a foundation (after Skempton and Bjerrum[34])

Lambe[7], and has been applied to settlement analyses in the London Clay[11]. It is necessary to estimate the initial *in situ* stresses and to calculate the increase in stresses caused by the applied loading. The major disadvantage of the method lies in the fact that very sophisticated and expensive laboratory techniques are involved.

CREEP SETTLEMENT

A complete settlement analysis should take into account the effect of creep, which is generally taken to be the settlement which occurs following dissipation of the excess pore water pressures.

Although the topic of creep has attracted the attention of a number of workers, no reliable method yet exists for calculating the magnitude and rate of settlement due to this cause. The analysis is inevitably complicated since it is known that creep is significantly affected by such factors as the stress increment ratio, the stress level, temperature and time effects. The effects of creep are undoubtedly significant, and a number of case records exist which demonstrate that a large proportion of settlement occurs after dissipation of the excess pore water pressures (see Foss[37]).

The prediction of creep settlements at present is unreliable, and no great accuracy can be expected. There is a clear need for further research work in this area of geotechnical engineering, preferably involving long-term field and laboratory studies.

METHODS OF SETTLEMENT PREDICTION IN SANDS

It has already been indicated that for sands, the allowable pressure which may be applied to a foundation is governed largely by considerations of settlement rather than rupture of the soil. For this reason, the estimation of settlement in sands is of considerable practical importance. In addition, the loading condition for these soils is generally drained, with the result that most of the settlement occurs during construction, and only minor creep effects occur after the end of construction provided that the foundation is not subjected to fluctuating loads nor that the foundation is very large.

A major problem in predicting the settlement of foundations on sand lies in the extreme difficulty in obtaining representative undisturbed samples. Thus whereas in clay, most of the methods used in prediction are based on laboratory tests on good quality samples, in sands the techniques are based on field tests. Some attempts have been made to use laboratory tests; for example, oedometer tests and stress-path triaxial tests. But even recompacting the samples back to the same relative density, does not guarantee that the results of stress-strain relationships obtained in the laboratory will be the same as those which apply in the field, partly because of the difficulty in reproducing the structural arrangement of the particles.

Plate loading tests have been widely used to assess the compressibility characteristics of granular soils. The major limitation of this type of test rests in the fact that the loaded area in the test is generally small compared with that loaded by the foundation. In addition, in variable ground a number of tests are required at different depths and preferably using plates of varying sizes. The settlement of the foundation is then obtained by extrapolation from the results of the plate loading test. Terzaghi and Peck[38] proposed that the extrapolation should be performed using the following relationship for footings loaded to the same intensity as the test plate:

$$\frac{s_B}{s_b} = \left(\frac{2B}{B + 0.3}\right)^2 \qquad (11.9)$$

ρ_B is the settlement of a footing of width B in metres, and
ρ_b is the settlement of the test plate of width 0.3 m.

This relationship implies that for large values of B, the ratio s_B/s_b will not exceed 4.

Based on a study of case records Bjerrum and Eggestad[39] found that there could be appreciable scatter from the correlation given in equation (11.9), and that settlement ratios considerably greater than 4 could occur. They suggested that density of the soil could have an influence on the correlations. Other workers have also found that the Terzaghi and Peck correlation may lead to considerable errors in predicting settlement from plate loading tests. At the present time it appears that no reliable correlation exists, and that the use of plate loading tests require extreme caution.

Probably the most widely used method for predicting settlements in granular deposits is the standard penetration test (see chapter 9). Terzaghi and Peck[38] proposed a correlation between the N value and allowable pressure. Their correlation relates the size of the footing, the N-value and the applied pressure to give a settlement of 25 mm (Figure 9.1). Terzaghi and Peck also suggested correction factors to take into account the depth of embedment of the foundation, and the position of the ground water table.

As discussed by Sutherland[40], the Terzaghi and Peck correlation was based on limited data, and probably did not include the results of any settlement observations on structures. It is now generally recognised that the Terzaghi and Peck correlation is conservative, and several modifications have been suggested.

Based on a laboratory investigation, Gibbs and Holtz[41], demonstrated that the penetration resistance of a soil is a function of both the *in situ* density and the effective stress. Subsequently a number of workers confirmed from field investigations that the effective overburden pressure had a significant influence on the penetration resistance of sands. The simple correction chart, shown in Figure 9.2, was proposed by Tomlinson[42] in order to allow for the effects of overburden pressure. It can be seen that for shallow foundations, a correction factor of up to 4 is indicated. The use of this chart to obtain corrected blow counts results in predictions of settlement which are more accurate than those obtained using the original Terzaghi and Peck method.

Meyerhof[43] used results obtained from a comparison of predicted and observed settlements to propose a revision of the Terzaghi and Peck correlation. He recommended:

(i) That the allowable pressures obtained from Figure 9.1 should be increased by 50%;
(ii) That the correction to allow for the position of the ground water table should be ignored since it should already be reflected in the N value.

Meyerhof's method also ignores the influence of the effective overburden stress. Meyerhof reviewed eight cases of settlement and found that even with these amendments, the predicted settlements were between 1.2 and 4 times larger than the observed settlements.

Peck and Bazaraa[44] recognised that the original Terzaghi and Peck correlation was over-cautious, and proposed three amendments. Firstly, the allowable pressure could be increased by 50%, as suggested by Meyerhof, secondly, they agreed that the effective overburden pressure did influence blow count, but thought that the Gibbs and Holtz[41] proposal was too high, and thirdly, they suggested a slightly different correction for the influence of the ground water table.

Among other methods which have been suggested is the simple empirical method proposed by Parry[45], in which it is assumed that settlement is a function of the width of the loaded area, the magnitude of the applied pressure, and the deformation modulus of the soil. Parry proposes an equation involving the N value and correction factors, one for the fact that excavation alters the stress system acting on the soil, another to allow for the influence of the water table, and a third correction factor for the thickness of the compressible layer.

De Beer and Martens[46] suggested an equation for predicting settlement from the results of Dutch cone tests. The relationship, which is based on a semi-empirical approach is:

$$s = \frac{2.3\,H}{C}\log_{10}\left[\frac{p' + \Delta p}{p'}\right] \tag{11.10}$$

where
s is the settlement,
H is the thickness of the compressible layer,
p' is the initial *in situ* effective stress at the centre of the compressible layer,
Δp is the increase in stress at the centre of the layer caused by the net foundation pressure (calculated from elastic stress distribution theory),
C is the compressibility coefficient, given by

$$C = \frac{1.5\,q_c}{p_0'}$$

where q_c is the static cone point resistance and p_0' is the effective overburden pressure at the depth where the test was performed. This method generally over-estimates settlement, and because of this, the following relationship for the compressibility coefficient has been suggested:

$$C = \frac{1.9\,q_c}{p_0'}$$

Schmertmann[47] used a different approach in using the results of cone tests to predict settlements, pointing out that the distribution of strain beneath loaded footings is not the same as the distribution of stress. Schmertmann points out that the greatest strain occurs at a depth of $B/2$ below a footing of width B, and then suggests an equation for calculating settlement which includes two correction factors, one for the effect of depth of embedment, and the second to allow for creep. Schmertmann's method is probably the only method in common use which includes a creep correction.

COMPARISON OF PREDICTED AND OBSERVED SETTLEMENT – COHESIVE SOILS

Butler[48] contains an excellent summary of predicted and observed settlements in over-consolidated clays. The use of the Skempton and Bjerrum[35] method appears to be reasonably satisfactory in such material. Thirteen case histories were considered, and a comparison between predicted and observed settlement ranged from +60% (i.e. over-estimate of actual settlement) to –12%. A similarly good measure of agreement was found for short term unloading, i.e. elastic heave.

Although several useful records of settlement exist, there is a clear need for further information, particularly for cases which record movement from the earliest stages of construction activity on a site. Equally important is to ensure that long-term measurements are taken in order to obtain the complete settlement picture. It is likely that only in this way, supplemented by thorough laboratory investigations of engineering properties and geotechnical studies, will information be obtained towards the understanding of creep in clays.

COMPARISON OF PREDICTED AND OBSERVED SETTLEMENT – COHESIONLESS SOILS

Because of the empirical nature of most of the methods for predicting settlements in granular soils, it is not surprising that the various methods give widely differing predictions. Simons, Rodrigues and Hornby[49] used a number of methods to predict the settlements of six structures where settlement observations had also been made. Their results are summarised in Table 11.1. It would

Table 11.1 RESULTS CALCULATED ON OBSERVED SETTLEMENTS
(after Simons, Rodrigues and Hornsby[49])

Method	*s calc / s obs average*	*s calc / s obs range*
Terzaghi and Peck	1.89	0.5–3.2
Terzaghi and Peck, modified by Meyerhof	0.70	0.2–0.1
Terzaghi and Peck, modified by Tomlinson	0.31	0.1–0.6
Peck and Bazaraa	0.63	0.3–1.4
Parry	0.72	0.1–1.3
de Beer and Martens	3.22	1.0–4.8
Schmertmann	1.48	0.2–4.0

appear from the table that the methods proposed by Meyerhof and Parry give the best agreements on average between predicted and observed settlements, but it must also be noted that the ranges of calculated to observed settlements are wide. It is therefore possible that in any single case, the calculated settlement may vary considerably from that predicted. Since many of the methods used are based on tests which do not measure compressibility of the soil directly, this poor agreement is understandable. The effect of past stress history (i.e. whether the soil is normally or over-consolidated) will significantly affect settlement of foundations, but would not be indicated by penetration tests.

In view of the wide scatter in results obtained from the various methods it is important that undue reliability is not placed on any one method. It would appear to be wise to calculate settlements based on as many different methods as possible, in order to assess the possible range of uncertainty.

APPARATUS FOR THE MEASUREMENT OF SETTLEMENT

Settlement observations on structures are important for two reasons: firstly, because very little is known about the actual performance of building and what differential settlement and distortions can be tolerated by the buildings, and secondly because the observations give important information about the ground. The Building Research Establishment has used surveyors levelling techniques for at least 20 years, and with careful installation, it is now possible to obtain records to an accuracy of 0.5 mm. The techniques pioneered by the BRE are described by Cheney[50].

It is important that observations of settlement are comprehensive, so that movements during construction, and subsequently, are monitored. In order to do this, Cheney[50] recommends the use of purpose-built levelling stations which are built into the structure. These units must not only enable accurate levelling to be undertaken, but also must be non-corroding and not liable to damage during construction or by vandals. The BRE levelling station consists of a stainless steel socket which is set into the structure, and a removable levelling plug, on which the base of the staff is placed when levels are taken.

It is also necessary to provide in the vicinity of the structure, a datum point or bench mark. Deep datum points may be excessive, but it may be possible to use nearby structures which have been in existence for long enough so that no

Figure 11.7 Settlement of building on clay at the quayside, Goteborg, Sweden. Note the window frames which are somewhat out of shape and the cracking in the brickwork (Courtesy of Dr. C.F. Moon)

further settlement is occurring. Cheney[50] gives details of a datum point used by BRE which consists of a rod installed in a 6 m deep borehole with a coupling and datum levelling dome protruding above ground surface. Care is essential to ensure that the levelling dome is adequately protected throughout the period of the settlement observations. Using the levelling stations and datum point, settlements can be recorded to an accuracy of 0.5 mm using a Wild N3 precision level and an invar staff.

Techniques and apparatus used in a wide variety of situations have been described by Hanna[51].

ALLOWABLE SETTLEMENTS

The foundation engineer may encounter considerable difficulties in predicting the magnitude of the settlement likely to occur beneath a particular structure. Of equal difficulty is the estimation of how much settlement the structure can tolerate. This is generally almost impossible in reality because of the complexity of the problem; the complexity arises from the number of factors which need to be taken into account. These include the interaction of the structural elements, the redistribution of loading which occurs during differential settlement, the effects of time and creep in the structure, and the difficulty in assessing the properties of the ground. A further problem is the question of assessing the criterion to be used for deciding that 'large' settlements have occurred.

In discussing 'tolerable' settlements, it should be emphasised that total settlements may not be too significant to the overall performance of the building, provided the settlement is uniform. Damage may be restricted only to the services to the structure. Of considerably greater importance is the relative, or differential, movement between adjacent footings. This differential settlement may cause tilt (noticeable in tall buildings) or distortion to the structure (Figure 11.7).

Skempton and MacDonald[52] suggested limits on total settlement and on angular distortion, which they defined as the ratio δ/l, where δ is the differential settlement between two points of a foundation, and l is the distance between these points. Skempton and MacDonald suggested the following limits on total settlements.

Isolated foundations on clay	65 mm
Isolated foundations on sand	40 mm
Rafts on clay	65–100 mm
Rafts on sand	50–65 mm

Bjerrum[53] developed the idea of Skempton and MacDonald and suggested the following limits for the ratio δ/l for the avoidance of damage:

δ/l	*Type of damage*
1/750	Difficulties with sensitive machinery.
1/600	Danger to frames with diagonals.
1/500	Limit for buildings where cracking is not permissible.
1/300	First cracking in panel walls; problems with overhead cranes.
1/250	Tilting of high buildings becomes noticeable.
1/150	Considerable cracking in brick and panel walls.

A significant contribution to the topic of allowable settlement and damage to buildings has been provided by Burland and Wroth[54]. They suggest that a fundamental understanding of the causes of damage is required, and approach the problem from the concept of critical tensile strain. They point out that the same limit for angular distortion (δ/l) of 1/500 suggested by Skempton and MacDonald is satisfactory for framed buildings of both traditional and modern construction. Damage has, however, been reported for very much smaller distortions from buildings with load-bearing walls. They particularly point out the significance of hogging, which can arise from a number of causes including mining subsidence, tunnelling, ground water lowering and movements around excavations. From both theoretical considerations and observation, it is clear that very small movements may cause damage if the mode of deformation is that of hogging.

Based on a critical tensile strain ϵ_{crit} = 0.075%, Burland and Wroth present data (given in Figure 11.8) relating δ/l to l/H (where H is the height of the

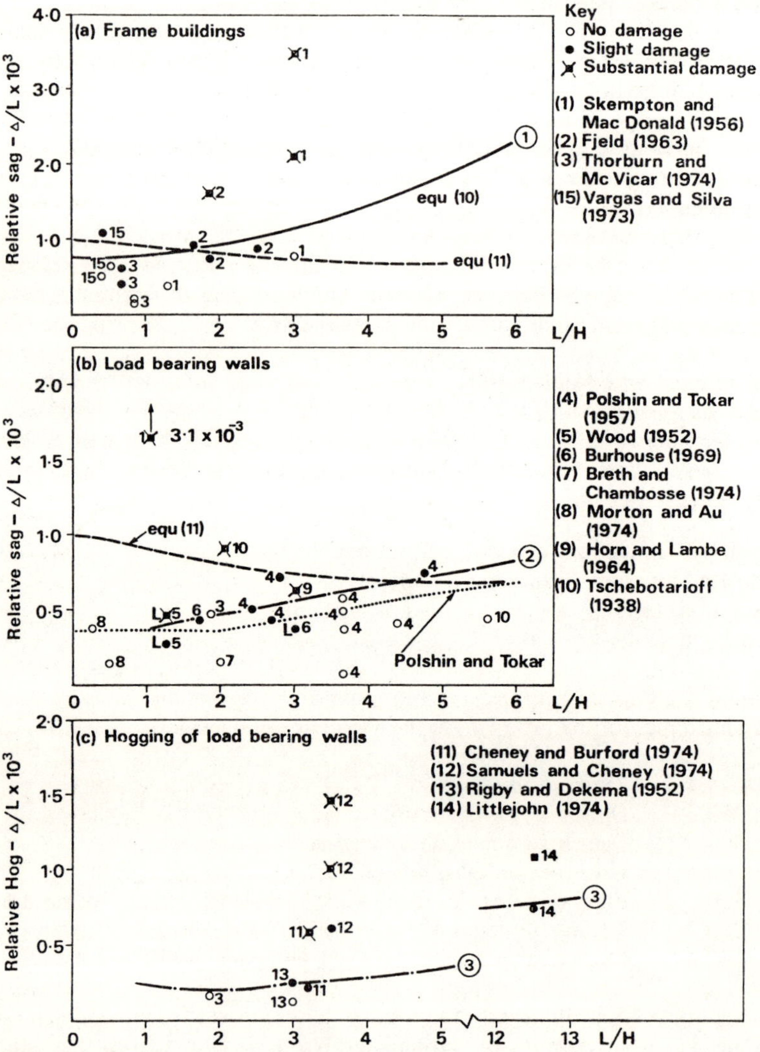

Figure 11.8 Relationship between Δ/l and l/h for buildings showing various degrees of damage

building), and indicate case histories where various degrees of damage have been observed. They suggest that predictions of relative deflections could be compared with the criteria of Figure 11.8 and if sufficiently large to suggest that damage can occur, then either the design of the structure or foundations can be modified, or provision made for remedial action.

Although some case histories exist comparing observed settlements to damage, there is a clear need for records which are comprehensive in reporting data on movements of the foundations and detailed observations of the structure and its elements.

A NOTE ON BEARING CAPACITY

Foundation design is primarily concerned with ensuring that movements of a foundation are within limits which can be tolerated by the proposed structure without adversely affecting its functional requirements. Structures vary widely in their capacity to accommodate movements of their foundations and so it is desirable that the design of the structure and the foundation be considered as inter-related problems. The design of a foundation structure requires an understanding of the local geological and ground water conditions and more particularly an appreciation of the various types of ground movement that can occur. Foundation movements may occur when the ground is excavated, when it is loaded, or independently of construction operations.

Movement of foundations under the influence of loading may occur as a result of overstressing the ground, which gives rise to plastic deformation in the ground beneath the foundation structure. In extreme cases shear failure may occur. In order to avoid shear failure or substantial shear deformation the foundation pressures used in design should have an adequate factor of safety when compared with the ultimate bearing capacity of the foundation. The ultimate bearing capacity is the value of loading intensity which causes the ground to fail suddenly in shear. The maximum safe bearing capacity is obtained when the ultimate bearing capacity is factored. In other words it is the maximum loading intensity which may be safely carried without the risk of shear failure, but there is a risk that excessive or differential settlement may occur in some circumstances. Thus the allowable bearing capacity is the value which is used in design, this taking into account all possibilities of failure, and so its value is frequently less than that of the safe bearing capacity. The value of ultimate bearing capacity depends on the type of foundation structure as well as the soil properties. For example, the dimensions, shape and depth at which a footing is placed all influence the bearing capacity. More specifically, in cohesionless soils the width of the foundation is important, the greater the width the larger the bearing capacity whilst in saturated clays the width of the foundation has little effect on the ultimate bearing capacity, density, frictional resistance and cohesion influence the shear strength of a soil, and its permeability, compressibility and consolidation must also be taken into consideration when the allowable bearing pressure is determined. With uniform soil conditions the ultimate bearing capacity increases with depth of installation of the foundation structure. This increase is associated with the confining effects of the soil, the decreased overburden pressure at foundation level and with the shear forces that can be mobilised between the sides of the foundation and the ground. The presumed bearing values for various types of soils and rocks are given in Table 11.2.

When a load is applied to a soil in gradually increasing amounts the soil deforms and a load-settlement curve can be plotted (Figure 11.9). When the failure load is reached the rate of deformation increases and the load-settlement curve goes through a point of maximum curvature which indicates that the soil

Table 11.2 PRESUMED BEARING VALUES UNDER VERTICAL STATIC LOADING (from CP 2004: 1972[55])

(These values are for preliminary design purposes only, and may need alteration upwards or downwards. No addition has been made for the depth of embedment of the foundation)

Group	*Class*	*Types of rocks and soils*	*Presumed bearing value* kN/m²	*Remarks*
I	1	Hard igneous and gneissic rocks in sound condition	10 000	These values are based on the assumption that the foundations are carried down to unweathered rock
Rocks	2	Hard limestones and hard sandstones	4000	
	3	Schists and slates	3000	
	4	Hard shales, hard mudstones and soft sandstones	2000†	
	5	Soft shales and soft mudstones	600 to 1000	
	6	Hard sound chalk, soft limestone	600	
	7	Thinly bedded limestones, sandstones, shales	To be assessed after inspection	
	8	Heavily shattered rocks	To be assessed after inspection	
II	9	Compact gravel, or compact sand and gravel	>600	Width of foundation (*B*) not less than 1 m. Groundwater level assumed to be a depth not less than *B* below the base of the foundation.
Non-cohesive	10	Medium dense gravel or medium dense sand and gravel	200 to 600	
soils	11	Loose gravel, or loose sand & gravel	<200	
	12	Compact sand	>300	
	13	Medium dense sand	100 to 300	
	14	Loose sand	<100	
III	15	Very stiff boulder clays & hard clays	300 to 600	Group III is susceptible to long-term consolidation settlement
	16	Stiff clays	150 to 300	
Cohesive	17	Firm clays	75 to 150	
soils	18	Soft clays and silts	<75	
	19	Very soft clays and silts	Not applicable	

has failed. The shape of the curve is influenced by the type of soil involved, for example, dense sand and insensitive clay show a more gradual transition, associated with progressive failure.

There are usually three stages in the development of a foundation failure. Firstly, the soil beneath the foundation is forced downwards in a wedge-shaped zone (Figure 11.10). Consequently the soil beneath the wedge is forced downwards and outwards, elastic bulging and distortion taking place within the soil mass. Secondly, the soil around the foundation perimeter pulls away from the foundation and the shear forces propagate outward from the apex of the wedge. This is the zone of radial shear in which plastic failure by shear occurs. If the soil is very compressible or can endure large strains without plastic flow the failure is confined to fan-shaped zones of local shear. The foundation will displace downwards with little load increase. On the other hand if the soil is more rigid, the shear zone propagates outward until a continuous surface of failure extends to ground surface and the surface heaves. This is termed general shear failure. The failure can be symmetrical, particularly if rotation is restricted, or it can tilt. A symmetrical bearing capacity failure is not common.

The weight of the material in the passive zone resists the lifting force and provides the reaction through the other two zones which counteract downward motion of the foundation structure. Thus the bearing capacity is a function of the resistance to uplift of the passive zone. This in turn varies with the size of

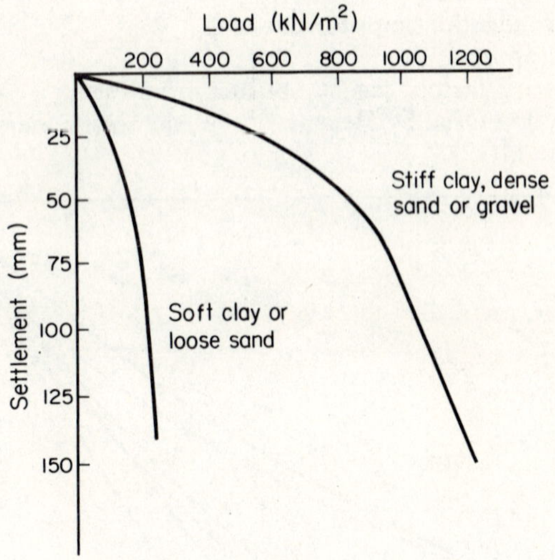

Figure 11.9 Load-settlement curve

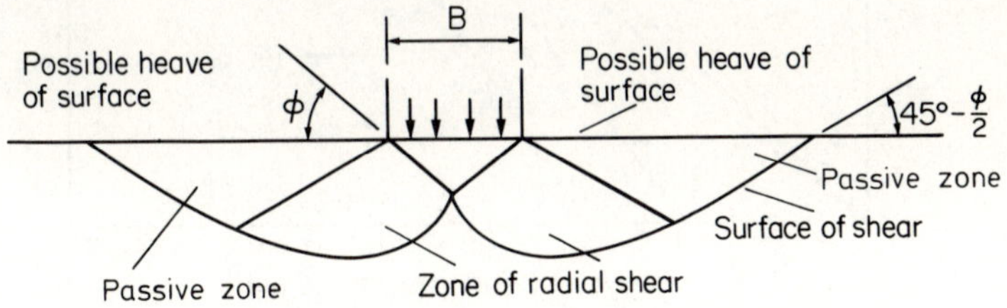

Figure 11.10 Foundation failure

the zone (which is a function of the internal angle of friction of the soil), with the unit weight of the soil and with the sliding resistance along the lower surface of the zone (which is a function of the cohesion, internal angle of friction and unit weight of the soil). A surcharge placed on the passive zone, or an increase in the depth of the foundation beneath ground surface, therefore increases the bearing capacity.

A number of bearing capacity factors are used to determine the influence of the various characteristics of the soil and foundation structure on the ultimate bearing capacity. The bearing capacity factors as suggested by Terzaghi[56] for shallow footings, that is, at a depth not exceeding the width of the footing, involve cohesion, density and overburden pressure, and the ultimate bearing capacity (q) for general shear is given by the expression:

$$q = cN_c + \rho z N_q + 0.5\rho B N_\gamma$$

where N_c, N_q and N_γ are bearing capacity factors respectively related to cohesion, overburden pressure and density which are functions of the angle of internal friction; c is the cohesion;
ρ is the density of the soil;
z is the depth of foundation emplacement, and
B is the width of footing.
The bearing capacity factors can be obtained from Figure 11.11. The analysis was subsequently improved by Meyerhof[57] but the results are expressed in the same form (Figure 11.12).

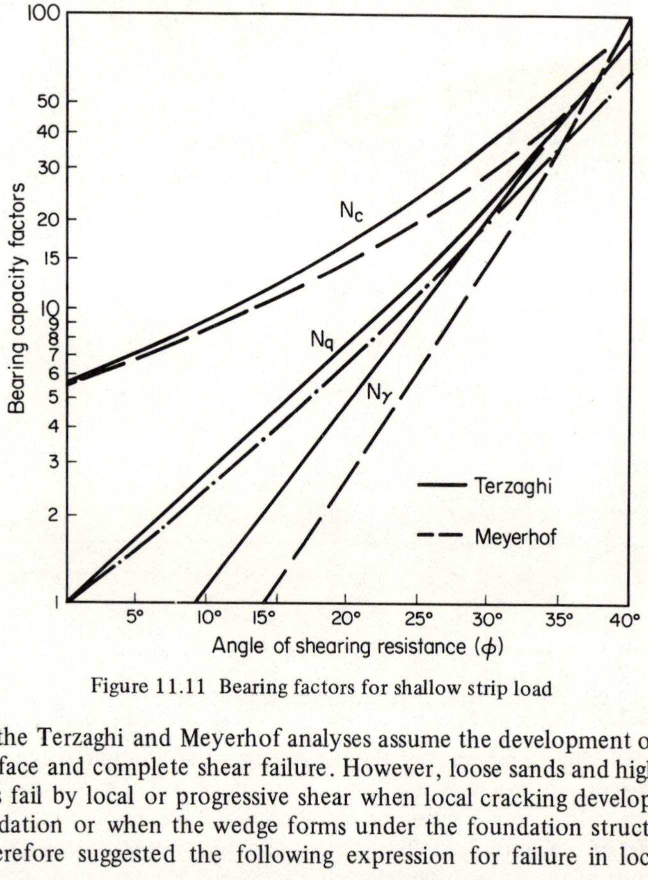

Figure 11.11 Bearing factors for shallow strip load

Both the Terzaghi and Meyerhof analyses assume the development of the full shear surface and complete shear failure. However, loose sands and highly sensitive clays fail by local or progressive shear when local cracking develops around the foundation or when the wedge forms under the foundation structure. Terzaghi therefore suggested the following expression for failure in local shear:

$$q = 2/3cN_c + \rho z N_q + 0.5\rho B N_\gamma$$

Meyerhof has developed other factors for deeply founded footings (Figure 11.12) and square footings (Figure 11.13). When a footing is rectangular or square the support given by the soil at the ends of the footing must be taken into account and the foundation has a higher ultimate bearing capacity. When the soil under a strip footing fails in shear the movement is laterally outwards. If a square, rectangular or circular footing fails the movement of the soil particles

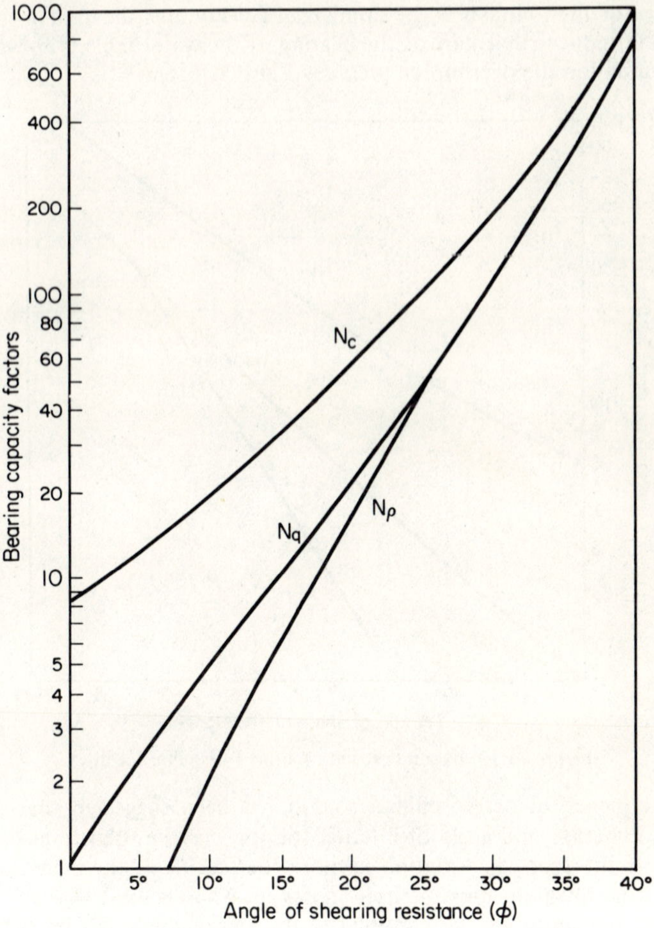

Figure 11.12 Bearing capacity factors for deep strip load

forms a radial pattern. According the Skempton[58] the ultimate bearing capacity for a rectangular footing is equal to the ultimate bearing capacity for a strip footing multiplied by:

$$1 + 0.2\frac{B}{L}$$

where B and L are breadth and length respectively. Terzaghi and Peck suggested the following expressions for square and circular footings:

(a) Square $q = 1.2cN_c + \rho zN_q + 0.4\rho BN_\gamma$

(b) Circular $q = 1.2cN_c + \rho zN_q + 0.6\rho rN_\gamma$

where r is the diameter

The angle of internal friction has the greatest influence upon the bearing capacity factors, all three increasing rapidly with only small increases in the angle. If the angle of internal friction is zero, as in saturated clay in undrained shear, only the cohesion contributes materially to the bearing capacity. When the water table is at the base of the foundation the submerged density must be used in the expressions which reduces that part of the bearing. If the water table is above the base of the foundation the overburden pressure is also affected.

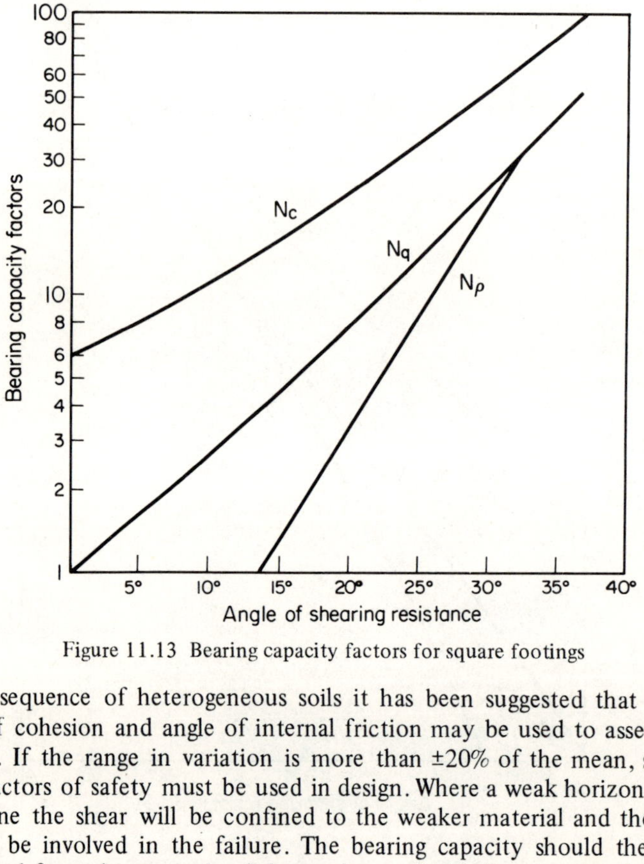

Figure 11.13 Bearing capacity factors for square footings

In a sequence of heterogeneous soils it has been suggested that the mean values of cohesion and angle of internal friction may be used to assess bearing capacity. If the range in variation is more than ±20% of the mean, somewhat higher factors of safety must be used in design. Where a weak horizon overlies a strong one the shear will be confined to the weaker material and the stronger will not be involved in the failure. The bearing capacity should therefore be determined from the strength of the weaker material. Because the shear zone is restricted the true bearing capacity will exceed that calculated. In the converse situation where a strong layer overlies a weak one then the former spreads the load thus reducing the pressure on the weaker horizon. Failure occurs by shear in the weaker material as the stronger bends under the load.

A reasonable approximation of how stress is distributed in soil upon loading can be obtained by assuming that the soil behaves in an elastic manner as if it was a homogeneous material. In such an instance the vertical pressure (P_V), produced at any point (N) in the soil by a load (Q) on the surface (Figure 11.14a) may be derived from the following expression:

$$P_V = \frac{3Q}{2\tau t^2}\left[\frac{1}{1+(r/t)2}\right]^{\delta\,½}$$

However, since it is assumed that the load applied by a foundation structure to the soil exerts a uniform pressure it is necessary to estimate the vertical pressure produced at some point on a horizontal point in the interior of the soil mass by a load uniformly distributed over an area of the ground surface. The Newmark chart provides a graphical method by which the solution to this problem can be determined (Figure 11.14b). In order to use this chart a tracing of the foundation is drawn to the same scale as the depth concerned, the line AB representing

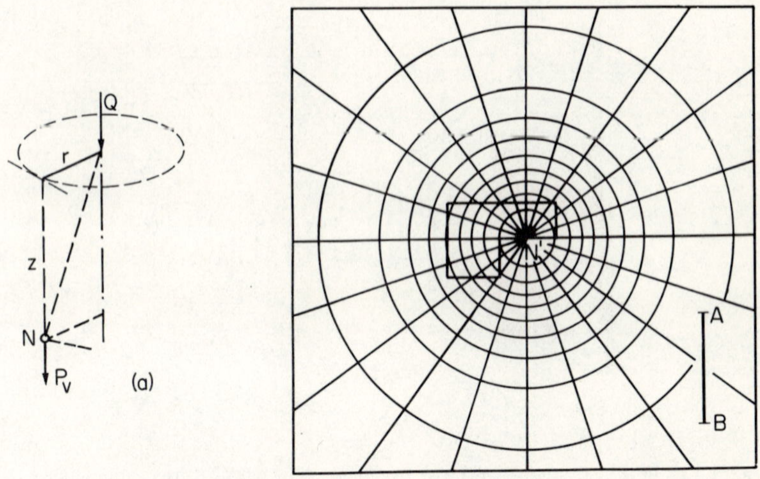

Figure 11.14 (a) intensity of vertical pressure at N when subjected to paint load, Q. r = radius; Z = depth; (b) Newmark chart

the depth, to scale, of the position at which the stress is to be evaluated. The scale drawing is then laid over the influence chart with the point vertically below which the stress is required being placed at the centre of the chart. Then the number of segments on the chart covered by the drawing are counted and multiplied by the influence factor of the diagram. This value is then multiplied by the loading to give the increase in pressure at the required depth. A bulb of pressure is a graphical illustration of the manner in which the loads imposed by a foundation structure are dissipated by the transfer of stress to increasing volumes of soil as the distance from the foundation structure increases (Figure 11.15). Closely spaced loads develop a cumulative pressure bulb as shown in Figure 11.15b.

The pressure acting between the bottom of a foundation structure and the soil is the contact pressure. The assumption that a uniformly loaded foundation structure will transmit the load uniformly so that the ground will be uniformly stressed is by no means always valid. For example, the intensity of the stresses at the edges of a rigid foundation structure on hard clay is theoretically infinite. In fact, of course, the clay yields slightly and so reduces the stress at the edges. As the load is increased more and more local yielding of the ground material takes place until, when the loading is close to that which would cause failure, the distribution is probably very nearly uniform. Therefore at working loads a uniformly loaded foundation structure on clay imposes a widely varying contact pressure. On the other hand a rigid footing on the surface of dry sand imposes a

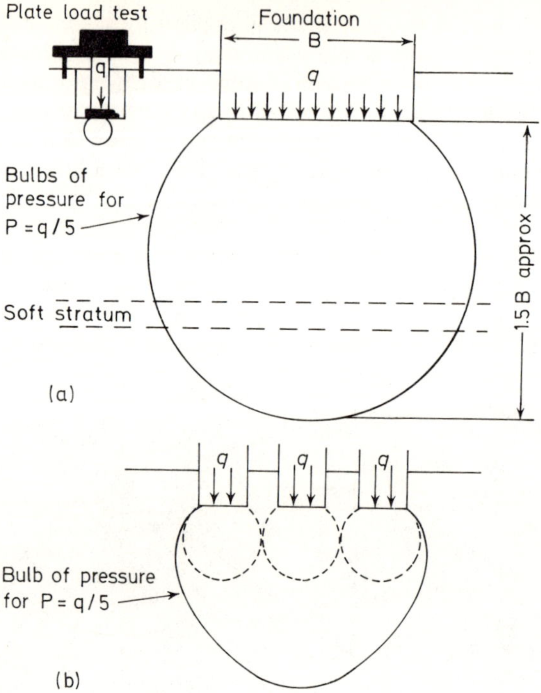

Figure 11.15 Bulbs of pressure. (a) For foundations of width B; (b) Cumulative bulb for three closely located loads. (*Note:* Bulb developed by plate load test does not affect soft stratum)

parabolic distribution of pressure. Since there is no cohesion in such material no stress can develop at the edge of the footing. If the footing is below the surface of the sand the pressure at the edge is no longer zero but increases with depth. The pressure distribution tends therefore to become nearly uniform as the depth increases. If a footing is perfectly flexible then it will distribute a uniform load over any type of foundation material. Also the foundation material may possess both cohesive and frictional properties to varying degrees. The mechanics of bearing capacity failure in large homogeneous rock masses are more complex than that of soil. Brittle failure, which involves a loss of strength after failure at low strains, is a major consideration in rock shear at low confining stresses. The bearing capacity of rock is controlled by local shear accompanied by cracking around the perimeter of the loaded area.

According to Hobbs[59] the stress at incipient failure depends upon the tensile strength of the rock and the magnitude of any *in situ* horizontal stresses in the rock mass. It is necessary for both effects to be overcome before the process can continue to ultimate failure, when the applied pressure is many times higher than that at incipient failure. In the absence of any horizontal stresses failure will begin when:

$$q_{fi} = 2.7q_c$$

where q_{fi} is the stress at incipient failure and q_c is the unconfined compressive strength of the rock. In a rock having a ϕ value of 40° the stress required to bring about ultimate failure is approximately eighteen times that needed to cause incipient failure. It is, however, impractical to use such high pressures, so that it is more convenient to work in terms of the stress to cause incipient failure. Consequently the safe bearing pressure approximates to the unconfined compressive strength.

It can be inferred from the foregoing paragraph that for ordinary concrete foundations on sound rock the bearing capacity is no limitation, indeed the rock is stronger than the concrete. But discontinuities usually traverse a rock mass and in such cases the mechanism of potential failure is different, depending on the size of the loaded area, and the discontinuity opening, spacing and orientation. Allowable contact pressures are also reduced when rocks are weathered, decomposed, heavily shattered or steeply dipping. The allowable contact pressure on unweathered rock should therefore not only be based on the inherent strength of the intact rock but also on the influence of such defects as joints, bedding planes, shear zones, solution features and the like.

Table 11.3 ALLOWABLE CONTACT PRESSURE FOR JOINTED ROCK*
(After Peck, Hanson and Thornburn[60])

RQD	*Allowable contact pressure (MN/m²)*
100	32.3
90	21.5
75	12.9
50	7.0
25	3.2
0	1.1

*If the value of the allowable contact pressure exceeds the unconfined compressive strength of intact samples then it should be taken as the unconfined compressive strength.
The RQD should be the average taken within a depth below foundation level equal to the width of the foundation, provided the RQD is fairly uniform within that depth. If the upper part of the rock, within a depth equal to about quarter the width of the foundation, is of low quality the value for this part should be used or the inferior rock should be removed. Since the values given in table are based on limiting settlement they should not be increased if the foundation is embedded into the rock

Table 11.3 represents values of allowable contact pressure for jointed rocks based on their RQD. If design is based on these values the settlement of foundations should not exceed 12.7 mm, even for large loaded areas.

The great variation in the physical properties of weathered rock and the nonuniformity of the extent of weathering even at a single site permit few generalisations concerning the design and construction of foundations. The depth to bedrock and the degree of weathering must be determined (see Chapter 1).

References

1. Bessey and Lea, 'The distribution of sulphates in clay soils and groundwater', *Proc ICE II, Part I*, p. 159 (1953)
2. Abelov and Askalonov. 'The stabilisation of foundations of structures on loess soils', *Proc 4th Int. Conf. Soil Mech. and Found. Eng.*, London, **I**, p. 259 (1957)
3. Davis, E.H. and Poulos, H.G., 'The use of elastic theory for settlement prediction under three-dimensional conditions', *Geotechnique*, **18**, pp. 67–91 (1968)
4. Poulos, M.G. and Davis, E.H., *Elastic solutions for soil and rock mechanics*, Wiley, New York (1974)

5. Janby, N., Bjerrum, L. and Kjaeonsli, B., *Veiledning ved Losning av Fundamentering-soppgaver*, N.G.I. Publication No. 16 (1956)
6. Gibson, R.E., 'Some results concerning displacements and stresses in a non-homogeneous elastic half-space', *Geotechnique*, **17**, pp. 58–67 (1967)
7. Lambe, T.W., 'Methods of estimating settlement', *ASCE, Div. Soil Mech. Found. Eng.*, **90**, No. SM5, pp. 43–67 (1964)
8. Gibson, R.E. and Sills, G.C., 'Some results concerning the plane deformation of a non-homogeneous half-space', *Proc. Roscoe Memorial Symposium on Stress-strain behaviour of Soils*, pp. 564–572 (1971)
9. Burland, J.B., Sills, G.C. and Gibson, R.E., 'A field and laboratory study of the influence of non-homogeneity on settlement', *Proc. 8th Int. Conf. Soil Mech. and Found. Eng.*, **I**, pp. 39–46 (1973)
10. Davis, E.H. and Poulos, H.G., 'Triaxial testing and three-dimensional settlement analysis', *Proc. 4th Aust. New Zealand Conf. on Soil Mech. and Found. Eng., Adelaide*, pp. 233–243 (1963)
11. Simons, N.E. and Som, N.N., 'Settlement of structures on clay with particular emphasis on London Clay', *CIRIA Report No. 22* (1970)
12. Marsland, A., 'Large *in situ* tests to measure the properties of stiff fissured clays', *Proc. 1st Aust-New Zealand Conf. on Geomechanics, Melbourne*, pp. 180–189 (1971)
13. Marsland, A., 'The shear strength of stiff fissured clays', *Proc. Roscoe Memorial Symposium on Stress-strain behaviour of soils', Cambridge*, pp. 59–68 (1971)
14. Marsland, A., 'Laboratory and *in situ* measurements of the deformation modulus of London Clay', *Proc. Symp. on Interaction of Structure and Foundation, Midlands S.M. and F.E. Society, Birmingham*, pp. 7–17 (1971)
15. Simons, N.E., 'Settlement studies on two structures in Norway', *Proc. 4th Int. Conf. on Soil Mech. and Found. Eng., London*, **1**, pp. 431–436 (1957)
16. Raymond, G.P., Townsend, D.L. and Lojkavek, M.J., 'The effect of sampling on the undrained soil properties of a Leda clay', *Canadian Geot. Jour.*, **8**, pp. 546–557 (1971)
17. Brooker, E.W. and Ireland, H.O., 'Earth pressure at rest related to stress history', *Canadian Geot. Jour.*, **2**, No. 1, pp. 1–15 (1965)
18. Skempton, A.W., 'Horizontal stresses in an over-consolidated Eocene clay', *Proc. 5th Int. Conf. on Soil Mech. and Found. Eng., Paris*, pp. 351–354 (1961)
19. Blight, G.E., 'Horizontal stresses in stiff and fissured lacustrine clays', *Proc. 4th Reg. Conf. African Soil Mech., Cape Town*, p. 95 (1967)
20. Poulos, H.G. and Davis, E.H., 'Laboratory determination of *in situ* horizontal stress in soil masses', *Geotechnique*, **22**, No. 1, pp. 177–182 (1972)
21. Bjerrum, L. and Andersen, K.H., '*In situ* measurement of the lateral pressures in clay', *Proc. 5th Inv. Conf. on Soil Mech. Found. Eng.*, **1**, pp. 11–20 (1972)
22. Ladd, C.C., 'Stress-strain behaviour of anisotropically consolidated clays during undrained shear', *Proc. 6th Int. Conf. on Soil Mech. and Found. Eng., Montreal*, **1**, pp. 282–286 (1965)
23. Wroth, C.P. and Hughes, J.M.O., 'An instrument for the *in situ* measurement of the properties of soft clays', *Proc. 8th Int. Conf. on Soil Mech. and Found. Eng., Moscow*, **1**, 2, pp. 487–494 (1973)
24. Bjerrum, L., 'Relasjon Mellom Malta og Beregnede Setninger av Byggverle pa Leire og Sand', *N.G.F. Foredraget*, 1964 N.G.I. Oslo
25. D'Appolonia, D.J., Poulos, H.G. and Ladd, C.C., 'Initial settlement of structures on clay', *ASCE, Soil Mech. Eng. Div.*, **97**, No S10, pp. 1359–1377 (1971)
26. Bjerrum, L., 'Embankments on soft ground', *ASCE Speciality Conference on Performance of Earth and Earth-supported Structures. Purdue University, Lafayette, Indiana*, pp. 1–54 (1972)
27. Simons, N.E., 'Normally consolidated and lightly over-consolidated cohesive materials', in *Settlement of Structures*, British Geotech. Soc., pp. 500–530, Pentech Press, London (1975)
28. Cole, K.W. and Burland, J.B., 'Observations of retaining wall movements associated with a large excavation', *Proc. 5th European Conference on Soil Mech. Found. Eng., Madrid*, **1**, pp. 445–453 (1972)
29. Ward, W.H., 'Displacements and strains in tunnels beneath a large excavation in London', *Proc. 5th Int. Conf. on Soil Mech. and Found. Eng., Paris*, **2**, pp. 749–753 (1961)
30. Atkinson, J., 'Discussion on Session 4; The interpretation of field observations', *Conf. on Field Instrumentation in Geotech. Eng.*, Part 2, pp. 677–679. Butterworths, London (1974)

31. Terzaghi, K., *Erdbaumechanik auf Bodenphysikalischer*, Grundlage, F. Deuticke, Vienna (1925)
32. Terzaghi, K., 'Undisturbed clay samples and undisturbed clays', *J. Boston Soc. Civ. Eng.*, **28**, No. 3, pp. 211–231 (1941)
33. Schmertmann, J.H., 'Estimating the true consolidation behaviour of clay from laboratory test results', *Proc ASCE*, **79**, No. 311 (1953)
35. Skempton, A.W. and Bjerrum, L., 'A contribution to the settlement analysis of foundations in clay', *Geotechnique*, **7**, No. 4, pp. 168–178 (1957)
36. Simons, N.E. and Menzies, B.K., 'A note on the principle of effective stress', *Geotechnique*, **24**, pp. 259–261 (1974)
37. Foss, I., 'Secondary settlements of buildings in Drammen, Norway', *Proc. 7th Int. Conf. on Soil Mech. and Found. Eng., Mexico*, **2**, pp. 99–106 (1969)
38. Terzaghi, K. and Peck, R.B., *Soil mechanics in engineering practice*, 1st edition, Wiley, New York (1948)
39. Bjerrum, L. and Eggestaad, A., 'Interpretation of loading tests on sand', *Europe Conf. on Soil Mech. and Found. Eng., Wiesbaden*, **1**, pp. 199–204 (1963)
40. Sutherland, H.B., 'Granular material' in *Settlement of Structures*, British Geotech. Soc., 473–499, Pentech Press, London (1975)
41. Gibbs, H.J. and Holtz, W.G., 'Research on determining the density of sands by spoon penetration testing', *Proc. 4th Int. Conf. on Soil Mech. and Found. Eng., London*, **1**, pp. 35–39 (1957)
42. Tomlinson, M.J., *Foundation design and construction*, p. 785, Pitman (1975)
43. Meyerhof, G.G., 'Shallow foundations', *Proc. Am. Soc. Civ. Eng., Soil Mech. Found. Div.*, **91**, No. SM2, pp. 21–31 (1965)
44. Peck, R.B. and Bazaraa, 'A.R.S.S. Discussion', *Proc. Am. Soc. Civ. Eng., J. Soil. Mech. Found. Div.*, **95**, No. SM3, pp. 305–309
45. Parry, R.H.G., 'A direct method of estimating settlements in sands from SPT values', *Proc. Symp. Internation of Struct. and Found. Midlands Soil Mech. Found. Soc.*, pp. 29–37 (1971)
46. de Beer, E.E. and Marteno, A., 'Method of computation of an upper limit for the influence of heterogeneity of sand layers in the settlement of bridges', *Proc. 4th Int. Conf. Soil Mech. and Found. Eng., London*, **1**, pp. 275–281 (1957)
47. Schmertmann, J.H., 'Static cone to compute static settlement over sand', *Proc. Am. Soc. Civ. Eng., J. Soil Mech. and Found. Div.*, **96**, No. SM3, pp. 1011–1043 (1970)
48. Butler, F.G., 'Heavily overconsolidated cohesive materials' in *Settlement of Structures*, Brit. Geotech. Soc., pp. 531–678, Pentech Press, London (1975)
49. Simons, N.E., Rodrigues, J. and Hornsby, P.A., 'Discussion, Conference', *Settlement of Structures*, British Geotech. Soc., pp. 658–660, Pentech Press, London (1975)
50. Cheney, J.E., 'Techniques and equipment using the surveyor's level for accurate measurement of building movement', *Proc. Symp. on Field Instrumentation, British Geot. Soc.*, Butterworths, London (1973)
51. Hanna, T.H., *Foundation instrumentation*, Trans. Tech. Publications (1973)
52. Skempton, A.W. and MacDonald, D.H., 'The allowable settlement of buildings', *Proc. Int. Civil Eng.*, **5**, No. 3, Pt. 3, pp. 737–784
53. Bjerrum, L., 'Discussion on Section 6, Europe', *Conf. on Soil Mech. Found. Eng., Wiesbaden*, **2**, pp. 135–137 (1963)
54. Burland, J.B. and Wroth, C.P., 'Allowable and differential settlements of structures including damage and soil-structure interaction' in *Settlement of Structures*, Pentech Press, London (1900)
55. CP 2004:1972, *Code of Practice on Foundations*, British Standards Institution, London
56. Terzaghi, K., *Theoretical soil mechanics*, Wiley, New York (1943)
57. Meyerhof, G.G., 'The influence of roughness of base and ground water on the ultimate bearing capacity of foundations', *Geotechnique*, **5**, pp. 227–242 (1955)
58. Skempton, A.W., 'The bearing capacity of clays', *Build. Res. Congr.*, Div. 1, pp. 180–189 (1951)
59. Hobbs, N.B., *Foundations on rock*, Training Pamphlet, Soil Mechanics Limited, Bracknell (1975)
60. Peck, R.B., Hanson, W.E. and Thornburn, T.H., *Foundation Engineering*, Wiley, New York (1974)
61. Terzaghi, K. and Peck, R.B., *Soil mechanics in engineering practice*, Wiley, New York (1967)

Chapter 12

Subsidence due to Mining Operations

INTRODUCTION

Subsidence at the surface can be regarded as ground movement which takes place due to the extraction of mineral resources. Subsidence damage has been defined by the Coal Mining (Subsidence) Act, 1957, as damage including an alteration of the level or gradient of the property or works such as to affect its fitness for use for the specified purpose. Subsidence is an inevitable consequence of mining activities and reflects the movements which occur in the mined out area, small deformations in the workings being associated with insignificant subsidence whilst their total closure can generate severe subsidence at ground level. Consequently subsidence can and does have serious effects on buildings, services and communications, can be responsible for flooding, lead to the sterilisation of land or call for extensive remedial measures or special constructional design in site development.

Coal is by far the most important mineral mined in the UK. It appears to have been worked from at least Roman times and the exploitation of many coalfields dates from the twelfth century. However, other materials have been, and some still are, won from the Coal Measures. These include gannister and fireclay for refractories; clays, shales and mudstones for bricks and coarse earthenware; sandstones for building purposes; and ironstones for ore. Such materials have been mined and quarried and as a result can also give rise to problems in ground engineering.

As far as coal mining is concerned most of the early workings were at surface outcrops. However, by the fourteenth century outcrop workings had largely given way to bell pits and drifting. The shafts of bell pits rarely exceeded 12.2 m in depth and their diameter was usually about 1.3 m. Extraction was carried on around the shaft until such times as roof support became impossible, another shaft was then sunk.

Increased demand for coal in the sixteenth century led to the development of the pillar and stall method of extraction. Underground workings were shallow and not extensive, for example, they rarely penetrated more than 40 m from the shaft. Indeed, when such limits had been reached, it was usually less costly to abandon a pit and sink another shaft nearby. Workings extending 200 m from the shaft were exceptional even at the end of the seventeenth century, the shaft itself usually being less than 60 m deep. With the coming of the Industrial

Revolution the demand for coal accelerated. Longwall working evolved more or less at the same time, probably first originating in the Shropshire coalfield. This method involves total extraction, panels being developed from an initial drivage within the seam. The working face is temporarily supported, the support being moved as the face advances leaving the roof from which support has been withdrawn to collapse. Today the vast majority of the coal mined in the United Kingdom is by panel working which is a development of the longwall system suited to mechanized extraction techniques.

In areas of productive Coal Measures, if no record of mining activities is available then it should not be assumed that mining has not taken place. Indeed old shallow workings close to outcrop should be expected in any urban area where exploitable beds were not covered by thick superficial deposits. Mining has gone on in many coalfields in Britain for several centuries but the first statutory obligation to keep mine records only dates from 1850 and it was not until 1872 that the production and retention of mine plans became compulsory. An appeal for the deposition of old working plans, launched by the Government, in 1925, allowed the Department of Mines to publish a catalogue of the plans of abandoned mines in 1931. Supplements to this catalogue were published annually until 1938. The catalogue and its supplements are now out of print, but most major libraries in coalfield areas have copies. A further appeal for old plans has been made recently by the Secretary of State for Energy. When the coal industry was nationalised in 1946 the mining companies handed over their abandonment plans to the Estates Mining Engineer of the National Coal Board. The abandonment plans of all other mines are stored by HM Mines Inspectorate.

Although the extraction of coal is the major cause of subsidence in the UK, subsidence has been and is brought about by the extraction of other materials from the ground, notably salt.

OLD WORKINGS IN COAL

Most of the large industrial centres of the UK, with the exception of greater London and parts of Merseyside, owe their development to the presence of accessible coal which was mined at shallow depth by partial extraction methods. In this way potentially unstable ground conditions may have been created and such workings can pose considerable problems in that the rational assessment of the risk of failure is difficult. This is because usually little or nothing may be known about the original mining geometry, hence old, shallow mine workings are not amenable to quantitative analysis and subsidence prediction. Surface collapse need have no regular pattern, it being specifically related to the collapse of strata overlying old mined out areas. Thus as far as foundations are concerned, the problem is not really one of producing a satisfactory design to withstand or accommodate the ground movements which are likely to occur, but one to prevent their occurrence by making, if possible, the ground on which the foundations are placed stable.

Although as a last resort measures can be taken to reduce the risk to a negligible level, they may be so costly as to render the development uneconomic. For example, the author was recently involved with a site where the presence of old workings was proved by a close pattern of rotary drill holes. Unfortunately the distribution of voids was irregular so that no real estimate of the extraction ratio could be made. On this basis the grouting contractor was only prepared to quote a minimum figure, refusing to specify a maximum cost, for filling the voids. This figure amounted to almost 28% of the contract price (the contract was for a two storey factory on an industrial estate). Furthermore the client felt that since the size of the void space requiring filling was not completely unknown, he could be signing an 'open cheque' if he allowed the ground treatment to go ahead. He decided to cut his losses and abandoned the site.

Little data is available which allows any reliable estimate of the frequency of incidence of damage to property resulting from old mine workings to be assessed. This perhaps gives some justification for concluding that such occurrences are not common in the UK. Nevertheless, it should be borne in mind that progressive deterioration of old workings could possibly give rise to an increasing number of failures in the future, and that in urban areas many modern buildings are often much larger and heavier and may be more prone to subsidence damage, than the buildings they replace.

In mediaeval times bell pits were used to exploit coal and ironstone and they are therefore a feature of coalfield areas where the drift cover is thin. As bell pits were very shallow workings, they tended to be concentrated near outcrops, indeed the outcrop of a seam in rural areas may be delineated by a series of bell pits and their associated mounds of spoil. If bell pits were backfilled, then the state of compaction of the backfill is generally unsatisfactory. This is to be expected considering the miscellaneous character of the backfill and the manner in which it was placed.

In the pillar and stall method of working, pillars are left to support the roof thus they have to sustain the redistributed weight of the overburden which means that they and the rocks immediately above and below are subjected to added compression. Stress concentrations tend to be located at the edges of pillars and the intervening roof beds tend to sag. The effects on ground level are normally insignificant unless the floor is unusually soft. In such cases the pillars may penetrate the floor and give rise to a general lowering of the surface. Obviously this becomes more significant with increasing depth or high extraction ratio since these impose greater loads upon the pillars. Although such subsidence is often slow it may be quickened by the ingress of water into the workings. In addition minor strain and tilt problems occur around the periphery of the basin thereby produced.

In very early mining, the remnant pillars were rather haphazard in size and arrangement but, with time, mining became more systematic and pillars of more or less uniform shape were formed. Generally the main galleries were driven in parallel lines, between which were left ribs of coal. Then narrow headways were cut at intervals through the ribs to connect adjacent galleries, thus leaving a series of pillars which were approximately square or rectangular in plan. The size of the rooms and pillars was often influenced by the thickness and quality of the seam, and the character and thickness of the overburden. The future working intentions also influenced their proportions, if the pillars were to remain then the object was to procure the greatest possible proportion of the coal at the first working, whilst if they were to be removed later it was usual to provide fairly adequate support to protect the workings and haulage roads until their removal on retreat.

There was a general tendency for the size of rooms to increase, for example, in the sixteenth century they were usually less than 2 m wide whereas in the eighteenth century they began to exceed 3 m in width. Wardell and Wood[1] noted that in the nineteenth century the normal width of room varied from 1.83 to 4.57 m, the extraction ratio varying from 30 to 70%. Several variations of the room and pillar method were devised, some to overcome particular local problems, others simply resulted from the prejudices of those directing the work. For example, Staffordshire squarework was developed to work the Ten Yards Coal seam, in the Sheffield area ribs of coal were generally left to support the roof rather than pillars. In present day workings the pillar support system can be designed for long term stability if it is important to protect the surface or to minimise damage to the roof (see Orchard[2], and Orchard and Allen[3]), as the miner can control the pillar dimensions and the percentage extraction.

Pillars often experience local failures whilst mining is taking place. If a pillar is highly jointed then its margin may fail and fall away under relatively low stress. Such action reduces and ultimately removes the constraint from the core thereby subjecting it to increasing stress. This can lead to pillar failure. Collapse in one pillar can bring about collapse in others in a sort of chain reaction because increasing loads are placed on those remaining. This has not been a significant problem in Britain. Where ground subsidence due to pillar collapse in shallow workings does occur it usually gives rise to a dish-shaped depression. This can cause some structural damage to buildings above, although the damage could not be regarded as catastrophic, the settlement being similar in rate and rather less in extent than that from limited longwall extraction. Collapse of a whole pillar district, however, has been known to happen, for example, in Coalbrook, South Africa (see Bryan, Bryan and Fouche[4]). In such cases the surface suffers substantial rapid subsidence, possible shock damage, often a zone of fracture and severe local strain and tilt damage. It must be admitted that the geological setting of the South African coalfields differs from that of British coalfields.

Slow deterioration and failure of pillars may take place years after mining operations have ceased, although observations at shallow depth and the resistance of coal to weathering suggests that this is a relatively uncommon feature at depths less than 30 m. However, old workings affect the pattern of ground water drainage which in turn may influence pillar deterioration. On the other hand the small pillars of earlier workings may be crushed out once the overburden exceeds 50 or 60 m. Old pillars at shallow depth have occasionally failed near faults and they may fail if they are subjected to the effects of subsequent longwall mining. The condition of pillars is difficult to assess unless access can be gained to old workings, although borehole cameras and closed circuit television can sometimes be employed.

Very often pillars were robbed on retreat. Extraction of pillars during the retreat phase simulates the longwall surface condition although it can never be assumed that all pillars have been removed. As noted above at moderate depths pillars, particularly pillar remnants, are probably crushed and the goaf compacted, but at shallow depths lower crushing pressures may mean that closure is variable. This causes foundation problems when large or sensitive structures are to be erected above.

Determination of pillar strength, pillar load and pillar stability is obviously important as far as assessment of possible ground movements is concerned. Some expressions which have been advanced in order to estimate these factors are given in Table 12.1. It can be seen from these expressions that, although the intrinsic strength of coal varies, the important factors relating to pillar stability

Table 12.1 SOME EXPRESSION FOR DETERMINING THE STRENGTH OF AND THE STRESS ON COAL PILLARS

Author(s)	*Expression*	*Remarks*
1. Greenwald, Howarth and Hartman[5]	Pillar strength $(\tau_p) = 2800 \times \frac{w^{0.5}}{h^{0.85}}$ psi w = width of pillar in inches h = height of seam in inches	1. Based on the results of *in situ* tests on square pillars in Pittsburg bed. Pillars varied in width from 1 to 5.3 ft and height from 2.4 to 5.3 ft.
2. Terzaghi[6]	$\sigma_V = \frac{\rho\ 0.5B}{K \tan \phi} (1 - \exp{-K \tan \phi\ z/B}) + q \exp{-K \tan \phi . z/B}$ σ_V = vertical pillar loading pressure K = constant representing ratio between horizontal and vertical pressure B = breadth of room ρ = density or unit weight of overburden z = depth ϕ = angle of internal friction q = uniform surcharge carried by soil per unit area	2. This expression was developed in relation to theories of arching and more particularly to the vertical load above a tunnel. It has been suggested that it could be used to calculate the total vertical load, including any proposed surface structure, above old mine workings.
3. Denkaus[7]	$\sigma_V = wz \frac{1}{1 - r}$ r = extension ratio $= \frac{2ab + b^2}{(a + b)^2}$ (see Figure 12.1)	3. This expression is based on two assumptions, namely, that each pillar supports the column of rock over an area which is the sum of the cross sectional area of the pillar plus a part of the room area; and that the load is vertical only and uniformly distributed across the cross sectional area. Steart[8] has pointed out, however, that the pressure is not evenly distributed, it frequently assuming a parabolic distribution with maximum pressure exerted on the central pillars of the mined out area. He further stated that the pressure gradient gradually increased with increasing area of development to a maximum given by Eq. 3, and that this maximum was reached when the development, if roughly circular, attained a radius equal to the depth, divided by the ratio $(a + b)^2/a^2$ (Figure 12.1)

Author(s)	Expression	Remarks
4. Coates[9]	$\sigma_V = \dfrac{1.1z}{1 - \dfrac{r}{100}}$ psi r = extraction ratio	4 & 5 According to Coates this expression gives valid results where the depth of the workings is less than half their extent, as is the case with most old mine workings. When a structure is to be built over old pillared workings the additional weight on the pillars can be estimated simply by adding the weight of the appropriate part of the structure to the overburden pressure supported by the pillar. The total load on a pillar determined in this manner is probably greater than the true load. Moreover the distribution of load over the area of the pillar is not uniform for stress is concentrated at the pillar edges. The transfer of the weight of a surface structure to residual pillars is normally calculated on the assumption that the additional weight acts vertically downwards and that there is no lateral spreading of the load. This is based on the fact that void migration causes dilation above the seam as well as joint opening so reducing or preventing such lateral distribution.
5. Wardell and Wood[1]	$\sigma_V = \dfrac{z}{1 - r}$ psi	
6. Salamon and Munro[10] and Salamon[11]	(a) $\tau_P = 1320 \dfrac{w^{0.46}}{h^{0.66}}$ psi (Dimensions in ft) (b) $\sigma_V = 1.1z \left(\dfrac{w + B}{w}\right)^2$ psi (c) Factor of safety (Fs) against pillar failure: $Fs = \dfrac{\text{strength of pillar}}{\text{pillar loading pressure}}$	6. These expressions are based upon a statistical analysis of a survey of pillar dimensions in both stable (98 cases) and collapsed (27 cases) mining areas in South Africa. *Summary of data* — *Stable* — *Collapsed* 1. Depth (m) — 20–220 — 21–192 2. Room height (m) — 1.2–4.9 — 1.5–5.5 3. Pillar width (m) — 2.7–21.3 — 3.4–15.9 4. Extraction ratio — 37–89 — 45–91 5. Pillar width:height ratio — 1.2–8.8 — 0.9–3.6 Salamon and Munro concluded that 99% of collapses occurred at safety factors lower than 1.48. Holland[12], however, had previously suggested that a factor of 1.8 was generally necessary and that in critically important areas it was 2.2.

Table 12.1 (continued)

Author(s)	*Expression*	*Remarks*
7. Bieniawski[13,14]	$\tau_p = 400 + 200 \frac{w}{h}$ psi	7. Based on tests carried out on coal pillars in South Africa, the pillars varying in height and width from 2 to 6.7 ft.
8. Wilson[15]	(a) Wide pillars, $w > 0.003\, mz$ ft (i) Square pillars $\sigma_v = 4\rho z\,(w^2 - 3wmz \times 10^{-3} + 3\,m^2 z^2 \times 10^{-6})$ tons (ii) Rectangular pillars $\sigma_v = 4\rho z\,(wl - 1.5\,(w + l)mz \times 10^{-3} + 3m^2 z^2 \times 10^{-6})$ tons (iii) Long pillars $\sigma_v = 4\rho z\,(w - 1.5mz \times 10^{-3})$ tons/ft run (b) Narrow pillars, $w < 0.003\, mz$ ft (i) Square pillars $\sigma_v = 444\rho \frac{w^3}{m}$ tons (ii) Rectangular pillars $\sigma_v = 1333\rho \frac{w^2}{m}\left(l - \frac{w + l}{2} + \frac{w}{3}\right)$ (iii) Long pillars $\sigma_v = 667\rho \frac{w^2}{m}$ tons/ft run	8. Wilson suggested that a pillar was surrounded by an outer yield zone in which the stress distribution varied in linear fashion from zero at the surface to the point of failure at the pillar core. The constraint given to coal in the pillar core can increase its strength appreciably. The six expressions take no account of the strength of the coal and it has been suggested that although they may be admissible for deep workings they could give underestimates of strength of pillars at shallow depths.
	where σ_v = load which pillar will carry, in tons ρ = average density of rock in tons/ft^3 l = length of pillar in ft m = height of roadway in ft w = width of pillar in ft z = depth of overburden in ft	(For further details concerning the estimation of the strength of pillars of coal (see Hustrulid, W.A., A Review of Coal Pillar Strength Formulae, *Rock Mechanics*, 8, 115–145 (1976))

are the ratio of seam thickness to pillar width and the depth below ground. Obviously the stress on a pillar increases as the extraction ratio increases, and it also depends upon the height/width ratio of the room. In addition the greatest stress occurs at the edges of pillars, between pillar and roof or floor. Pillars in the centre of the mined out area are under more stress than those at the sides. As remarked above subsequent undermining and deterioration may also adversely affect pillar stability.

The strength of coal pillars is affected by friction between coal and the rocks above and below. The width of pillars in relation to their height means that the rocks above and below exert a confining effect. Frequently the floor of an extraction area heaves, this being especially characteristic of mudstones, shales

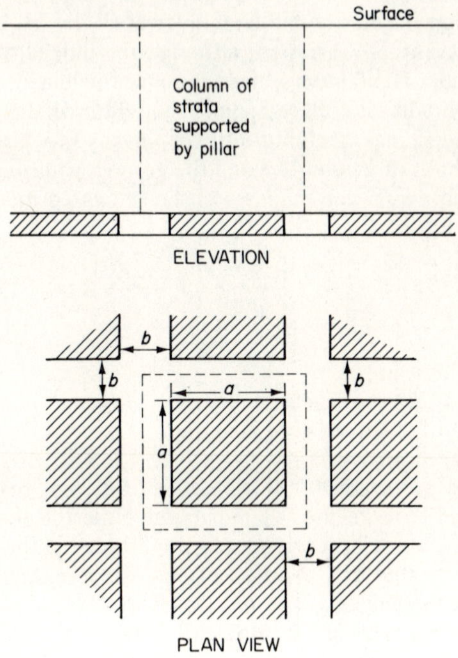

Figure 12.1 Plan view of loading on pillars in pillar and stall working

and fireclays. Floor heave can occur at shallow depths, for example, old roadways at less than 20 m depth have at times been completely closed. Floor heave places a constraint on pillars and reduces the void volume thus pillar spalling and void migration are reduced. Pillars in dipping seams tend to be less stable than those in horizontal seams since the overburden produces a shear force on the pillars.

Even if pillars in old shallow workings are relatively stable, the surface can be affected by void migration (Figure 12.2). This can take place within a few months of, or a very long period of years after mining, and in Britain this is a much more serious problem than pillar collapse. According to Wardell and

Eynon[16] the maximum width of a self-supporting span, S, can be derived from the following expression:

$$S = \sqrt{(2h\sigma_t)}$$

where h is the thickness of the seam and σ_t the ultimate tensile strength of the roof rock beam (this should make due allowance for discontinuities).

Generally the roof rock above a void consists of shale or mudstone which is respectively laminated or bedded, and transected by joints, so that it ultimately collapses, the collapsed zone forming a dome between the pillars. The process can continue upwards and occasionally may lead to the sudden appearance of a crown hole at the surface.

Void migration develops when roof rock falls into the worked out areas. When void migration occurs the material involved in the fall bulks, which means that migration is eventually arrested, although the bulked material never completely fills the voids. The factors which influence the height above the workings to which migration will take place include the width of the unsupported span, the height of the workings, the nature of the cover rocks, the thickness and dip of the seam, the depth of overburden and the ground water regime. But generally the height to which a void can migrate can be estimated by using the following expression:

$$H = t\left[\frac{\rho_1}{\rho} \div \left(1 - \frac{\rho_1}{\rho}\right)\right]$$

where

t, thickness of the seam;
ρ, bulk density of the roof rocks;
ρ_1, bulk density of collapsed roof materials;
H, height of migration.

The self-choking process may not be fulfilled if dipping seams are affected by water which can redistribute the fallen material. Nevertheless it has been suggested that void migration will not produce crown holes at the surface where the seam concerned is located at a depth in excess of approximately five times the width or ten times the height of the mine roadway. Even so voids just below the surface create just as awkward a problem. Void migration will be confined beneath a thick, competent rock unit.

TREATMENT OF OLD WORKINGS

Where a site which is proposed for development is underlain by shallow old mine workings there are a number of ways in which the problem can be dealt with. However, any decision regarding which is the most appropriate method must be based upon an investigation which carefully evaluates the site conditions and examines the economic implications of the alternatives (see Taylor[17]; Price, Malkin and Knill[18]; and Jennings[19]). One of the most difficult assessments to make is that related to the possible effects of progressive deterioration of the workings and associated potential subsidence risk. Also the placement of a new structure might alter a quasi-stable situation. Accordingly solutions tend to err on the conservative side.

A site investigation for an important structure(s) requires the exploration and sampling of all strata likely to be significantly affected by the structural loading. The location of subsurface voids due to mineral extraction is of prime importance in this context. In other words an attempt should be made to determine the extraction ratio, the pattern of the layout and the condition of the old pillar and stall workings. The sequence and type of roof rocks may provide some clue as to whether void migration has taken place and if so, its possible extent. The assessment of past and future collapse is obviously very important.

Figure 12.2 Void migration in shaley roof rocks above the Fire Quarter seam at Pethburn opencast site, Co. Durham (Courtesy of R.K. Taylor)

The location of old workings is generally done by exploratory drilling, the borehole locations being influenced by a pre-assessment of the possible pattern of the old workings i.e. by information revealed by the literature and map survey. For the most part drilling to prove the existence of old workings is done by open holes, which allows relatively quick probe drilling. In some cases once the basic pattern of the old workings has come to light, hand drills have been used to fill in the details (see Scott[20]). Detailed mapping of galleries is best made by driving a heading from the outcrop if this is close at hand or by sinking a shaft to the level of the coal seam to obtain access to the workings. Sometimes access has been gained via old shafts. Radial holes may be drilled from the shaft to establish the dimensions of the pillars and stalls. As old workings may prove dangerous it is advisable that exploration should be undertaken with the advice and aid of experts. Surface workings may be examined by using borehole cameras or closed circuit television, but illumination may be a problem. Occasionally smoke tests have been used to aid exploration. Except in special circumstances geophysical methods have not proved very successful in locating and revealing the layout of shallow old pillar and stall workings (see Maxwell[21]). As Burton and Paton[22] pointed out geophysical methods cannot be used without due reference to the geological situation and any survey must be planned by a geophysicist.

One of the most obvious methods of dealing with old workings is simply to avoid locating the proposed structure above them, placing it either on sound ground or over workings proved to be stable. After a recent site investigation for a hotel at Wideopen, Newcastle-upon-Tyne, it was found that part of the site was underlain by old pillar and stall workings and that coal was going to be extracted from beneath within the next few years. It was therefore recommended that the hotel complex should be relocated to avoid the area of potential subsidence, and that the design of the building should be altered. It is not generally sufficient to locate immediately outside the area undermined as the area of influence should be considered. In such cases the angle of influence or draw is usually taken as 25°, in other words the area of influence is defined by projecting an angle of 25° to the vertical from the periphery and depth of the workings. Such location is, of course, not always possible.

After a detailed investigation at a site in Airdrie underlain by old mine workings the concept of safe and unsafe zones was established (see Price[23]). In the safe zones the cover rocks were regarded as thick enough to preclude subsidence hazards (about 10 m of rock or 15 m of till was regarded as sufficient to ensure that crown holes did not occur at the surface) and normal foundations could be used for the two-storey dwellings which were to be erected.

If old mine workings are at very shallow depth then it might be feasible, by means of bulk excavation, to found on the strata beneath. This is an economic solution, particularly at depths of up to 7 m or on sloping sites, and is well suited to areas which were worked by means of bell pits. For example, Taylor[24] quoted an instance in Leeds where founding below bell pits in the Beeston Bed Coal meant that 21740 tonnes of coal was opencasted, providing a handsome return. Such excavations may be carried out rapidly if the overburden consists of clays, shales or fragmented and weathered rocks.

Where the allowable bearing capacity of the foundation materials has been reduced by mining, it may be possible to use a raft. Rafts can consist of massive concrete slabs or stiff slab and beam cellular rafts (Figure 12.2). The latter are suitable for the provision of jacking pockets in the upstand beams to permit the columns or walls to be relevelled if subsidence distorts the raft. A raft can span weaker and more deformable zones in the foundation, thus spreading the weight of the structure well outside the limits of the building. For example, the twelve storey block at Sheffield Polytechnic was provided with a reinforced raft foundation. This precaution was taken even though the old workings, which were dry and in relatively good condition, were overlain by the Silkstone Rock, a massive sandstone. This together with the estimated percentage extraction, 43%, allowed the engineers to judge that the ground would remain stable. However, rafts are expensive and therefore tend to be used where no alternative exists. For low buildings, up to four storeys in height, it is occasionally possible to use an external reinforced ring beam with a central lightly reinforced raft as a practical and more economic foundation.

Reinforced, bored pile foundations have also been resorted to. In such instances the piles bear on a competent stratum beneath the workings. The use of piles may be ruled out if a hard rock is located near the surface. For example, this occurred at a site in Attercliffe, Sheffield, where a strong mudstone overlay shallow old workings. The piles should either consist of pre-cast concrete units or be sleeved so that concrete is not lost into voids. To avoid the development of negative skin friction if overlying strata collapses and the consequential risk of serious overloading, piles can be coated with bituminous products. The sleeve also counteracts the effects of negative skin friction. Some authorities, however, have suggested that piling through old mine workings seems inadvisable because their emplacement may precipitate collapse. Subsequent collapse at seam level could possibly lead to piles being either buckled or sheared. There is also the question of the lateral stability of piles passing through collapsed zones above mine workings or through large remnant voids.

Where old mine workings are believed to pose an unacceptable hazard to development and it is impossible to use adequate measures in design, or found below their level, then the ground itself can be treated. Such treatment involves filling the voids in order to prevent void migration and pillar collapse. This is generally achieved by drilling holes from the surface into the mine workings, on a systematic basis, generally on a grid pattern, and filling the remnant voids with an appropriate grout mix. If it has been impossible to obtain accurate details of the layout and extent of the workings then the zone beneath the intended structure can be subjected to consolidation grouting. In exceptional cases, where the mine workings are readily accessible, barriers can be constructed underground and the workings filled hydraulically with sand or pneumatically with some other suitable material.

The choice of materials and composition of the grout mixes used in these operations us dependent, partly on the ultimate ground loadings and partly on the nature and extent of the voids. Surface subsidence resulting from the collapse of workings extends over a greater area than that of the workings themselves, and the extent of the area of treatment must ensure no measurable subsidence occurs within the bearing area of the foundation. This can be determined by using an angle of 25° to the vertical between the outer limits of the structure and the base of the workings. It has been suggested by Butler[25] that if the dip of a coal seam exceeds 10° then 5° should be added for every 10° of dip on the dip side whereas the angle can be reduced in the same manner on the rise side. He also suggested that in surface cover consisting of superficial deposits above the water table the angle of draw taken to define the safe limit should be $90^\circ - \phi$, whilst in superficial deposits below the water table this becomes $90^\circ - \phi/2$, where ϕ is the angle of internal friction.

It is not essential to fill voids with material of roughly equivalent strength to the host rocks and generally bulk filling by flushing via boreholes provides a satisfactory solution. This method cannot guarantee that every void will be completely filled but pressure grouting need not give better results. The grouts used in these operations commonly consist of cement, fly ash and sand mixes, economy and bulk being their important features. Pea gravel may be used as a bulk filler where the amount of grout required proves excessive. If the workings are still more or less continuous then there is a risk that grout will penetrate the bounds of the zone requiring treatment. In such instances dams can be built by placing pea gravel down large diameter boreholes around the periphery of the site. When the gravel mound has been formed it is grouted. The area within this barrier is then grouted. If the old workings contain water then a gap should be left in the dam through which the water can drain as the grout is emplaced. This minimises the risk of trapped water preventing the voids being filled.

The following case is provided as an example of ground treatment. As the Al approaches the Tyne bridge in Gateshead it is carried on a viaduct. This is a continuous pre-stressed concrete structure which is supported by piers founded on sandstone, the load on each pier being about 2000 tonnes. The area is underlain by the coal seams shown in Table 12.2. A further five seams, all thought to

Table 12.2

Seam	*Average depth* (m)	*Thickness* (m)	*Workings*
5 High Main	16.8	1.67	Old pillar and stall workings
4 Top Main	24.4	1.17	Presumed old workings
3 Bottom Main	30.5	0.66	No record of workings
2 Main	54.8	0.45–0.91	Thicker part of seam worked
1 Maudlin	82.3	1.07	Presumed old workings

have been worked, are found within 198 m of the surface. Sandstone generally occurs at 0.6 to 4.9 m below ground level, except where it was locally quarried for building stone, then it occurs at around 9.1 m, the quarry in question having been backfilled.

After the site investigation it was considered that the risk of void migration from the High Main seam causing surface trouble was low but it was thought that the stress on the pillars and the seat earth beneath could possibly give rise to subsidence damage, especially with the relatively high risk structure involved. Consequently it was decided to pressure grout the workings in the High Main seam. None of the other workings in the seams below the High Main were considered as potential sources of subsidence damage and therefore were not treated.

The perimeter wall about the grouted area totalled some 1675 m in length and consumed approximately 2295 m^3 of grout, the latter consisting of one part cement, 2.5 parts fly ash, 10.5 parts sand and 0.1 parts bentonite. The infill took 2850 m^3 of grout, the mix being 1 part cement, 5 parts fly ash and 18 parts sand. The grout holes were set out on a 3.04 m grid providing an 18.3 m wide treated area beneath each pier and an intervening zone 8.5 m wide into which grout was allowed to spread. After completion of the infill grouting the discontinuities in the overlying sandstone were grouted with a 1:1 mix of cement and fly ash at a pressure of 275 kN/m^2. It was calculated that for a 0.9 m worked seam (the complete thickness of the High Main coal was not worked) the total amount of grout consumed meant a grout take of about 39% of the seam volume. The grout take was assessed by injecting water into the grout holes and if a full return resulted then the grouting was assumed satisfactory.

Bulk grouting will result in a significant change in mass permeability and consideration must be given to the implications of such treatment to ground water flow.

Gray and Meyers[26] outlined a method of ground treatment used in the Pennsylvania coalfield. It involves the drilling of 150 mm boreholes down to the

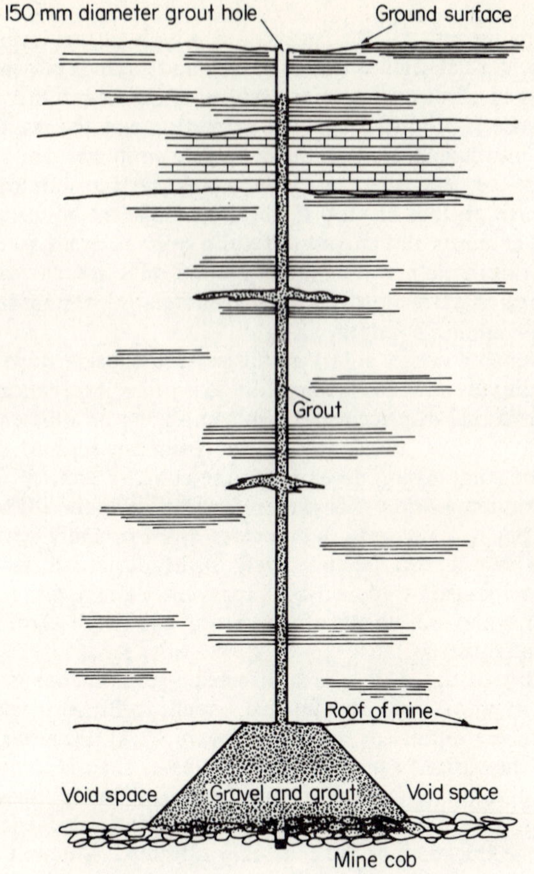

Figure 12.3 Grout column and ground movements

workings. Gravel is added if a void is encountered, thus a mound is formed which supports the roof, the gravel is then grouted. Grouting is then carried out from the borehole. The holes are simply grouted if no voids are encountered, the basic concept being to create a series of grouted columns, thereby strengthening the ground (Figure 12.3). The technique is possibly most successfully applied where roof subsidence has not occurred to any significant degree. The published costs indicate that there is little economic difference between this process and consolidation grouting based on closely spaced boreholes.

LONGWALL WORKINGS AND SUBSIDENCE

In longwall mining the coal is exposed at a face of 30 to 200 m between two parallel roadways. The roof is supported only in and near the roadways and at the working face. After the coal has been won and loaded the face supports are advanced leaving the rocks, in the areas from which coal has been removed, to collapse.

According to Lehman[27] the subsidence which occurs over a completely mined out area in a flat seam is trough-shaped and extends outwards beyond the limits of mining in all directions. For a review of the trough theory see Hilbig[28]. Trough subsidence is produced by total extraction and the resulting stratal and surface ground movements are regarded as largely contemporaneous with mining, producing more or less direct effects on any surface development. Trough shaped subsidence profiles develop tilt between adjacent points which have subsided different amounts and curvature results from adjacent sections which are tilted by differing amounts. Maximum ground tilts are developed above the limits of the area of extraction and may be cumulative if more than one seam is worked up to a common boundary.

Where movements occur points at the surface subside downwards and are displaced horizontally inwards towards the axis of the excavation (Figure 12.4). Differential horizontal displacements result in a zone of apparent extension on the convex part of the subsidence profile (over the edge of the excavation) whilst a zone of compression develops on the concave section over the excavation itself. Maximum tension occurs more or less inside the edge of the working and the transition to compression coincides approximately with the point of half maximum subsidence. Comparatively slight deviations in the subsidence profile are accompanied by appreciable variations in strain. Ground movement is three dimensional and movements of the vertical and two horizontal components may occur simultaneously.

The boundary of the surface area affected by subsidence is defined by the limit angle of draw or angle of influence, which, in British coalfields, approximates to 35° to the normal of the seam (Figure 12.4). However, Wardell[29] has suggested that the variation in limit angle is greater than is commonly believed.

A number of factors influence mining subsidence due to longwall extraction. Seam thickness is an obvious one. Maximum vertical subsidence can at times equal 90% of the thickness of the coal seam extracted. If more than one seam is worked simultaneously beneath the same area then the subsidence effects are cumulative. Usually there is an appreciable difference between the volume of mineral extracted and the amount of subsidence at the surface. For example, Orchard and Allen[3] showed that where maximum subsidence of 90% of seam thickness occurred, the volume of the subsided ground was only 70% of the coal extracted. This is mainly attributable to bulking.

One of the most important factors influencing the amount of subsidence is the width-depth relationship of the panel removed. In fact it has been shown that maximum subsidence generally begins at a width-depth ratio of 1.4:1 (see Orchard[30]). This is the critical condition above and below which maximum subsidence is and is not achieved respectively (Figure 12.5). Accordingly three stages have been distinguished, namely, sub-critical (width less than 1.4 × depth), critical (width equal to 1.4 × depth) and super-critical (width greater than 1.4 × depth). The concept of the area of influence or critical area of extraction has been explained by Wardell[31]. For a given point P on the surface the area of influence in a level seam is the circle at the base of an imaginary cone with its axis passing upwards through the overlying strata to P at its apex. The diameter of the area equals 1.4 × depth of seam. Any workings outside this area do not affect point P whereas all workings within the area do. However, all the coal from within this critical area must be extracted before point P undergoes maximum subsidence.

As long as mining is carried out within the area of influence of a surface point, that point continues to subside but once coal has been removed from the whole area the major part of the subsidence at the surface point will have taken place.

If only part of the critical area is worked, then point P only suffers partial subsidence.

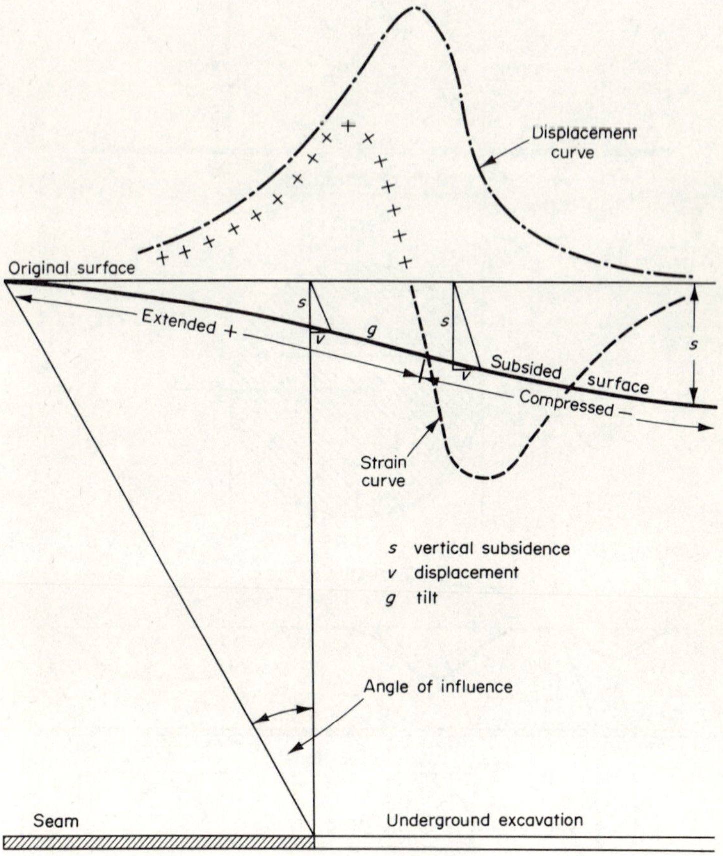

Figure 12.4 Curve of subsidence showing tensile and compressive strains, vertical subsidence and tilt, together with the angle of influence of draw (not to scale)

However, Orchard and Allen[3] subsequently pointed out that the large width-depth ratio necessary to cause 90% subsidence can usually only be achieved in shallow workings because with deeper workings the critical area of extraction is made up of a number of panels often with narrow pillars of coal left *in situ* to protect one or other of the roadways. These pillars reduce the subsidence and some coalfields, as a result, have no experience of subsidence in excess of 75% to 80%. They further pointed out that with shallow workings the supported or partially supported zones in a goaf, such as roadways and gateside packs, compose a much greater proportion of the depth than in deeper workings. Therefore deep panels can cause more subsidence than shallow ones with the same width-depth ratio.

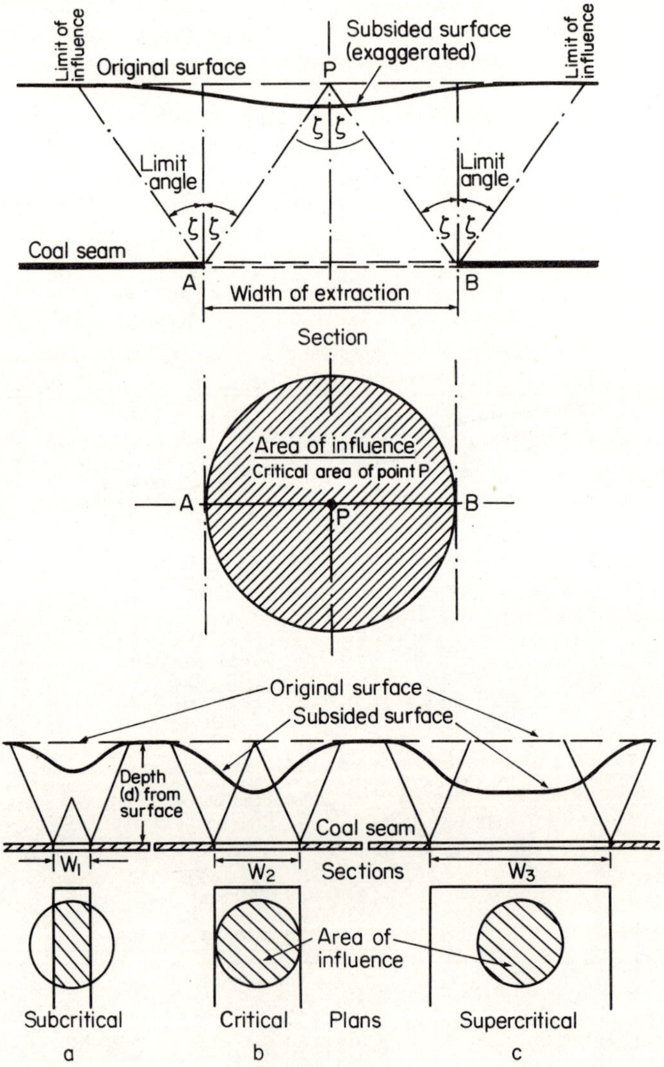

Figure 12.5 Subcritical, critical and supercritical conditions according to the depth-width ratio (After Orchard[30]). Depth width ratios (a) less than 1.4, (b) 1.4, (c) greater than 1.4

According to Orchard[32] a study of ground movements over shallow workings reveals many trends which are not so apparent over deep workings, mainly because of the masking effect of thick overburdens. For instance, for a given maximum subsidence the curvature of the ground surface is more marked over shallow workings than over deep workings owing to the smaller distance over which the subsidence curve is spread. The horizontal strains are proportional to subsidence and inversely proportional to the depth of workings. The maximum slope in the ground in a subsidence trough is also proportional to the subsidence at the bottom of the trough and inversely proportional to the depth of the workings. It follows that maximum slope is proportional to the maximum horizontal strain.

It has been suggested that if the area of influence in a seam is totally worked out then the maximum possible subsidence is a product of the seam thickness extracted and a 'subsidence factor' which depends upon whether or not the mined out area has been filled or packed. However, the character of any infill and the manner by which it is placed is so variable that really only qualitative assessments of the 'subsidence factor' can be made. In fact it could possibly be asserted that it is only in a single mining area where character and placement of infill are similar, that quantitative assessment of the effects of the 'subsidence factor' could be made properly (see Wardell and Webster[33]).

One of the most difficult problems of subsidence assessment involves the effect of older workings either above or below, or in the same seam as that being worked. Ground movements, in such instances, are usually intensified and increase the total surface effect.

Faults tend to be locations where strain is concentrated and unfortunately their exact location at the surface is not always easy to determine. What is more many coalfield areas are heavily faulted. If a fault is encountered during seam extraction and its throw is large, then the workings may terminate against the fault, thus permanent strains will be induced at the surface, probably accompanied by severe differential subsidence in the zone of influence of the fault. Indeed a subsidence step may occur at the outcrop of such a fault, sometimes with disastrous effects. Subsidence steps tend to occur when faults represent single, sharp stratal breaks. By contrast if a fault consists of a relatively wide shatter zone, then the surface subsidence effects are usually less pronounced but more unpredictable in terms of location and amount. The thickness and nature of the overlying unconsolidated deposits between bedrock and foundation level can influence the magnitude of such features.

The most notable steps occur when the coal is worked beneath the hade of a fault, faces in other positions being much less likely to cause differential movement. Steps are usually down towards the goaf but if old workings exist then steps may occasionally occur away from the face. The size of steps is on average one third of the maximum subsidence which takes place but this value varies appreciably (see Lee[34]). Their extent is very much limited to the area worked. Obviously building over faults in areas likely to be subjected to subsidence should be avoided wherever possible because of the uncertain nature of the surface ground movements and the fact that structures cannot normally be designed to withstand highly localised and attenuated differential subsidence. Indeed the Report on Mining Subsidence prepared by the Institution of Civil Engineers[35] recommends that structures be set back at least 16 m from their line of surface outcrop. On the other hand faults sometimes act as barriers to strains generated by subsidence.

Ground movements induced at the surface by mining activities are influenced by variations in the ground conditions, especially by the near surface rocks and superficial deposits. However, the reactions of surface deposits to ground movements are usually unpredictable. Indeed it has been suggested that 25% of all cases of mining subsidence undergo some measure of abnormal ground movement which, at least in part, is attributable to the near surface strata. Fissuring in brittle rocks may be due to the cumulative effects of tensile strain on the convex surface of a subsidence basin. Also joints may be caused to gape, anything up to a metre wide, at the surface. The necessary readjustment in weak strata can usually be accommodated by small movements along joints. However, as the strength of the surface rock and the joint spacing increases so the movement tends to become concentrated at fewer points. For instance, in massive limestones and sandstones movements may be restricted to master joints. Nevertheless fissuring and fracturing is more pronounced in sandstone and limestone because of their pre-existing discontinuities.

Rigid inclusions such as sandstone lenses in other rock materials such as marls and mudstones often present problems when subjected to compressive ground strains. These lenses tend to be forced upwards by the compressive forces with obvious consequences to foundations placed on or over them. It goes without saying, therefore, that it is not only necessary to examine the near surface ground conditions, but also to have some regard to the prevailing bedrock geology and the effects of past mining on any development in areas of current or future mining.

The massive sandstones of the Coal Measures and the Bunter Sandstone have been heavily fissured in places as a result of subsidence. The exact location and magnitude of these movements cannot be easily predicted. Cambering in the Magnesian Limestone, which is associated with valley bulging, has tended to open joints thereby forming gulls. This type of condition is likely to react in an unorthodox manner when affected by mining, differential displacement often being concentrated at particularly well-developed joints or fissures.

Moreover if fissures have developed then they may be partially closed or further opened by future mining. Thus if fissuring has occurred or is likely to take place, then foundations and superstructures should be designed to accommodate concentrated ground strains across those fissures. In this case friction between the underside of a raft and the founding horizon should be reduced, for example, by the inclusion of a granular layer and sliding membrane. The raft should be suitably reinforced in the event of contact being lost with the ground surface over and immediately adjacent to the fissure.

Fissuring or fracturing of the bedrock is normally accompanied by an increase in its permeability. As a consequence the hydrogeological regime may be considerably modified (see Mather, Gray and Jenkins[36]). The effects of these changes on the properties and characteristics of foundation materials must therefor be evaluated and foundations designed accordingly.

Wardell[37] demonstrated that the time taken for subsidence at a surface point to be completed is more or less inversely proportional to the rate of forward advance of the workings (Figure 12.6). However, it appears that the rate of advance does not affect the ultimate amplitude of subsidence. The transition of movement to the surface is almost instantaneous, commencing when the strata at seam horizon begin to relax. Measurable subsidence begins when a face is within three quarters of the depth from a surface point and reaches approximately 15% of maximum when the face is immediately beneath this point. Almost all subsidence at the surface point has occurred by the time the face has passed through its critical area of extraction. Residual subsidence then occurs, those points which were subsiding fastest experiencing the most residual subsidence.

Rarely does residual subsidence amount to more than 10% of the total subsidence. Depth influences the rate of subsidence, firstly, because the diameter of the area of influence and therefore the time taken for a working, with a given rate of advance, to traverse it, increases with depth and, secondly, because at greater depths several workings may be necessary before the area of influence is completely worked out. Consequently the time which elapses before subsidence is complete varies according to circumstances.

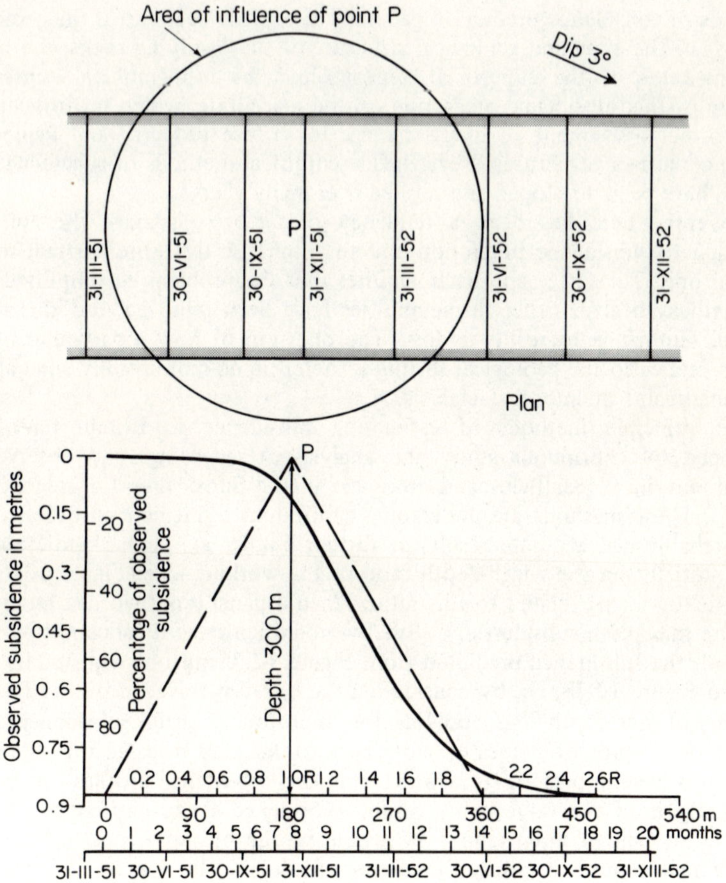

Figure 12.6 Subsidence-time relationship (After Wardell[37]). R = radius of critical area of influence

Drent[38] confirned the connection between the moment of maximum effect and depth of working. He suggested that in the Limburg coalfield in the Netherlands at the moment of maximum rate of subsidence, the surface point affected has undergone some 50% of the final subsidence to which it can be subjected. The thickness of non-Carboniferous strata appeared to influence the amount of maximum rate of subsidence more so than did the Carboniferous strata.

SUBSIDENCE PREDICTION

Unlike ground movements due to pillar and stall workings the amount of subsidence due to longwall mining can be predicted usually within ± 10%. Usually movements parallel and perpendicular to the direction of face advance are determined. Although adequate for many purposes such treatment does not consider the three dimensional nature of ground movement. For instance, recent observations have shown that individual points move on approximately helical paths, the pitch and radii change from point to point and the direction of rotation is different on opposite sides of the subsidence basin. Furthermore methods of subsidence prediction generally take little account of the geological setting i.e. the type and geological structure of the overlying rocks, the nature and thickness of the superficial deposits and the topographical expression. Because of this subsidence prediction can be inaccurate, which in turn can lead to incorrect assessment of likely damage to surface property and inadequate design of surface structures. Nevertheless empirical methods of subsidence prediction have been developed and refined over many years.

However, there has been a tendency of late to disparage the empirical approach to subsidence prediction and to emphasise the more abstract mathematical one. The latter approach requires that the problem be simplified until the mathematical or other intractabilities have been removed, but this means that all semblance to reality is lost. The question of how a particular mining system relates to the geological setting is therefore no more readily amenable to mathematical than empirical analysis.

The principle methods of estimating subsidence in Britain have been developed by continuous study and analysis of recordings and observations carried out in all coalfields and are given in the Subsidence Engineers Handbook[39]. These methods are undergoing continuous refinement in order to give more reliable and accurate results as further studies are made. Predictions are based initially on the width-depth ratio of the working face (Figure 12.7a), all other factors being related to this ratio. When a panel is not worked far enough to cause maximum subsidence, in other words when the critical condition is not attained, the subsidence predicted from Figure 12.7a must be adjusted by reference to Figure 12.7b, i.e. by considering the distance travelled by the face as a fraction of the depth and applying the appropriate partial subsidence curve.

The observance of shapes of subsidence troughs has revealed that they conform to a general pattern from which standard subsidence profiles can be produced. By itself the value of maximum subsidence over a particular extracted panel is of limited importance, the whole profile of subsidence having to be ascertained in order to study the effects of mining. The shape of a subsidence profile varies with the width-depth ratio of the extraction and, for any values, a curve can be plotted relating partial subsidence at various points to the distance from the centre of the panel as a proportion of depth. Figure 12.8 represents a series of profiles from which the subsidence values for any profile can be determined.

As remarked above, subsidence from shallow workings is usually less than that attributable to deep workings, other things being equal. However, if the extracted area is wide enough maximum subsidence will be produced in shallow workings. For example, the critical condition with a panel at 50 m depth should be 70 m but Figure 12.7 shows that maximum subsidence is not developed, subsidence approximating to 80% of the seam thickness. Maximum subsidence i.e. 90% seam thickness, is only obtained when the seam width reaches 170 m. Generally once the depth of workings exceeds 300 m, maximum subsidence is attained for the critical condition (however, refer to the remarks of Orchard and Allen above).

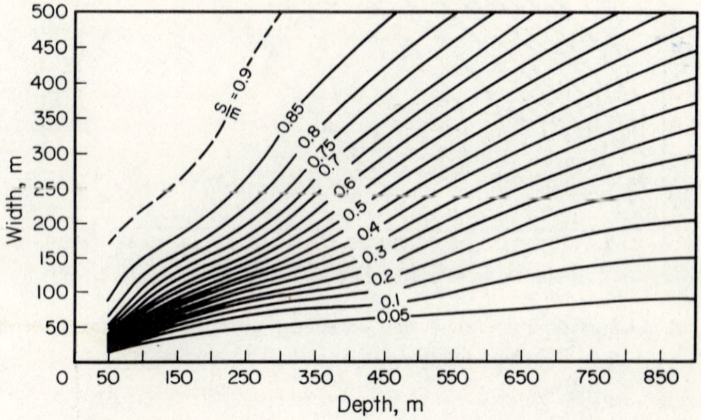

Figure 12.7(a) Relationship of subsidence to depth and width; prediction of maximum subsidence (S); m. thickness of seam extracted

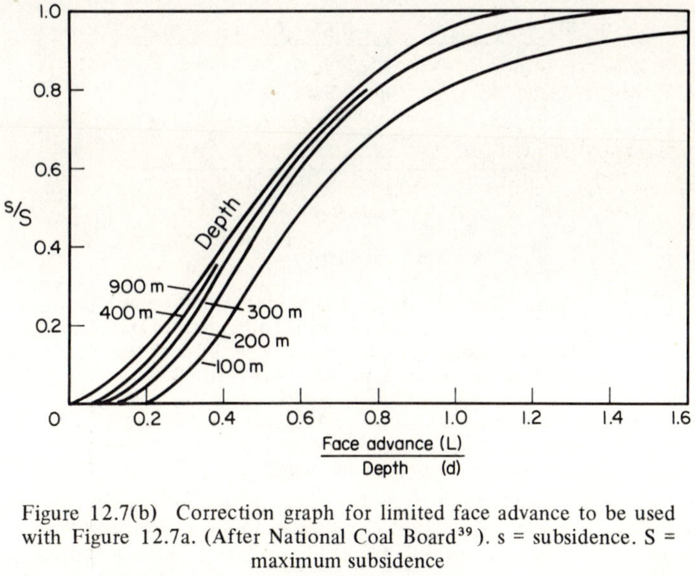

Figure 12.7(b) Correction graph for limited face advance to be used with Figure 12.7a. (After National Coal Board[39]). s = subsidence. S = maximum subsidence

The zone of maximum extension coincides with the position of the ribside where the width-depth ratio exceeds 1.35:1, lying outside the rib when the ratio is smaller. The maximum compression occurs in the centre of a panel in narrow panels (width less than 0.42 × depth) whilst at higher ratios the profile develops two compression zones (Figure 12.9). Figure 12.10 is a graph for predicting

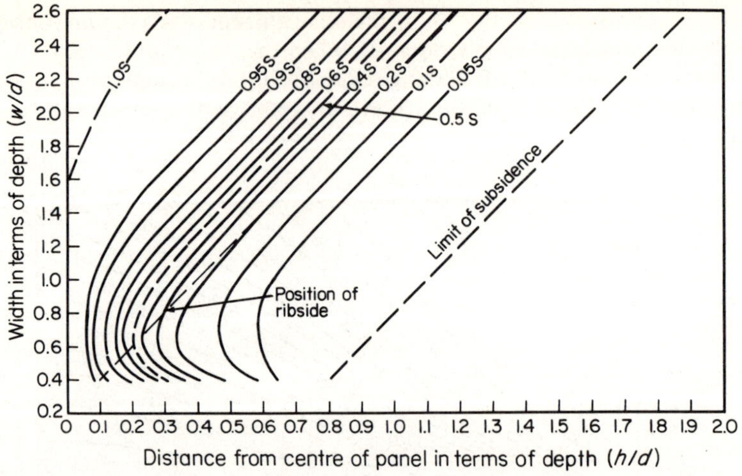

Figure 12.8 Graph for predicting subsidence profiles; w. width; d. depth; h. horizontal distance (After National Coal Board[39])

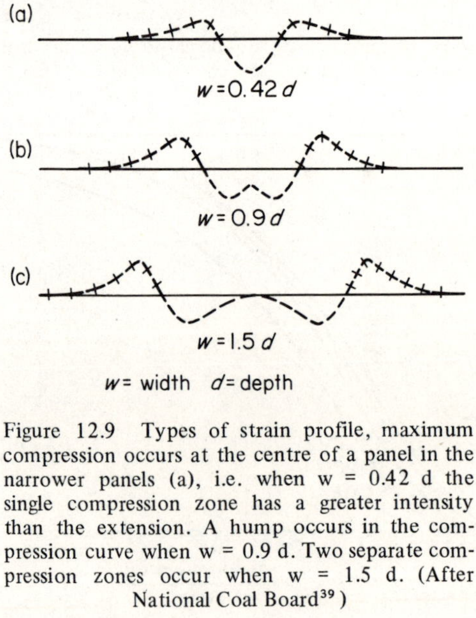

Figure 12.9 Types of strain profile, maximum compression occurs at the centre of a panel in the narrower panels (a), i.e. when w = 0.42 d the single compression zone has a greater intensity than the extension. A hump occurs in the compression curve when w = 0.9 d. Two separate compression zones occur when w = 1.5 d. (After National Coal Board[39])

strain profiles and the proportion of extension to compression for a given width-depth ratio can be derived from Figure 12.11 which shows that with narrow panels the intensity of compression greatly exceeds that of extension whereas that of extension is approximately 25% greater than compression when the critical condition is reached.

Because all principal factors are related to the width-depth ratio, this means that to be absolutely reliable this ratio has to remain constant throughout the production life of a working face. For various geological and operational reasons, for example, inclined seams or changes in the width of working face, this is not the case and consequently the above mentioned prediction methods must be modified.

Marr[40] has suggested that the zone area or circle method of prediction used in many coalfields of western Europe could be adopted. Basically surface subsidence is estimated by this method by constructing a number of concentric zones around the surface point, the radius of the outer zone being equal to the

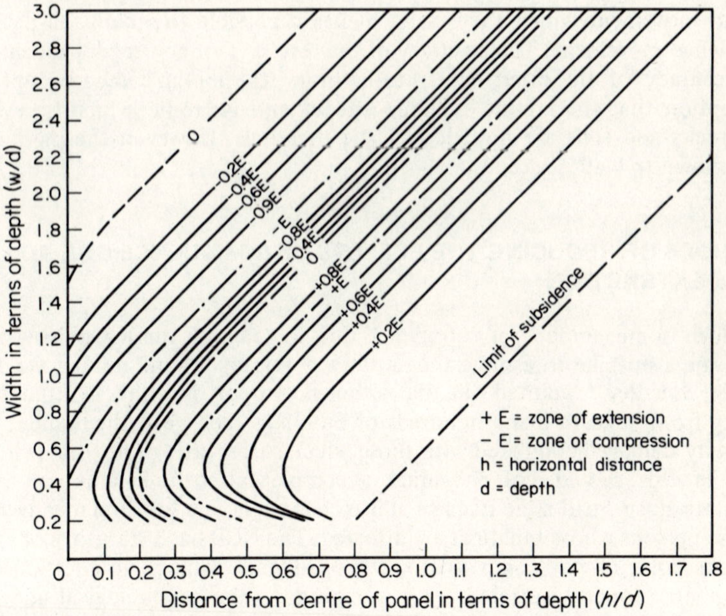

Figure 12.10 Graph for predicting strain profiles (After National Coal Board[39])

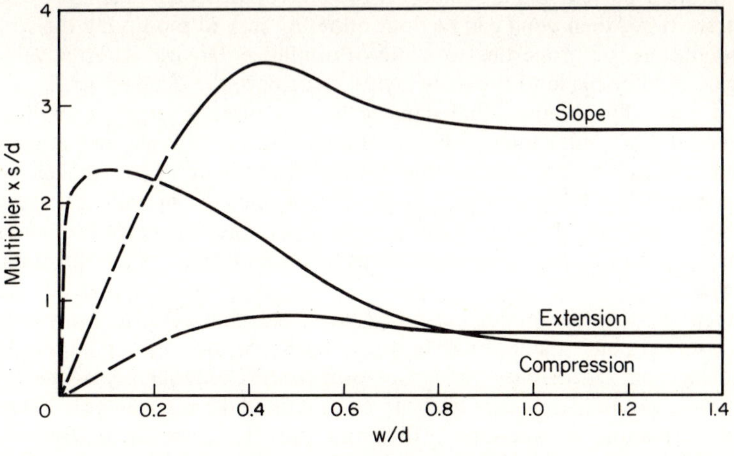

Figure 12.11 Graph for predicting maximum slope and strain for various width-depth ratios of a panel (After National Coal Board[39]) s = subsidence, d = depth, w = width

radius of the area of influence. The subsidence at the surface point is obtained from the summation of the proportions of each zone extracted, multiplied by its particular subsidence factor. In practice it has been found that 3, 5 or 7 zones are suitable for most estimations. The Western Area of the NCB have used this method of subsidence prediction for their computer programmes. These produce plans, to any required scale, which show subsidence contours and/or subsidence profiles of the mining area concerned.

Jones and Bellamy[41] have described a method of computer prediction of ground movement which they claim makes it possible to obtain full details of subsidence movement. The accuracy of the results is, of course, dependent upon the accuracy of the input data. Furthermore, the method is based upon the assumption that the ground is homogeneous and isotropic, which is never the case rocks and soils are not ideal elastic materials. However, the method has been shown to be 95% accurate.

METHODS OF REDUCING THE EFFECTS OF SUBSIDENCE DUE TO LONG-WALL EXTRACTION

Although some amount of subsidence due to longwall mining is inevitable it would be a mistake to assume that surface development and mining are incompatible. Smedley[42] pointed out that although support is withdrawn due to coal mining from beneath many hundreds of buildings each year, the number which are badly damaged compared with those which suffer little or no effects is small. Even so this should not encourage a complacent attitude amongst mineral operators since small scale damage still means economic loss and inconvenience for the persons whose buildings are affected. The NCB has a statutory obligation to make good or pay compensation for subsidence damage. In addition to the damage to property suffered by a community, the psychological and moral effects of subsidence must also be considered. Thus if subsidence is going to occur, the principal question which has to be answered is whether the anticipated ground movements are likely to produce unacceptable damage to surface property. If so, then what can be done either by way of modifying the structure or foundations of properties to avoid or minimize damage. Alternatively what changes could be made to the mine layout to achieve the same result.

It is generally accepted that damage to structures is largely caused by the differential horizontal movements of the ground (i.e. tensile and compressive ground strains) as opposed to the vertical movements. Nevertheless vertical subsidence in low lying areas can be disastrous since it can lead to flooding and drainage problems. Most buildings require approximately twice the amount of compression to develop comparable damage to that caused by a given amount of extension.

Except in the case of high rise buildings, tall chimneys, highways and rail tracks, services and drainage works, tilt in itself does not play an important part in causing physical damage to surface properties. Consequently where particularly severe ground strains are likely to be generated damage can only be avoided either by making foundations sufficiently rigid to resist these strains or, by constructing a structure sufficiently flexible to accommodate the future ground movements. However, usually it is not just a simple matter of examining the reaction of a structure to a particular value of tensile or compressive ground strain. For example, it is quite common for a building to be subjected to compressive strains in one direction and tensile strains in another direction. It may also be subjected to alternate phases of tensile and compressive ground movements so there will be a dynamic effect to consider. Thus any acceptable design for a structure situated in an active mining area must have regard to the nature, degree and periodicity of the ground movements likely to be caused by mining.

Table 12.3 NATIONAL COAL BOARD CLASSIFICATION OF SUBSIDENCE DAMAGE

Change in length of structure (mm)	*Class of damage*	*Description of typical damage*
Up to 30	1. Very slight or negligible	Hair cracks in plaster. Perhaps isolated slight fracture in the building not visible on outside
30–60	2. Slight	Several slight fractures showing inside the building. Doors and windows may stick slightly. Repairs to decoration probably necessary
60–120	3. Appreciable	Slight fracture showing on outside of building (or main fracture). Doors and windows sticking, service pipes may fracture
120–180	4. Severe	Service pipes disrupted. Open fractures requiring rebonding and allowing weather into the structure. Window and door frames distorted; floors sloping noticeably. Some loss of bearing in beams. If compressive damage, overlapping of roof joints and lifting of brickwork with open horizontal fractures
>180	5. Very severe	As above, but worse, and requiring partial or complete rebuilding. Roof and floor beams lose bearing and need shoring up. Windows broken with distortion. Severe slopes on floors. If compressive damage, severe buckling and bulging of the roof and walls

Investigations carried out by the National Coal Board have revealed that damage is not simply a function of the intensity of ground strain but that the shape and size of the structure play an important part as does the type and form of construction. Typical mining damage starts to appear in conventional structures when they are subjected to effective strains of 0.5 to 1.0 mm/m and damage can be classified as negligible, slight, appreciable, severe and very severe (Figure 12.12 and Table 12.3). Figure 12.12, of course, only serves as a crude guide to the likely damaging effects of ground strains induced by mining subsidence as it takes no account of the design of the structure or construction materials. Nevertheless it indicates that the larger a building, the more susceptible it is to differential vertical and horizontal ground movement. Furthermore there is no absolutely safe limit of ground strain.

Damage attributable to mining subsidence can be controlled and influenced by precautionary measures incorporated into new structures in mining areas, by preventative works applied to existing structures, and by mine design involving special underground layouts or any combination thereof. Several factors have to be considered when designing buildings for areas of active mining. Firstly, where high ground strains are anticipated, the cost of providing effective rigid foundations may be prohibitive. Secondly, experience suggests that buildings with deep foundations, on which thrust can be exerted, suffer more damage than those in which the foundations are more or less isolated from the ground.

Obviously because of the relationship between ground strain and size of structure, very long buildings should be avoided unless their long axis can be orientated normal to the direction of principal ground strain. Although tall buildings may be more susceptible to tilt rather than simply to the effects of horizontal ground strain, tilt can be corrected by using jacking devices (see Pryke[43]).

The most common method of mitigating subsidence damage is by the introduction of flexibility into design. This can be achieved by using specially designed rafts. Raft foundations should be as shallow as possible, preferably

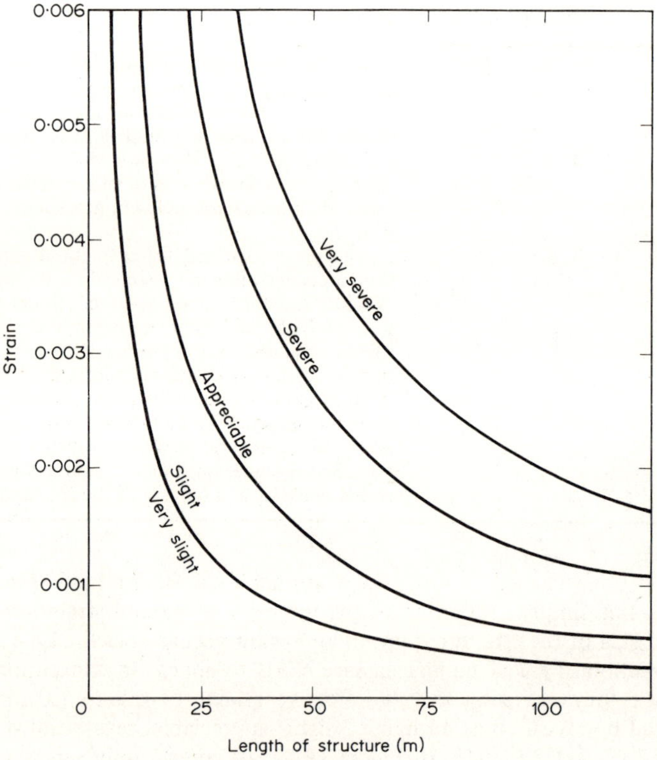

Figure 12.12 Relationship of damage to length of structure and horizontal strain (After Orchard[30])

above ground, so that compressive strains can take place beneath them instead of transmitting direct compressive forces to their edges and they should be constructed on a membrane so that they will slide as ground movements occur beneath them.

The use of reinforced concrete rafts, laid on granular material to reduce friction between the ground and the structure, is suggested by the NCB. The Ministry of Works (now the Department of the Environment[44]) have also suggested the use of a lightly reinforced concrete base slab bedded on a fine granular material. Wardell[45] proposed that where relatively small buildings (up to about 30 m in length) were concerned, they should be erected on a 'sandwich raft' foundation. The lower slab is separated along the two diagonals into four independent units, the upper slab is one unit and is separated from the lower by a layer of sand covered with building paper. In this type of foundation, the base of the raft is capable of movement relative to the upper reinforced section and friction between the two is reduced by the introduction of the sand layer. Usually high bearing pressures are adopted for heavy structures thus the plan dimensions of the raft are kept to a minimum and in this way the horizontal tensile and compressive forces acting on the underside of the raft are kept as low as possible. The length of the raft acting as a cantilever at the hogging stage or as a beam at the sagging stage is also minimised (see Mauntner[46]). Cellular rafts have often been used for multi-storey buildings.

The CLASP form of construction is also based on flexibility so that the structure can withstand surface ground movements induced by mining. This form of construction, which has been widely used by some Local Authorities, incorporates a structural framework which is sufficiently flexible to follow the differential vertical subsidence without cantilevering over the approaching subsidence depression. Accordingly it is pinjointed. In order to neutralise the elongation or shortening of the ground surface, the structure rests lightly on it and is free to slide over it, i.e. the building produces very light foundation loads. Walls, windows, floor and roof are designed to allow the structural frame to distort with the ground.

Flexible joints in service pipes and flexible pavements in highway construction are other examples which benefit by flexibility in design.

The incorporation of jacks to correct tilt in structures, machinery foundations and certain types of continuous processing plant has already been mentioned. The use of pivoting facilities, roller bearings etc, for bridge design in areas of mining subsidence are also commonly used.

The use of piled foundations in areas of mining subsidence presents its own problems. The lateral and vertical components of ground movement as mining progresses mean that the pile caps tend to move in a spiral fashion and that each cap moves at a different rate and in a different direction according to its position relative to the mining subsidence. Such differential movements and rotations would normally be transmitted to the structure with a corresponding readjustment to the loadings on the pile cap. In order to minimise the disturbing influence of these rotational and differential movements it is often necessary to allow the structure to move independently of the piles by the provision of a pin joint or roller bearing at the top of each pile cap. It may be necessary to include some provision for jacking the superstructure where severe dislevelment is likely to occur.

Paved or concreted areas surrounding buildings increase their effective surface area and hence their vulnerability to surface ground movements. For this reason gaps should be left between paved areas and the main structure. Similarly long brick or concrete walls, and continuous stone kerbs are badly affected even by small ground strains. This type of damage can often be avoided by leaving structural gaps of appropriate width at appropriate intervals.

Methods and materials aimed at flexibility do not necessarily mean higher costs in the long run, indeed as far as maintenance is concerned they may well prove cheaper. Emphasis is placed on flexibility because in most cases it is either too costly or it is impracticable to design rigid structures capable of completely resisting movements due to subsidence.

Preventative works can frequently be used to reduce the effects of movements on existing structures. Again one of the principal objects is to introduce greater flexibility. In the case of very large buildings, damage can be reduced by splitting the building into smaller, structurally independent units. In particular, such things as chimneys, lift shafts, machine beds etc, can be planned independently of the remaining, generally lighter, parts of a building and separated from the main structure by joints through foundations, walls, roof and floor which allow freedom of movement. Extensions and outbuildings should similarly be separated from the main structure by such joints.

The excavation of trenches around buildings subjected to compressive ground strains has significantly reduced the damaging effects of the ground movement (Wardell[45]). The trench is dug close to and around the structure down to foundation level which effectively breaks the continuity of the surface and hence the foundations are isolated from sidethrust. The trenches are backfilled with compressible material. Movement is allowed for by the provision of appropriate sliding joints.

Structures which are weak in tensile strength can be afforded support by strapping or tie bolting together where they are likely to undergo extension. Strapping consists of a system of steelwork placed around a building in an attempt to have it move as a whole. Frequently angle irons run the full height at each corner with two or more sets of tie rods around the building. In some cases rods may go through the building.

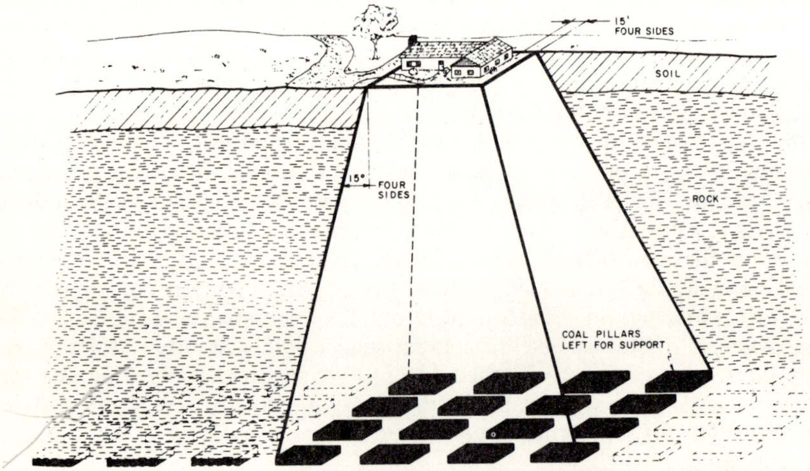

Figure 12.13 Method of supporting surface structures by leaving pillars of coal in place (After Gray and Meyers[26])

Damage to surface structures can be reduced by adopting a specially planned layout of underground workings which take account of the fact that surface damage to structures is primarily caused by ground strains. Thus to minimise the risk of damage, underground extraction must be planned so that surface strain is reduced or eliminated. Harmonious mining in Germany was described by Lehmann[47] as long ago as 1938 and in 1950 Grond[48] described a similar method used in Holland. This method of mining involves simultaneous working of two seams. The Lehmann method attempted to achieve a supercritical area of extraction by mining three or more seams simultaneously so that the area in the centre of the subsidence basin underwent uniform vertical subsidence and was more or less free from strain. Grond described how the anticipated tension from the workings in one seam were practically offset by compression produced by extraction in another seam.

Unfortunately British mining conditions seldom lend themselves to simultaneous extraction, however, Wardell and Beevers[49] referred to the stepped face layout which has been used to this end. Moreover they wrote that unlike the continental conception of harmonious mining this method allowed for the effects of the travelling movements which accompany an advancing face as well as the transverse movements. In the case of stepped face layout Wardell[50] indicated that the degree of cancellation is governed by the distance between the two faces and that if the distance is equivalent to 0.45 r (r = the radius of the area of influence about a surface point) then the tension from the leading face would not develop beyond 50% of its maximum. He, however, admitted that this was not always strictly valid. Unfortunately it is seldom possible to arrange the direction and location of underground workings to suit the need to minimise surface movement in the neighbourhood of a particular surface structure.

Subsidence can be much reduced by leaving pillars of coal between longwall panels. The individual pillars should be large enough to ensure that the overlapping subsidence from two adjacent panels does not exceed the required safe limit at the surface and does not cause humping and undulation in the total subsidence profile (see Orchard[2]; Orchard and Allen[3]).

Partial extraction by leaving pillars equivalent to approximately 0.25 depth of the workings has been successfully used in many cases in recent years. This has enabled large areas of coal to be partially worked (usually 50%) which hitherto would have been left, being considered commercially unworkable. A similar method has been used in the Pittsburg area and described by Vandale[51]. There an angle of 15° projected from the vertical outside a safety area 5 m wide around the foundation has been used to define the area of support for a property (Figure 12.13). Fifty per cent of the coal in the support area could be removed without ill effects.

Maximum subsidence is also influenced by the particular method used for packing the goaf. This subsidence factor, because it varies with method, needs to be determined from actual observations of individual cases in each mining district. Making allowance for variations in the percentage of actual packing, Wardell and Webster[30] noted the tendency for this subsidence factor to decrease with decreasing depth and added that a constant factor for a particular method of goaf support, when applied at varying depths, could not be adopted. Variations are due to different pressures at different depths, different types of packing materials and the difference in the quality and speed of pack construction. Pneumatic stowing is most effective and can reduce subsidence by up to 50%.

SUBSIDENCE AND EARTH TREMORS

In the past two years earth tremors have been recorded in the neighbourhood of Stoke-on-Trent, Staffordshire. For example, one of the largest tremors occurred at 7.15 on the morning of May 13th, 1976. It lasted for two seconds

Figure 12.14(a) A crown hole developed at the surface due to the collapse of a mine shaft in Swansea (Courtesy, Western Mail and Echo)

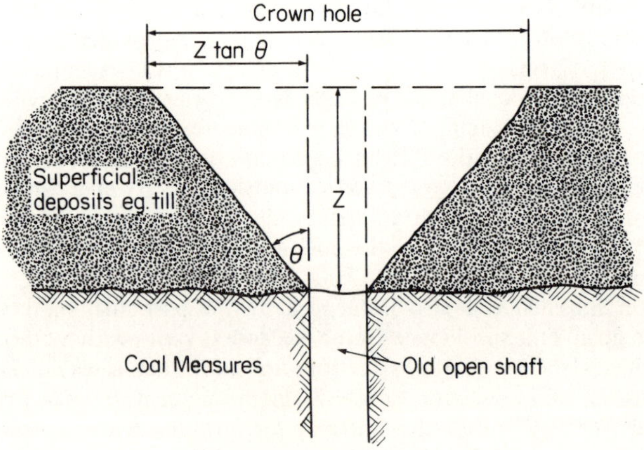

Figure 12.14(b) Sketch indicating the possible diameter of a crown hole developed in superficial deposits due to the collapse of a mine shaft. $\theta = 90^\circ - \phi$ (angle of internal friction)

and was centred on Trent Vale, affecting a wide area of the Lyme valley between Penkhull and Clayton. Exaggerated claims have been made regarding the 'violence' of this tremor, in fact one report of damage was recorded – a cracked kitchen ceiling. Nonetheless such movements are an obvious source of concern for local residents and continued tremors could weaken old or poorly constructed buildings. Moreover services could be damaged with undesirable consequences.

Over 100 low magnitude earth tremors, but less than a dozen large enough to be felt by the public, have been recorded since September 1975 when seismic equipment was installed for monitoring. At times as many as twelve tremors per week have been recorded. In a report by the Inspector of Mines and Quarries it has been suggested that there is a link between these earth tremors and the mining activity at Hem Heath Colliery. The foci of the tremors affecting the Hanford and Knutton districts are generally located above the level of these current mine workings. Retreat mining began at Hem Heath Colliery in the late 1960's and the area is interrupted by two major faults. However, the mechanics of this relationship and the relative importance of other factors such as geological setting, faults and old workings, which appear to combine in an exceptional way in this area, are still under investigation.

OLD MINE SHAFTS

Centuries of coal mining has meant that there are a great number of old shafts left behind. Unfortunately the location of many, if not most, is unknown and the location of those which are supposedly recorded is not always accurate. The location of a shaft is of great importance as far as the safety of a potential structure is concerned for although shaft collapse is fortunately an infrequent event, its occurrence can prove disastrous. Moreover from the economic point of view the sterilisation of land due to the suspected presence of a mine shaft is unrealistic.

According to Maxwell[52] the shape of the shafts and mode of lining the walls appear to have been largely governed by the mining custom in each coalfield. Hence in most parts of England, the shafts are circular whilst in Wales they are elliptical and where necessary the walls of the shafts are lined with stone or brickwork. In Scotland, the shafts are usually rectangular, and often lined with wood. Shaft diameter ranges from 2 to 5 m, and the maximum side of the rectangular shafts from 2 to 6 m. The number of shafts per mine varied, indeed some shallower mines had as many as ten shafts giving easy access to all parts of the workings. Air-shafts were generally much smaller in cross-section than winding-shafts.

Since a mine owner was not obliged by law to fill shafts he frequently failed to do so; it was supposedly made safe by capping it off. However, many cappings have been obliterated as surface features and therefore do not occur on present day Ordnance Survey maps. Shafts were capped with wrought iron domes and turfed over, or trees were dropped into the shaft to form a bridge, on top of which fill was placed; or more usually a wooden platform was laid across the buntons some 3 to 15 m below the surface and topped up with fill. With time the wood decayed to expose an open shaft. In the nineteenth century stronger wooden rafts were laid across the shaft void and projected outside its perimeter.

If shafts were filled it was frequently undertaken in a haphazard manner without regard for the future. Many shafts were filled with unsuitable material, that at hand being the most obvious, for example, rails, timbers, bogies, scrap metal and mine waste. As can be imagined this rarely meant that the shaft was properly filled; bridging and the formation of voids usually occurring. Such fills are capable of sudden collapse. Usually no attempt was made to seal shafts from the workings, which after the mine was abandoned became waterlogged. In such instances fine fill is likely to flow into the workings with the result that a cavity is produced in the shaft. With time the plug above the cavity is denuded until eventually the remnant collapses.

As the shaft lining, especially near the top, deteriorates a point is reached when its strength is so much reduced that it can no longer retain the surrounding material. This then craters into the open shaft to form a crown hole (Figure 12.14a). If the surface is covered by superficial deposits then the thicker these

Figure 12.15 Partial collapse of pillars in workings in Bath Stone (Courtesy of Kingston Minerals)

are the greater will be the dimensions of the crown hole, irrespective of the shaft diameters (Figure 12.14b). Hence it is generally recommended that the minimum distance for siting buildings from open or poorly filled shafts is twice the thickness of the superficial deposits up to a depth of 15 m, unless they are exceptionally weak. A consideration of the long term stability of shaft linings is required as well as the effects of increased lateral pressure due to the erection of structures nearby. Protective sheet piles or concrete walls can be constructed around shafts to counteract cratering, but such operations should be carried out with due care.

The location of shafts is obviously of paramount importance. If the position of a shaft is shown on a number of plans but not in the same location, then Taylor[24] has suggested that its position can be assessed by drawing equivalent circles of likely shaft dimensions about the plan locations. The overlap of the circles then provides a probability zone for detailed examination.

Confirmation of the existence of a shaft must always be done by excavation. This may be accomplished by exposing the top of the shaft by a drag-line securely anchored outside the area of probable position. If this is unsuccessful then the shaft will have to be exposed by boring, suitable safety precautions having been taken. A boring should go down the shaft into bedrock and another should be located next to the shaft to determine the thickness of overburden and succession if the latter is not available from mine plans.

The success of geophysical methods in locating an old shaft depends on the existence of a contrast between the physical properties of the shaft and its surroundings. If no contrast exists it will not be located. Most geophysical methods have had some success in locating shafts, where their position is known approximately, the proton magnetometer being an example (see Raybould and Price[53]). However, according to Maxwell[52] of the shafts which offer a magnetic contrast, and not all do, many do not provide an anomaly which differs significantly from surrounding anomalies over the site in question. Therefore all anomalies have to be investigated before a shaft is revealed. He further added that a resistivity survey was unlikely to produce vital evidence leading to shaft location.

The shaft void is only detectable if the depth of cover is less than the radius of the shaft and the surrounding material is homogeneous. This rarely occurs. Similarly if the infill on the shaft capping is more than 2 m thick a gravity survey will not detect an anomaly. McCann, Grainger and McCann[54] have claimed some success in locating cavities in limestone by using a direct recording seismic refraction technique between boreholes. Some success has also been claimed for seismic reflection in locating old shafts and subsurface cavities (see Howell and Amos[55]). Like Maxwell, Burton and Paton[22] concluded that because of the complexities involved, too much should not be expected of geophysical

Figure 12.16 Crown holes due to the presence of old workings in the Chalk which led to the abandonment of Jacqueline Close, Bury St. Edmunds (Courtesy of East Anglian Daily Times)

methods as far as the location of old mine shafts is concerned, the variations can be from complete failure to occasionally total success. It is therefore necessary to use all available information to estimate approximate location, as well as probable physical contrasts with surroundings, before choosing a particular geophysical technique. Furthermore two geophysical methods used in combination probably will give a better result.

If the depth is not excessive and the shaft is open then it can be filled with suitable granular material. If, as is more usual, the shaft is filled with debris in which there are voids then these should be filled with pea gravel. Dean[56] suggested that if the exact positions of the mouthings in an old shaft are known then these areas should be filled with gravel, the rest of the shaft being filled with mine waste, however, the latter will tend to consolidate much more than will gravel. Dean also supplied details concerning the concrete cappings needed to seal mine shafts. Where the depth of overburden is excessive then the problem of cratering exists if a shaft is open or badly filled. In such instances the shaft should be filled with gravel to the surface and this then grouted, with a cement-fly ash mix. The concrete capping should take the form of an inverted cone.

The easiest way of dealing with mine shafts is to avoid locating structures in their immediate vicinity. For instance, a suitably capped but unfilled mine shaft was located within a building site at Hoyland Common, near Barnsley. Unfortunately the shaft was near a fault and much of the area was going to be undermined at some future date. This possibly would have meant that the shaft would be affected and perhaps the cap would be dislodged. The overburden consisted of mine waste overlying till, approximately 6 m in total thickness. It was recommended that the shaft should be filled and re-capped, and that no building or structure should be located within 15 m of the shaft.

SALT MINING AND SUBSIDENCE

Salt has been worked in Cheshire, Lancashire, Shropshire, Somerset, Staffordshire, Worcestershire and Yorkshire (see Sherlock[57]), it being obtained either directly by mining or indirectly by brine pumping. Both these methods of extraction have involved subsidence, which in Cheshire was at times catastrophic. At the present day subsidence in Cheshire continues to be an inhibiting factor as far as major developments are concerned. This is chiefly because of its unpredictable nature.

A deep borehole sunk by the Geological Survey at Wilkesley revealed a concentration of rock salt into two groups which totalled 595 m in thickness (Pugh[58]). The succession was as follows:

	Thickness (m)	*Depth* (m)
Pleistocene: sands and clays	18	18
Lias: clay with bands of limestone	140	158
Rhaetic: mudstone and black shale	13	171
Keuper:		
Upper Keuper Marl	176	347
Upper Keuper Saliferous Beds	404	752
Middle Keuper Marl	327	1078
Lower Keuper Saliferous Beds	190	1268
Lower Keuper Marl	257	1525
Upper Keuper Sandstone	105	1630
Keuper Waterstones	54	1684

The area occupied by the lower group from its western and northern margins to the eastern boundary fault of the Cheshire Basin is approximately 1500 km^2 whilst that of the upper group is about 170 km^2. The old established salt industry of Northwich and Winsford is based on the lower group whilst the industry at Sandbach and Middlewich is based on the upper group.

The production of salt from land-brine in the UK dates back to pre-Roman times. Indeed until near the end of the seventeenth century most salt was obtained from natural brine springs. Then in 1670, during a search for coal, rock salt was discovered at Marbury, near Northwich, where the first mine was sunk in 1682. In the succeeding eighteenth and nineteenth centuries the salt industry in Cheshire expanded enormously. During this time numerous brine pits and mines were sunk, however, many of the mines had a short life, rarely more than 40 years. This was due to poor mining practices which led to roof collapses and the ingress of water, for example, in some mines the extraction ratios were as high as 90%. As adjacent mines collapsed subsidence spread until sets of flooded mines formed large underground reservoirs. Today the only salt mine still in operation in the UK is the Meadowbank Mine at Winsford which is run by ICI. The pillar and stall workings are in dry rock head and the extraction ratio is about 65%.

Natural brine springs were exploited in the valleys of the Weaver and Dane in mediaeval times and up to the seventeenth century brine was readily accessible from brine springs which overflowed at the surface. However, with increasing output of salt the brine level fell which meant that wells replaced the brine springs and as exploitation proceeded these had to be sunk deeper and deeper. These wells went into wet rock head, the aim of the pumper being to tap the brine runs flowing on the top of the salt beds. At the present time wild brine pumping takes place around Northwich, Middlewich, Sandbach and Lymm and it accounts for about 12% of the brine extracted in Cheshire.

Eighty-eight per cent of brine production in Cheshire is obtained by controlled solution mining. In this type of mining, salt is removed by solution from cavities, the salt surrounding the cavities providing support for the roof. These cavities are located in dry rock head so that there is no possibility of circulating ground waters dissolving the 'pillars' between them. The controlled field is centred about Holford, near Northwich, and is operated by ICI. The size and shape of each cavity is controlled by sonic methods and the borehole leading into the chamber is sealed by steel tubing grouted into place thus preventing the possibility of leakage. The roof of a cavity is protected from solution by a cushion of air or oil. Although the individual cavities are very large the extraction ratio is low. As a result no subsidence of any consequence has been recorded in the Holford field. Furthermore ICI use old cavities for the disposal of liquid waste, which does not dissolve the salt, and which can be later solidified.

Natural subsidence in wet rock head areas is believed to have been responsible for the formation of the Cheshire Meres but generally speaking this type of subsidence, although it operates over large areas, takes place extremely slowly, it is perhaps impercetible within the context of historic time. The Keuper marl which overlies the saliferous beds is incompetent which means that large solution cavities have not been able to form since the collapse of the marl was probably contemporaneous with salt removal. These collapses produced linear subsidence depressions at the surface. Where a brine run occurs beneath glacial sands cavities have presumably formed in the sands. These then collapse to produce rather circular shaped, small hollows at the surface. They make their appearance almost instantaneously but fortunately they occur only infrequently. It would seem that when the sand collapses it chokes the brine run thereby preventing further subsidence.

The most successful wild brine pumping was and is carried out on the major pre-existing natural brine runs. Such pumping accelerated the formation of solution channels. Active subsidence is normally concentrated at the head and sides of a brine run where fresh water first enters the system (see Bell[59]). Hence serious subsidence may occur at considerable distances, anything up to 8 km from pumping centres. The maximum strains developed by such subsidence can demolish buildings and it has proved at times impossible to allot the blame to any particular operator. As brine runs extend and widen the subsidence encroaches into new areas and deepens existing depressions. Obviously knowledge of the process of wild brine pumping is limited since the time factor involved in subsidence is unknown, as are the exact dimensions of the area undergoing solution. It is not really possible to predict when or where subsidence due to wild brine pumping will occur and it therefore remains a problem throughout mid-Cheshire. If any attempt is to be made at the prediction of the amount and location of subsidence then this can only be done by the extrapolation into the future of past trends, a method which cannot guarantee success.

Subsidence due to wild brine pumping has given rise to the formation of flashes. Flashes are water-filled, linear hollows which are developed in the tills of wet rock head areas. They may be about 10 m in depth by 70 or so metres wide. They form as a result of collapse above brine runs. Their sides are cambered which is a surface manifestation of the associated subsidence curves. The flanks may also be interrupted by tension scars along which movements may occur. Brine streams frequently merge and where they do larger flashes are developed.

The Brine Pumping (Compensation for Subsidence) Act, 1891, imposed a levy on wild brine operators in order to provide a compensation fund for subsidence damage. But these provisions only applied to a small district around Northwich. The Brine Subsidence Compensation Board was also to advise on the precautions to be taken in construction in subsidence area. The Cheshire Brine Pumping (Compensation for Subsidence) Act of 1952 largely followed the recommendations of the Weekes Report (1934), the chief of which was the establishment of a single compensation district covering all the areas of Cheshire where subsidence was a problem.

Since 1952 there has been a reduction in the number of wild brine pumpers and indeed the brine industry is now mainly concentrated in three major concerns. However, the reduction in the number of operators has meant that those remaining have generally expanded their production. Although no limit has been placed on the amount of wild brine pumped, since 1968, no permission has been given to extend wild brine pumping outside the demarkated area and now no permission is given to sink new wild brine wells.

Subsidence due to brine extraction will cease when all wild brine pumping has ceased, its place being taken by controlled pumping (see Collins[60]). Although there is a trend in this direction there are a number of difficulties involved in the cessation of wild brine operations. One of the most important of these is the higher operating costs of controlled pumping. The latter also depends upon the geological suitability of the site. For instance, the saliferous beds must be of considerable thickness and be protected by impermeable roof rocks. The Lower Keuper Saliferous Beds between Holford and Sandbach fulfil these conditions as do those between Winsford and Crewe, and the Upper Saliferous Beds at Wilkesley. Controlled pumping could also be employed at Middlewich of the Lower Keuper Saliferous Beds were tapped. In fact the only area where wild brine pumping goes on at present which seems unsuitable for controlled pumping is that about Lymm.

At the present it is almost impossible to delineate areas free from subsidence risk although every attempt is made to locate major developments on safe or relatively safe areas. Advice can always be obtained from the Brine Subsidence Compensation Board in Northwich, as to the likelihood of subsidence occurring and the types of construction to be used in subsidence areas. For example, liftable buildings have for a long while been erected in Northwich but steel framed buildings which can be jacked up are expensive and reinforced rafts are now being used. The Board can make a grant towards the cost of the construction of a liftable building but this is discretionary.

SOME FURTHER EXAMPLES OF MINING

The most important metal ore mined in Britain is iron. The sedimentary iron ores of the Jurassic system, as well as being open-casted, have been mined in several areas, chiefly in Lincolnshire, Northamptonshire and in the Cleveland Hills. The mines have been and are worked by the pillar and stall system. Little significant subsidence, however, has been associated with such mining, except for slight lowering of the ground surface, which, for example, has occurred to the east of Scunthorpe. The hematite deposits of Cumberland were worked by total extraction methods and these have given rise to subsidence, for example, subsidence due to extraction of ore from the Hodbarrow Mine, Millom, by 1922 was responsible for the formation of four large depressions each about 400 m wide and 15 m deep. Efforts to reduce subsidence included hydraulic storage of sand. Clay ironstones have been worked extensively in the past in coalfield areas. Frequently the workings took the form of bell pits and these sometimes present foundation problems.

Many other metals have been worked in Britain, notably lead and zinc in the Pennine orefield and the Lake District, and tin and copper in Cornwall. When these are vein deposits the workings follow the vein and are consequently narrow. In addition the host rock is invariably competent, for example, in the Pennine orefield it is generally the Carboniferous Limestone and in Cornwall, granite. Such workings do not as a rule produce subsidence problems. Old mine shafts, however, do constitute a problem. For example, it is believed that there are hundreds of unrecorded old mine shafts in north Derbyshire and these occasionally reveal themselves at the surface. Gangue minerals such as fluorspar, calcite and barytes have been and are also worked in the Pennine orefield. Some subsidence due to fluorspar working has occurred in north Derbyshire but no property has been damaged, as it affected agricultural land.

The Bath Stone (Great Oolite) has been worked in Wiltshire and Somerset, for example, within the Box-Corsham district of Wiltshire there are 96.6 km^2 of pillar and stall workings. Formerly the extraction ratio reached 85% but it has been reduced to 65%. Small subsidence depressions developed in the past, this being due to pillar collapse (Figure 12.15).

Shallow mine workings occur in the Chalk, particularly in East Anglia and Kent and have given rise to foundation problems in many towns such as Bury St. Edmunds, Norwich, Grays, Rochester and Gravesend. Some of these were made by Palaeolithic man in his quest for flint. These are usually similar to bell-pits, they may be about 3 m to 6 m in height, and have small tunnels, following the flints, running from the main chamber. Chalk has also been worked at later times for building and agricultural purposes. A notable series of subsidences in chalk occurred in Bury St. Edmunds in July 1966. The subsidences took the form of crown holes at the surface due to void migration from shallow workings. These holes appeared in Jacqueline Close with the result that the houses were eventually declared unsafe and therefore abandoned (Figure 12.16). As far as the workings are concerned the pavement is usually between 15 and 18 m below the surface and the galleries are 2 to 3.5 m high. Although the ground was potentially unstable, in that crown holes were likely to occur, the process was accelerated by site development which meant a change in the subsurface drainage regime. It appears that soakaway water led to extra solutioning and the rapid decay of the roofs above galleries. It was hoped that the workings could be filled and the site redeveloped but this proved too costly.

An interesting example of past mining which was not thought as extensive occurs in the Upper Greensand near Merstham, Surrey. Here the rock was worked for hearthstone by a crude pillar and stall method, the extraction ratio varying between 65 and 75%. The workings occur around and beneath the M23-M25 interchange and therefore required grouting. The grout mix consisted of 1 part cement, 4 parts sand and 5 parts fly ash by weight, with the minimum amount of water to render the grout workable. The grout holes were drilled on a 6.1 m grid, then infilled on a 3 m grid. Pea gravel was used for forming the curtain walls.

Fireclays and gannisters have been worked in most coalfields, by both quarrying and mining methods, for refractory purposes. These workings are widespread but gannister is of more importance in the Lower than Middle Coal Measures whereas fireclays are important in both, particularly the latter. Fireclays generally have been worked by pillar and stall methods and where these are old and at shallow depths they can present similar problems to those of old coal workings. But occasionally, as in some parts of Lancashire, the longwall system has been used, particularly where coal was also extracted. This obviously led to some subsidence.

References

1. Wardell, K. and Wood, J.C., 'Ground instability problems arising from old mine workings,' *Proc. Mid. Soc. Soil Mech. Found. Eng.*, 7, 5–30 (1965).
2. Orchard, R.J., 'Partial extraction and subsidence,' *Min. Engr.*, **123**, 417–427 (1964).
3. Orchard, R.J. and Allen, W.S., 'Longwall partial extraction systems,' *Min. Engr.*, **129**, 523–535 (1970).
4. Bryan, Sir A., Bryan, J.C. and Fouche, J., 'Some problems of strata control in pillar workings,' *Min. Engr.*, **123**, 238–266 (1964).
5. Greenwald, H.P., Howarth, H.C. and Hartman, I., 'Experiments on strength of small pillars of coal in the Pittsburg bed,' *U.S. Bur. Mines Tech. Paper 605* (1939), *Rep. Inv.* 3575 (1941).
6. Terzaghi, K., *Theoretical soil mechanics*, Wiley, New York (1943).
7. Denkaus, H.G., 'Critical review of the present state of scientific knowledge related to the strength of mine pillars,' *Jour. S. Af. Inst. Min. Met.*, **63**, 59–75 (1962).
8. Steart, F.A., 'Strength and stability of pillars in coal mines,' *Jour. Chem. Met. Min. Soc. S. Africa*, **54**, 309–325 (1954).
9. Coates, D.F., 'Pillar load. Part I. Literature survey and new hypothesis,' *Dept. Min. Tech. Surv. Ottawa, Mines Branch, Res. Rept. R168* (1965).
10. Salamon, M.D.G. and Munro, A.H., 'A study of the strength of coal pillars,' *Jour. S. Africa Inst. Min. Met.*, **68**, 53–67 (1967).
11. Salamon, M.D.G., 'A method of designing bord and pillar workings,' *Jour. S. Africa Inst. Min. Met.*, **68**, 68–78 (1967).
12. Holland, C.T., 'The strength of coal in mine pillars,' *Proc. 6th Symp. Rock Mech. Missouri Univ., Rolla, Pergamon Press*, 450–466 (1964).
13. Bieniawski, Z.T., 'An analysis of the results from underground tests aimed at determining the *in situ* strength of coal pillars,' *CSIR Rept. MEG 569*, Pretoria (1967).
14. Bieniawski, Z.T., '*In situ* large scale testing of coal,' *Proc. Conf. In Situ Testing of Rocks & Soils. Brit, Geot. Soc.*, 67–74 (1970).

15. Wilson, A.H., 'An hypothesis concerning pillar stability,' *Min. Engr.*, **131**, 409–417 (1972).
16. Wardell, K. and Eynon, P., 'Structural concept of strata control and mine design,' *Trans. Inst. Min. Met., Section A*, **77**, A125–150 (1968).
17. Taylor, R.K., 'Site investigations in coalfields: the problem of shallow mine workings,' *Quart. Jour. Eng. Geol.*, **1**, 115–133 (1968).
18. Price, D.G., Malkin, A.B. and Knill, J.L., 'Foundations of multi-storey blocks with special reference to old mine workings,' *Quart. Jour. Eng. Geol.*, **1**, 271–322 (1969).
19. Jennings, R.A.J., 'The problem below,' *Quart. Jour. Eng. Geol.*, **9**, 103–118 (1976).
20. Scott, A.C., 'Locating and filling old mine workings,' *Civ. Eng. Pub. Works. Rev.*, **52**, 1007–1011 (1957).
21. Maxwell, G.M., 'Some observations on the limitations of geophysical surveying in locating anomalies from buried cavities associated with mining in Scotland,' *Min. Engr.*, **134**, 277–285 (1975).
22. Burton, A.N. and Maten, P.I., 'Geophysical methods in site investigations in areas of mining subsidence,' (article in *Site Investigations in Areas of Mining Subsidence, Ed. by Bell, F.G.*), Newnes-Butterworths, London (1975).
23. Price, D.G., 'Engineering geology in the urban environment,' *Quart. Jour. Eng. Geol.*, **4**, 191–208 (1971).
24. Taylor, R.K., 'Characteristics of shallow coal mine workings and their implications in urban redevelopment areas,' (article in *Site Investigations in Areas of Mining Subsidence, Ed. by Bell, F.G.*) Newnes-Butterworths, London (1975).
25. Butier, F.G., *Personal Communication*.
26. Gray, R.E. and Meyers, J.F., 'Mine subsidence and support methods in the Pittsburg area,' *Am. Soc. Civ. Eng., Soil Mech. Found. Div.*, **96**, 1267–1287 (1970).
27. Lehmann, K., *Movements in open diggings and troughs*, Gluckauf (1919).
28. Hilbig, R., 'Lehmann's trough theory,' *Colliery Engng.*, 413–415 (1957).
29. Wardell, K., *Personal communication*.
30. Orchard, R.J., 'Recent developments in predicting the amplitude of subsidence,' *Jour. R.I.C.S.*, **86**, 864–876 (1954).
31. Wardell, K., 'Mining subsidence,' *Jour. R.I.C.S.*, **86**, 53–76 (1954).
32. Orchard, R.J., 'Prediction of the magnitude of surface movements,' *Colliery Engng.*, 455–462 (1957).
33. Wardell, K. and Webster, N.E., 'Surface observations and strata movement underground,' *Colliery Engng.*, 329–336 (1957).
34. Lee, A.J., 'The effect of faulting on mining subsidence,' *Min. Engr.*, **125**, 417–427 (1966).
35. Institution of Civil Engineers, *Report on mining subsidence*, Inst. Civ. Eng., London (1959).
36. Mather, J.D., Gray, D.A. and Jenkins, D.G., 'The use of tracers to investigate the relationship between mining subsidence and ground water occurrence at Aberfan, South Wales,' *Jour. Hydrology*, **9**, 136–154 (1969).
37. Wardell, K., 'Some observations on the relationship between time and mining subsidence,' *Trans. Inst. Min. Eng.*, **113**, 471–483; Discussion 799–814 (1954).
38. Drent, S., 'Time curves and the thickness of overlying strata,' *Colliery Engng.*, 271–278 (1957).
39. National Coal Board, *Subsidence engineers handbook*, NCB, London (1975).
40. Marr, J.E., 'The application of the zone area system to the prediction of mining subsidence,' *Min. Engr.*, **135**, 53–62 (1975).
41. Jones, C.J.F.P. and Bellamy, J.B., 'Computer prediction of ground movements due to mining subsidence,' *Geotechnique*, **23**, 515–530 (1973).
42. Smedley, N., 'Subsidence management in the North Derbyshire area,' *Min. Engr.*, **136**, 185–194 (1977).
43. Pryke, J.S.S., 'Eliminating the effects of subsidence,' *Colliery Engr.*, 501–508 (1954).
44. Ministry of Works, *Mining subsidence effects on small houses*, National Building Studies, Spec. Rept. 12, HMSO, London (1951).
45. Wardell, K., 'The protection of structures against subsidence,' *Jour. R.I.C.S.*, **90**, 573–579 (1958).
46. Mauntner, K.W., 'Structures in areas subjected to mining subsidence,' *Proc. 2nd Int. Conf. Soil Mech. Found. Eng., Rotterdam*, **1**, 167–177 (1948).
47. Lehmann, K., *Planmassige abbanfuhrung*, Gluckauf (1938).

48. Grond, G.J.A., 'Disturbance of coal measures strata due to mining,' *Iron and Coal Trades Review*, (1950).
49. Wardell, K. and Beevers, C., 'Recent research in mining subsidence,' *Trans. Inst. Min. Engrs.*, **114**, 233–244 (1954).
50. Wardell, K., 'The minimisation of surface damage,' *Colliery Engng.*, 361–367 (1957).
51. Vandale, A.E., 'Subsidence – a real or imaginary problem?' *Mining Engineering*, **19**, 86–88 (1967).
52. Maxwell, G.M., 'Old mine shafts and their location by geophysical surveying,' *Quart. Jour. Eng. Geol.*, **9**, 283–290 (1976).
53. Raybould, D.R. and Price, D.G., 'The use of the proton magnetometer in engineering geological investigations,' *Proc. 1st Int. Cong. Rock Mech., Lisbon,* **1**, 11–14 (1966).
54. McCann, D.M., Grainger, P. and McCann, C., 'Inter-borehole acoustic measurements and their use in engineering geology,' *Geophys. Pros.*, **23**, 50–57 (1975).
55. Howell, M. and Amos, G.W., 'Improved geophysical techniques for survey of disturbed ground,' (article in *Site Investigations in Areas of Mining Subsidence, Ed. by Bell, F.G.*) Newnes-Butterworths, London, 103–108 (1975).
56. Dean, J.W., 'Old mine shafts and their hazard,' *Min. Engr.*, **126**, 368–377 (1967).
57. Sherlock, R.L., *Rock salt and brine*, Mem. Geol. Surv. UK (1921).
58. Pugh, W., 'Triassic salt: Discoveries in the Cheshire-Shropshire basin,' *Nature*, **187**, 278–279 (1960).
59. Bell, F.G., 'Salt and subsidence in Cheshire, England,' *Eng. Geol.*, **9**, 237–247 (1975).
60. Collins, J.F.N., *Salt: A policy for the control of salt extraction in Cheshire*, Cheshire County Council Report (1971).

Chapter 13

Foundations and Earth Movements

INTRODUCTION

Earth movements which may affect foundations vary from slow, almost imperceptible creep to the sudden and dramatic effects of earthquake. There are so many types and causes of earth movements that it would be impracticable to discuss all in one chapter. Consequently this chapter concentrates chiefly on those mass movements which come under the heading 'tectonic adjustment'.

Geological processes are incessant. Consolidation of sediments and other rock forming agencies proceed simultaneously with weathering and erosion; and in the geological sense, no part of the Earth's surface is static; at the same time there are crustal movements. Every structure is subject to the influence of these factors to some degree, perhaps negligible, but sometimes a calculable certainty, or sometimes a risk difficult to define.

CREEP OF ROCK MASSES

As a result of tectonic movements and of rock degrading processes all rock masses sufficiently near the surface to be a foundation support contain joint systems. These processes also loosen the joints, which often become filled with clay, silt or sand. The tensile strength of the mass may be so impaired that gravity forces are sufficient to cause deformation at sloping faces, particularly the steep faces of an escarpment or in a gorge; often there is a continuing and relentless movement of the rock face, away from the rock mass, and with settlement. This slow gravity movement, referred to as creep, usually moves at rates up to 1 m per year, and affects both soil and weathered rock. Such gradual downhill movement leads to the accumulation of material at lower levels, particularly behind any obstacles in its path. A much higher component of horizontal movement is apparent compared with vertical.

The movement can be affected by diurnal temperature variations, or by cycles of freezing and thawing, and by seasonal rainfall, with exceptionally wet conditions causing marked acceleration. One example is the slow movement observed in the foothills of the Alps where vertical and lateral displacements of up to 1.4 m in 50 years have been recorded (see Kobold[1]). The influence and effects of mass movement must be investigated by boreholes and shafts. Inclinometers in the boreholes, pendulums and inverted pensulums in the shafts, will show the variation of movement with depth.

The effect of creep on structures depends very much on the size of the structure. It is impractical to stop the creep except by large expensive works. The general structure must therefore be designed to move. A small, independent building may perform satisfactorily throughout its useful life; a larger one may be subject to differential foundation movements which can be minimised by anchorage to rock at depth which moves less. A most difficult case is that of a tunnel portal which may be designed as a competent structure in itself, but which will inevitably move away from the tunnel it serves, and this must be taken into account especially if the tunnel carries water. A typical example is the portal to the main road access tunnel to Kariba underground power station. The separation between tunnel and portal has increased at a rate of 1 to 2 mm per year.

This same gorge side provides an excellent comparison between alternate designs of retaining wall in these circumstances. The walls are alongside the same road to the power station access, at the foot of the creeping section of the gorge. The first wall near the portal, was constructed early on when the access to the portal was first made. It consists of a near vertical reinforced concrete wall founded on rock with buttresses between the excavated rock face and the wall, the spaces between the buttresses being filled with rock debris. When a rock fall occurred just downstream of the wall, an extension was built. The profile was generally similar to the first section, but anchor bars, 6.4 m long, were provided through the buttresses, and were grouted into sound rock. The space between buttresses was filled with rock debris.

Records of movement have been kept at points at the top of each wall and within a metre of each other. The first wall has moved horizontally at a maximum rate of 20 mm per year, the second moving 10 mm per year. This indicates that the first wall moves more or less with the face of the rock. The second reflects the reduced movement at the depth of the anchor fixings. The natural rock face 20 m above the level of the top of the walls has moved horizontally at rates up to 60 mm per year.

It is clear that unless action is taken in time, the rock face where the bulge occurs will become steep to the point of instability and rock falls will occur. In the case quoted excavation of the more weathered surface material above the walls has recently been carried out, together with the installation of rock anchors. Where there is a slow creep, it must be accommodated. Structures of large area must be designed for differential settlements and footings tied or jointed to avoid or accommodate differential movements. Above all, the works must be designed to improve, and must not impair, the stability of the slope. The distribution of weight (i.e. reduction at higher levels and increase at lower levels), prevention of ingress of water, and the provision of drainage can do much to ensure a factor of safety for the future which is greater than for the past.

LANDSLIDES

Landsliding is one of the most effective and widespread mechanisms by which landscape is developed. It is of great interest to the engineer since an understanding of the causes of landsliding help provide answers relating to the control of slopes, either natural or man-made. An engineer faced with a landslide is primarily interested in curing the harmful effects of the slide. In many instances the principal cause cannot be removed so that it may be more economical to alleviate the effects continually. Indeed, in most landslides, a number of causes contributed towards movement and any attempt to decide which one finally produced the failure is not only difficult but incorrect. Often the final factor is nothing more than a trigger mechanism that set in motion a mass that was already on the verge of failure.

Varnes[2] defined landslides as the downward and outward movement of slope forming materials composed of natural rock, soils, or artificial fills. Movement may take place by falling, sliding or flowing or by some combination of these factors. Obviously this movement involves the development of a slip surface between the separating and remaining mass.

The common force tending to generate movements on slopes is, of course, gravity. Indeed the majority of stresses found in most slopes are the sum of the gravitational stress from the self weight of the material plus the residual stress. Over and above this a number of causes of landslides can be recognised. These were grouped into two categories by Terzaghi[3], namely, internal causes and external causes. The former included those mechanisms within the mass which brought about a reduction of its shear strength to a point below the stress due to the external forces imposed on the mass by its environment, thus inducing failure. External mechanisms were those outside the mass involved, which were responsible for overcoming its internal shear strength, thereby causing it to fail.

Internal slides are generally caused by an increase of pore pressures within the slope material which causes a reduction in the effective shear strength. An increase in water content also means an increase in the weight of the slope material or its bulk density, which can induce slope failure. Significant volume changes may occur in some materials, notably clays, on wetting and drying out. Not only does this weaken the clay by, for example, developing desiccation cracks within it, but the enclosing strata may also be adversely affected. Seepage forces within a granular soil can produce a reduction in strength by reducing the number of contacts between the grains. Water can also weaken slope material by causing minerals to alter or by solution.

An increase in the weight of slope material leads to a decrease in the stability of a slope which may ultimately give rise to a slide. This can be brought about by natural or artificial (man-made) processes. For example, removal of support from the toe of a slope either by erosion or excavation is a frequent cause of slides. Such a slide is an external slide in that an external factor caused failure. Other external mechanisms include earthquakes or other shocks and vibrations, particularly where slope material approximates to its angle of rest. Overloading the top of a slope can lead to its failure as can deafforestation.

Landslides have commonly been classified according to the type of movement undergone and the type of materials involved. Types of movement are generally grouped into falls, slides and flows; one can, of course, merge into another. The materials concerned can simply be grouped as rocks and soils (Figure 13.1).

Rock falls are very common. In such cases the moving mass travels mostly through the air by free fall, saltation or rolling, with little or no interaction between the moving fragments. Movements are very rapid and may or may not be preceded by minor movements. In rock falls the fragments are of various sizes and are generally broken in the fall. They accumulate at the bottom of a slope as a scree deposit. Toppling failure is a special type of rock fall which can involve considerable volumes of rock.

TYPE OF MOVEMENT	TYPE OF MATERIAL			
	BEDROCK		SOILS	
	ROCKFALL		SOILFALL	
SLIDES: Few units	rotational SLUMP	Planar BLOCK GLIDE	Planar BLOCK GLIDE	Rotational BLOCK SLUMP
SLIDES: Many units		ROCKSLIDE	DEBRIS SLIDE	FAILURE BY LATERAL SPREADING
FLOWS	ALL UNCONSOLIDATED			
	ROCK FRAGMENTS	SAND OR SILT	MIXED	MOSTLY PLASTIC
Dry	ROCK FRAGMENT FLOW	SAND RUN; LOESS FLOW		
		RAPID EARTHFLOW	DEBRIS AVALANCHE	SLOW EARTHFLOW
Wet		SAND OR SILT FLOW	DEBRIS FLOW	MUDFLOW
COMPLEX	COMBINATIONS OF MATERIALS OR TYPE OF MOVEMENT			

Figure 13.1 Classification of landslides (After Varnes[2])

In true slides the movement results from shear failure along one or several surfaces, this surface offering the least resistance to movement. The mass involved may or may not experience considerable deformation. One of the most common types of slide occurs in cohesive material where the slip surface is approximately spoon-shaped. Such slides are referred to as rotational slides. Rotational slides usually show tension scars in the upper part of a slope and the movement is more or less rotational about an axis located above the slope. When the scar at the head of a rotational slide is almost vertical and unsupported, then further failure is usually just a matter of time. As a consequence successive rotational slides occur until the slope is stabilised.

Translation slides occur in inclined stratified deposits, the movement occurring along a planar surface, frequently a bedding plane. The mass involved in the movement becomes dislodged because the force of gravity overcomes the frictional resistance along the potential slip surface, it having been detached from the parent rock by a prominent discontinuity such as a major joint.

Rock slides and debris slides are usually the result of a gradual weakening of the bonds within a rock mass and are generally translational in character. Most rock slides are controlled by the discontinuity patterns within the parent rock.

In a flow the movement resembles that of a viscous fluid. In other words slip surfaces are usually not visible or are short lived and the boundary between the flow and the material over which it moves may be sharp or may be represented by a zone of plastic flow. Some content of water is necessary for most types of flow movement but dry flows can occur. Debris flows are distinguished from mud flows on a basis of particle size, the former containing a high percentage of coarse fragments whilst the latter consist of at least 50% sand/silt/clay size material. Almost invariably debris flows follow unusually heavy rainfall or sudden thaw of frozen ground. These flows are of high density, perhaps 60 to 70% solids by weight, and are capable of carrying large boulders.

Mud flows may develop when a rapidly moving stream of storm water mixes with a sufficient quantity of debris to form a pasty mass.

An earth flow involves mostly cohesive or fine grained material which may move slowly or rapidly. The speed of movement is to some extent dependent on water content in that the higher the content, the faster the movement. Slowly moving earth flows may continue to move for several years. They are known as mudslides, and the movement is intermittent, totally some 5 to 25 m per year. If the material is saturated, a bulging frontal lobe is formed and this may split into a number of tongues which advance with a steady rolling motion. The term mudspate is used to describe the rapid transport of material which has a high water/soil ratio – the velocity may be greater than 1 m per second. Earth flows frequently form the spreading toes of rotational slides (Figure 13.2).

Figure 13.2 An earth flow forming the spreading toe of a rotational slide in deposits along the coast at Filey Bay

If landslides are to be prevented then areas of potential landsliding must first be identified, as must be their type and possible amount of movement. Then, if the hazard is sufficiently real, the engineer can devise a method of preventative treatment. Thus in any area of inherently low stability, especially where slides have occurred, a thorough investigation is called for before construction commences.

According to Root[4] landslide prevention may be brought about either by reducing the activating forces, or by increasing the forces resisting movement or by avoiding or eliminating the slide. In the first case reduction of the activating forces can be accomplished by removing material from that part of the slide which provides the force which will give rise to movement and, in the case of soils, by drainage which reduces pore pressures and the bulk density. Drainage also brings about an increase in shearing resistance. Indeed drainage is the most generally applicable preventative and corrective treatment for slides, regardless of type.

Restraining structures control landslides by increasing the resistance to movement. They include retaining walls, cribs, gabions, buttresses, piling and rock bolts. The following minimum information is required to determine the type and size of a restraining structure:

1. The boundaries and depth of the unstable area, its moisture content and its relative stability. For example, excessive pore pressures are likely to give difficulties in designing retaining walls.
2. The type of slide which is likely to develop or has occurred.
3. The foundation conditions since restraining structures require a satisfactory anchorage.

There are certain limitations which must be considered before retaining walls, cribs or piles are used as a method of landslide control. These involve the ability of the structure to resist shearing action, overturning and sliding on or below the base of the structure. Retaining structures should be designed for a predetermined load which they are to transmit to the foundation of known bearing capacity (see CP2:1951[4a] and Reimbert[4b]). They should include adequate provision for drainage.

EARTHQUAKE AND ITS ORIGIN

The advent of television has enabled us to see the effects of earthquake. The devastation has to be seen to be believed, and only those who share it can really appreciate the misery and suffering from injuries and loss of life, from destruction of property and loss of livelihood. But the engineer knows that most of the destruction, loss and injury could be avoided by suitable foundation and structure design.

Although earthquakes have been reported from all parts of the world they are primarily associated with areas of recent mountain building and with the global rift system, in other words with plate boundaries (Figure 13.3). The severity of ground motion at a particular point on the Earth's surface depends upon the total energy released in the form of seismic waves, modified somewhat by the exact mode and duration of the energy release, the distance from the source of the earthquake and the geological conditions. The principal motion of the ground is due to the shear component of the seismic wave. These motions, rather than the compressional waves, have greater amplitudes and thus cause the principal damage from ground shaking in earthquakes. In very small motions compressional waves frequently mask the slower travelling shear waves.

The focus of an earthquake is the name given to the location where it originates whilst the epicentre is located on the surface of the Earth immediately above the focus. Shock waves radiate from the focus in all directions. Earthquake foci are confined within a limited zone of the Earth, the lower boundary of which, occurs at approximately 700 km, and they rarely originate at the Earth's surface. In fact most earthquakes originate within the upper 25 km of the Earth. Because of its significance the depth of foci has been used as the basis of a threefold classification of earthquakes, those occurring within the upper 70 km are referred to as shallow, those located between 70 and 300 km as intermediate, and those between 300 and 700 km as deep.

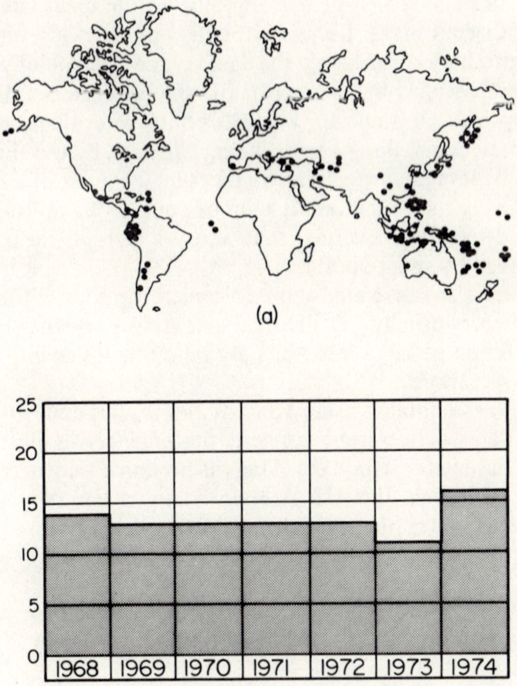

Figure 13.3 (a) Major earthquakes, magnitude 7.0 and above which have occurred throughout the world. (b) Frequency of major earthquakes of the world. Reported by the Center for Short-lived Phenomena. Both diagrams refer to the period 1968 to 1974

Large earthquakes cause shock waves which travel throughout the world. Near the epicentre, the shocks are only felt for a matter of seconds or minutes but with increasing distance from the epicentre the disturbance lasts for a progressively longer time. The vibrations of a shallow earthquake decrease rapidly in intensity away from the epicentre. Although the shocks of deep seated earthquakes are usually weakly felt when they reach the surface, they extend over a much wider area. The principal shocks of an earthquake is often preceded by foreshocks and followed by thousands of aftershocks.

It is assumed that the cause of most earthquakes is associated with Earth movements, in particular, faulting. However, Albee and Smith[5] noted that although the spatial association of larger earthquakes with faults suggests that strain accumulated along the faults, the spatial association of smaller earthquakes does not necessarily lead to the same conclusion, for example, smaller earthquakes occur almost randomly throughout southern California. Be that as it may, it has been maintained that initial movement may occur over a small area of a fault plane, being later followed by a slippage over a much larger surface. Such initial movements, which represent preliminary shattering of small obstructions along a fault zone, account for the foreshocks which precede an earthquake. When these have been overcome the main movement occurs but complete stability is not restored immediately.

The shift of strata involved in faulting relieves the main stress but develops new stresses in adjacent areas. Because stress is not relieved evenly everywhere, minor adjustments may arise along the fault plane and thereby generate aftershocks. The decrease in the strength of the aftershocks is irregular and occasionally they may continue for a year or more after the principal shocks. A second rupture may occur along a fault, being triggered by the first. Indeed there seems reason to believe that great earthquakes may consist of a rapid succession of such breaks, giving the overall impression of continuous rupturing. Some areas are affected by earthquake swarms, that is, by a series of nearly equally large shocks accompanied by small ones.

Earthquakes are also associated with volcanic eruptions, although the effects of such quakes are usually localised. Furthermore earthquakes have been induced by the filling of large reservoirs, by pumping fluids into the ground and by large mining operations.

The Earth's crust contains a network of faults, but only those regarded as active or potentially active need concern the engineer as possible sources of natural (i.e. not induced) earthquake. There is no universally accepted definition of active fault, although the US Atomic Energy Commission[6] has defined capable (or active) fault as one which has exhibited one or more of the following characteristics:

1. Movement at or near the ground surface at least once within the past 35 000 years, or movement of a recurring nature within the past 500 000 years.
2. Macroseismicity instrumentally determined with records of sufficient precision to demonstrate a direct relationship with the fault.
3. A structural relationship to an (active) fault according to characteristics 1 and 2 above such that movement on one could be reasonably expected to be accompanied by movement on the other.

In earthquake regions it is necessary for the engineer to establish the nature of the risk to new structures. Thus he has to assess the probability of earthquake occurrence, their intensity and magnitude, and the likelihood of earthquake damage to the structure concerned. As soon as the intensity detrimental to the planned structure is established, the probability of an earthquake of this intensity in the given region should be estimated.

The intensity of an earthquake can be measured by the amount of damage that it does. The severest earthquakes wreak destruction over areas of 2500 km^2 or more, however, most only affect tens of square kilometres. Earthquake intensity scales depend on human perceptability as well as the destructivity of the earthquake. Whereas the degree of damage may be estimated correctly and objectively, the perceptibility of an earthquake depends on the location of the observer and his sensibility. Observers at rest, or on the top floor of a high building or on loose ground feel slight earthquakes which may not be perceived by others.

The first earthquake intensity scale was formulated by Rossi in 1878. Ten years later this scale was revised by Mercalli and in 1904 Cancani attempted to give the scale a quantitative basis by defining the acceleration produced by the shocks. An elaboration of the Mercalli scale was published by Sieberg[7] in 1923. This form was in turn used as the basis for the Modified Mercalli Scale of 1931, commonly abbreviated MMO by Wood and Neumann[8], and was further slightly modified by Richter[9] in 1956 (see Table 13.1). This scale does not distinguish long period and short period effects. The latter are the more numerous and may be roughly correlated with acceleration. Long period effects represent large displacements which often go with comparatively moderate acceleration.

As noted above the intensity of an earthquake is characterised by its effects and is a qualitative concept, whereas its magnitude is an instrumentally measured quantity related to the total energy released during the earthquake. The magnitude assigned to a given earthquake therefore corresponds to the highest intensity of that earthquake. For example, a catastrophic earthquake spreads outwards from intensity 12 through all the lower intensities, but it has only one magnitude, which refers to the total energy set free by the shock. In physical terms intensity is determined partly by the duration and number of tremors, but mainly by the maximum rate of change of these movements of the ground, that is, by its maximum acceleration, which can be estimated from seismographic records.

In 1935 Richter[10] devised a logarithmic scale for comparing the magnitudes of Californian earthquakes. His method has since been widely extended and developed. He related the magnitude of a tectonic earthquake to the total amount of elastic energy released when overstrained rocks suddenly rebound and so generate shock waves, the relationship being given by the expression:

$$\log E = a + bM$$

where M is the magnitude and E the energy released expressed in ergs.

The values of the constants a and b have been modified several times as data has accumulated. One recently adopted form of the equation is:

$$\log E = 11.4 + 1.5M.$$

For example, for a magnitude of 8.6, only three times reached and twice exceeded during the present century, E amounts to $10^{24.3}$ ergs and an increase in magnitude of 2 represents at least a thousand fold increase in energy release. The largest earthquakes have had a magnitude of 8.9, and these release about 700 000 times as much energy as the earthquake at the threshold of damage. The average annual release of energy from all earthquakes ranges from about 10^{25} to 10^{27} ergs and about 80% or more is generally accounted for by a few really major shocks.

Table 13.1 MODIFIED MERCALLI SCALE (1956 VERSION) WITH CANCANI'S EQUIVALENT ACCELERATION

Degrees	*Description*	*Acceleration* mm/sec^2
I	Not felt. Only detected by seismographs	Less than 2.5
II	Feeble. Felt by persons at rest, on upper floors, or favourably placed.	2.5 to 5.0
III	Slightly felt indoors. Hanging objects swing. Vibration like passing of light trucks. Duration estimated. May not be recognised as earthquake.	5.0 to 10
IV	Moderate. Hanging objects swing. Vibration like passing of heavy trucks, or sensation of a jolt like a heavy ball striking the walls. Standing motor cars rock. Windows, dishes, doors rattle. Glasses clink. Crockery clashes. In the upper range of IV wooden walls and frame creak.	10 to 25
V	Rather strong. Felt outdoors, direction estimated. Sleepers wakened. Liquids disturbed, some spilled. Small unstable objects displaced or upset. Doors swing, close, open. Shutters and pictures move. Pendulum clocks stop, start, change rate.	25 to 50
VI	Strong. Felt by all. Many frightened and run outdoors. Persons walk unsteadily. Windows, dishes, glassware broken. Ornaments books, etc, fall of shelves. Pictures fall off walls. Furniture moved or overturned. Weak plaster and masonry cracked. Small bells ring (church, school). Trees, bushes shaken visibly or heard to rustle.	50 to 100
VII	Very strong. Difficult to stand. Noticed by drivers of motor cars. Hanging objects quiver. Furniture broken. Damage to masonry D, including cracks. Weak chimneys broken at roof line. Fall of plaster, loose bricks, stones, tiles, cornices, also unbraced parapets and architectural ornaments. Some cracks in masonry C. Waves on ponds, water turbid with mud. Small slides and caving in along sand or gravel banks. Large bells ring. Concrete irrigation ditches damaged.	100 to 250
VIII	Destructive. Steering of motor cars affected. Damage to masonry C, partial collapse. Some damage to masonry B, none to masonry A. Fall of stucco and some masonry walls. Twisting, fall of chimneys, factory stacks, monuments, towers, elevated tanks. Frame houses moved on foundations if not bolted down, loose panel walls thrown out. Decayed piling broken off. Branches broken from trees. Changes in flow or temperature of springs and wells. Cracks in wet ground and on steep slopes.	250 to 500
IX	Ruinous. General panic. Masonry D destroyed, masonry C heavily damaged, sometimes with complete collapse, masonry B seriously damaged. General damage to foundations. Frame structures, if not bolted, shifted off foundations. Frames cracked, serious damage to reservoirs. Underground pipes broken. Conspicuous cracks in ground. In alluviated areas sand and mud ejected, earthquake fountains, sand craters.	500 to 1000
X	Disastrous. Most masonry and frame structures destroyed with their foundations. Some well built wooden structures and bridges destroyed. Serious damage to dams, dikes, embankments. Large landslides. Water thrown on banks of canals, rivers, lakes, etc. Sand and mud shifted horizontally on beaches and flat land. Rails bent slightly.	1000 to 2500

Table 13.1 *(continued)*

Degrees	*Description*	*Acceleration* mm/sec^2
XI	Very disastrous. Rails bent greatly. Underground pipelines completely out of service.	2500 to 5000
XII	Catastrophic. Damage nearly total. Large rock masses displaced. Lines of sight and level distorted. Objects thrown into the air.	Over 5000

The engineer's problems associated with earthquake can be considered under three headings:

1. The continuing creep which takes place at active faults representing crustal adjustment usually leading to overstress in a section of the fault which then generates earthquake when sudden failure occurs.
2. The shear, thrust and tension movements which accompany the earthquake and which are seen where the fault breaks the surface.
3. The shaking resulting from seismic waves released from an earthquake, and its secondary effects, i.e., landslide, lurching (see under 'Secondary effects of earthquake' below), ground cracking and effects on underground water and drainage.

PRIMARY EFFECTS OF EARTHQUAKE

The shear, thrust and tension movements which accompany earthquake and the continuing creep at faults may be considered as the primary effects of earthquake.

In some parts of the world it is impossible to avoid building near, or even across potentially active faults. Therefore estimates of the amount and nature of earth movements which might occur in the lifetime of the structures involved must be made.

As yet no accurate method of earthquake prediction has been developed. Most attempts at prediction aim first at measuring the accumulation of strain within the Earth's crust and then at determining the instant at which the fault will begin to slip. According to Albee and Smith[5] an evaluation of the probability of faulting and the occurrence of earthquakes of various magnitudes in southern California can be made by noting the lengths of continuous, potentially active fault segments and the degree of activity. Evidence for strain accumulation along a fault zone can be inferred from previous earthquakes, recent surface faulting, strain measurements and regional geological relationships. Assuming that significant strain is accumulating along a fault zone, the question of when the strain will be released can be expressed only in terms of the probability of such an event occurring in a given time interval. Since the probability of faulting in an active system must increase progressively from one large earthquake toward the next, then the longer the time since the last movement, the more imminent is the next movement. This does not apply to a fault which is not active, hence it is important to determine at the onset whether or not a fault is active.

Rock bending at faults means that strain is accumulating in the rocks, possibly foreshadowing an earthquake and sudden fault displacement. Such tectonic movement can be measured with a geodimeter, by triangulation, by fault movement quadrilaterals and by tiltmeters. During the early 1960's several earthquakes in the range *M* 4–5 occurred in southern California in the vicinity of the geodimeter system set up by the Department of Water Resources[11] to monitor fault movements. Some of these events were preceded by anomalously changed distances between stations near the earthquake producing fault zone; the observed irregularities were relatively large. This, plus the coincidence of these anomalies with earthquake activity, led to the conclusion that an earthquake precursor may be observable. The precursors were observed 2 to 18 months prior to earthquakes. It therefore appears that sudden strain release may be more likely along a segment of a fault where the rate of movement differs from that of adjacent areas or from its past rate and it is this which may aid earthquake prediction.

The determination of earth movements is a subject for continued research, particularly field work to provide a basis for, and to check the theoretical estimates. An estimate of continuous creep is best based on actual measurements across the fault. For example, Ambraseys[12] reported that creep along the North Anatolian fault, Turkey, is progressing at the rate of 10 to 20 mm per year. At site it is first necessary to determine the width over which the movement is concentrated. This is best seen by geological examination in an adit crossing the fault and may also be an ideal location for the installation of studs for measuring creep, both horizontal and vertical movements being recorded.

The fault displacement due to an earthquake event is theoretically calculable from the source parameters derived from a distant seismic record. This can then be integrated over all events assumed to occur over the lifetime of the structure. Such estimates are likely to be far from accurate but they nevertheless give an indication of the order of magnitude of expected movement. These estimates must clearly be based on studies made jointly by the engineer, geologist and seismologist working together as a team.

At present, for most purposes the estimate is subject to engineering judgement and the provision in design is probably based on what it is reasonably possible to provide. Two cases are quoted below:

1. No suitable alternative site could be found when the Palmdale Dam was reconstructed across the San Andreas fault, California, in 1969 (see Sherard *et al*[13]). The dam is built across an active fault at which it was estimated that relative displacement of up to 6.2 m horizontal and 1.0 m vertical might occur in the life time of the dam. The design of the earth dam is conservative with interior zones wide enough to remain intact after being displaced by relative movements of 9.5 m horizontally and 1.6 m vertically. 'Self-healing' materials and proportions are specified in the design of the dam.

2. The Lar scheme now under construction in Iran includes an embankment dam 105 m high, a tunnel 20 km long, aqueducts 10 km long, hydroelectric power stations and other associated structures. Investigations of seismic activity have been in hand since 1972 including measurements from two sensitive seismometers installed in the site area. Many local events of magnitudes 1 to 3 have been recorded in this period, and it is concluded that every fault in the area must be considered potentially active.

The underground penstock leading to Kalan power station crosses the Mingun Musha fault. Tunnel excavation shows that the fault zone is less than 200 mm wide, containing calcite and highly altered green and red mudstones, but over a width of some 30 m the mudstone on one side and the sandstone on the other are highly sheared with dominant discontinuities sub-parallel to the fault. The tunnel will be 5.6 m internal diameter; concrete lined and steel sets have been provided for the immediate support of the rock. The penstock laid within this tunnel is a steel tube of 2.1 m internal diameter. Pivotted supports allow for angular movement and a series of expansion joints allow for change of length. These measures will provide for fault creep and for the small movements which would result from the local activity which is probable during the life of the structure. In the unlikely but possible event of a large earthquake causing major dislocation at the fault, the automatic valve provided at the head of the penstock will immediately close if the penstock breaks. Room and access are available for rapid repair of the pipe and the tunnel lining.

SECONDARY EFFECTS OF EARTHQUAKE

The most important aspect of earthquake is the shaking and any part of the Earth's surface may be affected and unfortunately large areas of the world are prone to severe shaking. It is the shaking and the secondary effects of shaking which cause nearly all the damage associated with any large event. The secondary effects include:

1. Landslide initiated where the ground slope or an old slide exists with a low factor of safety against slip.
2. Cracking of the ground, in loose alluvial and volcanic deposits.
3. Lurching, which is the cracking of the ground above or in slopes, and which is parallel to the escarpment or along the contours.

The engineer should avoid building structures in these sensitive situations which may be identified by thorough site investigations. It is more difficult to identify in advance, locations where groundwater and drainage problems may occur, but they will be associated with a high water table.

The energy of the earthquake is transmitted through the ground in the form of waves. At any location, seismometers will record waves in two horizontal directions at right angles to each other and in the vertical direction. The building and its foundation should therefore be designed to withstand such motions.

The foundation of the structure will move with the ground on which it is built. Whether it would be practical in special cases to isolate the building from the ground motions is a subject for study, but this would not necessarily simplify the foundation design. Apart from this hypothetical case, the structure above will respond to the shaking according to its several natural modes of vibration, and will oscillate in a manner which bears little resemblance to the ground vibrations. The foundation structure therefore acts as a buffer between opposing waveform forces. There is an added complication, that the foundation will tend to put back into the ground some of its own response to the motion of the structure above, and furthermore the interaction between foundation and ground results in radiation and dispersion of energy.

Ground motions produced by earthquakes enter a building via its foundation structure and during ground motion the foundation may be displaced horizontally and rotated at its base. According to Zeevaart[14], in a building with a raft foundation it may be assumed that ground motion is transmitted to the building by shear forces developed at the interface between foundation slab and soil, whereas if the foundation structures consist of shallow piles they take the shear force. The seismic response shear is transmitted by the passive earth pressure against the retaining walls in the case of pile foundations with basements. As remarked above the response forces in the building are related to the behaviour of the foundation. Rigid foundation structures are generally used in seismic areas for heavy and slender buildings.

In order to estimate the response of a foundation and building to strong ground motion it is necessary to know the free periods of vibration of the structure. Of particular significance is the rocking of a foundation due to overturning movements generated by the horizontal inertia forces in the building caused by the earthquake. Rocking of a foundation increases contact stresses in the soil at foundation grade level. It also increases the period of vibration of the structure. The stress increment has to be investigated to ensure that it does not exceed the elastic response behaviour of the soil, otherwise tilting will occur.

PSEUDO-STATIC METHOD OF DESIGN

The maximum violence in a destructive earthquake is generally reached within 10 sec of the first tremor although substantial motions sometimes continue for 70 sec or more. Except for the effects of fires, considerably more than half the property loss results from the failure of structures so weak that they should never have been erected (Figure 13.4). Some spectacular failures of old buildings in recent earthquakes are attributable to their progressive weakening by successive minor earthquakes. Buildings of frame construction fail because of lack of proper bracing, or inadequate joints; those of brick because of poor mortar, brick or connections. Cracking of walls and plaster is common. Chimneys frequently break at the roof line. Sometimes houses are thrown off their foundations.

Prevention of this type of damage is not difficult but unfortunately involves the use of untraditional materials. Largely on the basis of statistics and experience, building codes have been, and are still being, developed that may eventually solve this phase of the problem, hopefully without undue economic strain (see Nichols[15]). Usually, however, well built, simple structures can come through an earthquake unscathed. A reinforced concrete or steel framed building possesses both strength and flexibility and, although there may be superficial damage to partitions and curtain walls, if well constructed it usually remains substantially sound after even a large earthquake. For example, after the San Francisco earthquake of 1906 well-designed office buildings and factories remained in good condition (see Steinbrugge[16]). It can therefore be concluded that the object lesson to be learnt from destroyed and damaged buildings is, wherever possible, to avoid similar construction in the future.

The basis of the earthquake provisions in most building codes arose out of the intensive research done in Japan in the early part of this century, which introduced the concept of a seismic factor into building design. This means that the building is made strong enough to resist a horizontal force equal to a certain proportion of the weight of the building. One of the essential requirements is that at any floor level a building should be strong enough to withstand a horizontal force equal to 10% of the load above that level. This is generally referred to as the 0.1 g (gravity) formula. Depending largely on the seismic risk involved this value for structures or their component parts can be modified. While the

Figure 13.4 Earthquake damage at the Saada hotel, Turkey (American Iron and Steel Institute)

Japanese have found that buildings up to 30 m in height, designed to a seismic factor of 1/10, can be safely built even on soft ground, high rigidity is necessary to keep the building period below the dangerous range. This has generally involved an increase in costs of 10 to 15% (see 'Response spectrum method' below).

However, although experience has shown the above method to be a practical approach, Neumann[17] maintained that for certain types of structures it does not provide a satisfactory solution, and for many structures, especially high ones, a more scientific approach would result in greater economy in construction costs without loss of building strength. Generally a dynamic analysis will demonstrate that this pseudo-static method of design for earthquake is incorrect and that the horizontal force involved could be much greater.

Design regulations sometimes take partial account of this by providing for an increase in seismic coefficient with height of building. For example, in the USSR, the basic seismic coefficient for horizontal force applies to all parts of the structure up to ground level. The value of the coefficient is laid down in the regulations and depends on, firstly, site seismicity as shown on seismic zoning maps and, secondly, the economic value and consequential risk associated with the building. The regulations then provide for a linear increase in coefficient with height above ground, the mass at each level being taken into account according to its level (see Reference 18).

FULL DYNAMIC ANALYSIS

Advances in design methods making use of the computer, enables a full dynamic analysis. Theoretically, any specified ground motion can be applied to the structure in a step-by-step process in which the motion is digitised to give successive acceleration/time pairs, but this assumes that a time history of the ground motion can be specified. Typical records of earthquake show ground motions lasting from 2 to 70 sec. Experience has shown that the time interval for successive steps in the calculation should be about 0.01 sec and it will therefore be appreciated that the calculation is long and involved. The use of such a complex procedure for a range of possible ground motions would be impractical in most cases.

Research carried out at the Building Research Institute in Tokyo has involved the systematic analysis of the earthquake response of tall buildings in elastic and elasto-plastic ranges, this having been investigated by computer simulation (see Hisada, Izumi and Hirosawa[19]). The ultimate purpose of such investigations is to develop a design earthquake, i.e. to obtain an estimate of the severest earthquake which is likely to occur within a 50 or 100 year period.

RESPONSE SPECTRUM METHOD

All structures have a natural period at which they tend to vibrate. Hence the behaviour of a structure during an earthquake depends on its natural frequency and damping forces. If the earthquake waves have the same period they will shake a structure severely whereas if the period is very different the effect will be much less severe. In order to establish the earthquake design forces models have been vibrated at different variation periods and the corresponding accelerations determined. Thus an idealised spectrum may be obtained applicable to any structure and any earthquake. Indeed buildings have been vibrated by means of shaking machines to determine their performance. The period itself depends upon the manner in which the weight is distributed in a building and upon the stiffness of the construction materials. The addition or removal of interior partitions or the storage of heavy goods may change the period considerably.

Accordingly the method which, with present knowledge, seems appropriate is the response spectrum approach (see Reference 20). When the ground motions, i.e. acceleration, velocity and displacement are plotted against frequency (or period) it is found that the graphs can be smoothed to reveal approximately constant maximum values over wide ranges of frequency. From these maximum values a logarithmic tripartite chart of ground motion can be plotted. In considering the response of the structure to these motions account must be taken of the effects of damping and ductility, as these factors determine the energy absorption within the structure. This is effected in the design procedure by replotting the tripartite graph according to the amount of damping which is considered as a proportion of the critical damping factor.

The ductility factor is defined as the ratio of the useful displacement of a structure to the elastic limit displacement, and may vary from about 1.3 to 5. The response spectrum tripartite chart which has so far been based on a selected ground motion modified according to the selected damping factor, is then replotted to take account of ductility. In this way, the behaviour of the structural elements, their stresses, strains and displacements can be computed for the assumed earthquake parameters. This method has been used and is recommended for important structures such as nuclear power stations (see Newmark and Hall[21]). In a recent investigation of 11 buildings of varying type the extra costs of applying this design method were as follows:

Per cent increase in structure cost between 0.1% and 9.7%.
Per cent increase in engineering cost between 5% and 30%.

Those increases are the amounts greater than applying the Uniform Building Code.

However, the seismic data for the graphs is frequently inadequate and the method requires estimates of non-linear materials behaviour which are not yet confirmed. Earthquake engineering is a discipline in its infancy and therefore continual change can be expected in the next decades.

SITE SEISMICITY

Whatever design method is used, the parameters of a design earthquake must be stated which will apply at the site. A probability approach is adopted, and for structural design, two intensities may be specified, namely, the collapse threshold earthquake, defined as very low (about 10%) probability of exceedance in the life of the structure, which the structure should be designed to resist without collapse, and the damage threshold earthquake defined as moderately low probability (about 50%) of exceedance, which the structure should be designed to resist without significant structural damage (Note that the life of a structure is often assumed to be 70 years).

The estimation of site seismicity is a job for the expert geologist and seismologist. A seismic investigation prior to the design of a structure may provide the designer with some idea of the frequency and size of future earthquakes, as well as the spectrum of anticipated earthquake motion. This spectrum is an envelope of maximum motions at each specific period of frequency at each particular location. Such investigations include a survey of the geology and regional tectonics, with particular attention being paid to faulting over a wide area around the site.

As far as a seismic survey is concerned, teleseismic events are plotted from data available from the International Seismological Centre in Edinburgh or other institutions. In active zones this may be supplemented by records from a local array of seismometers. Reliable instrumental records of strong events, magnitude 4.5 and greater, are only available since 1960. Local stations will pick up the smaller events (magnitudes from 1 upwards) which occur with much greater frequency. A further value of the local array of stations is that the records of micro-earthquakes (magnitudes less than 1), although of no use in determining probability of large events, will locate fault activity and enable the plotting of faults which are not visible from the surface.

Determination of the physical properties of the soils and rocks at the site should also be included particularly density, stress-strain relationship and damping or energy-absorbing ability as this will assist in predicting the spectra of ground motions. The problem is not simple however, as next to the earthquake itself both general geological structure and local ground condition control the ground motion. Where depths of material of lower seismic velocity occur over the massive bedrock there will be wave reflections within the upper layers sometimes leading to reverberation (see Long[27]).

There are many snags in estimating site seismicity. A principal difficulty is that activity varies with time. A study of historical events has shown that periods of quiescence are followed by periods of activity. In some cases activity also moves along faults. The theory is that slip of a fault occurs progressively along its length; thus, at a given time, part of the fault will have slipped, another part being unchanged. Between the two the rock is dislocated and the propagation of slip along a fault may be considered as a dislocation flow. The dislocation is associated with a stress field, which, acting with the tectonic stress field, causes the slip movement.

When the total stress field exceeds the frictional stress resistance catastrophic slip failure occurs accompanied by earthquake. This causes further dislocations and so the process goes on. For example, referring again to the Lar scheme in Iran, this is in a notably active area. Research and site investigations have provided evidence of such periodic activity, and it appears that activity waves may cross the site, one per 17 years. It can therefore be concluded that at most sites, and with reliable records only since 1960, a simply probability study could result in a considerable underestimate of seismicity, especially if these recent years have been quiescent.

EFFECTS OF GEOLOGICAL SETTING

The response of structures on different ground materials has proved to be surprisingly varied. For instance, structures not specifically designed for earthquake loadings have generally fared far worse on soft, saturated alluvium than on hard rock. The effect of soils and alluvial deposits above bedrock during an earthquake is to increase the amplitude of the motion through the deposit. For example, the bedrock beneath Tokyo is Pliocene mudstone which is overlain by varying thicknesses of alluvial deposits and soils, which may be classified according to the velocity of shear wave transmission in them. It was found that there was an almost linear relationship between the amplification magnitude and the ratio of bedrock to surface shear wave velocity, the amplitude magnification reaches nearly 12 for a ratio of 7 (see Shima[22]). At a given site the increase in observed surface intensity over that found in the bedrock can be regarded as an intensity anomaly. Hence a history of such anomalies at a particular location should reveal a maximum value which can be used to assess the local seismic factor for use in structural design.

If there is a weak layer, there will be an upper bound to the motion, the amplitudes decreasing, but probably accompanied by settlement. This possibly explains why, for the same earthquake where the maximum acceleration is low (<0.2 g) the measured accelerations are greater on soil than on rock, but when the maximum acceleration is high (>0.2 g) the accelerations on rock appear to be greater than on soil. In most cases the measured velocities are nearly the same.

Loosely packed, saturated sands and silts tend to lose all strength and behave like fluids during strong earthquakes (see Seed[23]). When such materials are subjected to shock, a densification occurs. During the relatively short time of an earthquake, drainage cannot be achieved and such densification leads to excessive pore water pressures being developed, which cause the soil mass to act as a heavy fluid and can result in a complete loss of shear strength. A condition of initial liquefaction is then established. Deformation is probable and continued cyclic loading could result in actual liquefaction. An approximately linear relationship has been shown to occur between the relative density of sands and the stress required to cause initial liquefaction in a given number of stress cycles. Under repeated cycles of loading, clay soils can develop large deformation although their peak strength remains about the same. Nevertheless these deformations can reach a point where, for all practical purposes, the soil has failed.

From the remarks made above it can be concluded that it is possible to establish seismic intensity increments for different categories of ground. Indeed as long ago as 1908 Reid[24] introduced the concept of foundation coefficients for several major rock and soil types. The coefficient for the type of foundation material which produced the least vibrational force, as revealed by observed earthquake damage, was designated unity. Subsequently Medvedev[25] recognised a number of basic types of ground. He regarded granite as the standard type to which all others were referred, granite having a seismic intensity increment of unity. The other basic types, with the seismic intensity increments given in brackets, are sandstones and limestones (0.2 to 1.3), gypsum and marl (0.6 to 1.4), gravel (1 to 1.6), sandy ground (1.2 to 1.8), clayey ground (1.2 to 2.1) and fill (2.3 to 3). This scale emphasizes the fact, already stated, that in earthquake prone areas it is safer to build on a rock foundation and to avoid building different parts of a structure on different kinds of subsoil.

SEISMIC ZONING

In the procedure known as seismic zoning, a map of the country is divided into zones according to the estimated risk or severity of seismic effects. The basis of the zoning depends on the data available and on the purpose for which the map is intended, the maps being perhaps required to enable decisions relating to urban and regional planning to be made, or to give guidance for earthquake resistant design. It must be emphasised that seismic zoning maps generally do not take account of the influence of local soil conditions, nor the engineering problems of foundation-soil interaction.

There is at present no consistency in presentation of seismic zoning maps. Some summarise past earthquake effects only, while others endeavour to predict potential earthquake sources and effects. Some divide the territory into zones of macroseismic intensity, others into zones related to a code of practice for design. There is a need for maps which show quantities directly related to earthquake resistant design, such as peak acceleration, particle velocity, spectral density and period of shaking. These should be based on accurate topographic maps and should show the geology of the area. They should be accompanied by a report giving the data on which they are based, and the probable accuracy.

According to Medvedev[26] seismic zoning of a region in the Soviet Union is based upon a study of earthquakes occurring in the region concerned, an investigation of the laws governing the occurrence of earthquakes of different intensity, an analysis of the geological conditions under which earthquakes occur and the investigation of special features accompanying the occurrence of earthquakes. The production of maps of seismic zoning in the Soviet Union therefore takes account of the depth of earthquake foci, the relationship between energy at the focus and epicentral intensity, the correlation between magnitude and intensity, and information concerning attentuation of intensity with distance.

Specific conditions in each seismic region, peculiarities of the seimsic regime, forms of seismo-geological linking and the extent to which the area has been studied, mean that there can be no rigorously standard way of using the information for zoning in all regions. What is more in the compilation of these maps, all engineering data concerning the surface manifestations of earthquakes must refer to identical ground conditions as they can have a strong influence on intensity.

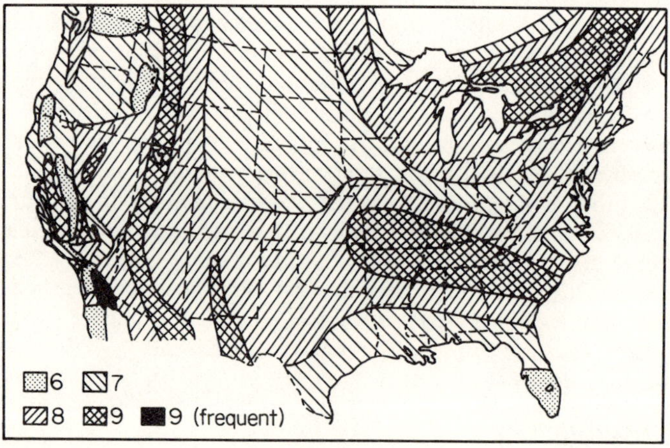

Figure 13.5 Map of the seismic zones of the United States (After Richter[9])

Historical and geological studies were used to distinguish regions at different levels of risk within the USA. A map zoning the USA at four levels was prepared by the US Coast and Geodetic Survey with the cooperation of seismologists in all parts of the country. This map was incorporated in the Uniform Building Code, of which it became legally a part. In 1959 Richter published a zoning map of the USA based on seismo-statistical and geological data. Zoning was according to the modified Mercalli scale. The map distinguishes regions of intensities 6, 7, 8 and 9, and it is assumed that earthquakes with intensities exceeding 9 are possible in the western zone (Figure 13.5). Much of the USA was considered highly seismic, about half occupying zones of intensity 8, and one quarter zones of intensities 9 and 10. However, Richter's map may be regarded as an overestimation. This is especially true of the eastern and central parts of the country, where earthquakes are comparatively rare and less strong than on the Pacific coast.

Seismic micro-zoning is a separate procedure which considers seismic effects at the scale of individual urban or industrial developments. The earthquake source mechanism, the relation of the site with respect to the source and the physical properties of the media through which the waves propagate must all be taken into account. Very often little is known about these factors and, as can be inferred from the foregoing, at present there remain many problems in the evaluation of seismic risk and in the estimation of earthquake engineering design criteria.

GUIDELINES FOR EARTHQUAKE RESISTANT DESIGN

To summarise, the architect and engineer dealing with a structure in a zone subject to earthquake should:

(a) Avoid active faults, areas of loosely packed deposits which may be subject to cracking, lurching or liquefaction, and areas with a high water table.

(b) Estimate the response of his structure to shaking and design the foundation to accept the additional forces resulting from this response.

(c) Make sure that in design and construction, the joints between structure and foundation (and in the structure) are able to take the forces due to oscillation of the structure.

(d) In general use materials and adopt a design able to deform beyond the elastic limit thus absorbing energy without collapse.

ACKNOWLEDGEMENTS

The authors acknowledge their indebtedness to Sir Alexander Gibb & Partners from whose works case histories are quoted, and to the Central African Power Corporation, Rhodesia and Zambia (Kariba Hydro-electric scheme) and the Tehran Regional Water Board (Shareholders Co.) Iran (The Lar scheme). Reference is also made to the discussion at the Intergovernmental Conference on the Assessment and Mitigation of Earthquake risk, UNESCO, Paris 1976.

References

1. Kobold, F., 'Alpine power stations and dams,' *RICS Jour.*, **95**, 9, 469–475 (1963).
2. Varnes, D.J., 'Landslide types and processes,' (article in *Landslides in engineering practice*, Ed. by Eckel, E.B.) Committee on Landslide Investigations, Highway Res. Board, Spec. Rept. 29, Washington, 20–47 (1958).

3. Terzaghi, K., 'Mechanisms of landslides,' (article in *Application of geology to engineering practice*, Ed. by Paige, S.) Berkey Vol., Am. Geol. Soc., 83–124 (1950).
4. Root, A.W., 'Prevention of landslides,' (article in *Landslides in engineering practice*, Ed. by Eckel, E.B.) Committee on Landslide Investigations, Highway Res. Board, Spec. Rept. 29, Washington, 131–149 (1958).
4a. Civil Engineering Code of Practice: 2, *Earth Retaining Structures*, Institution of Civil Engineers (1951)
4b. Reimbert, M. and A., Retaining walls, anchorages and sheet piling, *Trans.* Tech. Publications, Clausthal (1974)
5. Albee, A.L. and Smith, J.L. 'Earthquake characteristics and fault activity in Southern California,' (article in *Engineering geology in Southern California*, Ed. by Lang, R. and Proctor, R.) Ass. Eng. Geol., 9–35 (1969).
6. United States Atomic Energy Commission, 'Seismic and geologic siting criteria for nuclear power plants,' US Federal Register 38, No. 218 (23/11/73).
7. Sieberg, A., *Erdbebenkunde, Fischer, Jena*, (Sieberg's scale, 102–104) (1923).
8. Wood, H.O. and Neumann, F., 'The modified Mercalli intensity scale of 1931,' *Seismol. Soc. Am. Bull.*, **21**, 277–283 (1931).
9. Richter, C.F., *Elementary seismology*, W.H. Freeman & Co., San Francisco, 768p. (1958).
10. Richter, C.F., 'An instrumental earthquake scale,' *Seismol. Soc. Am. Bull.*, **25**, 1–32 (1935).
11. Department of Water Resources, 'Earthquake engineering programs,' *Bull. no. 116–4*, State of California, 139p. (1968).
12. Ambraseys, N.N., 'Some characteristic features of the anatolian fault zone,' *Tectonophysics*, **9**, 143–145 (1970).
13. Sherard, J.L., Cluff, L.S. and Allen, O.R., 'Potentially active faults in dam foundations,' *Geotechnique*, **24**, No 3, 367–429 (1974).
14. Zeevaart, L., *Foundation engineering for difficult subsoil conditions,* Van Nostrand, New York (1972).
15. Nichols, D.R., 'Seismic safety legislation and engineering geology: an example from California,' *Bull. Ass. Eng. Geol.*, **11**, 135–146 (1974).
16. Steinbrugge, K.V., *Earthquake damage and structural performance in the United States*, Ed. by Wiegel, R.L., 167–226, Prentice-Hall, Englewood Cliffs, N.J. (1970).
17. Neumann, F., 'Engineering seismology,' (article in *Reviews in engineering geology*, Ed. by Leggett, R.F. and Fluhr, T.) Geol. Soc. Am., **1**, 161–186 (1962).
18. 'A review of earthquake resistant design of dams,' *ICOLD Bulletin* 27 (March 1975).
19. Hisada, T., Izumi, M. and Hirosawa, M., 'Structural response of tall buildings to earthquake shocks,' *Proc. Int. Sem. Earthquake Eng.*, Unesco, 63–68 (1968).
20. *An evaluation of a response spectrum approach to seismic design of buildings*, Applied Technology Council, San Francisco (Sept 1974). A Study Report by Applied Technology Council, San Francisco, California. Sponsored by National Science Foundation Research Applied to National Needs Program, and National Bureau of Standards.
21. Newmark, N.M. and Hall, W.J., *Earthquake-resistant design of nuclear power plants,* Intergovernmental Conference on the Assessment and Mitigation of Earthquake Risk, Unesco, Paris (1976). (Unpublished).
22. Shima, E., (The University of Tokyo), *Japanese report on the seismic micro-zoning map of Tokyo*,' Unesco, Paris (1976). (Unpublished).
23. Seed, H.B., 'Soil problems and soil behaviour,' (article in *Earthquake engineering*, Ed. by Wiegel, R.L.) 227–252 (1970).
24. Reid, H.F., 'The Californian earthquake of April 18, 1906, Report of the State Earthquake Investigation Commission (Vol. 2)' *The Mechanics of the Earthquake*, Carnegie Inst. Washington (1908).
25. Medvedev, S.V., 'Measurement of ground motion and structural vibrations caused by earthquakes,' *Proc. Int. Sem. Earthquake Eng.*, Skopje, Unesco, 35–38 (1968).
26. Medvedev, S.V., *Engineering seismology*, Israel Programme for Scientific Translations, Jerusalem (1965).
27. Long, R.E., 'The problems of assessing effective ground motions from earthquakes,' *SECED 1977, Journal ICE*, Pt. 1 **64** (May 1978)

Chapter 14

Ground Treatment

INTRODUCTION

Although 'difficult' may be defined as hard to do, the term is relative; for example, it may be harder, in terms of energy expended, to drive a tunnel in Cornish granite than in London Clay, but it is not necessarily more difficult, indeed it may be easier since no support is usually required. When difficulties arise during construction, they may be due to a failure to suit the design or tools to the known prevailing ground conditions, or to lack of knowledge of these same conditions due to inadequate site investigation. Even though the latter may have been as thorough as could be expected, variations from the normal in certain parts of the work, whether known or not, may require more elaborate, and therefore expensive, procedures than elsewhere on the site.

When it is known that the ground conditions are likely to prove difficult over all or part of a new site, the engineer can either modify the structure, or he can make use of foundation engineering techniques such as piling, raft, etc, or he can modify the nature of the ground, either temporarily or permanently. For unexpected changes in ground conditions the third option is often the easiest to adopt, but combinations of these options may be possible. For instance, at Lower Drift Dam, Penzance, pneumatolytic alteration of the granite was found to be more extensive than had been disclosed by site investigation; the design of the spillway was altered to reduce the bearing capacity and the foundation grouting, already foreseen, was considerably extended and modified.

GROUND WATER LOWERING TECHNIQUES

All excavations below the water table encounter ground water, which, besides being a nuisance, can have undesirable effects. Piping, uplift, slope failures, and disintegration of weakly cemented rock under high pressure gradients are examples. Ground water lowering processes which depend on pumping are essentially temporary remedies, but if the terrain is suitable, permanent local control of the water table can be achieved by drainage ditches, counterfort drains, or drainage adits, the water being discharged into a stream at a lower elevation.

The method adopted for dewatering excavations depends upon the permeability of the soil and its variation within the stratal sequence, the depth of base level below the water table, piezometric conditions in underlying horizons, the method of providing support to the sides of the excavations and on safeguarding neighbouring structures. Surface water or shallow subsoil water can be dealt with by diversionary ditches or drains. However, the toes of ditch slopes in sandy soils may be suspect in that they develop quick conditions. In such instances a graded filter should be placed in the bottom of the ditch.

WELLPOINT SYSTEM

The simplest method of ground water lowering is to pump it from a sump within the excavation. This can generally be done in excavations in rock and in gravels where the rate of inflow does not cause instability of the sides or base of the excavation. However, in silts and sands the rate of inflow is often effective enough to cause erosion and slumping of the sides, and 'boiling' at the base. In these conditions a wellpoint system (Figure 14.1) or bored wells installed about

Figure 14.1 Wellpoint system (Courtesy of Ground Water Services Ltd)

the perimeter of the excavated face is required. Such ground water lowering techniques depend on excessive pumping which lowers the water table and thereby develops a cone of exhaustion (Figure 14.2). The radius of the cone of exhaustion at the withdrawal points depends upon the rates of pumping and recharge. The amount of discharge (Q) which is necessary to lower the water table through a given depth can be estimated by using the Dupuit equation, which for the gravity well condition is as follows:

$$Q = \frac{k(H^2 - h^2)}{\log_e (R/r)}$$

where
H is the elevation of the original water table above an impermeable horizon;
h is the elevation of the operating level of the pumping well above this horizon;

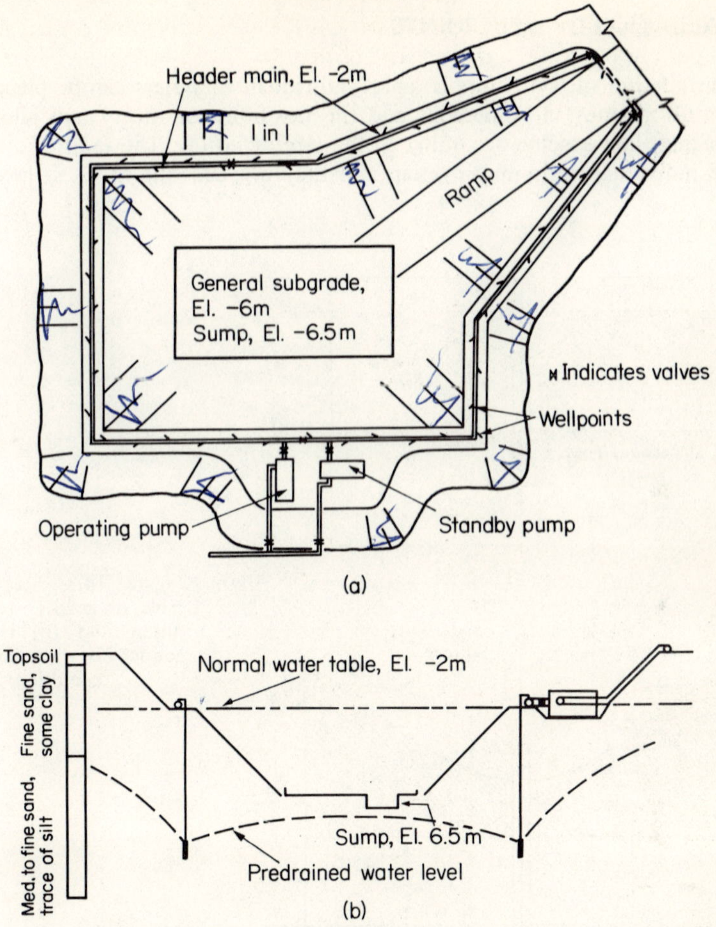

Figure 14.2 (a) Typical layout of a wellpoint system and (b) section through layout

R is the radius of the area of influence;
r is the radius of the well;
k is the coefficient of permeability (Figure 14.3a).

The equation for a confined aquifer under artesian pressures is:

$$Q = \frac{2\pi kb(H - h)}{\log_e (R/r)}$$

where b is the thickness of the confined layer (Figure 14.3b). Because drawdown is a linear function whereas soil volume is a cubic function, increased pumping from one point of withdrawal soon becomes inefficient.

A dewatering system not only aids construction operations by lowering the water table and intercepting seepage flow but also means that the stability of the excavated slopes and floor is enhanced. This in turn means that the load on any sheeting or bracing is reduced. Also, dewatering by 'drying' sandy soils improves their digging and hauling characteristics.

INSTALLATION OF WELLPOINTS

The installation of wellpoints is rapid. Individual wellpoints can be placed in a matter of minutes in some soils and the flexibility of the system allows for rearranging their spacing according to the rate of inflow. The radii of influence of the individual wellpoints overlap and they are laid out so as to lower the

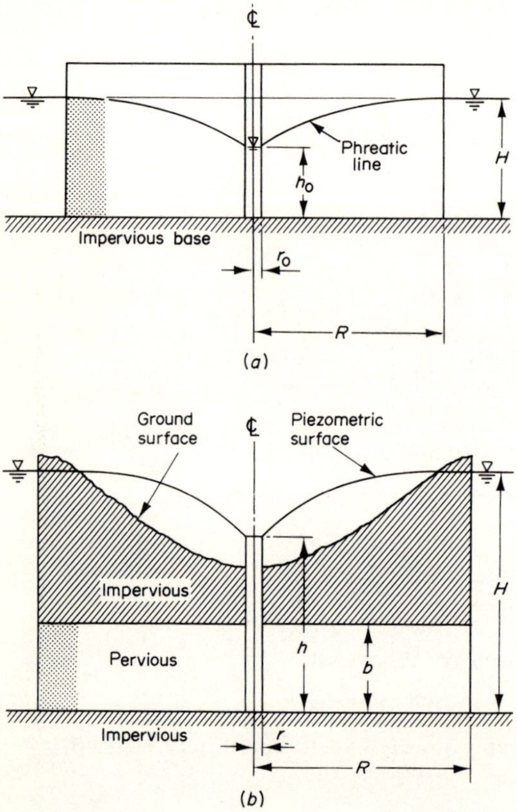

Figure 14.3 (a) Water table or gravity well condition and (b) confined aquifer or artesian well condition

water table by approximately a metre below the base level of the excavation. Some idea of the spacing between individual wellpoints can be gained from nomograms such as shown in Figure 14.4.

Wellpoints are generally installed using a high pressure water jetting pump although drilling methods have been used. The soil is penetrated by a filter casing from which the soil is removed by jetting. The wellpoint is positioned within the casing and a selected filter material placed about it. Then the casing is removed. The individual wellpoints are connected via a common suction header main to a high vacuum suction pump. In silty soils the mesh screen of the wellpoint may not prevent extraction of fines and so extra measures may be required. Wellpoint pumping is most useful where the required lowering is not more than about 4 to 5.5 m. In silty soils the achievable lowering averages about 30% less. In order to achieve greater lowering another tier of wellpoints, header main and pumps must be installed. The added excavation which this involves may prove a limiting factor as far as cost is concerned. Furthermore multi-stage operation may mean that the site becomes overburdened with wellpointing equipment thereby impeding construction work.

Cashman[1] referred to a new technique to place a continuous horizontal well-point at a depth of 6 m. Installation is carried out by a trenching machine

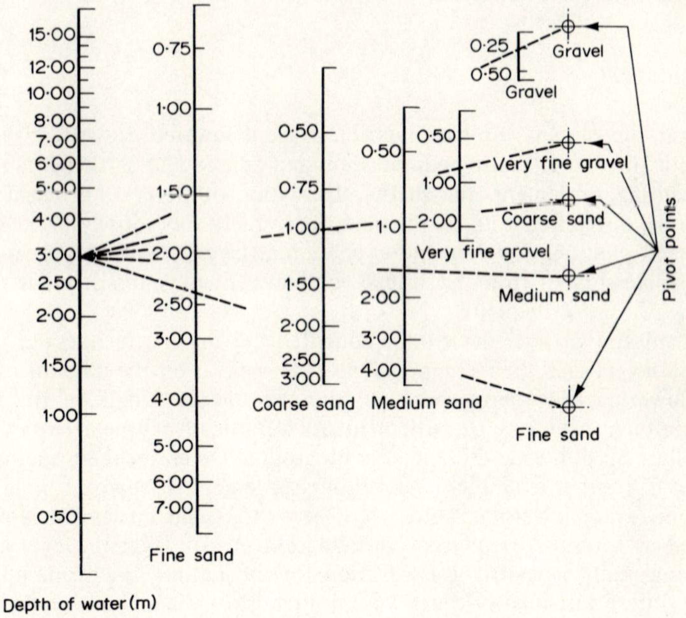

Figure 14.4 Nomogram for spacing (m) wellpoints in sand and gravel

which excavates the trench, lays the pipe, and then fills the trench in continuous operations. The advantages of this system is its speedy installation and lack of multiple risers and suction headers which means a clearer site.

BORED WELLS

A bored filter well consists of a perforated tube surrounded by an annulus of filter media and the operational depth may, in theory, be unlimited. Generally, wells are bored in a 600 mm diameter hole size. Having bored the hole to a sufficient depth within the aquifer (in the case of a thin aquifer the well bore is often taken some 1 to 2 m into the impervious stratum beneath the aquifer) the perforated tube (well screen) is lowered into place and the appropriate filter media are then placed. An electrical submersible pump is lowered into each well on its own riser pipe and connected to a common discharge main. The successful operation of deep bored filter wells is largely dependent on the grading of the filter. Bored wells are preferable to wellpointing for deep excavations where the area of the excavation is small in relation to the depth. They are also preferable in ground containing cobbles and boulders where wellpoint installation is difficult.

Deep wells are particularly suited to variable soils and multi-layer aquifers, as well as to the control of ground water under artesian or subartesian conditions.

le, deep wells are the appropriate method of dewatering an aquifer
eneath an impermeable stratum, in order to prevent a blow-out in an
n terminating with this impermeable material. A bored shallow well
can be used on a highly permeable site where pumping is required for
months, rather than wellpointing where risers at close centres could
r construction operations.

SETTLEMENT

Whenever the phreatic or piezometric surface is lowered, the effective load on the soil is increased, causing compression and consequent settlement. However, any resulting settlement due to the abstraction of water from clean sands is likely to be insignificant unless the sand was initially very loosely packed. On the other hand pumping from an aquifer containing layers of soft clay, peat or other compressible soils or from a confined aquifer overlain by compressible soils may cause significant settlements.

The amount of settlement undergone depends on the thickness of the compressible layers and their compressibility, as well as on the amount of ground water lowering. The permeability of the soil and the length of the pumping period influence the rate of settlement. As a result it may be necessary to limit the radius of influence by the use of ground water recharge methods (see Parsons[2]). Another example of preventing settlement of adjacent buildings due to ground water lowering is given by Zeevaart[3]. Such settlement would have occurred as a result of induced consolidation in clay layers brought about by dewatering sand beneath the excavation for the Latino Americana building in Mexico City. Consequently water was injected into the sand beneath the adjacent buildings so that seepage gradients were not developed in the clays.

ELECTRO-OSMOSIS AND ELECTROCHEMICAL STABILISATION

Electro-osmosis has been used as a dewatering method to stabilise soils of low permeability such as silts and clays which cannot be easily dewatered by the above mentioned techniques. The technique was first employed at Salzgitter in Germany in 1939 to stabilise an excavation for a long cutting in clayey silt for a railway and has since been promoted by Casagrande[4, 5].

Basically electro-osmosis consists of placing electrodes into the ground to a depth of about 2 m below excavation base level and passing a direct electric current between them. The electric current induces a flow of water from the anodes to the cathodes, the latter acting as wellpoints from which the water can be removed by pumping. The flow of water occurs because positive ions in solution are attracted towards the cathodes, the water molecules being attracted by the positive ions. Cathodes may be installed in one line, being spaced approximately 8 to 12 m apart, with anodes midway between them. If more than one line of cathodes is employed then the spacing between cathodes may be increased to 15 m. However, spacing is largely dependent upon the voltage available at the site.

According to Casagrande the coefficient of electro-osmotic permeability is similar for sands, silts and clays and therefore for practical purposes most soils may be assumed to have a value of about 0.5×10^{-3} mm/sec for a gradient of 0.1 v/mm. However, the mass permeability of stiff, fissured clays is frequently similar to their electro-osmotic permeability which means that electro-osmotic drainage methods are inappropriate since cheaper methods of stabilisation as, for example, sand drains, are more effective. Potential gradients in excess of 0.05 v/mm between electrodes should not be entertained for long term applications since they result in high energy losses in the form of ground heating.

According to Farmer[6] electro-osmosis suffers a fundamental drawback when used solely for dewatering in that it is a decelerating process, that is, as the soil dries out around the anode so its electrical resistance increases. In other words electro-osmosis becomes progressively less efficient as the soil water content is reduced. This in fact means that the water content in soils is rarely reduced sufficiently to achieve complete stabilisation and even where it is, subsequent re-hydration may often rapidly reverse the process. It also means that electro-osmosis only operates efficiently in saturated or almost saturated soils. Electro-osmosis also consumes a high amount of energy, it is therefore by no means a common method of ground treatment. For instance, on some large excavations where electro-osmosis has been employed the power required per cubic metre of excavated soil has amounted to 1 kW, whilst on small sites ten times this figure may be needed to excavate a cubic metre.

In the cases where electro-osmosis has been successful, this has generally been due to the introduction of a chemical into the soil, either through anode solution or by direct electrolyte replacement. This was discovered more or less accidentally. One of the initial difficulties with electro-osmosis was that the constant flow of electricity from the anodes meant that metal was continuously being lost from them and so new anodes had eventually to be installed. However, it was soon realised that this gave rise to ionic replacement in clay minerals or led to partial cementation of pore spaces, thereby enhancing soil stability. As Grim[7] pointed out, the engineering properties of a clay are related to its mineralogy and more particularly to the type of cations associated with the clay minerals. Montmorillonite provides the best example. For example, Na and Li montmorillonites can absorb large quantities of water and so have very low shearing strengths. On the other hand Ca, Mg, Fe or Al montmorillonites absorb appreciably less water and so have more stable structures. Because montmorillonite is prone to base exchange Na and Li may be replaced by Ca, Mg, Fe or Al, thereby improving the engineering performance of the soil. The technique of ionic replacement to improve soil properties is known as electrochemical stabilisation.

ELECTROCHEMICAL STABILISATION

Two methods are used for introducing stabilising electrolytes into a clay soil; firstly, this can be done by solution and dispersion of material from the anode by electrolysis or, secondly, it can be achieved by direct introduction of additives at the anode. Farmer indicated that chemical stabilisation brought about by anode solution depends on base exchange reactions occurring on the surfaces of clay minerals, thus replacing weak by strong cations, and soil cementing compounds being formed by reaction between the electrolyte and the dissolved anode. In the simplest cases these reactions can be achieved by hydrolysis and oxidation of anode material. For instance, when Fe is given off the ions react with water to form hydroxides and these act as cement binding soil particles together. Such treatment, Farmer recorded, could achieve increases in the shear strength of clay soils of up to 60%.

The addition of additives to the anode increases base exchange reaction thereby enhancing the quality and quantity of the cementitious material. Generally the additives used are organic and inorganic compounds of Al or Ca, the most frequently used being calcium chloride. Unfortunately, however, the use of additives leads to a reduction in flow with increase in electrolyte concentration, which in turn may limit stabilisation beyond a certain level.

EXCLUSION TECHNIQUES

In urban areas the presence of existing buildings and the high price of land frequently dictates that construction has to be carried out in the least possible space. Where surrounding property has to be safeguarded it will usually be more appropriate to provide a barrier about the excavation in order to prevent inflow of water, whilst maintaining the surrounding water table at its normal level, rather than to adopt a dewatering technique.

Methods of forming such a barrier, in a rough order of relative cost (cheapest first) include steel sheet piling, concrete diaphragm walls, contiguous bored pile walls, bentonite cut-off walls, cement or clay-cement grout curtains and frozen soil barriers (see Little[8]). The economy of providing a barrier to exclude ground water depends on the existence of an impermeable stratum beneath the excavation to form an effective cut-off for the barrier. If this stratum does not exist or if it lies at too great a depth to be practical to use as a cut-off then upward seepage occurs which may give rise to instability at excavation level. In these circumstances the barrier will not be effective unless it can be extended horizontally beneath the excavation. The only methods of forming a horizontal barrier are by grouting or freezing. Consequently it is necessary to carry out a site investigation before decisions on the type of wall to be adopted are made.

Perimeter exclusion walls should be capable of construction through soft, wet ground, should be economical in installation costs and should be capable of being installed without any adverse effects on the stability of adjacent buildings. If possible, they should eliminate the need for temporary work during construction of the permanent works. The consequences of any movement of the barriers can adversely affect both the structure to be incorporated within the excavation and the stability of existing buildings immediately adjacent to it. Accordingly measurements should be taken on the constructed walls and adjacent properties to check for any movements.

Steel sheet piling is perhaps the simplest and cheapest method of forming an impermeable barrier, particularly if the piles can be extracted for reuse, but it cannot be used in ground containing numerous boulders or other obstructions. On the other hand sheet piling is still attractive if it can be incorporated into the permanent structure without the necessity for a replacement reinforced concrete wall, for example, one of the major uses of sheet piling for permanent works is in quay walls. The usefulness and reliability of the Larssen and Frodingham sheet piles in all types of earth and water retaining structures have been proven over many years. Straight web steel sheet piles are used for cellular cofferdams.

The pile section to be selected depends on height of ground to be retaine and ground water conditions, and on whether the piling is tied back or st With all sections of pile or types of steel, higher stresses are permissibl used in temporary works rather than for permanent use. Most piling h used today are diesel driven and can be light, medium or heavy duty depending on the work involved. An important consideration is the noise level, and is the subject of a recent Building Research Establishment report[9].

Barriers can be formed by contiguous bored piles or secant piles, where a series of bored piles is first sunk at centres rather less than twice the pile diameter, intermediate piles then overlapping the first series to provide a wall which can be load-bearing. This type of wall can be economical in clays where mechanical augers can be used for drilling. In confined spaces this technique can conveniently be carried out using tripod rigs. In water bearing granular soils or silts the required impermeability of a bored pile wall can only be obtained by keying alternate secant piles after casting by a special grooving chisel (see Tomlinson[10]). This ensures that the infilling bored pile when cast effectively interlocks into the piles either side. In coarse granular soils it may be possible to seal the gaps by cement or chemical grouting. Contiguous piles can be designed to cantilever vertically, to be strutted or to be anchored back. It is usual to employ ground anchors, either temporary or as part of the final design, to reduce the necessity to resort to struts which limit the working space within an excavation and to reduce the diameter and amount of reinforcement within the pile.

A less popular walling technique is that of king post and sheeting. This method employs the use of rotary augers forming holes down which are lowered universal beams. These are secured at the lower end by concreting. As the excavation progresses, thin prestressed concrete planks spanning between the webs of the beams are progressively lowered, additional planks being inserted at the top of the beams. This system, again one which can make use of temporary ground anchors, is normally used as a temporary solution and although used extensively in the USA and on the continent has not been used on a large scale in Britain.

DIAPHRAGM WALLS

A diaphragm wall is the most economic form of exclusion barrier to construct in soils containing boulders since these can be removed by grabbing and chiselling. According to Fuchsberger[11] a minimum excess head of 1.25 m is required to exclude ground water. Artesian or semi-artesian ground water and sea water need special measures to ensure a satisfactory diaphragm wall. Organic compounds in ground water usually render the technique impracticable. Moving ground water may lead to one side of the trench being more stable than the other and so must be carefully monitored.

The major use of diaphragm walls in the UK has been in the construction of structural retaining walls in built up areas. The technique has several advantages, notably in providing support even when construction takes place immediately adjacent to multi-storey buildings, old foundations etc.

A diaphragm wall is constructed as a series of panels (the lengths usually varying between 4 and 8 m, the widths between 0.5 and 1 m), the trench being excavated, between shallow guide walls, by a grab attached to a crane (see

Sliwinski and Fleming[12]). The walls do not need to be straight in plan. Depths of walls range up to 60 m and a grouted cut-off can be used to extend the effectiveness of a diaphragm wall to greater depths or into bedrock. During excavation, the trench is kept filled with bentonite suspension mixed on site, which not only accounts for the impermeability but also stabilises the walls. For example, the suspension penetrates pervious soils, gels, and forms a filter cake on the surface of the excavation. The hydrostatic pressure of the bentonite in the trench supports the earth pressure without raising the pore water pressure within the soil.

There are a range of slurry types or modifications and these are able to support the excavation in most types of soil. According to Hutchinson *et al*[13] less than 3% sand can be added to bentonite suspensions, however, larger quantities spoil the bentonite and this may mean that the trench wall collapses. Their specification recommended a bentonite concentration in excess of 4.5%, a density varying between 1.034 and 1.25 Mg/m^2, a plastic viscosity greater than 20 cP, a 10 min gel strength between 50 and 200 dyn/cm^2 and a pH value of less than 11.7. The bentonite suspension must be able to suspend detritus and avoid sludgy layers being built up at the base of the trench. It also must allow clean displacement by concrete with no subsequent interference with the bond between reinforcement and set concrete. When the excavation is completed, reinforcement is positioned (Figure 14.5). Circular cross-section steel end stops are then lowered into the trench either side of the reinforcement cage. Concrete is tremied into position and the displaced bentonite recovered for subsequent use. The concrete must have a high workability and rounded aggregates are preferred. Concreting must be a continuous process and each panel should be completed in a single operation.

Reinforced diaphragm walls are used as part of structural load bearing elements, as retaining walls or foundations (see Keinberger[14]). The finish to the wall surface reflects the cutting action through the ground. If a good quality surface finish is required for visual purposes, then high strength precast wall elements with the required surface finish can be inserted into a modified bentonite and cement slurry which locks the panels together. Even mosaic tile finishes have been successfully applied to precast panels prior to insertion into the slurry trench. Unreinforced diaphragm walls are simply used to form cut-offs, for example below dams, and as such are not load bearing therefore reinforcement is not required. A recent development is to incorporate cement with a retarding agent into the suspension so as to form a self-setting mix.

Ground anchors have proved very successful as tie-backs for diaphragm walls, allowing interior support to be dispensed with. They have been combined with prefabricated panel walls on a number of projects in Europe according to Boyes[15]. The use of prefabricated panel construction means the elimination of site tremie concreting and reinforcement cage manufacture and placement problems, as well as stop-end removal difficulties. When permeable strata extend to greater depth, prefabricated panels can be used for the upper section of the wall whilst the lower part is filled with tremie concrete or self-hardening clay-cement slurry. Another recent innovation is the post-tensioned diaphragm wall method which enables placement in constricted conditions and the wall to maintain bending movements more than twice that permissible with a conventional diaphragm wall with the same thickness.

The use of self-hardening clay-cement slurries for the original excavation support of water cut-offs eliminates tremie placement. Speed of construction is

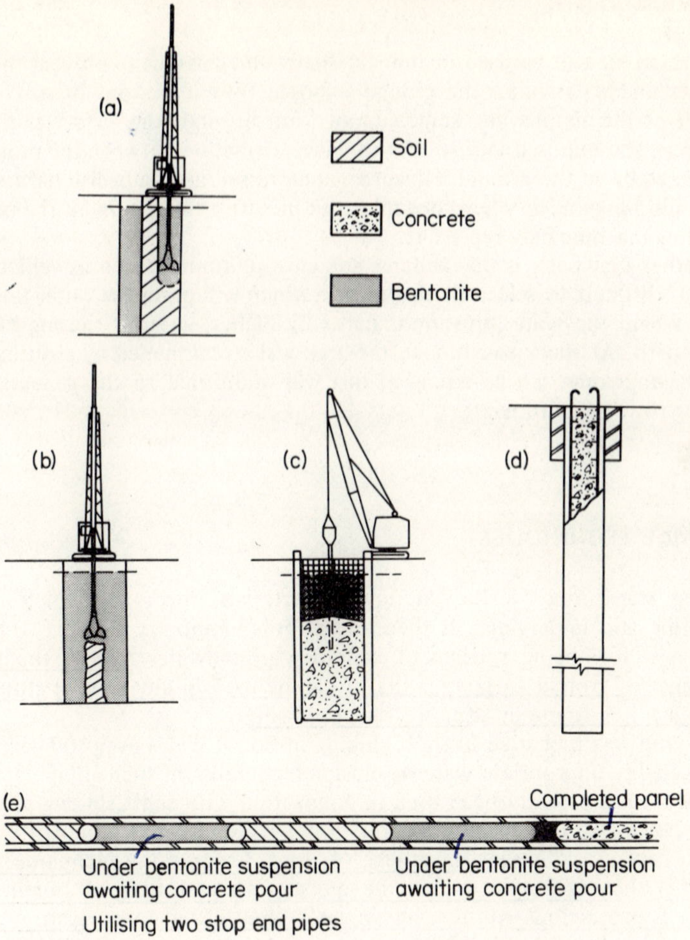

Figure 14.5 Diaphragm wall construction. General procedure for the excavation of panels during which time the excavation is kept filled with bentonite suspension. (a) First one end, then opposite end of panel is excavated to full depth. (b) Third and last stage is excavation of centre of panel. General procedure for concreting of panels when the bentonite is dsiplaced by the concrete. (c) Steel stop-end pipes and reinforcement cage positioned and concrete placed through tremie pipe. (d) Section through completed panel showing guide walls and steel. Series of panels on plan. (e) Illustrating variety of uses of steel stop-end pipes as required (Courtesy of Cementation Ltd)

increased as a number of alternate panels can be formed during the initial hardening period, the grab returning to dig intermediate panels of lesser width, keying them into the first panels to ensure watertightness. Cut-off walls can also be formed by driving and withdrawing steel piles, the space vacated on withdrawal being filled with a clay-cement slurry, thereby producing a thin plastic membrane.

COMPRESSED AIR

Compressed air can be used in tunnels, shafts and caissons to prevent the inflow of water and to stabilise the ground exposed by excavation. In soft clay the pressure of the air provides some support, but in sands the air drives the water back from the exposed face so that capillary attraction between the moist grains helps to stabilise the ground. However, compressed air is a health hazard and its use should be avoided where possible, and in soft ground tunnels the bentonite tunnelling machine may replace it.

Another drawback is that in large subaqueous tunnels such as vehicular tunnels it is difficult to select an air pressure which will minimise water flow at the invert, where the water pressure is naturally higher, without causing blow-outs at the soffit. At Blackwall tunnel, the arch was strengthened by grouting with a grout having some tensile strength; this was additional to the general annular grouting to reduce inflow.

THERMAL TECHNIQUES

Freezing transforms weak, waterlogged materials into ones which are self-supporting and impervious. It therefore affords temporary support to an excavation as well as being a means of excluding groundwater. One of the principal advantages of ground freezing is that it can be used in any type of ground. It is very well suited to use in saturated sands and silts.

Freezing was first used in shaft sinking in South Wales over 100 years ago, it being a highly appropriate water exclusion technique for such an operation. For example, it has recently been used in connection with shaft sinking in the Ely-Ouse scheme (see Collins and Deacon[16]) and at Boulby mine in north Yorkshire (see Cleasby *et al*[17]). One advantage of using the freezing technique in shaft sinking is that surface plant can be erected while drilling of boreholes and freezing is proceeding. Another advantage is that in a deep excavation of small diameter air is cooled appreciably and consequently the exposed surface of the frozen ground may not require insulation. By contrast where the freezing process is employed in a large, shallow foundation insulation of the exposed frozen surface should be carried out as soon as possible since no refrigeration plant is effective against solar radiation or open air convection. A temporary cover can be erected above large excavations which are likely to remain open for months.

FREEZE PROBES

Normally the freeze probes are laid out in linear fashion so that an adequate boundary wall encloses the future excavation when the radial development of ice about each probe unites to form a continuous section of frozen ground. If a very thick wall is required then more than one row of freeze probes will probably be necessary. The boreholes are usually sunk at approximately 1 to

1.2 m centres and are equipped with steel freeze tubes which range in external diameter from 100 to 200 mm. These are sealed at the lower end. Smaller internal tubes or feed pipes, with external diameters varying from 38 to 76 mm and open ended, are inserted within the larger tubes. The coolant is pumped through the ring main and down the feed pipes, returning to the refrigeration plant via the annular space between the tubes. Control of the coolant is obviously advisable. This can be accomplished by an outlet valve on each pipe and by short 'loops' in the system. Freezing is usually irregular but tell-tale thermocouples of other sensors focus attention on inefficient pipes or over or under cooled areas. Variation in flow of coolant may correct such situations. disadvantage ①

The classic coolant is supercooled brine although cryogenic liquids like liquid nitrogen, liquid propane and lithium chloride are now being used. These can reduce the freezing time, one of the usually quoted disadvantages of the technique, to about a third. For example, liquid nitrogen was used as the coolant when the elevator shaft was sunk at Oxford Circus for the new Victoria line. The freeze time was two months, it would have taken six months if supercooled brine had been used. Cryogenic liquids are, of course, more expensive. Ordinary steel pipes are quite satisfactory when the coolant used for freezing is brine but the problem of low temperature embrittlement must be considered if cryogenic liquids are used. Pipe fracture during excavation obviously must be avoided. Vertical heave and tensional stress due to adfreeze between the freeze probe and the ground generally do not give rise to any risk unless pipe joints are weak.

GROUND FREEZING

Ground freezing means heat transfer and, in the absence of moving ground water, this is brought about by conduction. Thus the thermal conductivities of the materials involved govern the rate at which freezing proceeds. According to Maishman[18], however, these values fall within quite narrow limits for all types of frozen ground. This is why ground freezing is such a versatile technique and can deal with a variety of soil and rock types in a stratal sequence, but a limitation is placed upon the freezing process by unidirectional flow of ground water. For a brine freezing project velocity exceeding 2 m per day will seriously affect and distort the growth of an ice wall. The tolerance is much wider when liquid nitrogen is used.

The success of the freezing process depends upon, firstly, a precise knowledge of the geological conditions, in particular ground water conditions should be investigated and, secondly, the attitude of each borehole within the ground. Permeability tests and the installation of piezometers in boreholes are required. Rupture of the frozen wall by high piezometric pressures is rare, when it has occurred freezing has not been uniform, or the suspect soil horizon has not been indicated on the borehole log. Ground water pressures should also be checked so that the instability of the floor is avoided during excavation.

Sanger[19] suggested that the centre line of a freeze pipe should not deviate more than 150 mm. However, even when the drilling is carried out with care and skill the holes may still deviate. Deviation may, in turn, mean that areas of ground are left unfrozen. These are referred to as windows and they can allow the penetration of water into the excavation. Freeze holes must therefore be surveyed to see whether any extra holes need to be drilled so as to avoid the development of windows. The inclination of individual boreholes can be determined by using an inclinometer. Large fissures in rock may also give rise to windows. On rare occasions a frozen wall, although 99% impervious, may still have small windows which may not close no matter what refrigeration power is used. Pumping must then be resorted to.

The salinity of water governs its freezing point therefore an analysis of ground water should be taken before freezing begins. Sea water presents no special problems in freezing since it solidifies at about –3°C. However, in some situations temperatures as low as –21°C have had to be achieved before salt saturated water has frozen. The ground water temperature should be recorded during the site investigation.

Because of the bond which exists between water and clay particles a significant proportion of the natural moisture content in a clay soil remains unfrozen even at temperatures as low as –25°C. However, Maishman maintained that this does not mean that clays cannot be effectively frozen for purposes of ground support. Indeed the unconfined compressive strength of clay as the temperature is lowered further below freezing point appears to increase exponentially. By contrast, the natural water content of sands and gravels is almost wholly converted to ice at a temperature of 0°C so that they exhibit a reasonably high compressive strength at only a few degrees below freezing. Consequently this parameter can be used as a design index of their performance provided a suitable factor of safety is used.

Ice creeps under uniform loading as do saturated soils in a frozen state. Hence the resistance to failure of such materials depends not only upon their temperature but also upon the duration of loading. It has been shown that the resistance to failure increases as the temperature below freezing is lowered. Indeed Sanger asserted that generally the strength of a frozen soil after an appreciable period of loading is reduced to about one third its normal strength. At ground freezing temperatures soils rich in silt-clay fractions may exhibit such behaviour to a significant extent and if excavations in them are to be open for an extended period of time their long term creep strength and plastic deformation should be estimated.

Although frost heave is frequently mentioned as a problem in silty ground, this rarely occurs when the ground is frozen artificially, rises in surface level of more than 75 mm being very exceptional. What is more the likelihood of such movement is reduced if freezing is rapid. After thawing the ground level slowly sinks towards its original level, complete recovery rarely being attained.

Grouting may be required when the ice wall has thawed out after completion of shaft excavation, either because incipient leakways have been concealed during the freeze period or because the temperature changes during thawing have cracked the lining. For this reason steel lining is usually installed.

HEAT TREATMENT

Clay soils can be stabilised by being subjected to heat treatment. Heating to 400°C brings about an appreciable reduction in the plasticity index of clays, and the moisture absorption capacity is significantly reduced when the temperature exceeds 600°C. Thermal treatment also influences permeability, it increases with heating up to 700°C, above which, due to the onset of fusion, there is a slight reduction. Furthermore it reduces the density, compressibility and sensitivity of clays. Thermal treatment has therefore been used to stabilise slopes in clays and loess in the Soviet Union and Rumania. The method consists essentially of driving exhaust gases from burning oil, at temperatures around 1000°C, into boreholes located 2 to 3 m apart. As a consequence the soil around the borehole is baked. Beles[20] described the stabilisation of a slide in clay by using this thermal technique.

GROUTS AND GROUTING

Grouting is the process of introducing a fluid, under pressure, into the ground in order to replace air or water in fissures, cracks, and pore spaces by a set product which will minimize the flow of water and, in some cases, strengthen the ground by a significant amount. In rocks grouting is usually required only to fill solution caverns and openings which occur along faults and joint planes. These represent the principal conduits by which water is conveyed through a rock mass and as such require sealing. By contrast, permeability attributable to the porosity of a rock mass is seldom significant enough to need reducing by grouting. However, during the excavation of deep mine shafts porous rocks such as the Bunter Sandstone may require treatment in order to reduce the cost of pumping from great depths (see Cleasby *et al*[17]). Loose granular materials have a much higher permeability than rocks of similar grain size and grouting serves to fill the pore spaces.

TYPES OF GROUT

Grouts may be divided into three classes, depending on the materials used:

(a) Suspension or particulate grouts
(b) Solution or non-particulate grouts which may be:
  (i) two-shot solutions
  (ii) one-shot solutions.

Solution grouts are sometimes termed 'chemical' grouts, a somewhat misleading term since 'chemicals' are frequently included in suspension grouts for one reason or another. A fourth class of grout has been recognised, the 'emulsion' grout, but these are seldom used nowadays and may be conveniently, if inaccurately, included in the other classes according to their constitutions.

SUSPENSION GROUTS

The basic suspension grout is a mixture of ordinary Portland cement and water, although other cementitious materials such as Trief cement (ground basic slag), fly-ash (pulverised fuel ash or PFA), or pozzolan may replace part or all of the cement. The initial water-cement ratio is far higher than is needed for hydration of the cement, in order to facilitate injection, and it is a requirement of the process that there must be adequate injection pressure to expel the surplus water, and somewhere for the water to go, either into fine non-groutable cracks or into pore spaces. Where considerable quantities of grout are accepted inert materials such as sand, clay, rock flour, or even gravel, corn husks, celluloid strip, or sawdust can be added as inert fillers if the hole size permits.

Natural or artificial bentonite can be added to cement grouts by up to about 6% of the dry weight of cement to minimise bleed and prevent segregation of the coarser inert fillers. The addition of bentonite ensures more complete filling of voids, but since the bleed water remains locked up in the set product the set strength can be considerably reduced. The effectiveness of bentonite depends on its swelling properties; since these are inhibited by the calcium liberated during hydration of cement they should be allowed to develop in a separate bentonite-water mix before being added to the main grout. Bentonite or processed clay can be combined with dispersing and peptising chemicals to form a clay suspension which can be injected into finer cracks and pore spaces than the coarser-grained cement grouts, and the thixotropic properties of such suspensions facilitate pressure injections and inhibit subsequent erosion when pumping ceases.

SOLUTION GROUTS

Two-shot solution grouts consist of two solutions, each of one or more chemicals, which will react almost instantaneously when they come into contact (see Glossop[21]). It is usual to inject the two fluids successively through the same injection rod. In the customary Joosten and Guttman processes a solution of sodium silicate, alone or with some other chemical, is injected as the pipe is advanced by short stages (Figure 14.6), each injection sufficing to fill voids within a calculated radius, usually about 300 mm. The pipe is then withdrawn while a second reagent, e.g. calcium chloride, is added to precipitate silica gel and thus fill the voids. The viscosity of the chemicals requires close hole-spacing and the method has been found particularly effective for increasing the strength of Thames gravels and thus facilitating many excavations for underground railway stations in London.

Single-shot solution grouts consist of two solutions, each of one or more chemicals, which when mixed together have a controllable delayed reaction ranging from a few minutes to several hours. Consequently they can be mixed at the surface and injected into the ground as one fluid which can penetrate to a distance varying with the viscosity of the grout, the permeability of the soil, and the pressure of injection before the reaction is completed. The effectiveness of the grout depends largely on the initial viscosity and the shape of the time-viscosity curve, and these have been widely discussed at the 1963 Grouting Symposium[22]. Many formulations exist, mostly patented, and a choice should be made after consideration of price, ground conditions, possible toxicity of components, and degree of permanence of the grout (some of the cheaper grouts based on silica gel will dry out if exposed above the water table in fluctuating ground water conditions).

Single-shot grouts are usually pumped in the required proportions by a proportioning pump before mixing at the head of the injection rod. This minimises the extent of piping choked by set grout in the event of power failure and

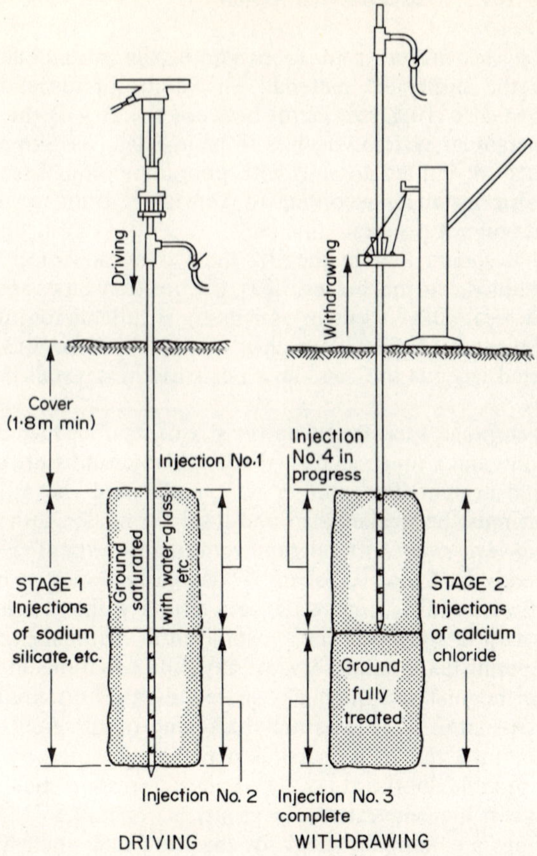

Figure 14.6 The Joosten process of grouting (from Little[8])

permits close control of the gelling time. While the mix proportions may be varied by altering the proportioning ratio it is simpler to alter the strength of either solution and keep the ratio constant.

HOT BITUMEN AND OTHER GROUTS

Other grouts which do not fall into the above classes include hot bitumen, heated by steam or electricity and sometimes used to grout into flowing water, often with the addition of cement or other solids, and bituminous emulsions but these are not popular today.

Mention should also be made of a Japanese grout which is claimed to need only ground water as a reagent and is therefore a true one-solution grout which can be stored and injected as required without further mixing. The possible effects of variations in ground water chemistry have not been publicised.

PENETRATION AND VISCOSITY OF GROUTS

The ability of a suspension grout to penetrate the ground depends on the particle size of the suspended material. The limits for suspension grouts are generally regarded at a 10:1 size factor between the D_{15} of the grout and the D_{15} size of the granular system which is to be injected (see Skipp[23]). Generally particulate grouts are limited to soils with pore dimensions exceeding 0.2 mm (Figure 14.7). For instance, according to Terzaghi[24] ordinary, coarse ground Portland cement will not penetrate fine sand.

Experience has shown that in the UK there is no advantage in using more expensive fine-grained cements but overseas, the position varies and it should not be assumed that local 'OPC' used for concreting is suitable for injection. At Ait Aadel Dam in Morocco (1968–70) 'scalped' cement was specified for injection into finely-fissured igneous rock and an air-classifier was installed to remove the coarser elements.

There has been much speculation on the size of fissure which can be injected with cement grout, and a rough guide is that a 100-micron fissure can just accept ordinary Portland cement. Put another way, in fissured rocks the D_{85} of the particulate grout must be smaller than one third the crack width. Estimation of fissure size is, however, very difficult, and various authorities (e.g. Cambefort[25], Londe and Sabarly[26], Snow[27], Sabarly[28], Lancaster-Jones[29]) have discussed estimation of fissure width from water-test results without agreement. Other factors such as fissure roughness and possible infill materials, acceptable pressures including permitted ground heave, if any, and variations in fissure spacing from the ideal orthogonal sets of equally spaced identical systems beloved of the theorists, affect penetration. Premature thickening of the grout by absorption of surplus water into the rock material may be prevented by a preliminary injection of a suitable lubricant, while various 'intrusion aids' are marketed which claim to assist the penetration of a grout.

Solution grouts are normally used for injecting finer-grained (sand or silt-sized) granular materials which will not accept the generally cheaper suspension grouts. An ideal grout would have a viscosity similar to that of water which would remain unchanged until a flash set occurred at the end of a pre-determined time, but although some grouts approach this ideal none reaches it. If high pressures are tolerable the cheaper more viscous grouts can be used, elsewhere only expensive grouts may be injectable at economic rates. Some advantage may be gained by clarification of the original grout solution to remove colloid-size particles which might reduce penetrability.

Permeability controls the time taken for a grout to migrate from its point of injection. Clay-cement mixes will not enter soils which have a permeability of less than approximately 10^{-5} m/sec. For permeabilities less than this silicate grouts are used, whilst resins may be used on the finest grained soils. By contrast where there is a large flow of underground water bitumen grouting may prove successful (see Moneymaker[30]).

HYDROFRACTURE

So far only permeation of the ground, either through fissures or pore spaces, without altering its basic physical structure has been discussed. However, in all

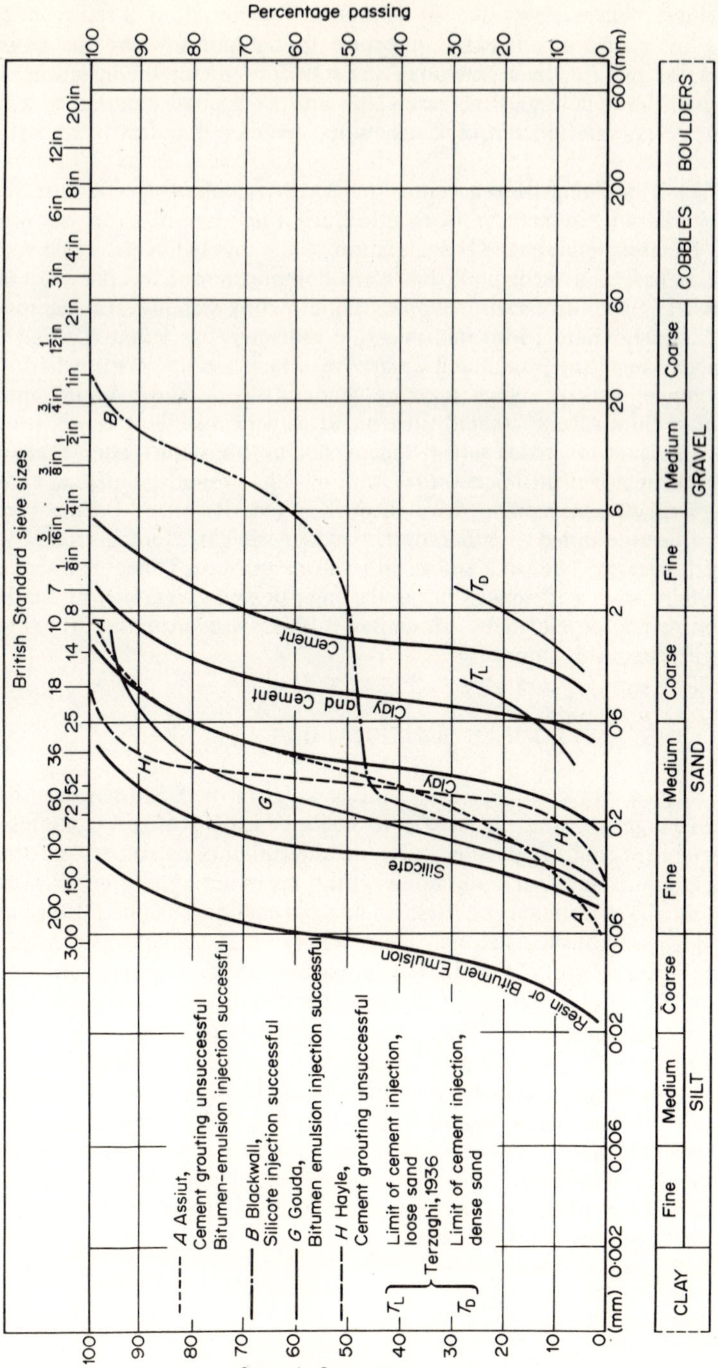

Figure 14.7 Limits of injection processes (from Little[8])

types of ground, penetration can be assisted by controlled hydrofracture. For many years it has been known that, in tight ground, penetration of the grout can be assisted by raising the injection pressure to the point where the ground commences to fracture, thus extending the source point for the injection outwards from the borehole and increasing the area of ground exposed to grout infiltration. Very careful control is necessary to keep the treatment within bounds.

In Europe it has long been accepted practice to deliberately fracture fine-grained material with cement grout to introduce a network of strong set grout which will facilitate subsequent consolidation of the fine sediments by chemical grouts and it has been recognised that some compression of the fine materials also occurs and this must have some beneficial effect on stability. An example is Malgovert Tunnel, where a four-fold injection sequence was followed when the heading penetrated soft, saturated quartzite which disintegrated when the advancing tunnel caused a high pressure gradient to develop. A preliminary injection of sodium silicate assisted subsequent cement injections which in turn facilitated injection of quick-setting chemicals to bind the grains together, followed by a final cement injection to compress the ground to such an extent that explosives were needed to drive through it (see Pelletier[31]). These techniques have been adopted at other tunnels where similar running ground was encountered, notably Roseland Tunnel in France and Awali Tunnel in Lebanon. However, when weak soft sandstones under high pressure were met at Seafield Colliery, Scotland, development of similar inflows was prevented by careful injection with chemicals alone.

'MUD-JACKING' AND COMPACTION GROUTING

For some years a process, known as 'mud-jacking' in the USA, has been used whereby thick grouts were injected into voids beneath structures to prevent further settlement, and, if possible, to raise the building or slab to its former level. Thick grouts reduced undesirable grout losses into the ground. It was observed that the surrounding soil was compacted and consolidated during this process and it has been suggested that sands or silty sands can be consolidated by deliberately injecting grouts too thick to permeate the soil, so that they form a series of 'bulbs' of set grout round the injection point (see Graf[32], and Brown and Warner[33]). This process has been named compaction grouting.

The alternative term squeeze grouting has been proposed by Hutchinson and Daw[34] who commented that it is in practice difficult to ensure that overpressuring does not induce small-scale hydrofracturing. The process is susceptible to quantitative planning if the coefficients of consolidation, compressibility and permeability of the soil are known or can be estimated. In silty sands the pressure has to be applied for many hours, typically 30 to 40, to expel the pore water and repeated injections are necessary.

USES OF GROUTING

The primary purpose of grouting is usually to reduce the permeability of the soil and thus minimise leakage, as when it is used around or below excavations, or to act as an extension to the cut-off of a water-retaining structure. While it is not possible to state with any certainty what will be the result of any given grouting programme, which depends on too many variables, it is fairly certain that a given degree of watertightness can be achieved if sufficient grouting is performed. In general terms a cut-off in alluvium of 10^{-6} m/sec and in rock of 10^{-7} m/sec is attainable although in individual cases an improvement of one or more orders over these figures has been claimed. It is easier to achieve these figures if the initial permeability is several orders higher.

To improve the effect, the spread of grout in alluvium may be restricted by sheet-pile walls, as at Shek Pik Dam, Hong Kong. Grouting may also be used to seal gaps in and below sheet-piles or to extend sheet-pile or diaphragm wall cut-offs into underlying alluvium or bedrock. At Great Island, Eire, circular sheet-pile cofferdams were driven to the sea bed with grout pipes welded to selected piles. Cement plugs in the grout pipes were then drilled out and the holes extended into the bedrock and grouted to complete watertight cylinders into which shafts were raised from the underlying intake tunnels.

Grouting can also be used to strengthen the rock, in order, for instance, to render the abutments of arch dams more homogeneous and to restore any shattering due to the use of explosives. In fact 'blanket' grouting to a depth of 5 to 10 m is almost standard practice below heavy structures such as gravity dams and their associated spillways and powerhouses. It is difficult to quantify the improvements achieved, but Lancaster-Jones[35] has listed sites where the changes in static or dynamic moduli and seismic velocity due to grouting have been measured. Knill[36] has shown that the variations in seismic velocity due to grouting can be correlated with the quantities of grout consumed, and has suggested that this may be a way of estimating grout take in advance. Granular material can also be consolidated, to strengths similar to concrete if cement grout can be injected but to appreciable strengths even with chemical grouts. Resin or polymer-based grouts are usually stronger than silicate.

Another use for grouts is to fill inaccessible voids which would prove a hazard to structures erected above or near them. Caverns in limestone; abandoned tunnels and sewers; old workings for coal, oil shale, and chalk, not always shown on mine maps; and known tunnels which have become flooded, are some of the hazards treated in this way. A cheap grout based on PFA alone, or with some cement added, will often suffice, but in flowing water more elaborate measures are needed. Bituminous emulsion and cement grout (as a strengthening medium and source of lime for emulsion cracking) may be injected through adjacent or co-axial pipes, and hot bitumen has been mentioned above. A cement-diesel oil mixture sets instantaneously on contact with water, and has been used with inert fillers such as cotton flock or coconut fibre (see Clarke[37]). Underground fires in tips or shallow mines have been quenched by injecting a grout containing limestone dust which gives off carbon dioxide on combustion; the voids can subsequently be stabilised with cement grout.

REINFORCEMENTS

Rock masses near enough to the surface to be involved in civil engineering works are usually intersected with discontinuities such as joints, faults and bedding planes even before they are shattered by excavation techniques. The rock material itself may be physically or chemically attacked by weathering so that the bond between the constituent grains is loosened. The very act of excavation leads to a change in stress pattern with consequent tendency to deform.

This tendency can be resisted by the use of tendons, of steel or other material, which are fastened into the rock mass at each end. Generally one fastening is on the surface, by means of a plate and nut, or concrete pad, while the other end is anchored into the rock mass by a cone and wedge, cement or resin grout, expanded shell, or by physical attachment to a plate or pad in another chamber or tunnel, either pre-existing or purpose-made.

The tendon may be unstressed, in which case deformation will be needed to stretch it until sufficient stress is developed to prevent further strain. Such a system is said to be one of rock support and joins other methods of passive support such as tunnel linings, pit props, retaining walls, and bridge abutments which permit limited deformation. These tendons are usually called anchor bars and will not be further considered here. Alternatively the tendon may be stressed after insertion, in which case movement of the rock mass is prevented unless the load exceeds the amount of prestressing. This system is entitled 'rock reinforcement' and its prime purpose is to increase the shear strength along planes of weakness and thus render stable an otherwise unstable rock mass. Alternatively thin strata may be locked together to form a rigid beam over an opening, or circumferential stresses may be developed so that a stable arch is formed. Stressed tendons may be rigid (rock bolts) or flexible (cable anchors), the latter usually but not always being the longer. Tendons, whether stressed or not, may be protected against corrosion by total immersion in grout or bitumen, by wrapping in Denso tape, or by using polypropylene-coated strands, which must be bared in the anchor length. Restressing may be facilitated by coating the strands with grease below the polypropylene cover or by heating the bitumen coating.

ROCK BOLTS

The latest theories and practices concerning rock anchors have been described at length recently by Littlejohn[38] while rock bolts have been reviewed by Lang[39]. The latter extends his review, which mainly concerns tunnelling, to cover the use of rock bolts alone or in conjunction with, successively, light wire mesh, steel reinforcement fabric welded to the bolt heads, and finally sprayed gunite or shotcrete. The rock to be supported becomes more fractured or liable to spalling and disintegration, until finally the thickness of gunite reaches a point where it can be considered a structural member with its own inherent strength rather than a means of mobilizing the strength of the rock.

Rock bolts are frequently used to stabilise road cuttings, as on the M5 in the Clevedon Hills in Somerset (see Eyre[40]) where the bolts are individually arranged to stabilise specific blocks of rock, while the joint pattern closely controlled the rock bolting pattern at Edinburgh Castle (see Price and Knill[41]). More regular patterns are usually adopted underground, perhaps the most extreme case being the Mont Blanc road tunnel, completed in 1962. On the French side up to fifty bolts had to be placed between each round to secure the 80 m^2 face before the next round could be drilled, while up to five bolts per square metre were needed in the roof. Although only about half this quantity was needed on the Italian side, nearly a million metres of rock bolts were installed before the face hol through.

CABLES

Stressed cables frequently fulfil a structural role, as when tying back a retaining wall, holding down a dock floor against uplift, or resisting overturning moments in dams after raising. Cables are, however, now finding a use in slope stability design in opencast mines or other large excavations. The cables can either be extended to a point behind the theoretical slip/circle or can be used in combination with a grillage of concrete beams to increase the coefficient of sliding friction where slab failure is expected, as at El Atazar Dam in Spain where more than one hundred cables, each stressed to 2.2 MN, held seven concrete beams against the fractured slates of the right bank. At Kawamata Dam in Japan, the left abutment was strengthened against tensile failure by a total prestressing force of 155 MN applied through pairs of opposing steel bars installed from galleries and stressed to 400 kN, while further downstream a concrete block was anchored against the rock with a total load of 8500 tonnes applied through stressed cables.

Although cables have usually served only a structural purpose in dams, at two recent sites dams have been constructed where the foundations were defective and cables were installed as a cure. Meadowbank Dam (Tasmania) and Muda Dam (Malaya) are buttress dams built on foundations consisting of alternating layers of sandstone and mudstone and there was a risk of sliding failure where the mudstone had been altered to a clayey material. Cables raking upstream increased frictional resistance to sliding and provided an upstream force to resist movement. In each case 2.7 MN cables were installed, 152 No. at Meadowbank (see Maddox[42]) and 205 No. at Muda (see Gosschalk and Taylor[43]).

GUNITE

Gunite is a pneumatically applied mortar which contains aggregates up to 25 mm in size (see Fox[44]). Because the material is sprayed into place it lends itself to bends and curves and rates of placement are high. Thicknesses of 200 to 300 mm may be achieved particularly when steel mesh is used; it strengthens very rapidly. At twenty-eight days flexural strengths between 550 and 700 kN/m^2 are normal and compressive strengths of about 4.2 MN/m^2 may be achieved after this period.

The modulus of elasticity of this material exceeds 35 000 MN/m^2 and shrinkage varies between 0.03 and 0.106%. Thus gunite can be used for stabilising poor ground, especially when combined with rock bolting or steel reinforcement, and has proved excellent as a temporary support for all qualities of rock. It has been used for treating slopes and tunnels, as channel and reservoir linings, for structural and repair work, and in combination with other techniques such as contiguous bored piles.

SYNTHETIC FABRICS

The advent of fibre technology for cementitious mixtures, whether they be glass, polypropylene or steel, has enabled guniting to become more versatile. The inclusion of fibres has allowed thinner sections to be formed, and the elimination of steel mesh reinforcement so that gunite follows the rock profile more closely. Moreover the surface of the gunite resists impact and abrasion more readily.

The use of fabric membranes in civil engineering has made rapid progress over the last ten years. Fabrics now have a wide range of different structures and properties but in civil engineering they can provide three basic functions, namely, reinforcement, separation and filtration.

According to Cannon[47] synthetic fabrics for use in civil engineering have many qualities which include high strength; ability to elongate without rupture under high loadings and repetitive stresses; flexibility; stable in contaminated ground waters; resistance to fungal attack; and high abrasion resistance.

The stress-strain behaviour of fabrics varies over a wide range. Some woven materials have strengths, elastic moduli and breaking strains, and creep properties approaching those of steel meshes. Therefore they could be substituted for steel or aluminium reinforcement in reinforced earth. On the other hand non-woven fabrics have much lower strengths and moduli and have high creep properties. They therefore cannot take off much load in a fabric reinforced system. Nevertheless McGown[48] has shown that strong low extension reinforcements restrict strains within a system by taking part of the load themselves. At the other end of the scale, the low strength and low modulus materials induce differential strains in the soil layers above and below them which may mobilise strains in a greater volume of the soil or at least modify the pattern of strains

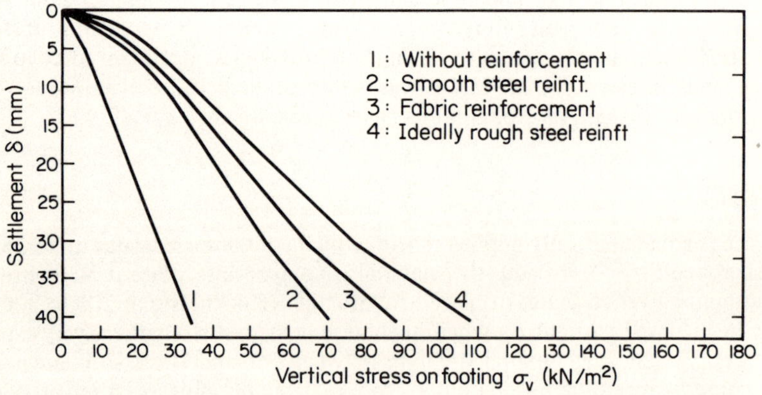

Figure 14.8 The figure shows load-settlement curves for an 80 mm footing (width, B) on loose sand with a single layer of reinforcement at a depth of 32 mm (depth, d). The displacement strain fields showed that strong low extensive reinforcements caused the sand to behave in a strain restricted manner whilst the fabric included ply-soil system exhibited a much more widely distributed strain pattern

over that in the unreinforced soil mass (Figures 14.8 and 14.9). In either case the system becomes more ductile and may increase its gross load bearing capacity. Such systems are known as ply-soil systems.

It appears that no particular values for fabric strength, modulus or breaking strain are of critical importance, but, as remarked, where a restriction of strain within a system is required then fabrics with the highest strengths and moduli and lowest breaking strains should be adopted. In systems which can accept a higher degree of strain, the low strength, low moduli, high breaking strain and lower cost fabrics can lead to more ductile, less brittle systems which in turn can lead to greater load bearing capacity. The former type of fabrics could be used in reinforced earth retaining walls whilst the latter could be used in granular soil embankments over highly compressible clays or peats.

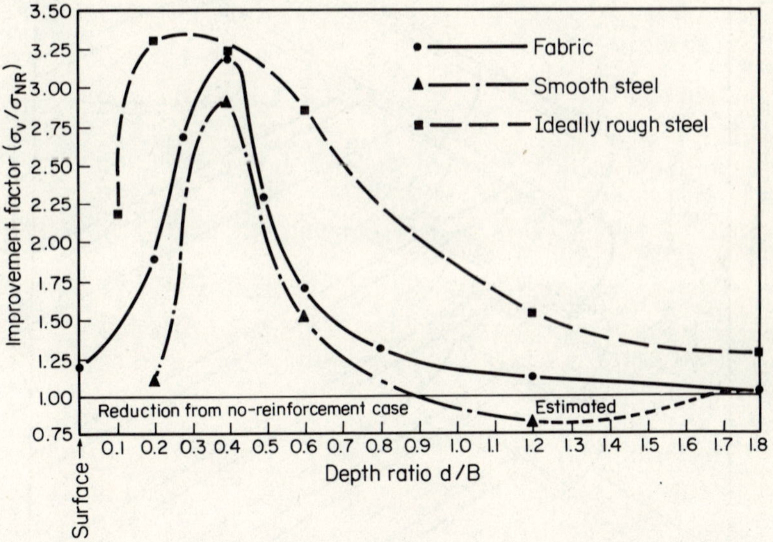

Figure 14.9 The figure shows the relative improvements of the footing bearing pressure on the reinforced system at a specific relative settlement. This clearly indicates that the depth of the reinforcement is critical. This improvement is not a constant with progressive straining of the system

The most commonly employed role of fabrics in roads or stabilisation of poor ground is as a separator/reinforcement layer at the sub-base/subgrade surface, thus preventing mixing of the two materials and increasing the load bearing capacity of the soil aggregate system. The savings in sub-base materials lost due to mixing with the subgrade can usually cover the cost of the fabric. Inclusion of fabric effects an improvement in pavement performance and it seems that sub-base depths can be reduced by approximately 30%, depending on soil and sub-base traffic loading conditions. Fabrics can also be used to advantage on reclaimed land and man-made fills. The positioning of a fabric layer within a soil system is critical, and the use of two discrete layers can give considerably greater improvement than simply twice the benefits of a single layer.

EARTH REINFORCEMENT

Ordinary construction fill can carry loads if the tendency for material to flow from beneath is counteracted. This can be achieved by using steel reinforcing strips, the soil being held in place by the frictional forces between the particles and the steel. Those particles not in direct contact with the steel are assumed to be held by arches of particles spanning between successive strips. The soil at the external surfaces of the structure is held by a facing skin.

Reinforced earth walls are constructed by erecting a thin front skin at the face of the wall whilst at the same time the earth is placed (see Baxter[45]). Thin steel strips are fixed to the back of the facing skin (Figure 14.10) at regular

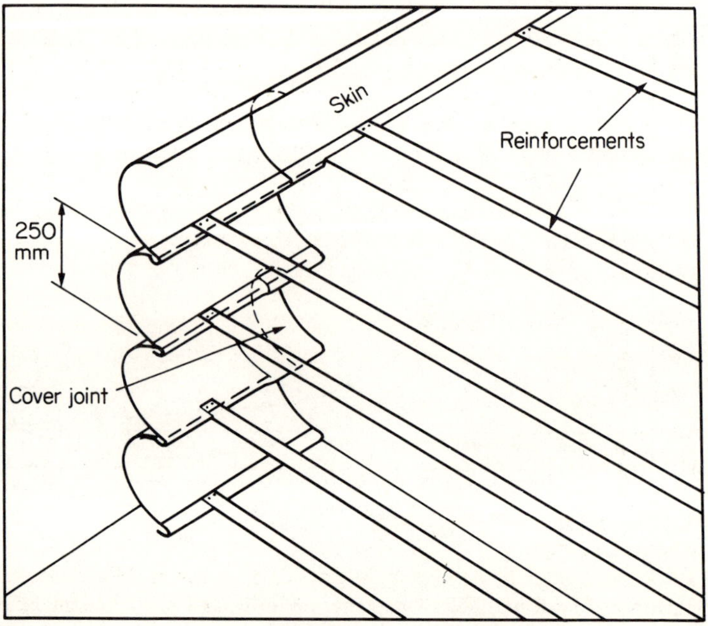

Figure 14.10 Earth reinforcement

intervals so that as each layer of compacted fill material is laid down, a number of strips are placed on top of it and buried by the next layer.

These structures are flexible and will settle and spread lightly. Consequently reinforcement must be flexible, strong in tension and, of course, corrosion resistant. Galvanised or stainless steel strips, only a few millimetres thick, are generally used. Glass fibre reinforcing strips were used in a trial embankment on the M62. The fill must give a good friction value with the reinforcement and have good drainage properties. This therefore excludes clay, fly ash and chalk. The facing material must also be flexible.

Reinforced earth can be used for retaining walls, vertically-sided embankments, dams, platforms on sloping ground or for cut-and-fill tunnels. Its adaptability to deformation means that it is particularly useful on poor ground where conventional alternatives would require expensive foundations.

Research carried out at the Laboratoire Central des Ponts et Chaussees (LCPC) in Paris, and quoted by Price[46], established that there were two modes of failure in reinforced earth. Firstly, the reinforcement broke (the tension mode), and secondly, there was a loss of adherence between the soil and the reinforcement (the bond mode). However, it was observed that if the length of reinforcement exceeded 0.8 H, where H is the height of the wall, failure due to loss of bond did not occur. As a result of this research a design method was proposed which involved calculating the maximum tension in the various levels of reinforcement and then calculating that length of reinforcement for which bond failure will not occur.

SOIL STABILISATION TECHNIQUES (see Bell[53])

Soil is a highly variable and complex material but because of its universal availability and low cost it is frequently used for engineering purposes. A soil at a particular locality, however, may be partly or wholly unsuitable from the point of view of construction. In such an instance it has to be decided whether to change the design, remove the soil, or to improve its engineering properties. The latter process is referred to as soil stabilisation. For example, soil stabilisation may be used to increase soil strength, to reduce its sensitivity, to alter its permeability, to reduce its compressibility or its frost susceptibility.

Although soil stabilisation may be brought about by a variety of means, because of the variability of soils no one method is ever successful in more than a limited number of soils. Consequently use of the correct method of stabilisation is dependent upon determining the character of the soil and which of its properties needs upgrading.

CEMENT STABILISATION

When the required stability cannot be obtained by mechanical means or where enhanced strength or resistance to water softening is required, stabilisation by the addition of cement or lime may provide the answer. Most inorganic soils can be stabilised by the addition of cement, although the amount of cement required and the resultant product vary with the character of the soil. The best results have been obtained with well graded soils in which the silt-clay fraction did not exceed 50% and the plasticity index was less than 20%. Clay in soil-cement can cause problems with pulverising, mixing and compacting the material. For satisfactory stabilisation, gravels generally require about 10% by weight of cement, sands 7 to 12%, silts 12 to 15% and clays 12 to 20%. Sands and gravels may develop compressive strengths between 3.5 and 10.5 MN/m^2 on being

Table 14.1 TYPICAL PROPERTIES OF CEMENT STABILIZED CLAY SOILS
(After Ingles and Metcalf, 1972[49])

Soil types	*Strength* kN/m^2	*Young's modulus* kN/m^2	*CBR*	*Permeability*	*Comments*
Sandy clays	1670–3500	690 × 10^4	600	Decreased by cement	Good material
Silty, sandy clays	690–1670	350–690 × 10^4	200		Sub-base for light traffic
Silty clays	350–1030	Less than 350 × 10^4	Less than 100	Increased by cement	Low-grade sub-base
Heavy clays	Less than 690	Less than 170 × 10^4	Less than 50	Increased by cement	Extreme difficulty in mixing Use of lime could be beneficial

stabilised with cement, whereas silty soils and clayey soils are usually below 2.0 MN/m^2 (Table 14.1).

Because the particles of non-cohesive soils are larger than those of cement, they are coated with cement paste during stabilisation. This binds the grains together when the cement hydrates. The improvement in strength depends upon the number of bonds between grains, the amount of cement used and the degree of compaction achieved. Mixing can either take place on site or materials can be premixed (see Lilley[50]).

Cement stabilisation in cohesive soils first requires that they are broken into small lumps, which are then coated by the addition of cement, the soil and the cement being well mixed. Next the material is watered and compacted. Clays are usually compacted slightly wet so that the moulding water content gives maximum density. According to Lilley, a secondary change occurs in clay soils which is due to the free lime from the cement penetrating the clay lumps, making them less cohesive and less prone to swelling or shrinkage with changes in moisture content.

When hydration of cement takes place in the cement stabilisation process it results in the formation of calcium hydroxide and the pH value of the aqueous phase is raised to approximately 12.2. The products formed after short periods of ageing are largely gelatinous and amorphous, but with further curing, poorly ordered varieties of hydrates calcium silicate and hydrated calcium aluminate develop. During the early stages of hydration gelatinous products form around anhydrous cement cores and are precipitated from solution in the interstices between particles. Hardening is due to the gradual desiccation of these gelatinous products and the crystallisation of new minerals. The process is promoted by the maintenance of an equilibrium condition between calcium hydroxide and the other components.

As the cement hydrates so the strength of the stabilised soil improves and it becomes water resistant. In fact the hydrated cement coatings form a skeletal structure of considerable strength, the actual strength depending on the lump size and the amount of cement used. The addition of as little as 2% cement to clay soil will generally modify the soil properties, the greater the proportion added the greater the change. Any cement may be used for stabilisation, Portland cement being most widely employed. In the UK soils subjected to cement stabilisation are designed to have a minimum 7-day compressive strength of 2.8 MN/m^2.

An indication of whether or not a clay is amenable to stabilisation by cement is provided by its Atteberg limits (see Croft[51]). Generally clays with plastic limits greater than 20 or liquid limits in excess of about 45 to 50 are considered uneconomic as far as this form of stabilisation is concerned. In other words such clays are difficult to mix and require a high addition of cement. It is often possible, however, to stabilise heavy clays after pre-treatment with a small amount of lime. This reduces the plasticity and thereby makes the soil more workable. The soil with lime is allowed to cure for one to three days and is then stabilised with cement.

The suitability of a clay soil for cement stabilisation is controlled by its texture and chemical and mineralogical composition. The results of several mineralogical examinations of clay-cement mixtures by certain authors, notably Croft[52], have suggested that certain of the clay minerals interfere with the stabilising action of cement. For instance, the expansive clay minerals may have a profound influence on the hardening of cement, notably montmorillonite, whose affinity for lime reduces the pH of the aqueous phase. The cementitious products developed during the curing of montmorillonite clay-cement mixtures are therefore inferior in their degrees of crystallinity to those of non-expansive kaolinitic or illitic clays. Hence the developed strengths are lower and unless cement addition is large enough to supply enough free lime to satisfy montmorillonite, as well as to promote hardening, the unsatisfactory properties of the clay may not be modified.

As much as 5% addition of cement may be necessary if noticeable changes are to be brought about in montmorillonitic clays whereas at least 15% is required to modify the clay significantly. Montmorillonitic clays, it would appear, respond to lime stabilisation, the early strengths of compacted mixtures often being higher than those developed in lime stabilised kaolinitic and illitic soils. Additional strength obtained after further ageing, however, is lower than that of the non-expansive clays. Addition of cement to montmorillonitic soils is therefore needed before the required strength is obtained.

Organic matter and excess salt content, especially sulphates, can retard or prevent the proper hydration of cement in soil-cement mixtures (see Bell[53]). One reason why organic matter retards the hydration of cement is because it preferentially absorbs calcium ions. The prior addition of hydrated lime may often enable such soils to be treated. The soil pH has little effect on the strengths developed with cement when large percentages of additive are present and expansive clay minerals and organic matter absent. Adequate strength, however, is difficult to achieve with economic amounts of additive in acid clay soils containing expansive clay minerals and, when more than 1% organic matter is present, stabilising reactions are inhibited. Sulphates have proved to be deleterious to soil-cement mixtures. It was assumed that sulphate interfered with cement hydrations and that crystallisation and expansion caused disruption of the soil-cement. Sherwood[54], however, demonstrated that the adverse effect of sulphates was due to a reaction with the clay content. In other words the disintegration of cement-stabilised soils only occurs when a soil has an appreciable clay fraction and when there is an increase in moisture content above that at the time of compaction. Addition of sulphate resisting cements is accordingly of little value.

SOIL STABILISATION USING LIME

Lime has been used to stabilise soils since Roman times. Slaked limes are most commonly used, but it may also be used in admixtures such as lime-cement, lime-fly-ash and lime-bitumen. In fact most soils with a plasticity index of less than 10 do not react with lime. In such cases the addition of a pozzolan such as fly ash will cause a satisfactory reaction. The addition of lime to a clay soil agglomerates the fine clay particles into coarse, friable particles due to base exchange. As the lime reacts with the clay, or additives such as fly ash, it causes hardening due to the formation of calcium aluminium silicates.

According to the lime stabilisation construction manual[55] the addition of lime to a clay soil:

(a) reduces its plasticity index sharply, thus making it more easy to handle;
(b) causes an increase in the plastic limit whereas the liquid limit falls;

(c) substantially decreases the soil binder content, thereby accelerating the break up of clay clods and making the soil more friable;
(d) appreciably reduces swelling and shrinkage;
(e) significantly increases the strength, for example, the unconfined compressive strength may be increased by as much as forty-fold;
(f) and substantially increases the Californian Bearing Ratio

The injection of lime slurry under pressure has been used to treat foundation soils in the USA. According to Wright[56] the method may be used to treat the foundation prior to building or as a remedial measure beneath an existing structure. Lime slurry pressure injection (LSPI) renders clay soils less active; the slurry forms a network of sheet-like seams. The seams also act as moisture barriers, stabilising the moisture content of the soil so that serious fluctuations are eliminated and therefore shrink-swell behaviour is reduced.

Recently Linden-Alimak[57] have introduced a method of stabilising cohesive soils by mixing quicklime into the soil by means of a giant steel whisk fitted to a rotary rig. As the whisk is screwed into the ground 3 to 8% of quicklime is added as required, injected by compressed air through a jet nozzle on the whisk. The rate of construction is rapid, producing 1 m of pillar, 1 m in diameter, per minute and the shear strength of the soil is increased more than ten-fold in an hour. The soil attains full supporting capacity after 30 to 90 days.

SOIL STABILISATION WITH BITUMEN

Bitumen, usually in the form of asphalt, has also been used for soil stabilisation. It stabilises the soil by acting as a binder for the particles and by protecting the soil from the deleterious effects of water. The density obtainable in a soil-bitumen mixture depends upon the volatile content of the bitumen and the amount and type of compaction undertaken. A low volatile content gives a higher strength. However, bitumen stabilisation does not improve the strength significantly. Asphalt, because it is too viscous to be mixed directly into the soil, is either heated, emulsified in water or thinned with petrol before being used. About 2 to 10% by weight of asphalt is incorporated into the soil and nearly every inorganic soil with which bitumen can be mixed can be so stabilised. The most satisfactory results are obtained in those soils which consist of less than 50% silt-clay fraction and have a liquid limit of less than 40% and a plasticity index of less than 20%. Organic matter is detrimental to soil-asphalt mixtures.

COMPACTION AND CONSOLIDATION

One of the most important aspects of any earthworks project is to ensure good compaction. Indeed compaction is probably the most important method of soil stabilisation. It involves packing the soil particles, by mechanical means, to bring about a decrease in the volume of air voids, thereby increasing the density of the soil. Compaction is extensively used in embankment construction, in the improvement of subgrades of roads and runways, and in fills. Consequently three properties need to be developed, namely, a high shear strength, low permeability and water absorption, and little likelihood of settlement under repeated loading.

It is almost impossible to recompact a disturbed soil to equal or greater density than its original state density without using heavy mechanical equipment, but weight alone is not the only requirement to effect compaction, of more relevance is the bearing pressure exerted at the soil surface. Moisture content is also an important factor in compaction, too much water is undesirable as the soil becomes sticky, while too little water makes compaction difficult or impossible. In fact the degree of compaction achieved is reflected by the dry density of the soil. As a result it is necessary to determine the bulk density (ρ_b) and its natural moisture content (m). The dry density (ρ_d) can then be derived from:

$$\rho_d = \frac{\rho_b}{1 + m}$$

Theoretically if a soil is compacted sufficiently a state will be reached when all the air expelled from the voids, the moisture remaining, so that the soil becomes effectively saturated. The saturation dry density for any given moisture content can be obtained once the specific gravity of the soil has been determined. However, it is impossible to achieve the saturation dry density during construction, this being an unattainably high standard of compaction, so that some arbitrary density must be chosen as the maximum state of compaction achievable.

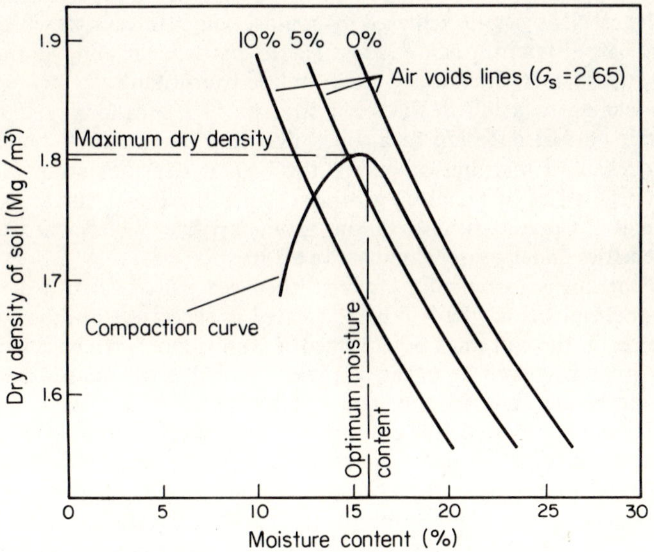

Figure 14.11 A compaction curve indicating maximum dry density and optimum moisture content

The relationship between dry density and moisture content, at increasing moisture contents, for a particular compactive effort can be assessed by a series of Proctor tests (see BS 1377:1975[58]). These two values, for each test, can be plotted against each other, the resultant curve having a peak value of dry density at an optimum moisture content (Figure14.11). Additional moisture content above the optimum results in a reduction of compacted density. This loss of density is caused by build up of pore water pressure which cannot dissipate. This is true in a confined situation but where little confining force exists the soil flows away from the compaction zone due to the loss of shear strength. When the moisture content is less than optimum the compacted density falls off due to the formation of air voids. These are present because of either the increasing shear resistance to the compaction force, or increasing resistance to grain fracture.

The relative compaction achieved on site should, if possible, come within 90 to 95% of the maximum dry density obtained in the Proctor test. For large earthworks, full scale tests will probably be necessary, a test section being compacted with the plant which is going to be employed. In this way it is possible to determine the number of passes the machine needs make in order to obtain the desired dry density. The Department of the Environment[59] specifies that compaction of cohesive soils should only be carried out when the moisture content of the soil is not more than 2% above the plastic limit. If it exceeds this figure then the soil must be allowed to dry. Put another way, as far as clayey soils are concerned it is better to compact at the maximum load which the equilibrium moisture content of the soil will sustain. As far as granular material is concerned it can be compacted at its natural moisture content. However, in defining requirements for compaction based on moisture content, it is important to keep in mind the purpose of the work. For example, earth dams should be constructed slightly wetter than optimum to keep down permeability.

The highest densities are achieved by compaction in those soils which contain a mixture of different particle sizes. Hence particle crushing during rolling can have an effect on the finished work. Indeed over-rolling can create an excess of fines, leading to a fall in the shear strength and occasionally it may even destroy the natural cohesion in a soil. Fuller and Thompson[60] demonstrated that in mechanical mixtures of gravels the highest densities attainable corresponded with a series of preferred gradings. Hence the Fuller curves express the size gradings for materials with various given top sizes which will pack to the highest densities under a given compactive effort.

If compaction is to be fully effective then the existence of a constraining layer around and below the soil being treated is necessary, in other words the shear stresses in the soil must be confined if compaction is to be achieved. This confining mass may well be in the soil itself, provided the applied compressive forces dissipate within a short distance of the point of application. However, where a layer of saturated soil or soil of a low bearing capacity underlies the layer being compacted, no real compaction can be achieved. This is because the lower layers cannot offer a reaction or resistance to the compactive load and deform so negating the compactive effort. If such a situation occurs then light rolling may be adopted to develop the soil strength, this being followed by heavier rolling to complete the compaction.

The soil to be compacted is spread in layers about 150 to 300 mm thick, each layer being compacted by a roller or rammer. For instance, natural subgrade to be built on is generally compacted to a depth of 150 mm. Fill is distributed in uniform layers, usually not exceeding 200 mm although 300 mm can be achieved with some materials and compaction plant. Compaction usually results in a layer depth of from two-thirds to three-quarters the loosely spread depth and is usually accomplished with a maximum of 6 to 10 passes of the compaction equipment. Increasing the number of passes beyond this usually proves uneconomic.

Heavier rollers achieve adequate compaction with fewer number of passes than do light rollers. However, sinkage, rutting and shoving of material can become a problem if roller weight is too large. Most of the surface compaction machinery must operate on soils at or below the optimum moisture content of the soil because they cannot cope with the compaction of loose or weak saturated soil in which they tend to become immobilised or bogged down.

DYNAMIC CONSOLIDATION

Dynamic consolidation by the Menard process is 'the improvement in the mechanical characteristics of a soil . . . by the repeated application of very high intensity impacts to the surface' (see Menard and Broise[61]). This is achieved by dropping a large weight, typically 10 to 20 tonnes, from a crawler crane, from heights of 15 to 40 m at regular intervals across the surface (Figure 14.12). The

Figure 14.12 Dynamic consolidation (Courtesy of Cementation Ltd)

high intensity stresses developed by tamping can penetrate deeply into the ground enabling compaction to take place to depths of as much as 12 m. Repeated passes are made over the site although several tampings may be made at each imprint during a pass. The first pass at widely-spaced centres improves the bottom layer of the treatment zone and subsequent passes then consolidate the upper layers. Care must be taken in establishing the treatment pattern, tamping energies and number of passes for a particular site and this should be accompanied by *in situ* testing as the work proceeds.

As far as coarse granular fill is concerned more energy is required to overcome the possibility of bridging action than is necessary for similar depths of finer material. The total energy imparted varies from 200 to 400 Tm/m^2.

Before subjecting sites which have previously been built over to dynamic consolidation underground services, cellars etc should be located. Pearce[6] suggested that old foundations should be demolished to approximately a metre depth below the proposed new foundation level prior to consolidation. Dynamic consolidation has proved a satisfactory method of deep compaction for coarse mine discard.

In finer materials the increased pore water pressures must be allowed to dissipate between passes, which may take several weeks. Menard and Broise[61] have pointed out that most saturated Quaternary soils contain 1 to 4% of gas in the form of micro-bubbles and it is the compression of this gas which allows imprints to occur. As the volume of the gas diminishes the soil approaches liquefaction, an undesirable state which is avoided in practice by careful timing combined with the use of a patented dynamic oedometer. However, according to West and Slocombe[63] the production of high pore pressures in low permeability soils due to tamping, leading to partial liquefaction also creates a new many-channelled drainage network. The latter allows consolidation to occur much more rapidly than would be the case under static loading. Regular tamping also initiates fissuring in the soil around the imprints which permits rapid dissipation of pore water. The effect of the consolidation can be checked by static penetrometers or pressuremeter tests.

Dynamic consolidation may normally be utilised up to 20 m from adjacent structures or services. Fortunately the transmitted vibration is low between 5 and 10 Hz. Indeed it is similar to that developed on quarry blasting where the peak particle velocity damage is 50 mm/sec (see chapter 17). The vibrations are usually appreciably below this limit.

COMPACTING NON-COHESIVE SOILS

Pile-driving at close centres can compact non-cohesive soils, the grains being physically displaced by timber or concrete piles, and also shaken into a closer packing by the vibration of the pile-driver. The method is an old one first described by Rankine and subsequently referred to by Glossop[64].

Explosive compaction to densify granular soils was described by Lynam[65] over thirty years ago. He found that loose sands lying under a zone of confining stable ground would compact by detonation of charges within the layer. The sand grains were shaken into a denser pattern and the released pore water expelled. Indeed the best results have been obtained in clean, loosely packed, coarse grained sands or fine grained gravels. The size and spacing of charges and use of the method in fill have been described more recently (see Prugh[66]) and in Russia charges exploded in water have been used to compact fill placed below sea-level (see Ivanov[67]). No applications of the method in Britain are known although the use of buried charges in cohesive soils ('camouflets') to make small chambers is standard military practice and has been considered for cable anchors. Small charges have been used in boreholes to fissure ground and facilitate grout penetration (the 'shoot 'n grout' technique proposed for the drifts at Selby, Yorkshire).

PRELOADING

Preloading consists of applying a load to the ground equal to or greater than the proposed foundation loading, so that settlement of the ground will be complete before the structure is erected, or in the case of a highway embankment so that settlement is completed prior to the pavement construction. The method is most suitable for loose granular soils or granular fills where settlement is rapid. It is also suitable for treating saturated clay soils. The surcharge weight used in both cases is normally derived from a local borrow pit. Preloading is most economical over a large area where the surcharge material can be provided in bulk and moved progressively across a site by using large earth moving equipment.

Care must be taken when applying the technique of preloading to saturated clays. For example, shear failure may occur if the application of preload is too rapid. This is because pore pressures are slowly dissipated in clay soils, hence consolidation is slow. Also preloading has to be sustained over a long period to be effective. To increase the effect of this surcharge and further accelerate consolidation it is necessary to reduce the drainage paths within the soil. Any reduction in the length of the drainage path is effective in reducing the time of settlement since the time varies with the square of the distance to a drainage channel.

SAND DRAINS AND SANDWICKS

The normal method of introducing drainage channels into a mass of soil is by the construction of sand drains. Indeed sand drains have been used for more than fifty years in order to dissipate pore water pressure and accelerate consolidation of silts and clays subjected to increased loads. The shear strength of the soil is also increased more quickly.

The traditional type of sand drain consists of a vertical borehole sunk through the compressible layer of soil to a relatively firm base and backfilled with a clean, coarse sand. There are instances where drains have exceeded 30 m in depth. The diameter of a sand drain varies from approximately 300 to 600 mm, in other words a sand drain need only be large enough to drain away expelled water and to allow for a margin of safety against clogging. The expelled water is usually drained into an overlying drainage blanket, however, if there is a permeable layer beneath the compressible soil which can act as a soakaway then the sand drains can be taken into it. The distance between individual sand drains is influenced by the permeability of the soil but generally they are spaced somewhere between 2 and 5 m centres. What is more their spacing should be less than their depth. The aim is that the spacing should effect most of the total consolidation during the construction period.

The method has now been refined to enable smaller diameter boreholes to be used, albeit at closer spacing, the sand fill being inserted when enclosed in a flexible, porous stocking of woven polypropylene. This is referred to as a sandwick. One of the reasons for this development is that it is difficult to construct sand drains of less than 250 mm diameter. It has been shown that the effect of diameter on the time of consolidation is much less than that of the drain spacing and that reduction in sand drain diameter only slightly lowers the efficiency. Various methods of boring are used to suit actual ground conditions and include auger drilling, jetting, driven casing and vibrated casing. The tube of sand is then lowered into the borehole. The usual diameter for this type of drain is 60 to 70 mm.

Sandwicks offer several advantages over sand drains. For example, they are flexible and therefore can withstand movement. Unlike the larger diameter sand drains they are less likely to act as piles and thereby restrict settlement. They are easier to install using lighter equipment than traditional methods and the flexible stocking acts as a filter reducing the tendency of the drain to clog with clay. Sandwicks have been used in embankments, particularly where the embankment has been built together with a temporary surcharge to speed consolidation, below oil tanks again with the tank preloaded to speed consolidation, in slopes in clays to lower the water table, and below proposed raft foundations, again to accelerate settlement.

An alternative to using sandwick drains is a more recent development which is not yet widely used in Britain. This involves the use of vertical paper drains. These usually consist of a central ribbed or castellated plastic core with thick paper either side, and measure approximately 100 mm × 4 mm. They act principally as capillary wicks transferring the soil moisture to the surface free draining layer. They can be punched into the ground to depths of approximately 20 m, although they require specialist installation equipment.

TRENCHES AND FABRICS

Yet another recent development in drainage techniques for speeding consolidation is the use of continuous deep, narrow aggregate filled trenches. The machines involved were developed by a Dutch company and are similar to those used for trencher type deep drainage installations. The trenches are excavated to approximately 6 m deep and are about 230 mm wide, at centres, depending on ground conditions, typically 3 m apart. The machine cuts the trench and fills it with free draining gravel simultaneously. The trenches are cut across the width of the embankment if used on a motorway project. After trench cutting a 0.5 m free draining blanket is laid on top. The machines are capable of installing between 160 to 400 m a day.

Fabrics can be used to help consolidate compressible soils by acting both as a filter and as a separator. Although fabrics cannot reduce the amount of overall settlement, this being a function of the soil properties, they can help to reduce some differential settlement by making it possible to form a stable platform of aggregate, resting on a fabric membrane, above suspect soil. Fabrics can also help overcome the variability within a non-homogeneous soil.

VIBROFLOTATION OR VIBROCOMPACTION

Vibroflotation or vibrocompaction is a technique used to improve poor ground below foundation structures. For example, it has been used to compact loose granular soils thereby enabling them to carry heavy loads (see D'Appolonia[68]). The treatment of such soils is intended to minimise total and differential settlement, as usually there is small risk of shear failure. Indeed the process may reduce settlement by more than 50%. Variants of vibroflotation can be applied to soft cohesive soils and fills. Normal spread footings can be used on ground treated by vibroflotation.

Vibrations of appropriate form can eliminate intergranular friction of cohesionless soils so that those that are initially loosely packed can be converted into a dense state. The basic tool used to accomplish this is the poker vibrator or vibroflot which is suspended from a crane (Figure 14.13). This is used to penetrate soil at depth and can also operate efficiently below ground water, so compacting soils inaccessible to surface plant without resorting to drainage. Such compaction reduces the soil volume so that similar material is required to make up levels.

Figure 14.13 The vibroflot suspended from a crawler crane (Courtesy Cementation Ltd.)

According to Thorburn[69] the vibrator develops a gyratory motion in the horizontal plane which in turn is responsible for the horizontal vibrations which agitate the granular soil particles, the vertical components of forced vibrations being unimportant. The frequency of the soil vibrations is the same as that of the vibroflot and in order to compact loosely packed granular soils the amplitude and lateral reaction at the end of the vibroflot must be maintained, the larger the amplitude the larger the radius of influence (see Greenwood[70]). In cohesionless soils most of the compaction occurs at any given depth within about 5 min from the time when it was first subjected to vibration.

As remarked above, vibrations nullify intergranular friction between soil particles. This causes liquefaction in an annular zone around the vibroflot about 300 to 500 mm thick. Liquefaction means that the soil particles are rearranged without stress by gravitational forces and, because the particles are unstressed, no release of stress can occur after compaction, thus the soil is permanently strengthened.

An adequately powered vibroflot can penetrate sandy soils under its own weight and power of vibration but frequently water, emitted from the tip of the poker, is used to facilitate penetration. The prime function of the water is to prevent the sides of the hole collapsing. The degree of compaction attainable depends upon the grading of the soil and the spacing of the compaction centres.

The permeability of the soil is also an important factor since some pore water needs to be expelled to achieve satisfactory compaction. Indeed low permeability means that the technique is uneconomic and it is ineffective in clayey soils. Furthermore the technique is not suited to coarse gravels since some, particularly those which are poorly graded, inhibit the ability of the vibroflot to penetrate. On the other hand uniformly graded cobble gravel can be penetrated to useful depths. The best results are obtained in fairly coarse sands which contain little or no silt or clay, since the damping effects of the latter two soil types reduce the effectiveness of the vibroflot. For example, it has been shown that the radius of influence decreases from around 2 m in clean sands to 0.7 m in sands containing more than 20% fines. According to West[71] area treatment of sand involves a grid spacing of between 1.5 and 3 m and Thorburn[69] maintained that no significant compaction occurs beyond a radius of 2.5 m from the vibration centre. Obviously the effectiveness of the vibrations in achieving compaction decreases with increasing distance from the vibroflot but this non-uniformity of compaction may be neglected for design purposes.

Loosely packed sands may have varying density, as may granular fills. Such variations in density may mean that differential settlement occurs on loading. Accordingly when such soils occur on site they should be treated so as to produce a zone of uniformly high density beneath the foundation structure. Greenwood[70] suggested that for unit loads between 100 and 430 kN/m^2 a compaction spacing or maximum radius of influence ranging between 1.8 and 2.5 m proves satisfactory. Usually under rafts or closely spaced footings treatment should extend to a depth of about 6 to 8 m.

Cohesive soils cannot be treated by vibroflotation since the vibrations are effectively damped within a small distance from the vibroflot. Such soils, however, can be strengthened by the formation of columns of coarse backfill such as crushed rock or slag, at individual compaction centres within the soil. The principal purpose of such treatment is to eliminate shear failure. The vibroflot is used for compacting these stone columns which, in turn, effect a reduction in settlement. The stone columns are sunk to a depth where the ground is firmer or, where the strength of the ground increases with depth, to a depth determined by elastic theory. They are distributed in a regular pattern over the foundation area, spacing being determined by the initial strength of the ground and the required load bearing capacity. These columns, like piles, carry the load of the structure they support and by dilating they mobilise the passive earth pressure of the surrounding ground thereby improving the bearing capacity. Since the granular backfill replaces the soil this process is sometimes known as vibroreplacement (see Thorburn and MacVicar[72]; Rathgeb and Kutzner[73]).

VIBROREPLACEMENT

Vibroreplacement does not have a significant destructive effect on the soil structure and is therefore commonly used in soft, normally consolidated compressible clays, soil with thin layers of peat, saturated silts, and alluvial and estuarine soils. Stone columns have been formed successfully in soils with undrained cohesive strengths as low as 7 kN/m^2 (see Greenwood[74]); although Thorburn[69] wrote that he had never used vibroreplacement in soils with undrained shear strengths of less than 14 kN/m^2 because of the low radial support afforded the stone columns. Any excess pore pressures developed as a result of loading are quickly dissipated through the columns which means that any settlement which may occur does so rapidly, as well as being appreciably reduced in amount, and the shear strength of the soil is enhanced. Columns need a closer spacing about the perimeter of a structure than they do in the centre.

In vibroreplacement a water jet is used to assist penetration wherever the borehole formed by the vibroflot is likely to be unstable and where ground water flows into the borehole. Water is allowed to overflow at the surface so that excess hydrostatic pressure and outward seepage forces support the uncased hole. The water is conveyed away in specially cut channels. It is sometimes possible to operate in fly ash without using water, but smaller columns are produced. Vibroreplacement has also been used dry to treat demolition debris on sites cleared of old houses. On reaching the required depth of treatment, coarse backfill is fed into the annular space surrounding the vibroflot and this is slowly withdrawn, surging up and down to effect compaction.

VIBRODISPLACEMENT

Vibrodisplacement is another related technique. This involves the vibroflot penetrating the ground by shearing and displacing the ground around it. The hole produced is infilled in stages with coarse backfill, the vibroflot compacting the material at each stage. The mode of formation of the generally irregular, roughly cylindrically shaped columns means that the passive resistance of the surrounding soil is fully mobilised at very small horizontal radial strains when the stone columns are subjected to vertical loading.

Water jetting is not used in vibrodisplacement but compressed air is used to break the suction which develops as the vibroflot is withdrawn, otherwise the sides of the hole are likely to collapse. If ground water is present within the hole great care must be taken to minimise the production of slurry due to the emission of compressed air, as the slurry will contaminate the stone column. This can reduce its shearing resistance significantly.

Vibrodisplacement is restricted to strengthening insensitive soils which have sufficient cohesion to maintain a stable borehole, that is, to those with over 20 kN/m^2 undrained strength. These soils are neither very weak not very compressible but require treatment primarily to boost their bearing capacity. In these soils displacement methods induce some measurable increase in the strength of the clay between these columns, this does not happen with vibroreplacement.

STONE COLUMNS

Hughes and Withers[75] outlined three loading situations for which stone columns could be used. Firstly, they could be used as single columns to support point loads such as small pad footings; in such cases the column has equal lateral restraint all round. Secondly, they can be used to support strip footings, the column being restrained more in the direction parallel to the strip. Thirdly, they can support widespread loads where the column again has equal lateral restraint all round. However, as the soil settles under a large load the lateral resistance of the ground increases.

Thorburn[69] recommended that for most structures, for which settlement is an important consideration, the loads should be supported entirely by the stone columns. Hughes and Withers[75] showed that the vertical displacement of the tops of stone columns within the range of working stresses was less than half the maximum radial strain. The magnitudes of radial strain required to develop working stresses are small because of the considerable radial displacements which occur in cohesive soils as stone columns are placed. Densely packed, clean stone columns may settle some 5 to 9 mm, the amount of settlement can be predicted by obtaining radial stress-strain data by using a pressuremeter (see Hughes, Withers and Greenwood[76]).

References

1. Cashman, P.M., *Control of ground water by ground water lowering*, article in 'Methods of treatment of unstable ground,' Ed. by Bell, F.G.) Newnes-Butterworths, London, pp. 12–25 (1975)
2. Parsons, J.D., 'Foundation installation requiring recharging of ground water', *Proc. Am. Soc. Civ. Eng. (Cons. Div)*, **85**, pp. 1–21 (1959)
3. Zeevaart, L., 'Foundation design and behaviour of the Tower Latino Americano in Mexico City,' *Geotechnique*, 7, pp. 115–133 (1957)
4. Casagrande, L., 'The application of electro-osmosis to practical problems in foundations and earthworks,' *Build. Res. Tech. Paper No. 30*, DSIR (1947)
5. Casagrande, L., 'Electro-osmotic stabilisation of soils,' *Jour. Boston Soc. Civ. Eng.*, **39**, pp. 51–83 (1952)
6. Farmer, I.W., *Electro-osmosis and electrochemical stabilisation* (article in 'Methods of treatment of unstable ground,' Ed. by Bell, F.G.) Newnes-Butterworths, London, pp. 26–36 (1975)
7. Grim, R.E., *Applied clay mineralogy*, McGraw-Hill, New York.
8. Little, A.L., *Ground-water control by exclusion* (article in 'Methods of treatment of unstable ground, Ed. by Bell, F.G.) Newnes-Butterworths, London, pp. 37–68 (1962)
9. Akam, E.A.A. and Lawson, P., *Construction site noise*, Building Research Establishment Report (1975)
10. Tomlinson, M.J., *The problems of unstable ground: a review of modern techniques of ground treatment* (article in 'Methods of treatment of unstable ground', Ed. by Bell, F.G.) Newnes-Butterworths, pp. 1–11 (1975)
11. Fuchsberger, M., 'Some practical aspects of diaphragm wall construction,' *Proc. Conf. on Diaphragm Walls and Anchorages, Inst. Civ. Engr.*, pp. 75–80 (1975)
12. Sliwinski, Z. and Fleming, W.G.K., 'Practical considerations affecting the construction of diaphragm walls,' *Proc. Conf. on Diaphragm Walls and Anchorages, Inst. Civ. Eng.*, pp. 1–10 (1975)
13. Hutchinson, M.T., Daw, G.P., Shotton, P.G. and James, A.N., 'The properties of bentonite slurries used in diaphragm walling and their control,' *Proc. Conf. on Diaphragm Walls and Anchorages, Inst. Civ. Engr.*, pp. 33–40 (1975)
14. Keinberger, H., 'Diaphragm walls as load bearing foundations,' *Proc. Conf. on Diaphragm Walls and Anchorages, Inst. Civ. Engr.*, pp. 19–22 (1975)
15. Boyes, R.G.H., 'The versatility of diaphragm walling,' *Civil Engineering*, pp. 57–59 (Dec 1975)
16. Collins, S.P. and Deacon, W.G., 'Shaft sinking by ground freezing: Ely-Ouse Essex scheme,' *Proc. Inst. Civ. Eng., Suppl., Paper 75065* (1972)
17. Cleasby, J.V., Pearse, G.E., Grieves, M. and Thorburn, G., 'Shaft sinking at Boulby Mine, Cleveland Potash Ltd,' *Trans. Inst. Min. Met., Section A*, 84, A7–28 (1975)

18. Maishman, D., *Ground freezing* (article in 'Methods of treatment of unstable ground,' Ed. by Bell, F.G.) Newnes-Butterworths, London, pp. 159–171 (1975)
19. Sanger, F.J., 'Ground freezing in construction,' *Jour. Am. Soc. Civ. Eng., (Div. Soil Mech. Found. Eng.)*, **94**, SM1, pp. 00–00 (1968)
20. Beles, A.A., 'Le traitment thermique du sol,' *Proc. 4th Int. Conf. Soil Mech. Found. Eng.*, **3**, pp. 266–267 (1957)
21. Glossop, R., 'The invention and development of the injection processes,' *Geotechnique*, **10**, 91–100; **11**, 255–279 (1960, 1961)
22. Grouting Symposium, *Grouts and drilling muds in engineering practice*, Butterworths London (1963)
23. Skipp, B.O., *Clay grouting and alluvial grouting* (article in 'Methods of treatment of unstable ground, Ed. by Bell, F.G.) Newnes-Butterworths, London, pp. 141–158 (1975)
24. Terzaghi, K., 'Discussion on methods for improving the physical properties of soils,' *1st Int. Conf. Soil Mech. Found. Eng.*, **3**, pp. 180–182 (1936)
25. Cambefort, H., 'Injection des sols,' *Eyrolles, Paris*, **1**, pp. 25–8 (1964)
26. Londe, P. and Sabarly, F., 'La distribution des permeabilites dans la foundation des barrages voutes en function du champ de contrainte,' *1st Cong. Int. Soc. Rock Mech., Lisbon*, **2**, pp. 517–521 (1966)
27. Snow, D.T., 'Rock fracture spacings, openings and porosities,' *Jour. ASCE (Soil Mech. Found. Div.)*, **93**, SM1, pp. 73–91 (1968)
28. Sabarly, F., 'Les injections et les drainages de foundations de barrages,' *Geotechnique*, **18**, pp. 229–249 (1968)
29. Lancaster-Jones, P.F.F., 'The interpretation of the Lugeon water test,' *Quart. Jour. Eng. Geol.*, **8**, pp. 151–154 (1975)
30. Moneymaker, B.C., 'Reservoir leakage in limestone terrains,' *Bull. Assoc. Eng. Geol.*, **6**, pp. 3–30 (1968)
31. Pelletier, J., 'The construction of Tignes Dam and Malgovert Tunnel,' *Proc. Inst. Civ. Eng.*, London (1951)
32. Graf, E.D., 'Compaction grouting techniques and observations,' *Jour. ASCE (Soil Mech. Found. Div.)*, **95**, SM3, p. 1151 (1969)
33. Brown, D.R. and Warner, J., 'Compaction grouting,' *Jour. ASCE, (Soil Mech. Found. Div.)*, **99**, SM8, p. 589 (1973)
34. Hutchinson, M.T. and Daw, G.P., *Squeeze grouting*, Cementation Research Ltd. Confidential Report R.15/75 (1976)
35. Lancaster-Jones, P.F.F., *Methods of improving the properties of rock masses* (article in 'Rock mechanics in engineering practice,' Ed. by Stagg, K.G. and Zienkiewicz, O.C.) Wiley, London, pp. 385–429 (1967)
36. Knill, J.L., *The application of seismic methods in the prediction of grout take in rocks* (article in '*In-situ* investigations into soils and rocks') *Brit. Geotech. Soc.*, London, pp. 93–100 (1970)
37. Clarke, J.F.F., *Discussion: grouts and drilling muds in engineering practice*, Butterworths, London, p. 112 (1963)
38. Littlejohn, G.S. and Bruce, D.A., 'Rock anchors – state of the art,' *Ground Engineering*, **8**, No. 3–9 (1975–1976)
39. Lang, T.A., Rock Reinforcement, *Bull. Ass. Eng. Geol.*, **11**, No. 3, pp. 215–240 (1972)
40. Eyre, W.A., 'The revetment of rock slopes in the Clevedon Hills for the M5 Motorway,' *Quart. Jour. Eng. Geol.*, **6**, pp. 223–229 (1973)
41. Price, D.G. and Knill, J.L., 'The engineering geology of Edinburgh Castle Rock,' *Geotechnique*, **17**, pp. 411–432 (1967)
42. Maddox, J.M. and others, 'Foundation studies for Meadowbrook buttress dam,' *Trans. 9th Int. Cong. Large Dams*, **1**, pp. 123–142 (1967)
43. Gosschalk, E.M. and Taylor, R.W., 'Strengthening of Muda Dam foundations using cable anchors,' *2nd Int. Cong. Int. Soc. Rock. Mech., Belgrade*, Paper 6–11 (1970)
44. Fox, K.H., *Use of shotcrete from the standpoint of the contractor*, (in use of shotcrete for underground structural support), *Am. Soc. Civ. Eng.* and *Am. Con. Inst.*, pp. 22–29 (1973)
45. Baxter, M., 'Reinforced earth,' *Civil Engineering*, pp. 30–33 (June 1974)
46. Price, D., 'Reinforced earth research,' *Civil Engineering*, 21–27 (Dec. 1976)
47. Cannon, E.W., 'Fabrics in civil engineering,' *Civil Engineering*, pp. 39–42 (March 1976)

48. McGown, A., *The properties and uses of permeable fabric membranes*, Univ. Strathclyde Res. Rept. (1976)
49. Ingles, O.G. and Metcalf, J.B., *Soil stabilisation: principles and practice*, Butterworths, Sydney (1972)
50. Lilley, A.A., *Stabilisation of soil with cement* (article in 'Methods and treatment of unstable ground, Ed. by Bell, F.G.) Newnes-Butterworths, London, pp. 172–181 (1975)
51. Croft, J.B., 'The problem of predicting the suitability of soils for cementitious stabilisation,' *Eng. Geol.*, **2**, pp. 397–424 (1968)
52. Croft, J.B., 'The influence of soil mineralogical composition on cement stabilisation,' *Geotechnique*, **17**, pp. 119–135 (1967)
53. Bell, F.G., 'The influence of the mineral content of clays upon their stabilisation by cement,' *Bull. Ass. Eng. Geol.*, **13**, 267–278 (1976)
54. Sherwood, P.T., 'The effect of sulphates on cement and lime stabilised soils,' *Roads and Road Construction*, Vol. 40, pp. 34–40 (1962)
55. American Road Builders' Association, 'Lime stabilisation construction manual,' *Tech. Bull. No. 243*, Washington, 35p (1959)
56. Wright, P.J., 'Lime slurry pressure injection Tames expansive clays,' *Civil Engineering, Am. Soc. Civ. Eng.* (1973)
57. Linden-Alimark, 'Quicklime eats into piling posts,' *Contract Jour.*, p. 28 (1976)
58. BS:1377:1975, *Methods of test for soils for civil engineering purposes*, British Standards Institution, London.
59. Dept. of the Environment, *Specification for road and bridge works*, HMSO, London (1969)
60. Fuller, W.B. and Thompson, S.E., 'The laws of proportioning concrete,' *Trans. ASCE*, **59**, pp. 67–143 (1907)
61. Menard, L. and Broise, Y., 'Theoretical and practical aspects of dynamic consolidation,' *Geotechnique*, **25**, 3–18 (1975)
62. Pearce, G.W., 'Dynamic consolidation new foundations for old,' *Civil Engineering*, pp. 45–49 (1975)
63. West, J.M. and Slocombe, B.C., 'Dynamic consolidation as an alternative foundation,' *Ground Engineering*, **6**, No. 6, pp. 52–54 (1973)
64. Glossop, R., 'Classification of geotechnical processes,' *Geotechnique*, **2**, pp. 3–12 (1950)
65. Lyman, A.K.B., 'Compaction of cohesionless foundation soils by explosives,' *Trans. ASCE*, **107**, 1330–1348 (1942)
66. Prugh, B.J., 'Densification of soils by explosive vibrations,' *Jour. ASCE (Const. Div.)*, **89**, CO1, pp. 79–100 (1963)
67. Ivanoff, P.L. and others, 'Binding loose soil by underwater explosions,' *Hydrotechnical Construction (Russia)*, No. 4 (1966)
68. D'Appolonia, E., 'Loose sands – their compaction by vibroflotation,' *ASTM Spec. Tech. Publ.*, No. 156, pp. 138–154 (1953)
69. Thorburn, S., 'Building structures supported by stabilised ground,' *Geotechnique*, **25**, pp. 83–94 (1975)
70. Greenwood, D.A., 'Mechanical improvement of soils below ground surface,' *Symp. on Ground Engineering, Proc. Inst. Civ. Eng.*, pp. 11–21 (1970)
71. West, J.M., 'The role of ground improvement in foundation engineering,' *Geotechnique*, **25**, pp. 71–78 (1975)
72. Thorburn, S. and MacVicar, R.S.L., 'Soil stabilisation by employing surface and depth vibrators,' *The Structural Engineer*, **46**, pp. 309–316 (1968)
73. Rathgeb, E. and Kutzner, C., 'Some applications of the vibroreplacement process,' *Geotechnique*, **25**, pp. 45–50 (1975)
74. Greenwood, D.A., 'Vibroflotation: rationale for design and practice' (article in 'Methods of treatment of unstable ground,' Ed. by Bell, F.G.) Newnes-Butterworths, London, pp. 189–210 (1975)
75. Hughes, J.M.O. and Withers, N.J., 'Reinforcing of soft cohesive soils with stone columns,' *Ground Engineering*, **7**, pp. 42–49 (May 1974)
76. Hughes, J.M.O., Withers, N.J. and Greenwood, D.A., 'A field trial of the reinforcing effect of a stone column in soil,' *Geotechnique*, **25**, pp. 31–44 (1975)

Chapter 15

The Stability of Slopes in Difficult Ground

Stability rather than deformation is the critical consideration in the design of slopes whereas in foundations, especially those in difficult ground, the opposite is usually true. Like all engineering structures, slopes must be designed to resist disturbing forces over long periods. When the slope is designed in strong and intact ground, the measure of the stability at the construction stage may, in some circumstances, also be the criterion for its stability at later periods. Even in good ground this statement cannot be of general application.

When soil is difficult e.g. showing high water content, fine particle grading, low shearing resistance, over-consolidation with fissures, stability achieved at the construction stage is not likely to be maintained indefinitely. This is the principal characteristic of difficult slopes when compared with slopes of sound material. There is overwhelming evidence from case studies that a simple analysis of stability is inadequate to ensure the absence of long-term failure within the expected life-span of the slope.

A description of some of the aspects of 'difficulty' which can be encountered in the design and construction of slopes is given in the previous paragraphs, but other features can also be found. Not only may the soil of which the slope is to be constructed have unsuitable properties tending to encourage instability, but its past geological history and possible future deterioration with time, may cause it to be a difficult material requiring study, planning and strengthening if long-term stability is to be achieved. Difficult soils do not conform to the simpler conditions of homogeneity and normal consolidation.

Both homogeneity and normal consolidation allow at least an assumption of performance in favourable circumstances. Difficult soils when encountered in an excavation or cutting, show diverse types of non-homogeneity, with perhaps stratified layers of soil; impermeable strata may support aquifers and soil water may flow at various levels. Soils can also be overconsolidated; the release of the original consolidation pressure, whether of sediments or of thick ice, allows of expansion and of the formation of fissures, both of which trends can be dangerous to long-term stability.

It is assumed in this chapter that embankments are not a subject of immediate concern. The construction of banks and dams of selected material is entirely under the control of the designer. Unless exceptional circumstances prevail, banks would not be built of difficult materials. Thus the stability of artificially constructed slopes is not considered here. The problems arising in slopes cut in natural strata cannot be solved by the controlled choice of materials, as is possible in dam construction. The design must be adapted to the materials and conditions prevailing. In this chapter, therefore, the emphasis is on slopes cut in naturally occurring and difficult strata.

MECHANISMS OF INSTABILITY

The engineer may satisfy himself that the ground into which a cutting is to be made is stable at the present time and has been so for as long as records are available. This assurance, however, especially if the soil is suspected of being 'difficult', is not enough. The excavation of a cutting, releases unbalanced forces which formed part of an equilibrated system up to that moment. These unbalanced forces can, and often do cause instability in an area where there have been no known episodes of slip or settlement. The engineer forming a slope in natural soil must be sensitive to the balance of forces in the soil. He must have an intuitive 'feel' for the pore water pressures which exist and be able to visualise how these pressures will be dissipated, redirected or intensified when structural operations commence.

Instability occurs when slip surfaces develop and movements are initiated in the mass of soil. Although there are many variations in the types of slip surface produced, a simplification which is acceptable in this short treatment of the subject, is to place examples of instability in two categories, namely *rotational slips* and *parallel slides* (see Chapter 13). Although individual failures of stability can show characteristics of both of these types, and this applies particularly in natural landslides, most failures in slopes in difficult soils can be attributed to one or other of the mechanisms of shear failure indicated by these two phenomena.

The analytical study of these mechanisms of instability must, inevitably, tend to consider idealised soils when the basic principles are being applied. Later adjustment is possible to take account of varying and more heterogeneous materials, complex geological histories or easily disturbed stability so often encountered on linear sites on sidelong ground. A short discussion is given below of rotational slips in difficult ground, the study of parallel slides being left until later in the chapter.

ROTATIONAL SLIPS

Readers of this book are likely to be familiar with the simpler analyses of circular slips in uniform soils. Disturbing moments are caused by the weight of the slipping mass around its centre of rotation and resisting moments by the shearing strength of the soil along the slip surface. This shearing strength is assumed to be equal to that determined by testing in the laboratory of typical samples.

In difficult soils such equality can be questioned for, in such soils, failure takes place with noticeable distortion; this distortion is not necessarily reflected

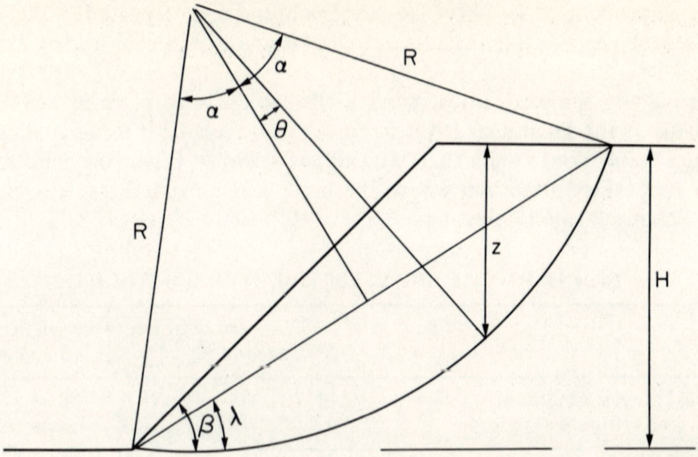

Figure 15.1 Geometrical conditions for circular slip

in the laboratory strengths from which the strength is deduced. However, to establish the basis of study of factors of safety, the geometrical relationships developed by Taylor[1] (for slopes in homogeneous clays) result in the following expressions:

$$M_D = (\gamma H^3/12)\ [1 - 2\cot^2\beta + 3\cot\lambda\cot\beta + 3\cot\alpha\cot\gamma - 3\cot\alpha\cot\beta]$$

where M_D is the disturbing moment and γ represents the bulk unit weight of the soil (in say kN/m^3) and the other parameters are shown in Figure 15.1. The restoring or resisting moment (M_R) is:

$$M_R = R\int_{-\alpha}^{+\alpha} c_z\, R d\theta$$

and the Factor of Safety is

$$M_R/M_D$$

The symbol c_z represents the cohesive or shearing strength of the soil at a depth z below the ground surface. This is where difficulties arise. Whereas it is simple to calculate the disturbing moment, the restoring moment depends on the value assumed for c_z. Is it uniform round the slip surface; how should its numerical value be determined?

Gibson and Morgenstern[2] and Hunter and Schuster[3] have translated the above calculation into charts of stability factors with differing regimes of the water table. In these charts it is assumed that the value of the shearing resistance (c_z) varies linearly with the depth z.

In difficult soils such a neat and simple relationship does not hold. It has been shown by back analysis of failures which have taken place in unstable soils, that simple analysis, using total stresses, as in Taylor's derivation, is inadequate for an accurate assessment of stability. This statement has been repeatedly shown to be true in recent years and the results of one of numerous careful studies illustrates this fact.

Kjaernsli and Simons[4] investigated a slip which took place in 1955 on the north bank of the Drammen River in Norway and related it to older slips in the same area since 1749. They then studied the stability of the clays both where the slip had taken place and where the bank had remained stable over a long period. The results are interesting:

Table 15.1 TOTAL AND EFFECTIVE-STRESS METHODS

Analysis by	*Factor of safety for soil in* *Slipped ground*	*Stable ground*
Total stress (Taylor's method)	0.47	0.71
Effective stress (Bishop's method)	1.01	1.26

From these figures it is evident that the total-stress analysis gives factors of safety which are too low. The slipped area would have failed long before the factor of safety could be reduced to 0.47 and a slope known to be stable could not have a factor of safety of 0.71.

The term *effective-stress analysis* in Table 15.1, refers to the use of normal pressures reduced by the values of the pore pressures known to exist at points along the line of the slip surface. The shearing resistance calculated from these reduced normal stresses are much closer to what might be expected, failure at a factor of safety of about unity, and stability at a factor above unity (Table 15.1). Rotational slips under effective stress are investigated by the methods of Bishop[5] or Spencer[6].

The values of the pore pressures which must be subtracted from the total normal stress on the slip surface can be found, but this still leaves the designer with the task of deciding on the relationship between the effective normal pressure exerted by difficult soils and the shearing resistance developed. This must be appropriate to the distortion and weakening which is known to occur near failure in such soils. For an answer to this question a study of shearing resistance of difficult soils under exaggerated strain is required.

STRENGTH OF DISTORTED CLAY

The vertical loading on an inclined slip surface, and hence the normal pressure, can be calculated. The variation of pore pressures along the assumed slip surface can be determined from flow nets or from piezometer readings, and effective normal pressures thus determined. The relationship between these effective normal stresses and the effective shearing resistance along the slip surface is the last of the links required for the analysis of possible rotational slips in difficult soils. To determine this relationship, the material most likely to slip, i.e. a clay, is studied in the circumstances of

(i) maintained pore pressure (total-stress analysis)
(ii) dissipated pore pressure (effective-stress analysis)
(iii) pronounced distortion (also under effective stress)

The first two of these have been briefly discussed above and are well-known. In the third type of test the apparatus for applying the shear is adapted to produce large strains. This is done by means of a multiple-reversal shear box or by a ring-shear apparatus. The shearing strength first reaches a peak value, which in the test of (ii) above, is looked upon as the shearing strength of the clay. As strain increases, the shearing strength begins to drop below the peak value as is shown in Figure 15.2. Eventually, continuing distortion causes the shear strength to reach a low value known as the *residual strength*. Although normally-consolidated clays show a relatively lower peak shear strength than overconsolidated clays, the pattern of Figure 15.2 is common to all clays.

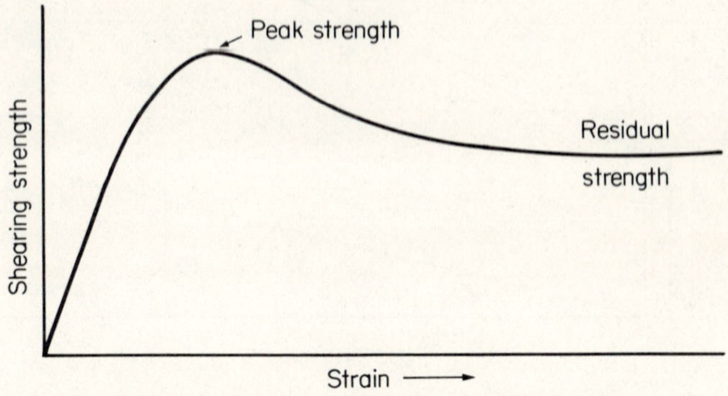

Figure 15.2 Peak and residual shearing strength of clay

In difficult soils there is a common history of distortions and slip movements. If slips are to be avoided and designs for stability prepared, then it must be accepted that the appropriate shearing strength to be used in assessing stability is that which can be sustained even in circumstances of some distortion. Since the greater the distortion, the closer does the available shear strength come to the irreducible minimum strength, failure in difficult soils must be accepted as occurring below the peak effective strength. Back analysis of slips which have taken place have established that the effective peak strength is not reached before failure and that instability occurs at or near the residual strength (see Symons and Cross[7]).

RELATIONSHIP BETWEEN PEAK AND RESIDUAL VALUES OF SHEARING STRENGTH

Suppose an analysis is made of the possibility of slip taking place in a weak clay slope. The stability is suspect because of the difficult nature of the clay, over consolidated, perhaps fissured and likely to deteriorate with time. By the usual procedures three differing 'strengths' can be determined experimentally (Figure 15.3).

Firstly, there is the cohesion measured in the undrained condition, the condition of total stress. This results in a constant value of shearing strength which is not increased or reduced by differences in the applied normal pressure. Any extra loading is carried by the pore water which has no shearing strength.

This type of loading in difficult soils is temporary, for the pore pressures are lowered under load and the effective strengths begin to show (c, $\phi = 0$) (CH).

Secondly, the strength can be measured under drained conditions, the excess pore pressure caused by the loading being allowed to dissipate so that the soil particles come into closer contact and develop a frictional resistance. In these conditions the line CH 'rotates' indicating that a higher normal loading develops

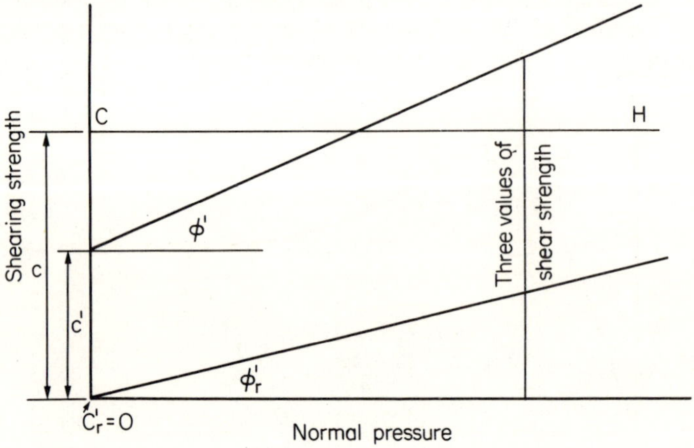

Figure 15.3 Three interpretations of the shearing strength of clay

a higher shearing strength. There is then less likelihood of the imposed shearing stress overtaking the strength which can be supplied. This occurs within the mass of soil on the site only if the pore pressure is dissipated there as well as in the laboratory test. The cohesion (shear at zero normal pressure) decreases while the material exhibits a frictional strength not shown in the undrained condition (c', ϕ'). The value of c' is normally small although still contributing a proportion of the strength of the clay.

Thirdly, it is not merely in the test sample that the residual strength is reached. In a slope, as distortion increases, and it need not be a large strain, the strength opposing the tendency to slip decreases and finally reaches the residual strength or some shearing resistance close to the lowest possible value. There may be a weakening of the soil, taking years to cause sufficient deterioration to reduce the shearing strength to instability level. Until the concept of the gradual weakening suffered by some soils, through time and weathering, was understood sudden failures of slopes which had stood safely for decades were unexplained. It was found that clays which have fissures or joints, for example, were likely to show slowly decreasing shear strength ultimately falling to the residual value. With weak or difficult clays it is thus imperative to investigate not only the value of the peak effective strength, but also that of the residual effective strength. Design for stability should be based on this figure unless there is evidence to show that a higher stress is applicable. Sometimes difficult soils may show an ultimate shearing strength at some value between the peak and the residual value. Whether this shearing strength is closer to the peak or to the residual depends on the type of clay. This point is discussed below.

The above discussion applies to studies under effective stress. The peak effective values are given the symbols c' and ϕ' and those at the residual strength level are shown as c'_r and ϕ'_r. The values of c'_r can safely be assumed to be zero, and the value of ϕ'_r is usually a few degrees less than the peak ϕ' (Figure 15.3).

THE RESIDUAL FACTOR

In his Rankine Lecture in 1964, Skempton[8] gave an illuminating treatment of the application of residual strength to the analysis of slips in banks. His concept of the *residual factor* is particularly helpful in visualising the proportion of the slip surface over which the strength has deteriorated to the residual value. In

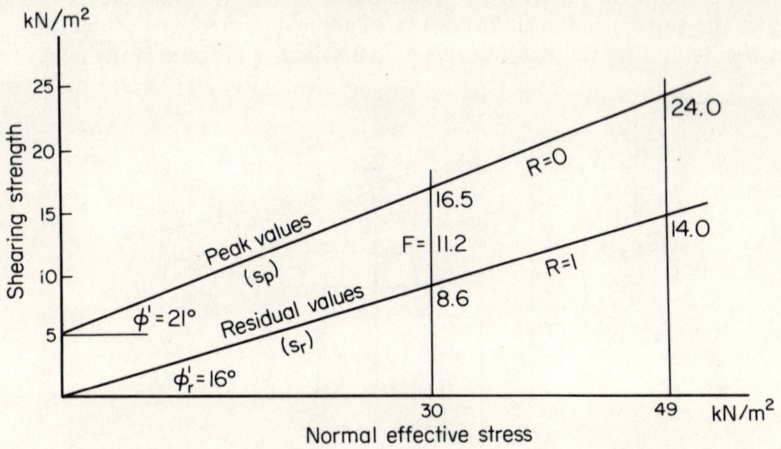

Figure 15.4 Numerical application of peak and residual shearing strengths

Figure 15.4 the test results obtained from laboratory tests of a soil in which a slip had taken place were:

$$\phi' = 21^\circ \qquad c' = 5 \text{ kN/m}^2$$

$$\phi'_r = 16^\circ \qquad c'_r = \text{zero}$$

In the back analysis of this slip, the mean normal stress on the length of the slip surface was found to be 30 kN/m² and the mean shearing stress at which slip had taken place was estimated at 11.2 kN/m² (shown as F in Figure 15.4) From Figure 15.4 the peak (s_p) and residual shearing (s_r) strengths can be assessed as follows:

$$s_p = c' + 30 \tan \phi' = 5 + 30 \tan 21^\circ = 16.5 \text{ kN/m}^2$$

and

$$s_r - c'_r + 30 \tan \phi'_r = \text{zero} + 30 \tan 16^\circ = 8.6 \text{ kN/m}^2$$

The residual factor (R) as defined by Skempton is

$$R = \frac{\text{Peak shear strength} - \text{mean shear stress at failure}}{\text{Peak shear strength} - \text{residual shear strength}}$$

which becomes

$$\frac{16.6 - 11.2}{16.5 - 8.6} = 0.67$$

The physical meaning of this is that R represents the proportion of the length of the slip surface on which the strength has fallen to the residual value. In this back analysis the figure is 67%.

FACTOR OF SAFETY AGAINST INSTABILITY OF AN INUNDATED SLOPE

The slope shown in Figure 15.5 is inundated so that the pore pressure is zero only at the surface and exhibits positive values in all other parts of the soil mass. This results in a substantial lowering of the effective pressure in the mass of the slope with a consequent risk of slipping. The analysis is for only one possible deep-seated slip, but other possibilities can readily be studied.

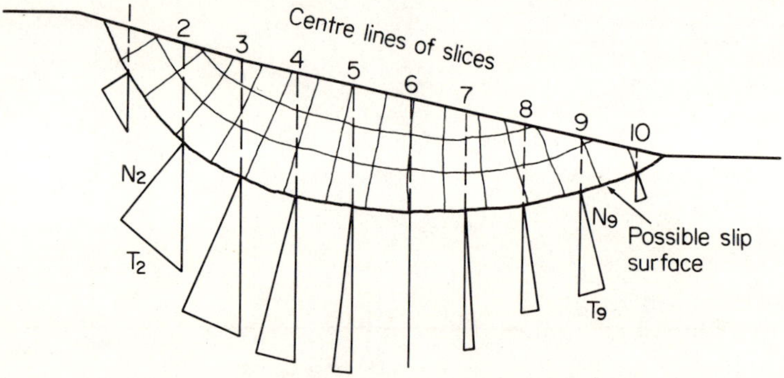

Figure 15.5 Factor of safety of inundated slope

Table 15.2 shows the physical conditions measured on each of the ten slices along the length of the slip surface (L). Each of these slices is 4 m wide. The slip surface is 45 m long, and the mean normal stress applied is thus

$$2212/45 = 49 \text{ kN/m}^2.$$

If it is to be assumed that the shearing strengths developed at failure were to be the peak effective strength, as found in the laboratory, then the factor of safety would be (using Figure 15.4 for data):

$$s_p = 5 + 49 \tan 21° = 24 \text{ kN/m}^2$$

$$\text{Factor of safety } (F) = \frac{c'L + s_pL}{\Sigma T} = \frac{5 \times 45 + 24 \times 45}{725} = 1.8$$

(T = tangential component of weight of strip)
This clearly indicates that the bank would be stable.

Table 15.2 TOTAL AND EFFECTIVE STRESS CONDITIONS ON THE SLIP SURFACE SHOWN IN FIGURE 15.5

Slice No.	*Mid height* (m)	*Volume* (m^3) ($19.4\ kN/m^3$)	*Weight* (kN)	*Normal load*, N (kN)	*Tangential load*, T (kN)	*Pore pressure head* (m)	*Pore pressure load* (kN)	*Effective normal* (N') *load* (kN)
1	3.0	12.0	233	129	202	2.1	82	47
2	6.6	26.4	512	399	326	5.5	216	183
3	8.0	32.0	621	578	247	7.3	287	291
4	8.4	33.6	652	634	163	7.9	310	324
5	8.3	33.2	644	640	73	8.1	318	322
6	7.8	31.2	605	605	0	7.8	306	299
7	7.0	28.0	543	539	−45	6.9	271	268
8	5.5	22.0	427	421	−73	5.7	224	197
9	3.8	15.2	295	393	−112	4.1	161	232
10	1.6	6.4	124	112	−56	1.6	63	49
				Total tangential force	725 kN	Total effective normal force		2212 kN

If, however, it is known that the material is difficult, and that shearing strength is likely to decrease in value with time, a lower factor of safety (at a mean value of strength along the slip surface equal to the residual strength), can be calculated. In those circumstances the factor of safety may be assessed as follows:

$$s_r = \text{zero} + 49 \tan 16° = 14\ \text{kN/m}^2 \text{ (Figure 15.4)}$$

and

$$F = \frac{\text{zero} + 45 \times 14}{725} = 0.9$$

This indicates a condition of instability, and failure would take place when the shear strength mobilised along the slip surface reached the value of the residual strength.

Some soils, however, show a *residual factor* which can be less than unity. If this soil is known to display this characteristic and the residual factor could possibly be say, 0.5, then a further factor of safety can be calculated. The mean strength at failure ($\bar{s}$) in these circumstances would be obtained from the expression for the residual factor given above:

$$R = 0.5 = \frac{24 - \bar{s}}{24 - 14} \quad \text{or} \quad \bar{s} = 19\ \text{kN/m}^2$$

$$F = \frac{\text{zero} + 45 \times 19}{725} = 1.2$$

If the residual factor does not increase above 0.5, the factor of safety indicates stability, and a possibility of a mean shearing resistance of 19 kN/m^2.

The problem which arises in the use of the residual factor for difficult soils is that of determining whether the soil in use can be credited with a low value of R. The lower the residual factor the greater is the average shearing resistance along the slip surface, and the less the proportion of the slip surface which would deteriorate to the residual shear strength if slip should take place.

Skempton[8] gives some figures for unfissured and unweathered till (not a difficult soil) and for fissured and weathered London Clay. In the first, the value of the residual factor was nearly zero (0.08) indicating that slip would take place only when the applied shear stress reaches the peak value for the clay. For the weathered London Clay the value varied from 0.56 to 0.80 showing that at slip from 56% to 80% of the length of the slip surface would have reached the low value of strength represented by the residual shear. Such figures illustrate considerable weakening over the years.

For any slip which has already taken place, the factor of safety is in the region of unity. When opportunity offers, analyses should be made to correlate the conditions of failure against the type of clay in question. This would add to the knowledge of slips in difficult clays.

GEOLOGICAL ORIGINS

Even a well-prepared and full site investigation often offers too little information on which to base design in difficult soils. The basic cause of possible instability can lie, not so much in the state of the strata at the present time, but on the geological history of the site. What is visible on first inspection of the area, and even what is described in the site investigation report, does not always give the critical information required to ensure stability in an excavation or cutting. If the soil is suspected of being 'difficult' or the region has a reputation for showing unstable engineering conditions, the advice of a professional engineering geologist is certainly required.

A site which appears to be in equilibrium may have reached that condition after a series of retrogressive slips, or movements obscuring more deep-seated failures. Such surface movements, and subsequent denudation by weathering. can cause the majority of the ground to be covered by soil creep material. The presence of such material is a danger signal. This weak material can be found even on flat slopes of a few degrees to the horizontal and may well be in a state of neutral equilibrium, ready to move if disturbed by excavation. Roads, which often require to follow the contours in a valley, are particularly likely to be the 'triggers' of movements on a geological scale. Excavation on sidelong ground to form a berm on which a road may be constructed may be sufficient to start a whole hillside moving. Even if this movement can be stopped at an agreed cost (this is not always possible) the choice of such a route will still prove to be a costly decision. Remedial measures are possible if the movement induced is not large or widespread (see Baker and Marshall[9] and Zaruba and Mencl[10]), but often the only solution is to re-align the road elsewhere.

The types of geological indicators of incipient instability vary according to the underlying formations and the subsequent geological pressures and denudations suffered. No exact rules can be laid down, but if the soil in which it is proposed to form a cutting is on a gentle slope which shows irregularities in surface, a closer investigation by a geologist is probably necessary. Small indicators like trees show a tendency to lean outwards over the slope, or the outer ends of dipping strata showing a downward bend or camber can ring warning bells.

That a geological as well as an engineering interpretation of the site investigation is often essential has been shown to be true on the lines of major highways even when a very full site investigation has been made. A geologist or an engineer with a thorough geological training in the field, may have acquired the 'eye' to observe when there is a danger of instability. An intuitive decision of this kind is of greater value than the results of many calculations.

Some of the signs which can act as danger signals even before the site investigation report is opened, are well described by Biczysko[11] in relation to the Jurassic soils of the east Midlands. Other formations, with different origins, show similar warnings. If these are ignored the slope will require elaborate and costly remedies when slipping eventually takes place. A successful geological investigation may take longer than was first expected. This is only one of the reasons for ensuring that a thorough site study can be made well ahead of the date of the proposed construction. Time must be available for geological and engineering consideration of all the possible effects of excavating slopes in difficult ground.

Even apparently sound and well-known formations can show instability when circumstances combine to give the necessary combination of gravitational and pore-pressure forces. An example of a successful study of instability, carried out in such conditions, is that made for Central Hill, Lambeth (see Buckthorpe *et al*[12]). The geological and analytical study on this site is a model of how natural slopes in difficult materials should be investigated. The successful conclusion of the constructional work was made possible only by an interlocking of the study and design procedures. The report again underlines the necessity of time being available for accurate assessment of conditions. The reading of the piezometers, for example, continued for at least two years.

The cost of such a thorough enquiry is not small. The cost of the remedial works required after failure of a natural slope or, worse still, of the structures erected on it, is much greater. Effective interpretation of all the facts available is of the highest importance if difficult slopes are to be used or constructed effectively. Early Ordnance Survey and other old topographical maps can often identify a sudden depth of silt in a single borehole as being on the site of an old pond, or an unexpected great depth of fill as an old filled-in quarry, long forgotten. Recent 'geological' changes are sometimes of significance in the design of excavations and slopes.

TIME-DEPENDENT FACTORS

In difficult soils, in particular, the designer must be conscious of the passage of time in its influence on the properties and stability of his construction. He must project his design to the end of its hoped-for lifetime, appreciate what conditions will then be operative, and design accordingly. He must, perhaps, incur capital costs, apparently not at present required, so that desirable conditions may eventually be developed and construction remain stable as long as it is required. Design and investigation should develop together with a convergence of information on properties of soil, geological forces, regime of ground water and proposed construction, resulting in permanent stability in slopes artificially cut or in natural slopes loaded by buildings or highways.

It cannot be too firmly asserted that, in difficult soils, time must be allowed for this interaction and cross-stimulation before decisions on construction are made. In the previous section emphasis is placed on the need to allow time for the geological investigation to develop. In difficult soils a specification for a complete site investigation cannot be written in advance. Such a step, although often attempted, is inconsistent with the aim of a study where the designer expects to meet problems of instability. In difficult soils inherent problems are numerous and often time-dependent. The roots of their occurrence can be identified only if the timing ensures adequate periods for exploratory and detailed investigations, for the study of time-dependent and seasonal changes and for later confirmation and checking of design proposals against continuing study of the site.

There are other time-dependent features which must be considered. One has already been studied. In an earlier section the deterioration which some clay soils suffer towards the residual strength of clay has been emphasised as a cause of unforseen instability. The changes which have taken place in geological conditions with the passage of time were also mentioned. The final time-dependent factor considered in this chapter is the regime of ground and surface water existing on the site. This has a particular significance in the determination of effective stresses, shown to be important in the stability of difficult soils. In granular soils there is no residual strength as it is measured in clays, but the presence and flow of ground water can reduce the strength of sands to a very low value if not controlled.

It is probably not an exaggeration to say that the establishment of the water-regime of the site is of prime concern in the study of the time-dependent factors in slopes in difficult soils. It is not sufficient in difficult ground to record only the levels at which water was struck or even, as is often done, the levels observed during the time the borehole is being driven. Seasonal variations in the state of the ground water may give clues to the possible behaviour of the site when it is subsequently excavated. The design of drainage systems depends on knowledge of changes which may require many months of recording to establish a sound conception of the conditions confronting the designer.

This search for time-dependent information requires an extended study of pore pressures, observation of the fluctuation in the level of the water-table, or, in more complex situations, the form taken by the phreatic surface where the pore pressures change from positive to negative values. The installation of at least standpipe piezometers in a proportion of the boreholes and their concomitant observation and recording is necessary in the collection of information. On some sites a more elaborate pattern of buried piezometers may be necessary. Whatever method is employed to obtain details of the pressures and movements of water in difficult soil, a period of months before the start of construction is needed to ensure the recording of even one cycle of seasonal variations.

The emphasis on soil moisture, which runs through this chapter, applies to any structure constructed on or of soil but is much more important when soils are difficult. False prediction of the state of ground water pressures which will exist after construction is complete may lead to unacceptable results. The site investigation must provide clear records of the liquid phase of the soil and of its changes if the slopes are to remain stable. Water is always the enemy: the disposition of its forces must be known.

SLIPS IN NATURAL CONDITIONS

The excavation of difficult soils to provide artificially constructed slopes is fraught with the possibility of producing instability. It is of some assistance, therefore, to observe, as warnings, the conditions which cause slip in naturally occurring soils. Such slips usually involve large volumes of soil and rock and widespread areas of land. The rotational slip, which is the commonest form of instability where artificial excavation of construction has disturbed equilibrium, also occurs naturally, but normally within the larger landslides. Rotational slips can form a pattern of smaller movements within a large slide which has reached a condition of at least temporary stability. They form retrogressive movements retreating up the shallow slope, each depositing debris below it. The upper edge of the whole often has a corrie-like shape, curved in plan with nearly vertical slip surfaces. These are relatively easy to recognise even if overgrown but the tumbled or sometimes smooth talus below, after long weathering, may not appear to be the delicately-balanced mass that it often is. Uncontrolled excavation in these detritus slopes have resulted in large and troublesome movements.

The chief noticeable difference between natural slides and rotational slips, apart from the contrast in the area of ground involved in the two types, lies in the depth to the slip surface. In rotational slips, the depth to the slip surface represents a substantial proportion of the length of the slip. In a slide the depth of disturbed ground is small when related to the extent of the sliding area. Landslides, particularly those under the generic term of *flow slides*, frequently take place in those disturbed soils deposited after the last glacial epoch and which were subsequently raised by isostatic uplift when the ice melted. Weathering can lead to such soil masses becoming unstable, especially in river valleys where material at the toe of the slope erodes more rapidly than the rest. A slide does not always develop into a massive failure but can show small and often seasonal movements. The designer of constructions in soil always hopes that these minor movements can be stopped and stability resumed. Such hopes are not always fulfilled and, when they are, the result is often costly in achievement.

If remedial measures are to be carried out on movements on natural hillsides or shallow slopes and it is clear that there are no indications of deep rotational movement, the relatively shallow depth to the slip surface should be determined before stabilizing measures are attempted. The technique used by engineers of British Rail[12] has been shown to be effective. Pliable plastic tubes are inserted into normal boreholes which are deep enough to pass through the estimated position of the slip surface. A long cylindrical plumb bob, in the form of a rod, is lowered into the tubes from time to time. When the plumb bob fails to penetrate to the full depth of a borehole, the inference is that the tube has been bent by lateral movement of the ground. Shorter cylindrical plumb bobs are then, in turn, lowered into the flexible tube, and the exact level of the slip surface located. Some areas of ground slip only at intervals, usually during the wetter times of the year. The process of determining the level of the slip surface can, therefore, be time-dependent. No movement may take place during dry seasons and evidence of the extent of the slip may appear only in the winter. This is one of the many reasons for programming site investigations of difficult ground a long time ahead of proposed construction. Long delays to major works have been experienced when it was discovered that the site showed a time-dependent instability.

FACTOR OF SAFETY OF A PARALLEL SLIDE

Features exhibited by natural slides or those initiated by uncontrolled excavation in unstable ground are as follows:

(a) Loose and often fine and cohesive material.

(b) Erosion, particularly at the toe of the slope.
(c) Leaching out of salts in the material.
(d) High water table; high pore pressure ratio above the slip surface.
(e) Retrogressive failure up the slope.
(f) Presence of small rotational slips contributing to general movement.
(g) A talus of soil creep material likely to move if disturbed.
(h) Temporary stability caused by piling up of debris from the slip.
(i) Because of remoulding, a reduction in shearing strength even below the residual strength, on occasion.
(j) Often shows a small slope to the horizontal.

If a natural slope shows signs of seasonal movement, perhaps because of excavation which has taken place, the existing factor of safety should be found before any decision can be made on what remedial measures are possible or likely to be effective. Meyerhof[14] suggests that, for flour sides at least, a factor of safety of 1.5 should be aimed at, in order to cover the effects of the uncertainties inherent in the study of slips in any natural slope. An assumed simple set of conditions allows of a theoretical study of a *parallel slide**. The use of geophysical methods can show whether these assumptions require modification for particular landslides (see Stewart and Armando Celis[15]).

It is assumed that, in a parallel slide, the following characteristics are shown:
(a) The slip plane lies close to the surface relative to the extent of the disturbed ground.
(b) The slip plane is parallel to the surface over the short length studied.
(c) The phreatic surface is parallel to the ground surface and to the slip plane over the short length studied.

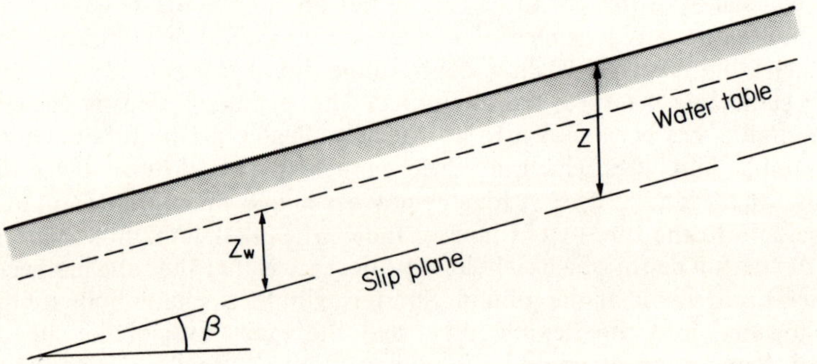

Figure 15.6 Factor of safety of parallel slide

In Figure 15.6 the depth to the slip surface is z and the height of the water-table above the slip surface is z_w. The bulk unit weight of the soil is γ and of water is γ_w. The pore-pressure ratio is thus:

$$\frac{\gamma_w z_w}{\gamma z} = r_u$$

The weight of the soil acting on a horizontal plane gives a principal stress of

$$\sigma_1 = \gamma z = OQ \text{ in Figure 15.7.}$$

*Frequently referred to as a translational slide

In the Mohr's circle of Figure 15.7 the angle *OSQ* (in a semi-circle) is 90° and thus:

$$OS = OQ \cos\beta = \gamma z \cos\beta$$

The values of the normal and shearing stresses on the slip surface are:

$$ON = OS \cos\beta = \gamma z \cos^2\beta$$

$$NS = OS \sin\beta = \gamma z \cos\beta \sin\beta$$

So far in this discussion no notice has been taken of the pore pressure at the level of the slip surface; only total stresses have been considered. When the

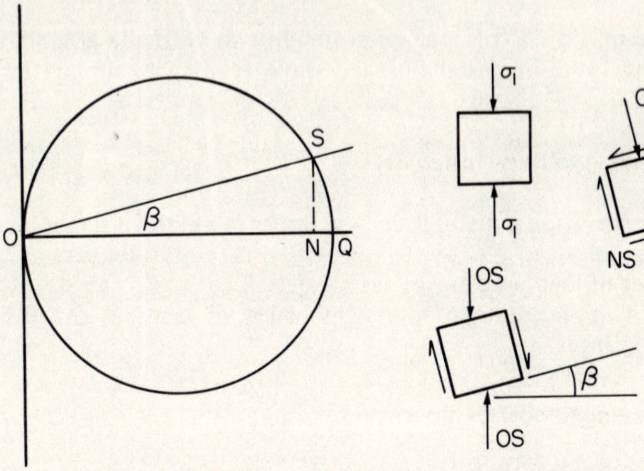

Figure 15.7 Shearing and normal stresses on slip plane

normal pressure on the slip surface is reduced by the pore pressure, which is hydraulic and acting equally in all directions, the effective normal stress is:

$$\gamma z \cos^2\beta - \gamma_w z_w$$

It may be assumed that at slip in these circumstances, the clay can offer only its residual strength. Thus, multiplying by the coefficient of effective residual friction, the frictional stress resisting movement is found:

$$(\gamma z \cos^2\beta - \gamma_w z_w) \tan\phi'_r$$

The force acting downward along a unit length of the slip surface is not affected by the pore pressure and remains

$$\gamma z \cos\beta \sin\beta$$

tending to cause movement.

The Factor of Safety is the ratio of the force on a unit length of the slip surface tending to resist movement, to that tending to cause it:

$$F = \frac{(\gamma z \cos^2 \beta - \gamma_w z_w) \tan \phi'_r}{\gamma z \cos \beta \sin \beta}$$

Simplifying this and substituting the pore-pressure ratio r_u:

$$F = \frac{(1 - r_u \sec^2 \beta) \tan \phi'_r}{\tan \beta}$$

DESIGN AND THE INFLUENCE OF WATER

In the design of slopes for long-term stability in difficult ground there are factors tending to cause instability and those tending to obviate movement:

(a) Factors tending to cause instability:

Ill-advised excavation in natural shallow slopes in a state of temporary stability.
The presence of high pore pressures.
The lack of appreciation of the significance of seasonal and other time-dependent changes.

(b) Factors tending to obviate movement:

Recognition that in cohesive soils, the strength of the material at slip should not be considered to be much above the residual value, which must be measured in laboratory tests.
The elimination of pore pressures by adequate drainage.
Appreciation of the influence of the geological history of the site on its sensitivity to movement, especially after disturbance.

From this selection of a few factors of particular significance, it is clear that, apart from the recognition of features of design to be taken into account or avoided, the most important practical step which can be taken to improve the stability of the slope is to study the methods of drainage available in difficult ground. By drainage, the pore pressure regime can be reduced in its capacity to produce instability. This section and the next one considers the increase in strength and stability which may be produced by the removal of water from the mass of the soil, and studies the means by which this desirable change may be made.

When a cutting is made in natural clays without previous site drainage, the very act of making the excavation has the effect of reducing the pore pressures. This, however, cannot be considered a permanent state. Later, when equilibrium pore pressures are again established by the passage of time, there could, in uninfluenced circumstances, be a rise in pressure in some parts of the mass of soil, with the effect of reducing the effective pressure and the strength. In difficult soils this must not be allowed to occur. If it is to retain permanent stability, the soil mass must be provided with adequate means of maintaining pore pressures at as low levels as can be arranged economically. Drainage of surface and ground water ensures that weakening influences, tending to increase pore pressures, rapidly revert to the state required for stability.

Water in difficult soils can cause problems of instability in three principal ways:

1. The ground may be wet and show a high water table in the natural state. It may show signs of instability before any excavation is commenced.

2. The soil may show no signs of natural slip, but have a water table at or close to the surface. This type of condition can lie between two extremes.

(a) when the soil is entirely granular and permeable on the one hand,
(b) when the soil is entirely cohesive and impermeable on the other (Figures 15.3 and 15.4).

The many variations of soil content between these two extremes show intermediate properties and may have intermediate treatment.

3. The soil water may exist and flow in permeable strata within an otherwise impermeable mass of soil. When the permeable strata are thick and clearly defined, the water they contain tends to pass over the surface of the next impermeable stratum below. This can cause a lubricated surface to develop, and the danger of slip to be intensified. When the permeable strata are very thin, high pore pressures can develop in these thin partings when they can become a part of a slip surface. Such thin partings may look insignificant in the boring records, but are possible paths of weakness within the mass of soil.

UNSTABLE GROUND IN THE NATURAL STATE

Some soft soils with a high water table show incipient or well-developed slipping before any excavation is made, or as a result of injudicious excavation, especially towards the toe of the slope. When a linear one-face cutting is to be excavated on sidelong ground of this type, it is very easy to start a slide movement inadvertently. Even if there is no sign of a slide, but the terrain is such as to be set in motion by disturbance, steps should be taken to prevent movement by physical support. Drainage for interception and to lower the water table should be considered before excavation is commenced. The loading of the slope by imported rock masses along the toe of any berm cut in sidelong ground should also be looked upon as one of the steps to be taken if there is possibility of movement.

For shallow, parallel slides, where excavation is required in sidelong ground, both support and drainage may be provided by the use of *counterfort-drains*. Deep trenches are dug into the slope of the bank and filled with drainage filter material and lined with artificial filter fabrics. The mass of granular fill in each trench acts both as resistant and supporting buttresses or counterforts and also as drainage channels removing ground water. For such a technique to be successful, the counterforts must penetrate through the slip surface, whether that of a shallow slide or of a deep-seated rotational slip. Unless the counterfort-drains key in to the mass of the hillside below the slip, they merely add unwelcome weight to the slipping mass. It is essential, therefore, for the designer to find the level at which slip is taking place by the method described in the previous section and by Waters and Bartlett[13].

The case shown in Figure 15.8 illustrates the type of problem which can be encountered in normal practice. Excavation A was made to provide a level recreational area for a school. The ground, being near the bank of a large river in

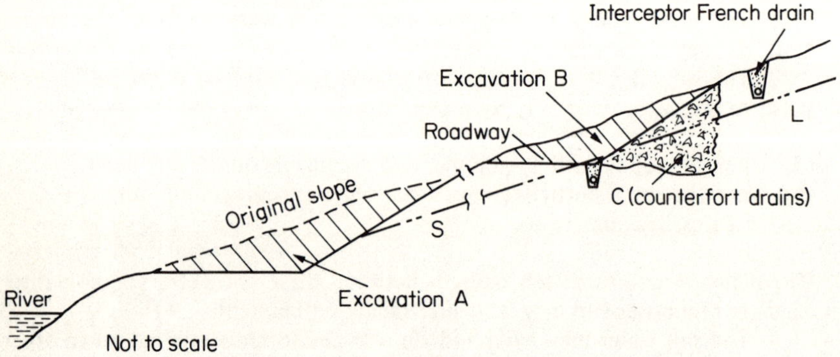

Figure 15.8 Case history of slide in alluvium

its lower reaches, consisted chiefly of alluvium. The resultant slide which started on slip line SL was unnoticed, for it was seasonal and not very active. Later, a road was to be constructed for access to the school; this work (excavation B) was found to be impossible until the hillside was stabilised. Counterfort drains C were constructed about every 3 m over the length of the moving section. The depth of the slip had been found to be about 2.5 m by using the flexible-tube technique, and it eventually became possible to build the roadway.

GRANULAR SOILS WITH A HIGH WATER TABLE

Even with a high original water table, granular soils are largely self-draining. It is necessary to control the drainage and to provide designed exits into the main drainage system. Pre-construction drainage is particularly desirable here to ensure water-free excavation. Free drainage down the slopes of the excavations is not so objectionable in this type of soil as it would be in softer or more cohesive materials. Drain channels of the familiar and traditional herringbone (Y) pattern are still recommended for the face of the excavation to remove both seepage and surface water. A development of this is to cover the whole of the face of the cutting with a layer of granular material resting on an artificial filter fabric and thick enough to carry all the drainage from surface and groundwater sources.

PERMEABLE AND IMPERMEABLE STRATA OUTCROPPING IN THE SAME CUTTING

Difficult soils often have a high water content and a high water table. Water may be flowing through channels forming aquifers in an otherwise impermeable mass.

The construction of a cutting or excavation in such circumstances, exposes both the permeable and impermeable layers on the face of the excavation. Water drains out as springs on the face of the slope along the lines of the upper levels of the impervious layers. These layers have trapped the water descending through the aquifers. The possibility of increased pore pressures developing on the upper levels of the impermeable strata is high, and slides might result.

The dip of the strata being towards the excavation is the worst situation, and therefore as much dewatering should take place as is possible before the water reaches the face of the slope. Flow down the face of the cutting cannot be freely permitted because of the danger of erosion of any exposed weak or cohesive strata. Although flow down the face cannot always be entirely prevented, it can be much reduced by the use of interceptor or cut-off drains, as considered in the next section.

When the site investigation shows the presence of thin layers of sand or silt in an otherwise dense impermeable mass, the danger presented by these layers is sometimes obscured by the apparently insignificant thickness of these 'partings'. However, if these thin layers are in continuous sheets, pore pressure can be transmitted along them with sometimes spectacular results, difficult to explain in terms of the main strata on the site.

In slopes, unexpected slips have sometimes been traced to a thin and almost un-noticed layer providing an area of low shear strength and high pore pressure. Reduction of this dangerous pore pressure held between impermeable layers can be assisted by thrusting porous pipes into the bank at intervals and at the relevant level. Neglect of thin partings carrying water under pressure can result in instability and movements of blocks of soil or rock on difficult slopes.

REMOVAL OF PORE-PRESSURE BY DRAINAGE

There are two aspects of drainage which can be discussed separately. The first is that of intercepting flows of ground water before they reach the site of the cutting. Such interception is easier and more effective than later attempts to remove water after it has penetrated the critical strata. Deep and near-surface flows should be intercepted, as far as is possible, before excavation of construction commences. The ground then becomes firmer because of the increase of effective stress; both men and machines can operate more effectively. This aspect of work on the site is sometimes overlooked in site investigation and in consideration of pre-construction drainage. Where the benefit of the interception of ground water is appreciated, it is unlikely to be neglected.

Interceptor drains must penetrate as deeply as possible into the upper strata behind the proposed crest of the slope. These drains, placed at a high level on the site, are not intended to draw down the water table but to prevent high pore pressures developing near the face of the cutting by preventing the penetration of ground and surface water. Cut-off drains intercept and collect the flow of water which would otherwise reach the main excavation unimpeded (Figure 15.8). Such drains are costly, since they must be deep to have their full effect as interceptors, but the savings effected, on excavation, are large. The foundations for twin hospital blocks, for example, built close together on sloping clay soil, with a perched water table, illustrate this. The first excavation was difficult because of the presence of large quantities of water which pumping alleviated but did not eliminate. The excavation for the second and similar block was carried out 'in the dry' easily and fast, after the whole site had been protected by a permanent interceptor drain at a convenient distance up the slope, clear of all permanent works.

It is not always essential that interceptor drains should be close to the area being protected. When they are used to form permanent protection to cuttings and run parallel to its length, they are better not placed close to the crest of the slope where they are not only less effective but where they might act as artificial 'cracks' and increase the possibility of a rotational slip being induced. Interceptor drains can also 'pick up' field drains, so protecting slopes and excavations in a secondary sense. It must be remembered that these cut-off drains require time to achieve their full effect. Cut-off drainage should be part of the pre-construction works as a device to ensure the removal of water and the firming of the soil before the main excavation commences.

Deep 'French' drains can be used to draw down a surrounding water table. The drains should be as close to the site to be dewatered as is possible and be of permanent construction. For instance, an excavation had to be made in very difficult ground i.e. consisting of fine, loose sands and silts which were low-lying, and inundated with the water table at the surface. Deep 'French' drains were inserted, ringing the site completely, and, in a short time, excavation which had been nearly impossible, was confidently carried out.

The Road Research Laboratory's Report LR(110)[16] gives details of valuable relationships existing between the dimensions of an excavation, the permeability of the soil and the volume of water which can be extracted by drainage. The use of the dimensionless ratios given in Figure 15.9 is of assistance where excavations

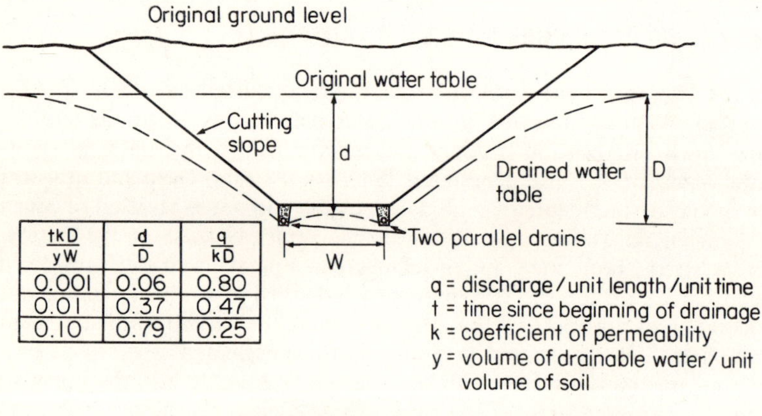

$\frac{tkD}{yW}$	$\frac{d}{D}$	$\frac{q}{kD}$
0.001	0.06	0.80
0.01	0.37	0.47
0.10	0.79	0.25

Figure 15.9 Lowering of water table by parallel drains

for buildings, or linear cuttings for roads are to be supplied with permanent deep drains to maintain a low level of water table under the construction. The soil forming the face of the cutting and the mass of soil lying behind it can be kept, by these means, in a state of uniform water content, with a sustained pattern of drainage flow. Subsoil drains are much assisted in their function by artificial filter fabrics which eliminate the need to provide filter gradings in 'French' drains.

SURFACE DRAINAGE

Although the subsoil drains intercepting underground flows and controlling them, are of prime importance, it is equally necessary to collect and remove surface water (from rain or snow). Permanent surface-drainage channels are part of the design to reduce pore pressures throughout the mass of soil behind the slope of the cutting.

On the upper surface, shallow drains discharging into the main interceptors give good protection. On the sloping faces, rubble drains, artificial filter fabrics weighted by porous material, and removal of spring water and seepage, all assist in maintaining that permanence in conditions which is the first step to stability.

MAINTAINED CONDITIONS

When it is remembered that the amount of softening and movement necessary to induce failure in wet clays is small, and need occur only over localised areas, the importance of maintaining a consistent water content is evident. Seasonal variations in moisture can be eliminated or decreased by adequate drainage. If variations are allowed to occur, and the drainage system does not over-ride the seasonal wetting and drying of the soil, the shearing strength can weaken and eventually produce instability.

It is important that the drains should be maintained as free-running to prevent pore-pressures building up in time, and causing instability. The Aberfan disaster of 1966 underlined the importance of free drainage channels in maintaining equilibrium. The Report[17] on the unstable tip emphasised the importance of the free passage of water and of the maintenance of permanent drainage.

SOME SALIENT POINTS

The study of the geological history of sloping sites, in particular, is of prime importance. Sabine[18] draws a composite and salutory sketch of a range of geological difficulties militating against the stability of slopes in difficult soils.

Reduction of pore pressure and its permanent maintenance at a low level helps to ensure the stability of slopes.

Adequate and carefully designed drainage systems form the best means of reducing pore pressures permanently. This reduction is particularly easily achieved in granular soils.

Slopes in cohesive soils are the most 'difficult'; their stability must be studied in effective stress conditions. In some clays, time-dependent deterioration may drop the ultimate shearing strength to the residual value or near it.

References

1. Taylor, D.W., 'Stability of earth slopes', *Boston Soc. of C.E.*, **24**, 197–246 (1937)
2. Gibson, R.E. and Morgenstern, N., 'A note on the stability of cuttings in normally consolidated clays,' *Geotechnique*, **12**, 212–216 (1962)

3. Hunter, J.H. and Schuster, R.L., 'Stability of simple cuttings in normally consolidated clays,' *Geotechnique*, **18**, 372–378 (1968)
4. Kjaernsli, B. and Simons, N., 'Stability investigations of the north bank of the Drammen River,' *Geotechnique*, **12**, 147–167 (1962)
5. Bishop, A.W., 'The use of the slip circle in the stability analysis of slopes,' *Geotechnique*, **5**, 7–17 (1955)
6. Spencer, E., 'A method of analysis of the stability of embankments assuming parallel inter-slice forces', *Geotechnique*, **17**, 11–26 (1967)
7. Symons, I.F. and Cross, M.R., 'The determination of shear-strength parameters along natural slip surfaces encountered during the Sevenoaks by-pass investigations,' *Transport Road Res. Lab. Report* LR 139 (1968)
8. Skempton, A.W., 'Long-term stability of clay slopes,' *Geotechnique*, **14**, 75–102 (1964)
9. Baker, R.F. and Marshall, H.E., 'Control and correction of landslides' (article in *Landslides in Engineering Practice*, Ed. by Eckel, E.B.) Committee on Landslide Investigation, Highway Res. Board, Spec. Rept. 29, Washington, 131–149 (1958)
10. Zaruba, Q. and Menac, V., *Landslides and their control*, Elsevier, Prague (1969)
11. Biczysko, S.J., 'East Midlands jurassic soils,' *Jour. Inst. Highway Engs.* **23** No. 11, 21–32 (1976)
12. Buckthorp, P., Butler, F., Dunican, P. and Morrison, J., 'Central Hill Lambeth,' *Struct. Eng.*, **52**, 395–407 (1974)
13. Waters, J.M. and Bartlett, D.L., 'A direct method for the location of slip planes,' *Civ. Eng. & Pub. Wks. Review* (Sept. 1956)
14. Meyerhof, G.G., 'The mechanism of flow slides in cohesive soil,' *Geotechnique*, **7**, 41–49 (1957)
15. Stewart, M. and Armando Celis, C., 'The use of geophysics in landslide studies,' *Transport & Road Res. Lab. Report* 703 (1976)
16. Russam, K., 'Sub-soil drainage and the structural design of roads,' *Transport Road Res. Lab. Report* LR 110 (1967)
17. *Report of* Tribunal appointed to inquire into the disaster at Aberfan (1966), HMSO (1967)
18. Sabine, P.A., 'The geological background to highway engineering,' *Jour. Inst. Highway Engs.* **24** No. 1, 23–27 (1977)

Chapter 16

Appraisal of Stability Conditions in Rock Slopes

GEOLOGICAL CONSIDERATIONS

The choice of a site for engineering works can rarely be decided solely on geological considerations. Once the site has been selected, however, considerable advantages will result if the precise position, orientation and layout of the works are designed so that the optimum relationship to the geology is achieved. In some cases, it will be possible to use geological features to advantage, so that construction problems are reduced. In the case of a rock slope, for example, minor changes in the orientation of steepness of the face may improve stability and reduce, or even obviate, the need for support. Such design decisions can only be made after a properly planned and executed site investigation. Both very cursory and very detailed, indiscriminate site investigations are usually ineffective. A successful investigation involves, firstly, deciding what data are relevant and, secondly, selecting appropriate techniques for their reliable measurement.

IMPORTANCE OF GEOLOGY TO ENGINEERING DESIGN

The importance of an accurate and complete definition of the geology is paramount: no design can be better than the quality of the geological input data. The investigation must be designed not only to provide a complete appraisal of local (i.e. site) geology but also to include regional and other factors which affect overall geological understanding. Consideration should include:

1. Regional factors which may affect the site, including the risk of landslides, flooding, earthquakes and vulcanicity.
2. Natural or man-made voids, including solution cavities, mine excavations and underground streams.
3. Underground services including sewers, electricity cables, drains, gas and water mains.
4. Degradation processes as evidenced by the examination of natural and man-made slopes and cuttings.
5. Regional hydrology, including the likely interaction between the proposed excavation and the groundwater system.

It is essential to make certain of the relationship, both vertically and laterally, between soils and rocks and between rocks of different types. The subsurface geology is frequently very different from its surface expression. It should never be assumed that geological conditions at one place are the same as those observed at another, particularly when comparing conditions at surface and at depth. The range of data that should, as a result of regional and site investigations, be available to the engineer responsible for rock slope design and the processes required to obtain it are summarised in Table 16.1.

Table 16.1 FLOWCHART FOR GEOLOGICAL DATA REQUIREMENTS RELATED TO ROCK SLOPE ENGINEERING

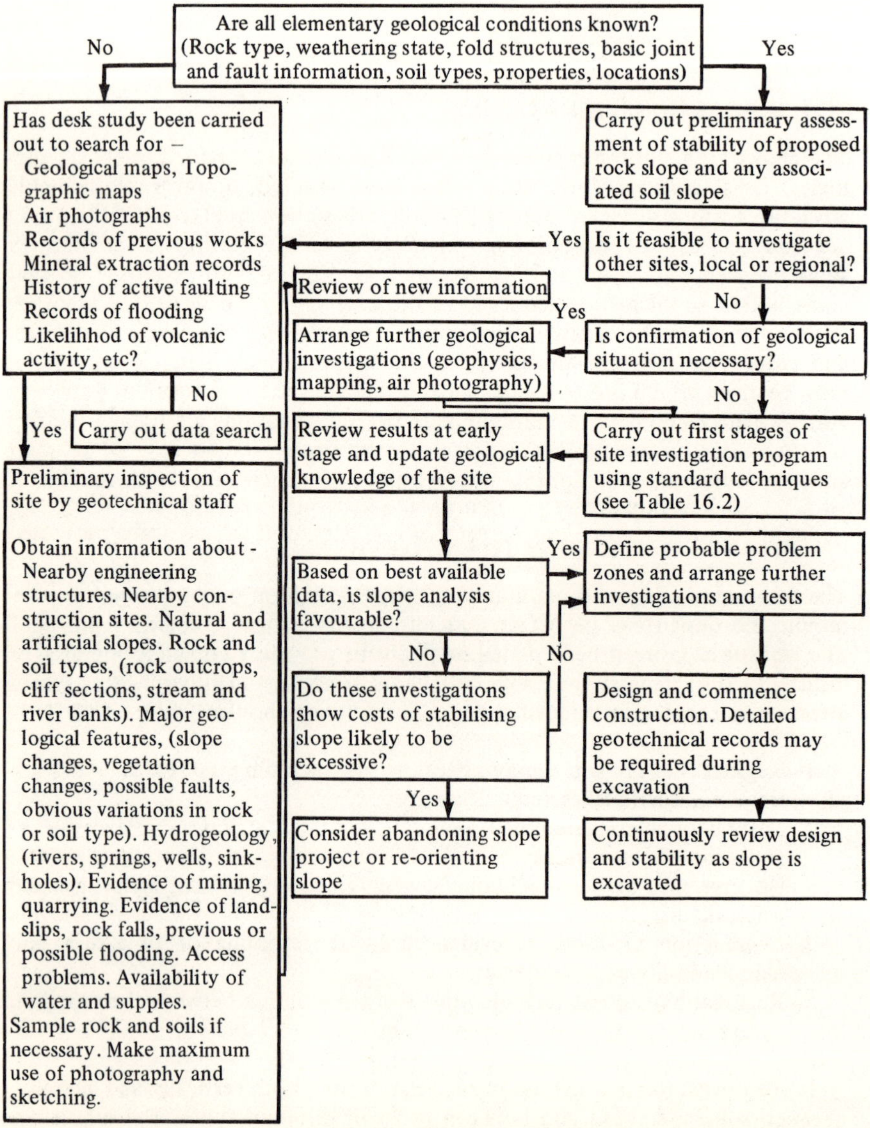

METHODS OF INVESTIGATION

The methods of investigation applicable to a given situation depend on the complexity and scale of the proposed excavation, the site conditions and geology and the amount of reliable investigation that has already been done. Often, when rock slopes form an ancillary part of a much larger project, a considerable amount of information will already be available and, after this has been evaluated, Table 16.1 can be used to indicate what additional investigation is appropriate.

Classification of the rock at an early stage in zones of similar geological, mechanical and weathering characteristics is advisable. For cost-effective investigation, it is preferable to build up data gradually, basing each new stage of investigation on a careful examination of the data already obtained.

Engineers involved in the design of surface constructions in rock should remember that natural processes progressively degrade all rock masses, eventually reducing the slopes of all land masses to near the horizontal. Degradation is usually by gradual erosion and occasionally by catastrophic failure. Most man-made slopes, particularly those in rock, must be thought of as being in conflict with the natural course of geological events. Engineers and geologists must be able to identify each of the factors which are likely to influence the stability of the excavation.

Methods of site investigation appropriate to rock slopes are summarised in Table 16.2. The dominant mechanical site investigation technique is diamond core drilling to confirm the positions and relationships between major discontinuities and to recover core for geological inspection and laboratory testing. Coring may be supplemented by percussion drilling to differentiate between features which can readily be identified from the chippings, particularly with extreme differences of hardness or colour.

Rock core is expensive to obtain, should be treated properly and transported and stored in properly designed coreboxes. Unless sheathed by a protective membrane, or sealed by a suitable process[2] it must be logged very quickly after drilling, prior to any deterioration. Logging should include a full geological and lithological description, with particular attention being paid to the position, orientation and description of major discontinuities, and the measurement of RQD, fracture spacing and point load strength. Colour photography is useful. Samples for laboratory testing of shear or intact strength, slake durability index, permeability etc. should be removed and effectively sealed immediately after logging; they should be selected to be representative and their position in the core string carefully recorded. The core should be resealed and stored carefully. Accurate and consistent recording of *all* data is essential.

Pits and trenches provide large scale information on the nature of overburden materials, the rock surface beneath and the weathering, infilling and continuity of joints. Trenches are preferable for defining the continuity of planar features. Because of the directional bias of a single trench, two series of trenches at right angle are advisable.

'Bulk sampling' techniques can be employed, stripping away overburden and clearing the rock surface with water and air hoses. The resulting rock exposure can be mapped in detail. This method is most applicable where there is a distinct boundary between rock and overburden.

During the advanced stages of investigation, large scale *in situ* shear tests or

Table 16.2 BASIC SITE INVESTIGATION METHODS APPLICABLE TO ROCK SLOPES (Based partly on Glossop[1])

METHOD	TECHNIQUE	APPLICATION
GEOPHYSICS	Resistivity	Backup for standard geological methods of investigation. Location of suspected geological features with simple structures. Expanding electrode method to find depth to near-horizontal changes in rock type, groundwater or rockhead. Constant separation method for lateral boundary location, e.g. buried channels, faults, igneous intrusions.
	Seismic	Backup for standard geological methods. Depth to bedrock or denser geological horizon in a simple, shallow-dipping rock sequence.
	Gravity	Regional geological surveys. Detection of buried channels or cavities.
	Induced Polarisation	Detection of clay-filled zones in rock mass. (Developed for finding disseminated metalliferous minerals but reacts to clay particles).
	Borehole Logging	Use of geophysical parameters to detect physical changes along the length of a borehole. Cost-effective in non-cored holes.
BOREHOLES	Rotary Percussion	Fast, cheap, unsatisfactory for precise work. Chip samples only. Penetration rate records suspect unless powerful, large diameter machine used. Identification of different, clearly distinguished rock types, large faults, cavities or soft horizons. Thickness of weathered rock. Attendance of geologist recommended, sampler essential.
	Rotary Coring	Quality of information varies depending on quality of driller and equipment used. Single-tube barrel not recommended for site investigation work. Double-tube barrel with rigid inner tube suited to massive rocks only. Core disturbed. Double-tube barrel with swivel inner tube helps prevent core disturbance, face discharge bits recommended for S.I. work. Plastic 'Mylar' film barrel liner available to improve core recovery and help preserve soft parts of core. Triple-tube barrel with split inner tube to aid core extraction and sealed sleeving to facilitate transport in 'as drilled' state. Wireline system allows recovery of core without withdrawing barrel and rods. Prevents caving of hole; cost-effective only in deep holes.
PITS	Hand or machine-excavated	Direct inspection of fresh rock through weathered material Limited discontinuity data for slope design.
TRENCHES	Ripping, blasting, or excavation through overburden to rockhead.	Inspection of rock section. Directional bias to linear or planar geological features. Ideally a number of cleaned trenches perpendicular to one another.
ADITS	Conventional tunnelling methods, mini-tunnelling machine if in use on site	First hand inspection of three-dimensional rock conditions. May prove ideal collaring site for investigatory drillholes. May be possible to use tunnel as future drainage or grouting gallery.
SHAFTS	Conventional sinking or large diameter boring	Similar information can usually be obtained from cheaper, large diameter cores. Gives access to rock mass for sampling or instrumentation.

plate bearing tests may be necessary; such tests on larger rock volumes will include the effects of discontinuities, strength, anisotropy etc, but are costly and time consuming. In very large civil works, such as dams, investigation by adits or tunnels may be required. These excavations, although costly, give excellent opportunities for detailed mapping, large scale examination of rock inside the zone of surface effects (weathering, etc) and monitoring of hydraulic behaviour of the rock mass. They may provide a suitable location for further exploratory drilling, or large scale tests or provide later use as drainage or grouting galleries.

Wherever possible, boreholes drilled to obtain core should be used for additional studies; these include measurements of ground water pressure and rock mass permeability, borehole logging techniques and visual inspection. Borehole logging (by sonic, dynamic or radiometric techniques) is a very specialised field but may provide useful data when properly applied. Visual inspection of boreholes by TV camera is sometimes useful, but problems arise with humidity, dirt and interpretation. Borehole cameras are particularly useful for investigating the extent and nature of cavities encountered during drilling.

No two geological features are identical and the differences, however subtle, may be important as far as stability is concerned. Unless the person doing the recording is also responsible for presentation and interpretation, he should record all such differences.

KEY DATA FOR SLOPE DESIGN

The site investigation must provide the fullest possible data on properties of the discontinuities present in a rock mass. These properties are generally the key elements of information required to design slopes successfully. The requirements are summarised in Table 16.3. Slope design requires, firstly, the determination of the spatial relationships between discontinuities so that possible modes of failure may be recognised and analysed and, secondly, the measurement of the factors affecting the shear strength of the rock mass so that an appropriate value may be used in the stability analysis. These two aspects are discussed fully in later sections.

CONSTRUCTION AND PERFORMANCE MONITORING

Geological and geotechnical data collection from all fresh exposures created during excavation should continue, and be used to confirm or amend the original assumptions made during design and, if necessary, as the basis for re-design. Monitoring of construction may be required for blasting vibrations, support performance or stress changes.

Long term monitoring of important structures is essential to ensure that stability is being maintained. Commonly applicable methods include the use of piezometer installations to monitor ground water pressure changes, borehole instruments (extensometers, etc) for measuring ground displacements and precise surveying methods for monitoring surface movement.

DATA PRESENTATION

Geological and geotechnical data obtained during site investigation and construction should be presented in a clear, unambiguous form and in adequate time to

Table 16.3 JOINT DATA REQUIREMENTS FOR ROCK SLOPE DESIGN

PROPERTY		METHOD OF MEASUREMENT
ORIENTATION		Direct measurement on surface of discontinuity, using specially designed compass/clinometer. Measurement of dip with clinometer, dip direction estimated by comparison with a known direction, preferably instrumentally, if magnetic compass cannot be used. Estimated by visual alignment with compass/clinometer. Measurement from oriented core in oriented hole. Strike directions determined from aerial photographs are of limited use in slope work. Determination of orientation from stereophotographs taken by phototheodolite. Ambiguous or involved coding should be avoided. Dip/dip direction is preferred to strike notations.
POSITION	ABSOLUTE	Measured position of feature with respect to surveyed or well-established reference on known global or site axes.
	RELATIVE	Measured position with respect to other, similar features such as members of the same joint set, to give apparent spacings, true spacings and intensity of discontinuities. Obtained by direct measurement or abstracted from geotechnical plans and sections. Measured position with respect to excavated faces.
CONTINUITY		Visual estimation or measurement of two-dimensional continuity in near-planar surface outcrops, with horizontal or vertical extrapolation to nearby exposures showing features with similar characteristics. Three-dimensional continuity assessed using evidence from vertical and horizontal exposures, boreholes and adits if available. Qualitative and quantitative descriptions related to extent of proposed excavations.
FACTORS AFFECTING SHEAR STRENGTH	INFILL	Details from freshly excavated outcrops, from drill core only if good quality recovery and if logged before deterioration. Chemical and mineralogical analysis may be necessary. Determination of physical properties of gouge by laboratory testing. Correct soils description of gouge desirable, e.g. as per Geological Society Working Party Report[3].
	WAVINESS	Description of the deviation of the surface from a true plane. Surface irregularities that are unlikely to shear off and that provide a keying effect. Quantitative description using approximation to wavelength and amplitude of the irregularity. Direct observation of large outcrops only.
	ROUGHNESS	Description of surface asperities that may shear off, but which affect the frictional properties of the surface. Obtained from fresh and occasionally weathered rock exposures and from drill core.
	FRICTIONAL PROPERTIES	Obtained by small-scale shear tests on specially selected rock samples containing joints, or from rock core. Difficult to avoid disturbing the surface of the discontinuity.

If the slope is very large and contains too many discontinuities to measure individually, it may be necessary to resort to statistical methods of analysis. These methods are becoming accepted in geotechnical work. If statistical or idealised methods are used it is essential that isolated, "non-systematic" or random variations are also examined to find out whether or not they affect stability. A "randomly oriented" discontinuity may provide the release surface necessary to allow failure. It is important to understand the limitations of any geological data collecting and assessing system.

Table 10.4 GRAPHICAL PRESENTATION OF GEOLOGICAL DATA

GEOLOGICAL MAPS, PLANS, SECTIONS (scale 1/10,000 to 1/1,000)	Representation of overall geological situation on a district to local scale. Indications of rock types, major structural features such as the larger faults and folds. May show extent and type of soil cover. Usually superimposed on a topographic map. Supplemented by annotated aerial or ground photographs if desirable.
GEOTECHNICAL PLANS, SECTIONS (Scale 1/1,000 to 1/1)	For showing the absolute and relative positions of geotechnical features (those geological features that could influence engineering constructions and excavations) in horizontal or vertical scaled form, occasionally obliquely. Non-planar surfaces should be divided into planar elements for representation purposes. Planar geological surfaces intersecting a roughly planar exposure at a low angle tend to give an irregular non-linear trace which can be misleading if represented in "sketch" form. Show as shaded surface or as a dot or cross with description codes appended. It is often possible to draw a "best fit" straight line through the intersection. Sections can be interpretative and predictive, constructed from a plan, or they can be scaled down representations of an exposed rock face. Full-scale plans and sections may be necessary at sites of specialised instrumentation or sampling locations. All maps, plans and sections are only complete when they state orientation, scale, title, location, geologist or engineer responsible, grid or magnetic compass bearings, elevation above a known datum, date and all other relevant details.
ISOMETRIC DRAWINGS	Block diagrams and isometric drawings of actual or predicted conditions are ideal ways of presenting geotechnical information. Drawings can show a slope at various stages of excavation and can include cut-away portions of rock to show internal structure. Isometric drawings can easily be amended, but several views may be required to show all detail.
SCALE MODELS	Physical models can be used to help geologists and engineers to understand relationships between geotechnical features and between those features and the proposed rock faces. Any method of presentation of information that removes ambiguity and helps to avoid confusion should be considered worthwhile. Scale models are the most taxing to construct and possibly the most time-consuming of all presentation methods. Amending and updating is difficult and the model itself cannot be constructed until plans and sections are available. Models can be used to investigate failure mechanisms and modes but it is not advisable to attempt to scale down physical properties such as rock strength, coefficients of friction etc. Waviness may be incorporated if significant. Two-dimensional models rarely represent the failure mechanisms of three-dimensional situations.
STEREONETS	Equal area (Lambert) stereonets usually used. Plots of great circles representing individual planes if not many involved. Plot in poles (preferably on a polar net) if a great number involved. Statistically significant numbers of poles can be contoured and representative sets determined. Sensitivity analyses can be incorporated into stability analysis for planes if known spread of values about each mean orientation of each set. Combination stereonet/polar net can simplify plotting. Stereonets essentially show angular relationships and cannot be used in a spacial sense. Curved surfaces must be divided into idealised planar elements or a representative plane used instead.

be used in the design process. Table 16.4 lists commonly used methods of data presentation. The list is not exhaustive, as many geologists and engineers have their own preferred methods.

Presentation methods must be considered at an early stage in data collection, as the choice of method may influence the collection and recording procedure, particularly if computer processing is intended. The practical aspects of geological data presentation are described elsewhere[3,4,5].

For larger volumes of data, it is usual to group discontinuities in 'sets', and some precautions are necessary. A set, strictly speaking, is a group of discontinuities having the same geological origin; the members of the set may have dissimilar orientations. For stability calculations, origin is not directly significant and a set consists of discontinuities having similar orientations. It is not unusual for the members of such a set to have very different surface characteristics, and the variations in shear strength resulting from these must be taken into account in stability analysis.

The orientation of discontinuities may vary in a gradual manner throughout the rock mass, i.e. a representative orientation of a set in one part of the rock mass may not be representative of that set in adjacent parts. For a large site therefore, it is important to sub-divide into areas having similar structural characteristics, and produce sets of discontinuities for each area; this information is then used for sensitivity analyses in stability calculations. Statistical methods are becoming increasingly useful in defining groupings of geological data.

Figure 16.1 is an example of the graphical presentation of data. It compares several discontinuity sets presented in the form of a stereographic projection (above) and an isometric projection (below). Figure 16.2 is an engineering geological map of the rock face from which the data shown in Figure 16.1 was obtained. The discontinuity surfaces are hatched in the same manner as the isometric projection. Information is given on the map about joint continuity and thickness of joint infill. Such presentations are of considerable assistance in depicting structural conditions, and for assessing areas of potential instability.

Figure 16.3 is an isometric view of a proposed excavation (not related to the previous data). This type of presentation, because of its clarity, is well-suited for reports and publications of a general nature.

BASIC MECHANISMS OF ROCK SLOPE FAILURE

The shear stresses generated in an excavated rock slope are usually low in comparison with the shear strength of the intact rock material. Pre-existing discontinuities in the rock mass may have sufficiently low strength for failure to occur along them. A common approach is to assume full joint continuity and to neglect any contribution to strength made by sections of intact rock along potential failure surfaces. This is conservative and is applicable to most civil engineering situations where permanent structures may be at risk and proven factors of safety are required.

By comparison, the economical design of mine excavations and temporary civil works imposes the condition that slopes should be closer to limiting equilibrium. For this case, multiple failure modes of rock and rock joints may be considered but are beyond the scope of this chapter.

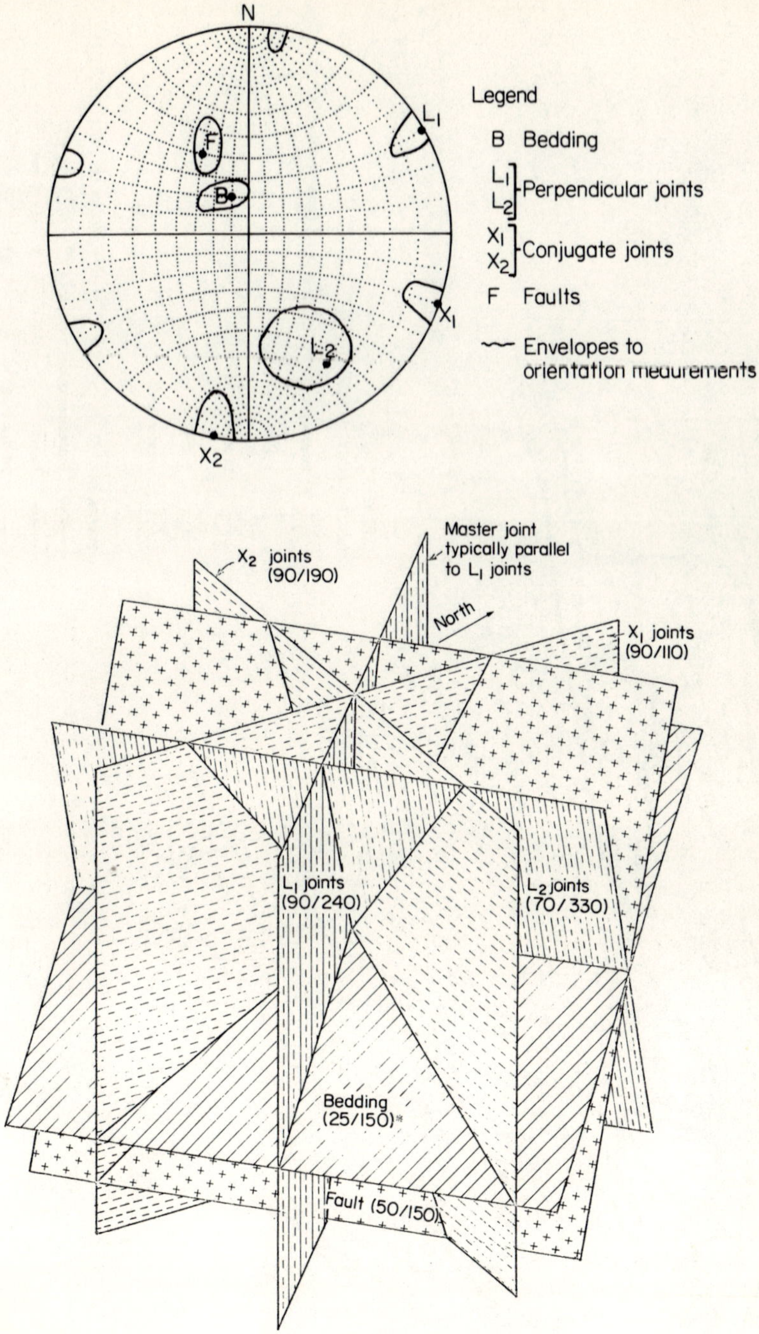

Dip and dip direction values used for plotting purposes
Isometric view

Figure 16.1 Methods of presentation of families of discontinuities

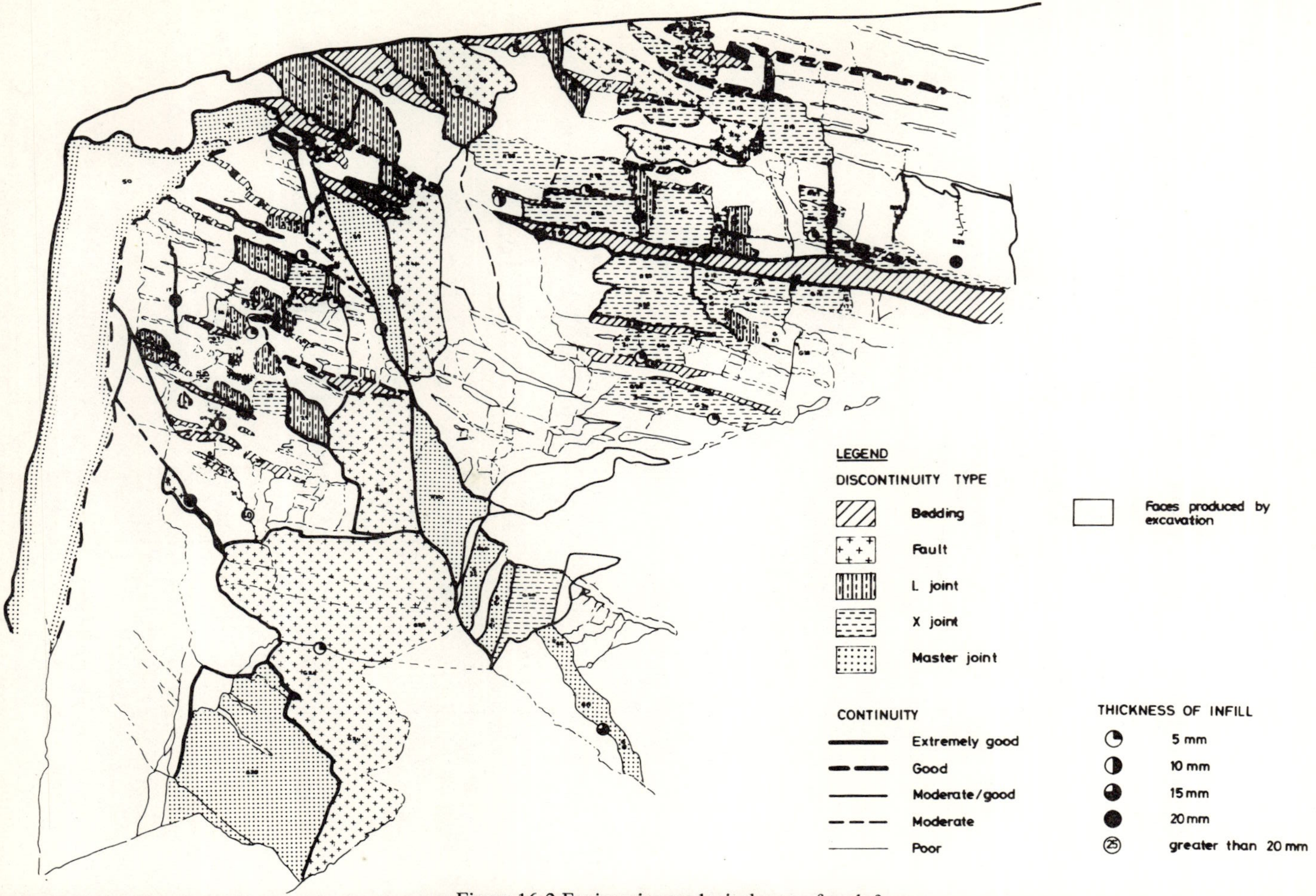

Figure 16.2 Engineering geological map of rock face

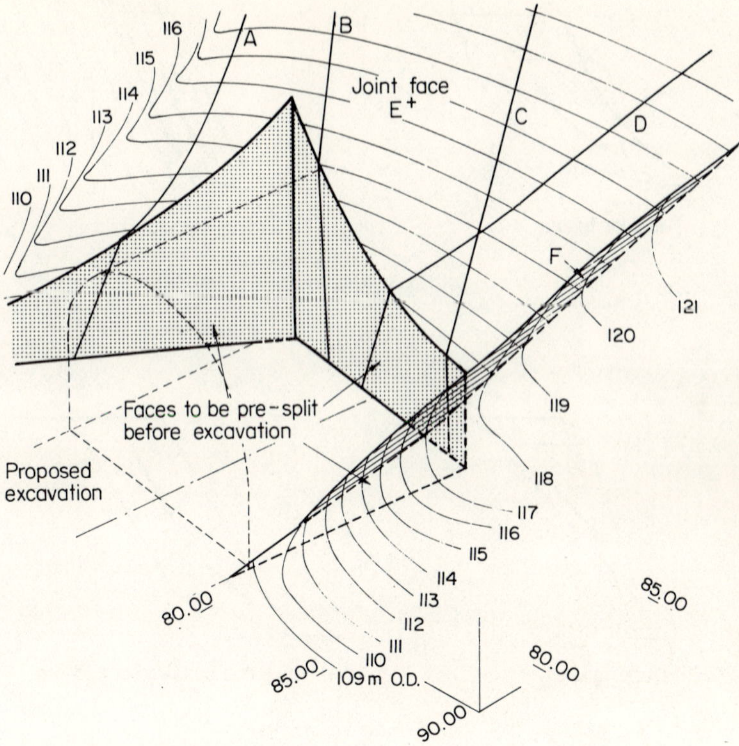

Figure 16.3 Isometric view of proposed excavation

The principal types of failure relevant to rock slopes are rotational, translational and toppling modes. These are idealised and combinations of any of these modes may be possible in any given situation. Figure 16.4 shows examples of the possible failure modes considered in this chapter.

ROTATIONAL FAILURES

Rotational failure modes in rock slopes are usually found only in structureless overburden material, highly weathered rock or very high slopes in closely jointed material. Either circular or non-circular failure surfaces may be generated.

Circular failures may occur in rock masses where the material is so intensely fractured that it may be considered as randomly jointed and isotropic. They may also occur where the scale is so large in relation to both the strength and jointing frequency of the rock that the stresses involved override the control of structural features in the rock mass.

In highly weathered materials discontinuities may persist in the form of relict jointing. Slides in weathered rock may occur along a combination of relict joints

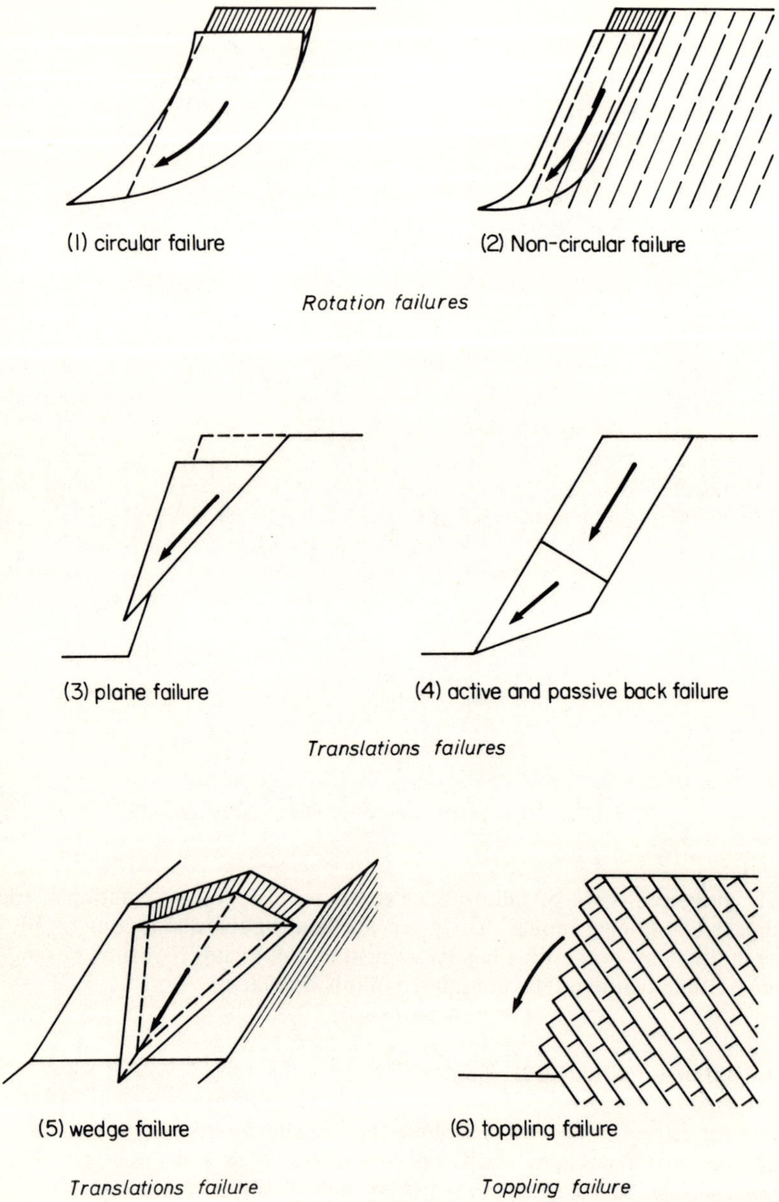

Figure 16.4 Examples of possible failure mechanisms for rock slopes

and failure planes through weak but previously intact material. These failure surfaces are often intermediate in geometry between planar and circular slides.

TRANSLATIONAL FAILURES

Plane failures frequently take place on rock slopes whereby sliding occurs along a single plane. In order for this to be kinematically possible, the following conditions must be satisfied.

(a) The failure plane must daylight in the slope face i.e. its dip must be less than the dip of the slope face.
(b) Release surfaces must be present with orientations defining the lateral boundaries of the potential sliding mass.

The presence of tension cracks is a major factor tending to produce instability of rock resting on a single plane. Water entering these cracks gives rise to substantial hydraulic driving forces behind the potential failure mass and in addition causes an increase in the uplift water pressure beneath the failure mass.

When considered in isolation, a single block near an excavated face may have adequate stability. Forces imposed by other, less stable blocks behind the face may produce instability of a system involving two or more blocks, thus giving rise to *active and passive block failures.*

Tetrahedral wedges may be formed in rock slopes by a combination of two planar discontinuities, the slope face, and the upper slope surface. In order for this wedge to be able to move, and therefore fail, it is necessary for the line of intersection of the two base planes to daylight in the slope face. Such failures are simply referred to as *wedge failures.*

TOPPLING FAILURES

Toppling failures involve the overturning of blocks of rock and are generally associated with steep slopes and sub-vertical jointing patterns. The overturning mechanism may lead to deep-seated rotational failures through loosening of the rock mass if allowed to progress.

SHEAR STRENGTH OF ROCK JOINTS

The shear strength of a rock joint is derived from two main components:

1. A basic frictional resistance for the joint surfaces. This basic friction angle is that obtained on a smooth, planar surface of the rock material. In general, it is equal to the residual strength of the joint.
2. An additional resistance caused by irregularities on the joint surface. These surface features give rise to an apparent cohesion or an increase in the friction angle over that of a planar surface.

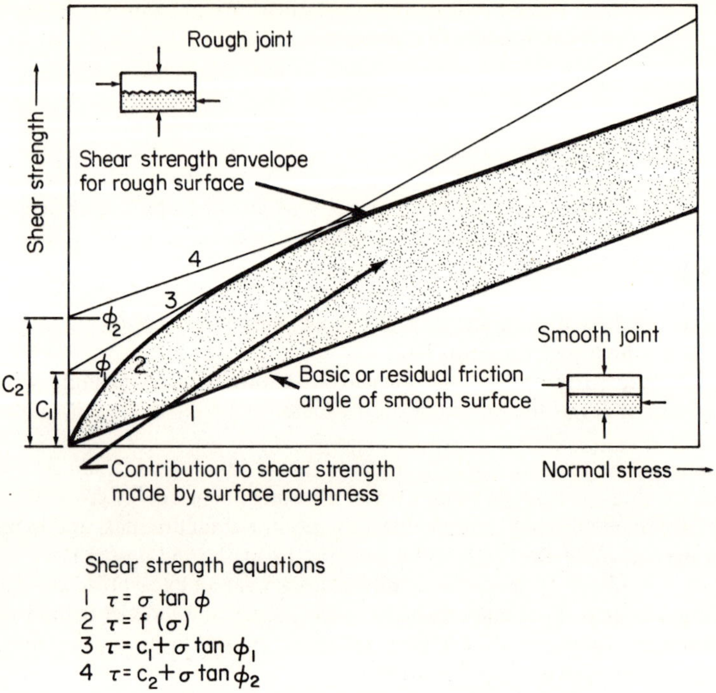

Figure 16.5 Simple models of joint shear strength

This simplistic model of joint strength (see Figure 16.5) may be greatly complicated if the effects of joint weathering, gouge infilling, lack of joint continuity and other factors are included.

BASIC LAWS OF ROCK FRICTION

The following relationships can be used to describe rock friction:

(a) If σ is the normal stress corresponding to a given shear strength τ, a coefficient of friction may be defined as:

$$\tan \phi = \tau / \sigma$$

(b) Better correlation can often be achieved with experimental work by assuming a value of shear strength at zero normal stress. This intercept is termed cohesion but does not necessarily imply any tensile strength normal to the joint. The relationship is expressed as:

$$\tau = c + \sigma \tan \phi$$

(c) Both the above criteria assume a linear relation between the shear strength of a joint and the normal stress level. A more general relationship is to assume a curved envelope in the form:

$$\tau = f(\sigma)$$

Various mathematical functions have been used to fit envelopes to experimental data.

FRICTIONAL CHARACTERISTICS OF MINERALS

The main influence on the basic friction angle of a rock surface is the mineralogical composition of the joint surface. The frictional behaviour of common minerals has been studied in the laboratory by Horne and Deere[6] and Kenney[7]. This data is summarised in Table 16.5 in which minerals are divided into two

Table 16.5 FRICTIONAL CHARACTERISTICS OF MINERALS

1. Data from Horn and Deere[6]

Mineral	Oven-dried $(\tan\phi)d$	Saturated $(\tan\phi)s$	$\dfrac{(\tan\phi)\,s}{(\tan\phi)\,d}$
Massive minerals			
Clear quartz	0.11	0.42	3.82
Milky quartz	0.14	0.51	3.64
Rose quartz	0.13	0.45	3.45
Microcline feldspar	0.11	0.76	6.90
Calcite	0.14	0.68	4.85
Layer lattice minerals			
Muscovite	0.43	0.23	0.54
Phlogopite	0.31	0.15	0.48
Biotite	0.31	0.13	0.42
Chlorite	0.53	0.29	0.47
Serpentine	0.62	0.48	0.63
Steatite	0.38	0.23	0.61
Talc	0.36	0.16	0.45

2. Data from Kenney[7]

Massive minerals			
Quartz-crushed	–	0.69	–
Feldspar-crushed	–	0.69	–
Calcite-crushed	–	0.69	–
Layer lattice minerals			
Muscovite-crushed	–	0.36–0.45	–
Na-Muscovite	–	0.31–0.34	–
Mica, various	–	0.29–0.49	–
Kaolinite	–	0.27	–
Montmorillonite	–	0.07–0.18	–

groups. Massive structured crystals are those with a three-dimensional structure, e.g. quartz, feldspar and calcite. Layer lattice crystals possess a two-dimensional crystal structure, e.g. muscovite, chlorite, biotite and talc. Table 16.5 shows that the crystal structure of a mineral has a very marked effect both on its frictional behaviour and on the way in which it is affected by water. The ratio of saturated friction to oven-dried friction is used as a measure of the lubricating or anti-lubricating action of water on the mineral. For massive minerals this ratio varies between 3.45 and 6.90, indicating that water has a strongly anti-lubricating effect on these minerals. This effect diminishes rapidly as the surface roughness of the mineral increases and does not appear to be relevant to natural rock surfaces. For lattice minerals the ratio is 0.42 to 0.63, showing that water is a lubricant for this type of crystal structure. This effect can be observed on natural joints.

SHEAR STRENGTH OF UNWEATHERED ROCK SURFACES

RESIDUAL SHEAR STRENGTH

Table 16.6 presents a summary (mainly adapted from Barton[8]) of basic friction angles for unweathered smooth rock surfaces. The rock types in this table have been grouped according to origin into sedimentary, igneous and metamorphic rocks. Sedimentary and igneous rocks generally have higher friction coefficients than metamorphic rocks. This is due to the relative proportions of massive and layer lattice minerals in the rock material.

The residual friction angle of a rock surface is generally assumed to be relatively independent of normal load, initial roughness and further shear displacement. Experimental evidence shows that with large displacements initially polished rock surfaces become scratched and gouged so that the friction angle increases to a residual value. Rough surfaces on the other hand become progressively smoother with displacement and the friction angle decreases to its residual value. The residual values for initially smooth and rough surfaces are usually identical provided that the roughness is not so pronounced that dilational behaviour is still possible after large displacements.

PEAK SHEAR STRENGTH

Roughness on joint surfaces causes an increase in shear strength. Firstly, there is the shear strength component caused by dilation and sliding at an angle to the applied shear force. Secondly, there is a strength component as a result of shearing through asperities. The dilational component of shear strength depends on the joint topography. The usual idealisation for joint roughness is to consider sliding taking place up a slope inclined at an angle i to the direction of shearing. In Figure 16.6 which considers a single asperity at limiting equilibrium, the relationship between shear load and normal load can be shown to be:

$$\tau/\sigma = \tan(\phi + i)$$

In other words, the effect of the inclination i on the shear strength of the joint is simply to increase the friction angle ϕ by an amount i. At high normal stresses, failure of the asperity becomes possible and a second failure criterion is introduced. The assumption of these two failure criteria, one at low normal stresses and another at higher stress levels, gives rise to a bilinear friction law.

Table 16.6 BASIC FRICTION VALUES OF VARIOUS UNWEATHERED ROCKS OBTAINED FROM FLAT AND RESIDUAL SURFACES

Rock type	*Moisture condition*	*Basic friction value* ϕ	$\tan \phi$	*Reference*
A. Sedimentary Rocks				
Sandstone	Dry	26–35	0.49–0.70	Patton 1966
Sandstone	Wet	25–33	0.47–0.65	Patton 1966
Sandstone	Wet	29	0.55	Ripley and Lee 1962
Sandstone	Dry	31–33	0.60–0.65	Krsmanovic 1967
Sandstone	Dry	32–34	0.62–0.67	Coulson 1972
Sandstone	Wet	31–34	0.60–0.67	Coulson 1972
Sandstone	Wet	33	0.65	Richards 1973
Shale	Wet	27	0.51	Ripley and Lee 1963
Siltstone	Wet	31	0.60	Ripley and Lee 1963
Siltstone	Dry	31–33	0.60–0.65	Coulson 1972
Siltstone	Wet	27–31	0.51–0.60	Coulson 1972
Conglomerate	Dry	35	0.70	Krsmanovic 1967
Chalk	Wet	30	0.58	Hutchinson 1972
Limestone	Dry	31–37	0.60–0.75	Coulson 1972
Limestone	Wet	27–35	0.51–0.70	Coulson 1972
B. Igneous Rocks				
Basalt	Dry	35–38	0.70–0.78	Coulson 1972
Basalt	Wet	31–36	0.60–0.73	Coulson 1972
Fine-grained granite	Dry	31–35	0.60–0.70	Coulson 1972
Fine-grained granite	Wet	29–31	0.55–0.60	Coulson 1972
Coarse-grained granite	Dry	31–35	0.60–0.70	Coulson 1972
Coarse-grained granite	Wet	31–33	0.60–0.65	Coulson 1972
Porphyry	Dry	31	0.60	Barton 1971
Porphyry	Wet	31	0.60	Barton 1971
Dolerite	Dry	36	0.72	Richards 1973
Dolerite	Wet	32	0.62	Richards 1973
C. Metamorphic Rocks				
Amphibolite	Dry	32	0.62	Wallace *et al* 1970
Gneiss	Dry	26–29	0.48–0.55	Coulson 1972
Gneiss	Wet	23–26	0.42–0.49	Coulson 1972
Slate	Dry	25–30	0.47–0.58	Barton 1971
Slate	Dry	30	0.58	Richards 1973
Slate	Wet	21	0.37	Richards 1973

Note This list of data has been compiled mainly from Barton[8].

A more general empirical description of peak shear strength has been suggested by Barton[8]:

$$\tau = \sigma \tan (JRC \log_{10} (JCS/\sigma) + \phi)$$

where *JRC* is the joint roughness coefficient which varies from 0 to 20 and is estimated visually. *JCS* is the joint compressive strength. An example of the shear strength envelopes predicted by this equation is given in Figure 16.7. The criterion is intended to apply between the limits of the basic friction angle and 70°. Values outside this range are invalid.

EFFECT OF WEATHERING ON JOINT SHEAR STRENGTH

Data on shear strength of weathered joints have been reviewed by Barton[8] and

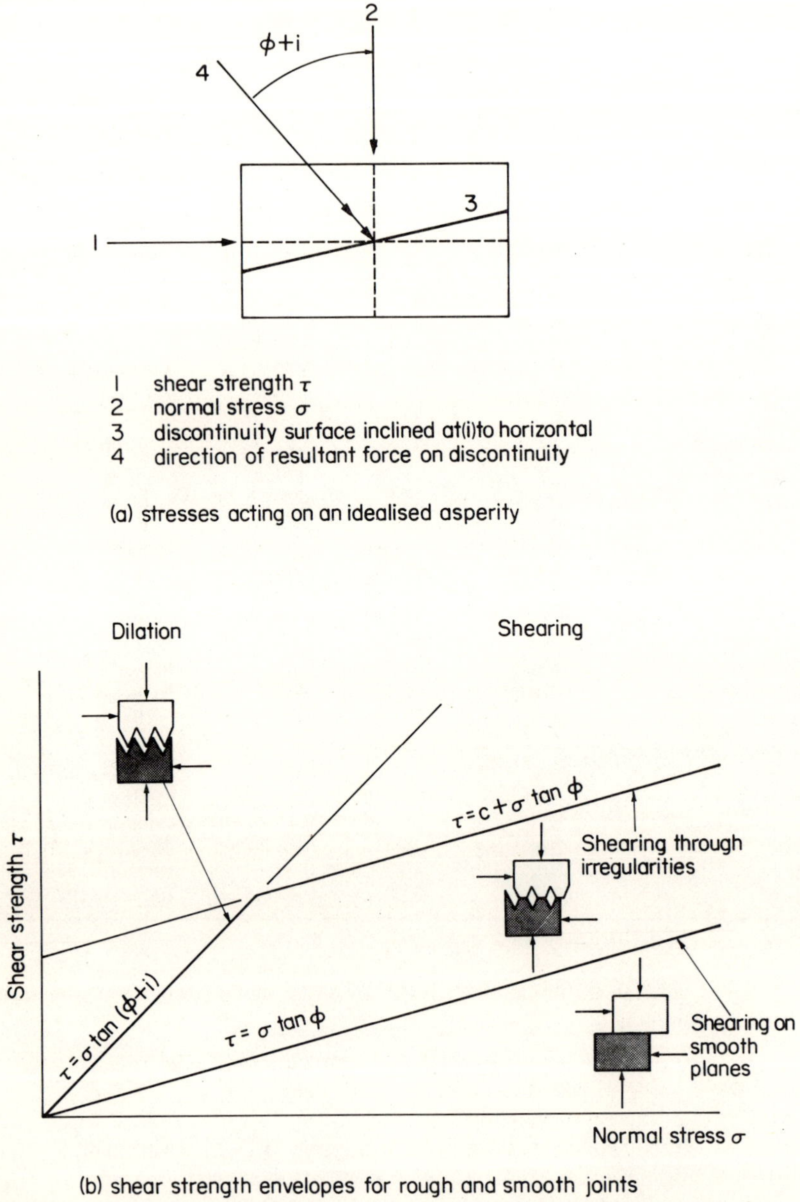

Figure 16.6 Shear strength behaviour of rough surfaces

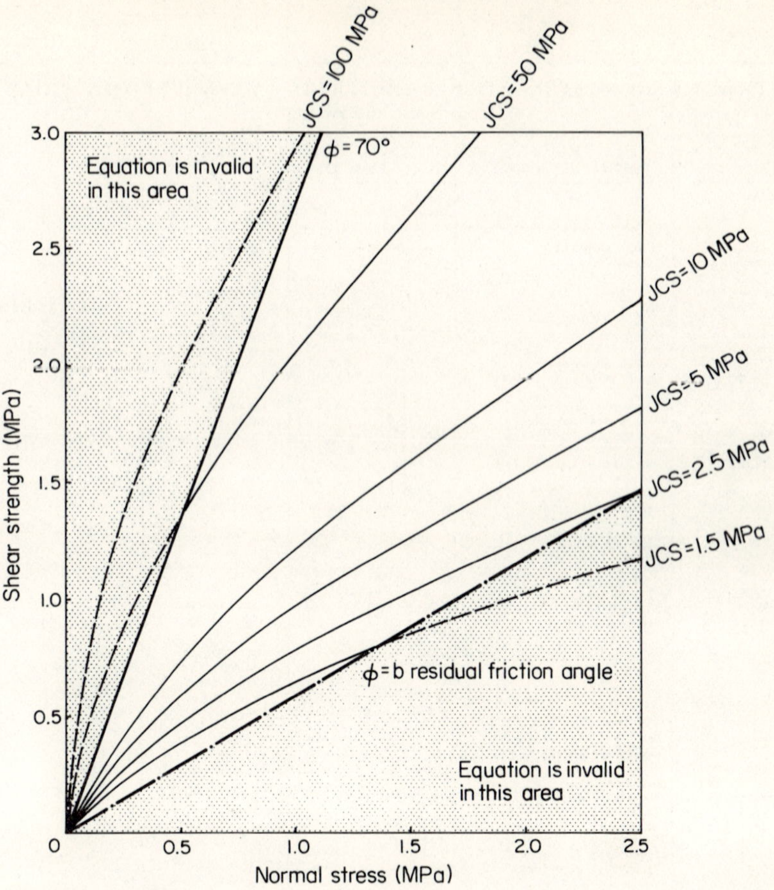

Figure 16.7 Peak shear strength for a rough joint as predicted by Barton's criterion

Deere and Patton[9]. Barton[8] has simplified the effects of weathering on rock joint shear strength to the following stages:

1. *Reduction of joint wall strength* to less than that of the compressive strength of the adjacent rock material.
2. *Common intermediate stage.* Weathered, water-conducting joints, impermeable rocks between. Joint wall strength some function of the rock material compressive strength.
3. *Penetration of joint weathering into rock blocks.* Progressive reduction of strength from the walls of the blocks inwards. Joint wall strength continues to reduce slowly.
4. *Advanced stage of weathering.* More uniformly reduced rock material, strength finally reduces to same level as joint wall strength, rock mass permeable throughout.

The major cause of shear strength reduction for a weathered joint is thus assumed to be a decrease in the compressive strength of the joint asperities.

Table 16.7 SHEAR STRENGTH PARAMETERS FOR WEATHERED MATERIAL
(from Deere and Patton[9])

ROCK TYPE	DEGREE OF WEATHERING	STRENGTH PARAMETERS	REMARKS	SOURCE
METAMORPHIC ROCKS		c,c', kgm/cm^2 ϕ,ϕ',ϕ_r degrees		
Gneiss (micaceous)	muram (zone IB) decomposed rock	c = 0.6 $\phi = 23^\circ$ c = 0.3 $\phi = 37^\circ$	direct shear tests	Gruner and Gruner (1953)
Gneiss	decomposed (zone IC)	$\phi = 18.5^\circ$	consol.undrained	St.John et al (1969)
Gneiss	decomposed (fault zone) much decomposed medium decomposed unweathered	c = 1.5 $\phi = 27^\circ$ c = 4.0 $\phi = 29^\circ$ c = 8.5 $\phi = 35^\circ$ c = 12.5 $\phi = 60^\circ$	direct shear tests on concrete-rock surfaces.	Evdokimov and Chiriaev (1966)
Schist	weathered (mica schist soil) partly weathered (mica and phyllites - broken)	- - - $\phi = 24.5^\circ$ c = 0.7 $\phi = 35^\circ$	from analysis of slides perp.to schistosity	De Fries (1971)
Schist	weathered - intermediate (zone IC)	c'=0.5 $\phi=15^\circ, \phi'=15^\circ$ c'=0.7 $\phi=18^\circ, \phi'=21^\circ$	S=50% Consol. S=100% Undrained	Sowers (1963)
Schist	weathered	$\phi = 26\text{-}30^\circ$	Rock fill	Wilson & Marano (1968)
Phyllite	residual soil (zone IC)	c = 0 $\phi = 24^\circ$ c = 0 $\phi = 18^\circ$	perp. to schistosity parallel to " " "	De Fries (1971)
IGNEOUS ROCKS				
Granite	decomposed granite	c = 0 $\phi = 27\text{-}31^\circ$	500 tests	Roberts (1970)
Granite	i = 15 i = 10 i = 7 i = 5 i = 3	c = 1 $\phi = 41^\circ$ c = 2 $\phi = 45\text{-}46^\circ$ c = 3 $\phi = 49\text{-}52^\circ$ c = 5 $\phi = 57^\circ$ c=6-13 $\phi = 62\text{-}63^\circ$	in-situ direct shear tests Alto Rabagao	Rocha (1964) Serafim and Lopez (1961)
Granite	weathered(IIB) i=7+ partly weathered (IIB) 3+ relatively sound 1-2	$\phi_r = 26\text{-}33^\circ$ $\phi_r = 27\text{-}31^\circ$ $\phi_r = 29\text{-}32^\circ$	lab direct shear tests Alto Lindoso	LNEC (1965)
Granite	red earth (zone IB) decomposed granite (IC)	$\phi' = 28^\circ$ $\phi' = 35^\circ$	average	Lumb (1965)
Granite	decomposed (finegrained) decomposed (coarse) decomposed(remoulded)	c = 0 if saturated $\phi = 25\text{-}34^\circ$ $\phi = 36\text{-}38^\circ$ $\phi = 22\text{-}40^\circ$		
Quartz diorite	sandy silt, decomposed	c = 0.1 $\phi = 30^\circ$+	lab,undisturbed	Li & Mejia (1967)
Diorite	weathered	c = 0.3 $\phi = 22^\circ$	consol. undrained	St John et al (1969)
Rhyolite	decomposed	$\phi = 30^\circ$		Lumb (1965)
SEDIMENTARY ROCKS				
Keuper Marl	highly weathered intermediate weathered unweathered	c'< 0.1 $\phi'=25\text{-}32^\circ, \phi_r=18\text{-}25^\circ$ c'< 0.1 $\phi'=32\text{-}42^\circ, \phi_r=22\text{-}29^\circ$ c'< 0.3 $\phi'>40^\circ$ $\phi_r=23\text{-}32^\circ$	2% carbonates 14% carbonates 20% carbonates	Chandler (1969)
London clay	weathered (brown) unweathered	c'=.1-.2 $\phi'=19\text{-}22^\circ, \phi_r=14^\circ$ c'=.9-1.8 $\phi'=23\text{-}30^\circ, \phi_r=15^\circ$		Chandler (1969) after Skempton.
JOINT FILLING				
Black seams	in zone IC	$\phi_r = 10.5^\circ$ $\phi_r = 14.5^\circ$	with slickensides without slickensides	St John et al (1969)
MINERALS COMMON IN RESIDUAL SOILS AND WEATHERED ROCKS				
Quartz sand		$\phi_r = 30^\circ$ $\phi_r = 35^\circ$	dune sand crushed sand	Kenney. (1967)
Kaolinite		$\phi_r = 22^\circ$	remoulded kaolinite-limited displacement	Kanji (1970)
		$\phi_r = 12^\circ$	remoulded kaolinite-limestone	Kanji (1970)
		$\phi_r = 12^\circ$	intact kaolinite - precut surface	Morgenstern (1967)
Illite		$\phi_r = 6.5^\circ$	remoulded illite-limestone	Kanji (1970)
		$\phi_r = 11.5^\circ$	remoulded illite	Kanji (1970)

Barton[8] found that a realistic approximation to weathered joint wall compressive strength could be obtained by dividing the compressive strength of the unweathered rock by a factor of four.

Deere and Patton[9] summarise the principal effect of weathering on the shear strength failure envelope as:

1. An appreciable decrease in shear strength at high stress levels.
2. A decrease in the normal stress level at which the initial mode of failure changes from sliding up to shearing through them.

At high normal stress levels, and for highly weathered rocks, the peak shear strength may be reduced nearly to the residual strength.

In addition to the deleterious effects on rock compressive strength, weathering may also change the basic friction angle of rock surfaces through changes in mineralogy. Data on weathered rock shear strength are summarised in Table 16.6 which is taken from Deere and Patton[9].

INFILLED JOINTS

If joint walls are separated by weak material, contact between the rock walls may not occur even after large shearing displacements. Such infilled joints have strength and deformability characteristics that are appreciably different from both the filling material and the rock joint. Goodman's[10] work on mica-filled plaster joints showed that the strength and stiffness of the infilled joint decreases with degree of joint filling. A summary of his results is given in Figure 16.8.

Where weathered layers are found as partings between rock joints, it is generally assumed that the strength of the infill represents the minimum strength present. The preceding results would indicate that this is true for peak strength conditions. However, there is occasionally field evidence to suggest that lower shear strengths exist along the interface between soil and rock walls. The following conclusions were drawn by Kanji[11], based on experimental work related to the shear strength at soil-rock interfaces:

1. The shear strength along soil-rock interfaces is considerably lower than that of the soil alone.
2. The reduction in strength at the soil-rock interface is greater at low normal stress levels than for higher normal stresses.
3. Polished rock surfaces produce greater reductions in shear strength than unpolished surfaces (hence the results are more applicable to slickensided rock features).
4. The shear strength versus displacement curves show differences (compared with those for soils alone) that are favourable to progressive failure.

All Kanji's results[11] showed loss of soil strength at the rock interfaces, the most significant loss being a 14.5° reduction in friction for a soil composed mainly of kaolinite sheared against a polished limestone surface. Simple test methods using modified soil shear machines are sufficient to test soil shear strengths at rock interfaces[12].

SHEAR STRENGTH PARAMETERS AND THE EFFECTIVE STRESS PRINCIPLE

The effective stress principle describes the decrease in the shear strength of discontinuities due to water pressure. Shear tests on rock joints are invariably carried out under drained conditions and effective friction angles determined. The 'cohesion' term often used to describe rock friction is not analogous with

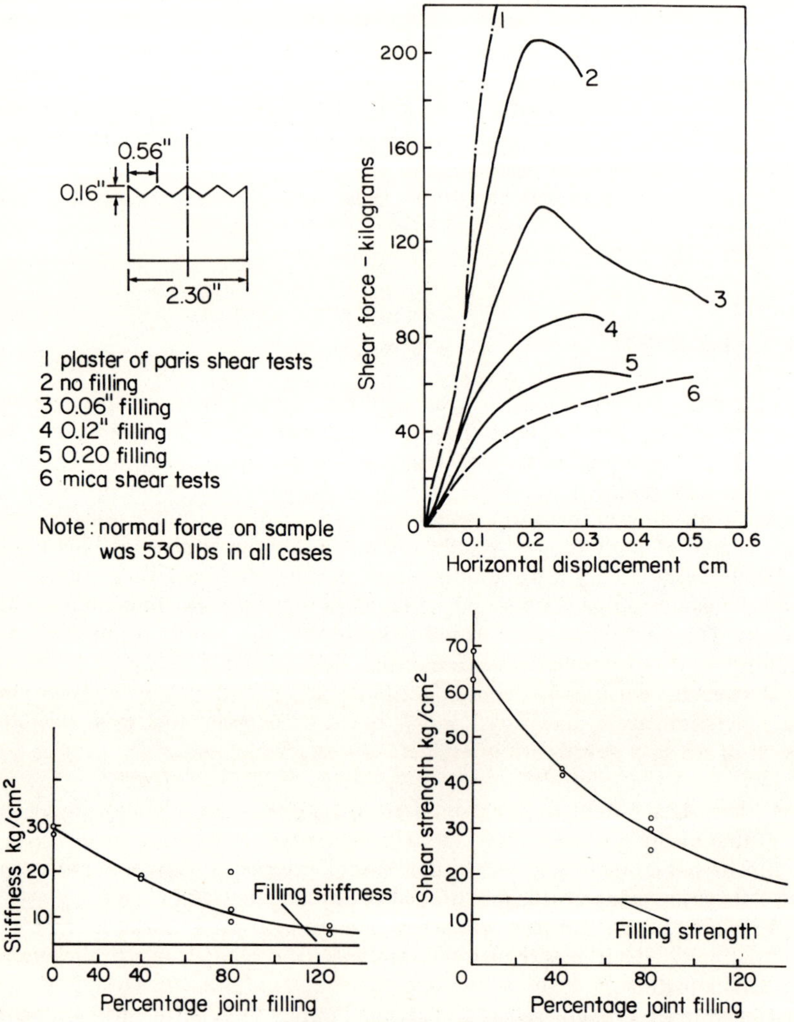

Figure 16.8 Test results for infilled joints

the 'c' parameter for soil material. If the shear strength envelope for a rock joint has a curved function $\tau = f(\sigma)$, it is a common practice to approximate this to a linear tangent at a given normal stress level (see Figure 16.5). This may lead to an over-estimation of the strength under conditions of water pressure as shown in Figure 16.9.

SENSITIVITY OF FACTOR OF SAFETY TO VARIOUS PARAMETERS

In a slope stability analysis, the relative sensitivity of the factor of safety to variations in the magnitudes of each function affecting stability will depend on the height of the slope being considered. The relative importance of the accuracy with which each item of input data has been measured is, therefore, different. This can be assessed by considering hypothetical 60° slopes with heights of 10 m, 100 m and 1000 m. These could be typical heights encountered in a road cutting, tunnel portal and an opencast mine respectively. A joint is assumed to strike parallel with the crest of the slope and to daylight at the toe of the slope.

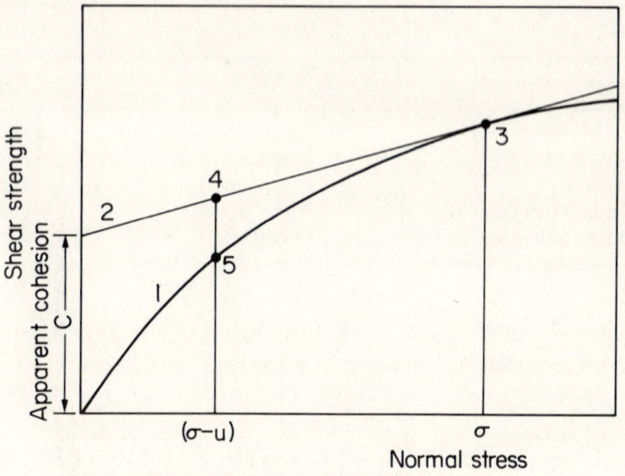

1. Non-linear shear strength envelope
2. Linear approximation to envelope at normal stress level
3. Shear strength at normal stress level
4. Shear strength based on linear approximation at effective stress level
5. Shear strength based on non-linear shear strength envelope at effective stress level

Shear strength is over-estimated by an amount (4−5)

Figure 16.9 Shear strength parameters and effective stresses

Table 16.8 (upper part) lists the functions which will affect the equilibrium of the wedge of rock above the joint for a given slope height and slope angle, and the range of magnitude likely to be encountered in each of them.

A sensitivity analysis involves varying in turn each function throughout its probable range while keeping the other functions constant at their mean values, for each of the three slope heights being considered, and calculating the effect on the factor of safety. This enables the functions to be graded in order of importance with respect to their effect on the factor of safety, and the results are shown in the lower part of Table 16.8. The results of the sensitivity analysis are given in Figure 16.10, which shows the relationship between factor of safety and each of the functions considered.

In all cases, the most significant parameter is that of joint inclination. On a small slope, density is more important than friction. Cohesion becomes less important with increasing slope height whereas the effects of water pressure

Table 16.8 SENSITIVITY OF FACTOR OF SAFETY TO VARIOUS PARAMETERS

Functions affecting the factor of safety (for a given value of slope height and angle)

Function	*Probable range of magnitude*
density, ρ	0–300 kPa
cohesion, c	15–30 kN/m^3
water pressure, hw/H	0–H metres
friction angle, θ	0–60°
joint inclination, ϕ	10–50°

Order of importance of functions

Rank	*Slope height*		
	10 m	100 m	1000 m
1	Joint inclination	Joint inclination	Joint inclination
2	Cohesion	Friction angle	Friction angle
3	Density	Cohesion	Water pressure
4	Friction angle	Water pressure	Cohesion
5	Water pressure	Density	Density

become increasingly more significant. For slopes of medium height and above, friction is the most important parameter after joint inclination.

DETERMINATION OF SHEAR STRENGTH PARAMETERS FOR SLOPE STABILITY CALCULATIONS

The methods for obtaining best estimates of shear strength parameters depend on the size, complexity and degree of risk associated with a given slope configuration. As shown in the preceding section, small non-critical slopes doe not require refined values of shear strength and it may often be sufficient to base designs on values obtained by inspection of the joint surfaces and a study of available data. Increasing slope heights and risk elements demand greater data refinement.

An attempt has been made in Table 16.9 to give general guidelines for the scope of work associated with shear strength determination. It must be emphasised that shear strength testing must be related to the geological investigations and preliminary design phases in order for the most relevant discontinuities to be identified and tested as necessary.

GROUND WATER CONSIDERATIONS

EFFECT OF GEOLOGY IN GROUND WATER BEHAVIOUR

The overall pattern of ground water flow is controlled primarily by structural geology because, in most rocks, flow occurs through the discontinuities rather than through the intact rock. Since discontinuities generally fall into 'sets'

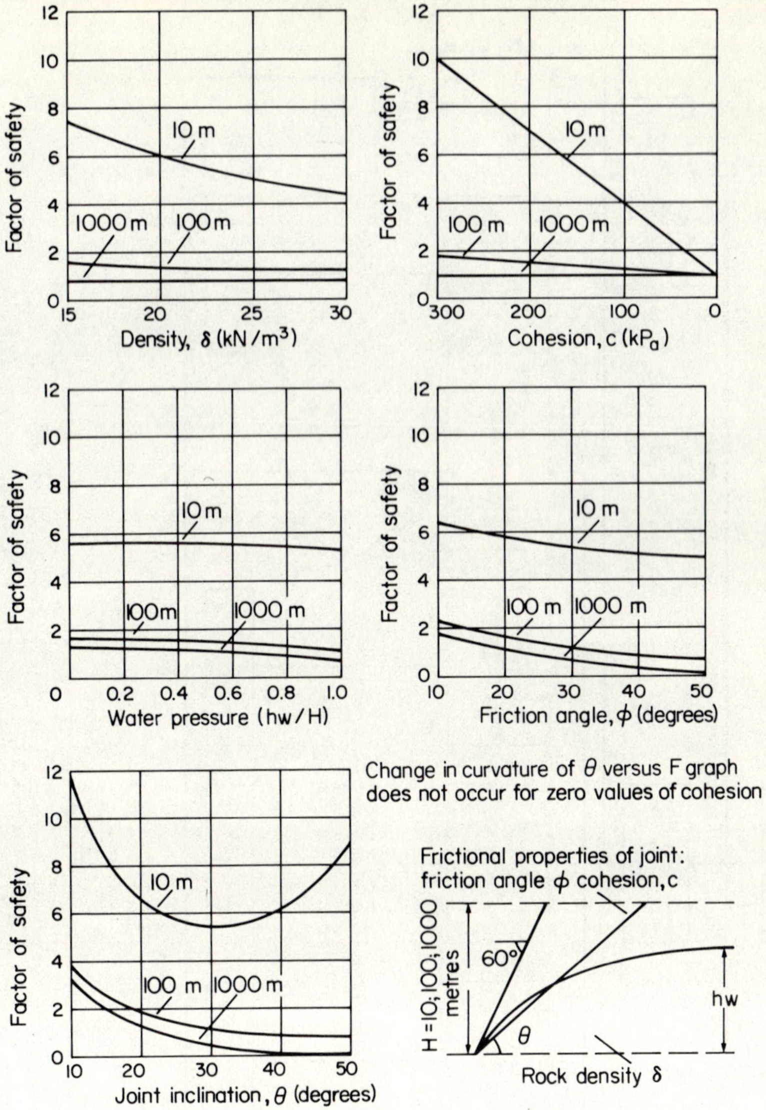

Figure 16.10 Sensitivity analysis for slope stability calculations

having a given orientation, the permeability of the rock will vary with direction, i.e. the rock will exhibit anisotropic permeability. In certain rock types, the permeability of the rock mass depends on the stress level, which influences the degree of joint opening. General permeability values for some rocks and rock masses are given in Table 16.10.

Most slopes are composed of a soil and a rock zone. Rock slopes may be composed of several different rock types or structural domains. Therefore zones of varying permeability will be present. The overall flow of water is controlled primarily by the distribution of these zones.

Table 16.9 DETERMINATION OF SHEAR STRENGTH PARAMETERS FOR SLOPE STABILITY CALCULATIONS

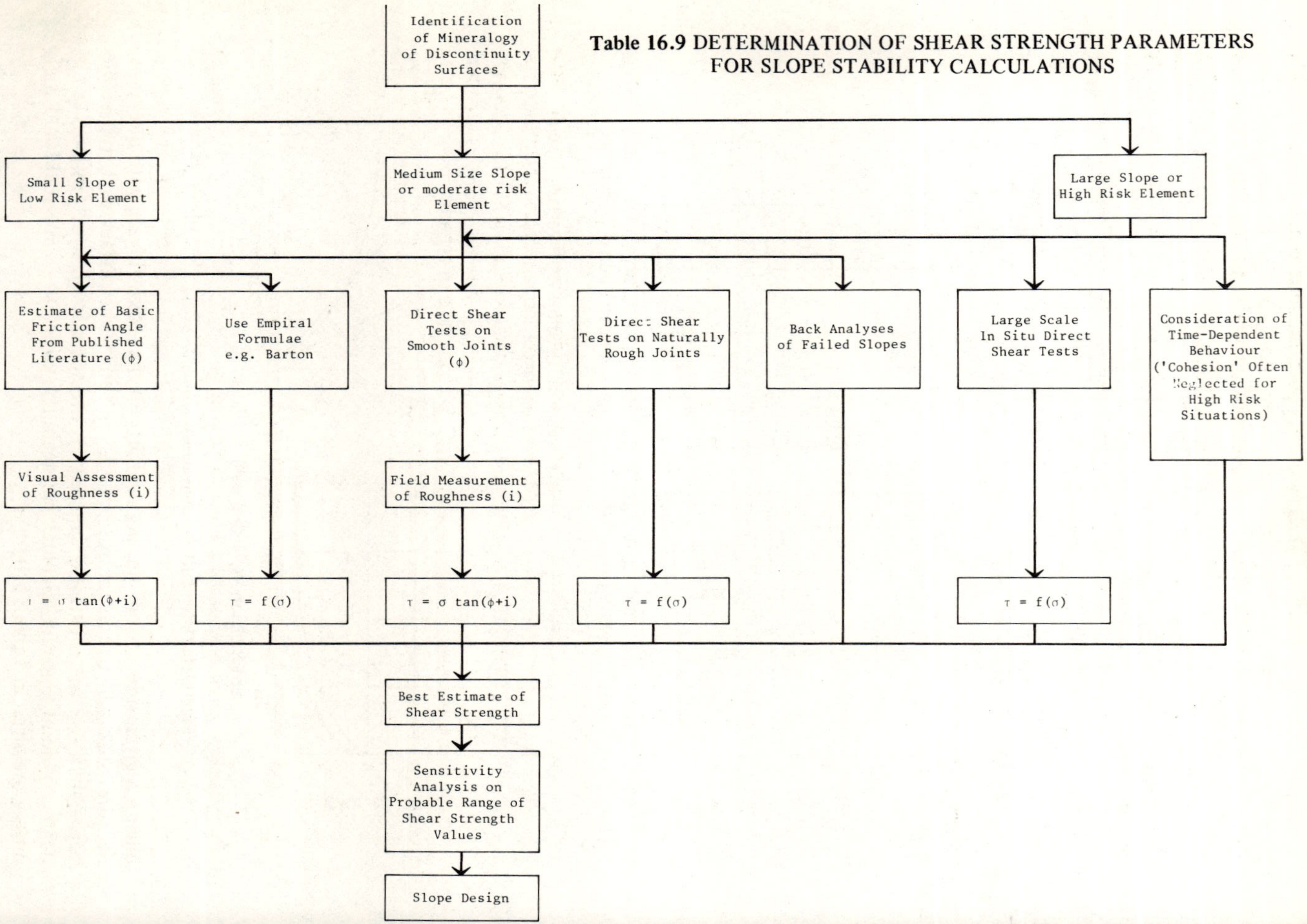

Since field permeability or ground water measurements represent spot values, an adequate description of the geology of the rock mass, including the identification of individual rock types, transgressive features and the local structural regime, is essential before an assessment of ground water conditions can be made or means of controlling ground water pressures designed. Certain rock types have predictable hydraulic properties. For example, shales can be expected to exhibit anisotropic permeability because of their laminated structure, limestones are likely to contain solution cavities and sandstones often have appreciable intact and bulk permeabilities.

Table 16.10 TYPICAL PERMEABILITY VALUES FOR SOME ROCKS AND ROCK MASSES

Rocks (intact permeability)	*k (m/sec)*
Limestone (2% porosity)	1×10^{-9}
Limestone (16% porosity)	1×10^{-6}
Silty sandstone	2×10^{-8}
Sandstone (29% porosity)	2×10^{-5}
Granite	5×10^{-11}
Slate	1×10^{-12}
Rock masses (bulk permeability)	*k (mm/sec)*
With clay filled joints	1×10^{-7}
Moderately jointed	$1 \times 10^{-6} - 1 \times 10^{-4}$
Well jointed	$1 \times 10^{-4} - 1 \times 10^{-1}$
Heavily fractured	$1 \times 10^{-1} - 1 \times 1$

Transgressive geological features may have a significant effect on ground water flow patterns. For example, faults containing gouge material impede flow and may cause high ground water pressures behind them. Alternatively, shatter zones or well-jointed dykes may form zones of high bulk permeability, concentrating ground water flow and effectively draining the rock on either side.

Regional structure and topography will indicate the potential for recharge of the rocks which are to be exposed in an excavation. Some of the effects of geological conditions are illustrated in Figure 16.11.

INFLUENCE OF GROUNDWATER ON ROCK SLOPE STABILITY

The primary influence of groundwater pressures on the stability of rock masses is to reduce the shear strength of discontinuities. Groundwater pressure can also reduce stability by producing hydraulic driving forces in steeply dipping discontinuities which are not subject to shear failure (e.g. tension cracks, near vertical faults). The presence of water may also have a deleterious effect on the shear strength of certain rock surfaces, as indicated in Tables 16.5 and 16.6.

The ground water pressure distribution in a slope must be known before the effect of ground water on stability can be calculated. Determination of the ground water pressure distribution requires direct measurements at representative locations using piezometers. Piezometric heads within rock slopes often vary considerably from point to point and a sufficient number of tips must be installed to define the overall conditions. Tips should be located with reference

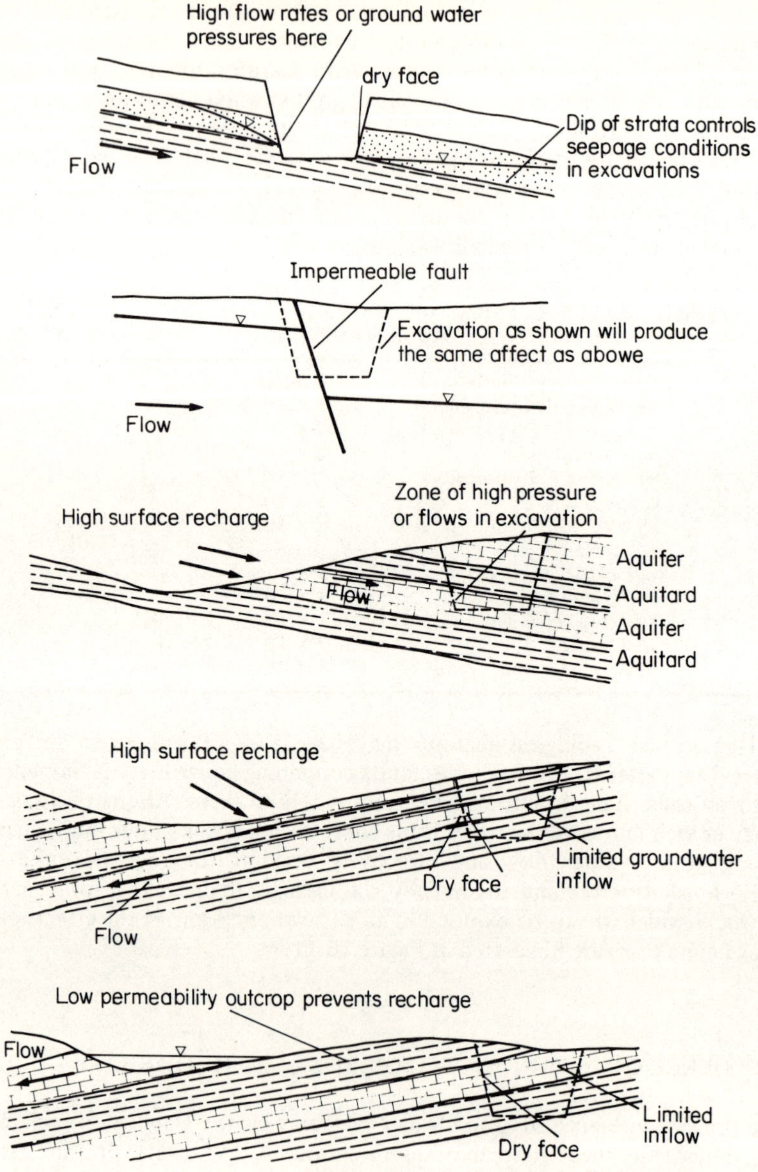

Figure 16.11 Some effects of geological structure on ground water conditions

to the geology, particularly with regard to intersecting major fissures, and must be properly sealed in the borehole above and below the tip so that measurements of piezometric head represent conditions only at the tip location. For this reason, open holes rarely give reliable indications. It is essential to provide a large enough number of piezometer installations to allow for malfunction of individual instruments and to provide a reasonable statistical assessment of piezometric conditions.

A general classification of piezometers, is shown in Table 16.11. Their methods of operation are illustrated in Figure 16.12. The most important physical properties required in a piezometer tip are ruggedness and long term reliability.

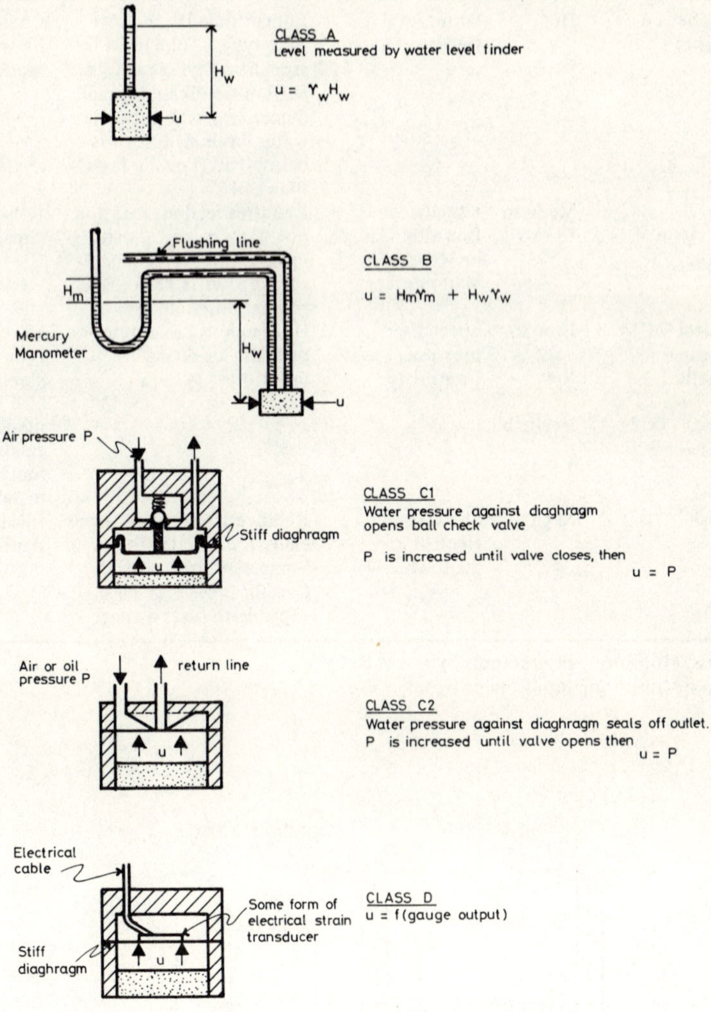

Figure 16.12 Piezometer classes showing principal characteristics

Piezometers must be selected with regard to the expected permeability of the rock mass. High volume demand piezometers (e.g. standpipes) are unsuitable for measuring rapid changes in ground water pressure in low permeability materials. Some installation methods for piezometers are shown in Figure 16.13. Full details of the selection and installation of piezometers are given elsewhere[13,14].

Rock mass permeability should be determined from correlation of detailed studies of the borehole core, information on water or air loss during drilling and

Table 16.11 PIEZOMETER TYPES: GENERAL CLASSIFICATION

Broad classification	*Class*	*Relative volume demand*	*Readout equipment*	*Major limitations*	*Major advantages*
Open Tube (standpipe)	A	High	Water level finder	Longer time lag in most rock types. Tube must be straight for whole length. Difficulties likely in small diameter plastic tubes if water levels significantly below 100 ft or dip less than c. 45°	Cheap. Simple to read
Closed tube (hydraulic)	B	Medium to low	Usually Bourdon Gauge or Mercury Manometer	Requires readout location not significantly above lowest water level. Therefore, not suitable for general borehole use	Relatively cheap
Mechanical diaphragm (Pneumatic or	C1*	Low to very low	Specialised pressure transmitter	Hydraulic types require periodic 'de-airing' of monitoring system	Excellent long term stability. Can be made very small. Simple to install
hydraulic)	C2*	Negligible			
Electrical	D	Negligible	Specialised electronic readout	Relatively expensive, especially if cable lengths large. Some zero drift possible. Certain types may be susceptible to blast damage	Ideal for remote monitoring

*C1 Class. Monitoring input closes bypass valve.
C2 Class. Monitoring input opens bypass valve.

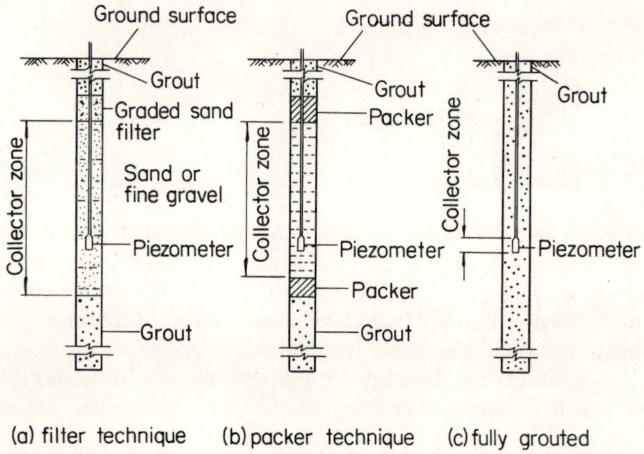

Figure 16.13 Piezometers: possible installation configuration

field permeability tests at representative locations. Cores should initially be logged for lithological changes and for changes in degree of fracturing (e.g. RQD). From this information, representative locations for permeability tests should be selected and the tests performed. Sufficient tests should be done to determine a correlation between permeability and geological/geotechnical variations. Two types of test are normally used.

In the *falling head test*, an uncased section of the borehole is subjected to a pressure head greater than static ground water pressure. This head is then allowed to dissipate to the static value and measurements of the loss of head with time are taken. A relationship between flow and pressure is determined, from which the permeability is calculated.

In the *constant head test*, an uncased section of the borehole is subjected to a steady excess pressure head (either by maintaining a constant excess water level in the hole above the test section or by pumping water at a steady pressure directly into the test section once it has been isolated by a mechanical or pneumatic packer unit). The resultant steady flow out of the test section is measured and the permeability is calculated from the relationship between flow and pressure. Details of the test methods and calculations are given elsewhere[13,15].

Suitable analytical methods to determine the ground water pressure distribution within a slope from a limited number of permeability and piezometric measurements include:

(a) Graphical flow net sketching
(b) Electrical resistance analogues
(c) Numerical (computer) analyses

Reference should be made to specialised texts[13,16,17,18] for details of these techniques. Using these methods of analysis, the existing ground water pressure distribution, geological structure and permeability measurements of representative rock or soil strata can be used to predict changes in ground water pressure conditions resulting from the intended excavation or construction of the proposed engineering works, from external changes (e.g. seasonal variations) and from drainage or other ground water control measures. When other input data (shear strengths etc) are known, stability calculations can be used to determine the most cost-effective slope design. Methods of incorporating ground water conditions into stability calculations are summarised in Table 16.12.

GROUND WATER CONTROL MEASURES

The most commonly used ground water control measure is drainage, which, by reducing the detrimental effects of ground water, can either increase the stability of existing slopes or allow slopes to be steepened while maintaining the same degree of safety.

The maximum benefit is gained from reducing water pressures within the slope prior to movements taking place during excavation and while the rock mass will still possess its undisturbed strength. To define the extent of the rock mass requiring drainage, the most probable failure zone (itself a function of the ground water conditions) must be determined. A typical zone which can be used for initial comparative studies is shown in Figure 16.14.

Table 16.12 METHODS OF USING GROUND WATER DATA IN STABILITY ANALYSIS

Rotational modes of failure (generally analysed as a series of elements of slices)	1.	CIRCULAR SHEAR FAILURE	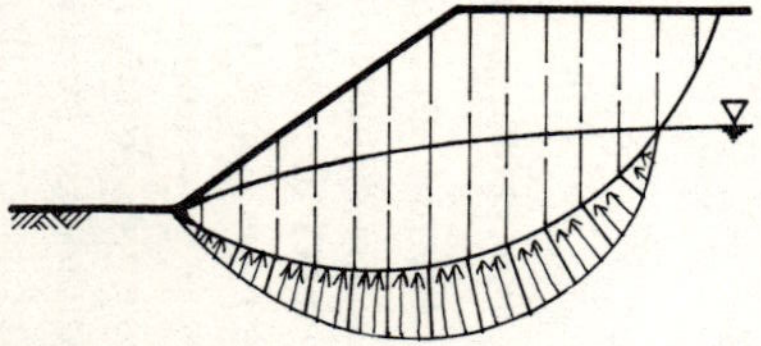	Consider individual elements of the failure surface. Calculate effective stress acting on each element. Then determine effective shear strength for each element and use as input to a stability analysis based on a method of slices. (19,20,21,22,23,24).
	2.	NON-CIRCULAR SHEAR FAILURE	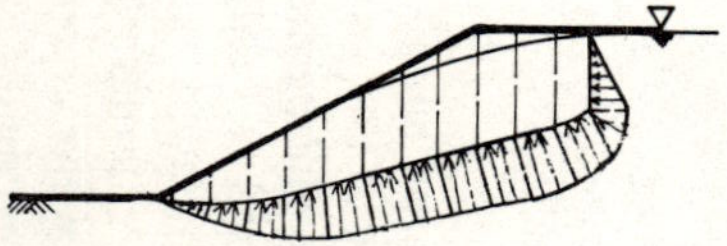	Steps as 1 above. Allowance must also be made for effect of tension cracks, if present, and hydraulic driving forces therein. (19,22,23,24)
Translational modes of failure (generally analysed as a rigid body)	3.	PLANAR FAILURE	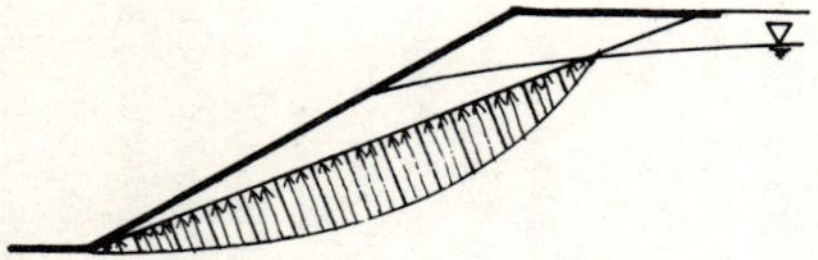	Define the groundwater pressure distribution acting on the failure plane. From this, calculate total force exerted by water. Calculate effective shear resistance of plane from effective stress. Consider tension cracks and release surfaces, if appropriate, and hydraulic driving forces therein. (21)

Translational modes of failure (generally analysed as a rigid body)	4. WEDGE FAILURE	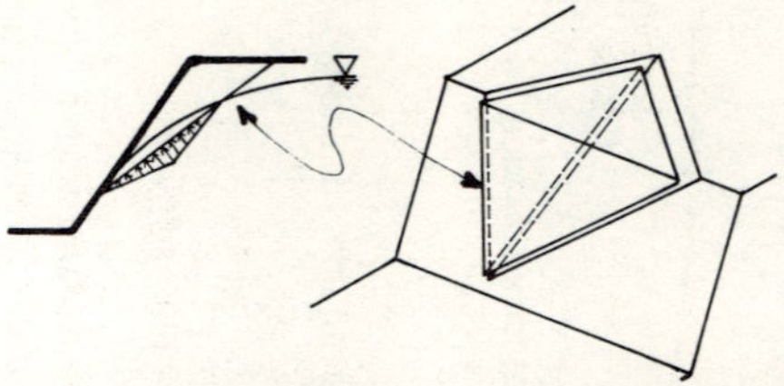	Steps as in 3 above but for two planes. (10,25,26)
	5. MULTIPLE FAILURE	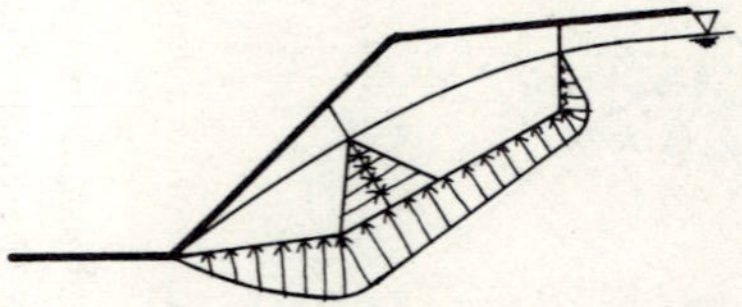	General principles as in 3 and 4 but analyse each block separately. It is particularly important to consider each discontinuity, to determine whether effective shear strength or hydraulic driving forces are applicable to the analysis.
Toppling mode of failure	6. TOPPLING FAILURE	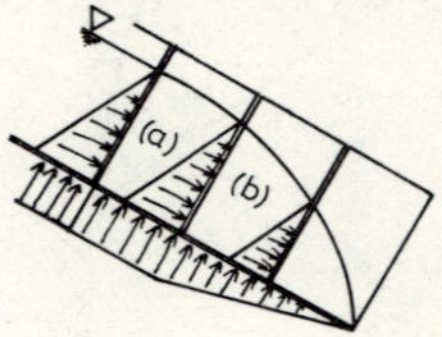	Effective shear strength not applicable. Hydraulic driving forces only are relevant to vertical discontinuities. Water pressure distribution on base of blocks needs to be considered to determine whether it is uniform (as in a) at left) or non-uniform (as in b)). Only in latter case will an additional **overturning force** need to be considered. (10,25,26)

Drainage implies the removal of water from an area which may be critical with respect to stability. In some cases, it may be more expedient to prevent the water from reaching the critical zone by creating a barrier (or cut-off) which retains the water behind it and allows natural drainage to occur within the slope. Cut-offs are advantageous where internal erosion of soft material may be caused by increased water flows induced by drainage measures. Methods of construction commonly used include:

(a) Injecting cement i.e. a grout curtain.
(b) Excavating a trench and backfilling with concrete or asphalt.
(c) Sheet piling.
(d) Sinking a row of pumped vertical wells i.e. a well curtain.

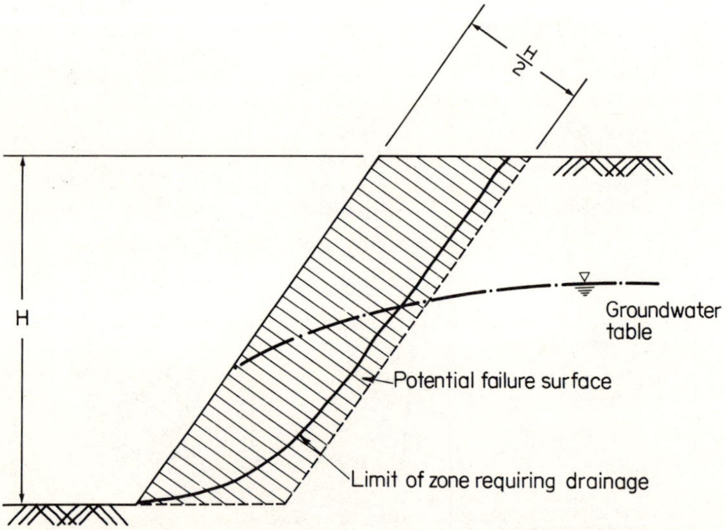

Figure 16.14 Definition of rock volume requiring drainage

The first three methods form physical barriers to ground water flow. The last method intercepts flow towards the excavation and provides 'dry' conditions on its downstream side. Slope drainage is the measure most frequently used.
Slope drainage systems currently in wide use are:

(a) Drainage trenches constructed down and/or along the slope face.
(b) Horizontal or near horizontal drain holes drilled into the slope face.
(c) Vertical pumped wells drilled behind the slope crest or on the slope face.
(d) Drainage galleries excavated in the rock mass behind the slope, with or without supplementary holes drilled from the gallery.

The method of slope drainage used will depend on slope geometry, permeability of the slope material, economic and operational considerations. The main features of the four systems are summarised in Table 16.13.

Effectiveness of drainage installations can be improved substantially by minimising infiltration into the slope as far as possible. This requires particular

Table 16.13 CHARACTERISTICS OF SLOPE DRAINAGE METHODS

	SCALE OF SLOPE (HEIGHT)	ROCK TYPES	INSTALLATION DETAILS	MAIN ADVANTAGES	MAIN DISADVANTAGES
a) DRAINAGE TRENCHES	SMALL (1-15 metres)	Soils and soft rocks	Used where thin impermeable strata overlie permeable strata and dip parallel to slope. Also used in floor of cut, to prevent "heave" in similar, but horizontally bedded, conditions. Must extend into permeable layer. Often backfilled with free draining material.	A cheap and direct method. Excavators usually readily available on site. Surface method, therefore conditions and performance can be visually monitored.	Limited to soft materials. Access problems on steep or high faces. Limited to slopes below 30°. Depth of penetration limited. High cost of backfill, in some situations.
b) HORIZONTAL DRAIN HOLES	MEDIUM (15-150 metres)	Hard and non-caving rocks	Usually percussive drills, but diamond drills for long holes. Inclined slightly upwards. Usually lined with finely perforated casing (except in very competent rock) except 2 m nearest slope face. Spacing and depth according to design study. Commonly depth equal to slope or bench height and spacing 2-20 m with 75 mm holes.	Quick and simple to install. Layout flexible and can be readily varied according to conditions. Drain by gravity. Can be installed on individual benches in high slopes.	Limited influence of drainage, particularly in large rock masses or low permeability rocks. Practical maximum depth about 150 m Lining difficulties in soft or caving ground. Periodic mantenance required.
c) PUMPED VERTICAL WELLS	MED-LARGE (>15 metres)	All rock types	Commonly 150 mm to 500 mm diameter, with individual submersible or turbine pumps installed. Diameter and spacing according to design study. Depth frequently 1.2 x slope height.	Prevent water from reaching face. Do not interfere with operations at face. Can improve stability prior to excavation of slope. Can be installed on each bench in high slopes	Comprehensive power supply required. Require discharge mains to remote point to avoid "re-cycling". Electrical or mechanical failure could cause instability. Non-gravity drainage therefore high cost. Regular skilled maintenance required
d) DRAINAGE GALLERIES	LARGE (>150 metres)	Self-supporting rock	Commonly small cross section (say 2 x 2 metres) with drain excavated in floor, and inclined slightly upwards. Can be parallel with or orthogonal to slope, depending on rock structure.	Large influence of drainage. Drainage can be augmented by drilling small diameter holes radially from tunnel, or driving cross-cuts. Give access to rock for geotechnical studies. Freezing of system inhibited. Low maintenance requirement.	High cost of excavation. Relatively inflexible layout.

attention to the control and disposal of surface water before it percolates into the slope. Under some circumstances, particularly where the slope is showing signs of distress and where open cracks are visible, specific waterproofing measures, such as laying impermeable membranes at the top of the slope, will be advantageous.

Control of water is particularly important, where individual benches are drained by horizontal holes, to prevent water from percolating into the bench below. Each row of drain outlet pipes should issue into a properly graded and maintained intercept or trench, lined with impermeable material if necessary.

Generally horizontal drain holes are self-cleaning. Periodically, it may be necessary to flush out each hole, using a high-pressure hose with a radial jetting attachment, to prevent loss of efficiency due to silting of the casing, algal growth or mineral precipitation.

If problems due to freezing of drain outlets are anticipated, an insulating barrier of free-draining rock fill can be constructed over the drain outlets to inhibit freezing. Properly installed wells should require little attention, other than mechanical and electrical maintenance of pumps.

Theoretically derived drainage designs should, except in very limited applications, be refined by drainage trials in the early stages of construction. An initial trial should be carried out, using carefully located piezometers and flow-rate measurement, to check the degree to which the design assumptions, particularly of reduction in ground water pressure, are being reflected by operating conditions. The results of the trial can be used to re-assess the performance of the drainage system and arrive at the most cost-effective drainage design for the remainder of the installation.

In critical stability conditions, or for important or long-term structures, a system of routine inspection and piezometric measurement should be instituted to monitor changing conditions due to the influence of the completed engineering structure or deterioration of the drainage system, so that any remedial works necessary can be designed and undertaken in good time. It must always be borne in mind that the magnitude of flow issuing from a drainage system gives an indication only of the permeability of the rock or soil – not of the adequacy of the drainage system.

METHODS OF ANALYSING THE STABILITY OF ROCK SLOPES

The methods used to analyse slopes in rock are generally based on the principle of limiting equilibrium. Rotational failures (circular or non-circular) are treated by the method of slices, as is used for soil problems. Structure-controlled failures are based on limit equilibrium analysis of one or more rigid masses. Detailed treatment of analytical methods for rock slopes is given in Hoek and Bray[21], and Goodman[10]. Reference should be made to these texts for a more comprehensive treatment of calculation techniques. For the analysis of general slip surfaces in soil material (see chapter 15) the methods used are equally applicable to the study of highly fractured or very weak rock slopes.

ROTATIONAL FAILURES

In homogeneous material, the friction circle method[22] is a convenient calculation technique for analysing circular failures. Slope stability charts for circular failure may be found in Taylor[20], and Hoek and Bray[21]. The latter charts are the more useful in that they include the influence of a critical tension crack and the effects of ground water in the slope. Design charts such as these should be used to separate potentially unstable slopes which will require further detailed treatment from stable slopes which require no further study. Circular failures in non-homogeneous materials are analysed by the more general methods described below for non-circular failures.

Non-circular failure surfaces may be clearly defined in rock material as a result of the presence of structural features such as joints. There may consequently be less need for a search for the critical failure surfaces. Depending on the degree of sophistication required, various calculation techniques are available to determine the stability of material above such failure surfaces. These include the methods of Bishop[19], Janbu[25], Morgenstern and Price[23], and modifications to these. The method described by Sarma[24] has been extremely useful in the study of large rock slopes associated with open cast mining operations. With the possible exception of the Janbu manual method[28], most of these techniques are suitable only for computer application.

Definition of strength parameters for rotational failures in rock is difficult, because laboratory tests may suffer from severe defects due to scale. Wherever possible, back analyses should be carried out on failures in similar conditions to establish a possible range of strength properties. Sensitivity analyses will almost always be necessary to check the effects of variations in the assumed strengths.

TRANSLATIONAL FAILURES

A block on an inclined plane is a simple model of rock slope failure along a discontinuity. Figure 16.15 gives the limiting slope angles for a block under various conditions of active and passive forces. For the examples given, the slope angles at limiting equilibrium range from 15° to 50°. Real situations are more complex but the principles involved in this simple case are applicable to the more comprehensive analyses of stability.

The Appendix to this chapter describes a method for identifying potential plane failures and for carrying out a simple assessment of stability based on gravity loading and frictional strength alone. Full analysis of plane failures requires consideration also of cohesion, water pressures, reinforcement loads and acceleration forces. Figure 16.16 shows a section through a potential failure surface along a discontinuity surface and details the various forces acting on the rock mass. An equation is given in this figure for factor of safety with respect to the shear strength on the plane. Hoek and Bray[21] give more comprehensive details.

The problem of active and passive block failure is dealt with in most textbooks on soil mechanics. The solution of the problem is simpler for rock than soil as the interface between active and passive blocks is defined by the geological structure and the rock friction angle may be used to determine the inclination of the forces across the interface. Iterative techniques are not therefore necessary. The input data required is similar to the simple sliding case except that consideration has to be given to the water pressure forces acting in the interfaces between the blocks.

The forces involved in two block sliding are shown in Figure 16.17. The

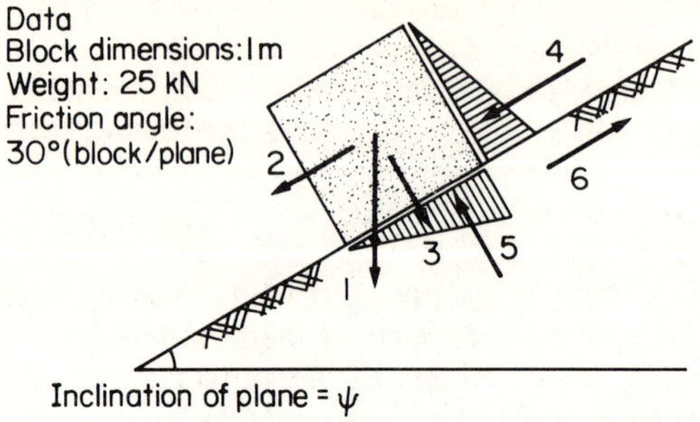

(a) general case of block on inclined plane

1. Weight of block, W
2. Component of block weight parallel to plane, $W\sin\psi$
3. Component of block weight normal to plane, $W\cos\phi$
4. Water force acting on back face of block, V
5. Uplift force acting on base of block, U
6. Shear resistance at base of block, T

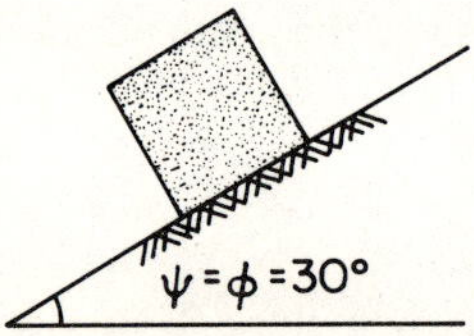

(b) frictional resistance only

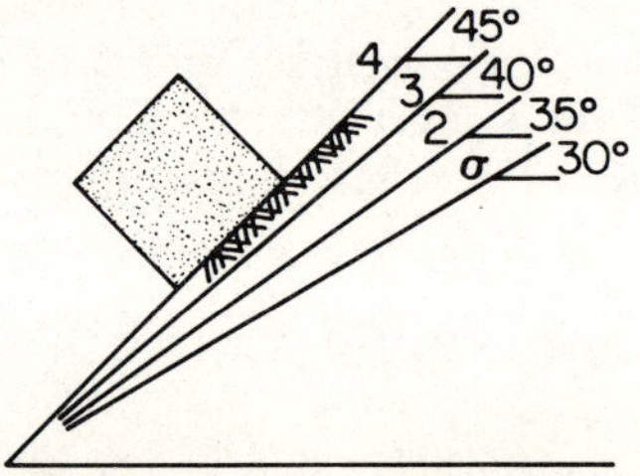

(c) 30° friction angle and cohesion

1. c = 0
2. c = 2.5 kPa
3. c = 5.0 kPa
4. c = 7.5 kPa

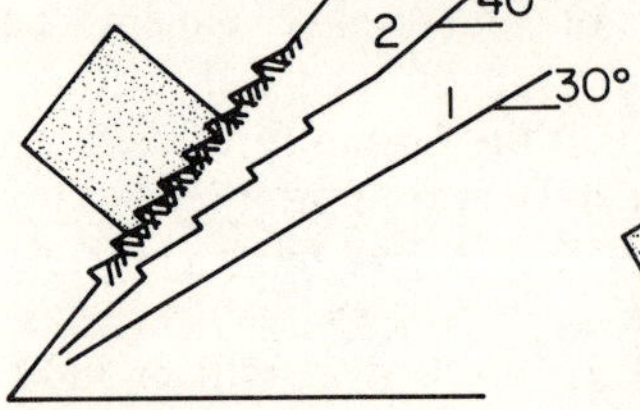

(d) stepped failure surface

1. i = 0
2. i = 10°
3. i = 20°

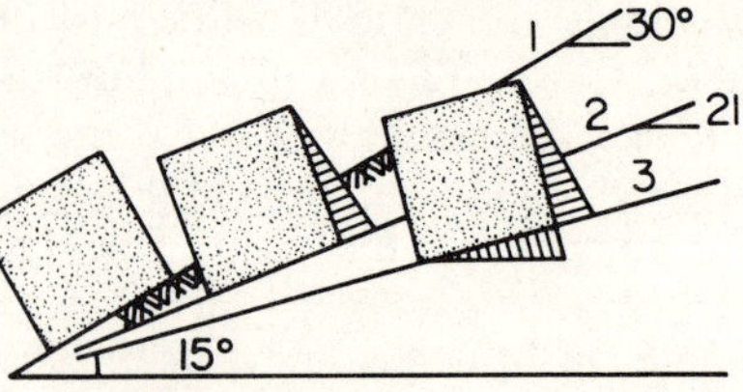

(d) water pressure

1. No water forces
2. Water force on rear face of block
3. Water forces on rear and base planes of block

Figure 16.15 Factors affecting the mechanism of simple sliding

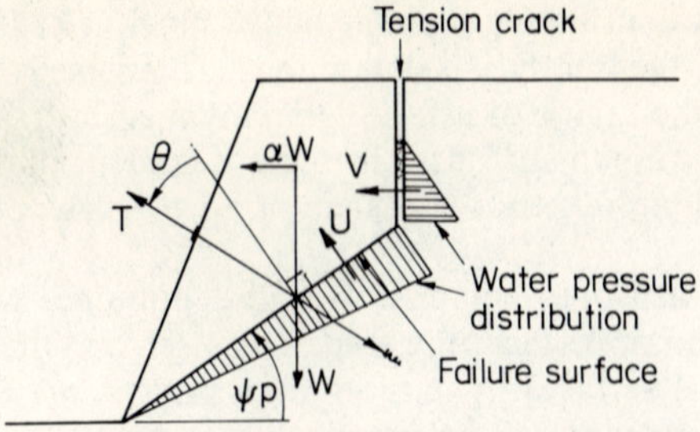

$$F = \frac{cA + \left[W(\cos\psi p - \alpha \sin\psi p) - u - V\sin\psi p + T\cos\theta\right]\tan\phi}{W(\sin\psi p + \alpha\cos\psi p) + V\cos\psi p - T\sin\theta}$$

Figure 16.16 Analysis of plane failures

T. total reinforcing force applied by rock anchors
θ. inclination of this force to the normal to the failure surface
c. cohesive strength of the failure surface
ϕ. frictional resistance of the failure surface
A. area of failure surface
W. weight of the sliding block
U_p. angle of the failure surface
a. coefficient of acceleration
U. water uplift force
V. force due to water in the tension crack

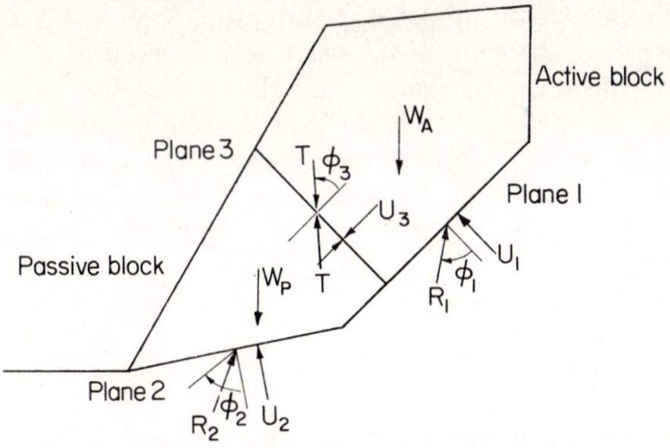

Figure 16.17 Forces involved in active and passive block slides

W_A, W_P. weights of the blocks
ϕ_1, ϕ_2, ϕ_3. friction resistances on Planes 1, 2 and 3
U_1, U_2, U_3. water pressure forces acting on Planes 1, 2 and 3
R_1, R_2. resultant forces of the active and passive blocks
T. interface force between the blocks

system can be analysed by considering that the active block is at limiting equilibrium and the force T at the interface is determined. This force is then applied to the passive block and the factor of safety determined against sliding on the base of the passive block. Goodman[10] describes the treatment of this problem by stereographic methods and extends the analysis to the case of two three-dimensional wedges.

Typical wedge failures in slopes comprise two discontinuities whose line of intersection daylights in the slope face, the slope face itself and the upper slope (which may be a horizontal surface). A tension crack may be present, in which case the upper apex of the wedge will be truncated. The general failure mode is by sliding parallel to the line of intersection of the two planes. The geometry of certain wedges is such that sliding occurs on one plane only under gravitational loading alone. This will show up during the kinematic check on planar sliding. Wedges involving a plane which has been shown to be unstable should be ignored as being less critical than the plane failure. There are also combinations of active forces which cause failure to occur on one plane only – these can generally be identified only during the calculation process. Where sliding occurs on one plane only, the problem can be treated by the methods for plane sliding. The calculation of stability for rock wedges may be handled by design charts, analytical methods or spherical projections. The Appendix to this chapter gives a simple method for identification and preliminary stability estimation. Detailed descriptions may be found in Hoek and Bray[21] and Goodman[10]. A full wedge analysis with water pressures, reinforcement and tension cracks is time-consuming and tedious if solved manually. Most programmable calculators of reasonable capacity can be used to carry out the calculations. In addition, there are commercially available programmes for the solution of wedge problems. A great many wedge problems may be simply disposed of during the process of checking kinematic admissibility and the simple stability check using the friction cone (as described in the Appendix).

For sliding on two planes with different frictional values, there are an infinite number of possibilities for producing limiting equilibrium. The factor of safety for wedges is often taken to be the value which produces the same factor of safety on both planes at the same time. This is obviously very arbitrary and possibly might be unrealistic. A more generally useful approach is to consider the sensitivity of the block to possible changes in shear strength characteristics and water pressure conditions (in effect a probabilistic study of stability with more emphasis placed on the sensitivity to change than the absolute value of safety calculated).

TOPPLING FAILURES

Toppling of individual blocks is governed by joint spacing and orientation. The criteria for toppling failure have been set down by John[26] for both two and three-dimensional modes of failure. The condition for stability is that the resultant force must be within the central two-thirds of the base of the block. Hydrostatic forces acting on the rear of near vertical joints will greatly affect the direction of the resultant force.

Figure 16.18 summarises the criteria required for stability against overturning for two dimensional conditions in terms of the mean spacing and orientation of

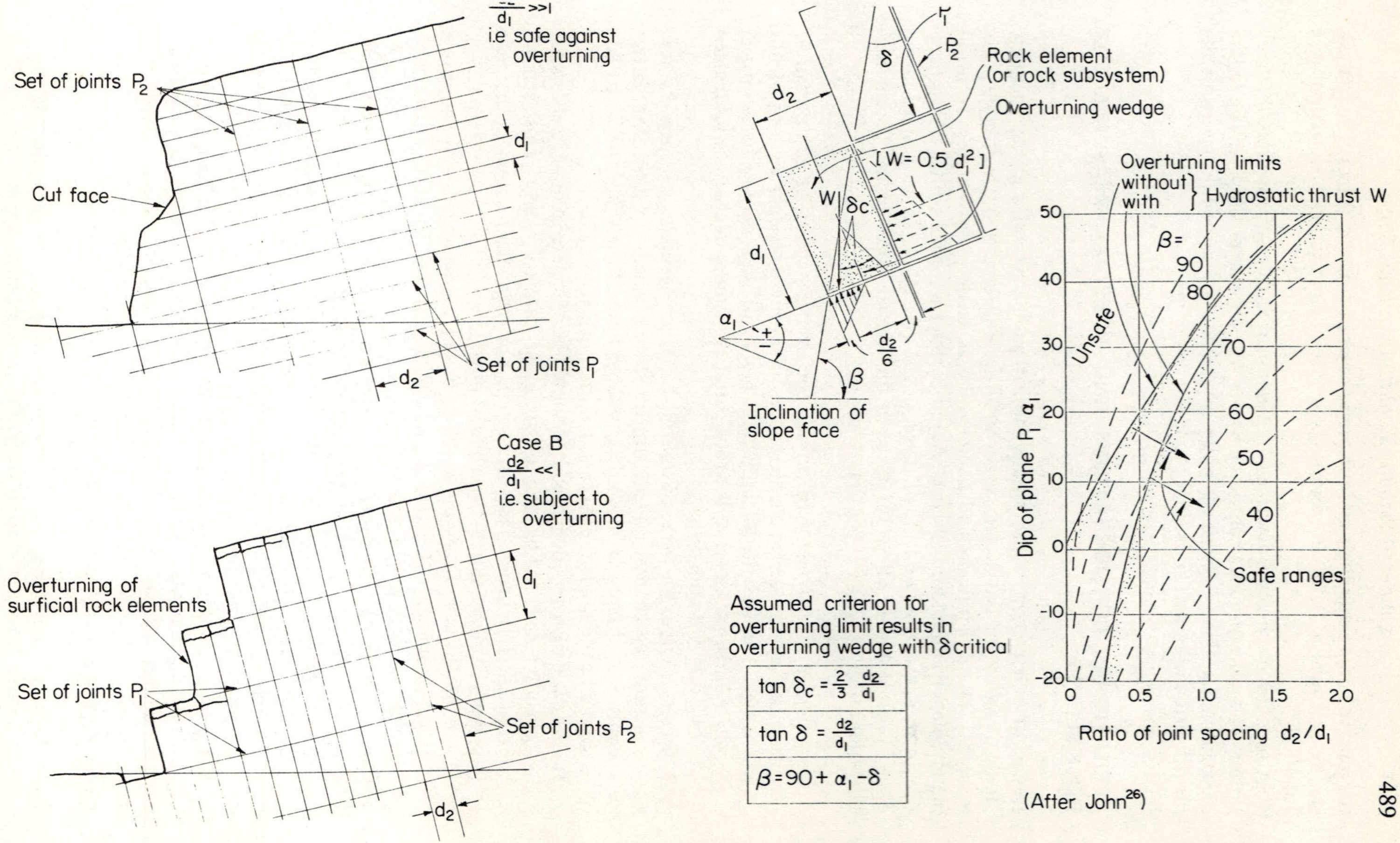

Figure 16.18 Criteria for toppling failure in two dimensions

the joint sets, the slope angle and the presence of hydrostatic forces. The analysis can be extended to three dimensions by the use of stereographic projections (see John[26]).

A toppling mechanism described by Goodman and Bray[27] involves the overturning of semicontinuous cantilever beams with interlayer slipping. The failure mechanism for this is described in the Appendix and a method given for the identification of potential flexural toppling.

OPERATIONAL CONSIDERATIONS AND REMEDIAL MEASURES FOR ROCK SLOPES

Design of stable rock slopes entails more than the provision of an adequate factor of safety for the overall slope. Small-scale instability or weathering problems may be as significant and troublesome as single major failures. To avoid inconvenience and additional costs caused by later remedial works, it is imperative to recognise the problems of ravelling and rockfall at an early stage in the design process and to allow for these in the overall slope design. Design considerations include:

1. *Adequate overall factor of safety*. The overall stability of a slope can be assessed by the design methods set out in the preceding sections of this chapter. The stability of individual blocks of rock may be more critical where these are closer to the rock surface and more likely to be influenced by blasting vibrations.
2. *Steep intermediate faces*. Controlled blasting techniques (especially presplitting, see chapter 17) are usually more efficient on steep face. Better control of drilling is ensured by steeper holes. Bench faces of 70 to 80° should be utilised if the geological structure is favourable.
3. *Adequate benches.* Bench widths should be adequate to provide access for inspection and scaling and provision of bench drainage. This is more economical if bench faces are steep. For permanent access to drainage systems, benches should be at least 5 m wide.
4. *Accessible faces.* Faces up to 10 in height can be scaled with conventional equipment. If access is not available, scaling by hand is necessary, which is expensive and time-consuming.
5. *Provision of catch benches.* It is rarely economical to design a rock slope so that no subsequent rockfalls occur. Restraint of individual blocks by reinforcement or mesh may be expensive and not completely effective. Except where absolute security is essential, therefore, slopes should be designed to allow small falls of rock under controlled conditions. Work by Ritchie[29] on rockfall protection has produced the data summarised in Figure 16.19. Slope designs should always be checked for compliance with these requirements, especially when the slopes are adjacent to permanent structures.

REMEDIAL MEASURES

Rock slopes are often designed for overall stability and an allowance made for redesign and stabilisation of local areas. For an economical design, about 10% of

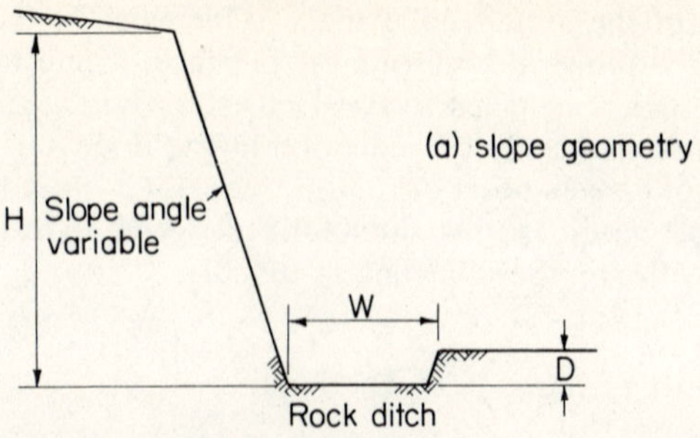

Variables in ditch design for rockfall areas

Rock slope (g) *angle*	*Height* (m) *H*	*Ditch width* (m) *W*	*Ditch depth* (m) *D*	*Notes*
90	5–10	3	1	
	10–20	5	1.5	1
	>20	7	1.5	1
75	5–10	3	1	
	10–20	5	1.5	1
	20–30	7	2	2
	>30	8	2	2
65	5–10	3	1.5	1
	10–20	5	2	2
	20–30	7	2	2
	>30	8	2.5	2
55	0–10	3	1	
	10–20	5	1.5	1
	>20			
45	0–10	3	1	
	10–20	3	2	1
	>20	5	2	2

Notes:
1. If dimension *D* is greater than 1, rock retaining fence should be used if ditch adjacent to highway.
2. Ditch dimension may be reduced to 1.5 if rock fence used.

Figure 16.19 Design of ditches for rockfall protection (After Ritchie[29])

the slope area may require some form of treatment. Fookes and Sweeney[30] have recently presented a useful review of slope stabilisation measures. Remedial works on rock slopes include the following.

1. Alteration of slope geometry
2. Structural support
3. Dental treatment
4. Drainage improvement

1 Alteration of slope geometry

(a) Scaling of loose blocks and clearing of benches is always a prerequisite to safe working conditions on rock slopes. It may be carried out mechanically or manually but must be done by skilled personnel or under the close scrutiny of the engineering geologist.

(b) Reduction of the overall slope angle to increase the factor of safety will involve further drilling and blasting on the face. Where trimming blasts are required, care should be taken to avoid excessive flyrock and blasting damage to the permanent slopes. Local reduction in the slope angle will often cause the formation of convex noses of rock. Slope stability is reduced in such areas and local slope modifications should be designed to cause as few abrupt changes of direction in the slope line as possible.

2. Structural support

(a) *Rock reinforcement*

Reinforcement can be used to provide additional stability to existing slopes or to allow steeper slopes to be excavated while maintaining the same degree of stability. It increases slope stability by providing additional strength on critical weakness planes within the rock. The design of reinforcement depends on prior knowledge of the potential failure mode of the slope.

Overall slope stabilisation requires an accurate assessment of the absolute stability of the slope before and after installation of anchors. The potential gains from large-scale anchorage programmes must be real and clearly defined. These conditions are generally met only in situations where:

(a) well-defined geological weaknesses exist and have been established by accurate exploration, and

(b) the relevant geological and strength parameters can be defined within narrow limits.

Various generalised examples of possible anchor applications are shown in Figure 16.20.

Anchor design is based on the results of stability analyses and yields such parameters as number, size and length of anchors, length of anchorage, bearing capafity required under head assembly, insertion angle, etc. The simplest approach is to determine the total anchor force required per unit length of slope. Once determined, this force can be distributed over the face according to practical anchor loadings and local geology. Where several alternative anchor spacing/individual load permutations are possible, the choice may depend on availability of materials and economic factors. An example of a slab resting on a single weakness plane is illustrated in Figure 16.21. It can be shown that the required anchor load per unit length of slope is given by:

$$T = \frac{W(F \sin \psi - \cos \psi \tan \phi)}{F \cos(\psi + \theta) + \sin(\psi + \theta) \tan \phi}$$

where W = weight of a unit length of slope section requiring support;
F = factor of safety of the anchored slope;
ψ = inclination of the assumed failure plane;
θ = inclination of the anchor load with respect to horizontal;
ρ = angle of shearing resistance on the failure plane.

The factor of safety, F, is applied to the shear strength of the weakness plane only and not to the anchor load T. The anchor load is generally specified as a working load. A typical design example is shown in Figure 16.22 which indicates the variation in factor of safety with anchor load. Also shown is the variation in anchor load for various anchor inclinations and a factor of safety of 1.1. The

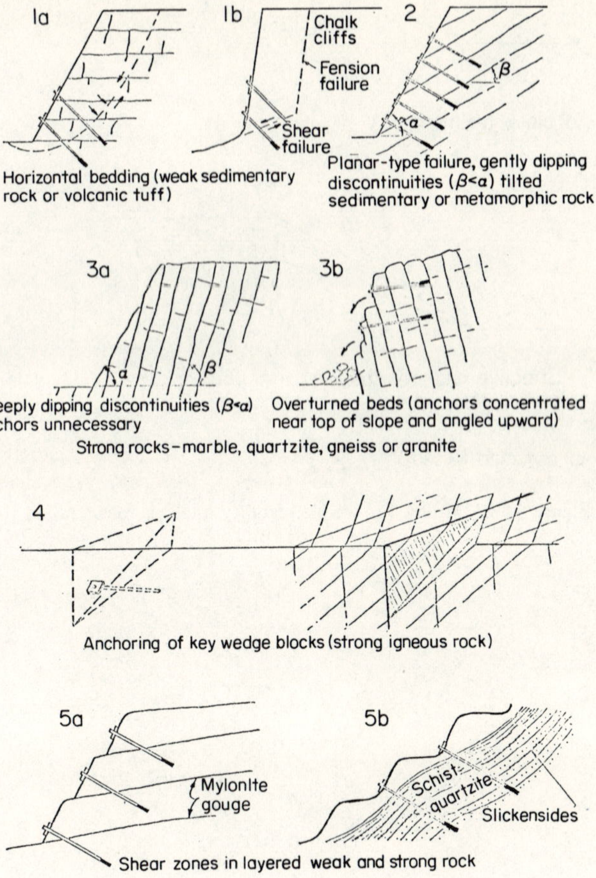

Type 1 Homogeneous or horizontally bedded structures. Commonly found with weak sedimentary rocks, volcanic tuffs and chalk. Anchoring near the toe of the slope would probably be most beneficial if the failure surface was as shown on (b).

Type 2 Tilted sedimentary or metamorphic beds with discontinuities dipping at an angle less than the slope angle. Stabilisation may require extensive use of relatively deep anchors.

Type 3 Strong, blocky or slabby rocks such as gneiss, marble or granite, where prominent discontinuities dip more steeply and generally into the slope face. Relatively shallow reinforcement may help to maintain stable slopes under conditions of ice or excessive water pressures behind the face. Toppling failures would be prevented by longer anchors.

Type 4 Strong igneous or other blocky masses with two or more intersecting discontinuities that could constitude wedge failures. Anchoring is best effected by securing key blocks.

Type 5 Shear zones and weak layers of sedimentary or metamorphic rock. Anchors could maintain the interlock of flutings and increase the normal force on shear planes.

Figure 16.20 Examples of slope structures and anchorage applications

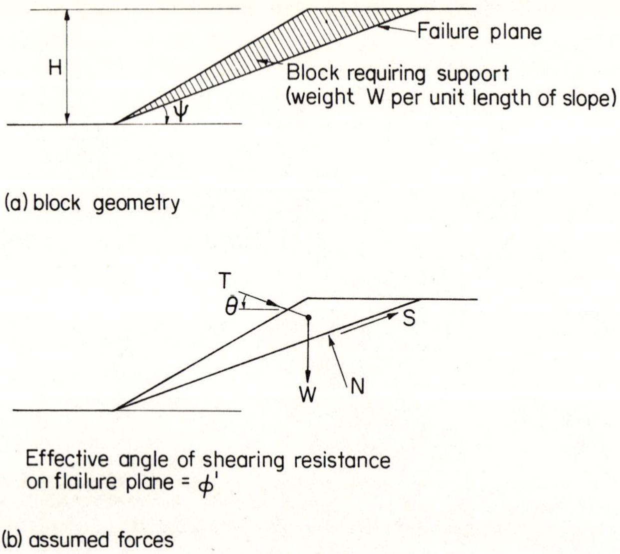

Figure 16.21 Model for simple stability analysis of anchored slope section

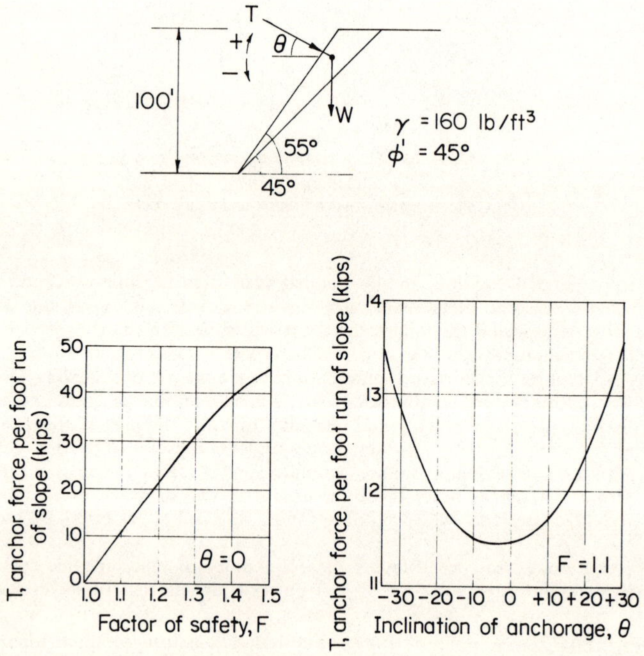

Figure 16.22 Typical anchor analysis

optimum inclination of the anchor (i.e. that which results in a minimum anchorage force) is:

$$\theta = \phi + \psi$$

The mechanical behaviour of *active (i.e. tensioned) reinforcement* when subjected to shear displacement is significantly different from the behaviour of *passive (i.e. untensioned) reinforcement*. A comparison for a smooth joint is shown in Figure 16.23. Untensioned reinforcement is applicable to situations where low

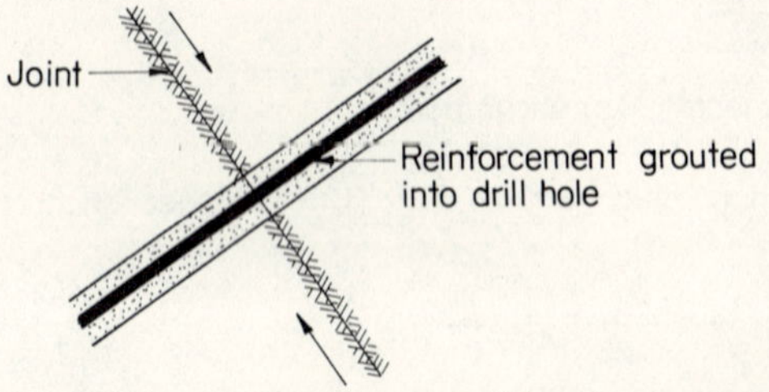

(a) *Untensioned reinforcement. Normal load acting across joint*

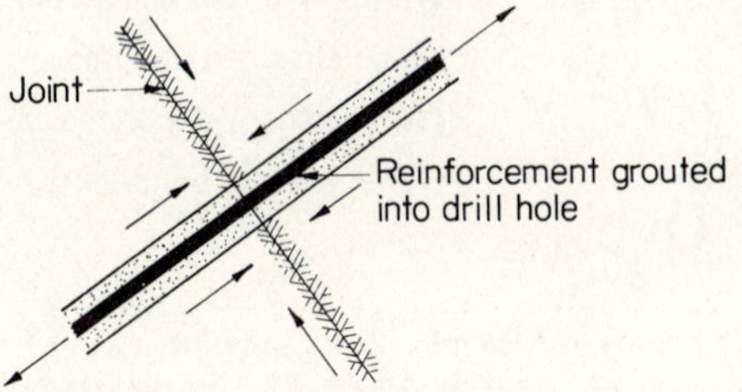

(b) *Tensioned reinforcement. Normal load always acts across joint*

(a) Untensioned reinforcement. Shear displacement along joint does not cause any significant normal dilation

1. No tension is induced in the reinforcement
2. No additional shearing resistance is mobilised along the joint
3. Hence, joints remain unreinforced (reinforcement acts as shear dowel only)

(b) Tensioned reinforcement. Shear displacement on joint is resisted by

1. Component of installed force in reinforcement acting parallel to joint
2. Frictional resistance due to normal component of reinforcement tension across joint. Hence, joint is always reinforced

Figure 16.23 Comparison between tensioned and untensioned reinforcement acting across smooth joint

loads only are required to increase stability and joint surfaces are at least moderately rough. Fully bonded reinforcement (whether active or passive) is always preferable to a single anchored point for reasons of reinforcement efficiency and corrosion protection. A comparison of the behaviour of bonded and unbonded reinforcement is given in Figure 16.24.

Simple design charts are useful to assist in the nomination of support requirements for spot bolting on large expanses of slopes. Figure 16.25 is an example of such a design chart. In this case, geological face maps were used to give information on potential failure surface inclinations and height of blocks. The

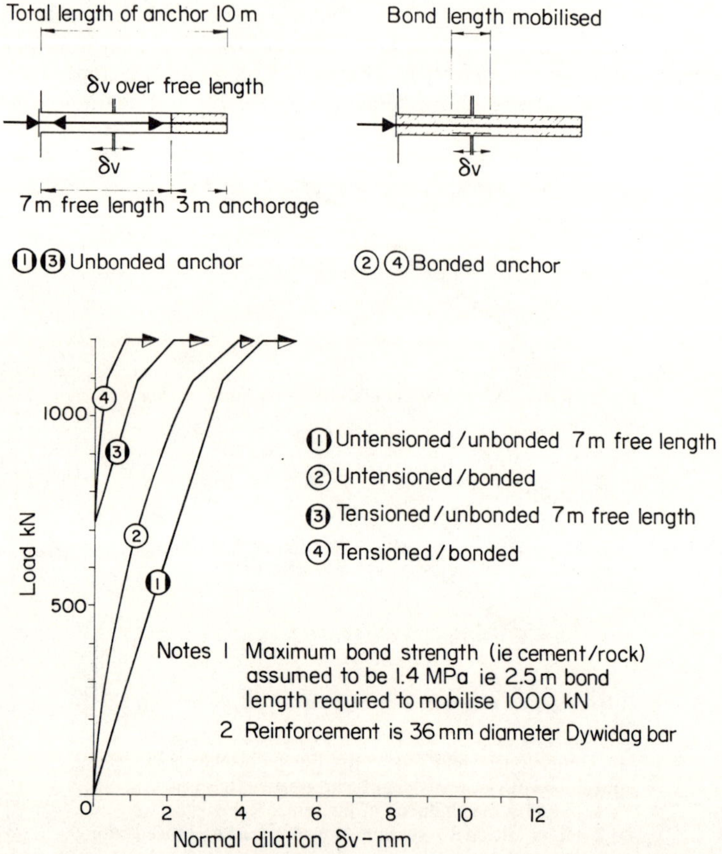

Figure 16.24 Comparison between behaviour of bonded and unbonded reinforcement

required bolt capacity for a factor of safety of 1.5 can be read directly off the chart. Details of anchors and anchorages are given in a series of review articles by Littlejohn and Bruce[31], which also provides extensive information on the testing and monitoring aspects of reinforcement.

(b) *Surface treatment*

Concrete mixtures applied pneumatically to the rock face (e.g. gunite and shotcrete) have been widely used to preserve the integrity of the rock face and prevent spalling. Such coatings also seal the rock surface and thus prevent weathering or erosion of the rock face, and may be reinforced with wire mesh. They can be badly affected by frost action and ground water pressures within the rock mass unless adequate drainage is provided. Generally such surface treatment is considered to offer negligible support to the overall slope structure.

Wire mesh tied to the face by short bolts can be used to retain spalling rock. Frequently, the mesh acts as a restraining curtain, behind which blocks may loosen and fall without endangering the working area below. This method is commonly used for the protection of roadways and important plant or working areas below slopes.

(c) *Buttresses*

Concrete or rock buttresses have been used occasionally to support major rock blocks, usually where overhanging rock bluffs have developed as a result of weathering or poor excavation techniques. The construction technique implies that the slope is stable without the presence of the buttress and therefore that such a structure will be used to provide added security against failure.

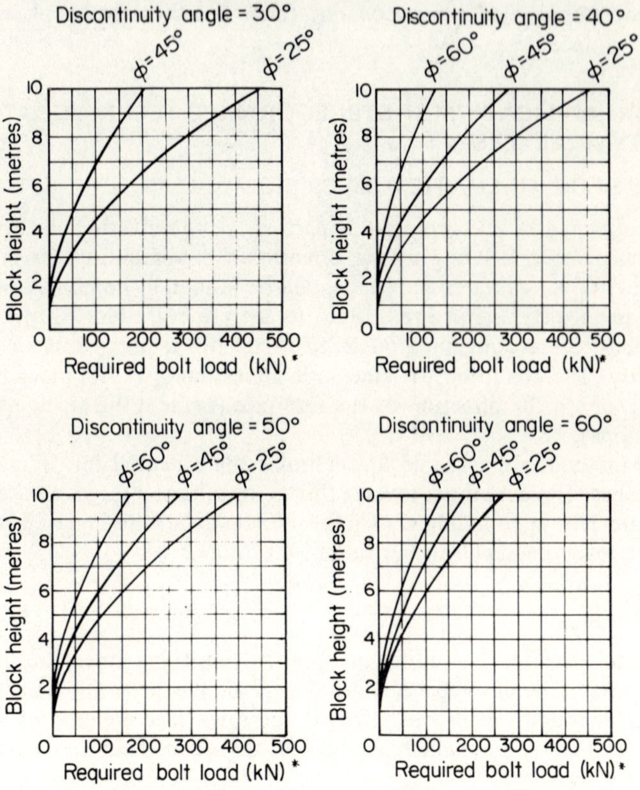

Figure 16.25 Example of rock bolt design chart

In general, for excavated slopes, it is more advantageous to improve the properties of the rock slope itself (by anchoring or bolting) than to remove the rock and replace it by concrete. Small buttresses may be very useful for retaining individual blocks of rock on a slope. Buttresses may be in the form of reinforced concrete retaining walls or gabions.

3. Dental treatment

Dental treatment consists of excavating bands of soft material and packing the slots with masonry or concrete. Careful attention to drainage by the provision of filter material and weep holes is imperative. Dental treatment may be used in conjunction with anchor installations. More details are provided in the paper by Fookes and Sweeney[30].

4. Drainage improvement

It is assumed that methods of drainage will have been considered as part of the overall design procedure. In addition to overall drainage requirements, attention should be given to the problem of water infiltrating fissures in the rock and causing near-surface instabilities. Open cracks can be sealed with cement pointing and horizontal surfaces graded to prevent ponding of water. Short holes can be drilled to prevent build up of water pressures behind individual blocks; they are essential where concrete (including shotcrete) is placed directly onto the rock face.

APPENDIX: APPLICATION OF STEREOGRAPHIC TECHNIQUES TO STABILITY ANALYSES

A. CHECKS FOR KINEMATIC ADMISSIBILITY

The data obtained by the geologist as part of the site investigation will include many measurements relating to the orientation of discontinuities in the rock mass. Blocks of various sizes and shapes will be formed by combinations of these planes. It is not practical or even useful to analyse every block formed. A convenient first-stage sorting method is to check the requirements of kinematic admissibility. The condition for kinematic admissibility is that block movement can occur only in the direction of the free face (i.e. that the block is not keyed in to the slope).

Simple methods are available to determine the potential for sliding and toppling in slopes. These methods involve the use of spherical projection techniques. Where extremely large numbers of joint measurements are involved, the use of computer programmes will further simplify sorting.

Daylight envelope

The daylight envelope is a useful method of identifying kinematically possible failure modes. This envelope is the locus of all the poles representing planes whose dip directions lie in the plane of the slope face. Poles lying within the envelope represent joints which daylight in the slope. This technique was first described by Markland[32].

The method of constructing this envelope is as follows:

(a) Draw the great circle representing the slope face (Figure 16.A1).
(b) Rotate the overlay so that strike line is at the 20° position and plot a further point at 90° to the intercept of the great circle and the W-E line. This point is the pole to a plane whose dip vector lies in the plane of the slope face and whose strike direction is at 20° to the strike of the slope face (Figure 16.A2(a)).

(c) Repeat step (b) at 20° intervals until overlay has been rotated by 180° from its position in (b). Join the points obtained to form the daylight envelope. Rotate the net to its original position with its north point coinciding with that on the net (Figure 16.A2(b)).

Application of daylight envelope to plane failures

Figure 16.A3 illustrates the conditions for potential movement of a mass of rock resting on a plane dipping towards the slope face. Poles which fall in the daylight envelope represent planes with kinematically possible failure modes.

Application of daylight envelope to wedge failures

Wedge failures are kinematically possible if the line of intersection of the two base planes daylights in the slope face. The line of intersection of two planes is found by drawing the two great circles on a stereographic projection and measuring the dip and direction of their intersection point (Figure 16.A4). An alternative method of establishing the direction of the line of intersection of two planes is by means of the poles to the planes. This technique is shown in Figure 16.A5.

The steps involved in checking kinematic admissibility for a wedge formed by two planes are illustrated in Figure 16.A6:

(a) Rotate overlay until the poles of both planes lie on the same great circle. Mark the point where this great circle cuts the W-E line. This point is at 90° to the line of intersection of the two planes and is thus analogous to the pole to a plane (Figure 16.A6a).
(b) Rotate overlay back to correct position on net. If the point found in (a) above falls within the daylight envelope, the line of intersection daylights in the slope face. If not, it is kinematically impossible for the wedge of rock to move out of the given slope face (Figure 16.A6(b)) and text (a) above.

Kinematic check for toppling

A simple kinematic test for toppling modes of failure has been described by Goodman and Bray[27]. This is based on the need for flexural slip prior to any large-scale movement of a closely layered columnar rock mass. Potential instability of isolated blocks of rock should be treated on the basis of geometrical considerations (e.g. see Hoek and Bray[21]). The condition for interlayer slip is derived as follows:

(a) The state of stress in the rock slope is uniaxial, with σ_1 parallel to the slope face (Figure 16.A7(a)).
(b) In order for the layers to slip, σ must be inclined at more than the angle of friction to the layers.

As shown in Figure 16.A7(b), this implies that

$$\theta > \phi + a$$

where θ is the slope angle;
ϕ is the friction angle of the layer surface;

α is the angle of the normal to discontinuities
i.e. for toppling to be able to occur, the pole to a joint set must be outside a great circle ϕ degrees below the rock cut (Figure 16.A8(a)).

(c) Toppling will only occur where the strike of discontinuities is approximately parallel to the strike of the cut slope. Goodman and Bray suggest $10°$ either side of the strike of the slope.

A toppling envelope can be drawn to define the above requirements and is described in Figure 16.A8.

B. FRICTION CONE CONCEPT

A convenient method for assessing the potential stability of kinematically permissible blocks is to use the concept of the friction cone. A block at rest on an inclined plane must have its normal at an angle of less than ϕ to the resultant forces acting on the block. For the simple case of gravity forces only, a stable block must have its normal within a circular cone described by the angle ϕ about the centre of the net (i.e. about the direction of the gravity force). This is shown in Figure 16.A9(a).

This friction cone may be constructed on a stereographic projection as shown in Figure 16.A9(b). A circle of radius ϕ is drawn around the origin. Poles falling within this cone represent planes which dip at less than the friction angle ϕ. Poles outside this area represent planes that dip at more than their friction angle.

More detailed analyses (incorporating water pressures, cohesion, etc) will be required to establish a factor of safety for failure modes involving these planes.

C. COMBINED KINEMATIC CHECK AND SIMPLE STABILITY ASSESSMENT

The kinematic checks for sliding and toppling can be combined with the friction cone technique to give a preliminary classification of potential failure modes.

The steps involved in this combined assessment are:

(a) Draw on a transparent overlay the great circle representing the slope face.
(b) Construct the daylight envelope for this face.
(c) Draw the friction cone around the origin of the net.
(d) Construct the toppling envelope.
(e) On another overlay, plot the poles to the joint planes under consideration.
(f) Place this on the first overlay and identify potential plane failures and toppling modes.
(g) Locate the 'poles' to the lines of intersection of all wedges formed by the joint planes. (For n joints, there will be $n(n-)1/2$ wedges possible).
(h) Place this plot on the first overlay and identify potential wedge failures.

Figure 16.A10 illustrates this process for a set of six hypothetical joint planes and a given slope face.

It must be emphasised that the above processes are not complete design calculations but are intended to be used only for preliminary sorting of data. Possible failure modes should be analysed using more rigorous analytical techniques which allow for the effects of water pressures, cohesion, reinforcement loads etc.

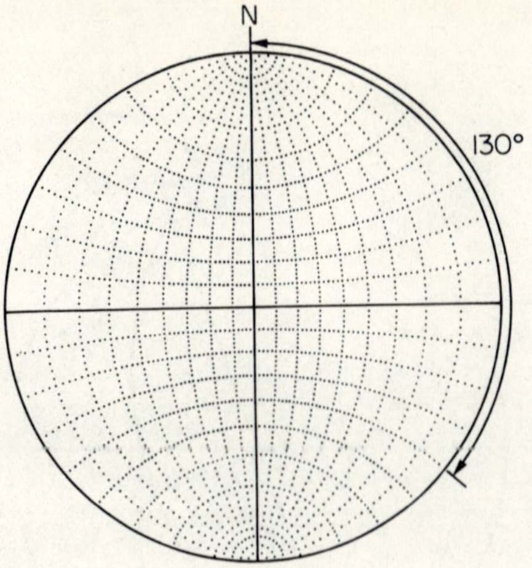

Slope Face Orientation
Dip 40°
Dip Direction 130°

(a) Mark off the dip direction of the plane on the circumference of the net as shown.

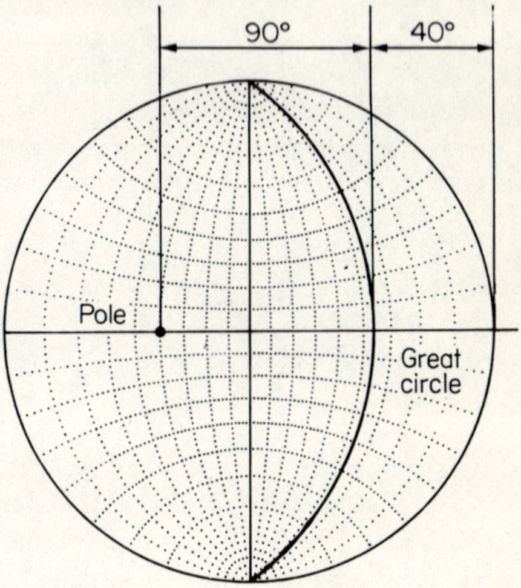

(b) Rotate the transparent overlay until the dip direction marked in (a) lies on the west-east axis of the stereonet. Measure off the dip angle from the outer circle of the net and trace the required great circle. The pole to the plane is at 90° to the strike of the plane, and the dip is measured along the W-E axis of the net from the centre of the net.

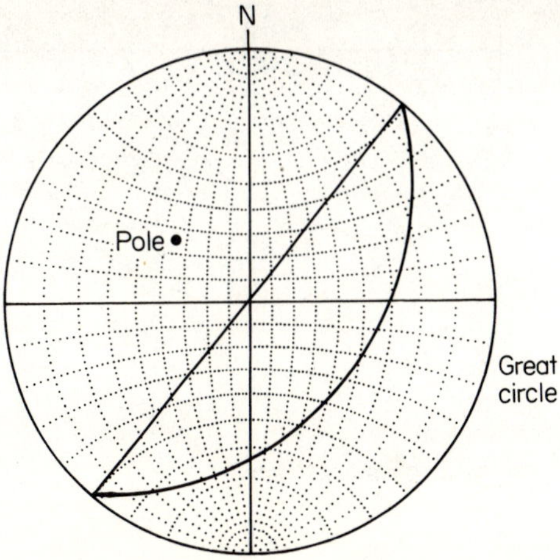

(c) Rotate the overlay to its original position so that the north mark on the tracing coincides once more with the north mark on the net.

Figure 16.A1 Construction of great circle representing slope face

Slope Face Orientation
Dip 40°
Dip direction 130°

(a) Rotate overlay so that strike line is at 20° and plot further point at 90° to intercept of great circle and E-W line.

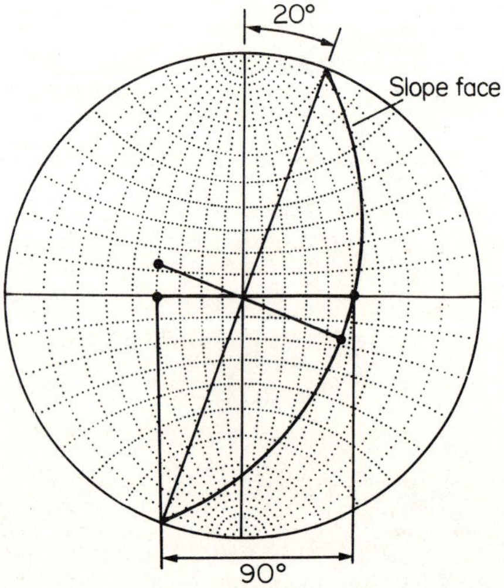

(b) Repeat above steps until complete envelope is drawn. Rotate north point on overlay so that it coincides with north point on net.

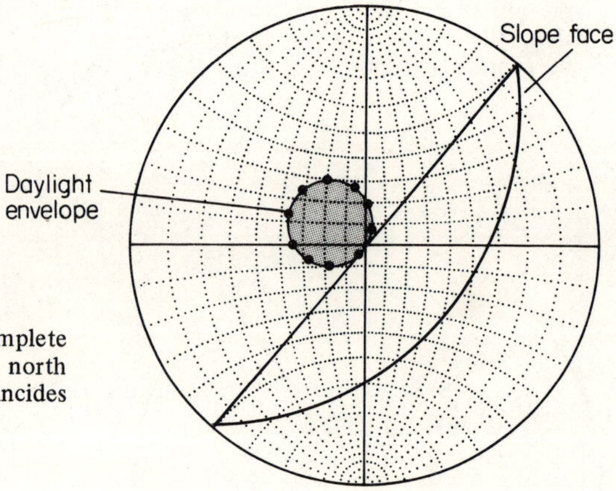

Figure 16.A2 Construction of daylight envelope

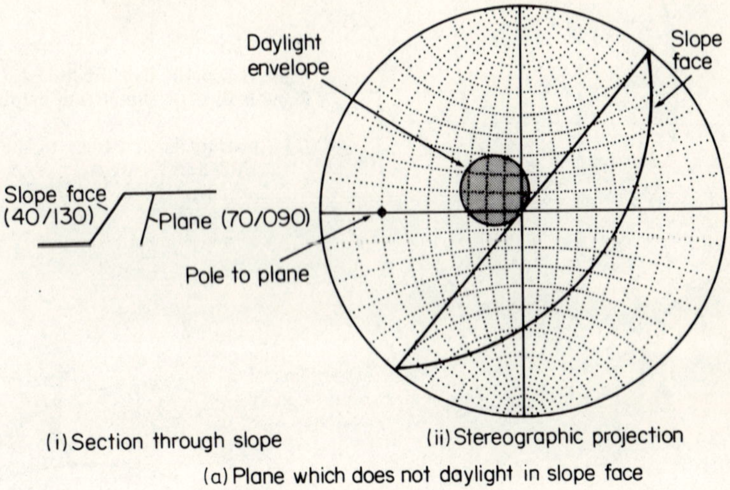

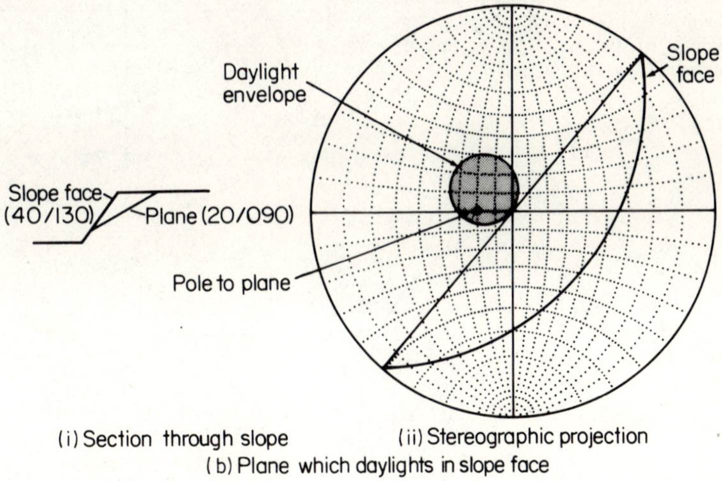

Figure 16.A3 Applications of daylight envelope to plane failures

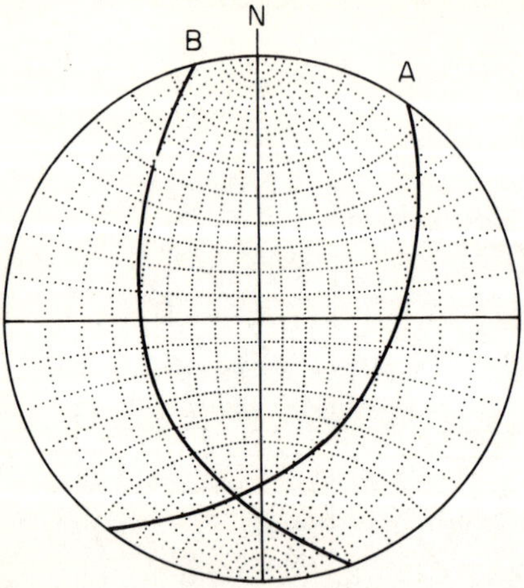

Plane A Dip 40° Dip direction 130°
Plane B Dip 30° Dip direction 250°

(a) Construct great circles to the planes as shown in Figure A.

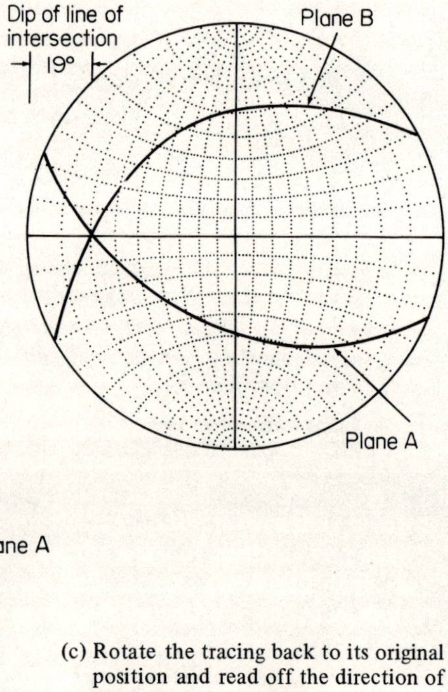

(b) Rotate the tracing until the intersection of the two great circles lies on the west-east axis of the stereonet, and measure the dip of the line of intersection from the circumference of the net. Dip of line of intersection = 19°.

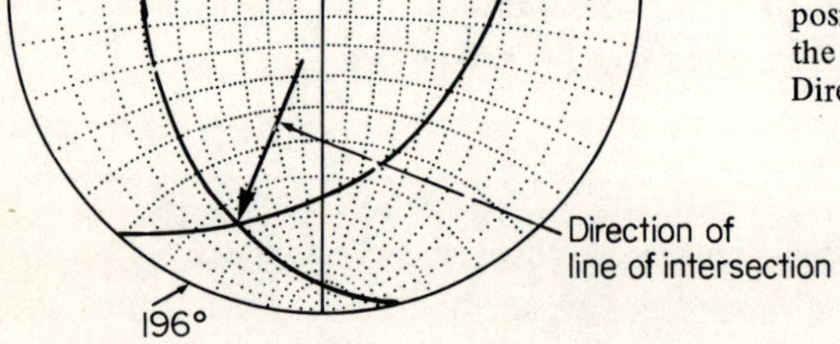

(c) Rotate the tracing back to its original position and read off the direction of the line of intersection.
Direction of line = 196°

Figure 16.A4 Determination of the orientation of the line of intersection of the two planes by the great circles method

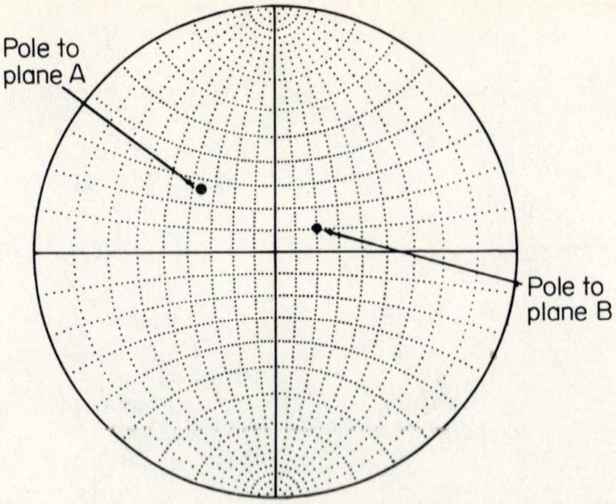

Plane A Dip 40° Dip direction 130°
Plane B Dip 30° Dip direction 250°

(a) Plot poles to great circles as shown in Figure A

(b) Rotate the tracing overlay until both poles lie on the same great circle. Locate the pole to this great circle – this is the line of intersection of the two planes. Measure the dip of this line.
Dip of line = 19°.

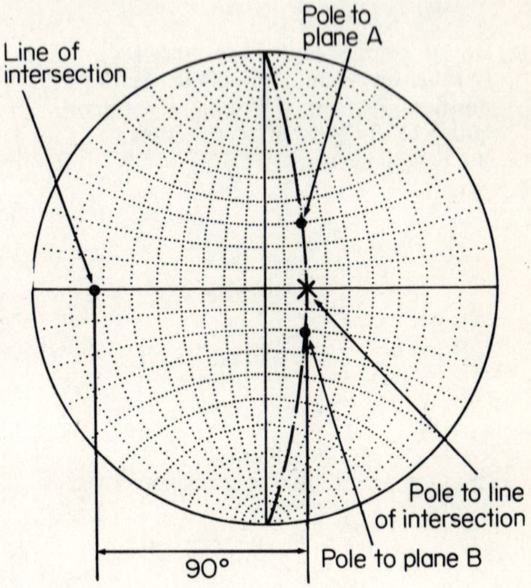

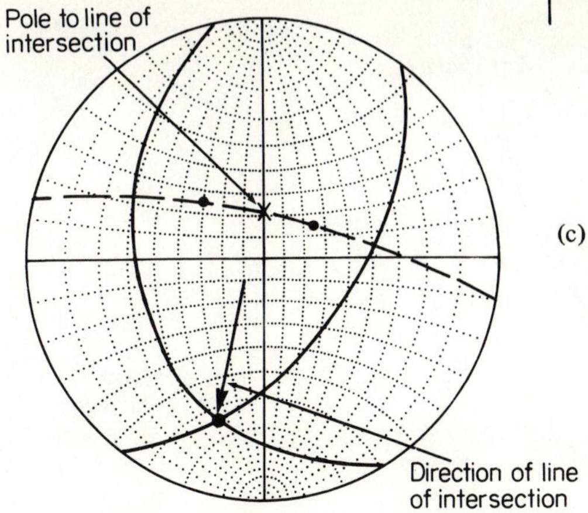

(c) Rotate the tracing to the original position and read off the dip direction of the line of intersection.
Dip direction = 196°.

Figure 16.A5 Determination of the orientation of the line of intersection of two planes by the poles method

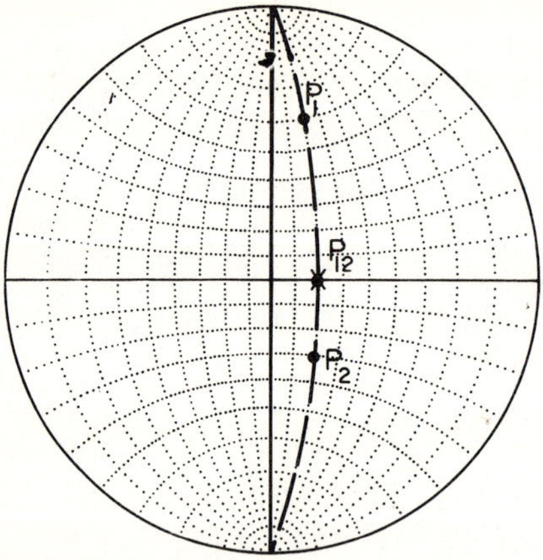

(a) Rotate overlay until both poles (P_1, P_2) lie on same great circle. Mark position P_{12} on E-W line, (ie the 'pole' to the line of intersection of the two planes).

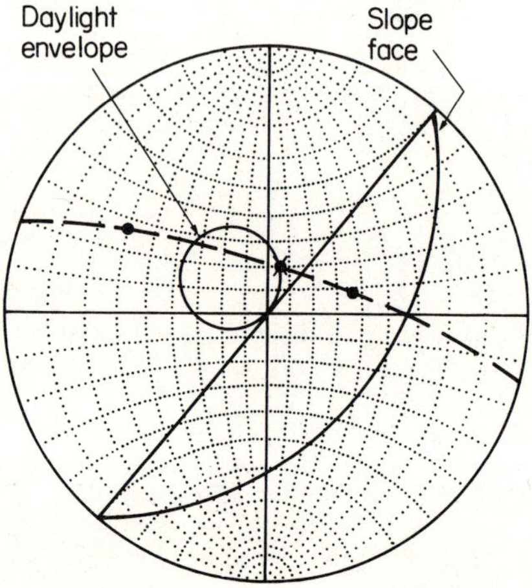

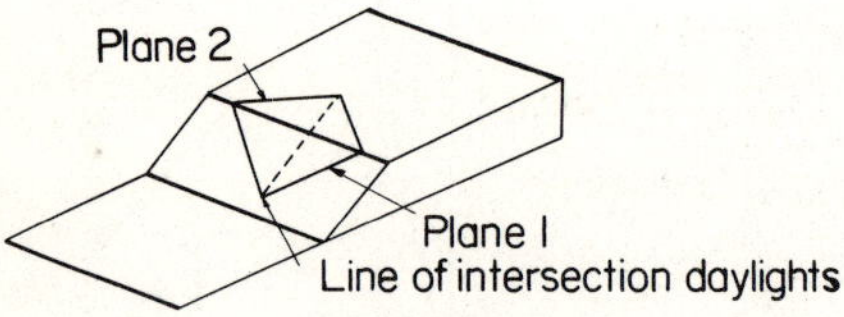

(b) Wedge which daylights in slope face.

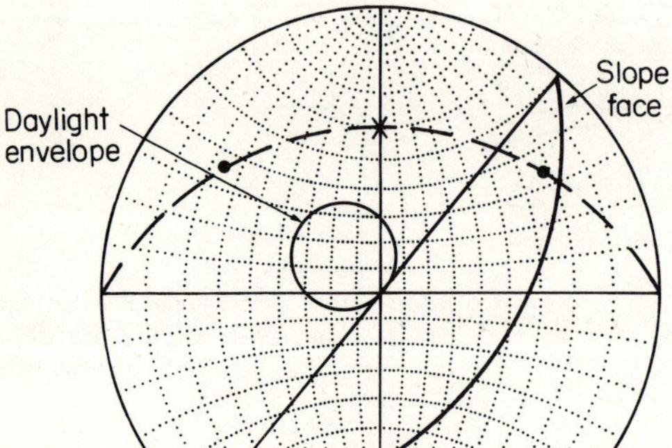

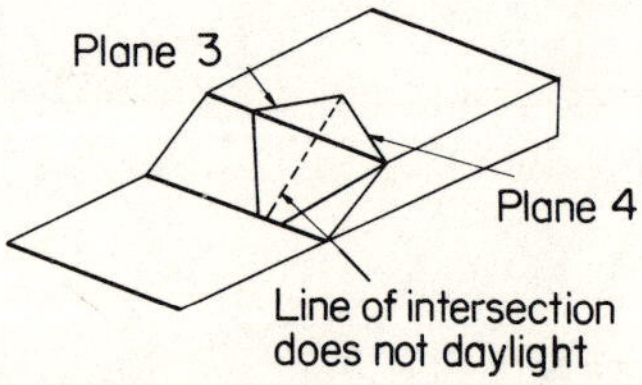

(c) Wedge which does not daylight in slope face.

Figure 16.A6 Application of daylight envelope to wedge failures

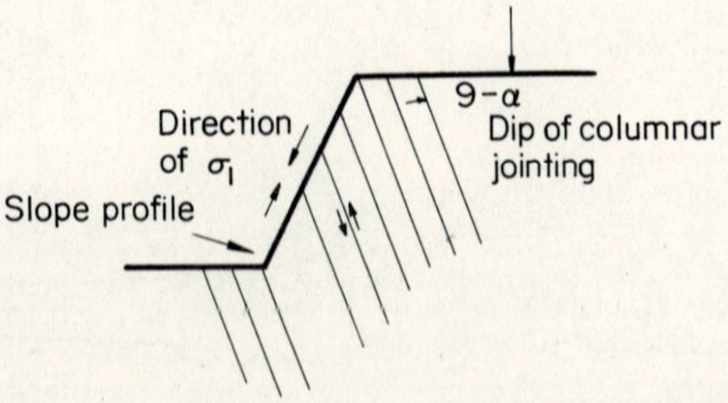

(a) Stresses and slip directions in the rock slope

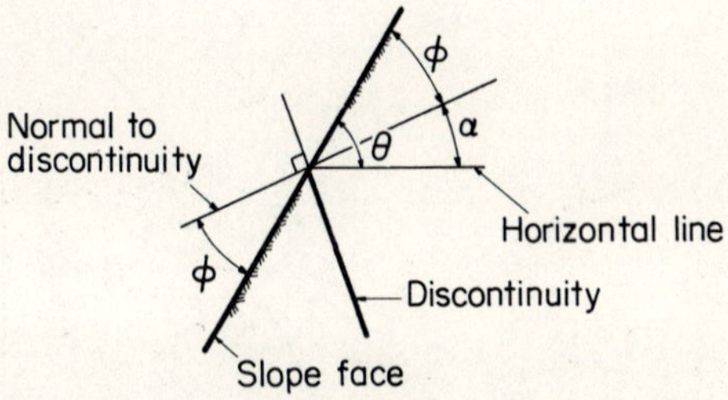

(b) Section through slope with a discontinuity at limiting equlibrium, i.e. $\theta = \phi + \alpha$

(c) Condition for interlayer slip is $\theta > \phi + \alpha$

Figure 16.A7 Relationship for toppling mechanism in layered slope (from Goodman and Bray[27])

Slope face orientation 70/130
Friction angle 30°

(a) 1 Mark 130° position on transparent overlay and rotate to W-E line. Draw slope face on great circle of 70° dip
   2 Draw great circle at angle of friction below slope face (i.e. 40° great circle)
   3 Draw small circle 10° either side of W-E line
   4 'Toppling envelope' is the area defined by the circumference of the net and lines 2 and 3

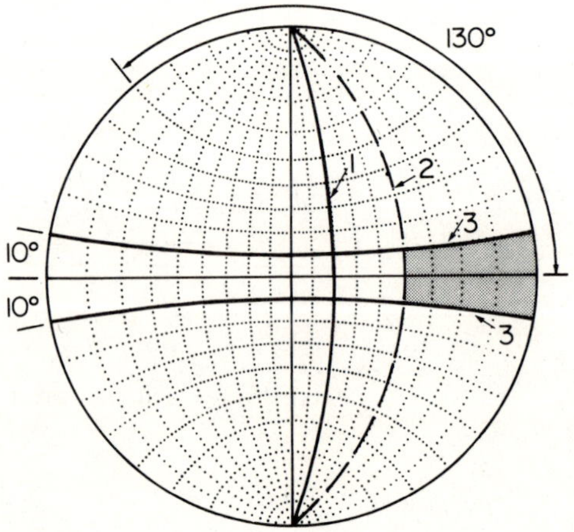

(b) Rotate overlay back to north position. Hatched area is toppling envelope. Poles falling within this area define planes with a potential for toppling

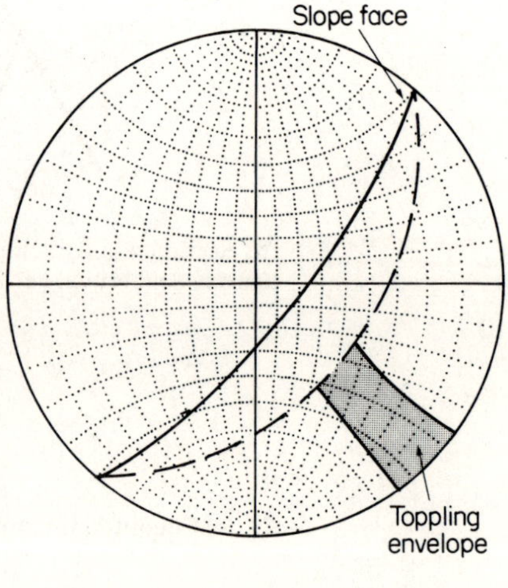

Figure 16.A8 Construction of toppling envelope

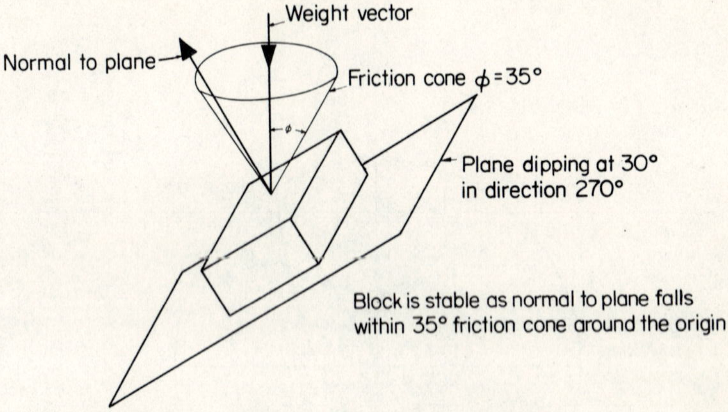

(a) Friction cone in relation to block on an inclined plane

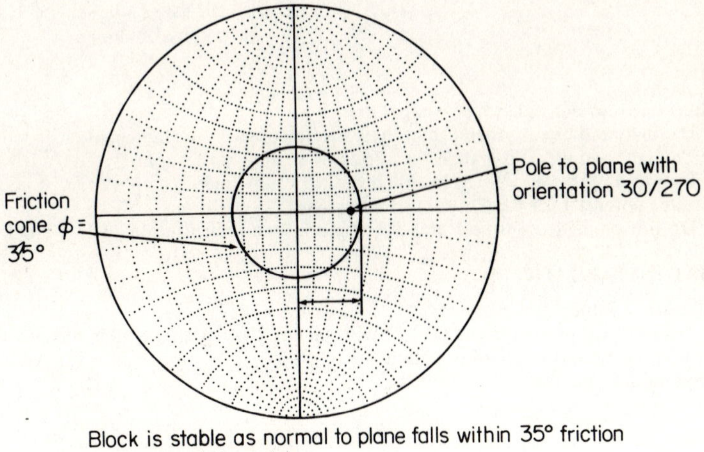

Block is stable as normal to plane falls within 35° friction cone around the origin

(b) Spherical projection of friction cone

Figure 16.A9 Friction cone concept

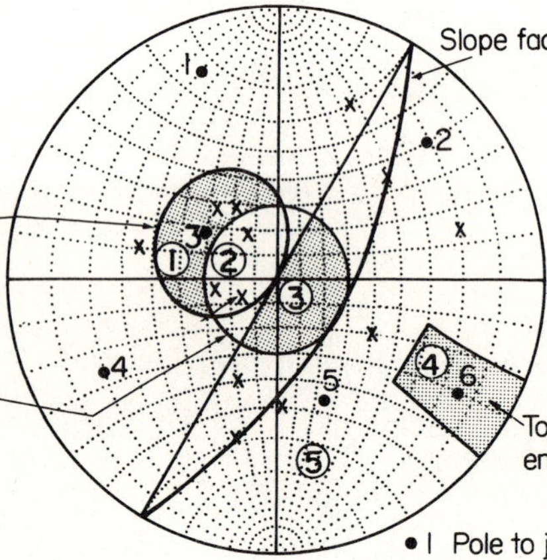

Poles in this zone represent planes or wedges that:
Zone ① Daylight and dip at greater than their friction angle – unstable zone
Zone ② Daylight and dip at less than their friction angle – stable zone
Zone ③ Do not daylight and dip at less than their friction angle – stable zone
Zone ④ Have potential for toppling – unstable zone
Zone ⑤ Do not daylight or topple and dip at more than their friction angle – stable zone

POTENTIAL INSTABILITIES

(a) Plane failures : Plane 3
(b) Wedge failures : Combinations of plane 3 with planes 2 and 4 are possible but will be less critical than plane failure on joint 3 alone
(c) Toppling failure : Plane 6

Figure 16.A10 Combined kinematic check and simple stability assessment

References

1. Glossop, R., 'The rise of geotechnology and its influence on engineering practice,' *Geotechnique,* **18**, 105–150 (1968)
2. Stimpson, B., Metcalfe, R.C. and Walton, G., 'A new technique for sealing and packing rock and soil samples,' *Quart. J. Eng. Geol.,* **3**, No. 2 (1970)
3. Geological Society of London, 'The preparation of maps and plans in terms of engineering geology,' *Quart. J. Eng. Geol.,* **5**, No. 4 (1972)
4. Badgley, P.C., *Structural methods for the exploration geologist,* Harper Row, New York (1959)
5. Ragan, D.M., *Structural geology: An introduction to geometrical techniques,* 2nd ed., Wiley (1973)
6. Horne, H.M. and Deere, D.U., 'Frictional characteristics of minerals,' *Geotechnique,* **12**, 319–335 (1962)
7. Kenney, T.C., 'The influence of mineral composition on the residual strength of natural soils,' *Proc. Geot. Conf., Oslo,* **1**, 123–130 (1967)
8. Barton, N.R., *The shear strength of rock joints in theory and practice, Norwegian Geotechnical Institute,* Oslo (1972)
9. Deere, D.U. and Patton, F.D., *Slope stability in residual soils, Proceedings Fourth Panamerican Conference on Soil Mechanics and Foundation Engineering,* San Juan, Puerto Rico (1971)
10. Goodman, R.E., *Methods of geological engineering,* West Publishing Co (1976)
11. Kanji, M.A., *Shear strength of soil-rock interfaces,* M.S. thesis Dept. of Geology, University of Illinois, Urbana (1976)
12. Richards, L.R., *The shear strength of joints in weathered rock*, Ph.D. thesis, University of London (1975)
13. Sage, R. (ed.), *Pit slope manual, chapter 4 – Groundwater,* Canada Centre for Mineral and Energy Technology (formerly Mines Branch, Energy, Mines and Resources Canada) (In preparation)
14. Hanna, T.H., *Foundation instrumentation,* Trans. Tech. Publication (1973)
15. Hvorslev, M.J., *Time lag and soil permeability in groundwater observations,* Bulletin No. 36, Waterways Experimental Station, Corps of Engineers, US Army (1951)
16. Cedergren, H.R., *Seepage, drainage and flow nets,* John Wiley & Sons, Inc. (1975)
17. Karplus, W.J., *Analog simulation,* McGraw-Hill Book Co., New York (1975)
18. Zienkiewicz, O.C., *The finite element method in engineering science,* McGraw-Hill Book Co., London (1971)
19. Bishop, A.W., 'The use of the slip circle in the stability of slopes,' *Geotechnique,* **5**, 7–17 (1955)
20. Taylor, D.W., 'Stability of earth slopes,' *Journal of the Boston Society of Civil Engineers,* **24**, 197 (1937)
21. Hoek, E. and Bray, J.W., *Rock slope engineering,* Institution of Mining and Metallurgy, London (1974)
22. Lambe, W.T. and Whitman, R.V., *Soil mechanics,* Wiley, New York (1969)
23. Morgenstern, N.R. and Price, V.E., 'The analysis of the stability of general slip surfaces,' *Geotechnique,* **13**, 121–131 (1963)
24. Sarma, S.K., 'Stability analysis of embankments and slopes,' *Geotechnique,* **23**, 423–433 (1973)
25. Janbu, N., *Stability analysis of slopes with dimensionless parameters,* D.Sc. Thesis, Harvard Soil Mechanics Series, No. 46 (1954)
26. John, K.W., *Three dimensional stability analyses of slopes in jointed rock,* Open pit mining symposium, Johannesburg (1970)
27. Goodman, R.E. and Bray, J.W., 'Toppling of rock slopes,' *Proc. of specialty conference on Rock Engineering for Foundations and Slopes.* Boulder, Colorado, Vol. 2 (1976)
28. Janbu, N., Bjerrum, L. and Kjaernsli, B., *Guidance in the solution of foundation problems,* (Chapter 2) *Norwegian Geotechnical Institute, Publication No. 16* (1956)
29. Ritchie, A.M., 'The evaluation of rockfall and its control,' Highway Record, 17, 13–28 (1963)
30. Fookes, P.G. and Sweeney, M., 'Stabilisation and control of local rock falls and degrading rock slopes,' *Quart. J. Eng. Geol.,* **9**, 1, 36–56 (1976)

31. Littlejohn, G.S. and Bruce, D.A., 'Rock anchors: state of the art,' *Ground Engineering* (1975)
32. Markland, J.T., *A useful technique for estimating the stability of rock slopes when the rigid wedge type of failure is expected, Imperial College Rock Mechanics Research Report No. 19* (1972)

Chapter 17

Open Excavation

Open excavation refers to the removal of material, within certain specified limits, for construction purposes. For this to be accomplished economically and without hazard the character of the rocks and soils involved and their geological setting must be investigated. Indeed the method of excavation and the rate of progress are very much influenced by the geology on site. Furthermore the position of the water table in relation to the base level of the excavation is all important, as are any possible effects of construction operations on the surrounding ground and buildings.

The cross section of an open excavation is influenced by the dimensions of the base, its depth and the profile of its slopes. Slopes must be as steep as possible, consistent with safety, in order to minimise the volume of material to be excavated. Allowance must be made for any drainage works as far as the dimensions of excavations are concerned.

EXCAVATIONS IN ROCKS AND SOILS

The maximum height which can be safely developed in a rock slope is roughly proportional to its shearing strength, i.e. the stronger the rock the steeper the slopes which may be cut into it (see chapter 16). For instance, excavations in fresh, massive, plutonic igneous rocks such as granite and gabbro can be left more or less vertical after removal of loose fragments. On the other hand volcanic rocks such as basalts and andesites are generally bedded and jointed, and may contain layers of ash or tuff, which are usually softer and weather more rapidly. Thus slope angles have to be reduced accordingly.

Gneiss, quartzite and hornfels are highly weather resisting and slopes in them may be left almost vertical. Some gneisses, however, may be severely contorted. Schists vary in character from talc schists to mica schists to sillimanite schists. The softer schists may be weathered and tend to slide along their planes of schistocity. Schists may also be severely contorted which may necessitate variations in the slope angle. Slate generally resists weathering. Slips may occur where cleavage or schistocity dips in the same direction as the slope of the cut face.

Vertical slopes can also be excavated in massive limestones and sandstones which are horizontally bedded. Both these rock types tend to weather slowly, although some Mesozoic limestones prove the exception. Safe slopes can be cut in chalk, which is softer than most limestones, with angles of 45° to 80°. In brittle, cemented shales slopes of 60° to 75° are usually safe but increasing fissibility and decreasing strength necessitate flatter slopes. Even in weak shales slopes are seldom flatter than 45°.

If the strata are horizontal then excavation is relatively straightforward and slopes can be determined with some degree of certainty. However, the dip in sedimentary formations is of major importance in excavation work. Slopes may have to be modified in accordance with the dip and strike directions as related to the excavation. The most stable excavation in dipping strata is one in which the face is orientated normal to the strike, since in such situations there is a low tendency for rocks to slide along their bedding planes. Conversely if the strike is parallel to the face then the strata dip into one slope. This is most critical where the rocks dip at angles varying between 30° and 70°. If the dip exceeds 70° and there is no alternative to working against the dip, then the face should be developed parallel to the bedding planes for safety reasons.

Sedimentary sequences in which thin layers of shale, marl or clay are present may have to be treated with caution, especially if the bedding planes are dipping at a critical angle. Weathering may reduce such material to an unstable state within a short period of time, which in turn can lead to slope failure. Therefore their presence should dictate that sufficient allowance be made in the design of slopes and the use of retaining walls should be considered. Argillaceous layers in sandstones and limestones, because of their low permeability, tend to cause ground water to move over their upper contacts, water emerging at these contacts along the sides of excavations. Furthermore such layers, particularly if less than 0.3 m thick, tend to be saturated and therefore weaker. Spring sapping may give rise to undermining and consequent rock fall. Indeed wherever weaker strata underlying stronger rocks are exposed upon excavation then undermining of the latter is likely to occur as the former are more readily removed by the agents of denudation. Ultimately this action will produce a rock fall or slide. Accordingly protection against weathering should be afforded to the softer beds, with adequate drainage being provided.

Faults which traverse the area in which excavation is to take place may cause serious trouble, especially if in soft ground. This is principally because of the greater freedom afforded rock masses to move along fault planes. In particular, if a fault intersects a prominent joint or bedding plane in such a way that it produces a wedge which opens into the excavation, then this is likely to slide.

Open joints in rocks facilitate weathering and generally aid slope failure. Frost action may be particularly significant in such situations. Fissure zones are usually zones of weakness along which rock may have been altered to appreciable depth by weathering. Indeed the degree of and depth to which weathering has occurred should be ascertained before excavation begins.

A slope of 1:1.5 is generally used when excavating dry sand, this more or less corresponding to the angle of repose, 30° to 40°. In other words a cutting in a non-cohesive soil will be stable, irrespective of its height, as long as the slope is equal to the lower limit of the angle of internal friction, provided that the slope is suitably drained. Slope failure in frictional soils is a surface phenomenon which is caused by the particles rolling over each other down the slope.

As far as sands are concerned their packing density is important. For example, densely packed sands exposed in sand pits may have almost vertical faces which are generally stable. The amount of cement and/or matrix is also significant in this context. Of lesser consequence are the size, sorting and shape of the grains.

The water content is of paramount importance in loosely packed sands, for if these sands are saturated they are likely to flow on excavation. On the other hand damp sands possess some cohesion and may be able to stand vertically for some time.

The most frequently used gradients in marls and many clays vary between 30° and 45°. In some clays, however, in order to achieve stability, the slope angle may have to be less than 20°. The stability of slopes in clay depends not only on its strength and the angle of the slope but also on the depth to which the excavation is taken and on the depth of a firm stratum if one exists not far below the base level of the excavation. Slope failure in a uniform clay takes place along a near 'circular' surface of slippage. The stability of slopes in such clays is amenable to analysis and reasonable estimates can be obtained (see chapter 15). For example, the critical height (H) to which a face of an open excavation in normally consolidated clay can stand vertically without support can be obtained from the following expression:

$$H = \frac{4c}{\rho \times 9.807}$$

where c is the cohesion of the clay and ρ its density.

In normally consolidated soft to firm clays, the required slope for a particular factor of safety can be estimated with reasonable reliability by the normal slip circle analysis. According to Tomlinson[1] the factor of safety depends on the risk involved in a major slip. If this would not damage property or risk life, then a low factor of safety such as 1.2 or 1.3 could be used whereas if slip must be avoided then a higher factor of 2 to 3 is necessary.

In stiff fissured clays the fissures appreciably reduce their strength below that of intact samples. For example, Skempton, Schuster and Petley[2] showed that the strength of the London Clay along the fissures was only slightly higher than its residual strength. Thus the reliable estimation of the stability of slopes in stiff fissured clays is difficult. Generally steep slopes can be excavated in such clays initially but their excavation means that fissures open due to the relief of residual stress, which also means a change from negative to positive pore pressures along the fissures, the former tend to hold the fissures together. This change can occur within a matter of days or hours. Not only does this weaken the clay but it also permits a more significant ingress of water, which means that the clay is softened. Irregular shaped blocks may begin to fall from the face and slippage may occur along well defined fissure surfaces which are by no means circular.

If there are no risks to property above the crests of slopes in stiff fissured clays then they can be excavated at about 35°. Although this will not prevent slips, those which occur are likely to be small.

Special attention should be paid to the effect of exposure to the atmosphere on any clays which are uncovered during excavation. The natural moisture content and consistency limits of clay are critical factors in such instances. Covering steep exposed faces of suspect clay may prevent slope failure and thereby result in a considerable saving in excavation.

Steep slopes can be excavated in tills which will stand unsupported for long periods, for example, several months, although lenses of waterlogged sand may cause instability.

Slopes in sand-clay mixtures, if well consolidated, should be stable at gradients of approximately 55° or possibly somewhat steeper.

Because of the presence of a vertical pattern of rootholes which characterises loess, and its carbonate cement, excavations can be made with vertical faces. These may remain stable for long periods, if failure does occur it generally takes the form of a vertical slice. By contrast inclined slopes are frequently subject to rapid erosion (see Clevenger[3]).

A high rock slope should be interrupted by berms or benches and should flatten out gradually towards the top. Berms effectively reduce the load on the lower part of the slope, help prevent rock fragments falling to the base of the cutting and intercept the flow of water over the surface of a slope. An adequate system of drainage should be installed on each berm (see chapter 16). The width of a berm depends upon the depth of the excavation, the character of the rocks involved and the topography above the top of the excavation. Ideally berms should be so constructed that they are accessible to maintenance equipment. Berms are also used to effectively reduce gradients of slopes in high cuttings in soils.

The stability of the floor of large excavations may be influenced by ground heave. The amount of heave and the rate at which it occurs depends on the degree of reduction in vertical stress during construction operations, on the type and succession of underlying strata, and on the surface and ground water conditions. It is generally greater in the centre of a level excavation in relatively homogeneous ground as for example, clays, marls and shales.

In particular in a deep excavation in soft clay there is a danger of failure due to bulging. In other words there is a downwards and inwards displacement of ground immediately surrounding the excavation while the central area is uplifted. If the lateral support to the sides of the excavation has been inadequately designed then plastic deformation and shear failures may occur. The weight of the soil at the sides of the excavation exceeding the shear strength of the clay is responsible for this. The zone of failure can be approximated by drawing a line at 45° to the bottom corner of the excavation and connecting it to a circular arc whose centre is the opposite bottom corner (Figure 17.1).

Uplift in homogeneous deposits may be divided into two major components. Firstly, there is an elastic rebound which occurs due to removal of overburden on excavation, which may pass unnoticed unless it is measured. Secondly, there is a slow, long-term swelling involving absorption of water from the ground surface or due to it migrating from below, which may continue for many years, or until adequate reloading restores the equilibrium.

The amount of elastic heave and the swelling of clays increases with the clay mineral content and the activity of the clay. The rate of swelling is generally slow, but the amount can be large (many tens of millimetres). Where the excavation is in overconsolidated clays or shales, swelling accompanied by absorption of water and softening takes place at the base on unloading. In such instances the process is quite rapid at the excavated surface and proceeds progressively in depth as time elapses.

In the case of clays with a low degree of saturation, swelling and softening take place very rapidly if surface water is admitted to the foundation area. The use of a thin blinding layer of concrete does not prevent swelling. The amount of swelling, however, can be reduced by excavating quickly and reloading the ground without delay. In a large excavation this is most easily done by excavating the final 2 to 3 m of ground in small areas and replacing it with a substantial thickness of concrete (which forms part of the foundation structure) and by subsequently imposing additional vertical loads as soon as possible. Swelling and softening at foundation level are augmented by freezing and thawing, such weakened material should be removed before concrete is placed. Heave and swelling are small in sands and gravels.

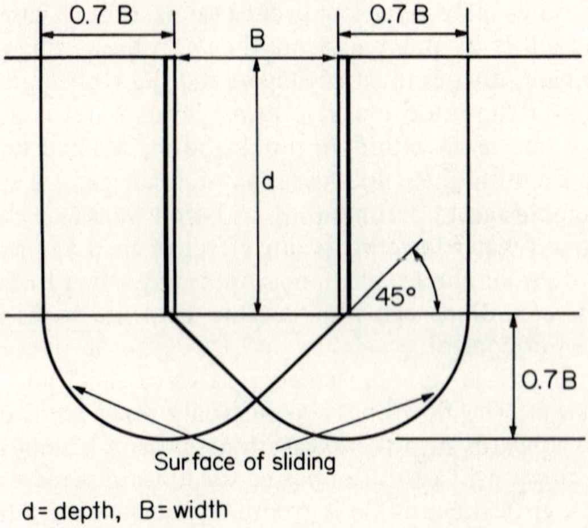

Figure 17.1 Stability of excavations in clay. The zone of clay contributing to failure has a width of approximately 0.7B. The downward force of this mass is resisted by the cohesion (c) along the depth (d) of the excavation so that the total loading (P) is P = 0.78pd – cd. The intensity of loading (P) is therefore $P = (\rho d - cd)/0.7B$ where ρ is density

Where an excavation is in the form of an open cut with sloping sides, the lateral forces on the foundation structure are largely determined by the properties and method of placement of the backfill. Backfill which is heavily compacted or likely to swell may exert a considerable lateral pressure on the foundation. In order to avoid this, lightly compacted well-drained granular material may be used since it exerts a small lateral pressure. Furthermore the placing of backfill into an open excavation with sloping sides may cause a settlement of the ground from its heaved position and hence settlement of the completed foundation structure. As far as foundation design is concerned it should be assumed that no skin friction develops around the vertical sides of the foundation to assist support. Indeed if highly compressible fill is used it may tend to drag the foundation structure downwards.

GROUND WATER AND EXCAVATION

Ground water frequently provides one of the most difficult problems during excavation work and its removal can prove costly. Not only does water make working conditions difficult, for example, vehicles can become bogged down, but an appreciable flow of water into an excavation can lead to erosion and failure of the sides. This, of course, increases costs since not only does the collapsed material have to be removed but the damage has to be made good. Subsurface water is normally under pressure, which increases with increasing depth below the water table to very high values. Hence data relating to the ground water conditions should be obtained prior to the commencement of operations.

Some of the worst conditions are met with in excavations which have to be taken below the water table in loosely packed sands. In such cases the water level must be lowered either by pumping from the excavation or by prior drainage of the site. If pumping, ditches must be dug in the floor of the excavation which lead the water to a sump located at a lower level. Water is pumped from the sump and the water level within it should be maintained below that in the ditches to ensure natural drainage. As far as predrainage of a site is concerned, this can be accomplished by wellpointing or bored wells (see chapter 14). If for some reason ground water lowering is not effective then a quick condition may develop in the floor of the excavation, slopes may begin to fail and the floor may rise. Quick conditions can also develop in dense sands if ground water conditions get out of control.

Quick conditions have sometimes occurred when sheet piling has been used about an excavation. The flow lines run vertically downwards outside the sheet piling then turn upwards into the excavation. As sand is removed from within the sheet piling the weight of soil opposing the upward seepage force is progressively reduced. A critical condition is eventually reached when the weight of the submerged sand is balanced by the upward acting seepage force. Further lowering means that the upward velocity of flow increases and a quick condition develops, and the sand may begin to boil. Hence the depth at which the critical hydraulic gradient becomes operative in such an excavation should be determined. The sheet piling should then be sunk deep enough to avoid a potential critical condition developing at the base level of the excavation. Relief wells can be used to reduce the hydrostatic head (see Ward[4]).

One of the most effective methods employed to avoid the development of quick conditions is to prolong the length of the seepage path (this, of course, is what is done by placing sheet piling to greater depth). This increases the frictional losses thereby reducing the seepage force, and can be accomplished by placing a clay blanket at the base of an excavation where the seepage lines converge. A quick condition may also be prevented by increasing the downward acting force. This may be brought about by loading the surface of the soil where seepage is discharging. Gravel filter beds may be used for this purpose. Suspect soils may also be grouted or frozen.

Silts can also prove troublesome in that they can also develop quick conditions but, because of their low permeability, they are usually difficult to dewater using a wellpointing system or bored wells. However, they have been dewatered by electro-osmosis (see Casagrande[5]). On the other hand unless a deposit of clay contains or is underlain by a horizon of permeable water bearing soil, ground water is not likely to give rise to difficulties during excavation. Where ground water is present it usually seeps from fissures and can be dealt with by pumping from sumps.

Usually pumping from sumps can deal with ground water in rock. The ground water generally emerges as springs or seepages, flowing from bedding planes and joints in sedimentary sequences or from fissure zones in igneous or metamorphic rocks. The influence of ground water on the stability of slopes in rocks is dealt with in chapter 16.

Artesian conditions can cause serious trouble in excavations and therefore if such conditions are suspected it is essential that both the position of the water table and the piezometric pressures should be determined before work commences. Otherwise excavations which extend close to strata under artesian pressure may be severely damaged due to blow-outs taking place in their floors. Such action may also cause slopes to fail and could lead to the abandonment of the site in question. Sites on which such problems are likely to be encountered should be dealt with, prior to and during excavation, by employing either dewatering or exclusion techniques. It must be remembered that the structures to be placed in these excavations will be acted upon by uplift pressures. If a particular structure is weak then uplift pressure may cause it to fail, for instance, a blow-out may occur in a basement floor. On the other hand if the structure is strong but light in weight, it may be subjected to heave. Uplift pressures can be taken care of by adequate drainage or by resisting the upward seepage force. Continuous drainage blankets are effective but should be designed with filters to function without clogging. The entire weight of a structure can be used to resist uplift pressures if a raft foundation is employed. Anchors grouted into bedrock can also provide resistance.

If slopes in soils are to remain stable for appreciable periods they should be adequately drained (see chapters 15 and 16, and CP 2003:1959[6]). Generally speaking interceptor drainage channels should be provided at the top and bottom of a slope, the surface of the slope also being drained. The open channels used in the drainage system should preferably be lined to prevent their enlargement by erosion. If this happens, especially at the top of a slope in cohesive soil, then the channel can act as a tension crack from which a slip may eventually develop.

METHODS OF EXCAVATION: DRILLING AND BLASTING

The method of excavation is very much determined by the geology of the site, however, consideration must also be given to the surroundings. For instance, drilling and blasting, although generally the most effective and economical method of excavating rock, may not be desirable in built up areas since damage to property or inconvenience may be caused, or where severe landslides or rock falls might result.

In any rock the drilling speed of tungsten carbide tipped bits is affected by the power of the rock drill, the air pressure at the rock drill, the shape of the cutting edge of the bit and the diameter of the borehole. Percussion type churn drills will penetrate the hardest rocks but they are slow and inefficient. In softer rocks rotary drills give faster penetration than the percussion types, but in hard to medium abrasive rocks the rotary drill is not economical because of the severity of bit wear. The best compromise according to Antill and Ryan[7] is the rotary-percussion drill.

According to McGregor[9] the rock properties which influence drillability include hardness, toughness, abrasiveness, grain size and discontinuities. The hardness of a rock depends not only upon the hardnesses of the individual minerals concerned but also upon their texture. The harder the rock the stronger the bit which is required for drilling since higher pressures need to be exerted.

Indeed Furby[10] suggested that rock drillability could be correlated with the results obtained by testing with a Schmidt hammer. Toughness is related to hardness and has been defined by Deere and Miller[11] as the ability of a material to absorb energy during plastic deformation and represents the work required to bring about fracture. It can be assessed by an impact test. With respect to drilling, abrasiveness may be regarded as the ability of a rock to wear away drill bits. This property is also closely related to hardness and in addition is influenced by particle shape and texture. The size of the fragments produced during drilling operations also influences abrasiveness. For example, large fragments may cause scratching but comparatively little wear whereas the production of dust in tougher but less abrasive rock causes polishing. This may lead to the development of high skin hardness on tungsten carbide bits which in turn may cause them to spall. Even diamonds lose their cutting ability upon polishing. Generally coarse grained rocks can be drilled more quickly than can fine grained varieties or those in which the grain size is variable.

The ease of drilling in rocks in which there are many discontinuities is influenced by their orientation in relation to the drill hole. Drilling over an open discontinuity means that part of the energy controlling drill penetration is lost. Where a drill hole crosses discontinuities at a low angle this may cause the bit to stick. It may also lead to excessive wear and to the hole going off line. If the ground is badly broken then the drill hole may require casing. Where discontinuities are filled with clay this may penetrate the flush holes in the bit, causing it to bind or deviate from alignment. As a consequence the rate of drilling is generally less difficult and therefore quicker if the hole runs at a high angle to the discontinuities.

The dip of a drill hole can be measured with an inclinometer. If the deviation is excessive then this will entail redrilling. Generally speaking the possibility of deviation tends to increase with distance from the top of the hole and its incidence tends to be greater in angled than vertical holes.

DRILLING AND BLASTING METHODS

Two of the principal methods of drilling and blasting used in foundation excavation are benching and well-hole blasting. Usually in blasting operations a free face is developed in front of the drill holes. As the name suggests, in benching the face is worked in a series of levels or benches, the maximum height of each bench generally not exceeding 10 m. Individual benches can be worked simultaneously, the drill holes being sunk by hand-held jackhammers or crawler-mounted, down-the-hole drills, depending on the size of the burden to be removed.

Single face with well-hole blasting is commonly used for working large excavations, indeed it is the standard method of working quarries. Here a line of holes is drilled parallel to the face, usually with down-the-hole drills. The holes may be inclined, often between 10° and 15°, for safety reasons. Spacing between shot-holes is determined on the one hand in relation to the strength, density and fracture pattern within the rock, and on the other in relation to the size of the charge. Sometimes two rows of holes may be used in blasting. The height of the face brought down is greater than that in benching, and tends to vary between 15 and 35 m.

Horizontal drilling is sometimes used when excavating steeply dipping rocks but is generally undesirable since it can lead to dangerous overhangs, and less stable faces, so that there is an increased danger of rock fall.

Under certain conditions heading blasting may be used. Small tunnels are driven into the base of the face and cross headings are driven from them. Chambers are excavated at intervals and filled with explosive. The headings are then backfilled. This is time consuming and there is little control over fragmentation.

The box cut is used for opening excavations on flat ground. Because there is no free face the blast must occur upward from the ground surface. For efficient operation the width of the cut should be at least twice the depth and the area broken by each charge should be in the form of a cone with its apex below the bottom of the cut. Hence there are a large number of holes drilled to overlap below grade level. The spacing of holes is often about half the depth to grade and nearly double the quantity of explosive used in blasting a free face in the same rock is required.

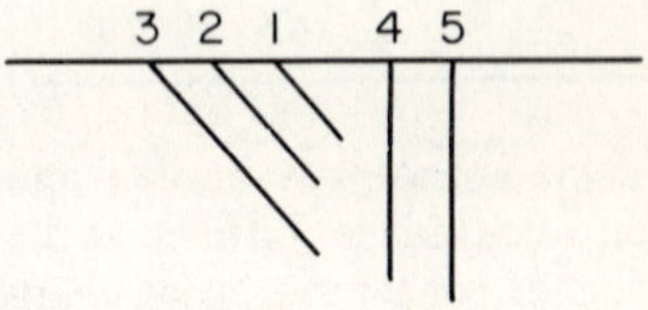

Figure 17.2 Drilling pattern for opening up a bench from a level floor, numbers show order of blasting

The drilling pattern adopted for excavating trenches in rock depends upon their width. In wide trenches the ground is first opened up with a pattern of angled holes drilled to successively greater depths, the object being to develop a free face for benching (Figure 17.2). Conversely, in narrow trenches, the rock may be blasted free with holes drilled along the edges of the excavation.

EXPLOSIVES FOR BLASTING

The quantity of a particular explosive required to blast a certain volume of rock is difficult to estimate since it depends upon its strength, toughness and incidence of discontinuities, but these vary throughout any individual rock mass. Consequently no rigid sequence for progressive resistance to blasting offered by different rock types can be formulated. Nevertheless it can be said that an explosive with the greatest energy and concentration is required for removing very hard rock; in medium-hard rocks a high velocity of detonation produces a shattering effect; a medium to high explosive can be used in medium to hard laminated rocks; and the greatest efficiency is obtained with fairly bulky explosive in soft to medium rocks.

Table 17.1 gives an idea of the amount of charge in relation to burdens for primary blasting. However, careful trials are the only method of determining the burden and blasting pattern in any rock. As a rule spacing of shot holes will vary between 0.75 to 1.25 times the burden. Generally 1 kg of high explosive will bring down about 8 to 12 tonnes of rock. Good fragmentation reduces or eliminates the amount of secondary breakage while minimising wear and tear on loading machinery.

Table 17.1 TYPICAL CHARGES AND BURDENS FOR PRIMARY BLASTING BY SHOT HOLE METHODS (Adapted from Sinclair[8])

Minimum finishing diameter of hole (mm)	*Cartridge diameter*	*Depth of hole* (m)	*Burden* (m)	*Spacing* (m)	*Explosive charge* (kg)	*Rock yield* (tonnes)	*Blasting ratio*	*Tonnes of rock per metre of drill hole*
25	22	1.5	0.9	0.9	0.3	3	10.0	2
35	32	3.0	1.5	1.5	1.8	19	10.5	6.3
57	50	6.1	2.4	2.4	9.5	97	10.2	15.9
75	64	9.1	2.7	2.7	18.0	180	10.0	19.8
75	64	12.2	2.7	2.7	25.0	245	9.8	20.1
75	64	18.3	2.7	2.7	36.3	365	10.1	20.0
100	83	12.2	3.7	3.7	43.0	430	10.0	35.3
100	83	30.5	3.7	3.7	104.3	1120	10.7	36.7
170	150	18.3	6.7	6.7	216.5	2235	10.3	122.1
170	150	30.5	6.7	6.7	363.0	3660	10.1	120.0
230	200	21.3	7.6	7.6	329.0	3350	10.2	157.3
230	200	30.5	7.6	7.6	476.3	4670	9.8	153.1

Calculation of the burden and charge must also take account of the toe section since this is confined at the back and base. In other words there needs to be a higher concentration of charge at the base than in the column whilst at the top of the hole there is no charging needed. This may be accomplished by increasing the charge density at the base, drilling below base level – usually not more than a metre, angling the face, or by horizontal drilling at the toe. Usually the concentration of explosive in the bottom of the hole is approximately 2.5 times greater than in the column. The basal charge may be regarded as extending from the bottom of the hole to a point above floor level equal to the burden, the column charge extends from there to a point below the crest equal to the burden.

If a clean floor is to be developed in an excavation stumps must be avoided at the foot of the face. These are less likely to occur with angled drill holes. In addition if the hole, as suggested above, is drilled a metre or so below floor level and an extra charge placed in the bottom then stumps are less likely to occur.

A drill hole is filled with alternating sections of explosive and stemming, the latter can be sand or dust from the drill hole. The top of the hole is also filled with stemming. Explosives used for primary blasting may be grouped into three categories, namely, conventional high explosives such as gelegnite, dynamite and trimonite; factory mixed blasting agents, such as nobelite; and compositions mixed on site, ammonium nitrate and fuel oil being the most common (AN-FO).

When the explosive in a drill hole is detonated it is transformed into a gas, the pressure of which may sometimes exceed 100 000 atmospheres (see Langefors and Kihlström[12]). The tremendous energy liberated shatters the zone around the drill hole and exposes the area beyond to enormous tensile stresses. This takes place under the influence of shock waves which radiate from the explosion, travelling at between 3000 and 5000 m/sec. A zone of intense deformation occurs about the hole, its thickness frequently approximating to the diameter of the hole, and radial cracking extends appreciably further (see Persson, Lundborg and Johansson[13]).

Under the influence of the pressure of the gases from the explosive the primary radial cracks expand and the free face yields and is moved forwards. When this occurs the pressure is unloaded and tension increases in the primary cracks which incline obliquely forward. If the burden is not too great several of these cracks expand to the exposed surface and the rock is completely loosened. In this way the burden is torn from the rock mass. The lateral pressure in the shock wave is initially positive but falls rapidly to negative values implying a change from compression to tension. Thus the area beyond the hole is exposed to vast tangential stresses. The shock wave itself is not responsible for the breakage of rock but it does provide the basic conditions for the process. Shock wave energy when using high explosive may only account for 5 to 15% of the total energy liberated.

If a rock is to be blasted efficiently it must be capable of transmitting the explosive energy for some distance from the shot holes. When this does not happen then the rock immediately surrounding the hole is pulverised and the between hole area is not fractured. The early movement of the rock face is most important for an efficient blast, i.e. one which achieves good fragmentation per drill hole with minimum quantity of explosive. Rock breakage in blasting, apart from the character of the rock itself, particularly its fracture index, depends largely on the relation between the burden and the hole spacing and also on the time of ignition between the holes. When ignition occurs simultaneously gas pressure in several holes cooperates and pushes the rock forward as the free face yields. If delays are used each hole must tear off a part of the burden. Blasting using milli-second delay intervals thus tends to increase the efficiency of explosives and improves fragmentation. The introduction of delays also means that the shock waves generated by individual blasts are mutually interfering which in turn means that vibration is reduced.

The discontinuities within a rock mass act as free surfaces which reflect shock waves generated by an explosion. They also provide paths of escape along which energy is dissipated. The geometry of the discontinuity pattern is very important since the greatest loss of energy occurs where most discontinuities intersect. At such locations the explosive energy opens existing breaks in the rock but generates few new ones. Secondary blasting is therefore required to break large masses of rock which have only been loosened.

DIFFICULT ROCKS

Bergh-Christensen and Selmer-Olsen[14] recognised three general types of rocks which caused difficulties in blasting. Firstly, there were those characterised by high specific gravity, and high intergranular cohesion with no preferred orientation of mineral grains. They had high tensile strength and very low brittleness values, cited examples being gabbros, greenstones and breccias. It was the high tensile strength which resisted crack initiation and propagation upon blasting.

The second group included those rocks, like certain brittle granites, gneisses and marbles, which were relatively weak and brittle with a low resistance to dynamic stresses. Blasting in such rocks gave rise to extensive pulverisation immediately about the shot holes, leaving the area between almost unfractured. These rocks were too weak to give an effective energy transfer from the detonated charge to the rock mass.

The third category of rocks giving rise to difficult blasting was that possessing marked preferred orientation. mica schist being a typical example. This difficulty arises from the influence of the mechanical anisotropy due to the preferred orientation of the flaky minerals. These rocks split easily along their lineation but crack propagation parallel to it is limited.

CONTROLLED BLASTING

In many excavations it is important to keep overbreak to a minimum. Apart from the cost of replacement, damage to the rock forming the walls or floor may lower the bearing capacity and necessitate further excavation. What is more smooth faces allow excavation closer to the payline and are more stable. There are three basic methods of controlled blasting which can be used for this purpose. They depend upon most of the energy released by blasting being absorbed by the burden so that little remains to damage the face produced.

Firstly, line drilling is most commonly used to improve the peripheral shaping of excavations. It consists of drilling alternate holes between the pattern shot holes forming the edge of the excavation. The quantity of explosive placed in each line hole is significantly smaller and indeed if these holes are closely spaced, from 150 to 250 mm, then explosive may be placed only in every second or third hole. The closeness of the holes is controlled by the type of rock being excavated and by the payline. These holes are timed to fire ahead, with or after the nearest normally charged holes of the blasting pattern. The time of firing is similarly largely dependent on the character of the rock involved. Line drilling is not always successful in preventing overbreak although it helps reduce it. Generally line holes in sedimentary and some metamorphic rocks are not as effective as in igneous rocks.

Fissile rocks such as slates, phyllites and schists tend to split along the planes of cleavage or schistosity if these run at a low angle to the required face. Paine, Holmes and Clark[15] suggested that the raggedness of faces in sedimentary and metamorphic rocks may occur more readily when shot holes are fired from 1 to 2 m away from the line-holes. Compressive strain pulses moving out from the shot holes are reflected from joint, cleavage, schistosity and bedding planes and the reflected tensile strain pulses cause rupturing of rock beyond the holes. On the other hand homogeneous igneous rocks sometimes split quite readily from hole to hole when lined drilled, because the major free surfaces are the holes and not a series of natural shear planes.

Cushion blasting is a type of line drilling where large holes, approximately 170 mm in diameter, are interspersed between the line holes to act as guides to the crack direction. Frequently one large hole to three small holes is used with either all three small holes or only the central one being charged. Cushion blasting often gives better results than line drilling.

Secondly, pre-splitting can be defined as the establishment of a free surface or shear plane in rock by the controlled usage of explosives in appropriately aligned and spaced drill holes. A line of trimming holes is charged and fired to produce a shear plane. This acts as a limiting plane for the blast proper and is carried out prior to the drilling and blasting of the main round inside the proposed break lines. The spacing of the trimming holes is governed by the type of rock and the diameter of the hole. Maximum effectiveness is achieved when the pre-split holes are fired simultaneously. Once pre-split the rock excavation can be blasted with a normal pattern of holes. If a rock mass is heavily fractured the pre-split and primary blast holes can be fired together, delays causing the latter to fire immediately after the former. In most rocks a shear plane can be induced to the bottom of the pre-split holes, that is, to base level, but in very tight, unfissured rocks difficulty may be experienced in breaking out the main blast to base level. In such instances the spacing between the outer shot holes and the crack may need to be reduced by 50 to 75%.

Thirdly, in ledge trimming the main excavation is taken to within a short distance, for example, 0.6 to 1.0 m, of the payline. The spacing between the line trimming holes (at about 0.4 to 0.6 m) is not more than approximately 80% of the burden. Light charges are fired simultaneously along a given section to give a smooth wall.

Franklin, Broch and Walton[16] noted that blasting and ripping could be related to point load strength. They quoted work by Bergh-Christensen and Selmer-Olsen which tended to show that in hard, massively jointed rocks the resistance to blasting decreased with increasing point load strength up to approximately

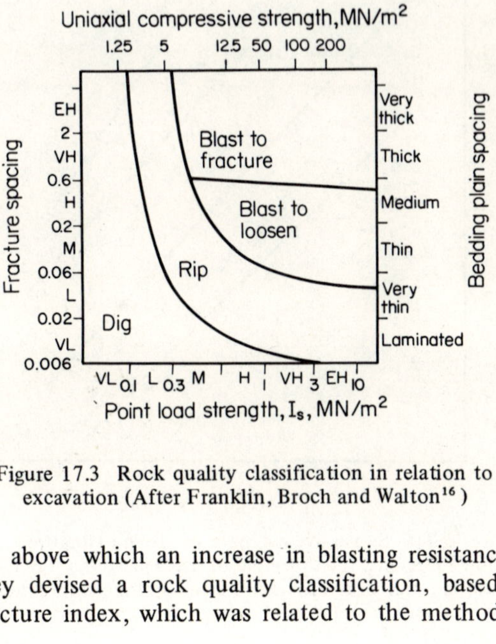

Figure 17.3 Rock quality classification in relation to excavation (After Franklin, Broch and Walton[16])

12–15 MN/m^2, above which an increase in blasting resistance was recorded. Accordingly they devised a rock quality classification, based on point load strength and fracture index, which was related to the methods of excavation (Figure 17.3).

DAMAGE TO STRUCTURES

When excavation in rock has to be carried out in urban areas the proximity of existing structures may determine whether or not blasting is used, since noise, air concussion and ground movement may cause inconvenience or even damage. Experience has shown that the great majority of complaints and even law suits against blasting operations are due to irritation, and subjective human response to vibrations normally leads a person to react strongly long before there is any likelihood of damage occurring to his property (Figure 17.4a and b).

As far as damage due to vibrations is concerned this can be estimated in terms of ground velocity, but it is extremely difficult to determine the limit values of ground velocity for varying degrees of damage. However, the conservative limit of 50 m/s suggested by Berger[17] seems to be commonly accepted as the limit below which no damage will be caused to internal renderings and plasterwork. Nevertheless low vibration levels may disturb sensitive machinery and in this case

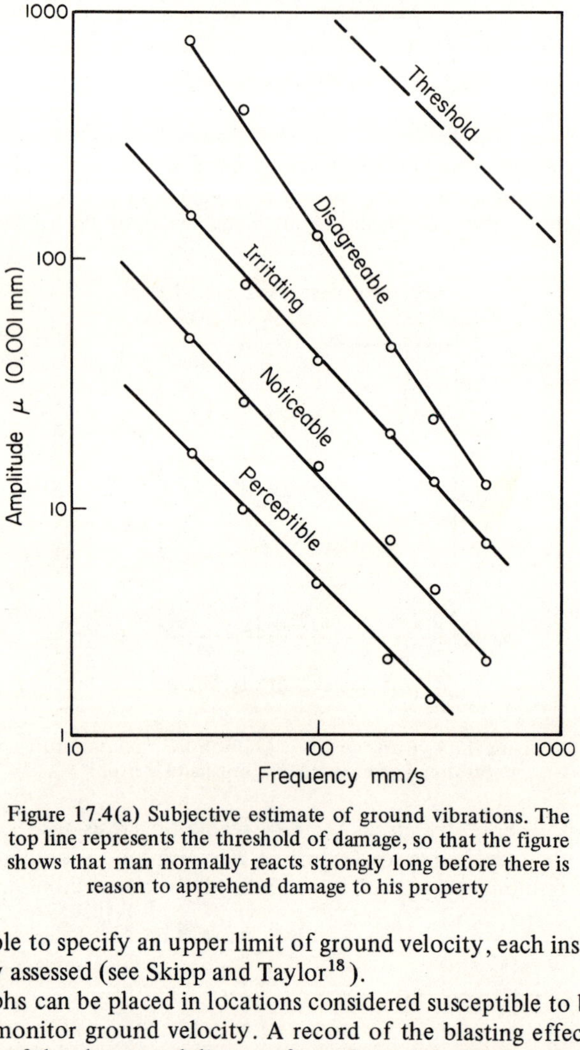

Figure 17.4(a) Subjective estimate of ground vibrations. The top line represents the threshold of damage, so that the figure shows that man normally reacts strongly long before there is reason to apprehend damage to his property

it is impossible to specify an upper limit of ground velocity, each instance should be separately assessed (see Skipp and Taylor[18]).

Vibrographs can be placed in locations considered susceptible to blast damage in order to monitor ground velocity. A record of the blasting effects compared with the size of the charge and distance from the point of detonation is normally sufficient to reduce the possibility of damage to a minimum (Figure 17.5). The use of a multiple-row blasting with short delay ignition reduces the effects of vibration due to blasting operations.

If the use of conventional explosives in urban areas is not permitted then alternative methods of rock breakage must be used. These include hand-held pneumatic breakers or lorry mounted diesel powered breakers. Unfortunately compressed air or diesel breakers are very noisy and also cause vibrations. Consequently in some situations the use of these tools is also excluded. In such instances the employment of hydraulic bursters has been found to be the most suitable method for producing large excavations. A burster is essentially a multiple hydraulic jack with several circular rams operating from one side. A series of holes is drilled parallel to the rock face and one or more bursters inserted.

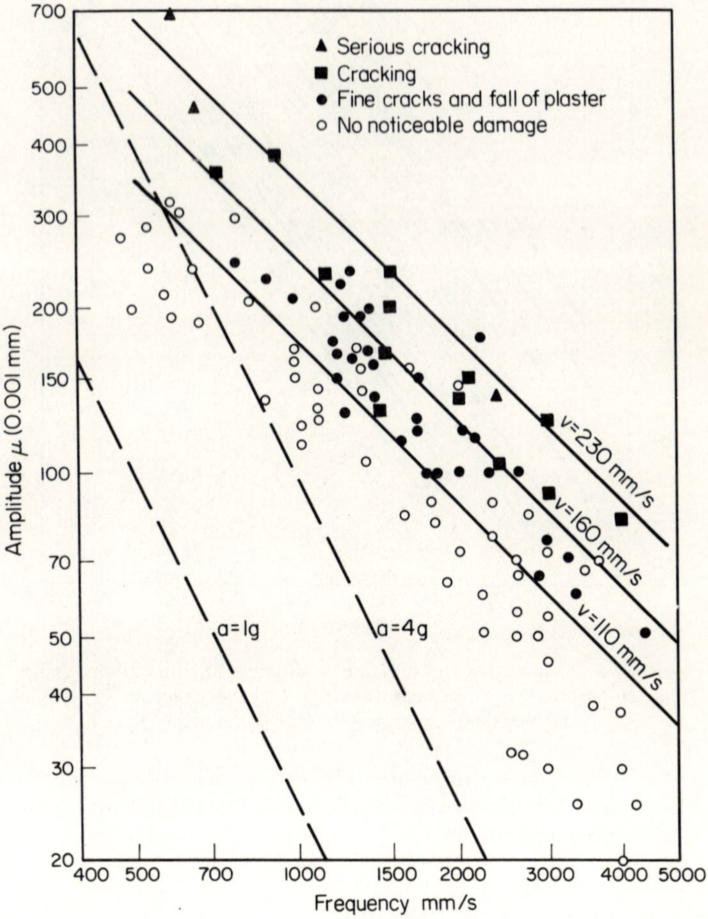

Figure 17.4(b) Connection between amplitude, frequency and damage (From Langefors and Kihlström[12])

Pressure is then exerted by the rams causing the rock to split along the line of holes. With holes of about 150 mm, going to a depth of 2 to 3 m, the burden and hole spacing should be about 1 m. Obviously use is made of discontinuities, especially bedding planes, when this method is used. The holes for the hydraulic burster can be drilled without much noise or vibration by rotary diamond core drills or by oxygen lance if no noise can be permitted. Splitting can also be accomplished by freezing water in the drill holes with the aid of liquid nitrogen.

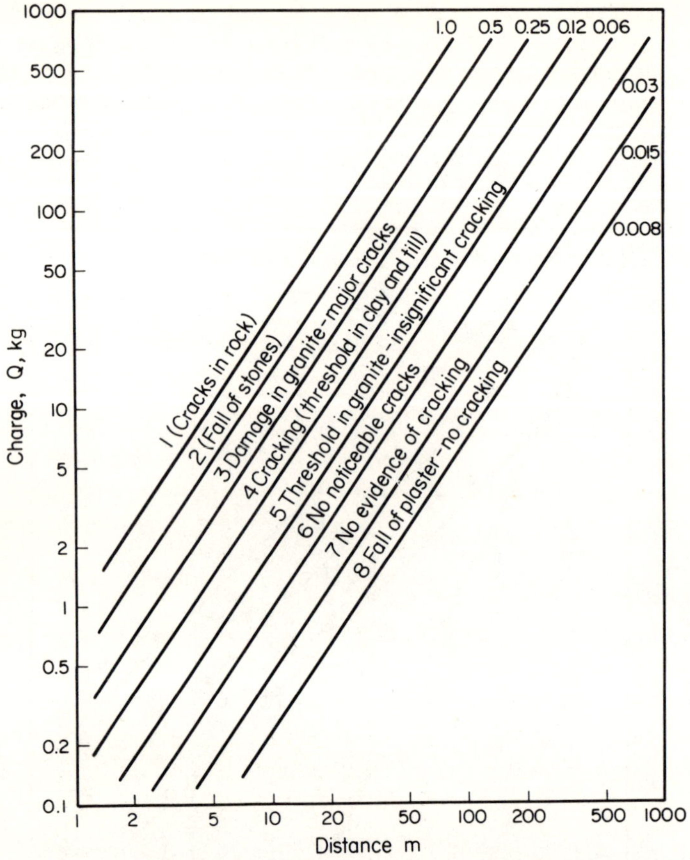

Figure 17.5 Charge (Q) as a function of distance (m) for various charge levels. Numbers 3 to 8 inclusive describe damage expected in normal houses. (After Langefors and Kihlström[12])

Chain saws or wire saws can be used for cutting rock and no doubt the flame torch method could be used in certain rock types. For example, it has been used successfully to quarry granite.

METHODS OF EXCAVATION: RIPPING

The major objective of ripping in construction practice is to break the ground just enough to enable economic loading to take place. Rippers used to break hard soil are pulled by tractors and consist of rugged frames into which are fitted one or more steel tynes. These are held down into the ground whilst being drawn through it, thus causing the ground to break. According to O'Callaghan and McCullen[19] the width of the cut in which pronounced fragmentation of the soil occurs under the action of a deep tyne decreases with depth along the tyne; the zone of disturbance being at a maximum at the surface and decreasing to the width of the tyne at the lower end. When the working depth of the tyne in the ground is more than three times its width, the width of the broken zone at the surface is about five times that of the tyne width.

Rippability in rock masses depends on their intact strength, fracture index and abrasiveness. In other words strong, massive and abrasive rocks do not lend themselves to ripping. On the other hand if sedimentary rocks such as sandstone and limestone are well-bedded and jointed or if strong and weak rocks are thinly interbedded then they can be excavated by ripping rather than by blasting. Indeed some of the weaker sedimentary rocks (less than 1 MN/m^2 point load strength; 15 MN/m^2 compressive strength) such as mudstones are not as easily removed by blasting as their low strength would suggest. The reason for this is that they are pulverised in the immediate vicinity of the hole. What is more, when blasted, mudstones may lift along bedding planes to fall back when the gas pressure has been dissipated. Such rocks, particularly if well-jointed, are more suited to ripping.

The run direction during ripping should be normal to any vertical joint planes, down-dip to any inclined strata, i.e., normal to the strike and, on sloping ground, down-hill. Atkinson[20] suggested that long ripping runs of 70 to 90 m usually gave the best results. He further suggested that where possible the ripping depth should be adjusted so that a forward speed of 3 km/hr could be maintained, since this is generally found to be the most productive, reduces track wear significantly and avoids impact shocks. Adequate breakage in rock depends on the spacing between ripper runs which is in turn governed by the fracture pattern in the rock.

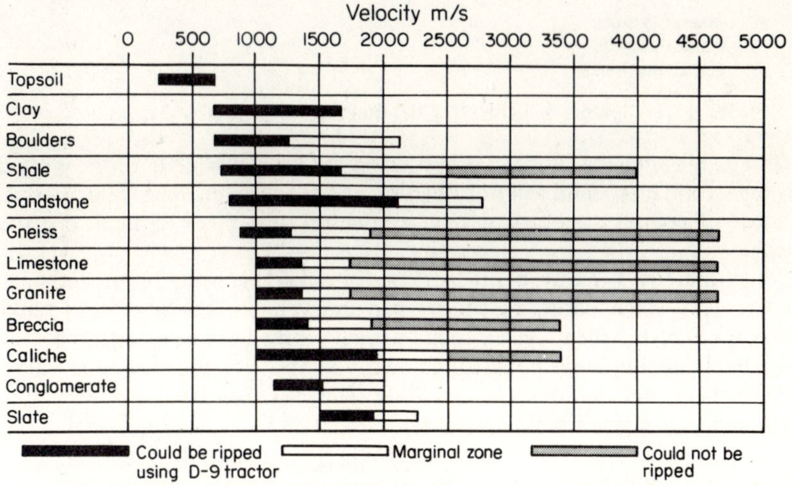

Figure 17.6 Seismic velocities in relation to rippability (After Soil Test Ltd)

The output of rippers varies very much according to the nature of the material involved but generally falls within the range 40 to 230 m^3 per hour.

According to Atkinson, the most common method for determining rippability is by seismic refraction. The seismic velocity of the rock concerned can then be compared with a chart of ripper performance based on ripping operations in a wide variety of rocks (Figure 17.6).

DIGGABILITY

The diggability of ground is of major importance in the selection of excavating equipment and depends primarily upon its intact strength, bulk density, bulking factor and natural water content. The latter influence adhesion or stickiness of soils, especially clay soils. At the present there is no generally acceptable quantitative measure of diggability, assessment usually being made according to the experience of the operators. However, a fairly reliable indication can be obtained from similar excavations in the area concerned, the behaviour of the ground excavated in trial pits or from tests on core samples.

Attempts have been made to evaluate the performance of excavating equipment in terms of seismic velocity (Figure 17.7). It would appear that most

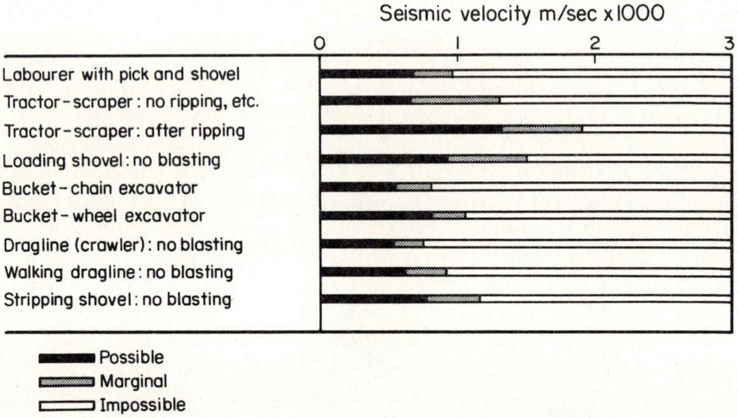

Figure 17.7 Seismic velocities for determining diggability (After Atkinson[21])

earth moving equipment operates best when the seismic velocity of the ground is less than 1000 m/sec and will not function above approximately 1800 m/sec, but in areas of complex geology seismic evaluation may prove difficult if not impossible.

When material is excavated it increases in bulk, this being brought about by the decrease which occurs in density per unit volume. The amount of bulking which takes place when a given rock or soil is worked can be ascertained experimentally by filling large boxes of known volume with the excavated material and averaging the results of several tests. This can then be compared with the *in situ* density to give a bulking factor:

$$\text{Bulking factor} = \frac{\text{Intact density per unit volume}}{\text{Disturbed density per unit volume}}$$

Some examples of typical bulking in rocks and soils are given in Table 17.2. The bulking factor is obviously important in relation to the loading and removal of material from the working face.

CHOICE OF PLANT FOR DIGGING

When choosing the type of plant required for digging, the quantity of the spoil, the means by which it is to be transported and the distance must be considered.

Table 17.2 DENSITY, BULKING FACTOR AND DIGGABILITY OF SOME COMMON ROCKS AND SOILS (From Atkinson[21])

Rock or soil type	*Density* (Mg/m^3)	*Bulking factor*	*Diggability*
1. Basalt	3.00	1.6	H
2. Granite	2.65	1.55	H
3. Sandstone, cemented	2.6	1.6	M–H
4. Sandstone, porous	2.5	1.6	M
5. Limestone, hard	2.7	1.6	M–H
6. Limestone, soft	2.2	1.5	M–H
7. Chalk	1.9	1.3	M
8. Shale	2.4	1.45	M–H
9. Gravel, dry	1.8	1.25	E
10. Sand, dry	1.7	1.15	E
11. Sand and gravel, dry	1.95	1.15	E
12. Clay, light	1.65	1.3	M
13. Clay, heavy	2.1	1.35	M–H
14. Clay, gravel and sand, dry	1.6	1.3	M

E. Easy digging, loose, free-running material such as sand and small gravel;
M. Medium digging, partially consolidated materials such as clayey gravel and clay;
M–H. Medium hard digging materials such as heavy wet clay, gravels with large boulders, well blasted limestones;
H. Hard digging, includes tough plastic clays and materials which require heavy blasting such as granite, basalt, strong limestones etc.

For example, if the tip area is within 100 m of the site then the earth can be moved by a bulldozer. The bulldozer, the blade of which is fixed at right angles to its line of advance, can only push its load in front of it, with an angle-dozer the blade can be angled to cast the spoil from the trailing adge of the blade. The dozer can produce high outputs on shallow excavation work provided, as noted, the haul is kept low, so minimising spillage from the sides.

Scrapers are most effective when employed for large earthmoving works such as road construction. Each type of scraper has an economic range of hauling distance. Under 100 m the bulldozer is more efficient than the scraper, between approximately 100 and 300 m the scraper towed by crawler tractor comes into its own, whilst between about 300 and 1000 m the scraper towed by rubber-tyred tractor is more effective because of its higher speed. Scrapers are unsuitable for deep excavations covering a small area for they must have ample room to turn, and access to such sites generally prohibits their use. They can excavate all types of soil except soft clays and silts.

Where the spoil haul exceeds 1 km, excavators which load into dump trucks are required. These include loading shovels, face shovels, backacters, draglines, grabs, drag-scrapers and continuous excavators.

Loading shovels are only suitable for cuts up to a few metres in depth. They are capable of handling hard, dense soils or badly fragmented ground but they have a poor sub-grade digging capability.

The face shovel is the most powerful of the front end excavators, with the highest output, and can operate in reasonably hard soils. It is designed primarily for bulk excavation above track level, working from the bottom of the excavation upwards. It is most efficient when operating against a high face in firm to hard clay, densely packed sand or gravel with clay binder, or soft chalk. The height of the working face which gives the most efficient output is that which allows the bucket to dig a full load in a single cut and is referred to as the optimum depth of cut (Table 17.3). Because of difficulties of access the face shovel is not suited to deep, confined excavations.

Although the depth to which backacters can excavate is limited to the length of the jib and bucket arm they can be used for deep excavation in small areas such as for trenches and column bases. Backacters can dig to close limits and make well controlled, even cuts. They operate satisfactorily in any type of soil which will stand up temporarily, they are therefore unsuitable for use in loose sand or soft clay.

Table 17.3 OPTIMUM DEPTHS OF CUT FOR FACE SHOVELS (After War Office[22])

		Optimum depth of cut (m)	
Size of bucket m^3	*Light soils e.g. sand, gravel, sandy clay*	*Medium soils e.g. dense sand, firm clay*	*Hard soils, e.g. stiff to hard clays, loess chalk*
0.3	1.2	1.4	1.8
0.4	1.4	1.7	2.1
0.6	1.6	2.1	2.4
0.8	1.8	2.4	2.7
1.0	2.0	2.6	3.0
1.2	2.2	2.8	3.3

With their long reach and dumping radius draglines are especially suited to bulk excavation below track level. Indeed they lose efficiency when employed on excavation above the level of their own tracks. Thus draglines can be used in loose soils or for soft ground and inundated areas, for they are designed to stand above the working area on firm ground and move backwards as the excavation proceeds. They can also be used to excavate badly fractured, weathered, and soft rocks such as chalk. Draglines can work to appreciable depths, although they cannot cut vertical faces. The most efficient digging area is immediately under the tip of the boom and this decreases as the bucket is drawn towards the base of the machine. Where access to the excavation is difficult the dragline has the advantage over the face shovel. One of the advantages of large draglines is that they can pile earth adjacent to the excavation when earth needs to be returned as backfill about the completed foundation structure.

Grabs or clam-shell buckets are the most suitable type of excavator for deep excavation in confined areas. For example, they are commonly used to excavate ground from within caissons.

Drag-scrapers are infrequently used in earthworks although they have been employed sometimes to excavate ground which is too soft to allow the operation of other types of excavators. They consist of a scraper bucket which travels along a wire cable suspended between two towers.

The two most important types of continuous excavators are the bucket-chain excavator and the bucket-wheel excavator. These are generally massive machines developed for use in open pit mining (see Atkinson[21]). However, smaller bucket-chain excavators can be used to dig trenches. These are capable of very high outputs when digging soils free from large stones.

Graders are designed for trimming, shaping and grading soils during road construction, they are not meant for heavy excavation and they cannot carry soil forward. They can be used for moving top soil, operating best in loose sand and gravel.

METHODS OF SUPPORTING THE SIDES OF EXCAVATIONS

The principal factors which influence the design of support required for the sides of an excavation are the strength and stability of the ground, which involves the ground conditions in relation to the depth of the excavation, as well as the length of time it remains open (see CP 2003:1959[6] and CP 2004:1972[23]).

Shallow excavations can be made in soil without providing support if there is adequate space to establish slopes at which the soil can safely stand. However, sites in urban areas frequently extend up to the boundaries of other properties and in such circumstances the sides of the excavation, because they will of necessity be excessively steep, require support.

Continuous support has to be afforded to faces in water-bearing sands or silts, this being accomplished by using timber runners or steel sheet piling. Under such conditions the face support is driven in advance of the excavation.

Continuous support must also be provided for faces excavated in loose sands or gravel or soft clays. In these instances close timbering can be used and it should be placed as soon as possible to avoid any slumping along the faces. Alternatively sheet piles can be driven, again ahead of excavation. Poling boards may be used to support the sides of an excavation in which the ground will stand vertically for a metre or so, for a long enough time to allow their installation.

Horizontal sheeting is used to support ground which will stand to a height of approximately 0.6 m while timbers are being positioned and strutted, or where the ground is heavily surcharged. It is also used in deep trenches. Because densely packed sands and gravels or stiff to firm clays can stand unsupported for varying lengths of time this means that timbering need only be provided at an open spacing sufficient to withstand inward yielding of the ground behind the faces. This also means that the risk of partial collapse of the faces is avoided. If the soil conditions vary then the type of support used can vary. However, special attention should be given to the soil around the bottom of the faces since it is there most heavily stressed.

Once the depth of an excavation exceeds 6 m the use of runners is generally less economical than steel sheet piles. The web types of steel sheet piling are used for the shallower excavations whilst the sides of deep excavations are protected by Z piling.

In most soils it is advisable to drive the sheet piles a metre or more below the floor of the excavation to avoid local heaves. In fact in some cases the embeddment may eliminate the need for installing struts at the base of the cut.

In narrow excavations horizontal struts can be placed between the faces. These struts are tightly wedged between the walings, or vertical beams (soldiers), placed at intervals. The walings run horizontally, supporting timber runners whilst the vertical beams support horizontally placed bracing boards or sheeting (Figure 17.8a). Conversely in wide excavations the bracing is usually vertical and is either supported by raking struts, connected to walings, or tied back by anchors (Figure 17.8b). In some sites it is possible to excavate the central area to its maximum depth and to place part of the permanent foundation structure

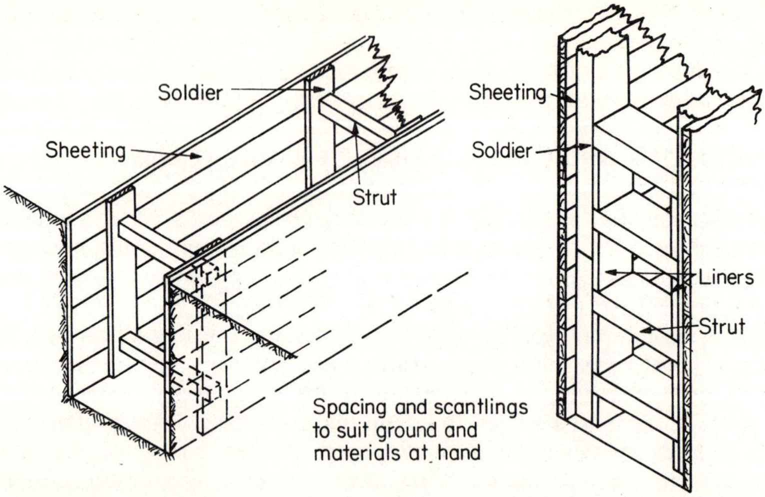

Figure 17.8(a) Timber support for the sides of excavations. Horizontal sheeting

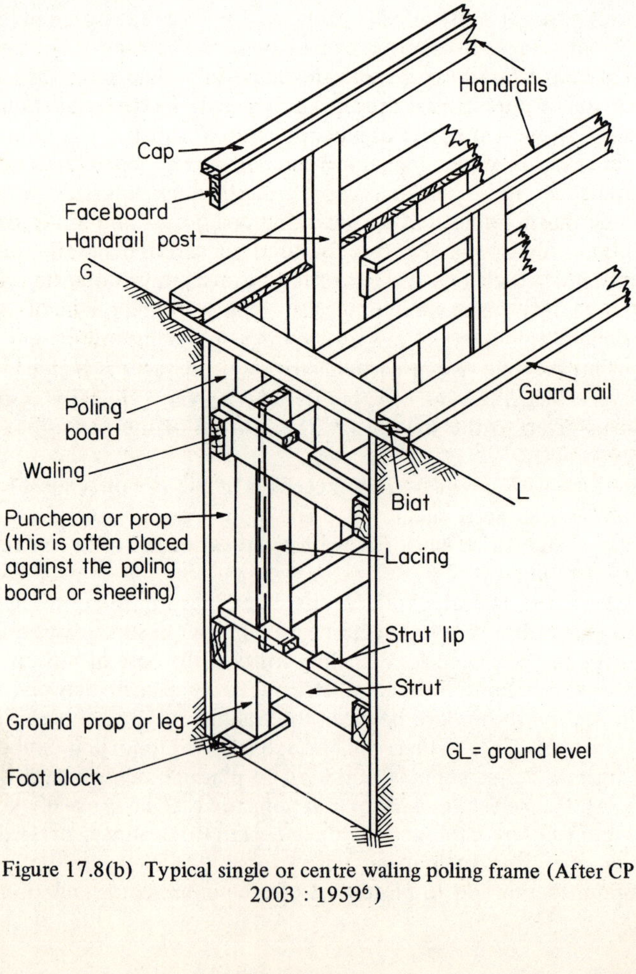

Figure 17.8(b) Typical single or centre waling poling frame (After CP 2003 : 1959[6])

so that this then offers support for the raking struts, required when the remainder of the site is excavated.

According to Tomlinson[1] it is unnecessary to calculate lateral pressures on supports to excavations less than 6 m unless hydrostatic pressure has to be taken into account. Lateral pressures at such depths can be highly variable, for instance, clays shrink from behind the timbering in dry weather and swell when wetted, indeed the swelling pressures may be high enough to deform or rupture the struts.

On the other hand economies in material can be achieved by calculating earth pressures in excavations deeper than 6 m although it is not always necessary to do so. For example, long experience of timbering in known ground conditions is often the best guide. However, when deep excavations have to be made in unfamiliar soils it is advisable to calculate earth pressures. It should be borne in mind that strutted excavations have some flexibility and that local concentrations of earth pressure can cause high loads on individual bracing members. If one strut fails it throws increased loads on to the adjoining members, thus initiating a general collapse of the system. Calculations of earth pressure should allow an economical spacing of walings and struts to be worked out and permit the development of the full flexural strength of the various components involved. The general method of earth pressure calculation is provided by Tomlinson[1] (pages 591 to 601).

Because excavation involves removal of material it causes a change in the state of stress in the rock or soil beneath and beside the excavated space, such a change occurring whether or not the sides are braced. Hence excavation is always associated with movements of the adjacent ground surface, which usually give rise to settlements. However, under certain rather rare conditions the ground surface may rise.

Properly designed and carefully placed bracing can materially reduce the change in lateral pressure in the material adjacent to the excavation and in this way is capable of reducing the settlements to a minimum value for a particular site. A common precaution for reducing movements of the adjacent ground to the practicable minimum is to prestress each strut as it is inserted.

Where the consequences of settlements of adjacent property or of failure are likely to be serious, field observations may be taken to provide early warning of unfavourable developments. These may include measurement of lateral and vertical movements of sheeting and bracing, of forces in raking or struts, of settlements of adjacent structures, and of piezometric levels below and adjacent to the excavation.

The use of slurry to stabilise the walls of a trench in which the foundation structure is subsequently placed is discussed in chapter 14 under 'Diaphragm walls'.

A NOTE ON COFFERDAMS AND CAISSONS

Cofferdams and caissons are used for construction work either in waterlogged or unstable ground which requires lateral support, or below water level (see White and Prentis[24], and White[25]). In other words they enclose an area down to foundation level from which water is sufficiently excluded to permit excavation and placement of the foundation structure to take place. A cofferdam is a temporary feature whereas a caisson is subsequently incorporated into the permanent structure. The principal factors influencing the choice between cofferdams and caissons are, firstly, the ground conditions and, secondly, the proposed depth of the foundation level.

Generally cofferdams are used up to depths of 18 m below water level, depending on the character of the soil. Because they need not exclude all water, indeed it is usually uneconomic to do so, then pumping may be necessary. In fact the water level may be lowered below excavation level by abstracting from wells inside the cofferdam. This is particularly necessary if there is a possibility of a quick condition developing. However, greater passive resistance will be offered at the cofferdam walls if the ground water level is appreciably lowered below excavation level. This accordingly affects the depth of the cut-off. In permeable soils the ground water level may be reduced by wells outside the cofferdam. The types of cofferdams are given in Figure 17.9.

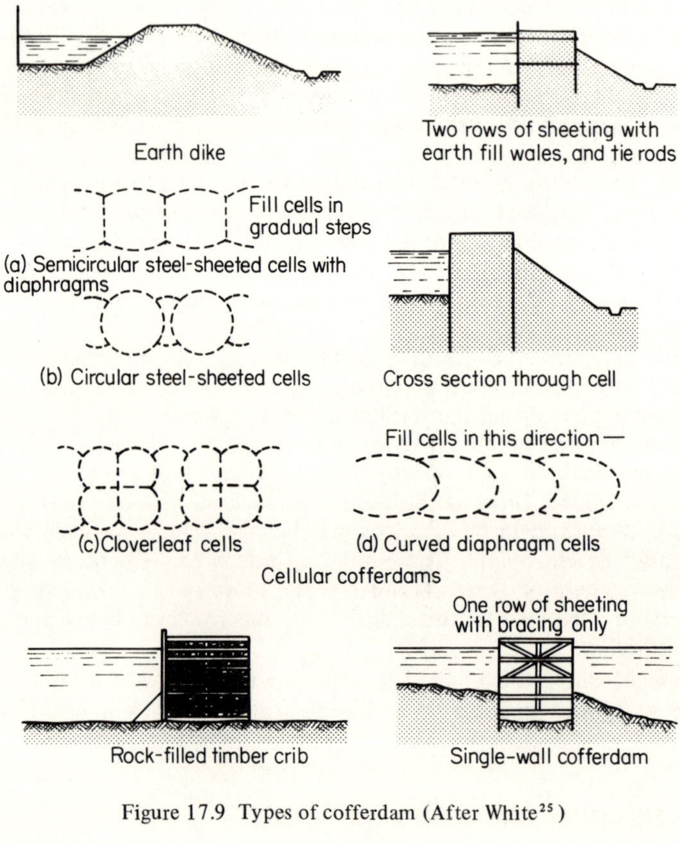

Figure 17.9 Types of cofferdam (After White[25])

Open caissons may be used for foundations which extend to and even exceed 45 m below surface water or ground level. However, they are difficult to pass through rock, or soil containing large boulders. What is more they are not usually employed in hard clays owing to the enormous weight needed to overcome skin friction. A box caisson is one which is closed at the bottom and open at the top, and a pneumatic caisson contains a working chamber in which compressed air excludes the entry of water and loose or soft ground. Caissons may be constructed of reinforced concrete, steel or cast iron segments.

In dense sand it may be possible to pump water from a caisson without destroying the stability of the floor since the deformation attributable to the seepage pressure does not bring about an increase in the pore pressure. But in loose sand pumping causes uplift pressure which in turn causes liquefaction. In order to prevent loss of ground or serious settlement immediately outside an open caisson in wet, loose sand, due to sand moving into the bottom of the caisson, the water level inside the caisson can be maintained a metre or more above the water level outside. This excess head means that water flows from the caisson and counteracts the tendency of the sand to rise. If pumping from sumps within a caisson is not possible then abstraction using deep wells or well-points should be considered or excavation can be carried out in a slurry filled hole or by using compressed air in a pneumatic caisson.

As a caisson is advanced into the ground, the downward movement is resisted by skin friction. Adding dead weights to the top of a caisson is a tedious procedure and so the weight of concrete caissons usually exceeds the resistance of skin friction during each stage of construction. Terzaghi and Peck[26] gave the following values of skin friction for different soil types:

Type of soil	*Skin friction* (kN/m^2)
Silt and soft clay	1035–4140
Very stiff clay	6900–27580
Loose sand	1725–4825
Dense sand	4825–9650
Dense gravel	6900–13790

The skin friction in cohesive soils can be reduced by painting the outside walls of caissons with a coating of oil. Clay grout can also be used as a lubricant.

References

1. Tomlinson, M.J., *Foundation design and construction,* Pitman, London (1975)
2. Skempton, A.W., Schuster, R.L. and Petley, D.J., 'Joints and fractures in the London Clay at Wraysbury and Edgware', *Geotechnique,* **19**, 205–217 (1969).
3. Clevenger, W.A., 'Experiences with loess as a foundation material', *Trans. Am. Soc. Civ. Eng.,* **123**, 151–180, Paper no. 2916 (1958).
4. Ward, W.H., 'The use of simple relief wells in reducing water pressure beneath a trench excavation,' *Geotechnique,* **7**, 134–139 (1957).
5. Casagrande, L., 'Electro-osmotic stabilisation of a high slope in loose saturated silt,' *Proc. 5th Int. Conf. Soil Mech. Found. Eng.,* **2**, 555–561, Paris (1961).
6. CP2003: *Code or Practice on earthworks,* British Standards Institution, London (1959)
7. Antill, J.M. and Ryan, P.W.S., *Civil engineering construction,* Angus & Robertson, Sydney (1967).
8. Sinclair, J., *Quarrying, opencast and alluvial mining*, Elsevier, London (1969).
9. McGregor, K., *The drilling of rock,* C.R. Books Ltd, (A Maclaren Co), London (1967).
10. Furby, J., 'Tests for rock drillability', *Mine and Quarry Eng.,* **30**, 292–298 (1964).
11. Deere, D.M. and Miller, R., *Engineering classification and index properties for intact rock,* Tech. Report. AFWL-TR-116, Air Force Weapons Lab., Kirtland Air Force Base, New Mexico (1966).
12. Langefors, U. and Kihlström, *The modern technique of rock blasting,* Wiley, New York (1963).

13. Persson, P., Lundborg, N. and Johansson, C.H., 'The basic mechanism in rock blasting', *Proc. 2nd Cong. Int. Soc. Rock. Mech.*, Belgrade, Paper 5–3, 19–33 (1970).
14. Bergh-Christensen, J. and Selmer-Olsen, R., 'On the resistance to blasting in tunnelling,' *Proc. 2nd Cong. Int. Soc. Rock Mech.*, Paper 5–7, 59–63 (1970).
15. Paine, R., Holmes, D. and Clark, H., 'Controlling overbreak by pre-splitting,' *Int. Symp. Min. Res. Univ. of Missouri,* **1**, 179–209, Pergamon, New York (1961).
16. Franklin, J.A., Broch, E. and Walton, G., 'Logging the mechanical character of rock,' *Trans. Inst. Min. Met.,* **80**, A1–9 (1971).
17. Berger, P.R., Blasting Seismology, 'Recent developments in criteria, regulations and instrumentation,' *Quarry Man. Jour.,* **55**, 187–190 (1971).
18. Skipp, B.O. and Tayton, J.W., 'Blasting vibrations – ground structure and response,' (article in *Dynamic Waves in Civil Engineering,* Ed. by Howell, D.A., Haigh, I.P. and Taylor, C.) Wiley-Interscience, New York (1970).
19. O'Callaghan, J.R. and McCullen, P.J., 'Soil mechanics in relation to earth moving machinery,' *Symp. Earth Moving Machinery*, Inst. Mech. Eng., 23–29 (1965).
20. Atkinson, T., 'Ground preparation by ripping in open pit mining', *Min. Mag.,* **122**, 458–469 (1970).
21. Atkinson, T., 'Selection of open pit excavation and loading equipment,' *Trans. Inst. Min. Met.,* **80**, A101–129 (1971).
22. War Office, *Engineer Construction Plant, Military Engineering,* 15, HMSO, London.
23. CP2004: *Code of Practice on foundations*, British Standards Institution, London (1972).
24. White, L. and Prentis, E.A., *Cofferdams,* Columbia Univ. Press, New York (1950).
25. White, R.E., 'Caissons and cofferdams', (article in *Foundation Engineering*, Ed. by Leonards, G.A.) McGraw-Hill, New York, 894–964 (1962).
26. Terzaghi, K. and Peck, R.B., *Soil mechanics in engineering practice,* Wiley, New York (1967).

Chapter 18

The Design of Foundation Structures for Difficult Ground

INTRODUCTION

The purpose of this chapter is to discuss the influence of difficult ground conditions on the design of foundation structures and to describe various procedures which can be followed in avoiding rather than overcoming difficulties at the construction stage.

The principal types of structure will be considered in relation to various conditions of difficult ground. The latter will generally be confined to experiences in the UK.

GENERAL PRINCIPLES OF DESIGN

Difficult ground conditions imply either a general weakness, or a high degree of variability, or a combination of these two, in the soil or rock affected by foundation pressures and by construction operations. Site investigations, even with advanced geophysical techniques, can rarely explore the ground conditions in sufficient detail to enable foundation designs to be tailored to suit a highly variable subsurface geology. The overall cost of a project places an economic limit to the amount of money which can be spent on a site investigation. Usually the best that can be done is to obtain an indication of the *range* of soil parameters affecting stability, settlement, and construction techniques.

Therefore the engineer must consider the foundation design in relation to the most unfavourable subsurface conditions. He must not be led into over-optimistic assumptions based on average conditions, with the idea that if weak ground is met the difficulties can in some way be overcome at the construction stage. The ideal foundation design is one for which a contractor can tender on fixed rates without having to cover for unforeseen difficulties, and which can be constructed to a firm programme without delays resulting from further site investigations, increases in excavation depths, or the adoption at a late stage in the contract of specialist geotechnical processes such as grout injections or ground water lowering. These processes can play a useful part in overcoming foundation difficulties and reducing construction costs, but if they are to be used effectively and economically, they must form part of the planned programme for design and construction and not be brought in as a last resort expedient when all other measures have failed.

The principle of 'design-as-you-go' or the observational technique in geotechnical engineering, which is sometimes employed in the design and construction of earth dams or highway embankments cannot be applied to foundations of structures. It is essential that superstructure and substructure should be designed as an integrated whole. The superstructure design cannot wait for the performance of the foundations to be monitored, or the ground studied in detail only at the time of foundation excavations.

The general aim should be to design a simple substructure which can be constructed as far as possible by mechanical plant. At one time it was easy to find operatives highly skilled in intricate timbering, shoring, and soft ground mining. These operations based on manual skills have been largely superseded by mechanical methods using plant operated from the ground surface or by remote control away from the working face. For low-cost structures such as housing or light industrial buildings it is particularly important to adopt simple foundation designs not dependent on accurate setting-out, intricate formwork, or sophistication in construction materials.

Where the foundation work in difficult ground is a critical factor in the overall cost and planned construction time of a project it can be advantageous to undertake trials in advance of the main contract to investigate possible construction difficulties or to confirm the efficacy of a particular specialist construction technique. For example, where diaphragm walling is proposed for a basement in ground containing large boulders, or a cemented soil layer, it can be helpful to make preliminary trials by excavating a trench under bentonite to explore the frequency of these obstructions and to develop methods of removing them.

Trial piling in advance of the main construction contract is almost always economically advantageous even for small projects. The object of the trial piling is not only to confirm the predicted ultimate pile loading capacity or to establish the settlement of the working load, but also to check that the proposed installation method is suitable for the ground conditions. On large projects it is advantageous to evaluate the installation problems, load-carrying performance and economics of several types of pile.

STRIP AND PAD FOUNDATIONS FOR VARIABLE GROUND CONDITIONS

Strip and pad foundations are suitable for lightly loaded structures in variable ground where total and differential settlements are not detrimental to the stability, or proper function of the superstructure. As far as possible the allowable bearing pressure should be selected with a view to avoiding any need for local deepening of the excavations. It can be tempting to design the foundations on the assumption that pockets of compressible soil below the nominal foundation level will be removed, thus obtaining uniform bearing conditions on a less compressible soil or rock.

While this procedure may appear to be economical, it means that the contractor cannot give a realistic price for excavation and concreting in foundations, since until the ground is opened up he has no idea of the extent, both laterally and in depth, of the weaker ground which has to be removed. If the extra excavation depth and concrete is billed and paid for on a measured basis there can be protracted delays while the exposed ground is inspected and tested, and the extra digging depths are agreed with the employing authority. It cannot be expected that the contractor will be allowed to make his own decisions on these extra depths. Assessment of 'uniform' bearing conditions is particularly difficult in weathered rocks in which there can be wide variations in the degree of weathering from a firm clay to a strong rock. The assessment of the foundation depth to obtain 'uniform bearing conditions' is often a matter of subjective judgment on which there can be quite wide differences of opinion.

Substantial extra foundation costs have been incurred on chalk sites where it has been specified that foundations should be taken below superficial drift deposits to uniform bearing on chalk which has been mistakenly assumed from trial pit excavations to exist at a reasonably uniform depth below the surface (Figure 18.1). It is preferable to base the foundation design on the allowable

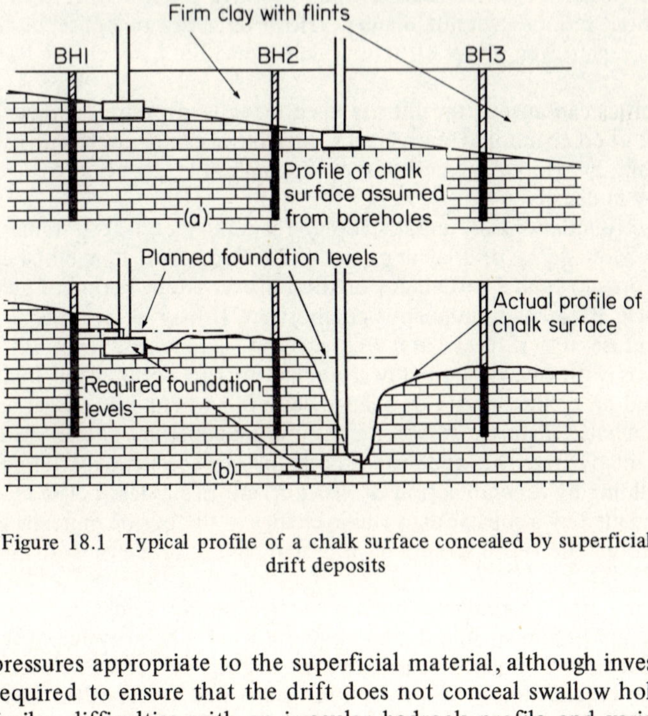

Figure 18.1 Typical profile of a chalk surface concealed by superficial drift deposits

bearing pressures appropriate to the superficial material, although investigations will be required to ensure that the drift does not conceal swallow holes in the chalk. Similar difficulties with an irregular bedrock profile and variable rock strength can occur with foundations in the Coal Measures where strong sandstones may be interbedded with weak clayey weathered shales or mudstones, often with steeply dipping beds.

The author has noted a preoccupation by some local authorities with the need to obtain uniform bearing conditions for foundations. In this respect they may be unduly influenced by Building Regulation D7 which is deemed to be satisfied if 'there is no made ground or wide variation in the type of subsoil within the loaded area'. The wording of this Regulation is not helpful, since it ignores the fact that soils which vary widely in type can have practically identical compressibility; the latter is the important consideration in foundation design.

BUILDING ON FILL

Shallow strip or pad foundations can be quite satisfactory on made ground, particularly on old tips of industrial waste material or quarry waste which have existed for many years. Penman and Godwin[1] described the behaviour of 2-storey houses constructed on deep fill consisting of clay and weak rock dumped by dragline and roughly levelled in a restoration operation on ironstone mine workings at Corby. The houses were built twelve years after placing the fill. Some were on standard strip foundations and some on reinforced concrete rafts. Although the houses on the strip foundations suffered the maximum differential settlement of 135 mm (total settlement 180 mm), compared with a maximum differential settlement for the rafts of 75 mm (total 135 mm) it was reported that cracking of the superstructure was relatively minor in extent and the degree of cracking was unrelated to the type of foundation. It is well worth noting the amount of total and differential settlement which brick built houses can withstand without serious structural damage. However, a private housebuilder may be obliged to repair even minor structural damage without any charge to the house purchaser.

Difficulties can arise if the fill has been recently deposited and varies widely in its state of compaction. However, the difficulties can be overcome by compacting the fill by vibratory methods or dynamically[2] to achieve a reasonable uniformity in density.

Shallow foundations are unsatisfactory for building on deep domestic refuse tips where slow decay of organic matter and collapse of rusting metal containers can result in substantial settlement of the fill over a long period of years. These tips are also liable to spontaneous combustion. It may be possible to build on shallow refuse tips if these are preloaded by surcharging with fill material.

Problems with heave of shallow foundations can occur when buildings are constructed on colliery waste or quarry waste fills where these materials contain pyritic minerals. Disturbances of the fill during construction can result in oxidation of the pyrites. The oxidation process may be accelerated after completion of the building by a combination of warmth and rising damp. The reaction can be accompanied by a substantial volume change in the pyritic minerals leading to heave of foundations and ground bearing slabs and in severe cases to combustion of the fill.

Waste materials containing pyrites are frequently acidic requiring special membrane protection of foundation concrete where the pH value of the ground water is less than 5. Foundations above the highest level of a fluctuating ground water table require no special protection against acid attack other than adopting a fairly rich, dense, well-compacted concrete made with ordinary Portland or sulphate resisting cement. A suitable protective scheme is shown in Figure 18.2.

BUILDING ON SHRINKABLE CLAYS

It is likely that foundation movements due to the shrinkage or swelling of heavy clays as a result of moisture content changes are responsible for more damage to buildings in south-eastern England than any other single cause. The need to take foundations in these soils to a depth of 0.9 m, to get below a depth subject to significant seasonal moisture changes, is well appreciated, as is the need for precautions against long-term shrinkage of clays where buildings are sited close to growing trees. These widely adopted precautionary measures appear to have worked quite satisfactorily for buildings sited in open ground.

The author has not examined any cases of damage to houses constructed on open ground away from trees or hedges within the last 30 years with a foundation depth of 0.9 m in shrinkable clays, even in the 1976 drought. However,

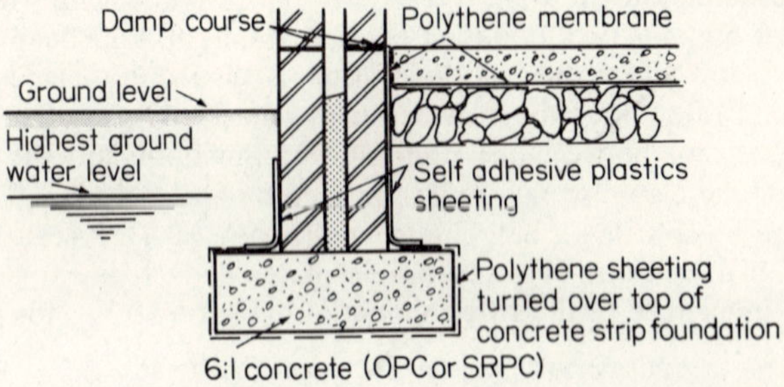

Figure 18.2 Method of protecting a strip foundation from attack by acid wastes in industrial fill

there have been cases of uplift on the strip foundations before erection of the superstructure. The uplift of the unloaded foundations has occurred due to a seasonal increase in moisture content causing swelling of the upper 0.5 m or so of clay which has adhered to the concrete in the 'trench-fill' type of construction commonly used for these deep strip foundations (Figure 18.3). The author has

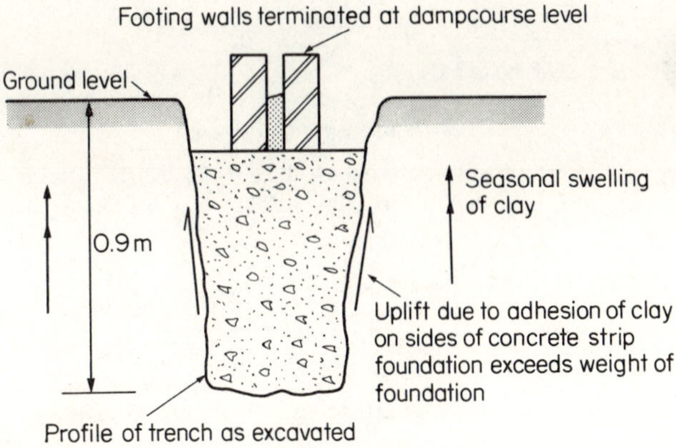

Figure 18.3 Uplift due to soil swelling adjacent to unloaded narrow strip foundation

also noticed quite widespread disregard of the need to restrict planting of new trees near to buildings on clay soils, which has caused damage to houses built over twenty years ago. A foundation depth of 0.9 m is, of course, quite inadequate to prevent damage caused by desiccation of clays by tree root action.

Swelling of clays which have been subjected to many years of desiccation due to growing vegetation or heating in buildings can cause severe damage to structures. The swelling takes place when the moisture content is allowed to increase, say after removing trees or hedges in advance of construction without allowing sufficient time for the moisture content to come to equilibrium with the surrounding soil. A period of one or two winter seasons for the clay to wet up is quite inadequate. Samuels and Cheney[3] have shown that a rise in moisture content, accompanied by a heave in the foundations of a house of up to 150 mm, occurred over a period of twenty years after cutting down the trees and commencement of building on a London Clay site. Merely to take deep strip or piled foundations below the zone of desiccation is also quite inadequate. Heave beneath ground floors must be provided for by suspending the floors with a gap of 150 to 200 mm between the underside of the floor and the soil surface.

Swelling of the clay also takes place in a horizontal direction. This can cause severe damage by splitting a building in two (Figure 18.4). Precautions can take the form of placing slabs of compressible material, such as expanded polystyrene, on the inner face of the strip foundations (Figure 18.5). However, because

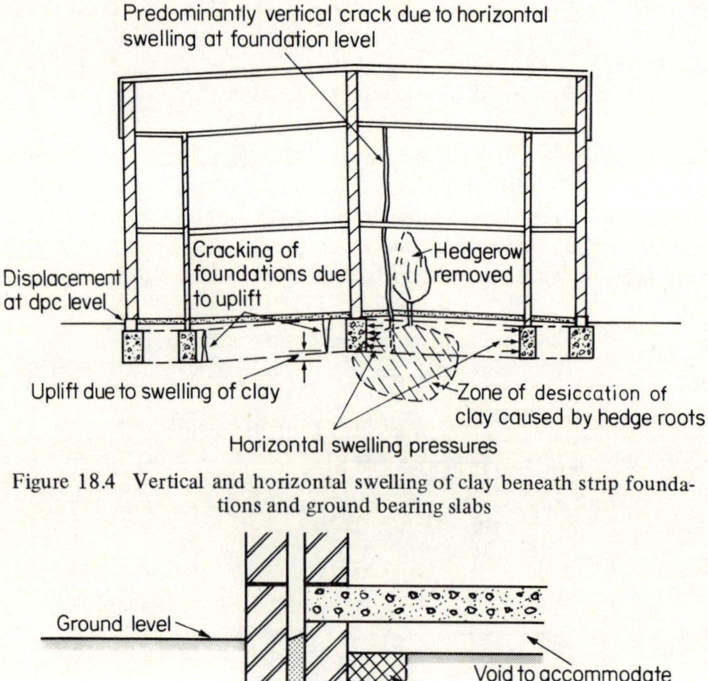

Figure 18.4 Vertical and horizontal swelling of clay beneath strip foundations and ground bearing slabs

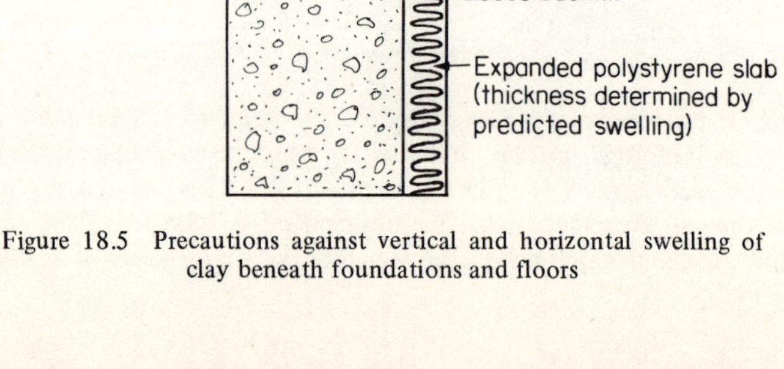

Figure 18.5 Precautions against vertical and horizontal swelling of clay beneath foundations and floors

of the cost of excavating to take the foundations below the desiccated zone together with the means to accommodate horizontal swelling, it will be found that the bored pile foundation with shallow ground beams will, in most cases, be a more economical solution.

It must be pointed out that unless a compressible layer such as expanded polystyrene, is placed beneath the ground beams, and a suspended floor slab is provided, the money spent on the bored pile foundations may be wasted.

RAFT FOUNDATIONS – VARIABLE SOILS AND MADE GROUND

A raft foundation is a useful expedient for building on soils or made ground of widely varying compressibility where differential settlements would be detrimental to the superstructure. Stiffness required to bridge over zones of weak ground is provided by incorporating beams in the substructure, placed either beneath the load bearing wall, or beneath rows of columns. Where the variation in ground compressibility or the variation in superstructure loading is not severe the raft can be designed essentially as a wide strip foundation without provision for load transfer to the ground bearing slab (Figure 18.6). In very weak ground

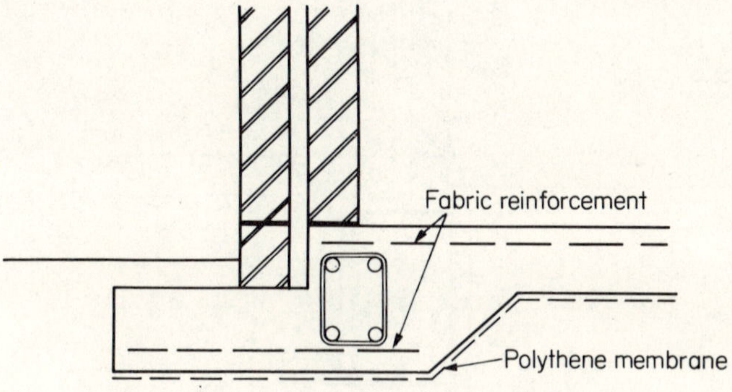

Figure 18.6 Wide toe foundation for lightly loaded building on variable soils of moderate compressibility

the superstructure loading must be distributed over the whole of the substructure which requires a rigid structural connection between the beam and the slab (Figure 18.7a) or for heavy structures, a stiff flat or cellular slab (Figure 18.7b).

Problems can arise in large loaded areas particularly if there are wide variations in the superstructure loading, as in the case of industrial buildings where heavy machinery bases can be placed near to lightly-loaded building columns. If possible large rafts should be loaded symmetrically to avoid tilting. Consideration can be given to articulating large rafts by means of dowelled joints. It may be necessary to continue these joints upwards to form movement joints in the superstructure.

Extending buildings on raft foundations can also cause problems of tilting. The ground beneath the existing raft may have consolidated fully, after which the extension is built on unconsolidated soil. If it is then rigidly attached to the existing raft it will drag the latter down causing it to tilt towards the extended portion. A preferred solution is to site the new raft a metre or more away and to link the existing and new structures by articulated slabs or beams.

Complications arise when rafts are constructed on sloping ground. Stepping the raft is a costly expedient and it is preferable to terrace the site so that the

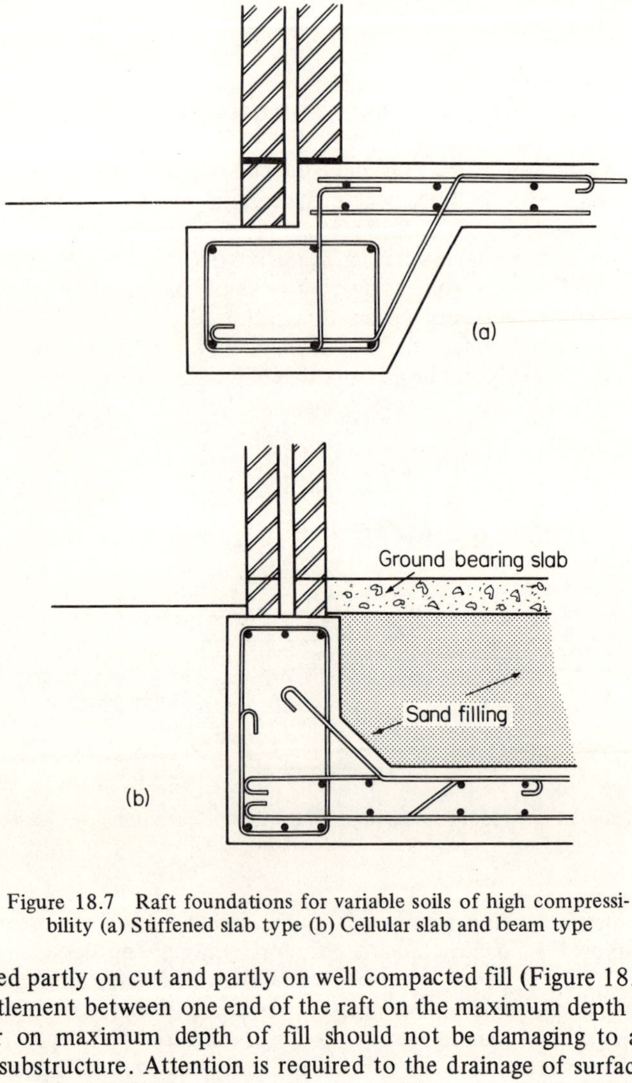

Figure 18.7 Raft foundations for variable soils of high compressibility (a) Stiffened slab type (b) Cellular slab and beam type

raft is sited partly on cut and partly on well compacted fill (Figure 18.8). Differential settlement between one end of the raft on the maximum depth of cut and the other on maximum depth of fill should not be damaging to a properly designed substructure. Attention is required to the drainage of surface and subsoil water on the uphill side of the raft.

RAFTS ON SHRINKABLE CLAY

Problems with shrinkable clays were described earlier. Although it was stated that bored piles were the most economical treatment for sites where moisture content changes extend to considerable depths, e.g. on sites subject to tree root action, or in arid countries, there are conditions where bored piling can be relatively costly for light structures. The clays may be stony or they may contain shaly bands difficult for the operation of a mobile earth auger. Pockets of water-bearing sand may exist above toe level of the piles requiring support by casing.

Although not widely used in UK, stiff raft foundations can be used to overcome the problem of differential swelling of a desiccated clay. They are used for this purpose on expansive clays in Australia. Walsh[4] has described the design of a

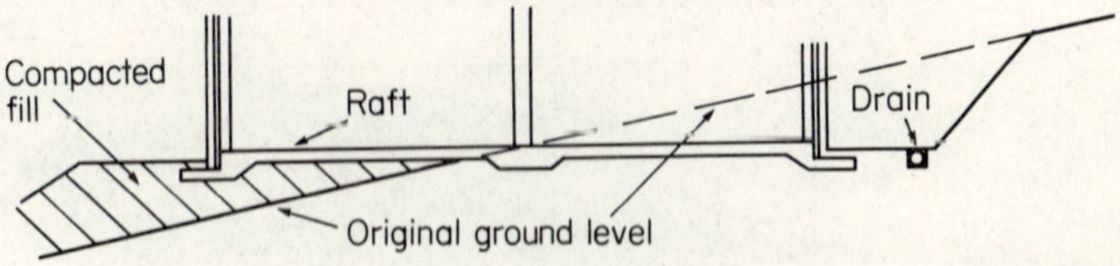

Figure 18.8 Raft foundation on sloping ground

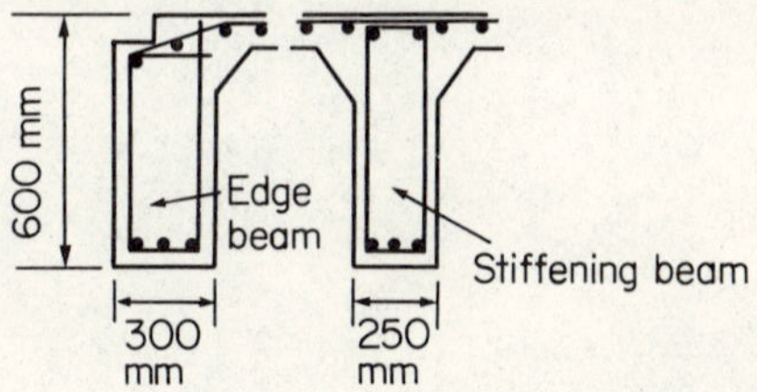

Figure 18.9 Raft foundation for a lightly loaded building on a swelling clay (after Walsh[4])

raft (Figure 18.9) suitable for single or two-storey housing on sites with a potential swelling of 53 mm. He states that deep lightly reinforced sections are preferable to shallow heavily reinforced sections for building on expansive clays.

BUOYANCY RAFTS

The principle of the buoyancy of a submerged hollow structure can be used to design the supports of heavy structures sited over deep highly compressible soils, where the alternative of long piles would be costly or would damage the adjacent structures. By adopting a hollow raft-type substructure sited below ground level the gross loading of the superstructure and foundations is reduced to a lower net value by the weight of the soil removed in excavation for the hollow raft.

Bjerrum[5] described the successful use of buoyancy rafts as foundations for multi-storey buildings sited over deep deposits of highly compressible marine clay in the Oslo area. He gave a comparison between the settlements of a building on a buoyancy raft, with those of an adjacent building on piled foundations. The piled basement reduced the gross loading of 65 kN/m^2 to a net value of

23 kN/m^2, the latter being transferred to friction piles. The building settled 120 mm in 10 years, and equally significant, the ground surface 15 m away from the building settled by 80 mm as a result of heave and reconsolidation of the soft clay due to pile driving displacements. By comparison the unpiled structure founded on a buoyancy raft which reduced the gross loading from 55 kN/m^2 to a zero net value settled by only 40 mm. This settlement was mainly reconsolidation of the soil which had heaved during excavation for the raft.

Somerville and Shelton[6] described the use of buoyancy rafts for 15-storey buildings in Glasgow sited over deep compressible glacial lake clays (Figure 18.10). Piling was not adopted because of the presence of abandoned coal mine

Figure 18.10 Buoyancy raft under construction for a 15-storey block of flats in Glasgow (Wimpey News photograph. Photographer W. Ralston)

workings below the bearing stratum for the piles. The gross loading of 135 kN/m^2 was reduced to a net value of 53 kN/m^2 and the buildings settled 40 to 60 mm in three years, which was an acceptable amount of movement.

At first sight, the buoyancy raft would appear to be an attractive and economical solution where buildings are sited on deep compressible soils. However, constructing such foundations can be difficult since the excavation can be accompanied by heave of the base. The latter movements can be catastrophic if excavations are taken below the critical depth for stability[7]. Earth pressure on the supports of a deep excavation in soft clay can be high resulting in yielding of the supports and settlement of the surrounding ground surface. A further disadvantage is that the hollow substructure cannot be put to a useful purpose. In order to act efficiently in balancing the superstructure loading the raft must be light yet stiff enough to prevent differential settlement. This requires a multi-cell structure with thin-walled cells, which is unsuitable for the storage of goods or for parking cars. There are further problems of preventing the accumulation

of explosive gas in the cells, and of preventing leakage of ground water i
structure thus reducing its buoyancy.

BASEMENTS

Basements are frequently used as foundation structures because of functional requirements in a building, say for storage, car parking, or the accommodation of industrial process plant. Deep excavation is involved which can cause construction problems in difficult ground. Examples of difficult conditions for basement construction are:

1. Soft clays extending below foundation level.
2. Water bearing granular soils and rocks.
3. Impermeable soils underlain by water bearing layers under artesian pressure.
4. Soil containing cemented layers, boulders, or other massive obstructions.
5. Existing structures, highways, or buried services close to the margins of the excavation.

Construction techniques are available to deal with all these difficulties and there is little that can be done in the design of a basement to reduce the severity of the problems they can cause[8]. Simplicity of construction is of course desirable, avoiding congested reinforcement, awkward shapes (involving intricate formwork), re-entrant corners and openings for pipework through external walls.

A circular basement is advantageous since the excavation support can consist of a simple ring beam waling at one or more levels, thus avoiding the need for cross-strutting which obstructs the working space. It is possible to adopt a circular form for many industrial structures such as pumping stations or conveyor pits. Ring beam walings can be designed in reinforced concrete and incorporated in the permanent substructure.

Diaphragm walling or contiguous bored pile walls combining support for the excavation with the permanent structural retaining walls can be advantageous in weak ground where existing structures are close to the excavation. However, they are not economical for open ground conditions compared to conventional retaining walls constructed within a sheet piled excavation, particularly if the sheet piles can be extracted for re-use[9]. Also, the construction of guide walls for the diaphragm wall excavation can be difficult in areas of existing buried foundations and services. The adoption of 'silent' methods of pile installation either by hydraulic rams or by enclosing the pile hammer in a sound-proof box has done much to alleviate problems of noise and vibration associated with sheet pile construction.

Difficulties with excavation in water bearing ground can be eased by provision in the design for drainage, e.g. by a layer of 'no-fines' blinding concrete over the base of the excavation, and by allowing adequate working space for a perimeter drain around the basement structure.

Possible damage to adjacent buildings can be avoided or minimised by utilising the floors[10] in a deep basement as supports for the excavation. The floor slabs are cast on the ground and excavation is taken out from beneath them and removed through an opening left for this purpose (Figure 18.11). If this method is to be used economically the floors should be at a uniform level across the

Figure 18.11 Excavating beneath the floor of a basement, Victoria Street, London (Wimpey News photograph)

width of the basement. Stepped floor levels can cause considerable problems of transferring horizontal thrust across the width of an excavation.

PILED FOUNDATIONS

There is a wide variety of types of pile or piling equipment which can be used to overcome difficult ground conditions[11]. Examples of these are given in the following paragraphs.

1. Loose water bearing granular soils

It may be impossible to drive solid section piles through these soils because of 'tightening' of the ground when piles are arranged in large groups. However, small displacement piles such as steel H-sections or tube piles can be adopted. Bored and cast-in-place piles can also be used but precautions are necessary against loss of ground causing settlement of adjacent structures. This can be avoided by drilling for the piles under a bentonite slurry or by filling the pile borehole up to ground level with water.

2. Ground containing boulders

Any form of driven pile may be unsuitable for ground containing large boulders or other buried obstructions. Large diameter bored piles of the type installed by grabbing inside a casing can be used. An oscillator is used to drill down the casing which can cut through concrete or weak rock and a chisel can be used in conjunction with the grab to break up the obstructions within the casing.

3. Ground containing layers of very dense or cemented soil

Problems of driving piles through dense or cemented soil layers can be overcome by using steel sections strengthened as necessary at the toe.

4. Very deep compressible soils

Problems with handling long piles can be avoided by adopting assemblies of jointed precast concrete sections. They can consist of solid hexagonal or square piles with a mechanically-locked joint, or cylindrical shells assembled over a mandrel which is withdrawn on completion of driving.

5. Piling close to adjacent structures

Problems caused by the effects of ground heave, or loss of ground, or vibrations on existing structures sited close to pile groups can be overcome by adopting bored and cast-in-place piles drilled under a bentonite slurry. One proprietary system employs a continuous flight auger with a hollow stem for relatively vibrationless drilling. The pile shaft is formed by pumping a sand-cement mix down the hollow stem while withdrawing the auger.

References

1. Penman, A.D.M. and Godwin, E.W., 'Settlement of experimental houses on land left by opencast mining at Corby,' *Proc. Conf. on Settlement of Structures, Cambridge,* Pentech Press, 53–61 (1975)
2. Various authors, 'Symposium on ground treatment by deep compaction,' *Geotechnique,* **25**, 1, 1–94 (March 1975)
3. Samuels, S.G. and Cheney, J.E., 'Long-term heave of a building on clay due to tree removal,' *Proc. Conf. on Settlement of Structures, Cambridge,* Pentech Press, 212–229 (1975)
4. Walsh, P.F., *The design of residential slabs-on-ground,* CSIRO, Australia, Tech. Paper No. 5, 2 ed (1975)
5. Bjerrum, L., 'Engineering geology of normally consolidated marine clays as related to the settlement of buildings,' *Geotechnique,* **17**, 2, 83–117 (June 1967)
6. Somerville, S.H. and Shelton, J.C., 'Observed settlement of multi-storey buildings on laminated clays and silts in Glasgow,' *Geotechnique,* **22**, 3, 513–519 (September 1972)
7. Bjerrum, L. and Eide, O., 'Stability of strutted excavations in clay,' *Geotechnique,* **6**, 1, 32–47 (March 1956)
8. *Design and construction of deep basements,* Inst. Struct. Eng. (1975)
9. Puller, M.J., 'Economics of basement construction,' *Proc. Conf. on Diaphragm Walls and Anchorages,* Inst. Civ. Eng. London, 171–179 (1975)
10. Hodgson, F.T., 'Design and construction of a diaphragm wall at Victoria Street, London,' *Proc. Conf. on Diaphragm Walls and Anchorages*, Inst. Civ. Eng. London, 51–56 (1975)
11. Tomlinson, M.J., *Pile design and construction practice*, Viewpoint Publications (1977)

Chapter 19

Piling

Piling is often associated with difficult ground conditions, but not exclusively so. With the increasing loads of modern buildings and engineering structures piling is often the most economic solution to the foundation problem, even in ground of relatively good bearing characteristics, where its primary function is to reduce settlement to within tolerable limits. There are cases when piling is conventional, and not needed. Despite important developments in theory applied to piling, which has occurred over the last two decades, it is still basically an art where Terzaghi's 'thinking with the hips' is essential. It is very much grounded in the knowledge and experience of numbers of specialist engineers, keymen and operatives employed in the piling industry.

TYPES OF PILE

Types of pile may be most rationally defined by the effect of the installation on the soil. From this standpoint there are three main types of pile in common use as defined in CP 2004: 1972[1].

1. Displacement piles

These consist of solid piles including timber and reinforced precast concrete, and steel and concrete tubes, closed at the lower end by a shoe or plug, which is either permanent or extruded to form an enlarged toe. They are normally driven. Displacement piles may be further sub-divided into preformed piles and cast-in-place piles. A preformed pile is either solid as in the normal precast concrete pile, or hollow, as in the case of concrete tubes driven with a permanent shoe. The latter are normally filled with cast-in-place concrete. For cast-in-place piles, a tubular void is formed by driving a closed end of tubular section into the soil, the tube being subsequently withdrawn as the void is filled with concrete. Displacement piles are normally referred to as driven piles.

2. Small displacement piles

These consist of driven steel sections including H piles, open-ended tubes, box piles and screw piles. Sometimes the ground is removed from open-ended tubes and replaced by concrete.

3. Non-displacement piles

These are formed by boring or excavation, and filling the resulting void with concrete. In many instances, either permanent or temporary support is given to the sides of the excavation, usually by casing, sometimes, for temporary support, by drilling mud. Such piles are normally referred to as bored cast-in-place piles.

The CP2004 classification has been somewhat amended (see Figure 19.1) in CIRIA/DOE Report PG1 (Weltman and Little[2]). Weltman and Little include small displacement piles with displacement piles.

There are many specialist systems of driven and bored cast-in-place piles, some of which are proprietary. While there are special features of particular systems which may argue in their favour, the proprietary features are often minimal and the experience and engineering judgment of the specialist contractor and his men are of primary importance. Reduced to the simplest terms in these systems a hole is formed in the ground by driving or boring, and subsequently filled with concrete. Many engineers tend to rely on a performance specification for driven piling systems, particularly those with enlarged bases, and on the contractor's judgment, integrity and guarantee. This is not surprising bearing in mind the shortcomings of almost every pile driving formula except in particular local conditions of which the specialist contractor may have knowledge. There are few proprietary features with most bored piling systems, and in this field the engineer more often than not designs the piles, leaving the contractor responsible for workmanship and materials. The essential features of most systems are described by Weltman and Little[2].

CHOICE OF PILE

There are three principal factors determining the type of pile to be used on a job; the environment, the ground conditions and economics.

ENVIRONMENT

When piles are to be constructed on an urban site surrounded by other buildings and structures there may be considerable objection to driven piling. This is because of the possibility of damage to neighbouring structures by vibration and the high noise which is unacceptable to occupants of buildings. Such circumstances may point to the use of bored cast-in-place piles, particularly augered piles, which may be constructed without appreciable vibration, and generally with a lower noise intensity. On open sites clear of other development, there are not the same objections to driven piles and the type of pile is dictated by the ground conditions and economics. The most intractable situations arise when the ground conditions demand a driven pile but environmental restrictions reject it. Nevertheless, the adverse effects of vibrations and noise have been exaggerated, and piles driven under carefully controlled conditions do not necessarily give rise to harmful vibrations. Displacement piles driven by internal drop hammer cause much less noise than piles driven by external hammer.

The use of vibrators and other proprietary devices to reduce the noise in driving sheet piles has done much in recent years to answer objections to driven

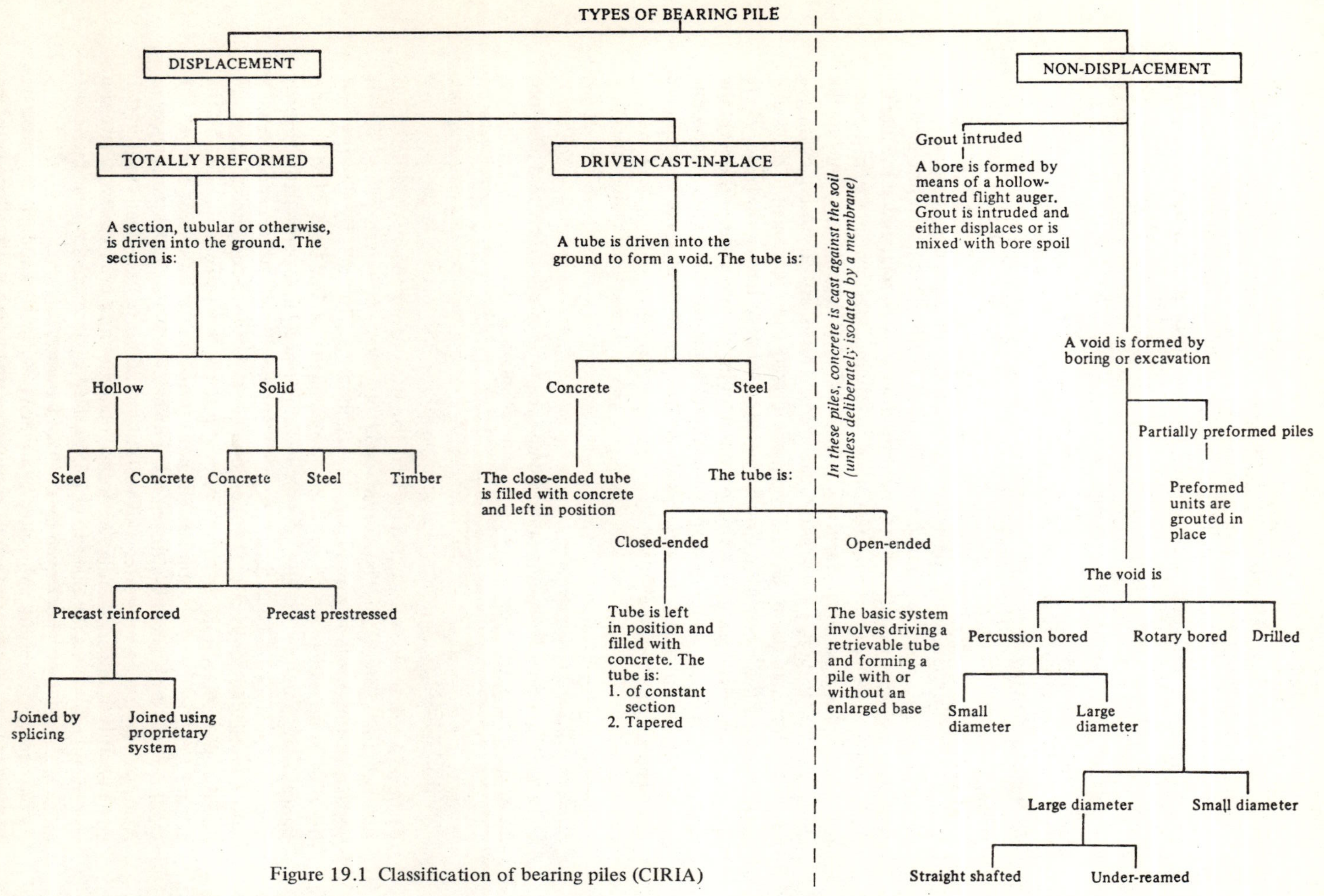

Figure 19.1 Classification of bearing piles (CIRIA)

piling on such sites. However, it must be admitted that many objections to noise are extremely subjective; cases have arisen where attention to lubrication on piling machines have eliminated unacceptable components of noise such as screeching and squealing to the residents' satisfaction but the noise level has been unaffected. Noise level determinations (see Beaman and Jones[3]) whilst useful are by no means the answer. Residents may prefer a noisier but quick piling job to a more protracted one satisfying Local Authority requirements. The community must decide; with the present trends the cost of piling work inevitably will be increased considerably as a result of present legislation.

In cases where piling gives rise to vibration in the ground communicated to neighbouring structures, great care is needed when resonant conditions are approached. However, in many instances it can be shown that vibration due to passing traffic is frequently greater than that due to pile driving. This is again another instance where subjective judgments are frequently made.

GROUND CONDITIONS

The ground conditions are a major factor in determining the choice of type of pile. Generally a predominance of clay strata favour construction of bored piles (non-displacement), granular water bearing strata favour some form of driven (displacement) pile. There are many intermediate conditions where the choice is not so obvious. The choice may be difficult when soil conditions dictate a driven pile and the environment a bored pile. On sites where such factors do not indicate one or the other, the choice will be made on economic grounds.

Where soft or loose alluvial water bearing deposits overlie a firm bearing stratum, such as sandy gravel, marl or rock, driven (displacement) piles are most frequently the best solution. Piles can be driven easily through soft or loose alluvial deposits, and driving may well increase the compactness (and hence bearing qualities) of loose or medium/dense granular strata into which they are driven.

Bored non-displacement piles require casing through soft or loose alluvial deposits and may cause a deterioration of the bearing qualities of granular strata in which they are founded. This is because of disturbance by the action of boring tools and upward seepage pressures. As a general principle bored piles should not be employed without special precautions where the toe level is in water bearing granular soils. Such special measures include drilling under bentonite drilling mud, which over the last decade in particular, has been increasingly and satisfactorily used in ground conditions which years ago would have seemed unsuitable for the installation of bored piles (q.v.).

Problems may occur when it is necessary to drive piles through a compact stratum, such as a cemented layer, or a stratum containing boulders, because the stratum beneath is weak or compressible. In such circumstances it may be necessary to drive piles to reach firm ground at much greater depth. Such problems may be accentuated where the piles are driven at very close centres through firm or stiff clays. This may result in ground heave because the ground cannot accommodate the volume of driven piles by increasing in density as with loose or medium/dense granular materials. The heavy driving necessary in these circumstances may result in breakage of piles. In some cases the difficulties may be overcome by the use of a bored cast-in-place pile when the ground conditions are suitable. In other cases where economics permit, small displacement piles, such as steel joists, open-ended tubes or fabricated box-piles may be employed. However, where none of these methods are suitable, it may be necessary to prebore through the compact or stiff strata and pitch the piles into the pre-bored holes before driving.

In firm to stiff clay deposits bored or augered non-displacement piles, where the boreholes are stable without casing, are usually most economic. They remain so, even with thicknesses of water bearing over burden up to 10 to 13 m in depth by using temporary casings to seal off the water-bearing over-burden from the boring.

ECONOMICS

Engineers must look at each site with an open mind, asking themselves if the proposed piling method is suited to the ground conditions, or indeed if piles are needed at all. Where on technical considerations a driven pile, for example, would appear preferable to a bored pile, say in deep overburden of soft and poorly compacted deposits, a cheaper tender from a bored piling contractor should be looked at critically. The cheapest tender, as many know to their cost, does not necessarily lead to the most economic job, and the incidence of possible claims and delays to the programme should be properly evaluated. Whether the employer or contractor foots the bill matters little; no credit is due to contractor or engineer on a job improperly conceived and poorly carried out. Unfortunately, however, economic pressures are such that technical considerations are often discounted because of an apparently favourable tender.

It is often in the best interests of the employer for the piling contractor to be nominated by the engineer from a limited list of tenderers known to be reliable by the engineer. Regrettably in the present economic climate, because of the alleged difficulties of nomination, piling contractors are selected by the main contractor by methods of selection more appropriate to the market place than to a soundly conceived engineering project.

At the same time careful consideration should be given to new methods being introduced such as bentonite piling permitting the extension of bored piling into the field formerly covered by the driven pile.

DISPLACEMENT PILES

The plant and techniques for driving piles are well described in the literature (Chellis[4], Weltman and Little[1]) and only incidental reference to these is made below.

TIMBER PILES

Timber piles have been used for centuries and, provided they are not exposed to tidal variation and the attack of marine borers, they are very durable as has been shown in a number of cases by the excavation of Roman piles. Protection against rot and attack from marine borers is usually given by the use of chemical preservatives and the choice of hardwoods, such as greenheart and ekki is also of assistance. Today, timber piles are chiefly used for marine structures, particularly for temporary works providing great resilience. Timber piles are generally used in square sawn sections, but in Scandinavia, for example, it is frequently the practice to drive whole tree trunks without preparation.

PRECAST CONCRETE PILES

Until the last decade the most widely used type of pile has been the reinforced concrete pile, usually of square sections. More recently prestressing wires have replaced bar reinforcement (see Figure 19.2). The purpose of reinforcement is

Figure 19.2 Prestressed pile casting yard, Anchor project, Scunthorpe (British Steel Corporation/John Mowlem & Co Ltd)

primarily to resist bending stresses when being lifted and transported. Except on the largest projects piles are made in a factory and transported to the site by road.

As with all preformed piles the chief advantages lie in the high degree of control that can be exercised in manufacture, and the avoidance of problems of bad workmanship which sometimes occur with in-place systems. However, piles can be damaged in driving without it being realised. There are also great advantages in the use of precast concrete, despite high initial cost, in parts of the world where it is difficult to exercise good control over workmanship of in-place systems. Precast concrete piles are most economical when the bearing stratum is at uniform depth below working level, and for marine works (Figure 19.3).

The chief disadvantages are the expense of cutting piles and wasting excess lengths, or having to lengthen them when the standard lengths are too short.

STEEL PILES

Steel piles are widely used in marine works, where their great strength permits heavy driving and is suitable for resisting horizontal impact loads on marine

Figure 19.3 Driving octagonal prestressed concrete piles with Menck frame, Beckton (Peter Lind & Co Ltd)

Figure 19.4 Rendhex and high tensile tubular piles, Coryton (Peter Lind & Co Ltd)

structures. A distinction should be made between heavy sections used for such works and lighter sections, often of spirally welded tubes hearted up with concrete for use as bearing piles. Common sections are octagonal box piles, Rendhex, tube piles and steel H piles (Figure 19.4).

Steel piles are often driven open-ended but sometimes they have a shoe or plug of dry concrete. The driving of open-ended piles frequently causes a plug to form and the driving is then similar to a displacement pile with a shoe or plug. Least resistance to driving occurs with H piles. Piles are either top-driven or bottom-driven with an internal hammer or mandrel. In some cases powerful electric or hydraulic vibrators are used to drive them.

The advantages of tubular steel piles include the possibility of internal inspection after driving. Steel piles also can be shortened or lengthened without waste and the heavier sections are well capable of acting as columns in compression, tension and bending. They can often be driven at closer centres than other preformed piles.

Whilst of high initial material cost, steel piles can often be the most satisfactory solution for piling in difficult ground conditions, with freedom from errors of workmanship in construction. For this reason thousands of tube piles are driven every year in areas overseas where adequate supervision of in-place work can be very difficult to ensure.

DRIVEN CAST-IN-PLACE PILES

In most systems an open-ended steel tube is driven into the ground with a plug or shoe at the toe. In the Franki or Dowsett systems aggregate between ½–1 m deep is placed at the bottom of the tube before driving and is then compacted by an internal drop hammer. Successive blows on the compacted plug because of the friction developed cause tube and plug to enter the ground until the required depth is reached. Further penetration of the tube is then effected by restraining the tube at ground level and the plug is driven into the ground by the internal hammer by adding successive charges of semi-dry low slump concrete. This tends to form a concrete bulb at the toe of the pile. Reinforcement is then added and the temporary driving tube is withdrawn as successive charges of concrete are placed (Figure 19.5).

In some systems detachable steel driving shoes are used in place of the plug. The tube is often driven with an internal mandrel and in the Alphapile system the shoe can be driven ahead and a bulb formed. Similar driving without bulb formation is adopted in the Simplex system.

In other systems the steel driving tube is left in place as a permanent casing after completion of the pile. In the BSP cased pile the driving tube is commonly 8 mm thick. In the Raymond system a ribbed light gauge permanent casing is driven using an internal expanding mandrel.

There are many variants offered by other specialist contractors and manufacturers including GKN Keller Foundations Ltd and BSP International Ltd, and the reader is referred to the literature, particularly Weltman and Little[2] for a more detailed survey.

SHELL CONCRETE PILES

These piles are formed by driving coupled lengths of precast concrete shells with a shoe. Typically in the West system shells are 915 mm long and 50 to 70 mm wall thickness, reinforced with polypropylene fibre. The shells are threaded onto a steel mandrel extending to the shoe and are jointed with plastic compound and recessed steel bands. They are top driven via the mandrel. After driving the core can be inspected and reinforcement may be added and the core hearted up with *in-situ* concrete. Surplus shells are removed after driving and used again. They are driven in diameters from 380 to 610 mm up to 75 m deep. For lighter loads a grouted segmental variant of the system is used. The system is well tried and

Figure 19.5 Crawler driven cast in place rig (Frankipile Ltd)

suitable for a variety of ground conditions, but it may not be suitable in very hard driving conditions, although this may be overcome to some extent by pre-boring (Figure 19.6).

SEGMENTAL CONCRETE PILES

Over the last decade a number of systems have been developed, particularly in Scandinavia, to overcome the problem of waste lengths of precast concrete piles. These use precast lengths of pile ranging from 1.3 to 4 m and in section width from 220 to 480 mm, with steel jointing pieces. Square or hexagonal cross sections are used. Joints consist either of a bayonet type or spigot and socket with locking pins. The precast segments are factory made using high grade concrete and high tensile reinforcement. Specially hardened rock shoes or pins are available for hard driving and lengths of up to 100 m can be driven. These systems are now widely used in the UK and throughout the world (Figure 19.7). They include systems operated by Balken Piling Ltd, GKN Keller and West's.

Figure 19.6 Shell piling rig (Wests Piling & Construction Co Ltd)

Figure 19.7 Segmental concrete piling rig (A. Johnson Construction Co Ltd)

HOLLOW CYLINDER PILES

Sometimes, especially in marine conditions where piles are not required at close spacing large diameter concrete shell piles, from 900 to 1400 mm diameter and 100 to 125 mm wall thickness, are used. Such piles are cast with longitudinal ducts for post tensioning the segments together. They are suitable in marine conditions, because of their great bending strength and resistance to corrosion. They have been driven up to 60 m depth to carry loads of 200 tonnes or more. Driving shoes are employed as necessary. They suffer to some extent when the bearing stratum is of variable depth and excess lengths have to be cut off.

NON-DISPLACEMENT PILES

BORED PILES

The distinction between large and small diameter piles is largely an artificial one; CP 2004 defines a large diameter pile as being 600 mm in diameter or over. However, tripod rigs employing percussive means of boring are used up to 600 mm in diameter and mechanical rotary auger rigs carry out much work in

Figure 19.8 Tripod rig (Soil Mechanics Ltd)

diameters of less than 600 mm diameter. The factors determining whether tripod or mechanical auger rigs are used are:

(a) ground conditions; the tripod rigs have the facility of using jointed casing which is readily adaptable to sudden changes in the depth of unstable ground;

(b) working space and headroom; tripod rigs are easily adapted to working in confined spaces;

(c) the size of the job; the cost of bringing heavy plant on site used by auger rig contractors tends to make the tripod method more economic on a small job.

In the large diameter range highly specialist plant such as the Bade & Benoto percussive rigs are occasionally used in special applications.

PLANT EMPLOYED AND CONSTRUCTION METHODS

Tripod rigs

Most rigs employed are basically the same, being a tripod about 4 m high with an air or diesel driven winch mounted on one leg and employing in the main cable-operated droop tools, clay-cutters, sand pumps or bailers, and heavy chisels to penetrate obstructions and rock (Figure 19.8). Square threaded short lengths of easily connected robust temporary casings are used to support unstable ground as boring proceeds, and heavy multi-reeved lifting tackle or jacks are used to extract casings. It is often the practice to pun the base of the hole where it is left rough by the boring tools.

Rotary small diameter rigs

During the 1960s a few contractors introduced mechanical rotary auger rigs specially designed for small diameter work, capable of handling up to 9 m of casing. Leaders are mounted on a short stub boom to a crawler crane, the drilling being by hydraulic motor driven from a power-pack mounted on the back of the base machine (Figure 19.9).

Figure 19.9 Rotary small diameter rig (Soil Mechanics Ltd)

Heavy duty rotary auger rigs

Although primarily introduced for large diameter work these are frequently used for small diameter work on big contracts. They are capable of superlative speeds of drilling in suitable ground, certainly of the order of 30/45 m per hour; or more than a 1.5 km of 600 mm diameter piling per rig per week in optimum conditions. Auger rigs are either mounted on a crane or a lorry. Lorry-mounted rigs have obvious advantages in being cheaply transported and set up on a site. However, their popularity has declined somewhat in favour of excavator crane (base machine) mounting because of ease of moving the latter on poor site surfaces, better output and robustness of crane rigs, and advantages of not tying-up capital in a piece of specialist machinery. An excavator crane can be used for a variety of purposes apart from piling.

The most usual power unit employed on crane rigs is a diesel engine mounted forwards of the crane, driving a ring gear through a torque converter, sometimes with a chain drive interposed. A kelly passing through the ring gear is used to transmit the torque to the drilling tool. The kelly is driven either through a yoke and drive-sleeve located within the ring gear (Calweld) through a roller nest (Watson, see Figure 19.10) or a combination of the two (Soil-Mec, see Figure 19.11). In the Watson the kelly passes freely through the rollers and the drilling

Figure 19.10 Watson rig drilling 2.5 m dia. piles

action is dependent solely on the weight of the kelly, whereas in the Calweld and Soil-Mec, the friction between the drive sleeve and the kelly is used to provide positive feed-down by means of hydraulic crowd rams. Similar rigs are manufactured by BSP International.

Kellys are normally telescoping, permitting drilling depths normally up to 30 m, or in the case of the quadruple Soil-Mec kelly to 60 m or more, depending on the size of the base machine. The Calweld uses a circular outer kelly and inner kelly of square cross section. The Watson employs square cross section kellys throughout and the Soil-Mec uses circular sections throughout.

In operation the kelly and auger are lowered into the hole, and the kelly is rotated to dig and fill the auger with spoil. The too is then raised to the surface and the spoil discharged, the process being repeated until the required depth is reached.

Figure 19.11 Soil-Mec rig

Another type of rig is the *McAlpine hydraulic rig.* The power unit for this is a diesel engine and hydraulic power pack mounted at the back of the base machine on a platform. It drives four hydraulic/hydrostatic motors mounted around the ring gear through rigid and flexible hydraulic pipework. A useful feature of the McAlpine rig is the ring gear which can be moved up and down the vertical boom incorporated in the rig, thus allowing considerable flexibility when using casings extending significantly above ground level.

The most used drilling tool is the flight auger. The diameters of flight augers vary from 375 mm to 2500 mm or more. The spoil is discharged from the auger by reversing the auger at a fast speed. The critical factor is the acceleration to the spin-off speed and the Watson, of all rigs, is pre-eminent in the speedy discharge of spoil. In drilling into weathered rocks or even tougher material, heavily-armoured augers are used, but in general the capability of such rigs is confined to drilling down to a firm seating on top of hard rock.

In bad ground when drilling under water or drilling mud, drilling buckets with retention flaps are often used to bring out the spoil. The buckets are then discharged by bottom opening which is automatically operated when the bucket reaches the proper elevation. Further reference should be made to the paper by Palmer and Holland[5].

Underreaming buckets

Underreaming tools which expand the base of a pile up to three times the shaft diameter are commonly used by auger rig contractors. The largest in use in the UK will enlarge an 1.8 m shaft to 5.5 m (Figure 19.12). This requires a heavy

Figure 19.12 5.5 m dia. under-reaming bucket (Lind Piling Ltd)

duty rig with the ring gear mounted some 6 m above the ground to accommodate the huge bucket. Such capability is at present limited to 2 or 3 contractors.

Casings for rotary rigs

Casings from 10 to 20 mm thickness of welded construction are normally used in poor ground in one continuous length. This necessitates cutting or welding when the depth of bad ground varies over the site. The length of casing employed is normally practically limited by the lifting crane capacity and extractive force necessary. Where the ground is stable it is only necessary to use a short length of casing at the top.

Benoto and Bade systems

The Benoto and Bade rigs are employed in Europe and elsewhere, employing specially jointed double-walled casings, cable-operated boring grabs, and hydraulic casing driving and extracting machines. They are essentially a me nised version of the tripod rig for large diameter work, and are useful in very ground. The hydraulic casing machines rotate the casings through several de in the horizontal plane and back again, and at the same time are capable of giving positive downthrust or uplift when extracting casings. Special cutting shoes are used for final penetration into the bearing stratum.

Figure 19.13 Bade rig (CET (Equipment) Ltd)

These machines have not found much favour in the UK for a number of reasons:

(a) the large amount of capital tied up in one rig with associated casings and tools;
(b) the relatively slow speed of drilling;
(c) the development of cheaper mud drilling techniques over the last decade.

The Benoto and Bade come into their own in very poor ground and in the construction of secant (interlocking) close bored piles (Figure 19.13).

VIBRO-DRIVERS

Development of electrical vibro-drivers for drilling and extracting casings has proceeded since the Russians used them in China on a bridge across the Yangste-Kiang at Hankow in 1955/57. In the main the German machines, Muller, Menck and Schenk have been employed using a system of rotating eccentric

Figure 19.14 Muller Vibro-driver

weights to import a downward thrust (Figure 19.14). The advantages of using vibro-drivers are:

(a) the casing is predriven and provides a stable support through which drilling may proceed rapidly, saving much time;
(b) overbreak, associated with other methods of installing casings, is avoided;
(c) internal vibration of concrete when extracting casings, ensuring sound construction.

The disadvantages are:

(a) the method cannot be used in stiff clays;
(b) the operating frequency of 20 to 30 Hz is very often in the range of the ground and structural frequencies, inducing resonant conditions which are at best psychologically and physically unacceptable, and at worst dangerous to nearby structures;
(c) at least a 40 tonne crane is needed, adding to congestion on site;
(d) electrical power demand of at least 125 to 250 kVA is needed. A preferred way of operating such vibro-drivers is to mount the generator on the back of the crane so that it can move readily from place to place on site.

BSP International have a hydraulic vibro-driver working hydraulically and permitting instantaneous variation of the driving frequency so that resonance with nearby structures can be avoided. Such a development is encouraging but it seems likely not to be used very much on piling for some time because of the high capital cost.

OTHER TYPES OF NON-DISPLACEMENT PILES

In the Prestcore system the boring is carried out by shell and auger using a tripod rig and casing as necessary. Precast concrete annular rings are assembled and are lowered into the borehole, longitudinal reinforcement added in the core, and the core and annulus between ground and precast rings grouted. Pile diameters range from 400 to 500 mm.

Proprietary systems of grout-intruded piles have been used of late. The holes are drilled by a hollow continuous flight auger in one pass. In one variant grout is pumped under pressure down the hollow stem of the auger to replace the load of spoil as the auger is withdrawn from the hole. Reinforcement is added after removal of the auger whilst the grout is still fluid. In another variant applicable to granular soils grout is mixed with the soil *in situ* by reversing the auger as it is withdrawn. Pile diameters vary from 300 to 450 mm and they can be constructed to a maximum depth of 25 m.

BORED PILES IN ROCK

Particularly in marine structures and in others where high bearing capacity and lateral resistance are required there is an increasing requirement for piles to be drilled many metres into hard rock. A limitation of the percussion and rotary bored piling systems previously described is that practical refusal to penetration occurs when rock is met. With tripod rigs chisels are used to obtain penetration but only to a limited degree. Auger rigs are often able to drill in weathered rock, especially chalk, but as soon as hard layers or sound rock is met then other methods are required to achieve penetration. Various methods have been tried with differing success. Rotary calyx boring can be used with perseverence but is slow and expensive.

Modern developments in bored piling follow, as in so many instances, the technology of the oil industry and various rigs are available for drilling by means of rock roller bits into hard rock. Water circulation is required to bring the rock cuttings back to the surface, either by direct circulation down the drill stem or by reverse circulation using airlift to bring the cuttings up the central stem. Vital to successful drilling is the design of the bit, often purpose-made for the job, and the deadweight of the tools and drill stem (Figure 19.15). Holes 2 m in diameter have been successfully and expeditiously carried out in UK by such methods.

A recent development from the Alaskan pipeline construction is a pneumatic down-the-hole rock drill marketed by Ingersoll-Rand. It is claimed that holes can be drilled up to 750 mm diameter at speeds of the order of 3 to 6 m per hour. The drill includes a button bit incorporating spherically-shaped tungsten carbide inserts. Air consumption is up to 70 m^3/min. The assembly is carried by

Figure 19.15 Rock roller bit (Peter Lind & Co. Ltd.)

leaders fitted to a crawler excavator machine, the drill being turned at 12 to 15 rpm by a hydraulic rotary head. Alternatively it can be operated in conjunction with a rotary auger rig ring gear.

CONSTRUCTION PROBLEMS

Plant breakdowns

Plant should be maintained in good working order. The construction of a pile may be endangered if the plant breaks down part way through the construction. It is not practicable to avoid plant breakdowns altogether, but well organised jobs will have contingency plans to meet many circumstances. For example, on a critical night possession of a railway line, a spare rig should be kept in readiness.

Clearing obstructions

One advantage of large diameter piling is that many obstructions can be cleared using a heavy duty auger. Also it is often possible to descend the hole in order to break up obstructions with pneumatic tools. In small diameter bored piling obstructions are usually cleared by percussion chisel and progress can be very slow, causing costly delays. In driven piling obstructions have to be broken out in advance of driving or on occasion piles are resited to avoid the obstructions.

CAST-IN-PLACE PILES (BORED AND DRIVEN)

A very comprehensive review of the problems in construction of cast-in-place piles has been given by Thorburn[6].

Concreting

This is the most critical operation in cast-in-place piling, particularly where temporary casings are used. It is essential to use concrete of high workability with a cement content of not less than 300 kg/m^3 and a slump of not less than 100 mm. Additives may be necessary to avoid bleeding and segregation. Particularly where ready mixed concrete is used a piling mix must be specified similar to that recommended by the FPS (Federation of Piling Specialists) and BRMCA (British Ready Mixed Concrete Association[7]). Close liaison is necessary to ensure the continuity of supplies. The time of mixing must be stated on the delivery ticket and the concrete must be placed and casings extracted within 2 hours of mixing, or even quicker where hot weather accelerates the setting time. Where circumstances demand a retarding additive should be used (Figure 19.16).

Figure 19.16 Concreting 2.5 m dia pile

In concreting under water the concrete should contain not less cement than 400 kg/m^3 and the slump should not be less than 150 mm. It is important that the concrete should not be allowed to mix with water and the operation should be carried through a tremie tube continuously until completed. The precautions noted in CP2004[1] for tremie work should be rigorously observed.

The use of concrete pumping has also been very successfully adapted to place concrete under water. The use of bottom-dump skips, except in very carefully controlled conditions, is not recommended. The skip can open early or late allowing the concrete to mix freely with the water in the borehole.

Segregation often occurs when concrete is carelessly tipped in a dry pile hole because the fall of the concrete is impeded by the reinforcement cage. In these circumstances it is good practice to use a hopper and a sufficiently long length of tube to direct the flow of concrete. Where there is any doubt it is an old bored piling trick of the trade to empty a bucket of grout down the hole before concreting as this is absorbed on the reinforcement bars preventing any further loss of grout.

Emphasis has been placed on designing a workable self-compacting mix of concrete. It is not normally advisable to compact the concrete using poker vibrators because this tends to fix temporary casing tubes in the ground, making extraction difficult or impossible.

Casing extraction

Before and during the casing extraction the head of concrete should be maintained well in excess of the ground water pressure. Casings should be maintained scrupulously clean and free from adhering concrete and clay so that arching does not occur. Extraction of casings should be at a steady controlled rate and without sudden shock loads such as are sometimes applied by snatching with a crane. The practice of 'standing a crane on its toes' is highly dangerous to the crane, the men on site and to the pile. The extraction of casings can be assisted using a mechanical extractor with pulley tackle, or using a vibrator. Checks should be made to ensure that the concrete is slumping properly as the casing is extracted.

Unless these precautions are rigorously observed it is certain that faulty construction will sooner or later occur, causing 'waisting' or complete breaks in the continuity of the pile.

Use of permanent casings

The use of permanent casings should be considered in works carried out through soft alluvium or near tidal rivers where there may be danger of loss of concrete into soft peaty layers or of cement being leached out of the concrete soon after placing. Permanent casings were specified at a large power station near the Thames where the piles are cased through the alluvium and river gravels to a depth of some 20 m, the toe of the piles being some 25 m below ground level, in chalk.

In another instance piles constructed through Coal Measures at Dudley, Worcs, were founded on hard rock about 13 m deep. However, it was known that unrecorded old mine adits existed at various levels above founding level, and there was a risk that the concrete placed in piles might flow into such adits leaving concrete placed at higher levels 'hung up', so that there would be unsuspected gaps in the piles. The risk was obviated by placing light gauge casings (18 SWG) as a permanent lining to the piles (Figure 19.17).

A further example occurred on a piling operation adjacent to the underground railway near Lord's cricket ground where it was desirable to ensure that no frictional loads was transferred from the large diameter piles to the tunnel. The

Figure 19.17 Permanent casing to 1050 mm dia pile

solution here was to install a light gauge permanent casing to the required depth covered with a 7 mm thick coat of special bitumen to allow slippage.

Reinforcement

Reinforcement for cast-in-place piles, except where necessary to resist lateral loads in bored piles, is generally nominal. Full length cages are often not necessary for large diameter piles, short splice bars at the top being sufficient. In cases where unstable overburden is cased, the cage, if not inserted after extraction of the casing, should extend a metre or two longer than the bottom of the casing to anchor it during extraction. Otherwise the cage may well be lifted as the casing is extracted.

Cages shown on engineers' drawings are often not structurally stable if built as shown. In such cases the contractor should ensure that sufficient extra steel is added to make the cage into a structurally handlable unit. If this is not done at worst the cage will collapse and injure, maim or kill somebody, or at best will 'snake' when lifted by the crane and be difficult to place in its required position in the hole.

The present practice not to measure strengthening steel is not only inequitable but is an encouragement to unworkmanlike and dangerous practice.

DRIVEN PILES

Heave

In granular soils the effect of driving displacement piles is for the soil to be compacted and to provide better bearing characteristics. In cohesive strata where the soil cannot change in volume the clay is displaced laterally and vertically and may lift or break piles already installed. Whereas friction piles may not be adversely effected, lifting of piles which are mainly end-bearing can give rise to excessive settlements.

Where ground heave occurs it will be necessary to check the levels of the pile tops before and after adjacent piles have been driven. Risen piles which depend largely on end-bearing will require to be re-driven. This means costly moving and re-setting of the piling rig.

Figure 19.18 Hands-England auger mounted on Delmag frame for preboring (British Steel Corporation/John Mowlem & Co Ltd)

It is often the practice in these conditions to prebore piles and pitch the piles into the prebored holes and drive to final refusal (Figure 19.18). This may also be necessary in granular soils, which are very compact before pile driving. Various rigs have been used for this purpose, rotary auger rigs commonly being employed, sometimes using very long flight augers.

Problems can be encountered in driving piles in closely spaced groups. The order in which piles are driven depends on the system of piling employed and may be affected by the requirement that the driving of an individual pile shall not damage an adjacent one. If the piles are cast-in-place then the age after driving is relevant. In some cases it is considered better to drive adjacent piles before the concrete has set rather than to risk fracture of newly set concrete.

Damage of preformed concrete piles

The tensile forces set up in driving can cause cracking of concrete piles, particularly where the hammer weighs less than the pile. Prestressing is of some help in reducing cracking but even so cracking can be a problem when driving prestressed piles through soft strata. Increased hammer weight and lower drop is of assistance in reducing cracking.

Driven-in-place piles

In addition to the problems of heave and lifting difficulties can arise when driving through soft sensitive soils which are remoulded and lose strength during driving. In these conditions there is little lateral resistance offered by the remoulded soil surrounding the pile and damage can result when driving other piles in the vicinity.

BORED PILES

Bad ground

Waterbearing ground unable to stand stably without support requires the adoption of various expedients. With a tripod rig the casing may be extended as boring proceeds but can give problems where the casing has to be taken to great depth involving turning and hammering, so that the subsequent extraction is rendered difficult or impossible.

With auger rigs the usual technique is to dig through the unstable ground to the full depth entering the casing subsequently. Sometimes the ground is sufficiently stable to permit this, but on other occasions it will collapse before the casing can be entered. In such cases a casing can be 'mudded in' or the hole drilled under reverse head of water or bentonite. 'Mudding in' means mixing either dry bentonite or premixed bentonite mud with the unstable ground without actually removing spoil from the hole. The casing is then entered through the resulting stiff fluid mix, driven into the underlying stable stratum and the fluid mix removed after adding, for example, lime to stiffen it, so that it can be handled into muck lorries.

Alternatively a head of bentonite fluid is maintained in the hole in excess of the ground water pressure and the spoil continuously removed by a drilling bucket with flaps maintaining the stability of the hole by the bentonite fluid until the desired depth for entering the casing is reached. In some cases, apart from a short length of casing at the top, a casing may not be used, bentonite mud being the sole stabiliser.

Where an appreciable amount of bentonite drilling is involved it is desirable to recover the bentonite when the concrete displaces it, both for economic reasons and to avoid contamination of the site, drains etc. with mud. The mud is passed into tanks and the silt-laden fluid first passes to vibrating screens and then to settling tanks. From there it is passed through hydrocylones before returning to a storage tank or to the next pile. Such a cleaning system is usually operated in conjunction with the bentonite mixing machinery (Figure 19.19). Control of

Figure 19.19 Bentonite desanding plant

the quality of bentonite and other valuable information is discussed in papers given to the ICE Conference on Diaphragm Walls and Anchorages[8] and published by CIRIA[12].

One disadvantage of digging under bentonite is the overbreak over the nominal size of pile. This adds to the cost of the piling because of the extra concrete required, and is avoided by the preferred technique applicable in soft and granular soils, namely, the driving of the casing by vibro-driver.

Cleaning the base of the pile

Because the auger tends to leave some loose spoil at the base it is normally essential to clean the base of an augered pile with a special cleaning bucket. In extreme cases, particularly in weathered rocks, it is possible to compact loose debris into a solid mass at the bottom of the hole, indistinguishable on ordinary inspection from a sound bottom. This can be overcome by an over-zealous cleaning-up first with an auger and then with the cleaning bucket.

The matter is of particular importance with underreamed piles, and although with well designed underreaming buckets it is possible to clean the hole fairly efficiently, it is essential to send men down to clean loose spoil and soft pockets. Little faith can be put on inspection by television camera. An unscrupulous rig driver can reverse the direction of the belling bucket and present to all remote viewers a seemingly perfect bottom.

Inability to seal off ground water with casing

On a site in London near the Thames considerable difficulty was experienced in penetrating the casings down to the London Clay, some 12 m deep, because of heavy obstructions. Because of the heavy flow it proved impracticable in 4 out of the 72 piles to pump the water out and thereby to allow access for men to break up the obstructions. The expedient of freezing with liquid nitrogen was adopted, the nitrogen being circulated in a ring main around the borehole to U-pipes inserted to the depth where the inflow was occurring. In optimum conditions the ground was sufficiently frozen in 48 hours and it was then possible to get access to break out the obstructions.

Close bored pile walls

The technique has been described in detail by North-Lewis and Lyons[9]. The construction of continuous bored pile retaining walls can be a very economic solution when the ground is sufficiently stable to allow construction of piles without substantial overbreak. Where there is unstable overburden, unless suitable expedients are adopted, the unhappy piling contractor finds himself having to break through the obstruction of his own concrete.

The largest bored pile wall in Europe, if not in the world, was carried out for the Liverpool approaches to the second Mersey Tunnel. These were 2.4 m in diameter, nearly 200 in all, constructed through stiff boulder clay and weathered Bunter Sandstone to seat on hard sandstone (Figure 19.20). The piles formed both a retaining structure and the seating for a 30 m span reinforced concrete arch. The reinforcement cage in each pile weighed 7 tonnes and 50 m^3 of concrete was placed in each pile.

Figure 19.20 2.5 m dia. close-bored pile wall, Mersey Tunnel approaches

PILES TO RESIST LATERAL LOAD

The construction of raking piles, up to rakes of 1 in 3, is quite common. The tripod rig has the advantage of being able to case as boring proceeds, although this problem can sometimes be overcome with auger rigs by vibro-driving the casing through unstable overburden. This expedient was adopted at the Anchor Project, Scunthorpe, where nearly 1000 raking piles, 900 mm in diameter were constructed (Figure 19.21). Particular care should be given to concreting and a chute may be necessary for the full length of hole to prevent segregation as the concrete falls through the reinforcement cage.

Figure 19.21 Test load on 900 mm dia. raking pile (British Steel Corporation/Lind Piling Ltd/Economic Foundations Ltd/ Sir Robert McAlpine & Sons Ltd)

It must be accepted that in the uncased part of the hole a certain tendency to 'banana' is present. Tight specifications requiring the pile to be within 1/100 of the required rake are out of place in this technology; 1/25 would be a more reasonable tolerance and the designer should allow for this. In some instances heavily reinforced vertical piles may be used to resist lateral loads. Lateral loads of up to 80 tonnes per pile were allowed for in the foundations of a bridge on the Burnley Inner Ring Road, using flat-bottomed rails as reinforcement. In another instance, as an emergency measure, heavily reinforced 1.2 m diameter vertical piles were constructed to give lateral support to a bridge abutment which was moving because raking piles had been omitted in the design.

SPECIFICATION OF PILING

Specification of piling is regrettably often confused by imperfect understanding of the construction requirements for workmanship and materials and by clauses which are essentially conditions of contract or methods of measurement. A good specification is such, irrespective of the form of contract of method of measurement under which the work, is carried out. Detailed guidance on piling specification may be obtained from 'Piling: Model procedures & specifications' published by the Institution of Civil Engineers[10].

SAFETY

As in the construction industry generally, risks are high where safety standards are low. Both CP 2004[1] and CP 2011[11] contain much vital information. However, the main risks on piling sites are those common to all construction work, falling objects, collapse of excavations and the like. The only serious accident resulting from descent into large diameter holes was when two engineers died through lack of air caused by freak conditions of low atmospheric pressure; carbon dioxide 'locked up' in the Woolwich and Reading Beds under normal conditions was released and displaced the respirable air in the underream.

Since ground conditions are many and varied, piling poses problems to the practitioner, both new and old. This demands a blend of art and science, and a humility of approach which we do not always achieve.

References

1. CP 2004:1972. *Code of Practice for Foundations*, BSI, London
2. Weltman, A.J. and Little, J.A., 'A review of bearing pile types', *DOE/CIRIA Report PG1* (1977)
3. Beaman, A.L. and Jones, R.D., 'Noise from construction and demolition sites – measured levels and their prediction', *CIRIA Report No 64* (April 1977)
4. Chellis, R.D., *Pile Foundations*, McGraw-Hill (1961).
5. Palmer, D.J. and Holland, G.R., 'The construction of large diameter bored piles with particular reference to the London Clay', *Proc. Sym. Large Bored Piles, ICE/RCA* (1966).
6. Thorburn, S. and Thorburn, J.Q., 'Review of problems associated with the construction of cast-in-place concrete piles', *DOE/CIRIA Report PG2* (1977)
7. Federation of Piling Specialists/British Ready Mixed Concrete Association, 'Recommendation for the specification and use of ready mixed concrete for cast-in-place piling', *Ground Engineering*, **10**, 4, pp 33–36 (May 1977).
8. Institution of Civil Engineers, *Proc. Conference on Diaphragm Walls and Anchorages*, ICE, London (1975)
9. North-Lewis, J.P. and Lyons, G.H.A., 'Contiguous bored piles', *Proc. Conf. Diaphragm Walls and Anchorages, ICE* (1974).
10. *Piling: Model procedures and specifications*, Inst. Civ. Eng., London (1978)
11. CP 2011:1969. Safety precautions in the construction of large diameter boreholes for piling and other purposes, BSI, London
12. Fleming, W.G.K. and Sliwenski, Z.J., 'The use and influence of bentonite in bored pile construction', *DOE/CIRIA Report PG3* (1977)

Chapter 20

Case Histories of Failures

INTRODUCTION

Every construction is a full scale experiment and, in ground engineering, model experiments are often difficult and costly. Much therefore can be learnt about the ground-structure interaction or the behaviour of soils by studying the case histories of failures. What should also be borne in mind is that it is very rare when there is only one clear cut cause of failure. Usually a number of causes are to blame and these may be totally unrelated to one another; by itself each one may have been quite harmless.

It must be stressed that the opinions expressed in the following account regarding the mechanism and causes of any mishap are those of the author alone. For obvious reasons the following case histories will be discussed with complete anonymity.

FAILURES DUE TO SHRINKAGE AND SWELLING IN CLAY

The first case history concerns a two-storey house near Wokingham which, as the result of settlements on one side, beneath a newly built single-storey extension, became structurally unsafe and had to be demolished. Figure 20.1 shows the extension, added a few years later on to the northern side of the two-storey house. The entire structure was of load-bearing brickwork, with traditional concrete footings which, at the new northern end, went to 1.52 m below ground level.

The extension generally settled as a unit on foundations in clay, which was subject to shrinking and swelling due to moisture content changes. Close to the north wall of the extension were three large trees, each absorbing up to 45500 litres of water per year. Test pit 1 showed that roots up to 50 mm thick, from the adjacent trees, had extended beneath the floor of the extension. Several very dry years had preceded the appearance of cracks, particularly between the new and the older part of the house. The natural moisture content of the soil was 32% in test pit 3, whilst in test pit 1 it was about 23%. This difference in moisture content could account for a reduction of some 16% in the original volume of the soil which is equivalent to reducing 1 m of soil below the footings by about 41 mm. This was a catastrophic settlement which had nothing to do with the bearing capacity of the ground (the factor of safety was over 6) and for which no one was really responsible. Numerous similar cases have occurred due to the dry summer of 1976.

However, the reverse effect must be borne in mind; structures built on such clays are, when the soil is dried out, subject to swelling on uplift pressures. Swelling pressures can amount to several hundreds of kilonewtons per square metre. Such swelling is not only effective beneath footings; larger areas, like floors or slabs of warehouses, etc, cause an upward capillary movement of

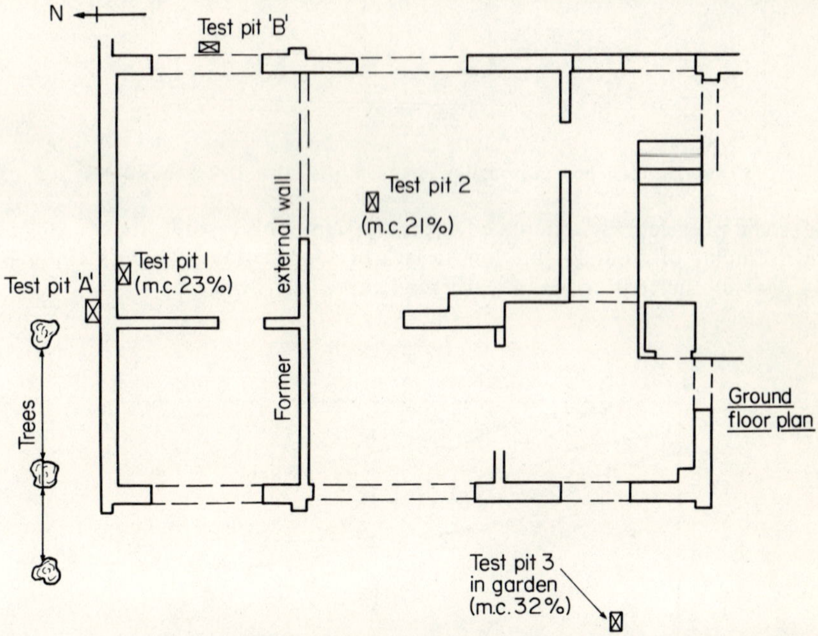

Figure 20.1 Two-storey house near Wokingham, Berks. Showing settlement which occurred beneath a newly-built extension

water which may result in considerable uplifts. In Cheltenham, for example, the central stairwell of a five-storey block of flats founded on Lias Clay was lifted 50 mm. A well-ventilated layer of granular material, 0.3 m thick, below the slab would have prevented this.

PILE FAILURES

It was decided to use piles as the foundation of a structure at an extensive site covering an old bed of the River Usk in South Wales. The site itself consisted of soft alluvium of very variable depths overlying Old Red to Sandstone marls, clays, sandstones and siltstones (Figure 20.2). The consulting engineers had accepted a tender for 7.6 m long Prestcore piles with a working load of 600 kN. Figure 20.3 shows the results of several pile tests, with several failures. Hundreds of piles were installed and had to be abandoned, the primary reason being that the 600 kN working load pile of 7.6 m length was only 0.3 m below the highest level at which good siltstone was encountered. As proved by a number of corings, percussion drilling had caused slurrying and softening of marl at toe level. Poor workmanship resulted in accumulation of loose material below the

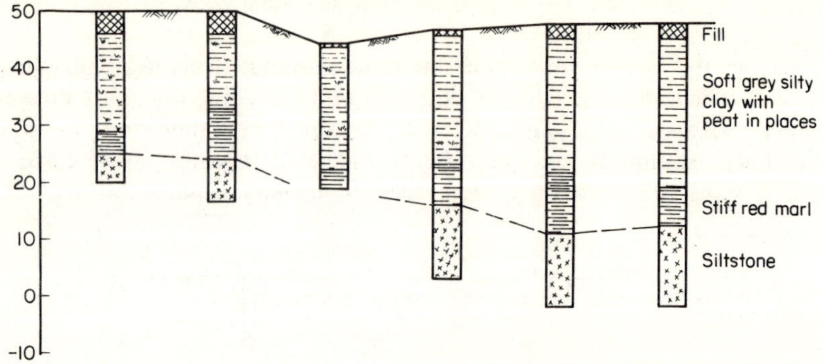

Figure 20.2 Borehole information obtained from River Usk, South Wales

toe; some piles actually terminated in the alluvium. This case illustrates a lack of understanding of ground conditions, a lack of appreciation of piling techniques and poor site supervision, all of which could have been avoided.

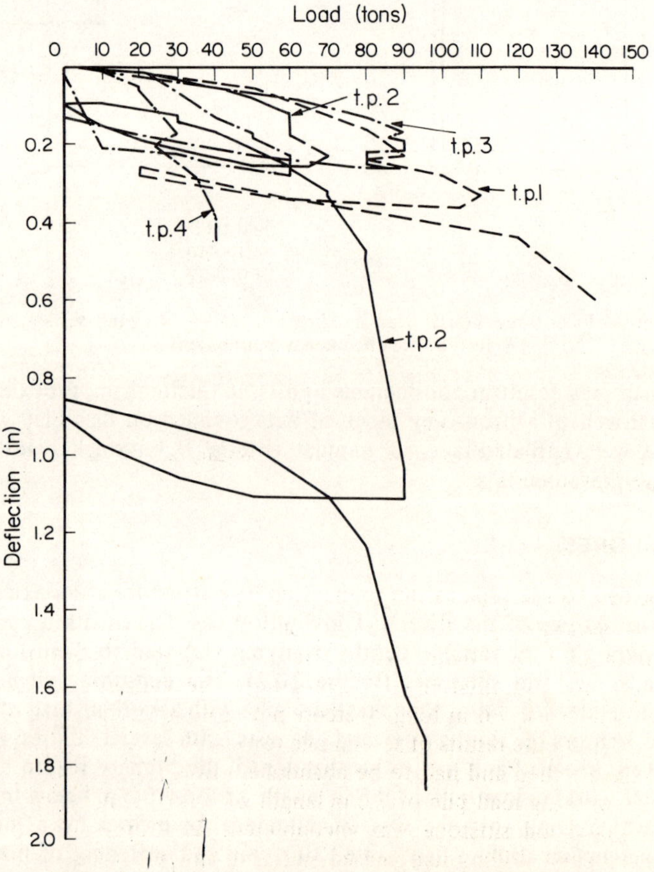

Figure 20.3 Test pile results for the settlement in Figure 20.2

Another interesting and controversial case of piling failure occurred on the west coast of Africa. The borehole data revealed 5.2 m of recently placed sand fill overlying soft compressible clay and peat, varying in thickness between 6.1 and 10.7 m from east to west across the site. A thick layer of medium dense sand was encountered at an approximate depth of 25.9 m. After the site investigation the use of 430 mm diameter, 27.4 m long vibro-piles, with enlarged toes, was recommended. Working loads were 500 kN which, supposedly, allowed for negative skin friction, but not for the varying amount of thickness of the compressible layer. Also the maximum concrete stress for a 430 mm pile is about

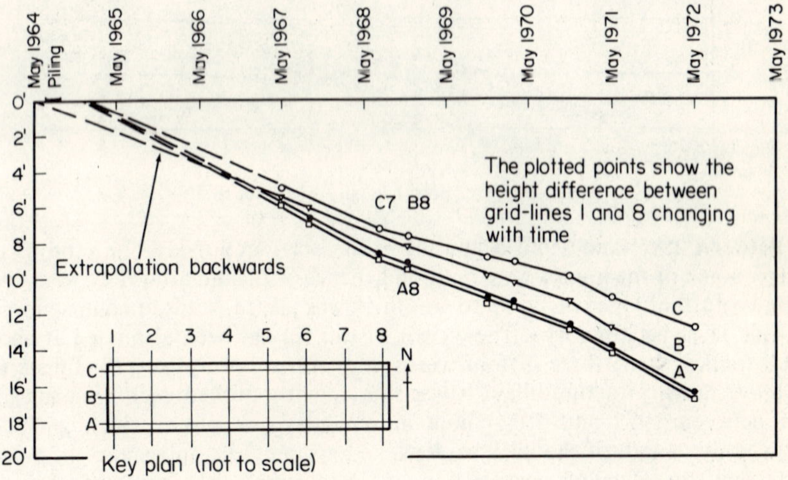

Figure 20.4 Progress of differential settlement with time of a piling failure which occurred on the west coast of Africa

500 kN, so there is some doubt about the factor of safety. Seventy six per cent of the piles were driven to only 26.5 m depth and to very variable sets. Settlement occurred and the amount increased with increasing thickness of the compressible layer (Figure 20.4). Whether the piles were too short, underdesigned or badly installed, or a combination of these factors, is never likely to be satisfactorily explained. Settlements were a direct function of the ground conditions and a raft would have behaved similarly. As it was the building had to be demolished. However, there was no consulting engineer on site and no client's representative.

SLOPES AND RETAINING WALLS

Figure 20.5 shows a geological section through the side of a valley in a Lancashire village from which it can be seen that it is carved out of sandstone and till. In the latter case the slopes frequently have been stepped by the use of berms. The presence of landslips and talus obscures some slopes. The valley side in question sloped at between 1:4 to 1:4.5 flattening off towards the top where the slope varied from 1:10 to 1:12. Houses, some over 100 years old, had been built at, or near to, the crest of the valley without any known distress occurring.

However, movements had started and were continuing. Aerial photographs taken in 1955 showed a number of springs issuing from the lower part of the slope and there were clear indications of some surface movements in the form of crescent-shaped terracettes, indicative of creep. A later aerial survey, taken in 1966, showed that the terracettes had moved slightly.

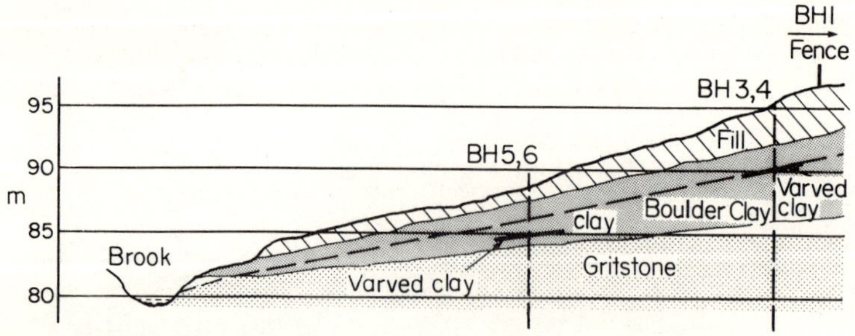

Figure 20.5 Section through slope after slide

Between 1969 and 1972 a housing estate was developed at the top of the valley. Some of the houses were built on benches of fill and after they were completed a further layer of fill up to 4 m thick was placed, without compacting, to provide level back gardens. The weight of this fill has been estimated at about 4000 tonnes. Some three of four weeks after it had been placed the fill started to move down into the valley. It was subsequently pushed back mechanically, but between 1972 and 1976 more movements took place, which gradually reduced the length of the back gardens, and assumed the appearance of a deep rotational slip. Detailed investigations in 1974 revealed the presence of varved clay in the till of the slope and at the toe (Figure 20.5). The initial movements probably represented the fill simply slipping down the valley side. However, once the fill had been replaced this meant that the top of the valley side was steepened to about 1:3, the weight of the fill then caused the deep-seated slide to develop within the till, the likely plane of failure probably lying along the varved clay.

Several points should be noted in connection with this case:

1. No previous site investigation had been carried out because no troubles had arisen on comparable sites in the past.
2. None of the several parties concerned had the experience or know-how to appreciate the danger due to placing fill on top of a slope. This is surprising because it is one of the commonest causes of slides on building sites of all sizes.
3. The failure plane followed a non-circular pre-determined line of weakness.
4. Very extensive and radical remedial measures will be necessary to stop movements completely, stabilise the valley side and safeguard the houses on top.

Another example of slope failure occurred in a railway cutting. Nearly 70 m of a retaining wall along the railway line were affected as the result of a rotating movement near to the rail level. Some 20 m of the wall were completely destroyed and large cracks appeared everywhere in the retaining wall. The wall was about 90 years old and essentially a gravity brickwork structure. It had weep-holes 4 to 5 m above track level, at about 5 m centres. No sizeable outward or translation movement and no significant heave of the track in front of the wall occurred.

One of the first danger signs was the appearance of a tension crack about one metre behind the wall (Figure 20.6). However, movements had also been reported four years earlier after which the amounts of fill and its slope was revised. A concrete toe block was installed after the earlier movements. Some 3 m of additional fill were placed before the failure.

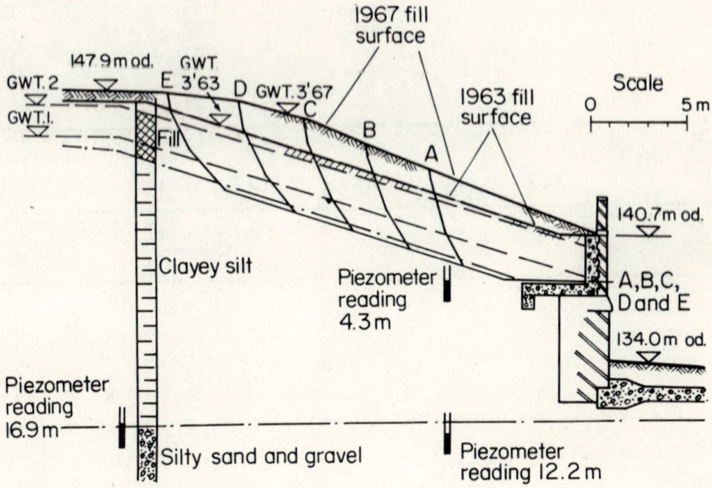

Figure 20.6 Retaining wall section showing three separate water tables and piezometer readings

A detailed site investigation, including continuous sampling and installation of piezometers, was carried out after the failure. It was found that rubble and clay fill overlay a laminated clayey silt, the surface of which sloped at 1:3 towards the railway cutting. This silt was considerably softened in the top 1 to 1.5 m directly below the fill. Frequent distortions in the silt observed in split undisturbed samples suggested solifluction movements of the slope during a post-glacial period. Therefore the residual shear strengths of the clayey silt were used in the stability analyses (Table 20.1). Piezometer readings showed the existence of three separate water tables, a varying one in the fill perched on the clayey silt, one in that stratum, close to the fill, and one in the underlying granular material (Figure 20.6).

Table 20.1 PARAMETERS USED IN STABILITY ANALYSIS

Material	*Wet density* Mg/m³	kN/m³	*Degrees*	*Remarks*
Fill	2.10	0	35	Assumed value
Clayey silt	2.10	10	23.5	Average of values measured in laboratory tests
Brickwork	2.10	0	0	Brickwork assumed to have no strength

In establishing the reasons for the failure, factual evidence ruled out a deep-seated circular slip. This view was strengthened by the existence of the zone of very much softened material directly below the fill, suggesting a preferential plane of failure. From the various analyses carried out the important results are given in Tables 20.2a and b.

Table 20.2 FACTORS OF SAFETY (1963)

Slip surface (see Figure 20.6)	GWT 1	*Ground water tables** GWT 2	GWT 3
A-A	2.32	1.71	1.42
B-B	2.04	1.55	1.33
C-C	2.03	1.53	1.36
D-D	2.02	1.53	1.37
E-E	2.02	1.55	1.36

*Assumed

Table 20.2b FACTORS OF SAFETY (1967)

Slip surface	GWT 1	*Ground water tables** GWT 2	GWT 3
A-A	1.96	1.53	1.07
B-B	1.75	1.41	1.05
C-C	1.72	1.40	1.01
D-D	1.73	1.42	1.02
E-E	1.83	1.50	1.13

*Assumed

These tables show that the reduction in the factor of safety due to a high water table is considerably greater than that due to the addition of fill. Normally a factor of safety of 1.5 is considered satisfactory. Because the weep-holes in the wall had not been functioning for many years and 100 mm of rain fell in the fortnight preceding failure, and also because of the absence of any drainage behind the wall other than in the granular material at toe level, it seemed reasonable to assume a very high water level in the fill. The addition of 3 m of fill alone could not have caused failure, but a high water table could have done so before 1967 and almost certainly did so in 1967. In this instance adequate drainage by means of vertical counterfort drains would have prevented a build-up of water pressures. The effect of vibrations caused by high speed electric trains on the structurally weak retaining wall, from which the overhead cables were suspended, is likely to have contributed towards the failure.

COFFERDAMS

A series of cofferdams were sunk on land and in the water of a tidal river to facilitate the construction of a water intake for an impounding station. The method of construction and strata concerned are shown in Figure 20.7. Steel sheet piling was used, driven 1 to 1.6 m into the chalk, beneath the superficial deposits. The river is tidal and mean high water is 32.7 m, spring tides reaching 34.0 m. However, when dredging had reached the level of the second frame on

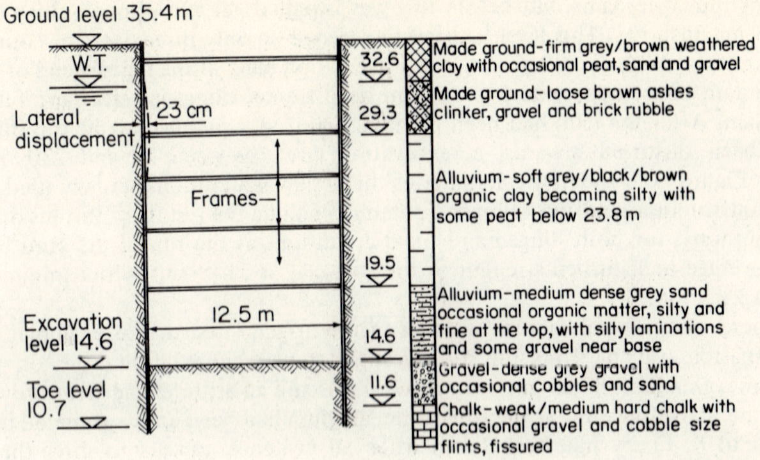

Figure 20.7 Section through cofferdam showing method of construction and strata

Description	Orig. design	Check calc. a	b	c	d	e	f
1. Soil strata Made ground	35.4-32.6 m	35.4-29.3	as a.				
γ_s	14.2 kN/m^3	18.1					
ϕ	35°	25°					
K_a	0.27	0.35					
Soft clay	32.6-19.5 m	29.3-19.5	as a.				
γ_s	14.9 kN/m^3	14.9					
ϕ	0°	0°					
c (active)	4.8 kN/m^2	28.8	28.8	28.8	28.8	22.8	22.8
(passive)	27.6 kN/m^2	5.8	28.8	28.8	17.3	22.8	22.8
Dense grey sand	below 19.5 m	below 19.5	as a				
γ_s	18.9 KN/m^3	18.9					
ϕ	35°	30°					
2. Water level	32.6 m	35.4	35.4	35.4	35.4	32.6	32.6
3. Structural condition	Fixed at top of sand at 19.5 m	Hinged at top of sand at 19.5	Fixed at 19.5	Hinged at 19.5	Hinged at 19.5	Fixed at 19.5	Hinged at 19.5
4. Theoretical deflection	25 mm	335	40	92.5	190	93.5	209
5. Max bending stress	5.72 kN/m^2 at 19.5 m	24.2 at 25.4	4.5 at 19.5	7.1 at 29.3	13.8 at 27.3	14.4 at 19.5	14.9 at 27.3

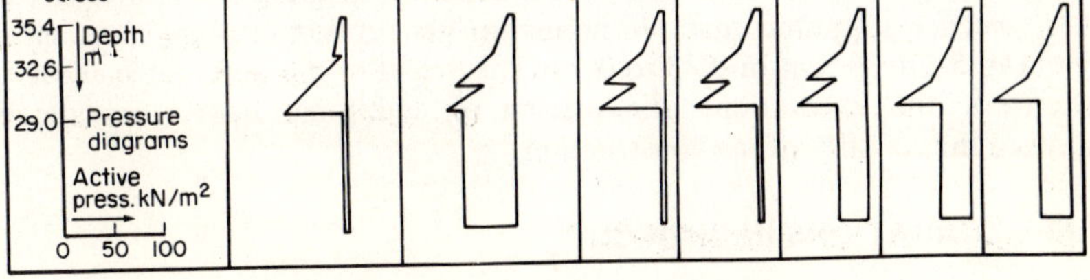

Figure 20.8 Design comparisons for cofferdams

one of the cofferdams, but before this was installed, an exceptionally high tide (35.5 m) occurred. This level had been exceeded on only three occasions during this century. Water broke through the river flood bank at the narrow end of the cofferdam concerned, scouring a 1.5 m deep trench along one side, and filling the dam. After the dam had been pumped dry it was found that the top frame had been distorted inwards, at excavation level, between 10 and 230 mm.

In Figure 20.8 the first column gives the design assumptions actually used for the construction, and the following columns include the results of various other assumptions, for both hinged and fixed conditions at the top of the sand bed. The average undisturbed strength of the soft clay is 31 kN/m^2 with a minimum of 20.3 kN/m^2.

There were several factual mistakes in the original design. For example, the information available from the site investigation was ignored in that the top 4 m of loose, granular fill had not been allowed for, and an arithmetical error showed that the active pressure at the top of the alluvium had been underestimated by a factor of 8. Fixity had been taken when all evidence appears to show that a hinge should have been assumed. What is more the active and passive earth pressures were unrealistic.

The subsequent analysis for different water levels, fixed and hinged conditions and various shear strength values showed that deflections of the order of magnitude of those measured could have been anticipated without an exceptional rise in the river once an unstrutted depth of excavation of about 6 m had been reached. It should be noted that the rule of thumb which gives the maximum depth for unstrutted excavations in normally consolidated clays is:

$$d = \frac{4c}{\rho \times 9.807}$$

where d = critical depth, c = shear strength, ρ = density.

Thus a very simple calculation would have given a warning. It is admitted, of course, that all earth pressure calculations are to a certain extent open to debate. In the present case, however, it would have been wise to place the second frame at a higher level. This example also illustrates the point that the depth of sheet piling below which failure occurs appears to be immaterial.

When excavation had reached 16.3 m, and 19.6 m, pools of water appeared in the bottom, indicating the occurrence of quick conditions. Sand and water entered the excavation at the places where the cofferdam joined another structure, and where clutches had been strained or split. In general, the problems which arose at these greater depths of excavation were due to insufficient penetration of the piles into the chalk, assisted, probably, by split clutches. The chalk is highly fissured, water bearing and has an uneven erosion surface. Therefore the flow path had not been sufficiently lengthened. Both the pressure diagram and other evidence suggested that the minimum penetration into the chalk should have been 5.5 to 6.5 m or 0.33 to 0.5 of the height of the water table above the bottom of the excavation. The placing of additional frames considerably improved the stability of the construction.

DEEP EXCAVATIONS IN OPEN CUT

The general arrangement of opencut excavations for an underground railway is shown in Figures 20.9a and b. The surface conditions consisted of up to 4 m of

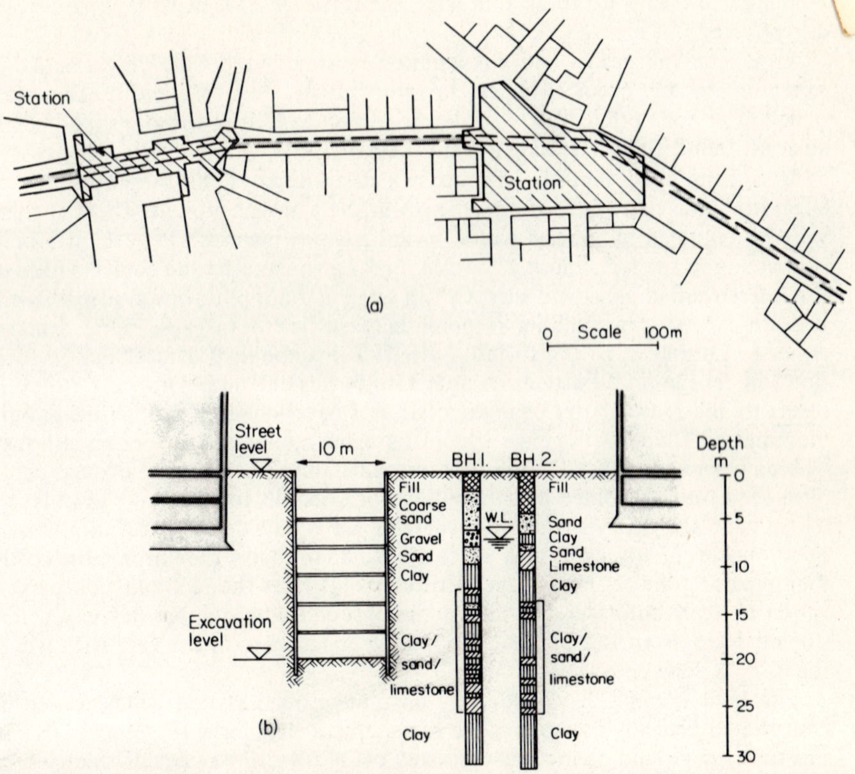

Figure 20.9 Open-cut excavations for an underground railway (a) General layout (b) Underground railway excavation

fill overlying between 2 and 10 m of alluvial sand and gravel with occasional bands of compressible silts and clays occurring at various horizons. This alluvium overlaid stiff clays, marls and limestones, with occasional lenses of silt, soft clays and sand, which were of Tertiary age. Ground water occurred at various levels in the alluvium and in the permeable Tertiary strata and was probably under artesian pressure.

The depth of the excavations was between 13 and 20 m. Diaphragm walls, continuous bored pile walls and the 'Berlin' and 'Hamburg' methods of construction were used. Ground water lowering techniques were extensively employed so that work could be carried out in dry conditions. All excavations were strutted with heavy girders set about 2 m horizontal and vertical centres. Struts at lower levels were pre-stressed. Measurements showed occasional inward movements of these retaining walls of only a few millimetres. Adjacent buildings had up to 2 basements (4 to 5 m in depth), were generally built around the turn of the century and 4 to 6 storeys high. They were not free-standing except for an old church and a monument. The excavations for the tunnels came as close as 1 m to existing structures. A very extensive amount of damage occurred along the excavations. It varied from minute cracks in basements, facades and inside the buildings, to cracks up to 50 mm wide at roof level. Pavements and roads were also affected.

In general the design and construction methods used were very sound. An exception to this occurred when, in a few places, steel piles and timbers were pulled after concreting. This was almost certain to lead to loss of ground below adjacent foundations, resulting in harmful settlements.

It is considered that the damage which occurred along the entire line of excavation was caused by stress changes in the adjacent ground. These changes were accentuated by ground water lowering which increased the effective pressure acting in the soil, causing compaction of granular soils and consolidation of the interstratified clays and silts. Only a small amount of downwards movement beneath an existing building is potentially sufficient to cause some damage, either to finishes or to the structure itself. The amount of damage and its location will depend on the structural design, since this determines how the structure reacts to and redistributes changes in stresses occasioned by differential ground movements. Therefore damage should have been expected to occur, although where, when and to what extent was unpredictable. Whether such damage could have been totally avoided by attempting, for example, to grout up known layers of granular soil or by installing recharging wells outside the excavation is also questionable. In any case, such measures would probably have been more costly than repairing the actual damage. This case illustrates the fact that a deep excavation is likely to affect the surrounding ground within approximately a line sloping 1:0.5 from the perimeter of its base regardless of the care with which the work is done.

The final example is provided by an underground station which was to be constructed beneath a large expanse of water in a city centre (Figure 20.10a). In this case the ground conditions consisted of fill overlying a variable thickness of recent alluvium consisting of interstratified, but not always continuous layers of sand, peat and organic clays. This is underlain by stiff till which contained sands and gravels (Figure 20.10b). All sandy strata were waterbearing. A recent deep valley, filled with similar alluvium, crossed the site from north east to south west. Buildings to the east and west of the lake consisted of 70 to 100 year old office blocks, up to seven storeys high and with 1 or 2 basements. They were either supported on wooden piles, driven into an alluvial sand, on piled rafts or on rafts. A previous underground construction in the vicinity, some fifty years ago, had caused some damage.

Construction in the water was done in a 120 m long by 40 m wide and 20 m deep sheet pile cofferdam, with the sheet piles keyed into impermeable till. A second 22 m deep cofferdam was installed 4 m from the nearest building. The tunnels were driven east and west by means of a shield. After excavation and dewatering commenced settlements varying between 2 and 200 mm were observed along the waterfront and occasionally away from it.

A building with two basements was located 90 to 100 m west of the cofferdam and was founded on endbearing wooden and concrete piles driven into a sand layer (3.5 to 4.5 m) with peat and organic silt above. Some shallow recharging wells had been installed in the upper and lower sand layers near the structure. After ground water lowering had commenced the water table beneath the structure went down by up to 4 m from just above basement level. Damage

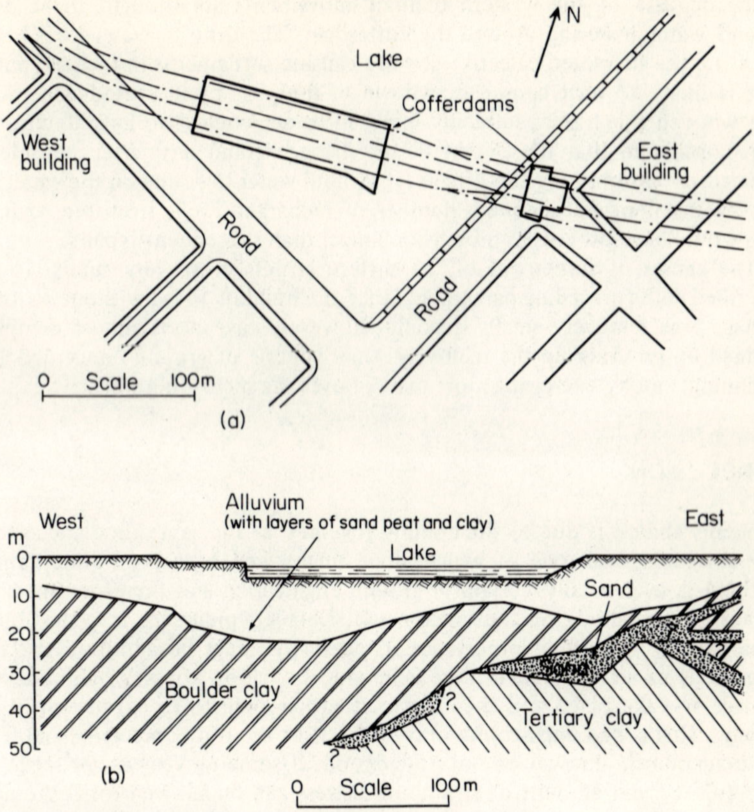

Figure 20.10 Underground station which was to be constructed beneath a large area of water beneath a city centre (a) Surface layout (b) Stata of stiff till which contained sand and gravels

to the building was considerable and amounted to cracks in the walls and on staircases up to 50 mm wide. Fears for its structural integrity were expressed. Further and deeper recharging wells were sunk into the sand.

The building near the landward cofferdam on the eastern side was a steel-framed structure on a 0.6 m thick raft (Figure 20.10a). The raft was supported by wooden piles driven into the sand layer. Here the thickness of the peat and silt layers varied between 0 and 2.5 m. Furthermore the water bearing sand in the till appears to rise steeply in the direction of this structure and it may be in communication with the upper, alluvial sand (Figure 20.10b). After the excavation of the cofferdam and during ground water lowering, the building tilted about 1:500 towards the excavation. The level of the water table fluctuated by 4 to 5 m despite recharging wells. When tunnelling started close to the corner of the building the tilt increased to 1:250. Cement was injected into the alluvial sand layers and this considerably decreased settlements. Two years later settlement differences across the building ranged from 30 to 200 mm. Despite the amount of movement damage was relatively small and no serious structural damage occurred.

In the case of the western building movements are thought to be due to ground water lowering around the cofferdam. The drop in water level beneath the structure increased effective stresses causing settlements in the peat and silt. This is likely to have caused drawdown of the piles by increased skin friction. The wooden pile-heads, suddenly exposed to air, could also have deteriorated. It is considered that the extent of the damage could have been considerably reduced by determining the effects of ground water lowering on the water table and by installing an adequate number of recharging wells from the beginning. This would have proved far more economical than subsequent repairs.

The causes of settlement of the eastern structure are very similar to those described in the preceding paragraph. Since the building tilted without distortion, damage was relatively small. It could, however, have been almost completely avoided by grouting up the relatively small volume of ground concerned at the beginning, and by exercising more control over the recharging wells.

CONCLUSION

Generally failure is due to the coming together, at the same time and place, of several causes, each one of which alone might well have been quite harmless. Nothing is ever totally certain in ground engineering and therefore one should always be prepared for the unexpected. Detailed predesign investigations are consequently of prime importance. However, it must be admitted that quite often it is easier and much more economical to allow for a certain amount of damage to take place and then repair it rather than to try to prevent it completely. This is bad engineering only if the risks are not appreciated and if facts are inadequately known or not understood. If some movement, or settlement, or even a certain amount of structural distress can be allowed for it should not be blindly ruled out.

In conclusion it may be stated that design is the preparation for a scheme to enable the construction of a structure best suited to its required function and safe under all foreseeable circumstances. If these conditions can be arrived at in several ways, then the most economical method may be deemed the best design.

ACKNOWLEDGEMENT

Some of the above cases were described by the author in a paper given at the Sixth European Conference on Soil Mechanics and Foundation Engineering, 1976, and their kind permission to quote is acknowledged. The illustrations in this article are the copyright of the author.

Index